MOLECULAR ANATOMY AND PHYSIOLOGY OF PROTEINS

BETA-LACTAMASES

MOLECULAR ANATOMY AND PHYSIOLOGY OF PROTEINS

VLADIMIR N. UVERSKY - SERIES EDITOR

INDIANA UNIVERSITY OF MEDICINE, US

Alpha -Lactalbumin
Permyakov, Eugene A. and Vladimir N. Uversky (Editors)
2005. 1-59454-107-8

Creatine Kinase
Christian Vial (Editor)
2006. 1-59454-715-7

Cellulosome
Irina A. Kataeva (Editor)
2006. 1-59454-950-8

Neuronal Calcium Sensor Proteins
Pavel P. Philippov and Karl-Wilhelm Koch (Editors)
2006. 1-59454-978-8

Human Stefins and Cystatins
Eva Zerovnik and Nataša Kopitar-Jerala (Editors)
Vladimir Uversky (series editor)
2006. 1-60021-233-6

Parvalbumin (Volume in Molecular Anatomy and Physiology of Proteins Series)
Eugene A. Permyakov and Vladimir Uversky (Editors)
2006. 1-60021-337-5

Methods in Protein Structure and Stability Analysis: \Luminescence Spectroscopy and Circular Dichroism
Vladimir N. Uversky and Eugene A. Permyakov (Editors)
2007. 1-60021-404-5

Methods in Protein Structure and Stability Analysis: Vibrational Spectroscopy
Vladimir N. Uversky and Eugene A. Permyakov (Editors)
2007. 1-60021-703-6

Methods in Protein Structure and Stability Analysis: NMR and EPR Spectroscopies, Mass-Spectrometry and Protein Imaging
Vladimir N. Uversky and Eugene A. Permyakov (Editors)
2008. 1-60021-705-2

Protein Adaptation in Extremophiles
Khawar Sohail Siddiqui, Torsten Thomas and Vladimir N. Uversky (Editors)
2008. 978-1-60456-019-0

Ribonucleotide Reductase
Prof. Kristoffer Andersson (Editor)
2008. 978-1-60456-199-9

Foldases: Enzymes Catalyzing Protein Folding
Natalya K. Nagradova (Editor)
2008. 978-1-60456-389-4

Biochemistry and Function of Antifreeze Proteins
Steffen P. Graether (Editor)
2010. 978-1-61668-265-1

Methods in Protein Structure and Stability Analysis: Conformational Stability, Size, Shape and Surface of Protein Molecules
Vladimir N. Uversky and Eugene A. Permyakov (Editors)
2011. 978-1-60021-704-3

Beta-Lactamases
Jean-Marie Frère (Editor)
2011. 978-1-61324-638-2

Biochemistry and Function of Antifreeze Proteins
Steffen P. Graether (Editor)
2012. 978-1-62100-435-6

MOLECULAR ANATOMY AND PHYSIOLOGY OF PROTEINS

BETA-LACTAMASES

JEAN-MARIE FRÈRE
EDITOR

Nova Science Publishers, Inc.
New York

For permission to use material from this book please contact us:
Telephone 631-231-7269; Fax 631-231-8175
Web Site: http://www.novapublishers.com

Library of Congress Cataloging-in-Publication Data

Beta-lactamases / editors, Jean-Marie Frhre.
p. cm.
Includes index.
ISBN 978-1-61324-638-2 (hardcover)
1. Beta lactamases. I. Frhre, Jean-Marie.
QP609.B46B472 2011
579.3'5--dc23

2011014373

Published by Nova Science Publishers, Inc. † New York

CONTENTS

PREFACE

The activity of a bacterial enzyme "able to degrade penicillin" had first been described in 1940, even before the exact structure of penicillin was elucidated and, by 1970, several enzymes had been purified to homogeneity, the amino acid sequence of a staphylococcal penicillinase was also known and that of its *Bacillus licheniformis* counterpart was well under way. By contrast, their catalytic mechanism remained quite mysterious. A Zn++ metallo-beta-lactamase (β-lactamase II, BcII or βII) had also been described as soon as 1967 and later purified. It was thus surprising that the first mechanistic information demonstrating the presence of a penicillin-binding serine residue was obtained with a penicillin-sensitive DD-peptidase rather than a β-lactamase. This seemed to open the floodgates and several class A β-lactamases were then rapidly shown to be active-site serine enzymes. This book presents current research in the study of beta-lactamases.

Chapter 1 - In the 3 decades from 1940 to 1970 the bewildering task of sorting out the properties and genetics of the penicillinases was done by a small number of biochemists and microbiologists. At first, even the chemical structure of the enzyme's substrate was unknown! Class A enzymes were the primary focus of these early studies, as little was known at the time of other classes. Here the authors salute these people, all of whom made the work of this book much easier.

Chapter 2 - Class A contains the largest number of β-lactamases. Despite a very little number of residues conserved in all these enzymes, the structures are strikingly similar. The first clinically relevant enzyme was produced by *Staphylococcus aureus*. It was essentially a penicillinase and never evolved to hydrolyse other compounds such as cephalosporins or even semi-synthetic penicillins such as methicillin. By contrast, the TEM and SHV enzymes produced by Gram-negative strains evolved to hydrolyse third-generation cephalosporins or to become more resistant to inactivation by clavulanate, sulbactam or tazobactam. A large number of 3D structures have been solved by X-ray crystallography. Some of them are described in this chapter where the authors also attempt to establish structure-function relationships.

Chapter 3 - The metallo-β-lactamase superfamily comprises a remarkable set of enzymes that catalyse the hydrolysis of a wide range of substrates such as peptides, nucleic acids, antibiotics of the penicillin family and organophosphorus derivatives. In the past ten years, X-Ray structures of representative enzymes from different families have been determined, with the metallo-β-lactamases being the most represented. The salient common structural feature is the presence of a catalytic metal centre embedded within a αβ/βα fold. The wealth of sequence and structural information on metallo-β-lactamases has allowed their classification into three subclasses. Structural information is now available for members of each subclass. Interestingly, these structures show the presence of either a mono or a di-nuclear metal centre in the active sites raising questions on the metal to protein stoichiometry under physiological conditions. In addition, the structures reveal a wide variability in the shape of the active site, which involves three variable loops lining the metal centre. For each enzyme a clear correlation is found between active site shape and substrate specificity.

Chapter 4 - Three-dimensional structures of a series of class C β-lactamases, alone or in complex with ligands, have been obtained by X-ray crystallography. In addition to their interest for studies on the mechanism of those serine-active hydrolases, structures of complexes between β-lactamases and synthetic compounds provide a rational basis for the design of original antibiotics. Indeed, as for other β-lactamases, hydrolytic properties of class C enzymes, and thus related antibiotic resistance, arise from specific molecular mechanisms driven by the structure of their active site. Therefore, the structure-function analysis of β-lactamases supplies critical information for the selection and development of antibacterial agents. These aspects are summarized and discussed on the basis of selected examples from the recent literature.

Chapter 5 - Class D β-lactamases were first considered as "honorary class A members" because their sizes and general properties seemed to make them close relatives of the members of this class. However, it rapidly appeared that, besides their ability to hydrolyse oxacillin, they exhibited very specific characteristics: the absence of the conserved Glu 166 residue and, more strikingly, a carboxylated Lys residue which was shown to act as the general base in the catalytic mechanism. Interestingly, although exclusively produced by Gram-negative strains, they exhibit a close structural similarity with the exocellular domains of the BlaR penicillin receptors produced by the Gram-positive *Staphylococcus aureus* and *Bacillus licheniformis*.

Chapter 6 - Beta-lactamases are probably the enzymes for which the largest quantity of kinetic parameters have been accumulated due to the large numbers of both available substrates and enzymes. Unfortunately, many of these parameters are not very reliable because the experimental conditions were rather poor. For instance, low K_m values are quite difficult to determine directly since the variation of the monitored property (e.g. absorbance or proton release) can become very small. In this chapter, the authors summarize the main features of β-lactamase kinetic analysis. If, in practice, the k_{cat} and K_m values are sufficient to correlate β-lactamase activity and MIC values (see chapter 16), the active-site serine and metallo-β-lactamases are analysed separately since they follow distinct kinetic pathways.

Chapter 7 - In this chapter, the catalytic mechanisms of β-lactamases are discussed in detail. After an introduction on the reactivity of compounds containing the the β-lactam ring, each class of β-lactamases is reviewed and analysed in terms of essential residues and active site chemistry. Special attention is given to class B enzymes (metallo-β-lactamases or MBLs) for which many questions remain presently unanswered and which seem to exhibit specific mechanisms depending on the subclass.

Chapter 8 - Nuclear Magnetic resonance (NMR) provides a substantial amount of structural and dynamic information short of a full protein structure. This technique has been successfully utilised to study the role of various residues in the catalytic mechanisms of the four classes of β-lactamases. In the active-site serine enzymes, NMR has been useful in elucidating the nature of the residue which acts as a general base in activating the active serine and for the metallo- β-lactamases, the focus has been on studies of the active-site metal ions environment.

Chapter 9 - The co-administration of a β-lactam antibiotic together with a β-lactamase inhibitor represents an efficacious treatment for infections caused by many β-lactamase producing strains. Most known inhibitors of β-lactamase, including the current commercial products, clavulanate, sulbactam, and tazobactam, are themselves β-lactams. These molecules are mechanism-based inhibitors which, subsequent to acylation of the active site serine, undergo a series of transformations leading to formation of a hydrolytically stabilized acyl-enzyme. Not surprisingly few display inhibition of the metallo-β-lactamases. This chapter discusses β-lactams which function as β-lactamase inhibitors from a mechanistic perspective. A review of inhibitor-resistance is also presented.

Chapter 10 - Besides the β-lactam containing inactivators, a variety of non-β-lactam containing compounds have been described as inactivators or inhibitors of β-lactamases. It is not surprising that very different compounds were found to be active against metallo-β-lactamases on the one hand and active-site serine enzymes on the other. Several of them were isolated from natural sources while many others are due to the imagination of chemists.

Chapter 11 - β-lactamase inhibitory proteins (BLIPs) are extracellular proteins produced by species of the soil bacteria *Streptomyces*. Three have been characterized and shown *in vitro* to selectively inhibit primarily class A β-lactamases. BLIP-0 and BLIP-I are homologous, 18 kDa proteins that are highly similar in function and with respect to their two-domain, tandemly repeated structure. BLIP-0 competitively inhibits TEM-1 β-lactamase using a large binding interface and loops that form specific contacts with key enzyme active site residues. BLIP-II has a molecular mass of 33 kDa and possesses a distinct fold and inhibitory mechanism, involving no direct contact with TEM-1 active site residues. Both BLIP inhibition mechanisms are reversible and show 1:1 stoichiometry. In addition to their structure and mechanism, the potential roles for BLIPs in nature and their implications for ß-lactamase inibitor design are discussed.

Chapter 12 - Theoretical models have become necessary to support experimental work. The molecular modelling field covers a wide range of theoretical tools and techniques. In this chapter, the methods of quantum chemistry (QM) are presented and the utilisation of QM in the study of β-lactamase reactivity is discussed in the light of publications by the authors' laboratory or from other sources.

Chapter 13 - The action of β-lactamases on β-lactam antibiotics is a major factor of bacterial resistance to these molecules, particularly for Gram-negative organisms. These enzymes are growing considerably in number and variety of catalytic properties associated to phenotypic expression(s). After the introduction of benzylpenicillin in human therapy of infectious diseases, emergence of "completely novel" β-lactamases occurred repetitively and usually step by step. This occurred within the four molecular classes of beta-lactamases defined on the basis of structural and mechanistic properties. In this chapter the authors report many of the novel sequences identified as naturally occurring or as sources of acquired resistance in Gram-negatives, speculate on the possible origin of "completely novel" β-lactamases and give some biochemical information about the most important leader type enzymes.

Chapter 14 - Substitutions at particular amino acids allow the common plasmid-mediated TEM, SHV, and OXA enzymes to become extended-spectrum β-lactamases able to hydrolyze oxyimino-β-lactams such as ceftazidime, cefotaxime, ceftriaxone, and aztreonam. Other amino acid changes in the TEM and SHV enzymes are associated with resistance to β-lactamase inhibitors such as clavulanic acid and sulbactam. Alterations at a few sites allow enzymes in the CTX-M family to attack ceftazidime or in the AmpC family to hydrolyze cefepime or cefpirome.

Chapter 15 - Multidrug resistant (MDR) bacterial strains present a challenging problem for the clinician. They are most taxing when belonging to the *Klebsiella pneumoniae*, *Pseudomonas aeruginosa* and *Acinetobacter baumanii* species. The clinical challenge, which can be very different depending upon the country and even the city, is analysed for each of these pathogens. Special attention is devoted to the KPC, CTX-M (both class A) and NDM-1 (class B) enzymes. In *Pseudomonas*, deregulated overproduction of the chromosomal class C enzyme can also be a major problem.

Chapter 16 - The resistance of a Gram-negative bacterial strain to penicillin is determined by the sensitivity of the essential penicillin-binding proteins (PBPs), the amount and properties of the produced β-lactamase(s), the permeability of the outer membrane and the efficiency of the efflux systems. The interplay between the last three mechanisms is discussed with examples stressing the importance of one or several of them in clinical isolates. Mathematical models are presented. In the relatively simple case of *Escherichia coli* (in which the efflux systems are often negligible when compared to the β-lactamase activity) the MICs can be predicted on the basis of these models. Similarly, correct MIC predictions could be made for an *Enterobacter cloacae* strain which overproduced a class C enzyme.

Chapter 17 - The production of a β-lactamase can be induced by the presence of a β-lactam in the growth medium. Three distinct mechanisms have been discovered. In *Bacillus licheniformis* and *Staphylococcus aureus*, a penicillin receptor whose properties are similar to those of a PBP detects the presence of the antibiotic. In *Citrobacter freundii*, the damage caused by β-lactams to the peptidoglycan is detected by an intracellular protein and a similar system might prevail in *Streptomyces cacaoi*. Finally, in *Aeromonas sp* regulation involves a two-component system.

Chapter 18 - Because of the ready selection for function, β-lactamases have been intensively studied by random mutagenesis methods. In particular, the TEM-1 β-lactamase has served as a model for methods development for random mutagenesis techniques and as a model for protein engineering and the directed evolution of new enzymatic functions. This chapter reviews the use of random mutagenesis methods for studies of β-lactamase structure, function, evolution and engineering.

Chapter 19 - Class A β-lactamases have proved useful as model proteins in studying a wide variety of aspects of protein folding. The authors review those features that have shed light on kinetic intermediates that take part in folding, including some insight into the molecular basis of the kinetic and stable molten-globule states that have been identified. The contrast between the folding behaviour of PC1 and the two lactamase mutants, P54 and P2, can be attributed in some detail to changes in molecular conformation. The early, very rapid stages of folding of β-lactamases have been shown to be multiphasic, and an interesting intermediate is described that has non-native contacts involving burial of the C-terminal tryptophan. A further feature is that the region around the disulphide bond in the TEM enzymes is formed very early in the folding reaction. The unusual feature of class A β-lactamases, i.e. a *cis* peptide bond critical in the stereochemistry of the active site that usually, but not always, involves a proline residue, has been shown to be important in accounting for the slow folding reactions. The effect of substrates on the stabilization of the enzymes and on their reversible deactivation is also reviewed.

Chapter 20- The activity of β-lactamases is easy to estimate, they often exhibit very high turn-over numbers and many of them are quite stable. For these reasons they have emerged as promising reporter enzymes to study gene expression and as models to analyse protein-ligand interactions. They also accept the insertion of rather large peptides without losing their activity. This has allowed the development of practically useful technologies which are described in this chapter.

Chapter 21 - With PBPs and active-site serine β-lactamases, β-lactams behave as acylating agents whereas the serine hydroxyl group acts as a nucleophile. In this chapter, the authors briefly discuss the sensitivity of β-lactams to nucleophilic agents, both small molecules and protein-borne and the β-lactamase-like activities of several other proteins, some of them unrelated to the penicillin-recognising enzymes families.

FOREWORD

In October 1970, I started working on a penicillin sensitive enzyme in Jean-Marie Ghuysen's laboratory. At that time, scientists in the field agreed on the fact that penicillin was inhibiting the DD-transpeptidase(s) responsible for catalysing of the last step of peptidoglycan biosynthesis but very little was known about the inhibition mechanism. No transpeptidase had been purified and, by contrast, β-lactamases already had a long history. The activity of a bacterial enzyme "able to degrade penicillin" had first been described in 1940, even before the exact structure of penicillin was elucidated and, by 1970, several enzymes had been purified to homogeneity, the amino acid sequence of a staphylococcal penicillinase was known and that of its *Bacillus licheniformis* counterpart was well under way. By contrast, their catalytic mechanism remained quite mysterious. A Zn^{++}metallo-beta-lactamase (β-lactamase II, BcII or βII) had also been described as soon as 1967 and later purified. It was thus surprising that the first mechanistic information demonstrating the presence of a penicillin-binding serine residue was obtained with a penicillin-sensitive DD-peptidase rather than a β-lactamase. This seemed to open the floodgates and several class A β-lactamases were then rapidly shown to be active-site serine enzymes.

Similarly, the first atomic resolution X-ray data for a penicillin-recognising enzyme were obtained for the *Streptomyces* R61 DD-peptidase. At that time, we thought that everything would be explained on the basis of detailed structures. As will be seen in this book, this act of faith was a little bit naive, although everybody agrees that the structures have significantly contributed to our understanding of the substrate profiles and catalytic mechanisms. But many major questions are still unanswered.

By 1970, it was generally recognised that β-lactamases were major factors in the resistance of Gram-negative bacteria to penicillins and first generation cephalosporins as demonstratedby the efforts of the pharmaceutical companies to develop compounds to circumvent resistance due to the TEM-2 and TEM-2 enzymes. In this context, it is quite ironic to note that Karen Bush's first β-lactamase paperwas rejected because it was dealing with an "unimportant enzyme". I hope that the referee who wrote this report has now changed his or her mind. More seriously, two approaches were devised to fight β-lactamase-induced resistance: the design of compounds which escaped their activity (the "camouflage" strategy) and the search for compounds which could inhibit or inactivate them (the "kill them" strategy). Both strategies yielded important results, as shown by the successes of third generation cephalosporins such as cefotaxime and ceftazidime (camouflage) and of clavulanate, sulbactam and tazobactam (kill them). But bacteria responded by selecting enzymes which could destroy the former compounds or exhibit increased resistance to the latter. Quite often, these new enzymes were plasmid-encoded and some of them managed to spread throughout the world (see for instance the numerous references to the CTX-M enzymes in chapters 2, 13 and 15).

In consequence, it appeared quite timely to produce a book devoted to β-lactamases. Our ambition was to collect contributions describing the important aspects of our knowledge of these enzymes, from structural to clinical problems. After a historical introduction, the first section deals with the structures of the 4 known classes of β-lactamases. The second section addresses the kinetic and mechanistic aspects and includes chapters on inhibitors and inactivators. In the third section, the clinical implications of β-lactamase production by pathogenic strains are discussed. But β-lactamases have also been used as models and tools, for instance in the study of folding and induction phenomena, in directed evolution experiments or as epitope carriers. This is the subject of the fourth section. Finally, in a last, short chapter we wonder if there might be other classes of β-lactamases and analyse the sensitivity of β-lactams to nucleophilic agents and to other enzymes.

One question has not been asked: do β-lactamases have another physiological activity in addition to their protective role? It is interesting to note that β-lactamases are produced by various species of the *Actinomycetes* family which also synthesise β-lactams. This could be the method by which these bacteria defend themselves against their own toxins. The genes could have been later transferred to other species, including pathogens.But what about the chromosome-encoded class C enzymes, found only in Gram-negative bacteria? It is now well accepted that active-site serine β-lactamases derive from DD-peptidases or Penicillin Binding Proteins (PBPs) or that the two families share a common ancestor. β-Lactamases also exhibit an esterase activity and a very low peptidase activity on typical DD-peptidase substrates. On this basis several attempts were made to mutate PBPs into β-lactamases, but no activity even remotely similar to those of the natural enzymes could be obtained. In fact, the highest activity against benzylpenicillin is that of a laboratory mutant of the *Streptomyces* R61 DD-peptidase, with a k_{cat} of $4x10^{-2}$ s^{-1} while the value of the same kinetic parameter can be larger than 1000 s^{-1} for some β-lactamases and their best substrates. Also worth mentioning are a native*Staphylococcus aureus* PBP (k_{cat} = 10^{-2} s^{-1}vsbenzylpenicillin) and a *Streptococcus peumoniae* PBP2x mutant (k_{cat} = 0.015 s^{-1}vscefotaxime, see chapter 18). It is also quite surprising that the best structural similarities between PBPs and β-lactamases are found over the Gram-positive/Gram-negative barrier, between class C β-lactamases and the *Streptomyces* R61 extracellular DD-peptidase on the one hand and between class D β-lactamasesand the extracellular domain of the *Bacillus licheniformis*BlaR penicillin receptor on the other. It can be hypothesised that both PBPs might represent primitive attempts to devise a protective mechanism, since the proteins could “mop up” β-lactam molecules present in the environment (R61 and *B. licheniformis* are soil bacteria).

The reader will certainly find some repetitions between the various chapters. It is difficult to ask a crystallographer to write about structures and to forbid him or her to include hypotheses about mechanisms in his (her) chapter. Similarly, if you write about mechanisms or inhibition, you must rely on the structures if they are available. This has the advantage of giving the reader different points of view, sometimes not in complete agreement or even contradictory. But it is interesting for the reader to realise that not everything is completely clear and straightforward in the field.

I asked colleagues and friends to contribute chapters or parts of chapters. I could not possibly ask all my friends to do so, because the book would have been several thousand pages thick! Some of these friends were very slow in doing their duty and I hope they are still my friends after the numerous annoying and even bullying messages I sent them. This explains why the preparation of the book took several years. Nonetheless, I hope that it will be a useful tool for those working in the field. As will be seen, bacterial resistance to antibiotics will remain with us and I strongly believe that the synthesis of peptidoglycan is probably one of the best, if not the best bacterial targets to try to interfere with. In this area, my feeling is that β-lactams still constitute a marvellous weapon in our fight against bacterial pathogens. An enormous number of molecules can be synthesised on the basis of the various available bicyclic or monocyclic nuclei. In this respect, countering the activity of β-lactamases (and of efflux pumps) should remain a priority (see chapter 16). Discovering non-β-lactam inhibitors of the transpeptidation reaction would be an alternative desirable goal but efforts in this direction have, up to now, remained rather unsuccessful. The "competition" between PBPs and β-lactamases to bind or destroy the same molecules demonstrates the pertinence of one of my favourite aphorisms: Life is a question of kinetics.

Tony Fink, one of the two co-authors of Chapter 7 died on March 3, 2008. Mike Page completed and updated their contribution. Tony's photograph is shown below and I would like to remind the readers of this book of Tony as both a good friend and an excellent scientist

Jean-Marie Frère

In: Beta-Lactamases
Editor: Jean-Marie Frère, pp. 1-18

ISBN: 978-1-61324-638-2

Chapter 1

BEFORE OUR TIME: EARLY ß-LACTAMASE PAPERS AND THE PEOPLE WHO WROTE THEM

James R. Knox*

Molecular and Cell Biology Department,
The University of Connecticut, Storrs, CT 06269 U.S.A.

ABSTRACT

In the 3 decades from 1940 to 1970 the bewildering task of sorting out the properties and genetics of the penicillinases was done by a small number of biochemists and microbiologists. At first, even the chemical structure of the enzyme's substrate was unknown! Class A enzymes were the primary focus of these early studies, as little was known at the time of other classes. Here we salute these people, all of whom made the work of this book's authors much easier.

INTRODUCTION

When writing a history, one has to specify beginning and ending dates of the period covered. The beginning date for the *recorded* history of our favorite enzyme is clear, December 1940, when Edward P. Abraham and Ernst B. Chain (Figures 1 and 2) in the Sir William Dunn School of Pathology at Oxford University published their one-page *Nature* paper about "An Enzyme from Bacteria Able to Destroy Penicillin" [1]. However, it can be claimed that this enzyme was first isolated 3 centuries earlier in 1689 by whomever collected, dried and cataloged certain plant specimens for the British Museum. The soil of these plants was later found by P. H. A. Sneath and Martin R. Pollock [2, 3] to contain spores of *Bacillus licheniformis* which, upon germination, yielded ß-lactamase-producing cells.

The ending date for this ß-lactamase history will arbitrarily be set to the early 1980s, before the enzyme was probed with newer techniques such as site-directed mutagenesis,

* E-mail: james.knox@uconn.edu; 860-429-8288

NMR, phylogeny analysis, and computational modeling. Another reason for chosing this ending date has to do with the way we now search the literature. The younger readers among us may not have read the early ß-lactamase literature nor have ventured to a library to find it. Rather, web-based search engines are used. Because many older publications have not been scanned into data bases, they are not found in web searches and therefore do not exist! It seems that only we older folks are aware of these foundation papers and perhaps have met the pioneering people who produced them. Therefore, this first chapter will attempt to bring some of these people and their classic labors to the attention of our younger colleagues.

FIRST ENCOUNTER

In 1939 Howard W. Florey's team at Oxford, while trying to produce large amounts of penicillin from the mold provided by Alexander Fleming in London, were often troubled by bacterial contamination which resulted in complete destruction of the penicillin as it formed [4]. These 'penicillin-destroying cultures' were examined by Edward Abraham and Ernst Chain, who set out to isolate and characterize the 'penicillin inhibitor'. They studied an extract from a crushed suspension of a '*B. coli*' (now *Escherichia coli*) strain not killed by penicillin. They believed the protective component in the cell extract was enzymic because it was non-dialysable, heat labile, papain inactivated, precipitated by alcohol, and had a pH-dependent penicillin-destroying activity.

Figure 1. Edward P. Abraham (1913-1999). Photo courtesy of Eric Sidebottom, Sir William Dunn School of Pathology, University of Oxford.

Figure 2. Ernst B. Chain (1906-1979). Photo courtesy of Eric Sidebottom, Sir William Dunn School of Pathology, University of Oxford.

Abraham and Chain reported that the enzyme was excreted by some bacteria, but not by others such as *B. coli*. They said it was absent from *Staphylococcus aureus*, yeast and *Penicillium notatum*. They found that some enzyme-producing bacteria were nevertheless sensitive to penicillin, and they concluded "..the presence or absence of the enzyme in a bacterium may not be the sole factor determining its insensitivity or sensitivity to penicillin." In the ending to this inaugural paper [1], Abraham and Chain state "The fact that a number of bacteria contain an enzyme acting on penicillin points to the possibility that this substance may have a function in their metabolism", an idea still being argued (See, for example, [5, 6]).

CHARACTERIZATION OF THE ENZYME - AND ITS SUBSTRATE!

Abraham and Chain named the new enzyme "penicillinase" rather than ß-lactamase because they had not yet established the chemical nature of its substrate.[1] Determination of the structure of penicillin was a considerable challenge to these biological chemists, who, like Fleming, at first believed the antibiotic was a protein. In any case, the penicillin-inactivating

[1] Their substrate was very impure. This 'penicillin' is now known to have been mostly penicillin F, the "Oxford" penicillin with a 2-pentenyl sidechain. Penicillin G, with the benzyl sidechain, was found later in the US from a different fermentation broth. Several years later in 1953 the cephalosporin series was found and developed by Abraham with Guy Newton. Not repeating the earlier British decision to forego the patenting of penicillin, only to see American firms do so on the basis that their penicillin G was different from Britain's penicillin F, Abraham obtained his own patents on several cephalosporins [7].

enzyme was not meant to be their primary focus in 1940. With a major war just beginning, and under pressure from their boss Howard Florey to produce larger amounts of the miracle antibiotic while he traveled to the US to seek help, the Oxford workers were supposed to work on penicillin, not penicillinase [7].

By 1943 Abraham and Chain had been able to deduce that penicillin was not a protein, but a small molecule containing a ß-lactam ring fused to a thiazolidine ring [8]. However, the prominent organic chemist at Oxford, Sir Robert Robinson, favored a more conventional unfused molecule of 2 separate rings, the so called oxazolone-thiazolidine structure. A third model, a tricyclic fused-ring, was proposed by F. Stokola at the Northern Regional Research Laboratory of the US Department of Agriculture in Peoria, Ill, where large scale fermentation was being tested [7, 9]. It was not until war's end in mid-1945 that the x-ray crystallographic structure of penicillin G by Dorothy Crowfoot, Charles Bunn and Barbara Rogers-Low at Oxford finally confirmed the fused bicyclic ß-lactam model [10]. These results foretold that total chemical synthesis of this unusual structure would be difficult. This task was taken on in the mid-50s by John Sheehan at the Massachusetts Institute of Technology [9]. His synthesis was successful, but was never used on a commercial scale.

With the chemical structure of penicillin finally established, many people set out to determine the nature of the breakdown catalysed by penicillinase. In December, 1943, Jackson Foster at the Merck Research Laboratory used a Warburg respirometer to detect CO_2 evolution from a bicarbonate solution and concluded that a new acid group appeared in the reaction product [11]. This finding was not published until 1945 and corrected the earlier negative finding by Abraham and Chain [1], who most likely used a low concentration of their very impure penicillin in their pH titrametry. Additional titrametic work in 1945 by Robert Benedict et al. in the USDA laboratory at Peoria, Illinois, showed that the new acid group in the product had a pK value 4.7, equivalent to that produced by chemical breakdown of penicillin by dilute alkali [12]. Thus, it was collectively established that penicilloic acid was the product of the inactivation by penicillinase.

Direct chemical assay methods could now be developed to replace the unwieldy microbiological methods based on the ability of a penicillinase-treated solution of penicillin to kill a standard bacterial culture. However, the chemical method used and the definition of activity units differed from lab to lab [13]. In addition to the manometric or titrimetric methods measuring the appearance of a new carboxylic group, other early methods included a microiodometric assay whereby the rate of penicilloic acid formation was determined by the decrease of iodine or iodine-starch absorption, a polarimetric assay measuring the change in optical rotation of penicilloic acid, or a colorimetric assay monitoring the formation of a complex of penicillin with hydroxamate and iron.

By the late 1940s, a penicillinase had not yet been purified to homogenetiy. Most bothersome was that many bacilli were found to produce both excreted and cell-bound forms of the enzyme (and in some cases both the serine and zinc-dependent enzymes). Another problem was the unusually strong adsorption to glass surfaces of some bacilli penicillinases, giving false indications of concentration, stability or activity unless gelatin was included. This adsorption to glass, however, provided a means of purifying and concentrating the enzyme from large volumes. Data on heat and pH stability, pH activity profile, and inactivation by reagents such as iodoacetic acid and azide were often conflicting [4]. Some preparations were contaminated with a penicillin amidase (or acylase) which produced 6-amino-penicillanic acid (6-APA) [14]. These factors made difficult the comparative evaluation of results from different laboratories, each using different bacterial sources, purification schemes, assays and activity units [15, 16].

For early quantitative studies of the enzyme, we are indebted to Riley D. Housewright (1913-2003) who, as a 30-year-old US Naval lieutenant with a PhD in microbiology, carried out studies of the enzyme at the US Army's Camp Detrick in Maryland in 1946.[2] After exploring and improving various production and purification methods for a *Bacillus cereus* enzyme, Housewright and Richard Henry made the first estimate of molecular mass, finding a value of 50,000 Da based on the diffusion coefficient measured by the porous diaphram method [17]. They confirmed the antigenic effects of the enzyme in rabbit serum, first demonstrated by Perlstein and Liebmann in 1945 [18], by using the resulting 'anti-penicillinase' to protect penicillin used therapeutically in animals [19].

In a series of experiments in 1947 on the physical and catalytic properties of the *B. cereus* enzyme, Henry and Housewright, using the manometry assay, found zero-order reaction kinetics in saturating penicillin concentrations [22] (Henri-Michaelis kinetics over a wide concentration range would not be demonstrated for another 10 years [23]). They found that the penicillinase had no metal dependence, that free amino or sulfhydryl groups are not required for activity, and that it had a pH optimum of 7.2.

Housewright's group probed the range of specificity of the enzyme by asking whether the N-CO bond in purines, pyrimidines, and similar naturally-occurring compounds might be attacked by the enzyme. They were the first to note the influence of the sidechain of penicillin on its rate of hydrolysis, finding that n-heptylpenicillin (penicillin K) was hydrolyzed at only 60-70% of the rate for either benzylpenicillin (penicillin G) or p-hydroxybenzylpenicillin (penicillin X). Another study of specificity a year later in 1948 by Otto Behrens and Mary Kingkade at the Lilly Laboratories in Indianapolis showed that 7 of Lilly's new biosynthetic penicillins were hydrolyzed much more slowly than penicillin G, but 11 were hydrolyzed up to 40% faster [24]. They also measured the effect of the new ß-lactams on a penicillin-resistant strain of *S. aureus* which did not produce a penicillinase.

Better purification procedures resulted in the first microcrystalline penicillinase from an unknown strain of *B. cereus* in Japan in 1950 [25] and of larger crystals from the 569 and 5B strains in 1956 [26]. This crystalline material was used for sedimentation and diffusion measurements by Hall and Ogston [27] to establish more accurate molecular masses and hydrodynamic parameters.

Into the 1950s no other compounds besides those containing the fused ß-lactam ring were found to be susceptible to the enzyme (but even this fact would be questioned in 1964). Neither dethiobenzylpenicillin nor the cyclic tripeptide N-phenylacetyl-L-cysteinyl-D-valine was hydrolyzed by the *B. cereus* enzyme [28]. In 1959 Batchelor et al. at Beecham

[2] Housewright also researched offensive and defensive biological warfare agents. During the war in 1943 Housewright was assigned to work on agriculture blights that could destroy German potatoes and Japanese rice, as well as anthrax and botulinum toxin for use against troops. Later in 1963, as scientific director of the Army's biological laboratories, he was consulted by the Kennedy Administration and Pentagon during the Cuban missile crisis about how biological agents might be used to incapacitate Soviet soldiers during an invasion of Cuba. In 1969, just before President Nixon banned offensive biological weapons work in the US, Nobel laureate Joshua Lederberg wrote to Housewright cautioning about work with biological agents. With foresight obvious to any microbiologist, he said "I am very worried about this, but hardly dare mention it for fear of putting an evil idea in someone's head." Housewright stepped down from the directorship in 1970 and in the early 1980s became the executive director of the American Society for Microbiology [20, 21].

Laboratories in the UK found that the just-isolated 6-APA nucleus was hydrolyzed, but at a rate slower than penicillin G [29].

As more ß-lactams with new sidechains were tested and found to be hydrolyzed to some extent, it was generally thought surprising that a single enzyme could have such a broad substrate specificity. Might there be more than one type of penicillinase in the preparations, or one enzyme species in different molecular states or forms? Nathan Citri in Israel concluded that a single *B. cereus* enzyme had both methicillin and benzylpenicillin activities, since the ratio of the 2 activities remained constant during purification, methicillin inhibited competitively the hydrolysis of benzylpenicillin, and partial inactivation by iodine or heat affected both activities equally [30].

Lee Sabath, working with Abraham, examined an induced *Pseudomonas pyocyanea* ß-lactamase and asked whether the penicillinase and cephalosporinase activities resided in a single enzyme [31]. Like Citri, he too found that the appearance of both activities followed a similar time course after induction, and that both activities could be inactivated in similar fashions. Mark Richmond at the National Institute for Medical Research at Mill Hill near London had drawn similar conclusions with the extracellular *S. aureus* enzyme [32]. By 1965 all studies conclusively showed that a single molecular species of ß-lactamase can hydrolyze a wide range of penicillins and cephalosporins. The early caution was well founded because it was known in 1956 that 2 types of ß-lactamase do in fact occur in the *B. cereus* 569/H strain. In 1966 Sabath established that 569/H produced a distinctive zinc-requiring ß-lactamase with very good activity on cephalosporin C [33]. This 'cephalosporinase' was later named ß-lactamase II to distinguish it from the serine-reactive ß-lactamase I. Kuwabara and Abraham characterized the substrate profile and properties of this metallocephalosporinase [34], the first member of what is now called class B.

Figure 3. Martin R. Pollock (1914-1999). Photo courtesy of Lindsay Sawyer and Graham Brown, Institute of Structural and Molecular Biology, University of Edinburgh.

We, as well as those outside the ß-lactamase field, owe much to the labors of Martin R. Pollock (Figure 3), who was active in the 1950s and 60s. While first at the National Institute for Medical Research in London, and from 1965 at the University of Edinburgh, he began studies of the immunological and catalytic properties of ß-lactamases. Pollock observed that "From a strictly enzymological standpoint, penicillinase has not been studied very intensively" prior to 1960 [14]. Recall that it was not until 1957 that Michealis-Menten kinetics was demonstrated for these enzymes [23].

To categorize the chemical makeup of ß-lactamases in the days before amino acid sequences, Pollock and his group determined amino acid compositions and the proportions of each type of amino acid in the various penicillinases. Chromatographic 'peptide maps' of trypsin digests were made and compared. Of great value in this analysis was immunological cross-reactivity. Antibodies raised in rabbits against a particular penicillinase were tested for their ability to precipitate other penicillinases. In this way, the relative affinity of one antibody for each penicillinase was used as a measure of molecular similarity. Another early method to distinquish and identify penicillinases was measurement of electrophoretic mobilities of the native enzymes on paper and in starch gels. Years later in the 1970s Roger Labia in France would identify penicillinases by exploiting isoelectric pH differences in high-resolution isoelectric focusing gels.

Pollock used ß-lactamase as the test enzyme in the first definitive studies of enzyme induction. Earlier experiments by others using impure ß-galactosidase were inconclusive as to the nature of the induced enzyme and its relation to the constitutive form. In the 1950s Pollock's group began by sorting out the confusing collection of the *B. cereus* enzymes, including the excreted and cell-bound forms, as well as the constitutive and inducible forms. Following upon Monod's 1951-53 papers on enzyme induction, Pollock used the highly purified 569 strain ß-lactamases, now available in large amounts, to demonstrate that induction with penicillin could increase enzyme levels by 300-fold. He emphasized that this induction was just as important as the enzyme activity itself in providing resistance, and he showed clearly that the induced 569 enzyme was equivalent to the constitutive 569/H enzyme [35]. In comparing specific activities, sedimentation and diffusion constants, electrophoretic mobility and salt solubility, he found the induced protein was equivalent to the one produced spontaneously. He concluded that induced and constitutive enzymes are formed by the same mechanism, in agreement with the hypothesis of Monod. In a remarkable 1957 paper, he shows considerable acumen by suggesting from his induction data that the "primary interaction between inducer and cell involves combination with a specific receptor". He first poses that "the specific combining groups of the penicillin receptor and of penicillinase may be structurally related", but considers it more likely that "the 'induction receptor' is identical with the 'fixation receptor' (or PBP in modern terms) [28].

P. H. A. Sneath in London in 1955 proposed another process by which penicillinase levels from Gram positives could be increased: In the presence of penicillin a spontaneous mutation leads to a derepression of penicillinase expression in some cells such that this population takes over the wildtype population [36]. Other early studies of the genetics of the Gram-positive enzymes have been described by Nathan Citri [13]. It was known since 1949 that the *Staph. aureus* ß-lactamase was usually lost during subculturing [37]. The use of transduction to restore production by means of bacteriophages was shown by Ritz and Baldwin in 1961 [38]. Richard P. Novick at Mill Hill showed that an unstable extra-chromosomal element or "plasmid" was the genetic determinant of ß-lactamase production in the *S. aureus* population [39]. This plasmid was lost every 1000 divisions and was said to be similar to the resistance-transfer factors (RTF) described in 1963 by Watanabe [40].

Only a few years later in 1965 the resistance-factor mediated production of penicillinase in Gram-negatives was found by Naomi Datta in London and Polyxeni Kontomichalou in Athens working with the *E. coli* TEM strain [41]. Its production was unlike the non-infective transduction process in the *Staph. aureas* strains. They found that penicillin resistance in the enterobacteriaceae is transferred by direct cell-to-cell conjugation, discovered earlier for other antibiotics in Japan in 1960. The R-factor periplasmic enzyme was characterized biochemically in 1966 by Datta and Richmond [42]. Their estimate of molecular mass (16,700 Da) was far from the correct value, however. The enzyme's 'physiological efficiency' (see below) on penicillin G turnover was similar to that of the bacillary enzymes, but very different from the (class C) enzymes from other constitutive *E. coli* strains.

Even into the 1960s the 'real' function of the enzyme was the subject of mild debate [3]. There was reluctance among some researchers to accept the hypothesis that antibiotic destruction was the primary job of the enzyme, despite its extremely high activity against penicillins. Might the ability to turnover hundreds of molecules of penicillin per minute just be an unfortunate happen-chance? The basis for this uncertainty was the claim by Arthur Saz at the US National Institutes of Health that the production of this enzyme in *S. aureus* could be induced by cyclic peptides and cell wall extracts (as well as penicillins) [43], and that the enzyme was inhibited by dipeptides containing D-valine [44]. Most of these disconcerting data, however, could not be reproduced later when highly purified material was used. In asking about the physiological roles of penicillins and penicillinase, Pollock introduced the idea of struggle for survival between the microfungi and bacteria [3], having himself demonstrated that a culturing together of *Penicillium chrysogenum* and *B. licheniformis* in the same soil sample produced considerably lower populations of each compared to their number if grown alone. He asked why penicillinase, if it has some other function in the cell, is found only in organisms having peptidoglycan, the synthesis of which is stopped by penicillin.

Catalytic comparison of two biochemically similar exocellular penicillinases from two strains of *B. licheniformis*, 749/C and 6346/C, led Pollock to question the surprising 6-fold difference in their specific activities against penicillin G . He noted that the affinities of the two for the antibiotic also differed by 6-fold, but in the opposite direction. Why would these 2 penicillinases, with undetectable differences in size, amino acid composition and antisera reactivity, have such different catalytic behavior against the same substrate? Believing that this reciprocal relationship between specific activity and affinity was unlikely to be coincidental, he concluded that at the very low antibiotic concentrations they were likely to encounter *in vivo*, below their Km, the 2 penicillinases would function against penicillin G at the same rate. To normalize the very different activities, Pollock introduced in 1965 the term "physiological efficiency" or Vmax/Km [45], and in this way he showed that under natural conditions the 2 penicillinases have the same ability to hydrolyse a common substrate.

Another anomaly addressed by Pollock was the much lower activities of the released cell-bound gram-negative enzymes compared to the excreted gram-positive enzymes. He realized that an intrinsic resistance in the gram-negatives, provided by the outer membrane and the enzyme's periplasmic location near the cell wall synthesizing enzymes, would explain this discrepancy. The very different *in vivo* and *in vitro* activities, if represented with the *ratio* of Vmax to Km, which took into account the low substrate levels normally existing *in vivo*, now showed how similar the cell-bound and excreted enzymes were. As current readers may know, Pollock's "physiological efficiency" morphed into "catalytic efficiency" with kcat replacing Vmax, and the ratio has become a part of the conventional terminology of enzymology.

Another pioneer examining the catalytic behavior of ß-lactamases was Stephen G. Waley in the Dunn School of Pathology at Oxford (Figure 4). As a graduate student at Oxford in 1943-45, he had studied in the organic chemistry group of Prof. Robinson, whom we mentioned above as one trying to establish the chemical structure of penicillin. The thesis project given Waley was to synthesize the two-ring oxazolone-thiazolidine structure proposed by Robinson for penicillin, incorrectly it turned out! Waley himself liked to describe his thesis as "a failed attempt to synthesize the wrong structure for penicillin".

By the 1970s and through the 80s, Waley was expanding upon the ß-lactamase enzymology started by Abraham and Pollock [46]. He developed a useful spectrophotometric assay for penicillin hydrolysis based on an extinction change in the UV region [47], not knowing a similar assay had been published 9 years earlier by Jansson for ampicillin hydrolysis [48]. Using primarily ß-lactamases I (serine) and II (zinc) from *B. cereus*, he studied catalytic properties with quantitative kinetics [49, 50], chemical modification, and the careful selection of substrates to reveal reaction paths and mechanisms.

Before complete amino acid sequences were known, chemical modification of reactive sidechains, either by chemical agents or by modified ß-lactam substrates, was one of the few tools available to Waley and others for probing the identity of active site groups. The early papers were often contradictory, and there were many false indications. For example, a 1964 hypothesis from Dupre et al. [51] said that a histidine may function as a proton donor in the *S. aureus* enzyme. The fact that penicillin sidechains with a positively charged group had a higher Km led them to propose that a similar charge may be present in the active site. But others showed that the histidine-binding reagent diethylpyrocarbonate did not reduce activity [52]. An arginine residue was implicated by phenylglyoxal inactivation of the *B. cereus* I enzyme by Waley [46], and a conserved tyrosine residue (105) was thought by David Thatcher to be important because modification by tetranitromethane and iodine reduced activity of the enzyme by 30% [53]. However, this arginine was shown to be nonessential for catalysis because the modified *S. aureus* enzyme showed no loss of activity [46, 54]. Waley used chemical modification and the analysis of the pH dependence of kinetic parameters to prove the participation of a carboxylic acid group in the *B. cereus* I enzyme [49]. The group of Ogawara was also active in the use of chemical modification and site-selective ß-lactams [55]. Reacting the *B. cereus* I enzyme with an irreversible inhibitor, 6-ß-bromopenicillanic acid, the mechanism of which had been studied by Rex Pratt and David Loosemore [56], Waley established its binding to a serine sidechain [57]. A few years later he also identified a reactive serine in the class C enzymes [58].

Stephen Waley examined the zinc-dependent 'cephalosporinase' (or ß-lactamase II) found in *B. cereus* by Abraham in the mid 1950s [34]. Waley's NMR and UV spectroscopy of the Co-substituted enzyme led to identification of 3 histidines and a cysteine around one of the two zinc centers. He used cryogenic cooling to extend the lifetime of non-covalent intermediates and proposed a branched pathway mechanism in these enzymes. This idea was confirmed at room temperatures using rapid-scanning stopped-flow spectrophotometry. The editor of this book, JMF, notes that Stephen Waley was an influential colleague who, through years of collaboration and personal friendship, helped and supported his graduate students at Liege in the study of enzyme kinetics and mechanism.

The Family Tree

The first data on the question of penicillinase evolution comes from Martin Pollock's (unpublished) 1967 analysis of 300 year old British Museum spores of *B. licheniformis*. Upon germination, the spores yielded cells which were found to produce a penicillinase with the same composition and properties found in the enzyme from modern bacilli. He was able to use sequential mutagen treatment to generate a mutant with an inverted substrate profile, hydrolyzing cephalosporin C better than penicillin G [3], 10 years before the Hall and Knowles chemical mutagenesis of *E. coli* TEM in 1976 [59]). Pollock postulated that a single amino acid substitution had resulted in the alteration in substrate specificity (not knowing the wildtype or mutant sequences).

Urgently needed sequence data were about to be generated by the University of Edinburgh group through the efforts of Richard P. Ambler (Figure 5). In the late 1960s Ambler was introduced to penicillinases by Pollock, now head of one of the first teaching departments of 'Molecular Biology'. Ambler and his students R. J. Meadway and David R. Thatcher began the tediuous work of determining complete amino acid sequences of 3 (class A) ß-lactamases [60].

Up to this time, only the sequence of short N-terminal peptides were determined and used for identification, and the detection of a single N-terminal residue in paper chromatography was commonly used as an indicator of purity. Ambler's team had to break the sequencing job down into several smaller ones by cleaving each penicillinase with CNBr or digesting each one with trypsin, chymotrypsin or pepsin into 4-5 overlapping blocks of peptide. Each block could then be sequenced by Edman degradation.

After obtaining most of the sequence of the *S. aureus* PC1 and *B. licheniformis* 749/C enzymes, they found a good match of the two by inspection. With the help of Ambler's wife, Pat, who devised a computerized alignment method, they optimized the alignment by incorporating 'conservative substitutions' in the matching. They also calculated match probabilities based on random sequences generated from the amino acid composition. In a prelude to what has become common methodology, they wrote: "We recommend the use of the computer technique early in the investigation of the structure of a protein if any sequences which may be homologous are already known".

This matching produced a similarity near 40%, but the significance of this percentage was unclear at a time when little or no data about sequence homologies were available for bacterial proteins from different genera. They pondered various hypotheses about the ancestry and evolution of the penicillinases in these genera. They later concluded that the penicillinases evolved in a single bacterial line and have reached the other lines by processes of intrageneric gene transfer [61].

With the *B. cereus* sequence, knowledge of 3 complete (class A) sequences simplified their search for catalytic groups to only those residues which were common to all sequences. Ambler found that a serine was conserved in a Ser-Xaa-Xaa-Lys motif. The absence of a conserved histidine ruled out an Asp-His-Ser triad like the charge relay system in the serine proteases.

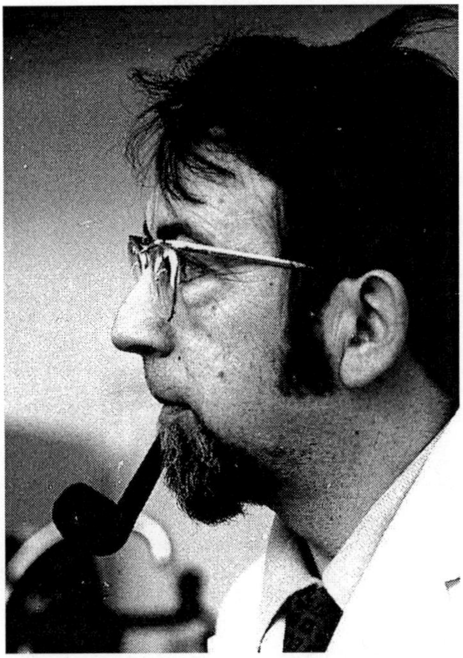

Figure 4. Stephen G. Waley (1922-1993). Photo courtesy of Eric Sidebottom, Sir William Dunn School of Pathology, University of Oxford.

The plasmid-encoded TEM ß-lactamase has the distinction of being the first enzyme sequenced from the DNA. Further, its hydrophobic 23-residue N-terminal sequence was identified as the first 'signal' sequence of a nascent protein. The background story of this TEM work is interesting.

Figure 5. Richard P. Ambler (1933-). Photo courtesy of Lindsay Sawyer, Institute of Structural and Molecular Biology, University of Edinburgh.

After Ambler and G. K. Scott had been struggling with the sequencing of the TEM enzyme from R6K (TEM2) by traditional methods for 3 years, Walter Gilbert at Harvard and his student J. G. Sutcliffe began using their new DNA sequencing methodolgy on the TEM gene in pBR322 (TEM1) in early 1977. The author, while on sabbatical in the same department at this time, was approached by Sutcliffe and asked how our crystallographic analysis of TEM was progressing without knowledge of the sequence [62, 63]. Sutcliffe believed he would have the TEM sequence for us within the year, and well before Ambler was finished in Edinburgh! (This boast was met with some skepticism.) Both sequencing groups agreed to carry on a friendly, blind competition in which their intermediate results would be held in Harvard's Chemistry Department by Jeremy R. Knowles, who had begun his own mechanistic studies of the class A ß-lactamases. In September 1977 Sutcliffe and Ambler met at the home of Prof. Knowles to compare the nucleotide and amino acid sequences. The results were found to be in broad agreement except for a few minor errors and unassigned regions in the Ambler sequence. However, a glaring difference at position 39 could not be resolved until it was realized that the Ambler group had used an *E. coli* producing a different isotype of the enzyme, and thus were discovered the TEM1 and TEM2 variants with glutamine and lysine, respectively. Sutcliffe and Ambler published their papers side-by-side in PNAS in 1978, though Ambler's sequence was not quite complete [64, 65]. The two TEM sequencing papers were significant in signaling the end of an old methodology and the beginning of a new one. Soon after, with more sequences, Ambler proposed the classification of the known ß-lactamases into 2 classes A and B [66]. A search for ancestral sequences would now begin.

ANCESTRY

The investigation of penicillin-binding proteins (PBPs) has its own history, which we will briefly relate to our ß-lactamase history. When probing the mode of action of penicillin in 1949, P. D. Cooper and D. Rowley at St. Mary's Hospital in London and Eleanor Maass at the University of Wisconsin independently found that growing bacteria would irreversibly bind S35-labelled penicillin [67, 68]. The London workers concluded that "Whatever the nature of the primary site of penicillin action, it is clear…that rapid growth exposes more centres in the bacteria with which the penicillin can react. The combination must either be by unusually strong adsorptive or by chemical forces." Cooper later said that penicillin must bind to a discrete 'penicillin-binding component' or PBC [69]. Although a very early 1942 study showed no adsorption or absorption of penicillin by bacterial cells, possibly not actively growing, the 1949 studies found that 500-1000 molecules of penicillin were irreversibly taken up by *S. aureus*. The first suggestion that penicillin inhibited the cross-linking of the 'mucopolymer' was due to Hans Martin in 1964 [70]. In the decade of the 1960s, Jack L. Strominger (Harvard) and Jean-Marie Ghuysen (Liège) were successful in demonstrating that the PBPs are involved in peptidoglycan synthesis and cross-linking [71-73]. Several workers proposed that penicillin is inhibitory because it may resemble a peptidoglycan component [74-76].

Martin Pollock's 1967 ideas about the origin of ß-lactamases had focused on these cell wall associated enzymes because both types of enzyme recognized and interacted with ß-lactams [3]. In 1976, penicillin was shown to inactivate the *Streptomyces* R61 penicillin-sensitive D-alanine-D-alanine carboxypeptidase-transpeptidase by modifying a serine residue [77], a finding which was later shown to be valid for all known PBPs. The identification of the class A ß-lactamases' active-site serine rapidly followed [57].

In 1979 short N-terminal amino acid sequences of penicillin-labelled PBPs became available for D-alanine-D-alanine carboxypeptidases from *B. subtilis* and *B. stearothermophilus* [78]. Alignment of these labelled PBP peptides with the 4 known class A ß-lactamase sequences exhibited a promising homology and a common serine, fortunately just within the forty-to-seventy-residue N-terminal segments obtainable by chemical sequencing [66, 79]. Shortly thereafter, our calculation of secondary structure potential for these few sequences predicted that tertiary structural homology exists between the PBPs and ß-lactamases [80, 81]. Our crystallization and x-ray analysis of J.-M. Ghuysen's and J.-M. Frère's PBP from *Streptomyces* R61 revealed for the first time a PBP folding motif, a ß-lactam binding site, and a second sequence motif K/H-T/S-G [82-84], all found 5 years later in the crystal structures of the serine ß-lactamases.

Conclusion

In May of 1979 a packed Discussion Meeting of the Royal Society on the subject 'Penicillin Fifty Years after Fleming' was held in London [85]. While many speakers dealt with the synthesis and activity spectrum of new ß-lactams, the more exciting talks (in the author's opinion) dealt with penicillin's mechanism of action, ß-lactamase diversity and inhibition, genetic aspects of ß-lactam resistance, and enzymes of cell wall synthesis. Unfortunately, Professors Abraham and Pollock gave only brief closing remarks. Ambler presented his sequence data and classification scheme. At the interval, I chatted with Professor Pollock and told him how his 1967 review paper [3] had sparked my interest in ß-lactamases in 1970. Ernst Chain was present at the meeting, but he appeared to be in poor health and did not speak. I vividly remember seeing the 73-year-old gentleman during a break in the meeting. Dressed in a dark suit, smiling but frail and unsteady, he slowly walked alone up a side isle of the crowded auditorium. Conversations quieted as he was followed by every eye in the room. He died several months later.

And finally, let us return to the late 1940s when there was some uncertainty about the enzyme's 'real' function and microbial distribution. Apparently, penicillinase was not considered by all to be a threat to the therapeutic use of penicillin, as shown by this comment from a clinician at the Boston University Medical School: "Fortunately, the development of so-called resistance or penicillin-resistant strains of bacteria has played little or no role in penicillin failures" [86]. Because many organisms which produced penicillinase were nevertheless still sensitive to certain ß-lactams, a 1949 review by Chain concluded "..there seems to be little evidence as to the relation between sensitivity and penicillinase production in any one strain." [15]. No doubt, with publication embargoes during the war years, poor understanding of induction effects, intrinsic resistance, substrate specificity, and the like, full comprehension of the growing biochemical and microbiological data base was lacking. It seems the enzyme was viewed by some with curiosity more than alarm, and only its practical applications for destroying residual penicillin in blood cultures or for treatment of penicillin allergic reactions were pursued.

But well before these statements were made, Alexander Fleming was quite alarmed. Though he failed to follow through on his 1928 discovery of penicillin, and was not an active participant in ß-lactamase history, he made a prescient comment during his Nobel lecture in 1945: "It is not difficult to make microbes resistant to penicillin in the laboratory… The time may come when penicillin can be bought by anyone in the shops. Then there is the danger that the ignorant man may easily underdose himself and, by exposing his microbes to nonlethal quantities of the drug, make them resistant" [87]. Most of the following chapters in this book will describe just how effectively these resistance mechanisms have developed in the 65 years since his lecture.

REFERENCES

[1] Abraham, E. P. and Chain, E. B. (1940) An enzyme from bacteria able to destroy penicillin. *Nature* 146, 837.

[2] Sneath, P. H. A. (1962) Longevity of microorganisms. *Nature* 195, 643-646.

[3] Pollock, M. R. (1967) Origin and function of penicillinase: a problem in biochemical evolution. *Brit. Med.* J. 4, 71-77.

[4] McQuarrie, E. B. and Liebmann, A. J. (1944) Studies on penicillinase. *Arch. Biochem.* 5, 307-315.

[5] Medeiros, A. A. (1997) Evolution and dissemination of ß-lactamases accelerated by generations of ß-lactam antibiotics. *Clinical Infectious Diseases* 24, S19-45.

[6] Hall, B. G. and Barlow, M. (2004) Evolution of the serine ß-lactamases: past, present and future. *Drug Resistance Updates* 7, 111-123.

[7] Lax, E. (2004) *The Mold in Dr. Florey's Coat.* Henry Holt and Co., New York, NY.

[8] Chain, E. B. (1946) T*he chemical structure of the penicillins.* Nobel Foundation, Stockholm, Sweden.

[9] Sheehan, J. C. (1982) *The Enchanted Ring: The Untold Story of Penicillin.* The MIT Press, Cambridge, MA.

[10] Crowfoot, D., Bunn, C. W., Rogers-Low, B. W. and Turner-Jones, A. (1949) The x-ray crystallographic investigation of the structure of penicillin. In *The Chemistry of Penicillin* (Clarke, H. T., Johnson, J. R. and Robinson, R., eds.), Princeton University Press, Princeton, NJ.

[11] Foster, J. W. (1945) Acid formation from penicillin during enzymatic inactivation. *Science* 101, 205.

[12] Benedict, R. G., Schmidt, W. H. and Coghill, R. D. (1945) Penicillin VII. Penicillinase. *Arch. Biochem.* 8, 377-384.

[13] Citri, N. and Pollock, M. R. (1966) The biochemistry and function of ß-lactamase (penicillinase). *Adv. Enzymol.* 28, 237-323.

[14] Pollock, M. R. (1960) *Penicillinase. In The Enzymes* (Boyer, P. D., ed.), pp. 269-278, Academic Press, New York.

[15] Chain, E. B., Florey, H. W., Heatley, N. G. and Jennings, M. A. (1949) Penicillinase. In *Antibiotics* (Florey, H. W., Chain, E. B., Heatley, N. G., Jennings, M. A., Sanders, A. G., Abraham, E. P. and Florey, M. E., eds.), pp. 1090-1110, Oxford University Press, London.

[16] Abraham, E. P. (1951) Penicillinase. In *The Enzymes* (Sumner, J. B. and Myrback, K., eds.), pp. 1170-1185, Academic Press, New York, NY.
[17] Housewright, R. D. and Henry, R. J. (1947) Studies on penicillinase I. The production, partial purification, and practical application of penicillinase. *J. Biol. Chem.* 167, 553-557.
[18] Perlstein, D. and Liebmann, A. J. (1945) The production of anti-penicillinase immune sera. *Science* 102, 197-199.
[19] Housewright, R. D. and Henry, R. J. (1947) Studies on penicillinase III. The effect of antipenicillinase on penicillin-resistant organisms. *J. Bacteriol.* 53, 241-247.
[20] Martin, D. (2003) Riley Housewright, Microbiologist, Dies at 89. *The New York Times,* New York, NY, Obituaries January 16, 2003.
[21] Hitt, E. (2003) Bio-warfare research pioneer dies. *The Scientist January* 24, 2003.
[22] Henry, R. J. and Housewright, R. D. (1947) Studies on penicillinase II. Manometric method of assaying penicillinase and penicillin. Kinetics of the penicillin-penicillinase reaction, and the effects of inhibitors on penicillinase. *J. Biol. Chem.* 167, 559-571.
[23] Banfield, J. E. (1957) Kinetics of penicillinase activity. Experientia 13, 403-404.
[24] Behrens, O. K. and Kingkade, M. J. (1948) Biosynthesis of penicillins VIII. Studies with new biosynthetic penicillins on penicillin resistance. *J. Biol. Chem.* 176, 1047-1050.
[25] Murao, S. (1950) *Symposia on Enzyme Chemistry* (Japan) 5, 81-94.
[26] Pollock, M. R., Torriani, A.-M. and Tridgell, E. J. (1956) Crystalline bacterial penicillinase. *Biochem.* J. 62, 387-391.
[27] Hall, J. R. and Ogston, A. G. (1956) Sedimentation and diffusion of samples of penicillinase. *Biochem. J.* 62, 401-403.
[28] Pollock, M. R. (1957) The activity and specificity of inducers of penicillinase production in *Bacillus cereus*, strain NRRL 569. *Biochem. J.* 66, 419-428.
[29] Batchelor, F. R., Doyle, F. P., Nayler, J. H. C. and Rolinson, G. N. (1959) Synthesis of penicillin: 6-aminopenicillanic acid in penicillin fermentations. *Nature* 183, 257-258.
[30] Ron-Zenziper, E. and Citri, N. (1963) Modified substrate specificity of cell-bound penicillinase. *Nature* 198, 887-888.
[31] Sabath, L. D., Jago, M. and Abraham, E. P. (1965) Cephalosporinase and penicillinase activities of a ß-lactamase from *Pseudomonas pyocyanea. Biochem. J.* 96, 739-743.
[32] Richmond, M. H. (1963) Purification and properties of the exopenicillinase from *Staph. aureus*. *Biochem. J.* 88, 452-459.
[33] Sabath, L. D. and Abraham, E. P. (1966) Zinc as a cofactor for cephalosporinase from *Bacillus cereus* 569. *Biochem. J.* 98, 11c-13c.
[34] Kuwabara, S. and Abraham, E. P. (1967) Some properties of two extracellular ß-lactamases from *Bacillus cereus* 569/H. *Biochem. J.* 103, 27c-30c.
[35] Kogut, M., Pollock, M. R. and Tridgell, E. J. (1956) Purification of penicillin-induced penicillinase of *Bacillus cereus* NRRL 569: A comparison of its properties with those of a similarly purified penicillinase produced spontaneously by a constitutive mutant strain. *Biochem. J.* 62, 391-401.
[36] Sneath, P. H. A. (1955) Proof of the spontaneity of a mutation to penicillinase production in *Bacillus cereus*. *J. Gen. Micro.* 13, 561-568.
[37] Barber, M. (1949) The incidence of penicillin-sensitive variant colonies in penicillinase-producing strains of *Staph. pyogenes*. *J. Gen. Micro.* 3, 274-281.

[38] Ritz, H. L. and Baldwin, J. N. (1961) Transduction of capacity to produce Staphylococcus penicillinase. *Proc. Soc. Exptl. Biol. Med.* 107, 678-680.

[39] Novick, R. P. (1963) Analysis by transduction of mutations affecting penicillinase formation in *Staph. aureus*. *J. Gen. Micro*. 33, 121-136.

[40] Watanabe, T. (1963) Infection heredity of multiple drug resistance in bacteria. *Bact. Rev.* 27, 87-115.

[41] Datta, N. and Kontomichalou, P. (1965) Penicillinase synthesis controlled by infectious R factors in enterobacteriaceae. *Nature* 208, 239-241.

[42] Datta, N. and Richmond, M. H. (1966) The purification and properties of a penicillinase whose synthesis is mediated by an R-factor in *Escherichia coli*. *Biochem. J.* 98, 204-209.

[43] Saz, A. K. and Lowery, D. L. (1964) Staphylococcal penicillinase. II Non-penicillin-like cyclic peptides as inducers of, and substrates for, the enzyme. *Biochem. Biophys. Res. Commun.* 15, 525-530.

[44] Saz, A. K., Lowery, D. L. and Jackson, L. J. (1961) Staphylococcal penicillinase. I Inhibition and stimulation of activity. *J. Bacteriol.* 82, 298-304.

[45] Pollock, M. R. (1965) Purification and properties of penicillinases from two strains of *Bacillus licheniformis*: a chemical, physiochemical and physiological comparison. *Biochem. J.* 94, 666-675.

[46] Abraham, E. P. and Waley, S. G. (1979) ß-Lactamases from *Bacillus cereus*. In *ß-Lactamases* (Hamilton-Miller, J. M. T. and Smith, J. T., eds.), pp. 311-338, Academic Press, London.

[47] Waley, S. G. (1974) A spectrophotometric assay of ß-lactamase action on penicillins. *Biochem. J.* 139, 789-790.

[48] Jansson, J. A. T. (1965) A direct spectrophotometric assay for penicillin ß-lactamase (penicillinase). *Biochim. Biophys. Acta* 99, 171-172.

[49] Waley, S. G. (1975) The pH-dependence and group modification of ß-lactamase I. *Biochem. J.* 149, 547-551.

[50] Waley, S. G. (1992) ß-Lactamase: Mechanism of action. In *The Chemistry of ß-Lactams* (Page, M. I., ed.), pp. 198-228, Blackie Academic & Professional, London.

[51] Depue, R. H., Moat, A. G. and Bondi, A. (1964) The relation between penicillin structure and penicillinase activity. *Arch. Biochem. Biophys.* 107, 374-381.

[52] Ferencz, I., Mile, I. and Czanyi, V. (1971) The role of the histidine side chains in the catalytic activity of *B. cereus* exopenicillinase. *Acta Biochim. Biophys. Acad. Sci. Hung.* 6, 5-7.

[53] Thatcher, D. R. (1975) ß-Lactamases (*Bacillus cereus*). Meth. Enzymol. 43, 640-652.

[54] Bristow, A. F. and Virden, R. (1978) Preferential nitration with tetranitromethane of a specific tyrosine residue in penicillinase from *Staphylococcus aureus* PC1. *Biochem. J.* 169, 381-388.

[55] Ogawara, H. and Umezawa, H. (1975) *Bacillus cereus* ß-lactamase. Reaction with N-bromosuccinamide and the properties of the product. *Biochim. Biophys. Acta* 391, 435-447.

[56] Pratt, R. F. and Loosemore, M. J. (1978) 6-ß-Bromopenicillanic acid, a potent ß-lactamase inhibitor. *Proc. Natl. Acad. Sci. U.S.A.* 75, 4145-4149.

[57] Knott-Hunziker, V., Waley, S. G., Orlek, B. S. and Sammes, P. G. (1979) Penicillinase active sites: Labelling of serine-44 in ß-lactamase I by 6ß-bromopenicillanic acid. *FEBS Lett.* 99, 59-61.

[58] Knott-Hunziker, V., Petursson, S., Jayatilake, G. S., Waley, S. G., Jaurin, B. and Grunstrom, T. (1982) Active sites of ß-lactamases. The chromosomal ß-lactamases of *Pseudomonas aeruginosa* and *Escherichia coli*. *Biochem. J.* 201, 621-627.

[59] Hall, A. and Knowles, J. R. (1976) Directed selective pressure on a ß-lactamase to analyse molecular changes involved in development of enzyme function. *Nature* 264, 803-804.

[60] Ambler, R. P. and Meadway, R. J. (1969) Chemical structure of bacterial penicillinases. *Nature* 222, 24-26.

[61] Ambler, R. P. (1979) Amino acid sequences of ß-lactamases. In *Beta-Lactamases* (Hamilton-Miller, J. M. T. and Smith, J. T., eds.), pp. 99-125, Academic Press, London.

[62] Knox, J. R., Zorsky, P. E. and Murthy, N. S. (1973) Preliminary crystallographic data for *Escherichia coli* TEM ß-lactamase. *J. Mol. Biol.* 79, 597-598.

[63] Knox, J. R., Kelly, J. A., Moews, P. C. and Murthy, N. S. (1976) 5.5 Å crystallographic structure of penicillin ß-lactamase and radius of gyration in solution. *J. Mol. Biol.* 104, 865-875.

[64] Ambler, R. P. and Scott, G. K. (1978) Partial amino acid sequence of penicillinase coded by *Escherichia coli* plasmid R6K. *Proc. Natl. Acad. Sci. U.S.A.* 75, 3732-3736.

[65] Sutcliffe, J. G. (1978) Nucleotide sequence of the ampicillin resistance gene of *Escherichia coli* plasmid pBR322. *Proc. Natl. Acad. Sci. U.S.A.* 75, 3737-3741.

[66] Ambler, R. P. (1980) The structure of ß-lactamases. *Phil. Trans. Roy. Soc. Lond.* B289, 321-331.

[67] Cooper, P. D. and Rowley, D. (1949) Investigations with radioactive penicillin. *Nature* 163, 480-481.

[68] Maass, E. A. and Johnson, M. J. (1949) Penicillin uptake by bacterial cells. *J. Bacteriol.* 57, 415-422.

[69] Cooper, P. D. (1956) Site of action of radiopenicillin. *Bact. Rev.* 20, 28-47.

[70] Martin, H. H. (1964) Composition of the mucopolymer in cell walls of the unstable and stable L-form of *Proteus mirabilis*. *J. Gen. Micro.* 36, 441-450.

[71] Izaki, K., Matsuhashi, M. and Strominger, J. L. (1966) Glycopeptide transpeptidase and D-alanine carboxypeptidase: penicillin-sensitive enzymatic reactions. *Proc. Natl. Acad. Sci. U.S.A.* 55, 656-663.

[72] Araki, Y., Shimada, A. and Ito, E. (1966) Effect of penicillin on cell wall mucopeptide synthesis in an *Escherichia coli* particulate system. *Biochem. Biophys. Res. Commun.* 23, 518-525.

[73] Ghuysen, J.-M. (1968) Use of bacteriolytic enzymes in determination of wall structure and their role in cell metabolism. Ba*ct. Rev.* 32, 425-464.

[74] Collins, J. F. and Richmond, M. H. (1962) A structural similariity between N-acetylmuramic acid and penicillin as a basis for antibiotic action. *Nature* 195, 142-143.

[75] Wise, E. M. and Park, J. T. (1965) Penicillin: Its basic site of action as an inhibitor of a peptide cross-linking reaction in cell wall mucopeptide synthesis. *Proc. Natl. Acad. Sci.* U.S.A. 54, 75-82.

[76] Tipper, D. J. and Strominger, J. L. (1965) Mechanism of action of penicillins: a proposal based on their structural similarity to acyl-D-alanyl-D-alanine. *Proc. Natl. Acad. Sci. U.S.A.* 54, 1133-1141.

[77] Frere, J.-M., Duez, C., Ghuysen, J.-M. and Vandekerkhove, J. (1976) Occurrence of a serine residue in the penicillin-binding site of the exocellular DD-carboxy-peptidase-transpeptidase from *Streptomyces* R61. *FEBS Lett.* 70, 257-260.

[78] Yocum, R. R., Waxman, D. J., Rasmussen, J. R. and Strominger, J. L. (1979) Mechanism of penicillin action: penicillin and substrate bind covalently to the same active site serine in two bacterial D-alanine carboxypeptidases. *Proc. Natl. Acad. Sci.* U.S.A. 76, 2730-2743.

[79] Waxman, D. J., Yocum, R. R. and Strominger, J. L. (1980) Penicillins and cephalosporins are active site-directed acylating agents: evidence in support of the substrate analogue hypothesis. *Phil. Trans. Roy. Soc. Lond.* B289, 257-271.

[80] Moews, P. C. and Knox, J. R. (1979) Predicted secondary structures of four penicillin ß-lactamases and a comparison with two lysozymes. *Int. J. Peptide Protein Res.* 13, 358-393.

[81] Moews, P. C., Knox, J. R., Waxman, D. J. and Strominger, J. L. (1981) Secondary structure relations between ß-lactamases and penicillin-sensitive D-alanine-carboxypeptidases. *Int. J. Peptide Protein Res.* 17, 211-218.

[82] Knox, J. R., DeLucia, M. L., Murthy, N. S., Kelly, J. A. and Moews, P. C. (1979) Crystallographic data for a penicillin receptor: Exocellular DD-carboxypeptidase-transpeptidase from *Streptomyces* R61. *J. Mol. Biol.* 127, 217-218.

[83] Kelly, J. A., Moews, P. C., Knox, J. R., Frere, J.-M. and Ghuysen, J.-M. (1982) Penicillin target enzyme and the antibiotic binding site. *Science* 218, 479-481.

[84] Kelly, J. A., Boyington, J. C., Moews, P. C., Knox, J. R., Dideberg, O., Charlier, P., Libert, M., Wery, J.-P., Duez, C., Joris, B., Dusart, J., Frere, J.-M. and Ghuysen, J.-M. (1987) Crystallographic comparisons of penicillin-binding enzymes and studies of antibiotic binding. In *Frontiers of Antibiotic Research,* pp. 327-337, Academic Press, Tokyo.

[85] Baddiley, J. and Abraham, E. P. (1980) *Penicillin Fifty Years After Fleming.* pp. 212, The Royal Society, London.

[86] Keefer, C. S. (1949) The clinical use of penicillin and streptomycin. In *Antibiotics* (Irving, G. W. and Herrick, H. T., eds.), pp. 119-128, Chemical Publishing Co., Brooklyn, NY.

[87] Fleming, A. (1945) *Penicillin. Nobel Foundation,* Stockholm, Sweden.

PART 1: STRUCTURAL DATA

In: Beta-Lactamases
Editor: Jean-Marie Frère, pp. 21-39

ISBN: 978-1-61324-638-2

Chapter 2

STRUCTURE OF CLASS A BETA-LACTAMASES

Wladimir Sougakoff
Laboratoire de Bactériologie-Hygiène (ER5, LRMA),
Faculté de Médecine Pitié-Salpêtrière, Université Pierre et Marie Curie,
UPMC, Sorbonne Universités, Paris, France.

ABSTRACT

Class A contains the largest number of β-lactamases. Despite a very little number of residues conserved in all these enzymes, the structures are strikingly similar. The first clinically relevant enzyme was produced by *Staphylococcus aureus*. It was essentially a penicillinase and never evolved to hydrolyse other compounds such as cephalosporins or even semi-synthetic penicillins such as methicillin. By contrast, the TEM and SHV enzymes produced by Gram-negative strains evolved to hydrolyse third-generation cephalosporins or to become more resistant to inactivation by clavulanate, sulbactam or tazobactam. A large number of 3D structures have been solved by X-ray crystallography. Some of them are described in this chapter where we also attempt to establish structure-function relationships.

INTRODUCTION

The class A beta-lactamases represent the major defence mechanism produced by gram-negative and gram-positive bacteria against beta-lactam antibiotics [1]. These enzymes have widely evolved in response to the massive use of beta-lactam antibiotics. To date, there is a plethora of class A beta-lactamases that can confer resistance to beta-lactam antibiotics with different substrate spectra. Understanding the molecular diversity of these enzymes requires a brief review of the functional classification of class A beta-lactamases that have been organized into various groups on the basis of their substrate spectrum and response to inhibitors [2].

The broad-spectrum beta-lactamases hydrolyse amino- (amoxicillin), carboxy- (ticarcillin), ureido penicillins and the narrow-spectrum cephalosporins (cephalothin), but not the oxymino- cephalosporins (cefotaxime, ceftazidime) nor the monobactams (aztreonam).

The most common enzymes in this subgroup are TEM-1, TEM-2 and SHV-1. To date, TEM-1 is still the most commonly encountered beta-lactamase in ampicillin resistant Gram negative bacteria. The extended-spectrum beta-lactamases (ESBLs) originally corresponded to mutant enzymes derived from the narrower spectrum TEM and SHV beta-lactamases and showing an expanded substrate spectrum. These enzymes are related by a few amino acid substitutions that change the geometry of the enzyme active site, allowing binding and hydrolysis of the oxymino-beta-lactams. More than 150 TEM enzymes are currently recognized in the TEM group, and more than 110 in the SHV family. While the majority of ESBLs are derived from TEM or SHV enzymes, a few others have also been reported that are not related to the TEM or SHV families of beta-lactamases. The most common ESBLs in this group are termed as CTX-M to highlight their greater activity against cefotaxime when compared to the other oxymino-cephalosporins. To date, more than 90 CTX-M enzymes have been described. Other class A ESBLs are known, such as PER, VEB and GES. PER is nowadays widely represented in gram-negative bacteria.

Beta-lactamase inhibitors were primarily developed to overcome resistance caused by the acquisition of TEM and SHV beta-lactamases. Amino acid substitutions, distinct from those leading to the ESBL phenotype, are associated with resistance to beta-lactamase inhibitors, thereby conferring clinical resistance to the beta-lactam-beta-lactamase inhibitor combinations (amoxicillin-clavulanate for instance).

Finally, carbapenems are currently regarded as the drugs of choice for treatment of infections caused by ESBL-producing organisms. The class A carbapenemases are therefore a source of considerable concern because they can be plasmid-encoded (KPC enzymes) and are very efficient against carbapenems.

General Fold of Class A Beta-Lactamases

Class A beta-lactamases share a very similar fold composed of two domains, one α/β and the other all α (Figure 1A). The α/β domain is formed by a five-stranded antiparallel beta sheet and three helices that pack against the sheet. The amino and carboxyl terminal helices are antiparallel. The binding site is in a crevice at the interface between the two domains, the catalytic site being limited by four highly conserved elements : Ser-Xaa-Xaa-Lys at position 70 where Ser70 is the active site serine ; Ser-Xaa-Asn at position 130, Ser130 being involved in the protonation of the substrate's living group ; Glu166, the general base located at the N terminus of the so-called omega-loop ; and finally the [Arg/Lys]-[Ser/Thr]-Gly sequence on the B3 strand facing the Ser_{130}-Xaa-Asn element on the other side of the active serine, with Arg/Lys234 acting as the cationic recognition site for the beta-lactam carboxylate (Figure 1B). The catalytic water molecule is sequestered by hydrogen bonding as a bridge between the general base Glu166 and active serine Ser70. Another important element in the catalytic site is the oxyanion hole formed by the backbone amide nitrogens of Ser70 and Ala237 and which sequesters by hydrogen bonding interactions either the beta-lactam carbonyl or a water molecule when the antibiotic is absent (Figure 1B).

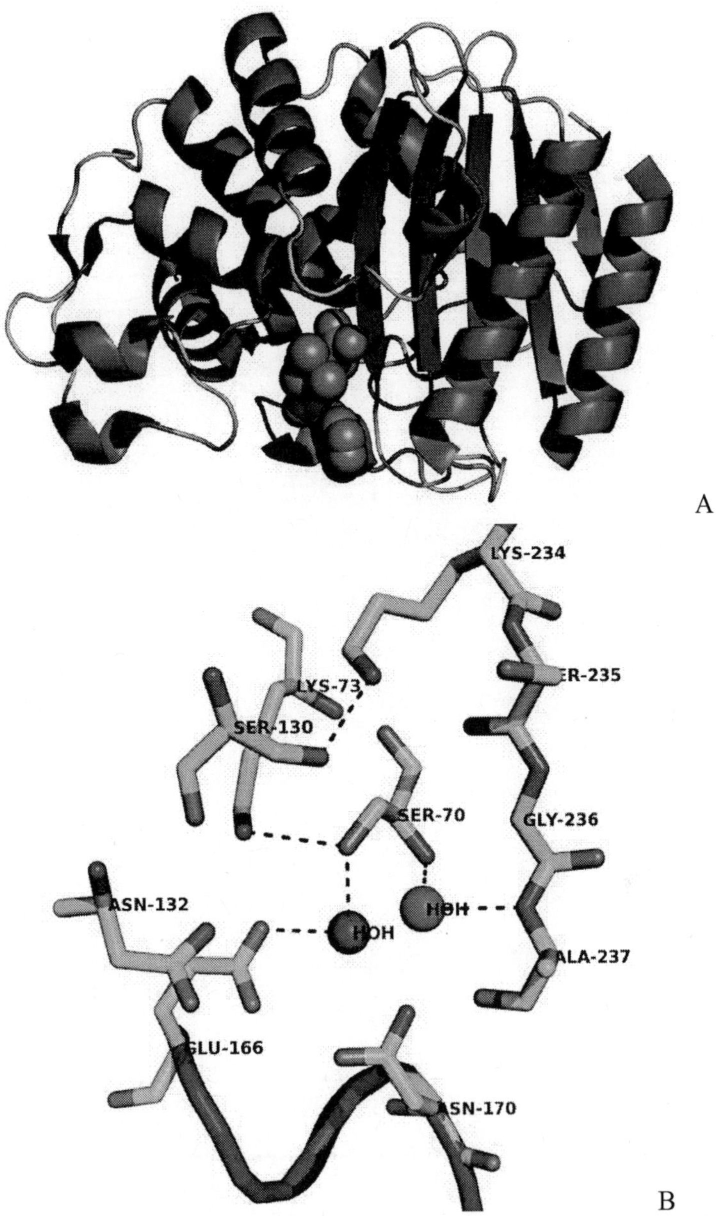

Figure 1. (A) general fold of class A beta-lactamases showing the secondary structure elements (α helices in red, β strands in blue) and the beta-lactam antibiotic in the catalytic cleft (represented by spheres). (B) key residues found in the four highly conserved elements in the catalytic pocket of class A beta-lactamases. The omega loop is shown in blue. The catalytic water molecule is indicated in red, the water molecule found in the oxyanion hole in magenta.

Class A beta-lactamases all have a mechanism (see chapters 7 and 8) that involves (i) the nucleophilic attack of the beta-lactam ring by the active site serine (Ser70) deprotonated by the general base Glu166, (ii) the formation of an acyl-enzyme intermediate (Figure 2) and (iii) hydrolysis of this intermediate by a general base activated water molecule (shown in Figure 2) [3].

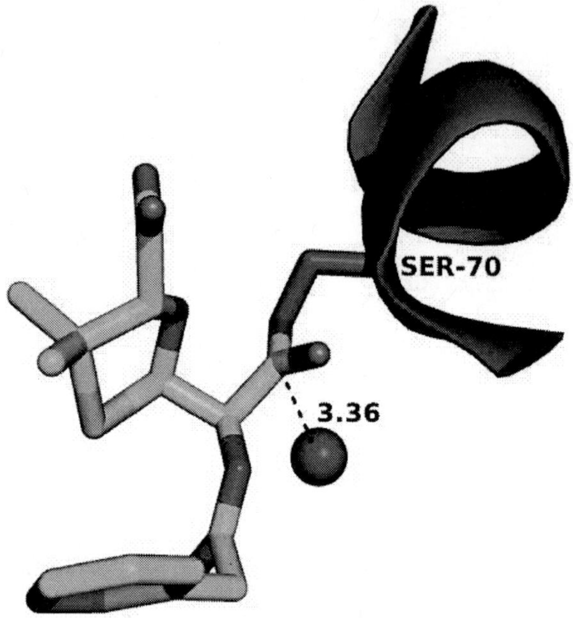

Figure 2. Acyl-enzyme intermediate formed after the nucleophilic attack of the beta-lactam antibiotic (in green) by the activated serine in position 70 (in red). The catalytic water molecule is represented by a red sphere.

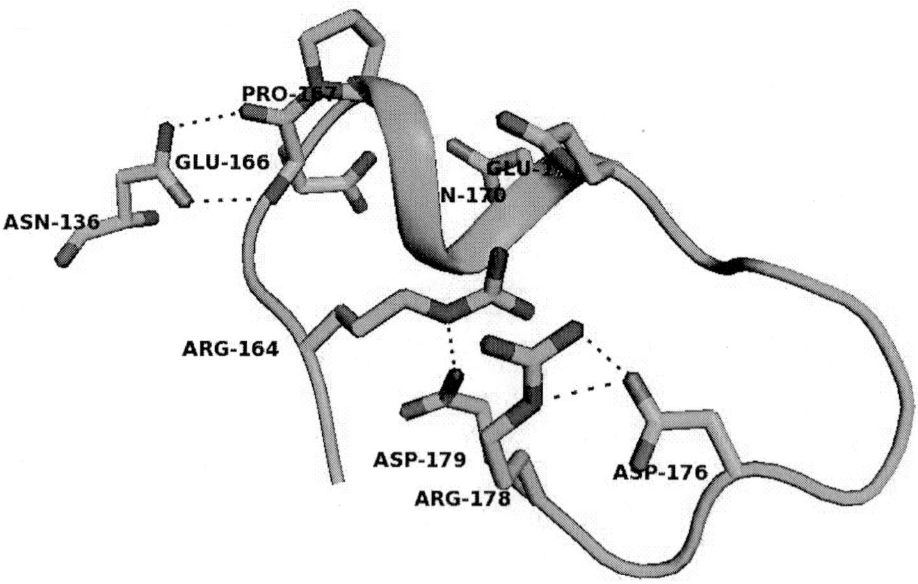

Figure 3. Structure of the omega loop in TEM-1. Hydrogen bonds are shown as dotted lines.

THE BETA-LACTAMASES FROM GRAM NEGATIVE BACTERIA

TEM and SHV Families

Overall Structures

The TEM and SHV class A beta-lactamases represent the best mechanistically characterized serine beta-lactamase group. The first structures were solved in the 1990s [4]. More recently, ultrahigh-resolution structures of TEM and SHV enzymes have been also determined [5,6]. When compared to the *S. aureus* PC1 beta-lactamase, the TEM-1 beta-lactamase crystal structure differs in the relative orientation of the two domains composing the enzyme (rms of 2.4 Å for the Cα coordinates), leading to a narrowed substrate binding cavity. As a result, these enzymes have reached a very high catalytic activity for their preferred substrates, the penicillins [7].

When compared, TEM-1 and SHV-1 share 68% sequence identity and very similar substrate profiles. Accordingly, all the α carbon atoms in SHV-1 and TEM-1 beta-lactamases overlay with an rms deviation of only 1.4 Å, and the atoms in the residues constituting the active site deviate only by 0.23 Å (rms). In both enzymes, the active serine Ser70 is at the N terminus of an α helix (H2). The side chain of the active serine is hydrogen bonded to Lys73 while Ser130 establishes a H-bond interaction with Lys234. The active sites of TEM-1 and SHV-1 contain the two important water molecules generally found in class A beta-lactamases (see Figure 1B). One is in the oxyanion hole with hydrogen bonds to the backbone amide groups of Ser70 and Ala237 and is displaced upon substrate binding by the beta-lactam carbonyl group. The other is the catalytic water molecule hydrogen bonded to the general base Glu166, Asn170 and the active Ser70.

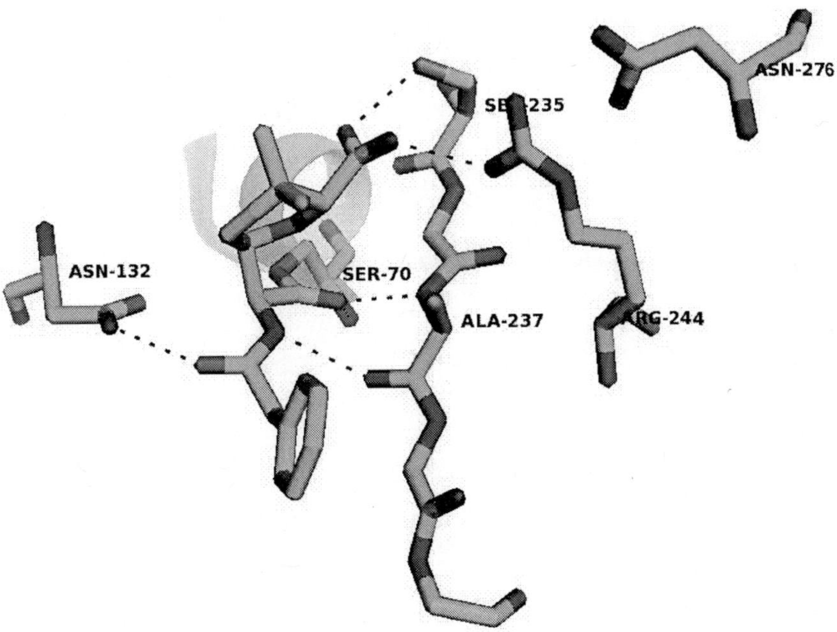

Figure 4. Details of the structure of the acyl-enzyme intermediate in TEM-1. Ser70 is covalently bound to the carbonyl-carbon atom of the beta-lactam antibiotic. H-bonds are shown by dotted lines.

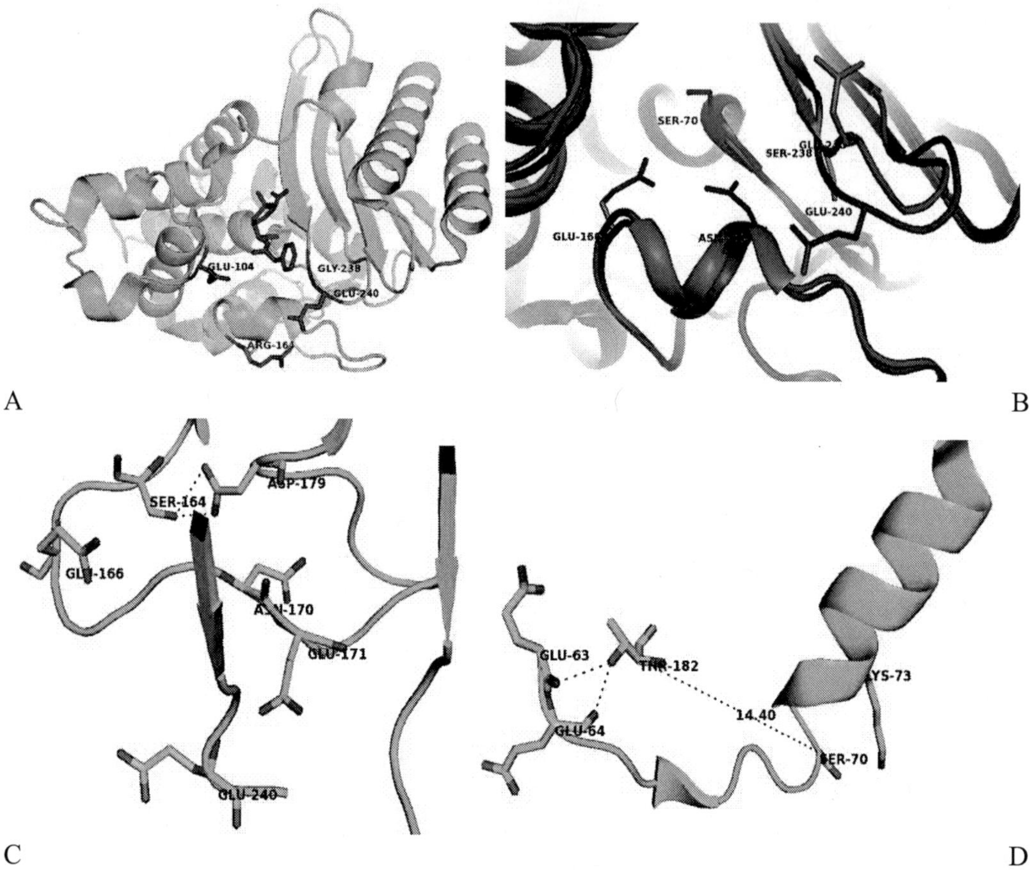

Figure 5. (A) Representation of the most commonly altered positions found in TEM-type ESBLs. The antibiotic in the binding pocket is shown in blue. (B) Structure-based alignment between the wild-type SHV-1 enzyme (in blue) and the Gly238Ser ESBL mutant (in red), showing the opening of the active site in the 238-240 region. (C) Effect of the mutation Arg164->Ser in the crystal structure of the ESBL TEM-64. (D) Effect of the suppressor mutation Met182->Thr in TEM-type ESBLs.

The omega loop (residues 161-179) which carries the essential residue Glu166, is stabilized by two salt bridges : Arg164-Asp179 and Asp176-Arg178 (Figure 3). In TEM-1, proper orientation of the general base Glu166 is provided by the cis peptide bond formed with Pro167 and stabilized by a H bond with the side chain of N170. In SHV-1, there is a difference at position 167 which is occupied by a Thr residue. Nevertheless, the Glu166-Thr167 peptide bond in SHV-1 was shown to adopt a cis conformation (as in nearly all crystal structures of class A beta-lactamases). Two hydrogen bonds between Glu166 and Asn136 stabilize the non-proline cis peptide and orient the omega loop so that Glu166 is correctly positioned to activate the catalytic water molecule [8].

Strand B3 forms one side of the substrate binding cavity and bears residues that are also important for enzyme activity. The 234-236 motive (see Figure 1B) is Lys-Ser-Gly in TEM-1 and Lys-Thr-Gly in SHV-1. This motive is highly conserved in beta-lactamases, the positive charge of Lys-234 being essential for transition-state stabilization [9]. Residue Ala237 represents the main steric constraint for the substrate positioning on this side of the cavity. Gly238, as mentioned hereafter, is a crucial residue to tune the substrate spectrum of

TEM/SHV variants. The absence of side chain at this position in the wild-type enzymes seems to be the best sequence adaptation to accommodate the presence of the side chain of Met69 which is very close in the three-dimensional arrangement [4].

Analysis of the crystal structures of TEM-1 in complex with beta-lactams antibiotics reveals how these antibiotics are recognized and accommodated in the acyl-enzyme intermediate [3,10]. In addition to the covalent bond between the carbonyl-carbon atom of the antibiotic and the side-chain oxygen atom of Ser70, the carbonyl-oxygen atom of the antibiotic in the complex forms two strong hydrogen bonds to the main-chain nitrogen atoms of Ser70 and Ala237 which are the two oxyanion hole nitrogens (Figure 4). The two oxygen and nitrogen atoms forming the R group peptide in the antibiotic make hydrogen bonds with the main-chain carbonyl-oxygen atom of Ala237 and with the amide-nitrogen atom of Asn132, respectively. Finally, the two carboxylate oxygen atoms strongly interact with the side-chain nitrogen atoms of Arg244 and the side-chain oxygen atom of Ser235.

The IRTs

This group of class A beta-lactamases includes inhibitor-resistant enzymes derived from the parental TEM or SHV enzymes by several amino acid substitutions. The variants resist inactivation by the mechanism-based inhibitors (clavulanate and sulbactam) and are characterized by single-point mutations at one of several key amino acids including in particular residues 69, 244 and 276. The Asn276Asp mutation of TEM-1 is representative of clavulanate-resistant IRT variants. Replacement of the neutral Asn-276 by a charged Asp residue results in a significant movement of the Asp side chain toward the positively-charged guanidinium of Arg-244 that is normally involved in substrate carboxylate recognition (see Figure 4) [11]. This results in the dissociation of a conserved water molecule involved in the critical fragmentation steps in inactivation of the TEM beta-lactamases by clavulanate [11].

Other TEM variants with mutations at position 69 also demonstrate clavulanate resistance. This could be related to the fact that residue 69 influences the positioning of the neighbouring strand B3 at the level of residue 237 which provides, together with Ser70, one of the two nitrogen atoms constituting the oxyanion hole. The molecular effects that could be related to beta-lactamase inhibitor resistance in the Met69 mutants have been investigated in detail for the Met69Val variant of SHV-1 [12]. Analysis of the X-ray structure of this variant suggests that the oxyanion hole in the mutant becomes smaller by 0.2 Å when compared to the wild-type enzyme, thereby reducing the acylation efficiency of clavulanic acid and, consequently, the level of inhibition of the enzyme [12].

Another point mutation, not directly related to resistance development but rather important to reinforce enzyme stability, can be found in some IRT variants, in particular M182T. This mutation appears to act as suppressor of certain deleterious mutations that promote misfolding and aggregation of the mutant protein, as discussed in the next paragraph and in Figure 5D.

The ESBLs

Following the introduction by pharmaceutical companies of extended-spectrum beta-lactam antibiotics such as cefotaxime and ceftazidime, resistance to these antibiotics has emerged, more commonly arising from point mutations in existing enzymes such as TEM-1 or SHV-1 and altering one to five amino acids near the catalytic centre. These changes are distributed widely in the enzymes and can have profound effects on the substrate specificity of the enzymes by altering the positioning of key groups within the active site. The resulting variants have an extended substrate spectrum responsible for resistance to the third- and fourth-generation cephalosporins (for example, cefotaxime, ceftazidime and cefepime) which are poorly recognized by the parental class A beta-lactamases TEM-1/-2 and SHV-1. To date, more than 150 TEM and 110 SHV beta-lactamase variants have been reported. The most commonly found mutations are located at positions 104, 164, 238 and 240 (Figure 5A).

The role played by the ESBL mutations in the transformation of a parental class A enzyme into an ESBL variant has been extensively studied for the Gly238Ser mutation. This single change results in a 4- to 6-fold increase of the MICs for oxymino-cephalosporins (cefotaxime and ceftazidime). The structural basis for this effect has been suggested from crystallographic investigations [6,13,14]. Gly238 is located on the B3 strand in proximity to the Asn170 residue of the omega loop element. Replacement by Ser results in a significant alteration of the conformation of the lower portion of the B3 strand, thereby widening the opening to the active site by as much as 2.8 Å (Figure 5B). It is generally admitted that the opening of the active site cavity induced by the Ser238 mutation facilitates the accommodation of the bulkier oxyimino side chain without altering the rest of the catalytic machinery. In addition, repositioning of the 238 region causes significant rearrangements at the level of the next residue, Glu240. The carboxylate moiety of this residue is thought to hinder binding of some cephalosporins by steric repulsion. Its repositioning in the ESBLs having Ser238 could reduce the interferences with substrate binding [13]. These hypotheses have been confirmed by determining the ultrahigh resolution structure (0.91 Å) of the ESBL SHV-2 which contains a single Gly238Ser change relative to the wild-type enzyme SHV-1 [6]. In this structure, a 1 to 3 Å displacement in the 238-242 beta-strand-turn segment was observed, and an important conformational modification in the 240 peptide region was seen that moves Glu240 far away from the binding pocket (see Figure 5B), making the beta-lactam binding site more open to third-generation cephalosporins with large R substituents and thereby expanding the substrate spectrum of the variant enzyme [6]. It must be noted here that a Glu240Lys mutation has also been observed in variants showing several combinations of mutations. In these mutants, a Lys at position 240 most likely interacts directly with the substrate at the level of its side chain [13].

In TEM-1, Glu-104 is at the entrance of the substrate binding cavity, facing the side chain of Glu240 but not establishing direct interactions with this residue. The Glu104Lys mutation acts synergistically with the Gly238Ser mutation discussed above, the double mutant displaying a >250-fold increase in MIC when compared to TEM-1. In the double mutant Glu104Lys-Gly238Ser, the side chain of the mutated residue Lys104 is oriented toward the reorganized Glu240 side chain at a distance closer than that seen in TEM-1, but its exact contribution to the extended-spectrum activity remains not clearly understood.

Regarding the alterations affecting Arg164, this residue is part of the omega loop and contributes to two of the four charge-charge interactions (with Glu171 and Asp179) that stabilize this important structural element (see Figure 3). The disruption of these interactions upon Arg164 mutations is likely to destabilize the omega loop region in TEM- and SHV-variants. Disrupting these salt bridges also increases the accessibility of the active site. In the crystal structure of TEM-64 [14], the Arg164Ser mutant loses two hydrogen bonds with Asp179 and Glu171, and introduces a packing defect. Consequently, Asn170, which normally points toward the centre of the binding site, rotates to fill the cavity created by the substitution of the bulky arginine by the small serine residue (Figure 5C), thus enlarging the binding site that can then accommodate the bulky oxymino side-chains of the third-generation cephalosporins [14].

Substitutions distant from the active site, such as Met182Thr located 14.4 Å from the active site, also occur in extended-spectrum TEM beta-lactamases. The hydroxyl group of Thr182 forms two H-bonds with the backbone carbonyls of Glu63 and Glu64 (Figure 5D). These new bonds likely increase the interaction between the two domains of the beta-lactamase and contribute to suppress the destabilization caused by the other ESBL mutations [15]. Therefore, the Met182Thr mutation is generally considered to be a global suppressor mutation that acts to stabilize the enzyme structure of the ESBL mutants [13].

The Non-TEM, non-SHV ESBLs of the CTX-M Group

Of all the ESBL variants reported so far, the CTX-M are of greatest concern. More than 40 CTX-M enzymes are currently known and these enzymes are found not only in nosocomial pathogens but also now in community strains [7]. These enzymes, which exhibit 40% or less identity with the TEM and SHV families, form an ESBL group with strong cefotaximase activity. Toho-1, which is plasmid-encoded, is classified into the CTX-M group and its crystal structure has been reported at 1.65 Å resolution [16,17]. The overall structure of Toho-1 is that seen for the other class A beta-lactamases, with a rmsd of 2.3 Å when compared to SHV-1 [16]. In comparison with other class A enzyme structures, the smallest Cα difference is observed with the beta-lactamase from *Proteus vulgaris* K1 (rmsd = 0.7 Å), an ESBL homologous to Toho-1 (70% amino acid identity). Within the active site cavity, an interesting feature in Toho-1 is that two alternative conformations of the Lys73 side chain are observed. In the first conformation (the conformation observed in most class A beta-lactamases), the ammonium group of Lys73 points toward the hydroxyl groups of Ser130, with H-bond interactions with Ser130, Ser70 and Asn132. In the other conformation, the side chain of Lys73 points toward Glu166 and establishes interactions with Glu166, Ser70 and Ans132. The existence of two conformations of Lys73 reinforces the idea that this residue plays an important role in the proton-relay pathway in the catalytic mechanism involving Glu166, Ser130 and Ser70 [16].

Extensive investigation of the crystal structures of the CTX-M Toho-1 enzyme acylated by cefotaxime (Figure 6A) clearly suggests that an expansion of the active site cavity operates to adjust the shape of the binding site to the bulkiness of the oxymino group of this antibiotic [16,18]. Indeed, a comparison between non-ESBLs and Toho-1 shows that the omega loop, which interacts at the level of the bulky side chain of oxymino-cephalosporins, adopts a specific position in Toho-1 with a c.a. 1 Å shift when compared to non-ESBLs (Figure 6B). This shift keeps the loop in Toho-1 away from the aminothiazole ring of cefotaxime, thus avoiding unfavourable steric contacts with the bulky side chain of cefotaxime and increasing the apparent affinity for this antibiotic.

When compared with the acyl-intermediate structures of non ESBLs, there are other key differences that can explain the specificity of Toho-1 towards oxymino-cephalosporins such as cefotaxime. First, Toho-1 does not contain Arg244 present in most non-ESBLs and which interacts with the carboxylate moiety of the substrate. In Toho-1, Arg276 was suggested to be a substitute for Arg244. However, this residue seems to have no interactions with the substrate in the acyl-intermediate structure of Toho-1 and instead, Ser237, which is conserved in all CTX-M type ESBLs, forms a hydrogen bond with the carboxylate oxygens of the antibiotic (Figure 6A), suggesting that this interaction plays a key role in the hydrolysis of cefotaxime [18]. One must note here that the position of residue 237 is affected by van der Waals contacts between the β strand B3 and the side chain of residue 69. In CTX-M-type ESBLs, a cysteine is conserved at position 69 which may play an important role to place Ser237 in the most adequate position [18].

Among the other interesting differences observed between the acyl-intermediate structures of Toho-1 and non-ESBL class A beta-lactamases, one can note in Toho-1 the presence of the conserved residue Asn104 with a unique peptide bond conformation between residues 104 and 105. In non-ESBLs, the position of Asn132, which interacts with the acylamide side chain of the substrate, is fixed in part by the interaction between its nitrogen side chain atom and the backbone oxygen of residue 104. In Toho-1, the side-chain of Asn132 still interacts with residue 104, but at the level of the side chain oxygen atom instead of the O-main chain (Figure 6A). Concomitantly, a unique hydrogen bond is formed between the side chain nitrogen atom of Asn104 and the carbonyl oxygen of the acylamide side chain of the beta-lactam antibiotic (Figure 6A). This interaction could contribute to increasing the binding efficiency of the bulky side chain of cefotaxime to the binding site.

Finally, it must be mentioned here that new CTX-M variants, such as D240G, show improved ability to hydrolyse ceftazidime in addition to the basic cefotaximase activity.

Other Non-TEM non-SHV ESBLs

The chromosomal class A beta-lactamase K1 from *Proteus vulgaris* is an ESBL exhibiting less efficient oxymino-cephalosporinase activity when compared to the CTX-M-type ESBLs. The structure of K1 was determined, showing a 3D organisation highly similar to that of Toho-1 (70% amino acid identity, rmsd = 0.7 Å, as mentioned above). Structural elements that could contribute to extended activity in K1 have been highlighted, including the Ser237 and the Lys276 amino acid residues which have been proposed to be involved in the cephalosporinase activity of the CTX-M-type ESBLs [18,19]. Conversely, several structural differences could be some of the reasons for the lower cefotaximase activity of K1 when compared to the CTX-M-type ESBLs. Among them, (i) the position of the omega loop in this enzyme shows higher similarity to those in the non ESBLs than that in Toho-1 [18], (ii) there is no Asn in position 104 in K1 (Ala104 is found instead) and (iii) in K1 an alanine residue replaces the Cys69 residue found in the CTX-M-type family.

Other class A ESBLs initially reported in *Pseudomonas aeruginosa*, such as PER-1, VEB-1, GES-1 are uncommon when compared to the TEM, SHV and CTX-M groups. The molecular mechanisms driving the ESBL activity in these enzymes are less understood. For instance, the ESBL PER-1 is a typical class A enzyme with respect to the conservation of the catalytic machinery. Nevertheless, resolution of its three-dimensional structure revealed that this family of ESBLs is not as homogeneous in structure as it could be thought. In PER-1, the fold of the omega loop is very different, with a trans 166-167 peptide bond instead of the cis peptide bond observed invariably between these two residues in other class A beta-lactamases. In addition, the PER-1 omega loop is highly organized in secondary structure elements, including a short helix between residues 166 and 171 (Figure 7A). Moreover, a histidine residue is specifically found at position 170 in PER-1 but this residue is displaced by

8 Å from the position of the homologous residue (Asn170) in the other class A beta-lactamases. Finally, there is no salt bridge in the PER-1 omega loop for which the stability of the fold stems mostly from the presence of secondary structure elements complemented by a H bond interaction between residues Asn179 (side chain) and Val163-Ala164 (main chain atoms) (not shown). The peculiar conformation of the omega loop in PER-1 is very likely involved in the ability of the enzyme to accommodate oxymino-cephalosporins [20]. This new fold of the omega loop, as well as the insertion of four residues (Gln-239, Ile-240, Lys-240A, Ala-240B) at the edge of strand B3, generates a broad cavity that is well adapted to accommodate the bulky R group of oxyiminocephalosporin substrates (14.3 Å between $Gln103_A$ and Lys240, Figure 7A). Another important residue in PER-1 is the Thr found in position 104. Replacement of this residue with Glu was shown to completely abolish the catalytic activity on penicillins and drastically reduce the kcat values for cephalosporins. As mentioned above for the ESBLs belonging to the CTX-M subgroup, an important interaction exists between residue 104 and Asn132. At the level of residue 104, this interaction involves the main chain oxygen atom of residue 104 in non-ESBLs such as TEM-1 while it is the side chain of the residue which is involved in the interaction in the CTX-M-type ESBLs (see Figure 6A). In PER-1, the interaction with Asn132 occurs through the gamma hydroxyl group of the side chain of Thr104 such that the 104 region is drastically relocated compared to other class A enzymes. The repositioning of the side chain of residue 104 and the peculiar fold of the omega loop in PER-1 seem in direct relationship with the oxymino-cephalosporinase activity of this enzyme by generating additional space in the region that is precisely located in the area expected to bind the bulky side chain of third and fourth generation cephalosporins.

Carbapenem-hydrolyzing Class a Beta-lactamases

The carbapenems (such as imipenem) resist hydrolysis by the classical class A enzymes such as TEM and SHV. However, a few class A enzymes, including the chromosomally-encoded SME-1 and NMC-A enzymes, and the more worrying plasmid-mediated KPC enzymes (KPC-2 to -10), are effective carbapenemases. The structural basis for the peculiar substrate spectrum of these enzymes was first investigated in NMC-A and SME-1 [21,22]. The overall structure of these enzymes is similar to those of other class A beta-lactamases and most of the active site residues are conserved in comparison with the non-carbapenemase class A enzymes (except at positions 104 and 238, as discussed below). Position 238 in all the class A carbapenemases studied so far is occupied by a cysteine forming a disulfide bridge with the other cysteine residue located at position 69. The presence of this 69-238 disulfide bridge has significant structural consequences in the binding pocket. In particular, it contributes to fix the position of Ser70 relative to that of the B3 strand by linking two important structural elements of the catalytic site : on the one hand, the helix bearing the catalytic serine residue Ser70 and, on the other, the β strand B3 harbouring the Lys_{234}-[Ser/Thr]-Gly_{236} motif (Figure 8A). The mechanism for carbapenem resistance is clearly associated with the presence of this disulfide bond (Cys69-Cys238) since a strain producing the mutant protein in which Cys69 was substituted by Ala was shown to be fully susceptible to imipenem (note that an effect on the *stability* of the mutated protein cannot be ruled out). However, the molecular effect by which this peculiar and specific element of class A carbapenemases contribute to the carbapenemase activity has not been elucidated to date. It must be stressed out here that the relative location of the active site serine residue Ser70 and the general base Glu166 is significantly different in SME-1 compared to the other class A enzymes since the Ser70-Glu166 distance is up to 1.4 Å shorter in this carbapenem-hydrolyzing enzyme. Consequently, the hydroxyl moiety of Ser70 is involved in direct hydrogen bonding to the carboxyl group of Glu166 (Figure 8B), an interaction that could strongly enhance the nucleophilicity of the hydroxyl group of Ser70 for the initial attack on the carbonyl C atom of the beta-lactam ring of carbapenems.

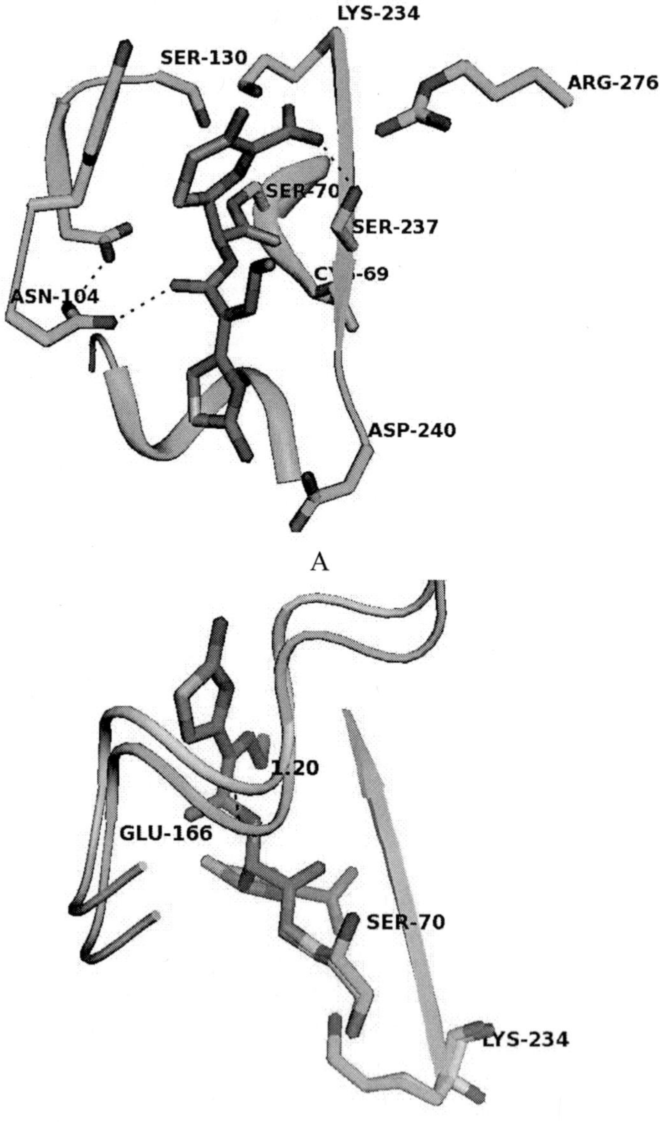

Figure 6. (A) Structure of the active-site region in the CTX-M Toho-1 enzyme (in green) after acylation by cefotaxime (in magenta) showing the omega loop region (at the bottom, visible between Asn104 and Asp240). (B) Superimposition of Toho-1 (in green) with a non-ESBL class A enzyme (in yellow) showing the shifted position of the omega loop in Toho-1.

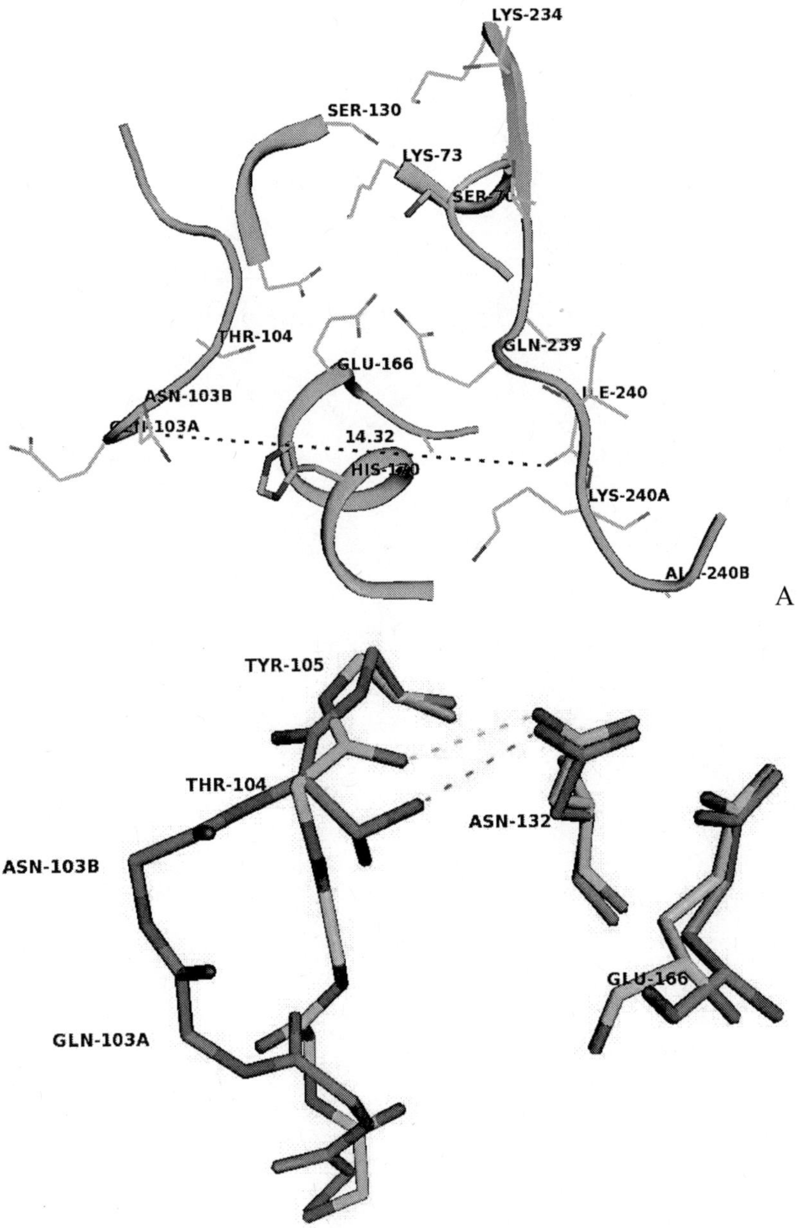

Figure 7. (A) Structure of the active-site region in PER-1. The distance between residues Gln103 and Lys240 (14.3 Å) is indicated by a dotted black line. (B) Superimposition of PER-1 (in pink) with TEM-1 (in green) at the level of residues 104 and 132.

Other elements could be of great importance, such as the positioning of Asn132 which is unusual in class A carbapenemases since it is displaced by approximately 1.25 Å compared to the TEM-1 enzyme (Figure 8A). This specific location might play a major role with respect to the carbapenemase activity because additional space is thus provided in a critical area for protein-imipenem interactions, which would permit accommodating the 6α-1R-hydroxyethyl substitutent of carbapenems. The 1.25 Å displacement of Asn132 away from strand B3 in class A carbapenemases would allow the hydroxymethyl substituent to be more easily accommodated and the substrate correctly oriented in the binding site cavity [22].

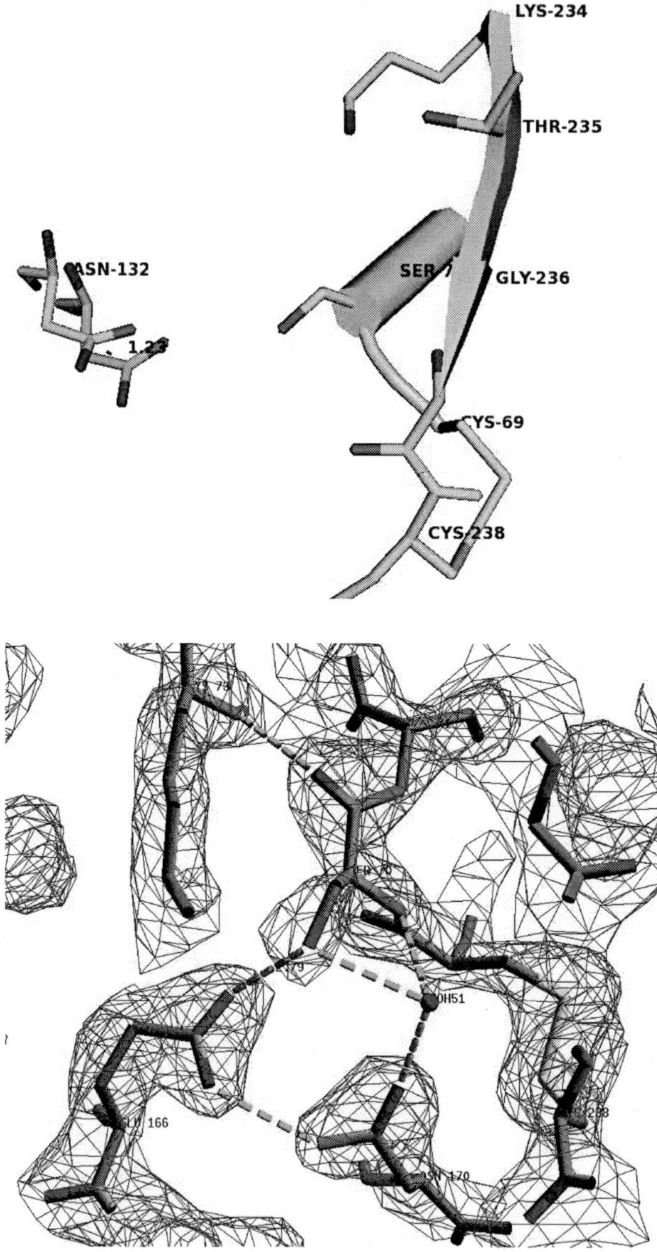

A

B

Figure 8. (A) Structural arrangement of the B3 strand in the class A carbapenemase SME-1 (in blue). The disulfide bridge between Cys69 and Cys238 is shown in yellow. The positioning of the Asn132 residue in TEM-1 is shown in green. (B) Electron density map around the active site of SME-1. The water molecule in the oxyanion hole is represented by a red sphere.

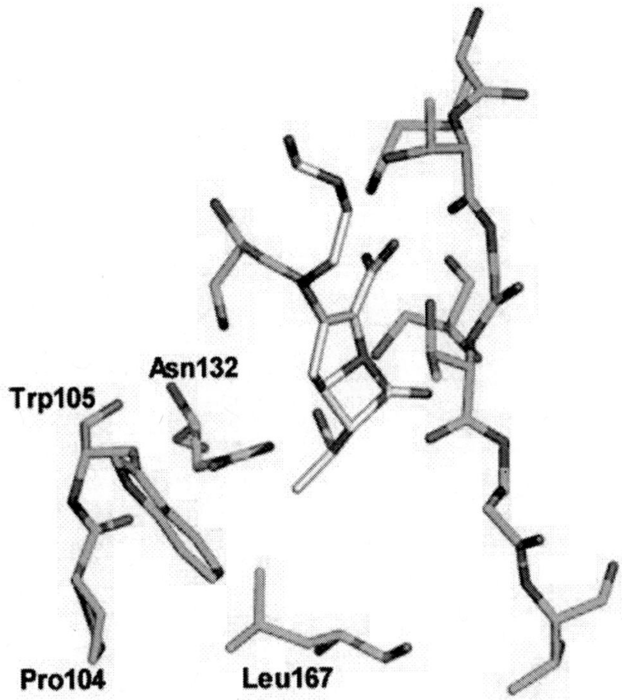

Figure 9. Structural organization of the binding site in the class A extremely broad-spectrum beta-lactamase KPC-2 (in green) in complex with imipenem (in grey).

The crystal structure of the extremely broad-spectrum beta-lactamase KPC-2 has been determined at resolutions of 1.85 Å [23] and 1.23 Å [24]. In addition to the Cys69-Cys238 disulfide bridge highly conserved in class A carbapenemases, both structures share more subtle structural changes likely involved in the ability of this class A enzyme to hydrolyze carbapenems, such as a catalytic serine in a more shallow position in the active site cleft, associated with a combination of active-site adjustments at the level of residues Asn132 and Asn170 [23,24]. In addition, analysis of the high resolution structure of KPC-2 suggests that residues Pro104, Trp105, Asn132 and Leu167 (see Figure 9) define a unique binding cavity capable of accommodating both the large R substituent of cefotaxime and the small 6α-hydroxyethyl group of imipenem, accounting for the extremely broad spectrum of activity characterizing the KPC enzymes that appear to be not only carbapenemases but also proficient cefotaximases [24].

Other Class A Enzymes: Carbenicillinases

Carbenicillinases are class A enzymes mostly produced by strains of *P. aeruginosa* and are highly active on carbenicillin, a penicillin derivative containing an additional carboxylate moiety at the α position. The structure of the carbenicillinase PSE-4 was determined, showing that the overall structure is similar to that of TEM-1 which shares 44% sequence identity with PSE-4 (rmsd = 1.3 Å for C_α atoms). At the level of the active site, the position and orientation of the catalytic residues in PSE-4 are conserved for Ser70 (the nucleophile), Lys73, Ser130,

Glu166 and Asn170. In contrast, most carbenicillinases are unique among class A beta-lactamases in that residue 234 is an arginine instead of a lysine in other class A enzymes. This specific residue seems to stabilize an alternate conformation of the Ser130 side chain that may be involved in accommodating a shift of the position of the bound carbenicillin necessary to avoid a steric clash between the additional carbenicillin α-carboxylate group of the antibiotic and the side chain of Asn170 [25].

THE BETA-LACTAMASES FROM GRAM POSITIVE BACTERIA

The crystal structures of the class A beta-lactamases from Gram-positive bacteria have been determined for *Staphylococcus aureus*, *Bacillus licheniformis*, *Streptomyces albus*, *Mycobacterium fortuitum* and *Mycobacterium tuberculosis*. [26-29]. The beta-lactamase from *S.aureus* is a narrow spectrum enzyme and exhibits the typical fold of class A beta-lactamases (when superposed to TEM-1, the Cα coordinates of the beta-lactamase from *S.aureus* diverge by 1.8 Å) [4]. When compared to *S. aureus*, the *Bacillus licheniformis* enzyme is significantly more efficient against oxyimino-cephalosporins such as cefotaxime [30]. Similarly, the specificity profile of the class A beta-lactamase of *M. fortuitum* has been reported to resemble that of the CTX-M type enzymes, including cefotaxime but not ceftazidime [31]. The overall structure of the *M. fortuitum* beta-lactamase is similar to those of other class A beta-lactamases, in particular for the residues responsible for the catalytic activity (Ser70, Lys73, Ser130, Asn132, Glu166 and Lys234). Following the $Lys_{234}ThrGly_{236}$ motif, residue 237 is a serine in the *M. fortuitum* beta-lactamase, with the side-chain hydroxyl group lying near the active site and interacting with the substrate carboxylate together with the Arg278 residue corresponding to Arg276 in the CTX-M type enzymes. Thus, the *M. fortuitum* enzyme possesses two of the specific features of the CTX-M-type enzymes (Ser237 and Arg278) that have been shown to favour cephalosporinase activity. By contrast, the *M. fortuitum* beta-lactamase lacks the $Asn_{104}Tyr$ motif typical of CTX-M-type ESBLs, which is thought to play a significant role in the cefotaximase activity of enzymes such as Toho-1 (CTX-M) or PER-1 [18,20].

The crystal structure of the class A beta-lactamase from *M. tuberculosis*, an enzyme also showing a broad spectrum of activity, has also been determined, showing interesting differences when compared to that of *M. fortuitum*. Indeed, although the overall structure of the *M. tuberculosis* beta-lactamase is similar to those of other class A beta-lactamases, several of the highly conserved residues involved in the binding site pocket of the *M. tuberculosis* beta-lactamase are unusual, particularly Gly132, Ala164, Ala244 and Glu276 (replacing the generally conserved residues Asn132, Arg164, Arg244 and Arg276). The replacement of Asn132 by Gly in the *M. tuberculosis* beta-lactamase makes the substrate binding pocket significantly larger, a point that deserves to be underlined in regard to the presence in this enzyme of amino acid residues such as Thr237 and Arg220 that are generally regarded as important for hydrolytic activity against third-generation cephalosporins such as cefotaxime and that could also account for the broad spectrum of activity of the *M. tuberculosis* beta-lactamase [32]. Another unusual residue in this enzyme is Ser104 which is typically an Asp or an Asn in other class A beta-lactamases. Like the Asn132Gly modification, Ser104 creates more space in the active site. Finally, a structural feature that could also contribute to the broad substrate specificity of the *M. tuberculosis* beta-lactamase is the presence of Ala164 (instead of Arg). This notable modification is located in the omega loop and is well known to play an important role in the expansion of activity against third-generation cephalosporins in TEM and SHV ESBLs.

CONCLUSIONS

Although the class A beta-lactamases display a highly conserved fold, this group of enzymes is characterized by a very wide degree of amino acid and functional diversity, as illustrated in this chapter by the description of the various subclasses of class A variants. This underlines the ability of bacteria to modify protein structure to acquire new functions under selection pressure.

REFERENCES

[1] Wilke MS, Lovering AL, Strynadka NC: Beta-lactam antibiotic resistance: a current structural perspective. *Curr Opin Microbiol* 2005, 8:525-533.

[2] Shah AA, Hasan F, Ahmed S, Hameed A: Characteristics, epidemiology and clinical importance of emerging strains of Gram-negative bacilli producing extended-spectrum beta-lactamases. *Res Microbiol* 2004, 155:409-421.

[3] Strynadka NC, Adachi H, Jensen SE, Johns K, Sielecki A, Betzel C, Sutoh K, James MN: Molecular structure of the acyl-enzyme intermediate in beta-lactam hydrolysis at 1.7 A resolution. *Nature* 1992, 359:700-705.

[4] Jelsch C, Mourey L, Masson JM, Samama JP: Crystal structure of Escherichia coli TEM1 beta-lactamase at 1.8 A resolution. *Proteins* 1993, 16:364-383.

[5] Minasov G, Wang X, Shoichet BK: An ultrahigh resolution structure of TEM-1 beta-lactamase suggests a role for Glu166 as the general base in acylation. *J Am Chem Soc* 2002, 124:5333-5340.

[6] Nukaga M, Mayama K, Hujer AM, Bonomo RA, Knox JR: Ultrahigh resolution structure of a class A beta-lactamase: on the mechanism and specificity of the extended-spectrum SHV-2 enzyme. *J Mol Biol* 2003, 328:289-301.

[7] Fisher JF, Meroueh SO, Mobashery S: Bacterial resistance to beta-lactam antibiotics: compelling opportunism, compelling opportunity. *Chem Rev* 2005, 105:395-424.

[8] Kuzin AP, Nukaga M, Nukaga Y, Hujer AM, Bonomo RA, Knox JR: Structure of the SHV-1 beta-lactamase. *Biochemistry* 1999, 38:5720-5727.

[9] Brannigan J, Matagne A, Jacob F, Damblon C, Joris B, Klein D, Spratt BG, Frere JM: The mutation Lys234His yields a class A beta-lactamase with a novel pH-dependence. *Biochem J* 1991, 278 (Pt 3):673-678.

[10] Maveyraud L, Pratt RF, Samama JP: Crystal structure of an acylation transition-state analog of the TEM-1 beta-lactamase. Mechanistic implications for class A beta-lactamases. *Biochemistry* 1998, 37:2622-2628.

[11] Swaren P, Golemi D, Cabantous S, Bulychev A, Maveyraud L, Mobashery S, Samama JP: X-ray structure of the Asn276Asp variant of the Escherichia coli TEM-1 beta-lactamase: direct observation of electrostatic modulation in resistance to inactivation by clavulanic acid. *Biochemistry* 1999, 38:9570-9576.

[12] Totir MA, Padayatti PS, Helfand MS, Carey MP, Bonomo RA, Carey PR, van den Akker F: Effect of the inhibitor-resistant M69V substitution on the structures and populations of trans-enamine beta-lactamase intermediates. *Biochemistry* 2006, 45:11895-11904.

[13] Orencia MC, Yoon JS, Ness JE, Stemmer WP, Stevens RC: Predicting the emergence of antibiotic resistance by directed evolution and structural analysis. *Nat Struct Biol* 2001, 8:238-242.

[14] Wang X, Minasov G, Shoichet BK: Evolution of an antibiotic resistance enzyme constrained by stability and activity trade-offs. *J Mol Biol* 2002, 320:85-95.

[15] Raquet X, Vanhove M, Lamotte-Brasseur J, Goussard S, Courvalin P, Frere JM: Stability of TEM beta-lactamase mutants hydrolyzing third generation cephalosporins. *Proteins* 1995, 23:63-72.

[16] Ibuka AS, Ishii Y, Galleni M, Ishiguro M, Yamaguchi K, Frere JM, Matsuzawa H, Sakai H: Crystal structure of extended-spectrum beta-lactamase Toho-1: insights into the molecular mechanism for catalytic reaction and substrate specificity expansion. *Biochemistry* 2003, 42:10634-10643.

[17] Ibuka A, Taguchi A, Ishiguro M, Fushinobu S, Ishii Y, Kamitori S, Okuyama K, Yamaguchi K, Konno M, Matsuzawa H: Crystal structure of the E166A mutant of extended-spectrum beta-lactamase Toho-1 at 1.8 A resolution. *J Mol Biol* 1999, 285:2079-2087.

[18] Shimamura T, Ibuka A, Fushinobu S, Wakagi T, Ishiguro M, Ishii Y, Matsuzawa H: Acyl-intermediate structures of the extended-spectrum class A beta-lactamase, Toho-1, in complex with cefotaxime, cephalothin, and benzylpenicillin. *J Biol Chem* 2002, 277:46601-46608.

[19] Nukaga M, Mayama K, Crichlow GV, Knox JR: Structure of an extended-spectrum class A beta-lactamase from Proteus vulgaris K1. *J Mol Biol* 2002, 317:109-117.

[20] Tranier S, Bouthors AT, Maveyraud L, Guillet V, Sougakoff W, Samama JP: The high resolution crystal structure for class A beta-lactamase PER-1 reveals the bases for its increase in breadth of activity. *J Biol Chem* 2000, 275:28075-28082.

[21] Sougakoff W, L'Hermite G, Pernot L, Naas T, Guillet V, Nordmann P, Jarlier V, Delettre J: Structure of the imipenem-hydrolyzing class A beta-lactamase SME-1 from Serratia marcescens. *Acta Crystallogr D Biol Crystallogr* 2002, 58:267-274.

[22] Swaren P, Maveyraud L, Raquet X, Cabantous S, Duez C, Pedelacq JD, Mariotte-Boyer S, Mourey L, Labia R, Nicolas-Chanoine MH, et al.: X-ray analysis of the NMC-A beta-lactamase at 1.64-A resolution, a class A carbapenemase with broad substrate specificity. *J Biol Chem* 1998, 273:26714-26721.

[23] Ke W, Bethel CR, Thomson JM, Bonomo RA, van den Akker F: Crystal structure of KPC-2: insights into carbapenemase activity in class A beta-lactamases. *Biochemistry* 2007, 46:5732-5740.

[24] Petrella S, Ziental-Gelus N, Mayer C, Renard M, Jarlier V, Sougakoff W: Genetic and structural insights into the dissemination potential of the extremely broad-spectrum class A beta-lactamase KPC-2 identified in an Escherichia coli strain and an Enterobacter cloacae strain isolated from the same patient in France. *Antimicrob Agents Chemother* 2008, 52:3725-3736.

[25] Lim D, Sanschagrin F, Passmore L, De Castro L, Levesque RC, Strynadka NC: Insights into the molecular basis for the carbenicillinase activity of PSE-4 beta-lactamase from crystallographic and kinetic studies. *Biochemistry* 2001, 40:395-402.

[26] Knox JR, Moews PC: Beta-lactamase of Bacillus licheniformis 749/C. Refinement at 2 A resolution and analysis of hydration. *J Mol Biol* 1991, 220:435-455.

[27] Herzberg O: Refined crystal structure of beta-lactamase from Staphylococcus aureus PC1 at 2.0 A resolution. *J Mol Biol* 1991, 217:701-719.

[28] Lamotte-Brasseur J, Dive G, Dideberg O, Charlier P, Frere JM, Ghuysen JM: Mechanism of acyl transfer by the class A serine beta-lactamase of Streptomyces albus G. *Biochem J* 1991, 279 (Pt 1):213-221.

[29] Dideberg O, Charlier P, Wery JP, Dehottay P, Dusart J, Erpicum T, Frere JM, Ghuysen JM: The crystal structure of the beta-lactamase of Streptomyces albus G at 0.3 nm resolution. *Biochem J* 1987, 245:911-913.

[30] Matagne A, Misselyn-Bauduin AM, Joris B, Erpicum T, Granier B, Frere JM: The diversity of the catalytic properties of class A beta-lactamases. *Biochem J* 1990, 265:131-146.

[31] Sauvage E, Fonze E, Quinting B, Galleni M, Frere JM, Charlier P: Crystal structure of the Mycobacterium fortuitum class A beta-lactamase: structural basis for broad substrate specificity. *Antimicrob Agents Chemother* 2006, 50:2516-2521.

[32] Wang F, Cassidy C, Sacchettini JC: Crystal structure and activity studies of the Mycobacterium tuberculosis beta-lactamase reveal its critical role in resistance to beta-lactam antibiotics. *Antimicrob Agents Chemother* 2006, 50:2762-2771.

In: Beta-Lactamases
Editor: Jean-Marie Frère, pp. 41-77

ISBN : 978-1-61324-638-2

Chapter 3

X-RAY STRUCTURES AND MECHANISMS OF METALLO-β-LACTAMASES

Carine Bebrone[1,*], Gianpiero Garau[2,*], Isabel Garcia-Saez[2], Laurent Chantalat[3], Andrea Carfi[4] and Otto Dideberg[2†]

[1] Centre d'Ingénierie des Protéines, Université de Liège, Allée du 6 Août B6, Sart-Tilman, 4000 Liège, Belgique.

[2] Institut de Biologie Structurale *Jean-Pierre Ebel*, 41, rue Jules Horowitz, F-38027 Grenoble Cedex 1, France.

[3] Structural Biology GALDERMA R&D Sophia Antipolis 635, route des Lucioles BP 87 06902 Sophia Antipolis Cedex, France.

[4] Instituto di Ricerche di Biologia Molecolare *P. Angeletti*, Viale Pontina Km 30,600, I-00040 Pomezia, Rome, Italy.

ABSTRACT

The metallo-β-lactamase superfamily comprises a remarkable set of enzymes that catalyse the hydrolysis of a wide range of substrates such as peptides, nucleic acids, antibiotics of the penicillin family and organophosphorus derivatives. In the past ten years, X-Ray structures of representative enzymes from different families have been determined, with the metallo-β-lactamases being the most represented. The salient common structural feature is the presence of a catalytic metal centre embedded within a αβ/βα fold. The wealth of sequence and structural information on metallo-β-lactamases has allowed their classification into three subclasses. Structural information is now available for members of each subclass. Interestingly, these structures show the presence of either a mono or a di-nuclear metal centre in the active sites raising questions on the metal to protein stoichiometry under physiological conditions. In addition, the structures reveal a wide variability in the shape of the active site, which involves three variable

---

* The first 2 authors contributed equally to this work.

† Corresponding author, Institut de Biologie Structurale *Jean-Pierre Ebel* (CNRS-CEA-UJF), 41, rue Jules Horowitz, F-38027 Grenoble Cedex 1, France. Tel.: +33 438 78 56 09; Fax: +33 438 78 54 94; E-mail: otto@ibs.fr

loops lining the metal centre. For each enzyme a clear correlation is found between active site shape and substrate specificity.

Abbreviations

ABL: standard numbering scheme for class A β-lactamases
BBL: standard numbering scheme for class B β-lactamases
DTT: dithio-threitol
PBP: Penicillin Binding Protein
TCEP:Tri (2-carboxyethyl) phosphine

Introduction

β-lactamases are bacterial enzymes that catalyse the hydrolysis of the β-lactam ring of antibiotics of the penicillin and cephalosporin families. After hydrolysis, these compounds are no longer efficient against their targets, the membrane-bound transpeptidase enzymes (Penicillin Binding Proteins-PBPs). Thus, β-lactamases represent the major causes of bacterial resistance to β-lactam based antibiotics. β-lactamases have been divided into four classes, A to D on the basis of their amino-acid sequences [1,2]. In enzymes of classes A, C and D, a serine residue is responsible for the nucleophilic attack on the carbonyl carbon atom of the β-lactam ring. The serine β-lactamases and the PBPs are structurally related and the former probably evolved from the latter or from a common ancestor [3-5]. Class B enzymes have no structural similarity with the active-site serine enzymes and require a bivalent metal ion (Zn^{2+}) for activity.

The first class B β-lactamase identified was found in a relatively innocuous *Bacillus cereus* strain [6] and initially received little attention. Thereafter class B enzymes have been found in at least 20 species of bacteria, many of which include pathogenic strains. The observations that class B enzymes can hydrolyse almost all β-lactam antibiotics, including carbapenems, and that they are not sensitive to the classical β-lactamase inhibitors underline their clinical relevance. Moreover, class B enzymes are often plasmid-encoded and their spreading to other pathogenic bacteria is a frightening possibility that further highlights the importance and urgency of finding effective inhibitors. Several recent reviews have discussed various aspects of class B enzymes [7-13]. The present contribution is focused on the results obtained by X-ray crystallography and their implications for metallo-β-lactamase evolution, mechanism of catalysis and design of inhibitors.

1. Metallo-β-lactamase Superfamily Structure: Historical Aspects

Metallo-β-lactamase sequences share a small number of conserved motifs related to the zinc ligands. They display significant sequence diversity and have been classified into three subclasses: B1, B2 and B3 [14]. All the metallo-β-lactamases have similar sizes (220-280 amino acids but the L1 enzyme from *Stenotrophomonas maltophilia* is a homotetramer) and within each subclass share more than 50% sequence identity. The first structure of a metallo-β-lactamase was reported in 1995 for BcII, a subclass B1 enzyme, from *Bacillus cereus* [15]. The structures of other subclass B1 enzymes were later solved (Table 1): CcrA from *Bacteroides fragilis* [16], IMP-1 from *Pseudomonas aeruginosa* and *Serratia marcescens* [17]; BlaB from *Chryseobacterium meningosepticum* [18]; VIM-2 and VIM-4 from *Pseudomonas aeruginosa* [19, 20]; SPM-1 from *Pseudomonas aeruginosa* [21] and IND-7 from *Chryseobacterium indologenes* [22]. The structure of the first subclass B3 protein, L1 from *Stenotrophomonas maltophilia* was reported in 1998 [23], and was followed by FEZ-1 from *Fluoribacter gormanii* [24] and BJP-1 from *Bradyrhizobium japonicum* [25] (Table 1). Finally, the first structure of a subclass B2, CphA from *Aeromonas hydrophila*, was solved in 2005 [26]. Additional structures were published for subclass B1 enzymes: BcII [27, 28], BcII mutants [29-33], cadmium- and cobalt-substituted BcII (Garcia-Saez, unpublished data, PDB code 1MQO) [34]; CcrA [35], CcrA mutants [36], cadmium- and mercury-substituted CcrA [37], CcrA in complex with small molecules [38-40], IMP-1 mutants [41], IMP-1 and VIM-2 in complex with small molecules [17, 42-45]. For subclass B2, an intermediate of the enzymatic reaction was trapped into the active site of CphA [26] and structures of CphA in complex with small molecules were recently reported [46-48]. Finally, L1 mutants [49], apo-, mono-zinc, native and Cu-substituted forms of L1 and L1 in complex with D-captopril [50] were reported as well as a FEZ-1 mutant [51]. (Table1).

Metallo-β-lactamases belong to a structural family [15] with an αβ/βα fold of novel topology (Figure 1). In the structure, helices are exposed to the solvent and surround the compact core of the protein consisting of two central β-sheets. The polypeptide chain folds into the first β-sheet then into the second β-sheet without intermediate crossing-over between them. The large amount of structural information available has allowed a structural alignment of all available metallo-β-lactamase sequences [52]. Moreover, in order to facilitate the comparative analysis of the 3D structures, a standard numbering scheme has been proposed [52], the class **B** **b**eta-**l**actamase numbering (BBL), by analogy with the ABL numbering which has been widely used for class A β-lactamases [53]. The structural alignment highlights two highly conserved motifs: His/Asn/Gln116-Xaa-His118-Xaa-Asp120 and Gly/Ala195-His196-Ser/Thr197. These motifs, along with amino acids conserved in each subclass, are located on the upper part of the two β-sheets (Figure 2) and define two potential zinc binding sites (namely Zn1 and Zn2) in the enzyme active site. These two sites are close to each other and, when both occupied, form a binuclear centre (the zinc ligands for each subclass are listed in Table 2). The metal centre is located at the bottom of a cleft formed mainly by amino acid residues in the L_1 (56-67), L_2 (151-168) or L_3 (223-238) loops (Figure 2).

The loops vary in length, secondary structure and flexibility in each subclass. All the metallo-β-lactamases discovered until now have been reported to function *in vivo* in the presence of Zn^{2+} although *in vitro* they can also be active in the presence of Cd^{2+} or Co^{2+}. However, the number of Zn^{2+} ions required for optimal activity varies within the metallo-β-lactamase family; how these differences in metal requirement translate into differences in the catalytic mechanisms and what is the real metal to protein stoichiometry under physiological conditions is still an open question [10].

Table 1. Metallo-β-Lactamase structures available in 2010

Depo Year	Sub class	PDB ID	Name	Source	Protein	Reso (Å)	Ref.	Comments
1995	B1	1BMC	BcII	*BACILLUS CEREUS*	Strain 569/H/9	2.50	[15]	First X-ray structure of the superfamily
1996	B1	1ZNB	CcrA	*BACTEROIDES FRAGILIS*	Wt	1.85	[16]	First di-zinc structure of the superfamily
1997	B1	1A8T	CcrA	*BACTEROIDES FRAGILIS*	Mutant A171T, D208N	2.55	[39]	L-159,061 complex
1997	B1	2BC2	BcII	*BACILLUS CEREUS*	Strain 569/H/9	1.70	[27]	C221-cysteine-S-dioxide (CSD)
1997	B1	2ZNB	CcrA	*BACTEROIDES FRAGILIS*		2.15	[37]	Cadmium derivative
1997	B1	3ZNB	CcrA	*BACTEROIDES FRAGILIS*		2.70	[37]	Mercury derivative
1998	B1	1A7T	CcrA	*BACTEROIDES FRAGILIS*	Mutant A171T, D208N	1.85	[38]	Mes complex
1998	B1	1BC2	BcII	*BACILLUS CEREUS*	Strain 569/H/9	1.90	[27]	500μM Zn, pH 4.5, 2 Zn
1998	B1	1BVT	BcII	*BACILLUS CEREUS*	Strain 569/H/9	1.85	[28]	100μM Zn, Zn2 site partial occupancy
1998	B1	2BMI	CcrA	*BACTEROIDES FRAGILIS*		2.00	[35]	New crystal form
1998	B1	3BC2	BcII	*BACILLUS CEREUS*	Strain 569/H/9	1.70	[27]	C221-cysteine-S-dioxide (CSD)
1998	B1	4ZNB	CcrA	*BACTEROIDES FRAGILIS*	Mutant C181S	2.65	[36]	C221S mutant
1999	B1	1DD6	IMP-1	*PSEUDOMONAS AERUGINOSA*	Wt	2.00	[17]	Mercaptocarboxylate complex
1999	B1	1DDK	IMP-1	*PSEUDOMONAS AERUGINOSA*	Wt	3.10	[17]	Acetic acid complex
2000	B1	1DXK	BcII	*BACILLUS CEREUS*	Strain 569/H/9; Mutant C168S	1.85	[29]	Role of C221 in the catalytic mechanism
2000	B1	1HLK	Cfia	*BACTEROIDES FRAGILIS*	Wt	2.50	[40]	Tricyclic inhibitor complex
2000	B1	1KR3	CcrA	*BACTEROIDES FRAGILIS*	Wt	2.50	[40]	Tricyclic inhibitor complex
2001	B1	1JJE	IMP-1	*PSEUDOMONAS AERUGINOSA*	Wt	1.80	[42]	Biaryl succinic acid complex
2001	B1	1JJT	IMP-1	*PSEUDOMONAS AERUGINOSA*	Wt	1.80	[42]	Biaryl succinic acid complex
2001	B1	1KO2	VIM-2	*PSEUDOMONAS AERUGINOSA*	Wt	2.20	[19]	C221-cysteine sulfonic acid (OCS)

Depo Year	Sub class	PDB ID	Name	Source	Protein	Reso (Å)	Ref.	Comments
2001	B1	1KO3	VIM-2	*PSEUDOMONAS AERUGINOSA*	Wt	1.91	[19]	Reduced form
2001	B1	1MQO	BcII	*BACILLUS CEREUS 569/H/9*	Strain 569/H/9	1.35	n/a	Cadmium derivative
2002	B1	1M2X	BlaB	*CHRYSEOBACT E- RIUM MENINGOSEPTI - CUM*	Wt	1.50	[18]	D-captopril complex
2004	B1	1WUO	IMP-1	*SERRATIA MARCESCENS*	Mutant D81A	2.01	n/a	Role of D120 in the catalytic mechanism
2004	B1	1WUP	IMP-1	*SERRATIA MARCESCENS*	Mutant D81E	3.00	n/a	Role of D120 in the catalytic mechanism
2004	B1	2BFK	BcII	*BACILLUS CEREUS*	569/H/9Mutant R121C	2.00	[30]	20mM Zn, pH 7.0, 1mM DTT, C221-S-hydroxy-cysteine (CEA)
2004	B1	2BFL	BcII	*BACILLUS CEREUS*	569/H/9Mutant R121C	1.80	[30]	20mM Zn, pH 5.0, 1mM DTT,
2004	B1	2BFZ	BcII	*BACILLUS CEREUS*	569/H/9Mutant R121C	2.30	[30]	20mM Zn, pH 4.5, 1mM DTT, C221-cysteine-S-dioxide (CSD)
2004	B1	2BG2	BcII	*BACILLUS CEREUS*	569/H/9Mutant R121C	2.40	[30]	20mM Zn, pH 4.5, 1mM DTT-TCEP, C221 reduced
2004	B1	2BG6	BcII	*BACILLUS CEREUS*	569/H/9Mutant R121C	2.30	[30]	20μM Zn, pH 5.0 1mM DTT, C221-cysteine-S-dioxide (CSD)
2004	B1	2BG7	BcII	*BACILLUS CEREUS*	569/H/9Mutant R121C	2.10	[30]	20μM Zn, pH 4.5 1mM DTT, C221-S-hydroxy-cysteine (CEA)
2004	B1	2BG8	BcII	*BACILLUS CEREUS*	569/H/9Mutant R121C	2.50	[30]	20μM Zn, pH 4.5 1mM DTT-TCEP, C221-S-hydroxy-cysteine (CEA)
2004	B1	2BGA	BcII	*BACILLUS CEREUS*	569/H/9Mutant R121C	2.70	[30]	20μM Zn, pH 7.0 1mM DTT, C221-cysteine-S-dioxide (CSD)
2005	B1	1VGN	IMP-1	*SERRATIA MARCESCENS*	Wt	2.63	[43]	Inhibitor complex
2005	B1	2FHX	SPM-1	*PSEUDOMONAS AERUGINOSA*	Wt	1.90	[21]	
2006	B1	2DOO	IMP-1	*SERRATIA MARCESCENS*	Wt	2.43	[44]	Complex with DansylC4SH

Table 1. (Continued).

Depo Year	Sub class	PDB ID	Name	*Source*	Protein	Reso (Å)	Ref.	Comments
2006	B1	2NXA	BcII	*BACILLUS CEREUS*	569/H/9 Double mutant R121H-C221D	2.29	[32]	Mono-zinc
2006	B1	2NYP	BcII	*BACILLUS CEREUS*	569/H/9 Double mutant R121H-C221D	1.84	[32]	di-zinc
2006	B1	2NZF	BcII	*BACILLUS CEREUS*	569/H/9 Double mutant R121H-C221S	2.28	[32]	Space group C2
2006	B1	2NZE	BcII	*BACILLUS CEREUS*	569/H/9 Double mutant R121H-C221S	1.80	[32]	Space group P3121
2007	B1	2UYX	BcII	*BACILLUS CEREUS*	569/H/9 Mutant D120S	1.95	[31]	Role of Asp120 in the catalytic mechanism
2008	B1	2YZ3	VIM-2	*PSEUDOMONAS AERUGINOSA*	Wt	2.30	[45]	Complex with mercaptocarboxylate inhibitor
2008	B1	3FCZ	BcII	*BACILLUS CEREUS*	569/H/9 Mutant N70S-V112A-L250S-G262S	2.80	[33]	
2009	B1	3I0V	BcII	*BACILLUS CEREUS*	Wt	1.60	n/a	Apo-form
2009	B1	3I11	BcII	*BACILLUS CEREUS*	Wt	1.45	n/a	Di-cobalt form
2009	B1	3I13	BcII	*BACILLUS CEREUS*	Wt	1.74	n/a	At pH 5.8, 2 Zn
2009	B1	3I14	BcII	*BACILLUS CEREUS*	Wt	1.55	n/a	Di-cobalt form, Cys221 partially oxidized
2009	B1	3I15	BcII	*BACILLUS CEREUS*	Wt	1.55	n/a	Mono-cobalt form, Cys221 fully oxidized
2010	B1	2WHG	VIM-4	*PSEUDOMONAS AERUGINOSA*	Wt	1.90	[20]	Di-zinc form
2010	B1	2WRS	VIM-4	*PSEUDOMONAS AERUGINOSA*	Wt	2.60	n/a	Mono-zinc form
2010	B1	3L6N	IND-7	*CHRYSEOBACTERIUM INDOLOGENES*	Wt	1.65	[22]	Di-zinc form

Depo Year	Sub class	PDB ID	Name	*Source*	Protein	Reso (Å)	Ref.	Comments
2004	B2	1X8G	CphA	*AEROMONAS HYDROPHILA*	Wt	1.70	[26]	First structure of the subclass B2
2004	B2	1X8H	CphA	*AEROMONAS HYDROPHILA*	Mutant N220G	1.60	[26]	Mutant, Zn-Arg121 bond
2004	B2	1X8I	CphA	*AEROMONAS HYDROPHILA*	Wt	1.90	[26]	biapenem complex
2006	B2	2GKL	CphA	*AEROMONAS HYDROPHILA*	Wt	1.86	[46]	complex with the inhibitor pyridine-2,4-dicarboxylate
2007	B2	2QDS	CphA	*AEROMONAS HYDROPHILA*	Wt	1.66	[47]	Complex with the inhibitor D-captopril
2009	B2	3F90	CphA	*AEROMONAS HYDROPHILA*	Wt	1.70	[100]	Structure of the di-zinc form of CphA
2009	B2	3FAI	CphA	*AEROMONAS HYDROPHILA*	Mutant N220G	2.03	[100]	Structure of the di-zinc form of the N220G CphA mutant
2009	B2	3IOF	CphA	*AEROMONAS HYDROPHILA*	Mutant N220G	1.44	[48]	Complex with the mercaptophosphonate inhibitor 10a
2009	B2	3IOG	CphA	*AEROMONAS HYDROPHILA*	Mutant N220G	1.41	[48]	Complex with themercaptophosphonate inhibitor 18
1998	B3	1SML	L1	*XANTHOMONAS MALTOPHILIA*	Wt	1.70	[23]	First structure of the subclass B3
2001	B3	1JT1	FEZ-1	*FLUORIBACTER GORMANII*	Wt	1.78	[24]	D-captopril complex
2001	B3	1K07	FEZ-1	*FLUORIBACTER GORMANII*	Wt	1.65	[24]	First structure of a monomeric subclass B3 enzyme
2003	B3	1L9Y	FEZ-1	*FLUORIBACTER GORMANII*	Mutant Y228A	2.01	[51]	Inhibition by the product formed upon hydrolysis of cephalosporins
2005	B3	2AIO	L1	*STENOTROPHO-MONAS MALTOPHILIA*	Wt	1.70	[110]	Hydrolyzed moxalactam complex
2006	B3	2FM6	L1	*STENOTROPHO-MONAS MALTOPHILIA*	Wt	1.75	[50]	Native form
2006	B3	2FU6	L1	*STENOTROPHO-MONAS MALTOPHILIA*	Wt	2.05	[50]	Apo-form
2006	B3	2FU7	L1	*STENOTROPHO-MONAS MALTOPHILIA*	Wt	1.85	[50]	Cu-substituted form
2006	B3	2FU8	L1	*STENOTROPHO-MONAS MALTOPHILIA*	Wt	1.80	[50]	d-captopril complex
2006	B3	2FU9	L1	*STENOTROPHO-MONAS MALTOPHILIA*	Wt	1.80	[50]	Mp2 inhibitor complex

Table 1. (Continued).

Depo Year	Sub class	PDB ID	Name	*Source*	Protein	Reso (Å)	Ref.	Comments
2006	B3	2H6A	L1	*STENOTROPHO-MONAS MALTOPHILIA*	Wt	1.80	[50]	Mono-zinc form
2006	B3	2GFJ	L1	*STENOTROPHO-MONAS MALTOPHILIA*	Wt	1.80	[50]	Inhibitor 1
2006	B3	2GFK	L1	*STENOTROPHO-MONAS MALTOPHILIA*	Wt	1.90	[50]	Inhibitor 2
2006	B3	2HB9	L1	*STENOTROPHO-MONAS MALTOPHILIA*	Wt	1.75	[50]	Inhibitor 3
2006	B3	2GM N	BJP-1	*BRADYRHIZO-BIUM JAPONICUM*	Wt	1.40	[25]	
2007	B3	2QDT	L1	*STENOTROPHO-MONAS MALTOPHILIA*	Wt	1.78	[47]	L1-IS38 inhibitor complex
2007	B3	2QJS	L1	*STENOTROPHO-MONAS MALTOPHILIA*	Mutant D120N	2.25	[49]	Role of Asp120 in the catalytic mechanism
2007	B3	2QIN	L1	*STENOTROPHO-MONAS MALTOPHILIA*	Mutant D120C	1.76	[49]	Role of Asp120 in the catalytic mechanism

Accurate determination of Zn^{2+} affinity in absence and presence of substrate is available for four representative metallo-β-lactamases: BcII, BlaB, CphA and L1 [54]. For the formation of the mono-zinc enzymes the dissociation constants are in the nanomolar and picomolar range, in the absence and presence of substrate, respectively (but see the conclusion section below). Notably, the free Zn^{2+} concentration has been proposed to be in the femtomolar range [55].

Table 2. Putative zinc ligands in class B β-lactamases

	"Histidine" site (Zn1)		"Cysteine" site (Zn2)
B_1	His116 His196	His118	Asp120 Cys221 His263
B_2	[Asn116][a] His196	His118	Asp120 Cys221 His263
B_3	His/Gln116[b] His196	His118	Asp120 His121 His263

[a] In CphA, Asn116 is not involved in the binding of the second zinc (the distance is more than 4 Å) [100].

[b] GOB enzymes (B3) have a Gln116 instead of a histidine in the "histidine" but the presence of a zinc ion in this site remains controversial (see text).

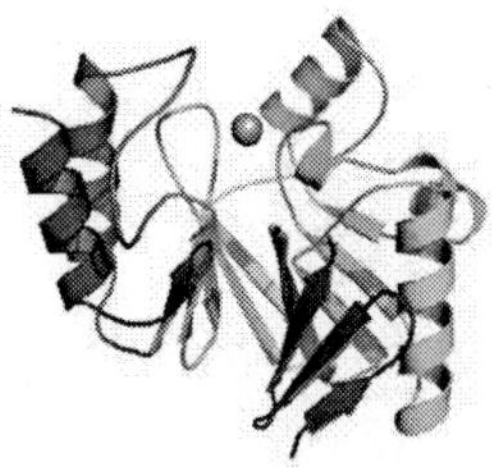

(A)

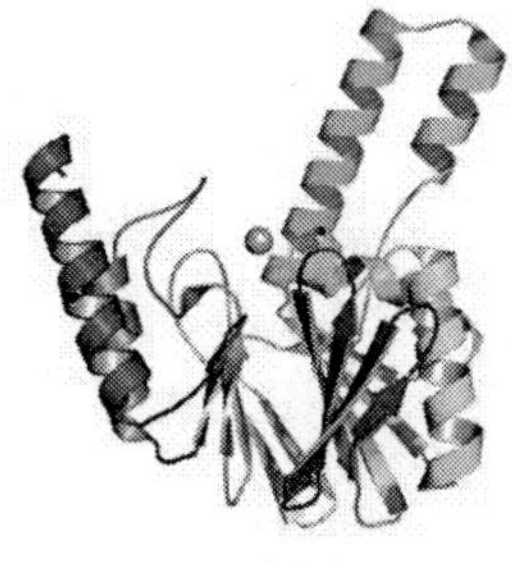

(B)

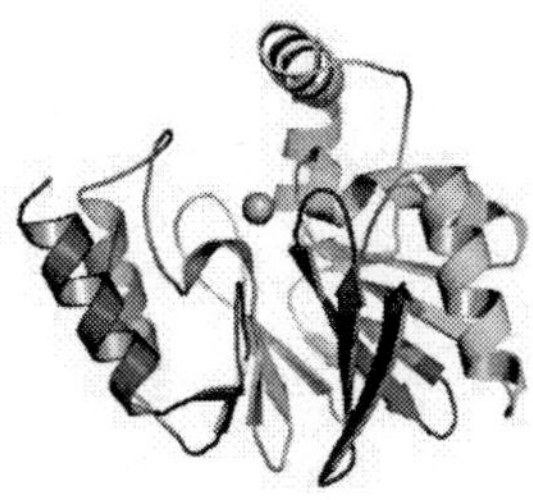

(C)

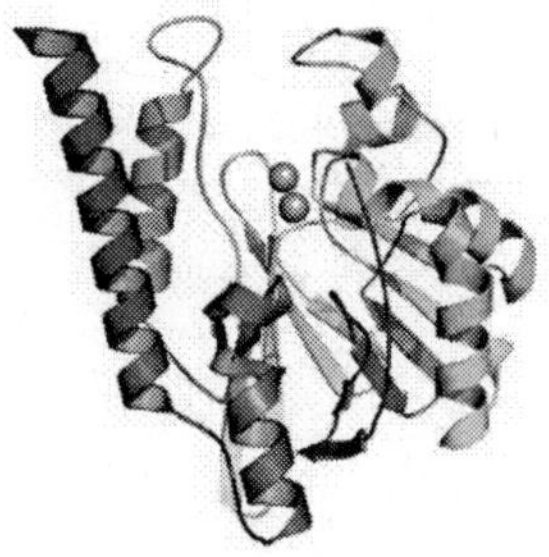

(D)

Figure 1. Rainbow cartoons of representatives of the three subclasses of metallo-β-lactamases.
(A) Subclass B1: BcII from *B. cereus* (PDB code 1BMC)
(B) Subclass B1: SPM-1 from *P. aeruginosa* (PDB code 2FHX)
(C) Subclass B2: CphA from *A. hydrophila* (PDB code 1X8G)
(D) Subclass B3: FEZ-1 from *F. gormanii* (PDB code 1K07)
Zinc ions are represented as grey spheres.

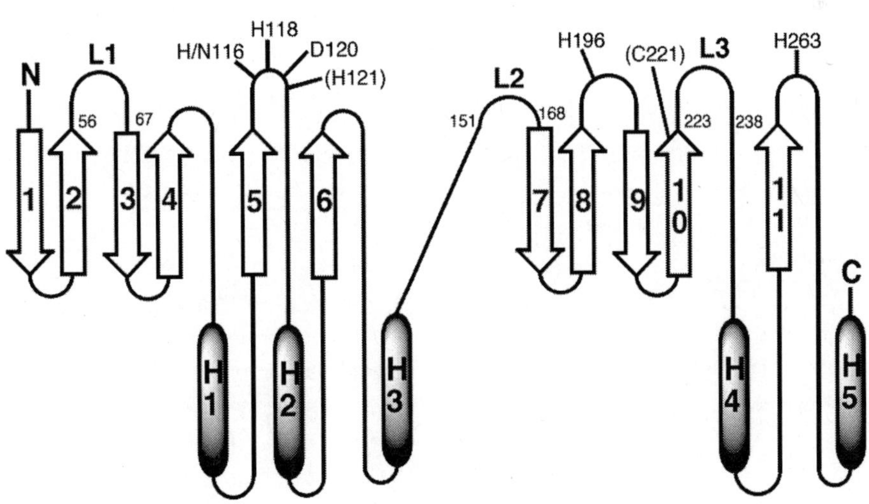

Figure 2. General metallo-β-lactamase topology.
Strictly conserved secondary structures in all subclasses are represented as arrows and cartouches for β-strands and α-helices, respectively. Corresponding β-strands in the two sheets are labelled in red. L1 loops line the active-site cavity. Zn-ligands are labelled.

1.1. β-lactamases: Evolutionary Aspects

Based on their three dimensional structures, β-lactamases can be classified into two distinct structural superfamilies. The first superfamily is essentially bacterial and contains the active-site serine β-lactamases and penicillin-binding proteins (PBPs). Their common ancestor is more likely an essential PBP involved in peptidoglycan synthesis. Presumably the active site and function were slowly modified by evolutionary pressure to produce the three presently known enzyme classes A, C and D. The founding member of the second structural superfamily is the metallo-β-lactamase BcII whose structure was solved in 1995 [15]. The members of this superfamily have very different functions related to hydrolysis and redox reactions, DNA repair and uptake, and RNA processing [56-58]. The metallo-β-lactamase superfamily contains more than 18500 proteins spanning all natural kingdoms. The largest group, composed of more 3200 sequences, is ubiquitous and the corresponding enzymes are involved in an essential function, t-RNA maturation. These two features might suggest that t-RNA maturase is the common ancestor of the superfamily. The structures of four of these enzymes have recently been described (tRNAses Z from *Thermotoga maritima*, *Bacillus subtilis*, *Escherichia coli* and *Bacillus anthracis*) [59-61] (Brunzelle *et al.*, unpublished results, PDB code 1ZKP). The 3D structures of other members of the metallo-β-lactamase superfamily have also been reported:

- Glyoxalase II from *Arabidopsis thaliana*, *Salmonella typhimurium*, *Leishmania infantum* and human glyoxalase II [62-67],
- ETHE1 from *Arabidopsis thaliana* At1g53580 [68]
- Cytosolic redox proteins such as *Desulfovibrio gigas* rubredoxin oxygen:oxidoreductase flavoprotein [69], FprA from *Moorella thermoacetica* [70] and Tflp from *Thermoanaerobacter tengcongensis* [71],
- *Pseudomonas* sp. strain WBC-3 methyl parathion hydrolase [72]
- the *Streptococcus pneumoniae* teichoic acid phosphorylcholine esterase (CbpE) [73, 74],
- SdsA1 from *Pseudomonas aeruginosa* [75],
- the *Bacillus thuringiensis* N-acyl-L-homoserine lactone hydrolase [76] and the quorum-quenching lactonase (AiiB) from *Agrobacterium tumefaciens* [77, 78],
- the *E. coli* l-ascorbate-6-P lactonase [79],
- Proteins acting on nucleic acid substrates such as TTHA0252 from *Thermus thermophilus* HB8 [80] (Ishikawa *et al.*, unpublished results, PDB codes 3IEK, 3IEL, 3IEM) and its mutants (Ishikawa *et al.*, unpublished results, PDB codes 3A4Y, 3IDZ, 3IE0, 3IE1, 3IE2), TTHA1623 from *Thermus thermophilus* HB8 [81], TTHA1429 from *Thermus thermophilus* HB8 [82], the 73-kDa and 100-kDa subunits of cleavage and polyadenylation specificity factor (human CPSF-73 and yeast CPSF-100) [83], the RNase J of *Thermus thermophilus* [84] and subunits of archeal cleavage and polyadenylation specificity factors (Nishida *et al.*, unpublished results, PDB codes 3AF5, 3AF6) [85]
- two proteins of unknown function whose structures were solved in a structural genomics joint effort, TM0207 and TM0894 from *Thermotoga maritima* (unpublished results, PDB codes 1VJN and 1ZTC),

- an uncharacterized metallo-protein from *Vibrio cholerae* (Minasov *et al.*, unpublished results, PDB code 3BV6).

In all of these proteins most of the metal ligands are structurally conserved and are located, as in metallo-β-lactamases, on loops connecting secondary structure elements (Figure 2). However, differently from the metallo-β-lactamases, in all of the above mentioned structures the carboxylate group of an aspartic acid bridges the two active-site metal ions, with the exception of SdsA1 in which Glu221 only binds Zn2. Interestingly, within the metallo-β-lactamase superfamily the metal requirements vary dramatically among the different enzymes, with mono-zinc [27], di-zinc [23], zinc-iron [62], di-iron [69], mono-manganese [79] or di-manganese metal centers [62]. On the basis of a structure based phylogeny analysis of the metallo-β-lactamase superfamily it was suggested that subclass B1 and B2 enzymes arose 2.2 billion years ago [86]. Interestingly, subclass B3 homologues were identified in both *Eubacteria* and *Archaebacteria* suggesting than subclass B3 appeared earlier than 2.2 billion years ago [86].

2. Metallo-β-lactamase Protein Structures: 2010 Status

X-ray structure determinations can yield very valuable information regarding an entire protein family. However, one should take into consideration that X-ray structures can be affected by some of the experimental conditions. In this review, we would like to emphasize this point. X-ray structures of metallo-β-lactamase from all three subclasses have been deposited in the PDB [87]. The seventy-eight coordinate data sets now available (Table 1) were obtained from X-ray diffraction experiments at various resolutions (3.1-1.35 Å). Subclass B1 is well represented with 51 structures whereas only 9 and 18 structures are present for subclasses B2 and B3, respectively. X-ray structures were solved for native and mutant proteins and in some cases for protein-inhibitor complexes. In addition to the zinc ions located in the active-site, and sometimes more than 150 water molecules per monomer, other ligands are very often identified: most of them are present in the crystallisation (sulphate, carbonate and acetate ions) or in cryo-cooling (glycerol) solutions. Unfortunately, these molecules are present at non-physiological concentrations and disturb the active-site structures, preventing a detailed comparison between native, mutants and/or inhibitor-enzyme complex structures.

Other experimental conditions can also affect protein structure, such as pH of the buffer, presence of Zn^{2+} ions and/or reducing agent in the crystallisation medium. A wide range of pH buffers have been employed for metallo-β-lactamase crystallisation from 4.5 (BcII R121C mutant, [30]) to 9 (CcrA, [35]). In 2005, crystal structures of the R121C BcII mutant enzyme were solved at pH 7.0, 5.0 and 4.5, each in the presence of either 20 μM or 20 mM Zn^{2+}. Surprisingly, although Arg121 is buried and proximal to the active site, its mutation into cysteine has little effect on the active-site structure, in agreement with a small decrease in k_{cat}/K_M values for both mono and di-zinc R121C BcII enzymes when compared to the wild-type enzymes [88]. A network of ordered water molecules replaced the interactions made by the buried arginine side chain. However, at pH 4.5 and in the di-zinc form of the enzyme, the

Zn^{2+} in the Zn2 site moved away from Zn1 increasing the Zn1-Zn2 distance from 3.6 Å to 4.1 Å. In addition, the bridging water/hydroxide molecule was absent from the active site. The observed structures suggested that at low pH, protonation of the active-site Asp120 and of the bridging water/hydroxide molecule is likely to occur. The buffer pH clearly has an effect on the binuclear centre structure, but other factors are also important. A parameter to take into consideration is the Zn^{2+} concentration which in turn affects the zinc occupancy of sites Zn1 or Zn2. Variation in Zn^{2+} occupancy is observed in most of the subclass B1 enzymes: in BcII (PBP code 1BVT), BlaB (PBP code 1M2X) and VIM-2 (PBP code 1KO3), the zinc ion in site Zn2 was refined in the range of 60% to 80% occupancy.

B1 and B2 enzymes have a cysteine (i.e. Cys221) as one of the metal ligands in the Zn2 site that is susceptible to oxidation. Cysteine oxidation could lead to three types of chemical modification: generation of a S-hydroxy-cysteine (PDB acronym CEA, formula $C_3H_7NO_3S$), a cysteine-S-dioxide (PDB acronym CSD, formula $C_3H_7NO_4S$) or a cysteinesulfonic acid (PDB acronym OCS, formula $C_3H_7NO_5S$). To date, these modifications were observed in 44, 73 and 27 X-ray structures, respectively, for 31,434 deposited structures. The best resolutions were reached for the *E.coli* β-galactosidase in complex with galactose, with a CEA modification (1.5 Å; PDB code 1JZ4), the methylsulfoxide reductase from *Rhodobacter sphaeroides*, with a CSD residue (1.3 Å resolution; PDB code 1EV1) and for the quorum-sensing protein LuxS from *Bacillus subtilis*, with a OCS residue (1.2 Å resolution; PDB code 1J98). These high-resolution structures constituted a strong experimental support to the nature of the chemical modifications. In metallo-β-lactamase structures, the first observed modification was a cysteine-s-dioxide (CSD) in the BcII wild type enzyme (PDB code 2BC2 and 3BC2) and more recently in the R121C mutant (PDB code 2BFZ, 2BG6, 2BGA). For the R121C mutant in some conditions a S-hydroxy-cysteine (CEA) is observed (PDB code 2BFK, 2BG7, 2BG8). In some cases, an oxidized cysteine is observed even in presence of 1mM DTT and/or 1mM TCEP. When crystals of the R121C mutant of the BcII enzyme containing an oxidized Cys221 were soaked in a solution containing the chromophoric substrate nitrocefin, they turned red, suggesting the presence of an active enzyme [30]. Interestingly, in the crystals grown at pH 4.5, in presence of 20 μM $ZnSO_4$, 1mM DTT and 1mM TCEP, Cys221 is oxidized (CEA) in one molecule of the asymmetric unit but not in the other, and both molecules contain only one Zn^{2+} ion located in Zn1 site [30]. SPM-1 presents a CSD modification in one molecule of the asymmetric unit and a CEA modification in the second one (PDB code 2FHX). The third modification (OCS) was observed in the VIM-2 metallo-β-lactamase (PDB code 1KO2), another subclass B1 enzyme, for which the presence of a cysteinesulfonic acid was unexpected. To try to discern at which step of the structural study Cys221 became oxidised, electrospray mass-spectrometry analysis of samples from either freshly-purified samples, or protein stored at - 20°C or after crystallisation were performed. It appears that oxidation occurred during the crystallisation process even in presence of 10 mM DTT. Only the presence of 1 mM TCEP prevented the oxidation process [19]. However, oxidized VIM-2 is an inactive enzyme [19]. Recently, an OCS modification was also observed in the apo- and cobalt-forms of BcII [34].

The presence of Zn^{2+} ions raises another problem in terms of structure-function analysis. To grow crystals or to improve their diffraction quality, cofactors, ligands or inhibitors are often added to the crystallization buffer. In many cases, Zn^{2+} salts were added in the crystallisation buffers at concentrations far above the physiological ones. The resulting active-site structures might be misleading from a functional point of view. The best example is for the BcII enzyme, for which crystals were grown in presence of 100 μM [15] [28] or 500 μM Zn^{2+} [27]. The corresponding active-site structures displayed either a mononuclear [15] or a binuclear centre [27, 28], with an increasing occupancy by zinc in site Zn2.

Most of the metallo-β-lactamase crystals contain one molecule of protein in their asymmetric unit. In a few cases, two molecules were observed (BcII, PDB code 1BC2) (CcrA, PDB code 1A7T) (SPM-1, PDB code 2FHX).

Finally, BlaB from *C. meningosepticum*, a subclass B1 enzyme, was crystallised with four molecules in the asymmetric unit (PDB code 1M2X). The monomers are basically identical, with a few local differences due to crystal packing effects. In the crystal, 12 monomers form a particle with a cubic symmetry. However, BlaB is monomeric in solution. The only oligomeric metallo-β-lactamase, L1 from *S. maltophilia*, a subclass B3 enzyme, is a tetramer in solution but crystallised with one molecule in the asymmetric unit. The other monomers are generated by crystallographic two-fold axis.

An interesting aspect in protein crystallography is the possibility to diffuse small molecules into the crystal.

The preferred method to prepare a protein-inhibitor complex is by soaking the protein crystal into a solution of the inhibitor. The success of the soaking procedure strongly depends on the presence of channels in the crystal that are wide enough to allow the inhibitor to diffuse and to reach the active-site of the enzyme. Such crystals are very useful for X-ray crystallographic screening of novel inhibitors [89]. Based on reported X-ray structures of protein-inhibitor complex CcrA, IMP-1, BlaB, CphA, L1 and FEZ-1 crystals could be useful in the screening method.

3. Metallo-β-lactamases Protein Structures: Examples for Each Subclass

3.1. Overall Structure of a Subclass B1 Metallo-β-lactamase

Structures from eight different subclass B1 enzymes have been deposited in the PDB. The identities among their sequences aligned with the help of 3D-structural data vary between 28 and 39%. Figure 3A shows a comparison of representative structures of these enzymes. The main differences are observed in the conformation of the L_1 loop that forms the left side of the active-site cleft. The flexible L_1 loop has been shown to interact directly with bound inhibitors. The SPM-1 enzyme lacks this L1 loop.

3.1.1. BcII from Bacillus cereus

BcII is one of two extracellular β-lactamases produced by the soil bacterium *Bacillus cereus* 569/H [6]. Unlike β-lactamase I (a class A β-lactamase), which is active mainly against penicillins, BcII has a very broad spectrum of activity encompassing most bicyclic β-lactams. Twenty-six crystal structures of native, mutant and metal-substituted BcII are available, representing both mononuclear and binuclear species [90].

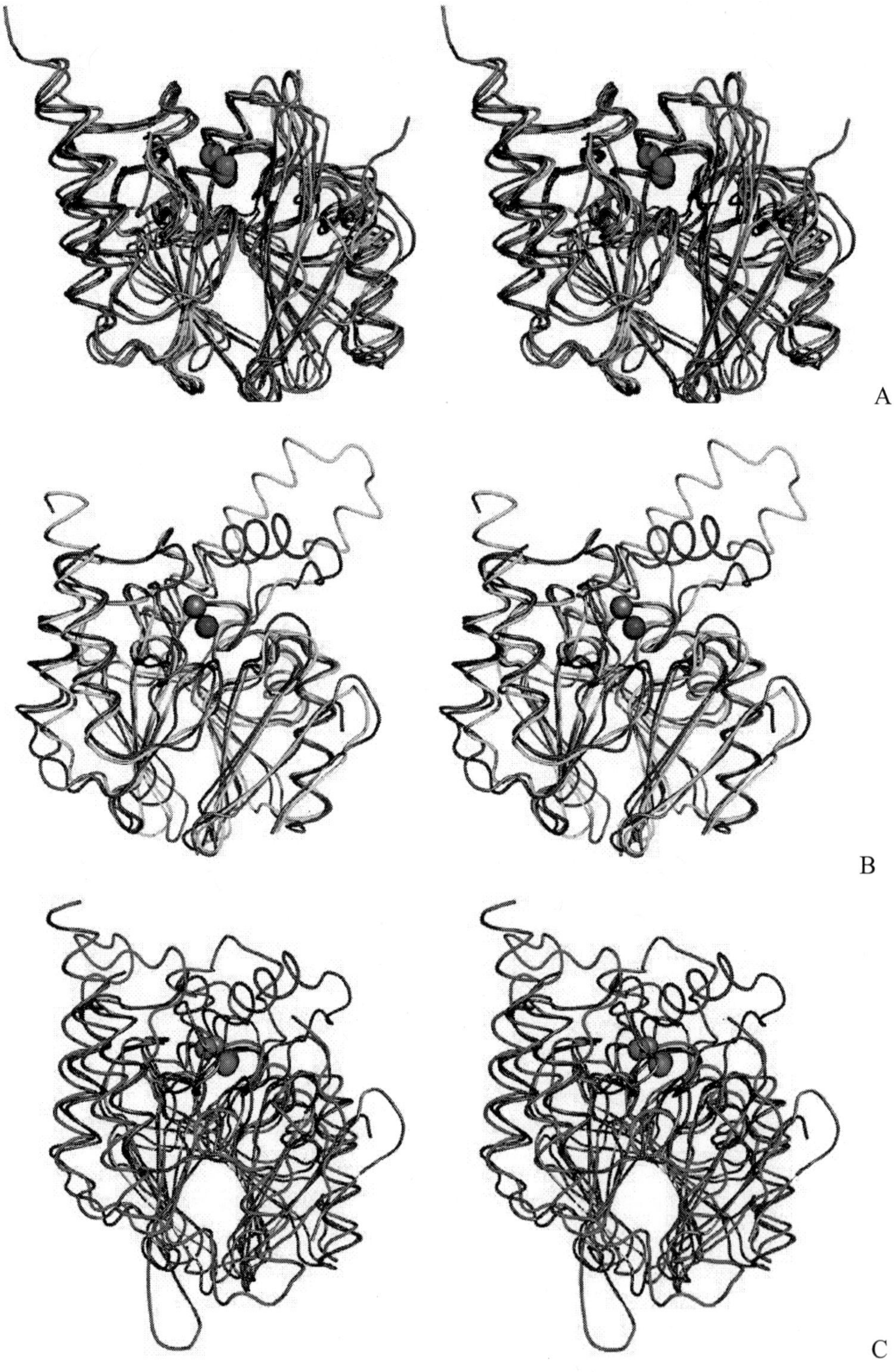
A
B
C

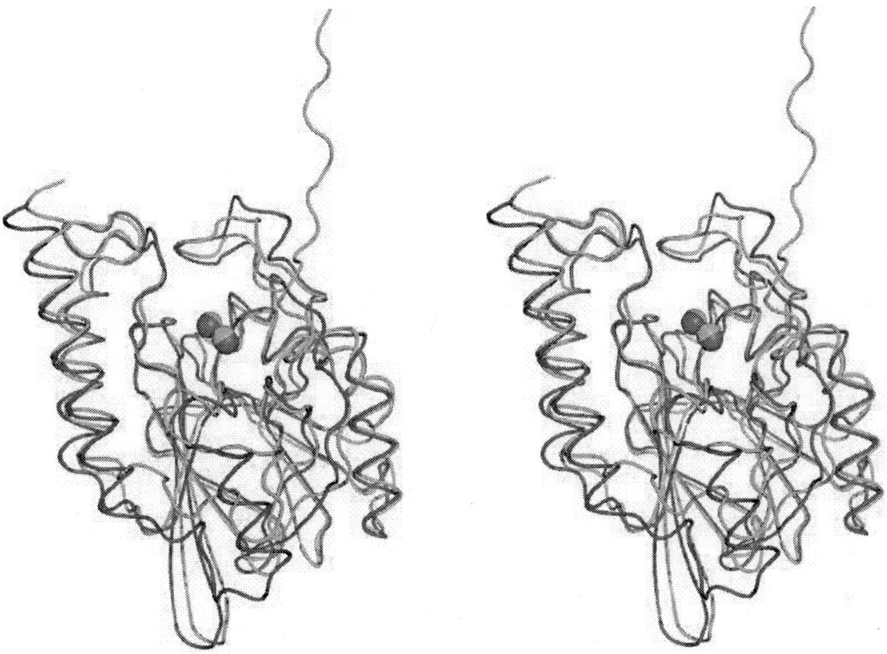

Figure 3. Stereo view of the superimposed Cα backbones for each subclass.
(A) BcII (grey), CcrA (blue), IMP-1 (magenta), BlaB (orange) and Vim-2 (green) enzymes
(B) BcII (grey), SPM-1 (cyan) and CphA (blue)
(C) BcII (grey), CphA (blue) and FEZ-1 (magenta)
(D) FEZ-1 (magenta) and L1 (green)
(E) Zinc ions are represented as spheres.

Structure of the Active Site

The X-ray structures reveal that the Zn1 site is always fully occupied and that the Zn2 site is variably occupied. The occupancy of the Zn2 site seems to depend on the pH of the crystallization solution, cysteine oxydation and concentration of Zn^{2+} ions [29]. Accordingly, the BcII enzyme likely acts as a mononuclear species in physiological conditions [10]. Two mononuclear Zn^{2+} structures were reported [15] (PDB code 1BMC) and [30] (PDB code 2BG2), however only at 2.5 Å and 2.4 Å resolution, respectively. In order to perform a detailed theoretical study of the mechanism of β-lactam hydrolysis, a model was generated based on the mono-zinc–1BMC structure at the highest available resolution [91]. In this theoretical BcII model, Zn^{2+} in site Zn1 is coordinated by three histidines and a hydroxide ion in a tetrahedral arrangement (Figure 4A). For comparison purposes the stereo view of the mononuclear structure (1BMC) is also represented (Figure 5A). The theoretical model slightly differs from the 1BMC structure. Crystals were grown at pH 5.6 which may explain the observed longer Zn-water/hydroxide distance (3.3 Å) in the crystal. Addition of free Zn^{2+} ions and increasing the pH buffer drives the enzyme from a mononuclear to a binuclear species [30].

The active site is located at the bottom of a long groove running on the surface of the protein (Figure 6A). The shape of the cleft is modulated by conformational changes of loops L_1 and L_3 (Figure 2).

Structure of the Binuclear Active Site

The X-ray structure of the BcII R121C mutant at pH 7.0 and in the presence of 20 mM zinc sulphate was solved in 2004 [30]. Superposition of the mutant and native enzyme structures revealed no major structural changes (rmsd for 250 Cα is 0.38 Å). In the active site, both zinc binding sites are occupied. Importantly, the presence of the second Zn^{2+} ion does not affect the geometry of amino acids delimiting the active site. In the Zn1 site, the zinc ion is coordinated by four ligands in a tetrahedral geometry: the nitrogen atom of the imidazole rings from three different histidine residues (Table 2) and an oxygen atom that was refined as a water/hydroxide ion (Wat1). In the Zn2 site, the zinc ion is coordinated to five ligands: three amino acid side chains and two water molecules (Wat1 and Wat2) (Figure 4B), adopting a distorted trigonal bipyramidal geometry. A stereo view of the active site is also represented (Figure 5B).

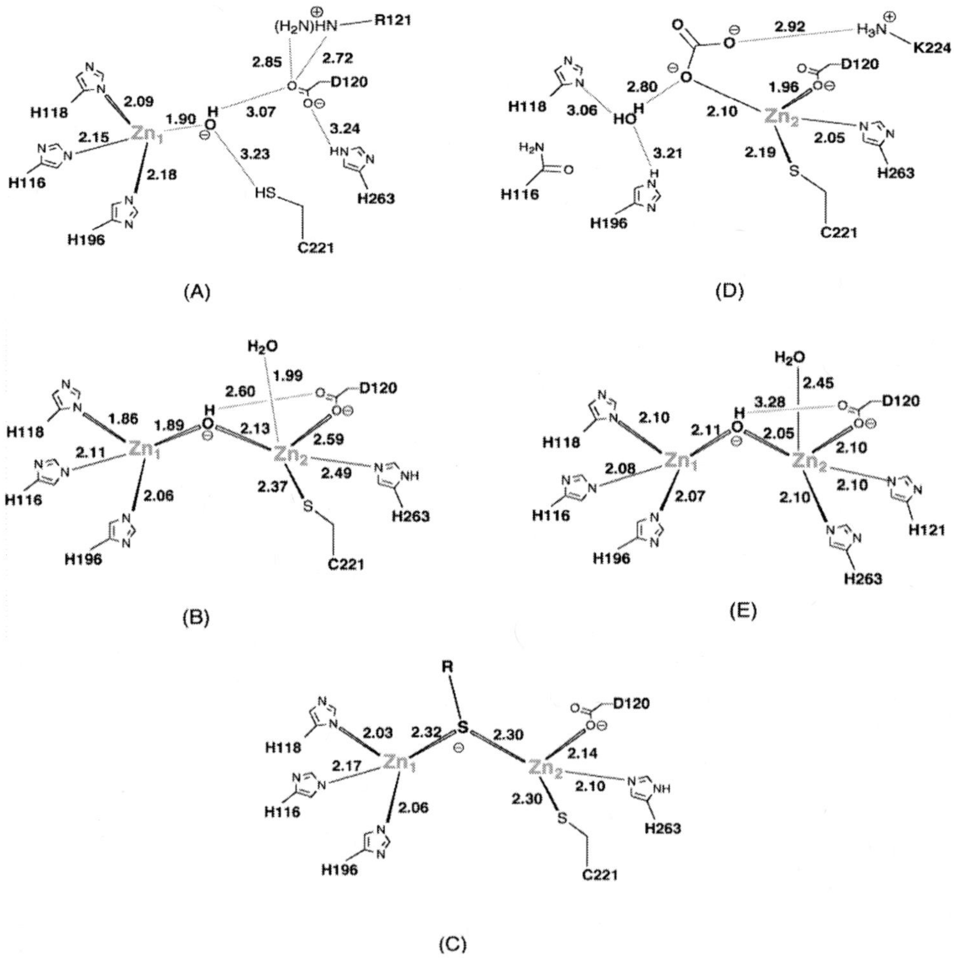

Figure 4. Schematic diagram of the Zn interactions.
(A) Mono-zinc enzyme BcII model [91]
(B) Di-zinc enzyme BcII R121C mutant (PDB code 2BFK)
(C) Di-zinc enzyme BlaB (PDB code 1M2X)
(D) Mono-zinc enzyme CphA (PDB code 1PX1)
(E) Di-zinc enzyme L1 (PDB code 1SML).

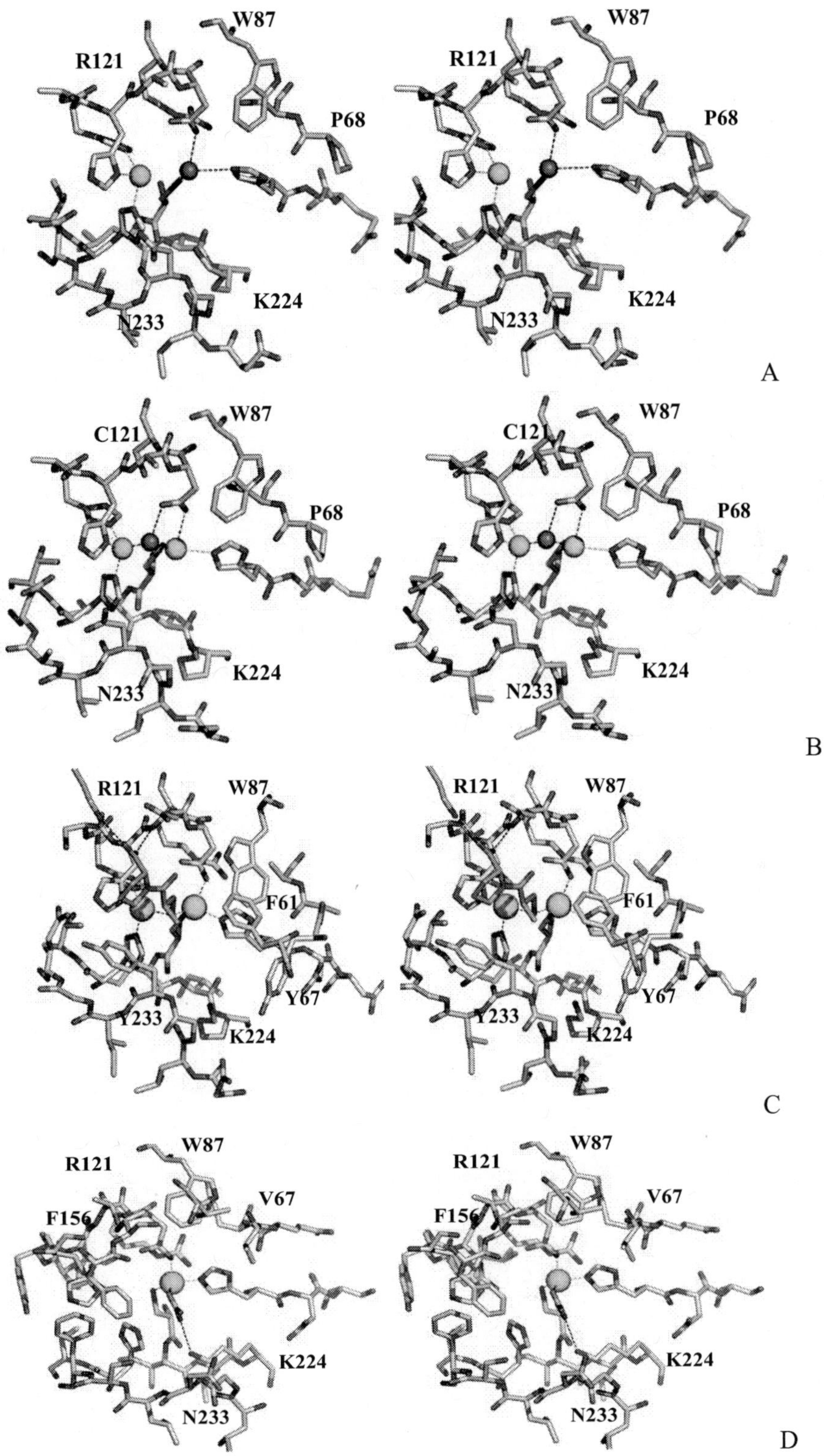

Figure 5. (Continued)

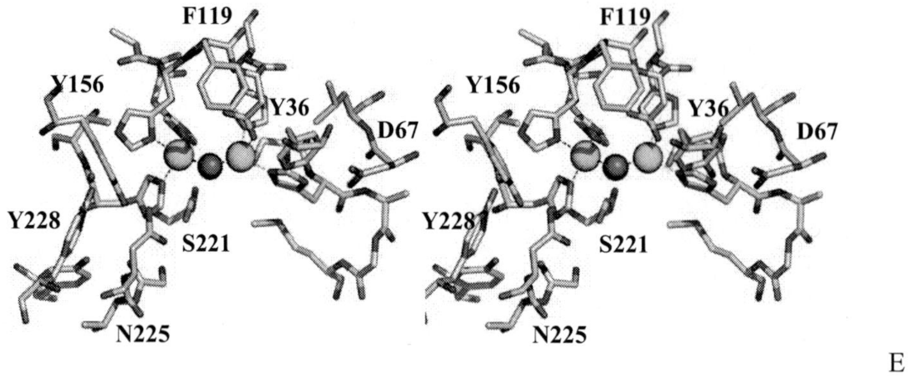

Figure 5. Stereo views of the active-site of representative enzymes.
(A) BcII (PDB code 1BMC)
(B) BcII R121C mutant (PDB code 2BFK)
(C) BlaB (PDB code 1M2X)
(D) CphA (PDB code 1PX1)
(E) FEZ-1 (PDB code 1K07)
Zinc ions and water molecules are represented as blue and red spheres, respectively.

The structures of BcII displaying modified environments at the Zn2 site and reproducing the metal coordination environments of structurally related metallohydrolases, have also been solved [32]. Mono-zinc R121H-C221D exhibits a single zinc ion located in the Zn1 site (i.e. His116, His118 and His196). Zn1 exhibits a tetrahedral geometry, resembling the Zn1 site of mono-zinc-BcII (PDB code 1BMC). Instead, the Zn1 site of di-zinc R121H-C221D is analogous to that of wild-type BcII on C2 crystals (PDB code 1BVT), exhibiting a distorted trigonal-bipyramidal geometry, coordinated to three His residues and two solvent molecules. In contrast, the Zn2 site is markedly altered compared to wild-type BcII, since both engineered residues act as metal ligands and define the position of this metal ion. The His121 residue drags Zn2 deeper into the protein, making it less solvent-accessible and increasing the Zn1–Zn2 distance from 3.82 Å in wild-type BcII to 4.73 Å. Remarkably, the His121 side-chain displays the same orientation in both mono- and di-zinc forms, suggesting that this residue provides a rigid, more buried anchoring site for Zn2. The engineered Asp221 also binds Zn2, resulting in a tighter ligand set that disfavors coordination to His263, whose Nε2 atom is 3.23 Å away. Two water molecules complete the vacant coordination positions of Zn2 in a distorted octahedral geometry. This R121H-C221D BcII double mutant is a catalytically impaired β-lactamase even as a di-zinc enzyme, due to the modified position of Zn2 which is thus unable to provide a productive interaction with the substrate.

3.1.2. BlaB from Chryseobacterium meningosepticum

C. meningosepticum is a ubiquitous, rod shaped Gram-negative bacterium. This bacterium is clinically relevant because it can cause neonatal meningitis, adult septicaemia and nosocomial infections. Moreover, *C. meningosepticum* is resistant to most β-lactams antibiotics, including carbapenems [92-94] and constitutively expresses a chromosome-encoded metallo-β-lactamase, BlaB [95].

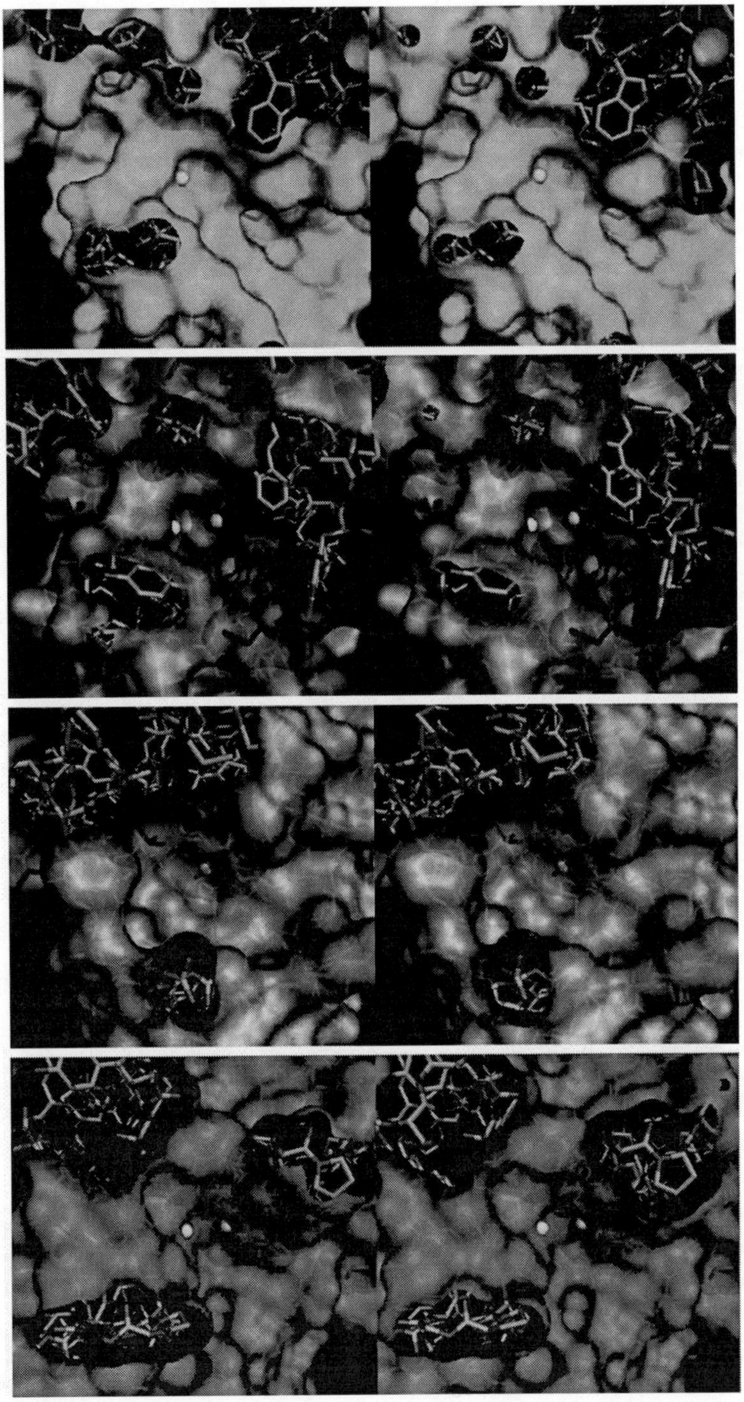

(A)

(B)

(C)

(D)

Figure 6. Solvent-accessible surfaces of representative enzymes.

(A) BcII (PDB code 1BMC)
(B) BlaB (PDB code 1M2X)
(C) CphA (PDB code 1PX1)
(D) FEZ-1 (PDB code 1KO7)

The same slab was used for the four figures, displaying internal regions of the protein.

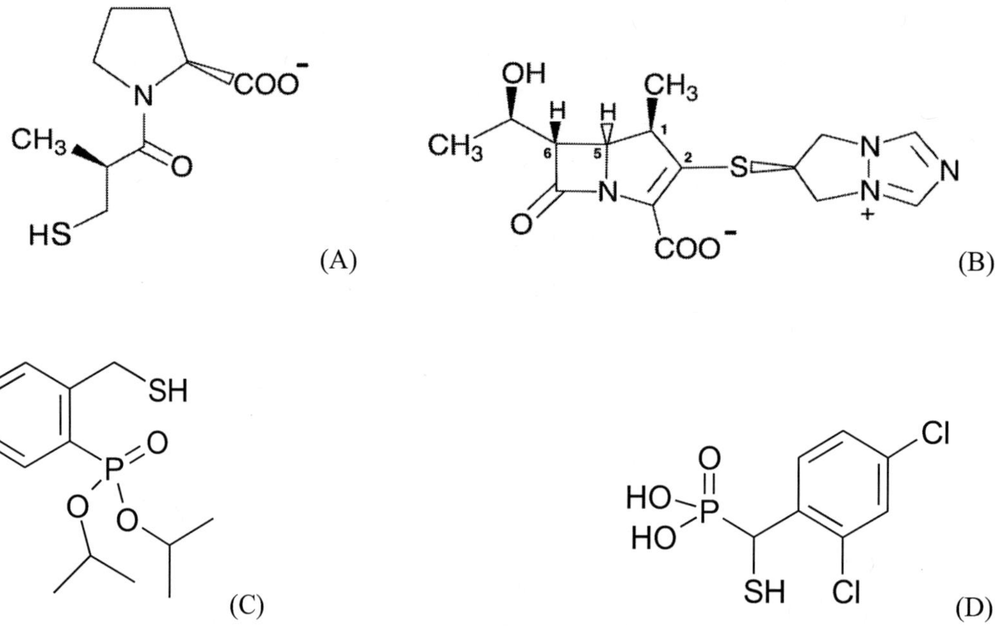

Figure 7. Structures of D-captopril (A), biapenem (B), diisopropyl [2-(sulfanylmethyl)phenyl] phosphonate (C) and [(2,4-dichlorophenyl)(sulfanyl)methyl] phosphonic acid (D).

Active-site Structure

BlaB crystals were grown at pH 8.4 in a buffer containing 100 μM zinc acetate [18]. Two Zn^{2+} ions are present in the active site, but the Zn^{2+} occupancy in site Zn2 was refined to 80%. Unfortunately, no data are available on the BlaB-Zn^{2+} dissociation constants. In addition the crystal was soaked for five weeks in a solution containing 2mM D-captopril (Figure 7A) a good competitive inhibitor of BlaB (Ki = 70-100μM) [18]. The D-captopril presented a well defined electron density and was refined with full occupancy. The average thermal B-factor for the inhibitor was 15.6 $Å^2$, close to the average value observed for the protein (15.0 $Å^2$), suggesting a tight binding of the D-captopril moiety. Interestingly, the bridging hydroxide ion is replaced by the D-captopril sulphur group (Figure 4C). The inhibitor is located on the left part of the active site (Figure 5C). As for BcII, the cleft is very long and can be modulated by conformational changes of the L_1 and L_3 loops (Figure 6B).

3.1.3. SPM-1 from Pseudomonas Aeruginosa

Pseudomonas aeruginosa is a leading cause of nosocomial infections, ranking first as a cause of nosocomial pneumonia in Brazil and the USA. A new transferable metallo-β-lactamase, designated SPM-1, has been identified from a carbapenem-resistant *Pseudomonas aeruginosa* strain [96].

The sequence shows that SPM-1 can be distinguished from the other B1 metallo-β-lactamases thanks to a central insertion of 24 amino acid residues and substitutions at two highly conserved positions, Asp84(Ser) and Arg121(Gly). SPM-1 possesses two metal sites of very different affinities, its Zn2 site is potentially the weakest that has been identified so far [21].

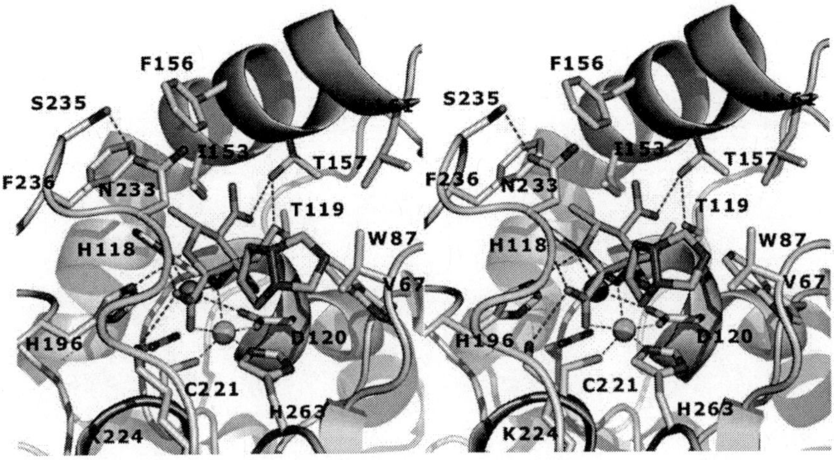

Figure 8. Interactions in biapenem-CphA complex.
The product of the reaction is represented in sticks with carbon, oxygen and nitrogen coloured in orange, red and blue, respectively.

Active-site Structure

SPM-1 crystals were grown at pH 8.0 with 100 μM $ZnCl_2$ [21]. As crystallised, SPM-1 contains a single Zn^{2+}. Both the active site hydrogen-bonding network and main-chain geometry at Asp120, a key component of the binding site for the second zinc ion, differ significantly from previous B1 metallo-beta-lactamase structures. This decreases the zinc affinity for the SPM-1 Zn2 site. It has been proposed that variable interactions made by the Asp120 carboxyl group modulate affinity for a second Zn^{2+} equivalent in the subfamily B1 metallo-beta-lactamases. The structure of the SPM-1 enzyme shows that it lacks the 61-65 mobile loop and that strands β3 and β4 are separated by a stretch of only five amino acid residues that form a tight turn. Moreover, SPM-1 accommodates a central insertion of 24 amino acids in an extended α3-α4 region (residues 150b-165). Deleting this insertion has only marginal effects upon binding and hydrolysis of a range of β-lactam antibiotics [21]. Both these structural features of SPM-1 are reminiscent of those found in the subclass B2 CphA enzyme (see below). By these aspects, SPM-1 thus represents a structural hybrid between subclass B1 and subclass B2. Figure 3B shows a superposition of BcII (B1), SPM-1 and CphA (B2).

3.2. Overall Structure of Subclass B2 Metallo-β-lactamases

In 2005, we solved the first X-ray structure of CphA, an enzyme produced by various species of *Aeromonas* [26]. The sequence identities between CphA and members of subclass B1 enzymes vary between 24 and 32%. The putative zinc ligands for the two subclasses are identical, except for position 116 which is an asparagine in subclass B2 and a histidine in subclasses B1 and B3 (Table 2). Figure 3C shows a superposition of one representative structure for each subclass: BcII, CphA and FEZ-1 (a subclass B3 enzyme see below). The main differences in the Cα fold are the presence of a long helix near the active-site groove in CphA (upper right) and an additional helix in FEZ-1 (lower middle).

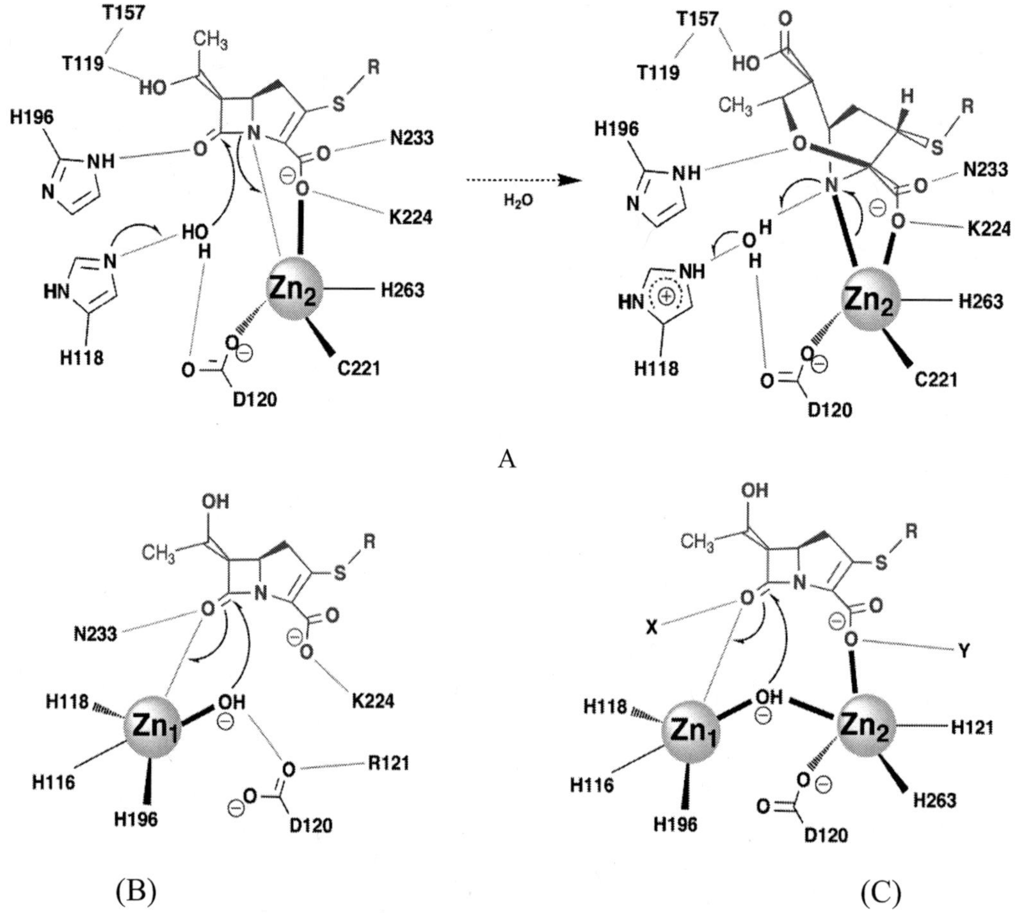

Figure 9. Schematic diagram of the hydrolysis of carbapenem.

(A) for the mononuclear CphA enzyme.
(B) for the mononuclear BcII enzyme.
(C) for a binuclear enzyme.

3.2.1. CphA from Aeromonas hydrophila

The first paper on CphA was published in 1991 [97]. The authors cloned the *Aeromonas hydrophila* gene, named CphA, coding for a carbapenem-hydrolyzing metallo-β-lactamase, in *E. coli* by screening an *Aeromonas* genomic library for clones able to grow on an imipenem-containing medium. Later, the same gene was found in *Aeromas veronii, Aeromonas caviae, Aeromonas jandaci,* and other strains of *A. hydrophila* [98]. CphA has a narrow substrate profile limited to carbapenems, is active only in the monozinc form and is non-competitively inhibited by added zinc with a K_i value of 50 μM [99].

Active-site Structure of the Mono-zinc wild Type Enzyme

Crystals of CphA were obtained at pH 6.5 using N220G mutant micro-crystal seeds [26]. Before data collection, the crystals were transferred to a solution containing 10 mM $ZnCl_2$. Only one zinc ion is observed in the active site at the Zn2 site suggesting that the monozinc species represents the native form of CphA. An oxygen from a carbonate ion is bonded to the Zn^{2+} and together with Asp120, Cys221, His263 generate a tetrahedral geometry around the Zn^{2+} ion (Figure 4D). One of the most striking features of the CphA structure is the presence of a "hydrophobic wall" formed mainly by an elongated α_3 helix [26]. The wall closes the left side of the active site (Figures 5D and 6C). Unlike subclass B1 enzymes, the active site cleft is almost rigid and can accommodate a carbapenem molecule but not a larger substrate, a structural feature in good agreement with the narrow activity profile of the CphA enzyme.

Active-site Structure of the Biapenem-CphA Complex

The biapenem-CphA complex (Figure 7B) was obtained by adding an excess of biapenem powder directly to a drop containing crystals [26]. A well-defined electron density in the active site could be interpreted as neither a biapenem nor a hydrolyzed biapenem molecule. Moreover, the presence of two fused rings was unexpected [26]. In order to clarify the chemical nature of the bound species that would account for the active site electron density, substrate and product inhibition studies were performed. The kinetic data clearly reveal a model of product inhibition with a Ki value of 320 μM. X-ray results and kinetic data agree with the presence of a hydrolysis reaction intermediate (Figure 8).

Active-site Structure of CphA in Complex with Inhibitors

Some pyridine carboxylates are competitive inhibitors of the subclass B2 CphA enzyme. The structure of the CphA-Zn(II)-2,4-pyridine dicarboxylic acid (PDCA) complex shows that the 2,4-PDCA inhibitor binds to the catalytic zinc ion, acts as an NO-chelating ligand, and interacts with the enzyme active site with both electrostatic and hydrophobic contacts [46]. The binding of the inhibitor leads to the movement of the Gly232-Asn233 loop located at the active site entrance so that it closes the active site. The comparison of the CphA-Zn(II)-2,4-PDCA complex structure with that of the reported CphA-Zn(II)-hydrolyzed biapenem complex (PDB 1X8I) reveals two interesting analogies: first, a similar Gly232-Asn233 loop closure upon binding of 2,4-PDCA and hydrolyzed biapenem, and analogous positions for the zinc-bound N-1 and 2-carboxylate group of 2,4-PDCA and for the zinc-bound N-4 and 3-carboxylate group of hydrolyzed biapenem.

The diisopropyl [2-(sulfanylmethyl)phenyl] phosphonate (*10a*) and [(2,4-dichlorophenyl)(sulfanyl)methyl] phosphonic acid (*18*) (figure 7) turned out to be quite potent broad spectrum MBL inhibitors [48]. The crystallographic structures of CphA in complex with *10a* and *18* were solved (PDB 3IOF and 3IOG). Both inhibitor-enzyme complexes are stabilized by a network of hydrophobic contacts and hydrogen bonds. Interestingly, these complexes point to two types of interactions with the enzyme depending on the complexed inhibitor. Indeed, in the case of *10a*, an interaction of the sulphur atom with the active site Zn^{2+} ion is observed while *18* interacts with the Zn^{2+} ion *via* its phosphonato group. A similar situation is observed in the CphA/D-captopril complex (2QDS) where the main interaction between D-captopril and the metal occurs *via* the inhibitor carboxylate group and not *via* the thiolate group of D-captopril [47]. In contrast, D-captopril acts as an inhibitor for B1 and B3 metallo-β-lactamases with interchelation of its thiolate group between the two zinc ions [18, 50]. However, the comparison of the CphA-*10a* and CphA-*18* complex structures with that previously reported for the carbapenem substrate [26] or for other inhibitors (2QDS and 2GKL) [46, 47] reveals a general pattern composed of three important contacts: (i) contact to the Zn^{2+} ion, (ii) contact to the conserved Lys224 residue (directly or *via* a water molecule) and (iii) contact to Asn233 that stabilizes the Gly232-Asn233 loop closure.

Active-site Structure of the Di-zinc Inhibited Wild Type Enzyme

More recently, X-ray data from zinc-saturated crystals yielded the crystal structures of the di-zinc form of the CphA enzyme [100]. Three zinc ions are present in the structure with two observed in the active site and one at the surface. The first zinc ion binds in the Asp120-Cys221-His263 site, as previously determined for the mono-zinc form of the enzyme. The tetrahedral coordination sphere is completed by a sulphate ion from the crystallization solution at a distance of 2.04 Å. The second zinc ion occupies a slightly modified “histidine” site, where the conserved His118 and His196 residues act as metal ligands. The distances between the zinc ion and the His118 ND1 and the His196 NE2 are 2.01 and 2.03 Å, respectively. In CphA, the third histidine ligand (His116) is not conserved and is replaced by an asparagine residue, which is not involved in the binding of the second zinc (the distance is more than 4 Å). A sulphate ion (1.95 Å) and a water molecule (2.01Å) complete the vacant coordination positions of Zn2 ion, forming a tetrahedral coordination sphere. The sulphate ion acts as a bridging agent between the zinc ions. The distances between the sulphate ion and Zn1 and Zn2 are 2.04 and 1.95 Å, respectively. The distance between Zn1 and Zn2 is 4.08 Å. This atypical coordination sphere probably explains the rather high dissociation constant for the second zinc ion compared to those observed in enzymes of subclasses B1 and B3. The third observed zinc ion is found on the surface, away from the active site (17.43 Å). It is coordinated by His289 (2.09 Å from His NE) and the coordination sphere is completed by three chloride ions (2.19, 2.25 and 3.17 Å).

The differences between the mono- and di-zinc structures are limited to the Gly232–Asn233 loop region. This loop, located at the entrance to the active site, is known to be stabilized upon the binding of the substrate [26] or inhibitors [46-48]. For the wild-type di-zinc structure, this loop can exist in two conformations (both half occupied), corresponding, respectively, to the open form (as in the mono-zinc structures 1X8G and 1X8H) or the closed form (as in the structures with substrate or inhibitor in the active site, 1X8I, 2QDS, 2GKL, 3IOF, 3IOG). The closed conformation is stabilized by additional H-bonds between the Asn233 side chain and the Ser235 side chain oxygen and backbone nitrogen. Inhibition by the second zinc ion results from immobilization of the catalytically-important His118 and His196 residues, as well as the folding of the Gly232–Asn233 loop into a position that covers the active site.

Extended-Spectrum Mutant

Among class B β-lactamases, the subclass B2 CphA enzyme is characterized by a unique specificity profile. CphA efficiently hydrolyzes only carbapenems. A site-directed mutant that exhibits a strongly broadened activity spectrum when compared with the WT CphA has been generated [101]. Strikingly, the substrate profile of the N116H/N220G double mutant is close to that observed for the broad spectrum subclass B1 enzymes. The double mutant is significantly activated by the binding of a second zinc ion under conditions where the WT enzyme is non-competitively inhibited by the same ion.

3.3. Overall Structure of Subclass B3 Metallo-β-lactamases

The structure of tree subclass B3 enzymes (L1, FEZ-1 and BJP-1) have been deposited in the PDB. The identity of the structurally aligned sequences is 33%. Sequence identities between subclass B3 and both B1 or B2 are low and range from 12 to 20%. L1 is a tetrameric enzyme and, compared to the other subclass B3 members, has an N-terminal extension that is involved in oligomerization. Figure 3D displays a comparison between the L1 and FEZ-1 enzymes. The main differences are that FEZ-1 lacks of the L1 N-terminal extension and exhibits two additional α-helices in the C-terminal region.

3.3.1. FEZ-1 from Fluoribacter gormanii

FEZ-1, a subclass B3 enzyme, was identified in legionella *Fluoribacter gormanii* [102], overproduced in *E. coli* and biochemically characterised [103]. FEZ-1 exhibits a broad-spectrum activity profile, with a strong preference for cephalosporins such as cephalothin, cefuroxime and cefotaxime.

Structure of the Binuclear Active Site

FEZ-1 crystals were obtained at pH 6.0, in the presence of 20 μM $ZnCl_2$ [24]. In the active site the two Zn^{2+} ions are separated by a distance of 3.4 Å with the Zn1 ligands adopting a tetrahedral geometry and the Zn2 ligands a distorted trigonal bipyramidal geometry (Figure 5E). Unfortunately, in the FEZ-1 structure a disordered glycerol molecule, from the buffer used for cryo-cooling the crystals, is located in the active site. The glycerol molecule seems to have displaced the expected bridging hydroxide ion. However, the protein ligands are in the same positions as in the L1 enzyme [23] (Figure 4E). When compared to enzymes of subclasses B1 and B2, the Zn2 site has a histidine ligand, His121, instead of Cys221 (Table 2).

Coordination of the metal ion in the Zn1 site is made by the conserved His116, His118, His196 and a bridging water/hydroxide (Wat1), whilst the Zn^{2+} in the Zn2 site is bound to Wat1, Asp120, His121, His263 and an additional water molecule (Wat2). Interestingly, the same Zn coordination was also found in other subclass B3 enzymes, with a notable exception for GOB-1, which has the H116Q mutation in site Zn1.

Compared to the other subclasses enzymes, the active site cleft also has a different shape. In subclass B1, there is a long L_1 loop (56-67), which is disordered in the native structure, but is stabilized upon inhibitor binding. The L_1 loop is missing in subclasses B2 and B3. The L_2 loop (151-168) is absent in subclass B1 and forms the elongated α3 helix in subclass B2 and a short mobile loop in subclass B3 (Figure 2). Finally, the L_3 loop (223-238) is always present and modulates the shape of the active site. In FEZ-1, the active site has a V-shape with the binuclear metal centre at the apex (Figure 6D).

4. Mechanism of Action (See also Chapters 7 and 12)

In the metallo-β-lactamase superfamily, seventeen groups of enzymes have been identified based on the biological functions [58]. X-ray structures have been established for representatives of ten groups. Most of the structures revealed a binuclear centre with metal ligands located on loops connecting secondary structure elements (Figure 2). The rubredoxin oxygen: oxidoreductase flavoprotein from *Desulfovibrio gigas*, a strict anaerobe, contains a di-iron centre [69]. For the teichoic acid phosphorylcholine esterase, the situation is less clear; the full-length enzyme has a di-zinc centre [74] while the catalytic domain structure displays a di-iron centre [73]. Differences also appear between species: for instance, human and *Arabidopsis* glyoxalases II contain two zincs or one zinc and one iron, respectively, in the active-sites [62, 64]. All characterized metallo-β-lactamases are active as mono or di-zinc states. Zn^{2+} is preferred to other ions because of its flexible coordination geometry and the possibility to have different coordination numbers (4, 5 or 6). In addition, Zn^{2+} is an electron lone pair acceptor and it coordinates well with donors such as nitrogen of the imidazole moiety in histidine residues, the sulphur of cysteines and oxygen of water/hydroxide or of aspartic or glutamic acid residues. Its Lewis acid strength, intermediate polarizability and lack of redox activity are also very important.

In the crystal structures of metallo-β-lactamases examined thus far, Zn1 is tetrahedrally coordinated to three histidines and a water molecule. In the di-zinc state, Zn2 is coordinated to three residues (a cysteine, a histidine, and an aspartate) and two water molecules in a trigonal bipyramid geometry in subclass B1 or to two histidines, an aspartate and two water molecules in subclass B3. Since mono and di-zinc centres are observed, the catalytic mechanisms of the two species must be considered separately.

In subclass B1, the active-site cleft is wide and lined by mobile loops. In contrast, in CphA, a subclass B2 enzyme, the active site is narrow. Finally, in B3 enzymes, the active site is more open and flexible, like in B1 enzymes, but has a V-shape (Figure 6). Early attempts to obtain substrate complexes in crystals always failed. Docking procedures were used but with limited success due to the flexibility of the active site, [17, 23, 24, 104]. Using a high concentration of biapenem in the crystallisation drop, we succeeded in trapping an intermediate filling the narrow active-site pocket of CphA (Figure 8) [26]. On the basis of the X-ray structures of native and biapenem-complexed proteins, structures of intermediates of the reaction pathway were modelled (Figure 9A). The presence of a biapenem molecule in the active site leads to the closure of a small external loop (Gly232 – Asn233). The proposed reaction pathway proceeds by proton transfer, water molecule displacements and amino acid side chain rotations.

In 2005, a report was published describing active-site dynamic simulations of native and biapenem-complexed CphA enzymes, using a quantum mechanical / molecular mechanical method and density functional theory [105]. The proposed mechanism is consistent with the observed enzyme-intermediate in the crystal and the previously proposed pathway of reaction [26].

For B1 and B3 enzymes, in absence of experimental information on the substrate enzyme complex, two catalytic mechanisms have been proposed. In the mononuclear enzymes, the zinc-bound water / hydroxide ion is expected to be the attacking nucleophile [15]; this mechanism resembles those of other metallo-enzymes such as carboxypeptidase A and thermolysin [106-108]. In this mechanism, Zn^{2+} has a dual role: firstly, it activates the water molecule and, secondly, it polarizes the carbonyl group of the β-lactam (Figure 9B). In addition, Asp120 also participates in the activation of the water molecule. After the hydroxide attack, a tetrahedral intermediate is formed. Finally, the proton on Asp120 is transferred on the β-lactam nitrogen and the β-lactam C-N bond is cleaved (Figure 9B). Based on pH-activity dependence studies another catalytic mechanism was proposed [88, 109]. In this model, Zn^{2+} acts as a Lewis acid to stabilize the dianionic form of the tetrahedral intermediate and to provide a hydroxide ion bound nucleophile. Asp120 acts as a general base to form the dianion and to later render the proton to the leaving β-lactam nitrogen. However, the Asp120Asn mutant, whose activity decreases 70-200-fold when compared to the WT, [90] is not significantly less active than mutants of several zinc-liganding residues (His118, 196 and 263), a result which seems to be inconsistent with a central role for Asp120 in catalysis [11]. On the basis of a study of Asp120 BcII mutants, Llarull *et al* suggest that Asp120 is neither the proton donor nor an essential actor in nucleophilic activation. It is proposed that the role of Asp120 is to act as a strong Zn2 ligand, locating this ion optimally for substrate binding, stabilization of the development of a partial negative charge on the β-lactam nitrogen and protonation of this atom by a zinc-bound water molecule [31].

In the binuclear enzymes, the water / hydroxide ion that bridges the two Zn^{2+} ions is expected to be the attacking nucleophile [8, 16, 23]. In the proposed mechanism, two binding interactions between the β-lactam and the enzyme were postulated (Figure 9C). First, the side-chain labelled X in the figure and Zn1 form an oxyanion hole, which could accommodate the carbonyl of the β-lactam. X is not conserved in di-zinc enzymes. Enzymes of subclass B1 such as IMP-1, CcrA and VIM-2 have an Asn233, and BlaB a Tyr233 (Figure 4). FEZ-1 and L1, subclass B3 enzymes, also have an Asn233 but 14 Å away from the postulated oxyanion hole. Asn225 in FEZ-1 and Tyr228 in L1 are in better positions to fulfil this role [23, 24]. Second, a strong interaction occurs between the side-chain labelled Y and the substrate carboxyl group (Figure 9C). Y is not conserved in di-zinc enzymes. All subclass B1 enzymes have a Lys224, except VIM-2, which has a tyrosine. L1 and FEZ-1 have a serine and glycine, respectively at this position. In subclass B3, a movement of the 223-229 loop upon substrate binding cannot be ruled out, which can bring other side-chain residues into the active-site [24]. Another catalytic mechanism for the di-zinc enzyme was proposed for the di-zinc form of BcII [23]. In this model, the role of X is conserved but the apical water at Zn2 is replaced by one oxygen atom of the carboxyl and the lone pair of the β-lactam nitrogen which interacts with Zn^{2+} in the Zn2 site. In addition, as for the mono-zinc mechanism, Asp120 plays the role of general base. More recently, a study of the Asp120Asn and Asp120Ser L1 mutants concluded that while interactions of Asp120 with active site water molecules are important to proton transfer and possibly nucleophilic attack by the water molecule, its primary role is to position Zn2 optimally for catalytically important interactions with the negatively charged nitrogen of a reaction intermediate [49]. It should however be noted that the existence of this intermediate has only been documented with the chromogenic substrate nitrocefin (see chapters 6 and 7).

Time-dependent absorption and rapid-freeze-quench EPR spectra were obtained using a Co(II)-substituted form of L1 and four substrates: nitrocefin, cephalothin, meropenem and penicillin G [110]. The authors suggest the presence of a metal-bound intermediate in the hydrolytic reaction pathway, more likely an N5 nitrogen anion interacting with Zn^{2+} in the

Zn2 site. The same intermediate was proposed based on mechanistic studies on a binuclear Zn^{2+} model complex with nitrocefin [111]. The proposed intermediate is reminiscent of what is observed in the CphA-biapenem complex (Figure 9A).

Finally, in the crystal structure of the L1 enzyme in complex with the hydrolysis product of moxalactam (PDB code 2AIO) [112], the apical water molecule on Zn^{2+} in the Zn2 site is displaced by the C4 carboxylate and, in the Henri-Michaelis complex, moxalactam is interacting with the binuclear centre as shown in Figure 9C. There is no amino-acid side-chain at the X site and Ser221 occupies the Y site. Interestingly, Ser221 is conserved in subclass B3 enzymes, except in GOB-1 and GOB-18 which have a methionine. Moreover, in these GOB enzymes, His 116 is replaced by a Gln, a mutation which strongly modifies the properties of the Zn1 site, making it similar to that of B2 enzymes which have an Asn in the same position. This structural feature suggested that GOB-18 might be active in the mono-Zn^{2+} form, with the metal ion in the Zn2 site as in the *A. hydrophila* B2 enzyme. Accordingly, GOB-18 is fully active against a broad range of β-lactam substrates as a mono-zinc species. Different spectroscopic techniques, 3D modeling, and mutagenesis experiments, reveal that the zinc ion is bound to Asp120, His121, His263, and a solvent molecule, i.e., in the canonical site 2 of dinuclear B3 metallo-β-lactamases [113]. However, in a study of GOB-1, two zinc ions were found per GOB-1 molecule after a purification procedure during which no Zn^{++} had been added to the buffers [114]. These two GOB enzymes differ by a few point mutations, apparently far from the active site (Leu94Phe, Ala137Val and Asp282Asn).

CONCLUSION

After 15 years of investigation of the structures of metallo-β-lactamases, we are starting to have a better understanding of the main features of each subclass for both fold and active site. This information combined with biochemical and biophysical studies performed on specific enzymes can shed light on the catalytic mechanisms. However, the central question remains the protein-zinc stoechiometry under physiological conditions [10]. Until now, most of the structural and biochemical studies used an excess of zinc in the buffer, which forced the binuclear state. More likely, the mononuclear state might be predominant at low zinc concentration. Metal-free states may exist in absence of zinc but the apo-enzyme is inactive, suggesting that the mononuclear state might be the relevant species under physiological conditions, if the free zinc ion concentration is in the femtomolar range. Since the zinc is present in site Zn1 in BcII and in site Zn2 in CphA, two distinct catalytic mechanisms could exist even for a mono-zinc species. The number of Zn^{2+} ions necessary for optimal activity of B1 enzymes remains controversial. For the BcII 569/H enzyme, an anticooperative binding of the two Zn^{2+} ions has been suggested, with a slightly higher activity of the di-Zn^{2+} species [115]. Accordingly, X-ray data have indicated a "looser binding" of the metal ion in the Zn2 site [27]. However recent results obtained by mass spectrometry and NMR [116] indicate cooperative binding with a K_{d1}/K_{d2} ratio equal to or larger than 5. By contrast, the activity of the mono-Zn^{2+} VIM-4 appears to be very low, binding of a second Zinc ion (K_{d2} = 10 μM) yielding the fully active enzyme [20] while the SPM-1 enzyme remains mononuclear even in the presence of 100 μM Zn^{2+} [21]. Such a heterogeneity is surprising for a set of closely related proteins. As seen above in the case of the GOB enzymes, the "ideal" number of Zn^{2+} ions is also controversial in subclass B3 and many important questions have not yet been answered. Such a binding site heterogeneity, if confirmed, might constitute a significant obstacle to the design of general drug inhibitors of metallo-β-lactamases.

Finally, in a more optimistic tone, the metallo-β-lactamase superfamily reveals a nice example of protein evolution from an ancient protein. Nature has generated enzymes with a large variety of biological functions from the same fold.

ACKNOWLEDGMENTS

We would like to thank all co-workers from the metallo-β-lactamase group who contributed to these studies, in particular Jean-Marie Frère, Emile Duez, Moreno Galleni, Gian Maria Rossolini and Gianfranco Amicosante.

We would like to thank Andrea Dessen for critical reading of the manuscript.

REFERENCES

[1] Ambler, R.P. (1980) *The structure of β-lactamases.* Philos.Trans.R.Soc.London Ser.B 289, 321-331.

[2] Joris, B., Ledent, P., Dideberg, O., Fonze, E., Lamotte-Brasseur, J., Kelly, J.A., Ghuysen, J.M. and Frère, J.M. (1991) Comparison of the sequences of class A β-lactamases and of the secondary structure elements of penicillin-recognizing proteins. Antimicrob. *Agents Chemother.* 35, 2294-2301.

[3] Kelly, J.A., Dideberg, O., Charlier, P., Wery, J.P., Libert, M., Moews, P.C., Knox, J.R., Duez, C., Fraipont, C., Joris, B., Dusart, J., Frère, J.M. and Ghuysen, J.M. (1986) On the origin of bacterial resistance to penicillin. Comparison of an active-site serine β-lactamase and an active-site serine D-alanyl-D-alanine-cleaving peptidase. *Science,* 231, 1429-1431.

[4] Samraoui, B., Sutton, B.J., Toddy, R.J., Artymiuk, P.J., Waley, S.G. and Phillips, D.C. (1986) Tertiary structural similarity between a class A Beta-lactamase and a penicillin-sensitive D-alanyl carboxypeptidase-transpeptidase. *Nature,* 320, 378-380.

[5] Parès, S., Mouz, N., Pétillot, Y., Hakenbeck, R. and Dideberg, O. (1996) X-ray structure of *Streptococcus pneumoniae* PBP2x, a primary penicillin target enzyme. *Nature Struct.Biol.,* 3, 284-289.

[6] Sabath, L.D. and Abraham, E.P. (1966) Zinc as a cofactor for cephalosporinase from *Bacillus cereus* 569. *Biochem. J.,* 98, 11C-13C.

[7] Bush, K. (1998) Metallo-β-lactamases: a class apart. *Clin. Infect. Dis.,* 27, S48-53.

[8] Wang, Z., Fast, W., Valentine, A.M. and Benkovic, S.J. (1999) Metallo-beta-lactamase: Structure and Function. *Curr. Opin. Chem. Biol.,* 3, 614-622.

[9] Page, M.I. (1999) The reactivity of beta-lactams, the mechanism of catalysis and the inhibition of beta-lactamases. *Curr. Pharm. Des.,* 5, 895-913.

[10] Heinz, U. and Adolph, H.-W. (2004) Metallo-β-lactamases: two binding sites for one catalytic metal ion? *Cell. Mol. Life Sci.,* 61, 2827-2839.

[11] Herzberg, O. and Fitzgerald, M.D. (2004) *Handbook of Metalloproteins. Ed. Albrecht Messerschmidt,* Wolfgrame Bode, Mirek Gygler - John Wiley & Sons, LTD., 3, 217-234.

[12] Walsh, T.R., Toleman, M.A., Poirel, L. and Nordmann, P. (2005) Metallo-beta-lactamases: the Quiet before the Storm? *Clin. Microbiol. Rev.*, 18, 306-325.

[13] Bebrone, C. (2007) Metallo-β-lactamases (classification, activity, genetic organization, structure, zinc coordination) and their superfamily. *Biochem. Pharmacol.,* 74, 1686-701.

[14] Rasmussen, B.A. and Bush, K. (1997) Carbapenem-hydrolyzing beta-lactamases. Antimicrob. *Agents Chemother.*, 41, 223-232.

[15] Carfi, A., Parès, S., Duée, E., Galleni, M., Duez, C., Frère, J.M. and Dideberg, O. (1995) The 3-D structure of a zinc metallo-β-lactamase from *Bacillus cereus* reveals a new type of protein fold. *EMBO J.,* 14, 4914-4921.

[16] Concha, N.O., Rasmussen, B.A., Bush, K. and Herzberg, O. (1996) Crystal structure of the wide-spectrum binuclear zinc beta-lactamase from *Bacteroides fragilis*. *Structure,* 4, 823-836.

[17] Concha, N.O., Janson, C.A., Rowling, P., Pearson, S., Cheever, C.A., Clarke, B.P., Lewis, C., Galleni, M., Frère, J.M., Payne, D.J., Bateson, J.H. and Abdel-Meguid, S.S. (2000) Crystal Structure of the IMP-1 Metallo β-Lactamase from *Pseudomonas aeruginosa* and its Complex with a Mercaptocarboxylate Inhibitor : Binding Determinants of a Potent, Broad-Spectrum Inhibitor. *Biochemistry*, 39, 4288-4298.

[18] García-Saez, I., Hopkins, J., Papamicael, C., Franceschini, N., Amicosante, G., Rossolini, G.M., Galleni, M., Frère, J.M. and Dideberg, O. (2003) The 1.5 A structure of *Chryseobacterium meningosepticum* Zn-β-lactamase in complex with the inhibitor D-captopril. *J. Biol. Chem.,* 278, 23868-23873.

[19] Garcia-Saez, I., Docquier, J.D., Rossolini, G.M. and Dideberg, O. (2008) The three-dimensional structure of VIM-2, a Zn-β-lactamase from *Pseudomonas aeruginosa* in its reduced and oxidised form. *J. Mol. Biol.*, 375, 604-611.

[20] Lassaux, P., Traoré, D. A.K., Loisel, E., Favier, A., Docquier, J.D., Bebrone, C., Frère, J.M., Ferrer, J.L. and Galleni, M. (2010) Biochemical and Structural Characterization of the Subclass B1 Metallo-β-Lactamase VIM-4. Antimicrob. *Agents Chemother., in press*

[21] Murphy TA, Catto LE, Halford SE, Hadfield AT, Minor W, Walsh TR and Spencer J. (2006) Crystal structure of *Pseudomonas aeruginosa* SPM-1 provides insights into variable zinc affinity of metallo-beta-lactamases. *J.Mol.Biol.,* 357, 890-903.

[22] Yamaguchi, Y., Takashio, N., Wachino, J., Yamagata, Y., Arakawa, Y., Matsuda, K. and Kurosaki, H. (2010) Structure of metallo-beta-lactamase IND-7 from a *Chryseobacterium indologenes* clinical isolate at 1.65 Å resolution. *J. Biochem.,* 147, 905-15.

[23] Ullah, J.H., Walsh, T.R., Taylor, I.A., Emery, D.C., Verma, C.S., Gamblin, S.J. and Spencer, J. (1998) The crystal structure of the L1 metallo-beta-lactamase from *Stenotrophomonas maltophilia* at 1.7 Å resolution. *J. Mol. Biol.,* 284, 125-136.

[24] Garcia-Saez, I., Mercuri, P.S., Papamicael, C., Kahn, R., Frère, J.-M., Galleni, M., Rossolini, G.M. and Dideberg, O. (2003) Three-dimensional Structure of FEZ-1, a Monomeric Subclass B3 Metallo-β-lactamase from *Fluoribacter gormanii*, in Native Form and in Complex with D-Captopril. *J. Mol. Biol.,* 325, 651-660.

[25] Stoczko, M., Frère, J.M., Rossolini, G.M. and Docquier, J.D. (2006) Postgenomic scan of metallo-β-lactamase homologues in rhizobacteria : identification and characterization

of BJP-1, a subclass B3 ortholog from *Bradyrhizobium japonicum*. Antimicrob. *Agents Chemother.,* 50, 1973-81.

[26] Garau, G., Bebrone, C., Anne, C., Galleni, M., Frère, J.M. and Dideberg, O. (2005) A metallo-β-lactamase enzyme in action: crystal structure of the monozinc carbapenemase CphA and its complex with biapenem. *J. Mol. Biol.,* 345, 785-795.

[27] Fabiane, S.M., Sohi, M.K., Wan, T., Payne, D.J., Bateson, J.H., Mitchell, T. and Sutton, B.J. (1998) Crystal structure of the zinc-dependent beta-lactamase from *Bacillus cereus* at 1.9 Å resolution: binuclear active site with features of a mononuclear enzyme. *Biochemistry,* 37, 12404-12411.

[28] Carfi, A., Duée, E., Galleni, M., Frère, J.M. and Dideberg, O. (1998) 1.85 Å resolution structure of the Zinc^{II} β-lactamase from *Bacillus cereus. Acta Crystallogr.,* D54, 313-323.

[29] Chantalat, L., Duée, E., Galleni, M., Frère, J.M. and Dideberg, O. (2000) Structural effects of the active site mutation Cysteine to Serine in *Bacillus cereus* Zinc-β-lactamase. *Protein Sci.,* 9, 1402-1406.

[30] Davies, A.M., Rasia, R.M., Vila, A.J., Sutton, B.J. and Fabiane, S.M. (2005) Effect of pH on the active site of an Arg121Cys mutant of the metallo-beta-lactamase from *Bacillus cereus*: implications for the enzyme mechanism. *Biochemistry*, 44, 4841-4849.

[31] Llarrull, L.I., Fabiane, S.M., Kowalski, J.M., Bennett, B., Sutton, B.J. and Vila A.J. (2007) Asp-120 Locates Zn2 for Optimal Metallo-beta-lactamase Activity. *J. Biol. Chem.,* 282, 18276-85

[32] González, J.M., Medrano Martín, F.J., Costello, A.L., Tierney, D.L. and Vila, A.J. (2007) The Zn2 position in metallo-beta-lactamases is critical for activity: a study on chimeric metal sites on a conserved protein scaffold. *J. Mol. Bio*l., 373, 1141-56.

[33] Tomatis, P.E., Fabiane, S.M., Simona, F., Carloni, P., Sutton, B.J. and Vila, A.J. (2008) Adaptive protein evolution grants organismal fitness by improving catalysis and flexibility. *Proc Natl Acad Sci U S A.,* 105, 20605-10.

[34] González, J.M., Buschiazzo, A. and Vila, A.J. (2010) Evidence of adaptability in metal coordination geometry and active-site loop conformation among B1 metallo-beta-lactamases. *Biochemistry.,* 49, 7930-8.

[35] Carfi, A., Duée, E., Paul-Soto, R., Galleni, M., Frère, J.-M. and Dideberg, O. (1998) X-ray structure of the Zn^{II} β-lactamase from *Bacteroides fragilis* in an orthorhombic crystal form. *Acta Crystallogr.,* D54, 45-57.

[36] Li, Z., Rasmussen, B.A. and Herzberg, O. (1999) Structural consequences of the active site substitution Cys181 ==> Ser in metallo-beta-lactamase from *Bacteroides fragilis.* Protein Sci., 8, 249-252.

[37] Concha, N.O., Rasmussen, B.A., Bush, K. and Herzberg, O. (1997) Crystal structures of the cadmium- and mercury-substituted metallo-β-lactamase from *Bacteroides fragilis*. Protein Sci., 6, 2671-2676.

[38] Fitzgerald, P.M.D., Wu, J.K. and Toney, J.H. (1998) Unanticipated Inhibition of the Metallo-β-Lactamase from *Bacteroides fragilis* by 4-Morpholineethanesulfonic Acid (MES): A Crystallographic Study at 1.85 Å resolution. *Biochemistry,* 37, 6791-6800.

[39] Toney,J.H., Fitzgerald, P.M., Grover-Sharma, N., Olsonn S.H., May, W.J., Sundelof, J.G., Vanderwall, D.E., Cleary, K.A., Grant, S.K., Wu, J.K., Kozarich, J.W., Pompliano, D.L., Hammond, G.G. (1998) Antibiotic sensitization using biphenyl

tetrazoles as potent inhibitors of *Bacteroides fragilis* metallo-beta-lactamase. *Chem.Biol.,* 5, 185-196.

[40] Payne, D.J., Hueso-Rodríguez, J.A., Boyd, H., Concha, N.O., Janson, C.A., Gilpin, M., Bateson, J.H., Cheever, C., Niconovich, N.L., Pearson, S., Rittenhouse, S., Tew, D., Díez, E., Pérez, P., De La Fuente, J., Rees, M. and Rivera-Sagredo, A. (2002) Identification of a series of tricyclic natural products as potent broad-spectrum inhibitors of metallo-beta-lactamases. *Antimicrob. Agents Chemother.,* 46, 1880-1886.

[41] Yamaguchi, Y., Kuroki, T., Yasuzawa, H., Higashi, T., Jin, W., Kawanami, A., Yamagata, Y., Arakawa, Y., Goto, M. and Kurosaki H. (2005) Probing the role of Asp-120(81) of metallo-beta-lactamase (IMP-1) by site-directed mutagenesis, kinetic studies, and X-ray crystallography. *J. Biol. Chem.,* 280, 20824-20832.

[42] Toney, J.H., Hammond, G.G., Fitzgerald, P.M., Sharma, N., Balkovec, J.M., Rouen, G.P., Olson, S.H., Hammond, M.L., Greenlee, M.L. and Gao, Y.D. (2001) Succinic acids as potent inhibitors of plasmid-borne IMP-1 metallo-beta-lactamase. *J. Biol. Chem.,* 276, 31913-31918.

[43] Kurosaki, H., Yamaguchi, Y., Higashi, T., Soga, K., Matsueda, S., Yumoto, H., Misumi, S., Yamagata, Y., Arakawa, Y. and Goto, M. (2005) Irreversible inhibition of metallo-beta-lactamase (IMP-1) by 3-(3-mercaptopropionylsulfanyl)propionic acid pentafluorophenyl ester. *Angew. Chem. Int. Ed. Engl.,* 44, 3861-3864.

[44] Kurosaki, H., Yamaguchi, Y., Yasuzawa, H., Jin, W., Yamagata, Y. and Arakawa, Y. (2006) Probing, inhibition, and crystallographic characterization of metallo-beta-lactamase (IMP-1) with fluorescent agents containing dansyl and thiol groups. *Chem. Med. Chem.,* 1, 969-72.

[45] Yamaguchi, Y., Jin, W., Matsunaga, K., Ikemizu, S., Yamagata, Y., Wachino, J., Shibata, N., Arakawa, Y. and Kurosaki, H. (2007) Crystallographic investigation of the inhibition mode of a VIM-2 metallo-beta-lactamase from *Pseudomonas aeruginosa* by a mercaptocarboxylate inhibitor. *J. Med. Chem.,* 50, 6647-53.

[46] Horsfall, L.E., Garau, G., Liénard, B.M., Dideberg, O., Schofield, C.J., Frère, J.M. and Galleni, M. (2007) Competitive inhibitors of the CphA metallo-beta-lactamase from *Aeromonas hydrophila. Antimicrob. Agents Chemother.,* 51, 2136-2142 .

[47] Liénard, B.M.R., Garau, G., Horsfall., L., Karsisiotis, A.I., Damblon, C., Lassaux, P., Papamicael, C., Roberts, G.C.K., Galleni, M., Dideberg, O., Frère, J.M. and Schofield, C.J. (2008) Structural basis for the broad-spectrum inhibition of metallo-beta-lactamase by thiols. *Org. Biomol. Chem.,* 6, 2282-94.

[48] Lassaux, P., Hamel, M., Gulea, M., Delbrück, H., Mercuri, P.S., Horsfall, L., Dehareng, D., Kupper, M., Frère, J.M., Hoffmann, K., Galleni, M. and Bebrone, C. (2010) Mercaptophosphonate compounds as broad-spectrum inhibitors of the metallo-beta-lactamases. *J. Med. Chem.,* 53, 4862-76.

[49] Crisp, J., Conners, R., Garrity, J.D., Carenbauer, A.L., Crowder, M.W. and Spencer J. (2007) Structural basis for the role of Asp-120 in metallo-beta-lactamases. *Biochemistry,* 46, 10664-74.

[50] Nauton, L., Kahn, R., Garau, G., Hernandez, J.S. and Dideberg, O. (2008) Structural insights into the design of inhibitors for the L1 metallo-beta-lactamase from *Stenotrophomonas maltophilia. J. Mol. Biol.,* 375, 257-69.

[51] Mercuri, P.S., García-Sáez, I., De Vriendt, K., Thamm, I., Devreese, B., Van Beeumen, J., Dideberg, O., Rossolini, G.M., Frère, J.M. and Galleni, M. (2004) Probing the

specificity of the subclass B3 FEZ-1 metallo-β-lactamase by site-directed mutagenesis. *J. Biol. Chem.,* 279, 33630-33638.

[52] Garau, G., Garcia-Saez, I., Bebrone, C., Anne, C., Mercuri, P., Galleni, M., Frère, J.-M. and Dideberg, O. (2004) Update of the standard numbering scheme for class B β-lactamases. Antimicrob. *Agents Chemother.,* 48, 2347-2349.

[53] Ambler, R.P., Coulson, A.F., Frère, J.M., Ghuysen, J.M., Joris, B., Forsman, M., Levesque, R.C., Tiraby, G. and Waley, S.G. (1991) A standard numbering scheme for the Class A β-lactamases. *Biochem. J.,* 276, 269-270.

[54] Wommer, S., Rival, S., Heinz, U., Galleni, M., Frère, J.M., Franceschini, N., Amicosante, G., Rasmussen, B., Bauer, R. and Adolph, H.W. (2002) Substrate-activated Zinc binding of metallo-β-lactamases, *J. Biol. Chem.,* 277, 24142-24147.

[55] Outten, C.E. and O'Halloran, T.V. (2001) Femtomolar sensitivity of metalloregulatory proteins controlling zinc homeostasis. *Science,* 292, 2488-2492.

[56] Neuwald, A.F., Liu, J.S., Lipman, D.J. and Lawrence, C.E. (1997) Extracting protein alignment models from the sequence database. *Nucleic Acids Res.,* 25, 1665-1677.

[57] Aravind, L. (1999) An evolutionary classification of the metallo-β-lactamase fold proteins. *In Silico Biol.,* 1, 69-91.

[58] Daiyasu, H., Osaka, K., Ishino, Y. and Toh, H. (2001) Expansion of the zinc metallo-hydrolase family of the beta-lactamase fold. *FEBS Lett.,* 503, 1-6.

[59] de la Sierra-Gallay, I.L., Pellegrini, O. and Condon, C. (2005) Structural basis for substrate binding, cleavage and allostery in the tRNA maturase RNase Z. *Nature,* 433, 657-61.

[60] Ishii, R., Minagawa, A., Takaku, H., Takagi, M., Nashimoto, M. and Yokoyama, S. (2005) Crystal structure of the tRNA 3' processing endoribonuclease tRNase Z from *Thermotoga maritima. J. Biol. Chem.,* 280, 14138-44.

[61] Kostelecky, B., Pohl, E., Vogel, A., Schilling, O. and Meyer-Klaucke W. (2006) The crystal structure of the zinc phosphodiesterase from *Escherichia coli* provides insight into function and cooperativity of tRNase Z-family proteins. *J. Bacteriol.,* 188, 1607-14

[62] Schilling, O., Wenzel, N., Naylor, M., Vogel, A., Crowder, M., Makaroff, C. and Meyer-Klaucke, W. (2003) Flexible metal binding of the metallo-beta-lactamase domain: glyoxalase II incorporates iron, manganese, and zinc *in vivo.* B*iochemistry,* 42, 11777-11786.

[63] Marasinghe, G.P., Sander, I.M., Bennett, B., Periyannan, G., Yang, K.W., Makaroff, C.A. and Crowder, M.W. (2005) Structural studies on a mitochondrial glyoxalase II. J. *Biol. Chem.,* 280, 40668-75

[64] Cameron, A.D., Ridderstrom, M., Olin, B. and Mannervik, B. (1999) Crystal structure of human glyoxalase II and its complex with a glutathione thiolester substrate analogue. *Structure,* 7, 1067-1078.

[65] Campos-Bermudez, V.A., Leite, N.R., Krog, R., Costa-Filho, A.J., Soncini, F.C., Oliva, G. and Vila, A.J. (2007) Biochemical and structural characterization of *Salmonella typhimurium* glyoxalase II: new insights into metal ion selectivity. *Biochemistry,* 46, 11069-79.

[66] Stamp, A.L., Owen, P., Omari, K.E., Nichols, C.E., Lockyer, M., Lamb, H.K., Charles, I.G., Hawkins, A.R. and Stammers, D.K. (2010) Structural and functional

characterization of *Salmonella enterica serovar Typhimurium* YcbL: an unusual Type II glyoxalase. *Protein Sci.,* 19, 1897-905.

[67] Silva, M.S., Barata, L., Ferreira, A.E., Romão, S., Tomás, A.M., Freire, A.P. and Cordeiro, C. (2010) Catalysis and structural properties of *Leishmania infantum* glyoxalase II: trypanothione specificity and phylogeny. *Biochemistry,* 47, 195-204.

[68] Mc Coy, J.G., Bingman, C.A., Bitto, E., Holdorf, M.M., Makaroff, C.A. and Phillips G.N. Jr (2006) Structure of an ETHE1-like protein from *Arabidopsis thaliana. Acta Crystallogr., sect.* D, 62, 964-70.

[69] Frazão, C., Silva, G., Gomes, C.M., Matias, P., Coelho, R., Sieker, L., Macedo, S., Liu, M.Y., Oliveira, S., Teixeira, M., Xavier, A.V., Rodrigues-Pousada, C., Carrondo, M.A. and Le Gall, J. (1999) Crystallization and preliminary diffraction data analysis of both single and pseudo-merohedrally twinned crystals of rubredoxin oxygen oxidoreductase from *Desulfovibrio gigas. Acta Crystallogr. D Biol. Crystallogr.,* 55, 1465-1467.

[70] Silaghi-Dumitrescu, R., Kurtz, D.M. Jr., Ljungdahl, L.G. and Lanzilotta, W.N. (2005) X-ray crystal structures of *Moorella thermoacetica* FprA. Novel diiron site structure and mechanistic insights into a scavenging nitric oxide reductase. *Biochemistry,* 44, 6492-501.

[71] Gu, S.Y., Yan, X.X. and Liang, D.C. (2008) Crystal structure of Tflp: a ferredoxin-like metallo-beta-lactamase superfamily protein from *Thermoanaerobacter tengcongensis. Proteins,* 72, 531-6.

[72] Dong, Y.J., Bartlam, M., Sun, L., Zhou, Y.F., Zhang, Z.P., Zhang, C.G., Rao, Z. and Zhang, X.E. (2005) Crystal structure of methyl parathion hydrolase from *Pseudomonas sp.* WBC-3. *J. Mol. Biol.,* 353, 655-63.

[73] Garau, G., Lemaire, D., Vernet, T., Dideberg, O. and Di Guilmi, A.M. (2005) Crystal structure of phosphorylcholine esterase domain of the virulence factor choline-binding protein e from *Streptococcus pneumoniae*: new structural features among the metallo-beta-lactamase superfamily. *J. Biol. Chem.,* 280, 28591-600.

[74] Hermoso, J.A., Lagartera, L., Gonzalez, A., Stelter, M., Garcia, P., Martinez-Ripoll, M., Garcia, J.L. and Menendez, M. (2005) Insights into pneumococcal pathogenesis from the crystal structure of the modular teichoic acid phosphorylcholine esterase Pce. *Nat. Struct. Mol. Biol.,* 12, 533-538.

[75] Hagelueken, G., Adams, T.M., Wiehlmann, L., Widow, U., Kolmar, H., Tümmler, B., Heinz, D.W. and Schubert, W.D. (2006) The crystal structure of SdsA1, an alkylsulfatase from *Pseudomonas aeruginosa*, defines a third class of sulfatases. *Proc. Natl. Acad. Sci. USA.*, 103, 7631-6.

[76] Liu, D., Lepore, B.W., Petsko, G.A., Thomas, P.W., Stone, E.M., Fast, W. and Ringe, D. (2005) Three-dimensional structure of the quorum-quenching N-acyl homoserine lactone hydrolase from *Bacillus thuringiensis. Proc. Natl. Acad. Sci.* USA, 102, 11882-11887.

[77] Liu, D., Thomas, P.W., Momb, J., Hoang, Q.Q., Petsko, G.A., Ringe, D. and Fast, W. (2007) Structure and specificity of a quorum-quenching lactonase (AiiB) from *Agrobacterium tumefaciens. Biochemistry,* 46, 11789-99.

[78] Liu, D., Momb, J., Thomas, P.W., Petsko, G.A., Guo, H., Ringe, D. and Fast, W. (2008) Mechanism of the quorum-quenching lactonase (AiiA) from *Bacillus thuringiensis*. 2. Substrate modeling and active site mutations. *Biochemistry,* 47, 7715-25.

[79] Garces, F., Fernández, F.J., Montellà, C., Penya-Soler, E., Prohens, R., Aguilar, J., Baldomà, L., Coll, M., Badia, J. and Vega, M.C. (2010) Molecular architecture of the Mn^{2+}-dependent lactonase UlaG reveals an RNase-like metallo-beta-lactamase fold and a novel quaternary structure. *J. Mol. Biol.,* 398, 715-29.

[80] Ishikawa, H., Nakagawa, N., Kuramitsu, S. and Masui, R. (2006) Crystal structure of TTHA0252 from *Thermus thermophilus* HB8, a RNA degradation protein of the metallo-beta-lactamase superfamily. *J. Biochem.,* 140, 535-42.

[81] Yamamura, A., Okada, A., Kameda, Y., Ohtsuka, J., Nakagawa, N., Ebihara, A., Nagata, K. and Tanokura, M. (2009) Structure of TTHA1623, a novel metallo-beta-lactamase superfamily protein from *Thermus thermophilus* HB8. *Acta Crystallogr Sect F Struct Biol Cryst Commun., 65, 455-9.*

[82] Yamamura, A., Ohtsuka, J., Kubota, K., Agari, Y., Ebihara, A., Nakagawa, N., Nagata, K. and Tanokura, M. (2008) Crystal structure of TTHA1429, a novel metallo-beta-lactamase superfamily protein from *Thermus thermophilus* HB8. *Proteins,* 73, 1053-7.

[83] Mandel, C.R., Kaneko, S., Zhang, H., Gebauer, D., Vethantham, V., Manley, J.L. and Tong, L. (2006) Polyadenylation factor CPSF-73 is the pre-mRNA 3'-end-processing endonuclease. *Nature,* 444, 953-6.

[84] de la Sierra-Gallay, I.L., Zig, L., Jamalli, A. and Putzer, H. (2008) Structural insights into the dual activity of RNase J. *Nat. Struct. Mol. Biol.*, 15, 206-12.

[85] Mir-Montazeri, B., Ammelburg, M., Forouzan, D., Lupas, A.N. and Hartmann, M.D. (2010) Crystal structure of a dimeric archaeal Cleavage and Polyadenylation Specificity Factor. *J. Struct. Biol., in press.*

[86] Garau, G., Di Guilmi, A.M. and Hall, B.G. (2005) Structure-based phylogeny of the metallo-beta-lactamases. *Antimicrob. Agents Chemother.,* 49, 2778-2784.

[87] Berman, H.M., Westbrook, J., Feng, Z., Gilliland, G., Bhat, T.N., Weissig, H., Shindyalov, I.N. and Bourne, P.E. (2000) The Protein Data Bank. *Nucleic Acids Res.,* 28, 235-242.

[88] Rasia, R.M. and Vila, A.J. (2002) Exploring the role and the binding affinity of a second zinc equivalent in *B. cereus* metallo-beta-lactamase. *Biochemistry,* 41, 1853-1860.

[89] Nienaber, V.L., Richardson, P.L., Klighofer, V., Bouska, J.J., Giranda, V.L. and Greer, J. (2000) Discovering novel ligands for macromolecules using X-ray crystallographic screening. *Nat. Biotechnol.,* 18, 1105-1108.

[90] de Seny, D., Prosperi-Meys, C., Bebrone, C., Rossolini, G.M., Page, M.I., Noel, P., Frère, J.M. and Galleni, M. (2002) Mutational analysis of the two zinc-binding sites of the *Bacillus cereus* 569/H/9 metalloβlactamase. *Biochem. J.,* 363, 687-696.

[91] Dal Peraro, M., Vila, A.J. and Carloni, P. (2002) Structural determinants and hydrogen-bond network of the mononuclear zinc(II)-beta-lactamase active site. *J. Biol. Inorg. Chem.* 7, 704-712.

[92] Fraser, S.L. and Jorgensen, J.H. (1997) Reappraisal of the antimicrobial susceptibilities of *Chryseobacterium* and *Flavobacterium* species and methods for reliable susceptibility testing. *Antimicrob. Agents Chemother.,* 41, 2738-2741.

[93] Rossolini, G.M., Franceschini, N., Riccio, M.L., Mercuri, P.S., Perilli, M., Galleni, M., Frère, J.M. and Amicosante, G. (1998) Characterization and sequence of the *Chryseobacterium (Flavobacterium) meningosepticum* carbapenemase: a new

molecular class B β-lactamase showing a broad substrate profile. *Biochem. J.,* 332, 145-152.

[94] Vessillier, S., Docquier, J.D., Rival, S., Frère, J.M., Galleni, M., Amicosante, G., Rossolini, G.M. and Franceschini, N. (2002) Overproduction and biochemical characterization of the *Chryseobacterium meningosepticum* BlaB metallo-β-lactamase. Antimicrob. *Agents Chemother.,* 46, 1921-1927.

[95] Bloch, K.C., Nadarajah, R. and Jacobs, R. (1997) *Chryseobacterium meningosepticum*: an emerging pathogen among immunocompromised adults. Report of 6 cases and literature review. *Medicine (Baltimore),* 76, 30-41.

[96] Pellegrino, F.L., Casali, N., Dos Santos, K.R., Nouér, S.A., Scheidegger, E.M., Riley, L.W. and Moreira, B.M. (2006) *Pseudomonas aeruginosa* epidemic strain carrying bla(SPM) metallo-beta-lactamase detected in Rio de Janeiro, Brazil. *J. Chemother.,* 18, 151-6.

[97] Massidda, O., Rossolini, G.M. and Satta, G. (1991) The *Aeromonas hydrophila* cphA gene: molecular heterogeneity among class B metallo-beta-lactamases. *J.Bacteriol.,* 173, 4611-4617.

[98] Rossolini, G.M., Zanchi, A., Chiesurin, A., Amicosante, G., Satta, G. and Guglielmetti, P. (1995) Distribution of cphA or related carbapenemase-encoding genes and production of carbapenemase activity in members of the genus *Aeromonas.* Antimicrob.*Agents Chemother.,* 39, 346-349.

[99] Hernandez Valladares, M., Felici, A., Weber, G., Adolph, H.W., Zeppezauer, M., Rossolini, G.M., Amicosante, G., Frère, J.M. and Galleni M. (1997) Zn(II) dependence of the *Aeromonas hydrophila* AE036 metallo-β-lactamase activity and stability. *Biochemistry,* 36, 11534-11541.

[100] Bebrone, C., Delbrück, H., Kupper, M.B., Schlömer, P., Willmann, C., Frère, J.M., Fischer, R., Galleni, M. and Hoffmann, K.M.V. (2009) The structure of the dizinc subclass B2 metallo-beta-lactamase CphA reveals that the second inhibitory zinc ion binds in the histidine site. *Antimicrob. agents and chemother.,* 53, 4464-71.

[101] Bebrone, C., Anne, C., De Vriendt, K., Devreese, B., Rossolini, G.M., Van Beeumen, J., Frère, J.M. and Galleni, M. (2005) Dramatic broadening of the substrate profile of the *Aeromonas hydrophila* CphA metallo-beta-lactamase by site-directed mutagenesis. *J. Biol. Chem.,* 290, 28195–28202.

[102] Boschi, L., Mercuri, P.S., Riccio, M.L., Amicosante, G., Galleni, M., Frère, J.M. and Rossolini, G.M. (2000) The *Legionella (Fluoribacter) gormanii* metallo-β-lactamase : a new member of the highly divergent lineage of molecular-subclass B3 β-lactamases. Antimicrob. *Agents Chemother.,* 44, 1538-1543.

[103] Mercuri, P.S., Bouillenne, F., Boschi, L., Lamotte-Brasseur, J., Amicosante, G., Devreese, B., van Beeumen, J., Frère, J.M., Rossolini, G.M. and Galleni, M. (2001) Biochemical characterization of the FEZ-1 metallo-β-lactamase of *Legionella gormanii* ATCC33297^{T} produced in *Escherichia coli. Antimicrob. Agents Chemother.,* 45, 1254-1262.

[104] Dal Peraro, M., Vila, A.J. and Carloni, P. (2004) Substrate binding to mononuclear metallo-beta-lactamase from *Bacillus cereus. Proteins,* 54, 412-423.

[105] Xu, D., Zhou, Y., Xie, D. and Guo, H. (2005) Antibiotic binding to monozinc CphA beta-lactamase from *Aeromonas hydrophila*: quantum mechanical/molecular mechanical and density functional theory studies. *J. Med. Chem.,* 48, 6679-89.

[106] Lipscomb, W.N. (1983) Structure and catalysis of enzymes. *Annu.Rev.Biochem.* 52, 17-34.

[107] Lipscomb, W.N. and Sträter, N. (1996) Recent Advances in Zinc Enzymology. *Chem.Rev.,* 96, 2375-2434.

[108] Matthews, B.W. (1988) Structural basis of the action of thermolysin and related zinc peptidases. *Acc.Chem.Res.,* 21, 333-340.

[109] Bounaga, S., Laws, A.P., Galleni, M. and Page, M.I. (1998) The mechanism of catalysis and the inhibition of the *Bacillus cereus* zinc-dependent beta-lactamase. *Biochem. J.,* 331, 703-711.

[110] Garrity, J.D., Bennett, B. and Crowder, M.W. (2005) Direct evidence that the reaction intermediate of metallo-beta-lactamase L1 is metal bound. *Biochemistry,* 44, 1078-1087.

[111] Kaminskaia, N.V., Spingler, B. and Lippard, S.J. (2001) Intermediate in beta-lactam hydrolysis catalyzed by a dinuclear zinc(II) complex: relevance to the mechanism of metallo-beta-lactamase. *J. Am. Chem. Soc.,* 123, 6555-6563.

[112] Spencer, J., Read, J., Sessions, R.B., Howell, S., Blackburn, G.M. and Gamblin, S.J. (2005) Antibiotic recognition by binuclear metallo-beta-lactamases revealed by X-ray crystallography. *J. Am. Chem. Soc.,* 127, 14439-44.

[113] Morán-Barrio, J., González, J.M., Lisa, M.N., Costello, A.L., Peraro, M.D., Carloni, P., Bennett, B., Tierney, D.L., Limansky, A.S., Viale, A.M. and Vila, A.J. (2007) The metallo-beta-lactamase GOB is a mono-Zn(II) enzyme with a novel active site. *J. Biol. Chem.,* 282, 18286-93.

[114] Horsfall L., Izougarhane, Y., Lassaux, P., Selevsek, N., Liénard, B.M.R., Poirel, L., Kupper, M.B., Hoffmann, K.M., Frère, J.M., Galleni, M. and Bebrone, C. (2010) Broad antibiotic resistance profile of the subclass B3 Metallo-β-Lactamase GOB-1, a dizinc enzyme. *FEBS J.,* submitted.

[115] R. Paul-Soto, R. Bauer, J.-M. Frère, M. Galleni, W. Meyer-Klaucke, H. Nolting, G.M. Rossolini, D. de Seny, M. Hernandez-Valladares, M. Zeppezauer and Adolph, H.W. (1999) Mono- and binuclear Zn^{2+}-β-lactamase. Role of the conserved cysteine in the catalytic mechanism. *J. Biol. Chem.,* 274, 13242-13249.

[116] Jacquin O, Balbeur D, Damblon C, Marchot P, De Pauw E, Roberts GC, Frère JM and Matagne, A. (2009), 392, 1278-91 Positively cooperative binding of zinc ions to 569/H/9 beta-lactamase II suggests that the binuclear enzyme is the only relevant form for catalysis. *J. Mol. Biol.,* 392, 1278-91.

In: Beta-Lactamases
Editor: Jean-Marie Frère, pp. 79-101

ISBN: 978-1-61324-638-2

Chapter 4

STRUCTURES OF CLASS C β-LACTAMASES AND PERSPECTIVES IN DRUG DESIGN

***Johan Wouters*[1,*] *and Cédric Bauvois*[2]**

[1] Department of Chemistry, University of Namur,
61 Rue de Bruxelles, B-5000 Namur, Belgium.
[2] Institut de Recherches Microbiologiques Wiame, Campus Ceria,
1 Av E. Gryzon, B-1070 Brussels, Belgium.

ABSTRACT

Three-dimensional structures of a series of class C β-lactamases, alone or in complex with ligands, have been obtained by X-ray crystallography. In addition to their interest for studies on the mechanism of those serine-active hydrolases, structures of complexes between β-lactamases and synthetic compounds provide a rational basis for the design of original antibiotics. Indeed, as for other β-lactamases, hydrolytic properties of class C enzymes, and thus related antibiotic resistance, arise from specific molecular mechanisms driven by the structure of their active site. Therefore, the structure-function analysis of β-lactamases supplies critical information for the selection and development of antibacterial agents. These aspects are summarized and discussed on the basis of selected examples from the recent literature.

1. INTRODUCTION

β-lactamases (3.5.2.6) are enzymes which catalyse the hydrolysis of an amide bond in the β-lactam ring (Scheme 1) of antibiotics belonging to the penicillin/cephalosporin family [1]. Four kinds of β-lactamases have been identified: class-B enzymes are zinc containing proteins whilst enzymes of classes A, C and D are serine hydrolases. The three classes of serine β-lactamases are evolutionary related and belong to a superfamily that also includes DD-peptidases and a variety of other penicillin-binding proteins (PBP's). All these proteins

* Corresponding author. Email : johan wouters johan.wouters@fundp.ac.be

contain a Ser-Xaa-Xaa-Lys motif, where the serine is the active site residue. Although sharing some similarities, the sequences of the three classes of serine β-lactamases exhibit a large degree of variability and only a small number of residues are conserved in addition to the catalytic serine.

β-lactamases are the main cause of bacterial resistance to penicillins, cephalosporins and related β-lactam compounds. Among the four classes of β-lactamases, classes A and C are the most common clinically and have been the subjects of extensive biochemical and structural studies. In particular, it has become possible to analyse how specific structural changes in the enzyme structures might modify their catalytic properties on the basis of their X-ray crystallography structures. The detailed understanding of these phenomena constitutes the cornerstone for the design of future generations of antibiotics.

Class C β-lactamases (e.g. AmpC from *Escherichia coli* K12 or *Pseudomonas aeruginosa*) hydrolyze cephalosporins at high rate and the hosts that overproduce them become β-lactam-resistant. A particularity of this class of hydrolases is that some molecules (*e.g.* cloxacillin) that are substrates for many β-lactamases of other classes, behave as transient inactivators for AmpCs [3, 4]. Similarly, unlike a large majority of class A enzymes [5], class C β-lactamases are able to efficiently hydrolyse cephamycins (such as cefoxitin) characterized by a methoxy group on the β-lactam ring. As will be seen later, this is a direct consequence of the geometry of their active sites. Knott-Hunziker *et al.* [6] showed that clavulanic acid was inefficient for inactivating class C β-lactamases. Indeed, this compound was increasingly used in combination with another β-lactam antibiotic (e.g. Augmentin®, which combines clavulanic acid and amoxicillin) to counter emergence of bacteria producing β-lactamases. This probably contributed to the subsequent propagation of strains which produce an AmpC-type enzyme.

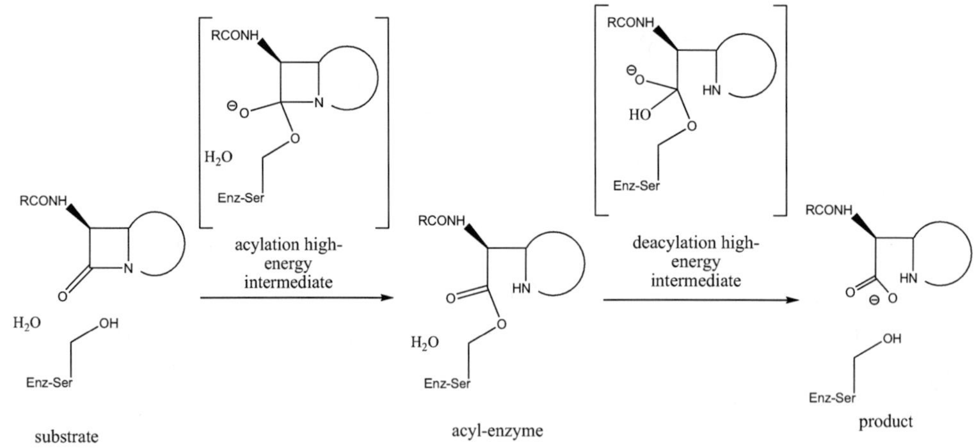

Scheme 1. Reaction proceeds through a pre-covalent complex, a tetrahedral high-energy acylation intermediate, a low-energy acyl-enzyme complex, a tetrahedral high-energy deacylation intermediate following the attack of the catalytic water, and finally the hydrolyzed product, which is subsequently released from the enzyme complex (adapted from [2]).

More generally speaking, AmpC's are not very sensitive to 'classical' β-lactamases inhibitors such as clavulanic acid, sulbactam or tazobactam. On the contrary, they are transiently inactivated by isoxazoyl penicillin derivatives (such as oxacillin or cloxacillin), monobactams (azthreonam) and reversibly inhibited by boronic acids. This latter property was exploited for purification by affinity chromatography and mechanistic studies. More significantly, it has allowed the design of novel inhibitors as will be discussed in the next sections of this chapter [7, 8] and in chapter 7.

Also of clinical importance, AmpC-type variants have been identified in many bacterial species, especially in the Enterobacteriaceae family including pathogens such as *Citrobacter*, *Salmonella* and *Shigella*. Although first exclusively found on bacterial chromosomes, the ampC genes can also be located on transferable plasmids (for reviews, see Philippon *et al.* [9] and Jacoby [10]). Consequently they can appear in strains which did not express class C enzymes and enhance the emergence of new resistant strains. However, in many cases AmpC-related resistance ismainly due to deregulation of expression leading to AmùpC overproduction. This is frequently associated with other resistance factors like membrane permeability alteration and/or co-production of several different β-lactamases types (see chapter 16). There are not many characterized genuine extended spectrum AmpC, one exception being the GC1 mutant which presents an unusually high hydrolytic activity towards ceftazidime. Nevertheless, class C β-lactamases should not be underestimated. Firstly, since these enzymes can evolve, as shown by the GC1 mutant and, like the other classes of enzymes, can change their substrates profiles against clinically used compounds. Secondly and, as already observed, transfers of *ampC* genes on plasmids will undoubtedly result in the appearance of new variants among different bacterial species. Inevitably, this will increase the group diversity and enhance the emergence of novel extended spectrum AmpC's. Thirdly, AmpC producers remain difficult to identify this may lead to the choice of inappropriate antibiotic treatment. Design of more specific new inhibitors would certainly circumvent these problems and this justifiesy the interest for this class of enzymes both from a biochemical point of view and as important drug targets for the design of antibiotics.

2. STUCTURAL FEATURES OF CLASS C β-LACTAMASES

Crystallographic structures of different AmpC enzymes have been resolved, including the structures of the chromosomal AmpC's from *Enterobacter cloacae* P99, *E. coli* and *Citrobacter freundii,* P*seudomonas aeruginosa* 2WZZ, psychrophilic strain 2QZ6 and of plasmid-mediated enzymes (ACT-1, CMY-2, CMY-10). The later two enzymes are close to the equivalent chromosome-encoded proteins (AmpC from *E. coli* and *C. freundii* respectively) and their structures present no significant differences as observed for clas a and class D enzymes. The overall structure of class C β-lactamases is formed by two distinct domains: the first contains only α-helices (α domain) and the other contains a β sheet and α - helices (β/α domain) (Figure 1a). All class C β-lactamases share several sequence motifs separated in the primary structures by constant intervals : S_{64}XXK, Y_{150}(A/S)N and K_{315}TG (Figure 1b). These motifs are located in close vicinity in the tertiary structures and form the active site. They are all involved in the catalytic mechanism which explains why these residues are well conserved in the sequences. Ser-64 is the active serine performs the nucleophilic attack onto the β-lactam carbonyl and where the acylation takes place. Tyr-150, as shown by directed mutagenesis, is an essential residue and is generally considered as the general base for increasing serine and water nucleophilicity during the acylation and deacylation steps, respectively. The role of this residue has recently been revisited in the light of recent ultrahigh resolution crystallographic data, as will be detailed in section 3.

Beside the 'classical' AmpC class C β-lactamases, the extended spectrum class C β-lactamase deserves special attention from a structural point of view, to understand both its activity and to suggest possible inhibitors. This class C chromosomal extended-spectrum (ES) β-lactamase (ESBL) was discovered in 1995 in *E. cloacae* GC1 [11]. Kinetic and mutagenesis experiments with the GC1 ESBL revealed an important deacylation activity that is somehow due to a unique three-residue repeat insertion [12]. Its crystal structure (PDB code: 1GCE) shows that the insertion makes the Ω loop at the bottom of the binding site more flexible than that of the parental wild type *E.cloacae* P99 enzyme. It was suggested that this structural feature permits the GC1 enzyme to accommodate cephalosporins with large side chains and allows more conformational freedom for the acyl intermediate, properties which should increase the hydrolysis rate. Indeed, a first crystallographic analysis of the unliganded GC1 β-lactamase (PDB code: 1GCE) revealed disorder at the level of residues 189-226 below the binding site in the Ω-loop. The disordered segment included part of an unusual tandem tripeptide insertion, with residues 213-215 invisible in the electron density map. The insertion moved the Ω-loop 1-2 Å away from the reactive Ser64 and appeared to enlarge the binding cavity for β-lactam entry. These findings supported the hypothesis described above. This was further studied by the determination of the structures of various complexes (PDB entries 1GA0, 1ONH, 1RGZ), as will be presented later.

More recently, a second class C ESBL, CMY-10, was isolated from the virulent clinical strain, *Enterobacter aerogenes*. CMY-10 is a plasmid encoded enzyme which confers resistance to cephamycins and presents some carbapenemase properties. According to the crystal structure, a three-amino-acid deletion appears to be responsible for the ESBL activity. Indeed, this deletion is accompanied by the broadening of the part of the active site cavity which accommodates the R2 side-chains of β-lactam antibiotics. This allows the active site to accommodate β-lactams with long R2 substituents, such as imipenem, thereby expanding the substrate spectrum of the variant enzyme [13, 14]

X-ray structures of mutated extended-spectrum AmpC's have also been obtained (e.g. PDB entries 3IXG, 3IXH) that further provide structural bases for understading antibiotic resistance [15, 16].

3. Stuctural Features of the Active Sites of Class C β-Lactamases and Mechanistical Considerations

Structures of complexes with class C β-lactamases were used to approach the mechanism of these enzymes. First studies on *Citrobacter freundii* AmpC suggested the importance of Ser 64 and Tyr 150 in the catalytic mechanism of class C β-lactamases. This was confirmed by the crystal structure of the *E. cloacae* enzyme. In these early works, the structures of the AmpC β-lactamase from *E. coli*, alone and in complex with a transition-state analogue, m-aminophenylboronic acid, (*1*, Table 1) were also determined(Figure 2) [17].

The structure of AmpC from *E. coli* resembles those previously determined for the class C enzymes from *Enterobacter cloacae* and *Citrobacter freundii*. The structure in complex with *1* shows that the inhibitor is covalently bond to the active site Ser 64 (Figure 2 a, c). The transition-state boronic acid analogue makes several interactions with AmpC that were unexpected. In particular, the putative "oxyanion" of the boronic acid forms a hydrogen bond with the backbone carbonyl oxygen of Ala 318 suggesting that the high-energy intermediate for amide hydrolysis by β-lactamases and related enzymes could involve a hydroxyl and not an oxyanion. The involvement of the main-chain carbonyl in ligand and transition-state stabilization would thus be a distinguishing feature between serine β-lactamases and serine proteases, to which they are often compared. The structure of the m-aminophenylboronic acid adduct also suggested several ways to improve the affinity of this class of inhibitor as discussed in the next sections (point 4).

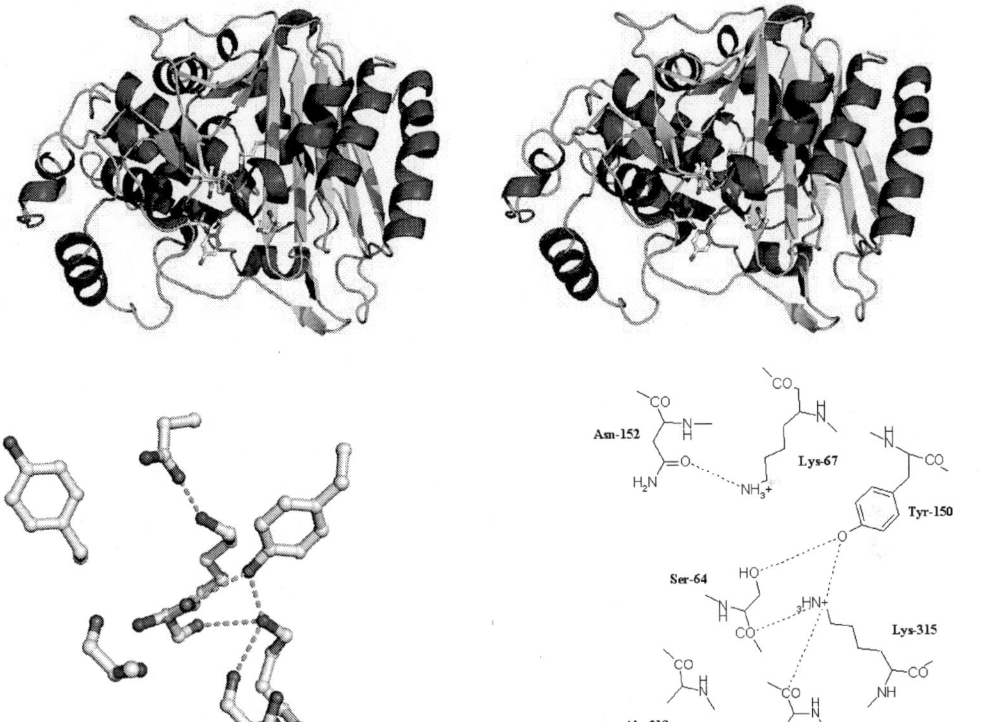

a) b)

Figure 1. Stereoview (top) of the overall fold of class C β-lactamases composed of two distinct domains: an all α domain (left of the molecule) and an α /β domain (right); a) view of the active site showing the conserved sequence motifs : S_{64}XXK, Y_{150}(A/S)N and K_{315}TG; b) schematic view of the active site of class C β-lactamases. Coordinates are taken from PDB entries 2BLS and 3BLS.

The mechanism of the transient inactivation of AmpC by imipenem has also been approached by crystallography. The X-ray crystal structure of the acyl-enzyme complex was determined to a resolution of 1.80 Å (PDB code : 1LL5) [18]. In the complex, the lactam carbonyl oxygen of imipenem has flipped by approximately 180 degrees compared to its expected position; the electrophilic acyl center is thus displaced from the point of hydrolytic attack. This conformation ressembles that of imipenem bound to the class A enzyme TEM-1 but is different from that of moxalactam bound to AmpC.

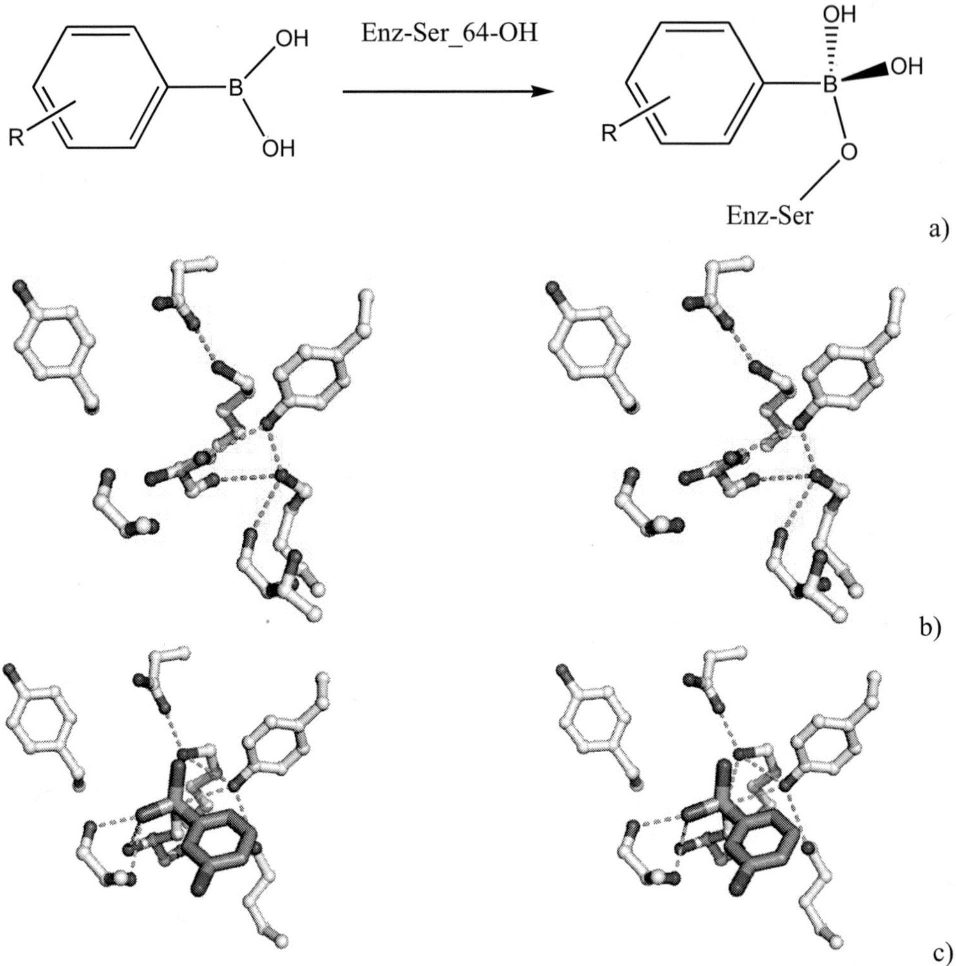

Figure 2. Inhibition of serine active class C β-lactamases by aryl boronic acid inhibitors. General mechanism of inhibition (a). Stereoviews of the active site showing the conserved sequence motifs (shown as ball-and stick) : $S_{64}XXK$, $Y_{150}(A/S)N$ and $K_{315}TG$; active site of Ampc C without inhibitor (b) and in complex with m-aminophenylboronic acid (shown as sticks) **1** (c). Coordinates are taken from PDB entries 2BLS and 3BLS.

The structure of the S64G mutant AmpC in complex with the β-lactam cephalothin [*2*] in its substrate and product forms was also determined by X-ray crystallography to 1.53 Å resolution 1KVL [19]. The acyl-enzyme intermediate between the wild-type AmpC and cephalothin was determined to 2.06 Å resolution (PDB code : 1KVM). The ligand undergoes a dramatic conformational change as the reaction progresses, with the characteristic six-membered dihydrothiazine ring of cephalothin rotating by 109 degrees. These structures correspond to all three intermediates along the reaction pathway and provide insights into substrate recognition, catalysis, and product expulsion.

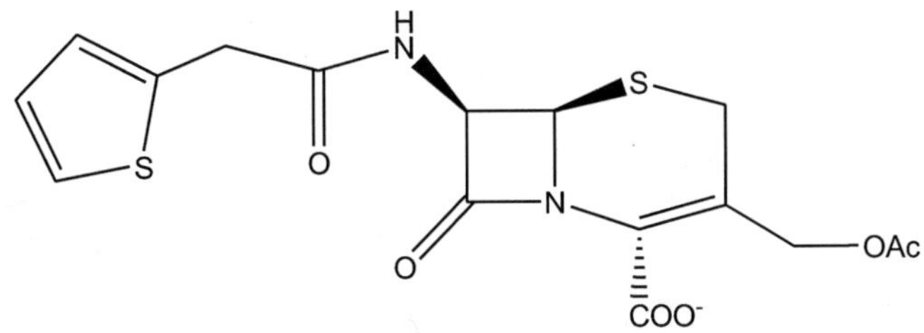

2

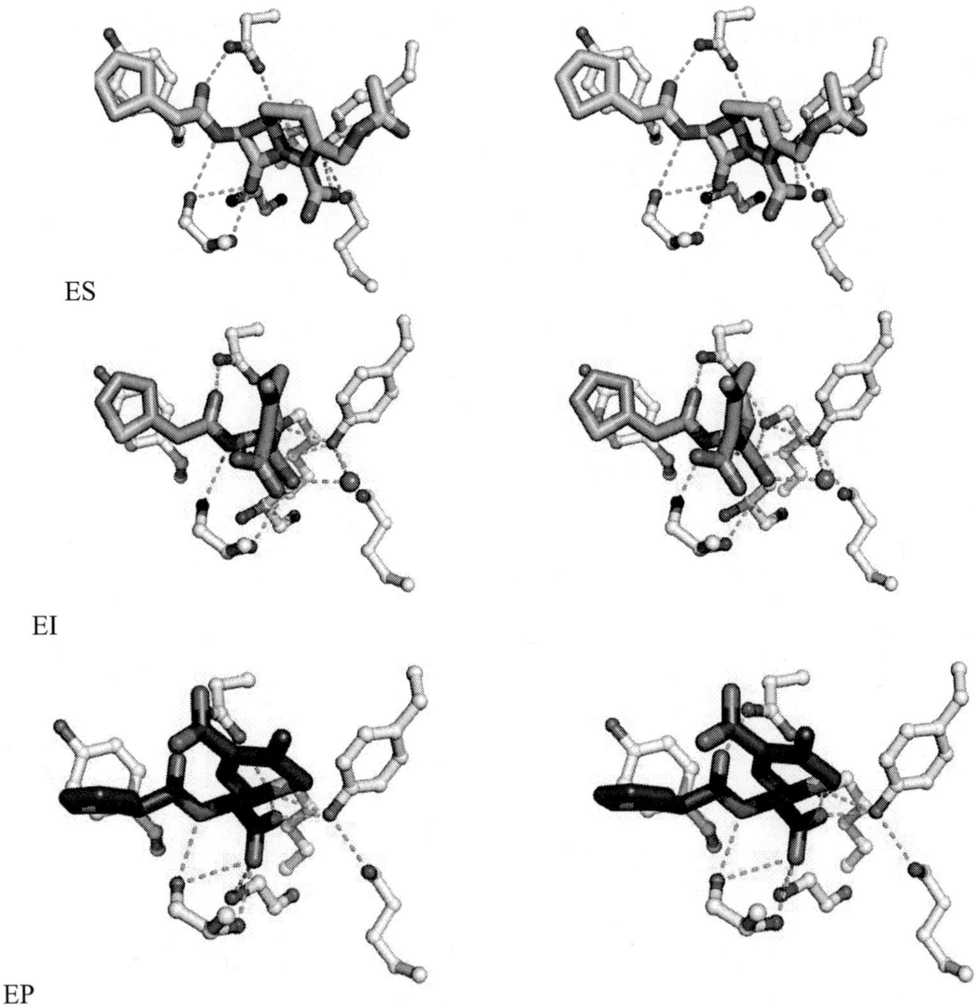

Figure 3. Stereoviews of the active sites of *E .coli* AmpC in complex with cephalothin (*2*) showing snapshots along the catalytic pathway of hydrolysis by the β-lactamase : ES (S64G enzyme-substrate complex), EI (acyl-enzyme intermediate with WT) and EP (enzyme (S64G mutant) product complex). Coordinates are taken from crystalloghraphic structures 1KVL and 1KVM.

More recently, the mechanism of AmpC deacylation was further studied by crystallography at ultrahigh resolution. Shoichet *et al.* [2] described a 1.07 Å X-ray crystallographic structure of AmpC β-lactamase in complex with boronic acid mimicking the deacylation transition-state analogue (*3*, Table 1) (PDB code: 2FFY).

The high quality of the electron density map allows the determination of many proton positions. The proton on the Tyr150 hydroxyl group is clearly visible and is donated to the boronic oxygen mimicking the deacylation water. Meanwhile, Lys67 hydrogen bonds with Ser64 Oγ, Asn152 Oδ1, and the backbone oxygen of Ala220. This suggests that this residue is positively charged and has relinquished the hydrogen bond with Tyr150 observed in acyl-enzyme complex structures. Together with previous biochemical and NMR studies [20], these observations indicate that Tyr150 is protonated throughout the reaction coordinate, disfavoring mechanisms that involve a stable tyrosinate as the general base for deacylation. Rather, the hydroxyl of Tyr150 appears to be well positioned to electrostatically stabilize the negative charge buildup in the tetrahedral high-energy intermediate. This structure, in itself, appears consistent with a mechanism involving either Tyr150 acting as a transient catalytic base in conjunction with a neutral Lys67 or the lactam nitrogen as the general base. Whereas mutagenesis studies suggest that Lys67 may be replaced by an arginine, disfavoring the conjugate base mechanism, distinguishing between these two hypotheses may ultimately depend on direct determination of the pKa of Lys67 along the reaction coordinate. However, these results have recently been contradicted by Shoichet et al. who reexamined the K67R AmpC mutant. Indeed, these authors have observed that the Lys67Arg mutation greatly affects both acylation and deacylation steps. Consequently, they concluded that their results do not disfavour the conjugate base catalysis mechanism [21].

These experimental details disagree with computed values of the pKa values of titratable residues in a set of class C β-lactamases [22]. In these calculations, the pKa of the phenolic group of residue Tyr150 is the only one to stand out with an abnormally low value of 8.3, more than one pKa unit lower than the measured reference value for tyrosine in solution. Comparisons between the relative kcat values and the Tyr150 pKa value in mutants revealed a striking correlation supporting the hypothesis that the phenolate form of Tyr150 is responsible for the activation of the nucleophilic serine.

Extended-Spectrum Class C β-lactamase

As already mentioned, bacterial resistance to the third-generation cephalosporins is an issue of great concern in current antibiotic therapeutics. An important source of this resistance comes from production of ESBL's by bacteria. The *Enterobacter cloacae* GC1 enzyme is an example of a class C ES β-lactamase. Unlike wild-type (WT) forms, such as the *E. cloacae* P99 and *Citrobacter freundii* enzymes, the GC1 ESBL rapidly hydrolyzes third-generation cephalosporins such as cefotaxime and ceftazidime.

The crystallographic structure of the *Enterobacter cloacae* GC1 class C ESBL inhibited by *4*, an alkylidenecephalosporin sulfone, has been determined by X-ray diffraction at 100 K to a resolution of 1.6 Å (1GA0 [23]), revealing the structure of a reaction intermediate (*4**) with a cephem sulfone (Scheme 2). The crystal structure was solved by molecular replacement using the unliganded structure (1GCE) [24]. Cryoquenching of the reaction of the sulfone with the enzyme produced an intermediate where the sulfone is covalently bound to Ser64 (Figure 4).

After acylation by the β-lactam ring, the dihydrothiazine dioxide ring opens with departure of the sulfinate. Nucleophilic attack of a side chain pyridine nitrogen atom on the C6 atom of the resultant imine yields a bicyclic aromatic system which helps to stabilize the acyl enzyme. Positioning of the anionic sulfinate group close to Tyr 150 and the acyl ester bond could block the approach of a potentially deacylating water molecule and explain observation of intermediate *4**.

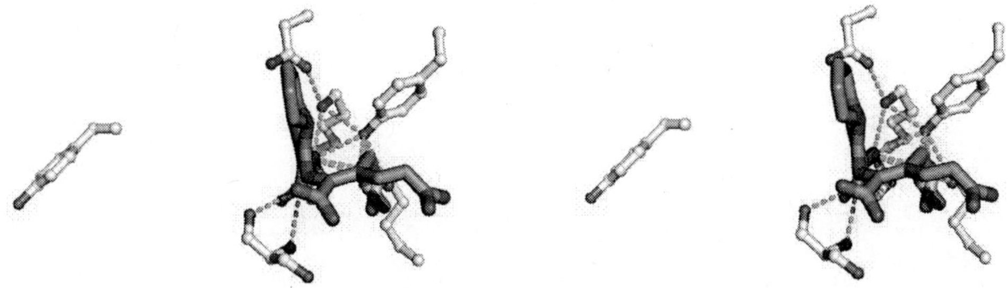

Figure 4. Active site (stereoview) of *E. cloacae* GC1 extended-spectrum β-lactamase inhibited by cephem *4* leading to intermediate *4** after acylation by reactive serine Ser64 and intra-molecular rearrangement. Coordinates are taken from crystallographic structure 1GA0.

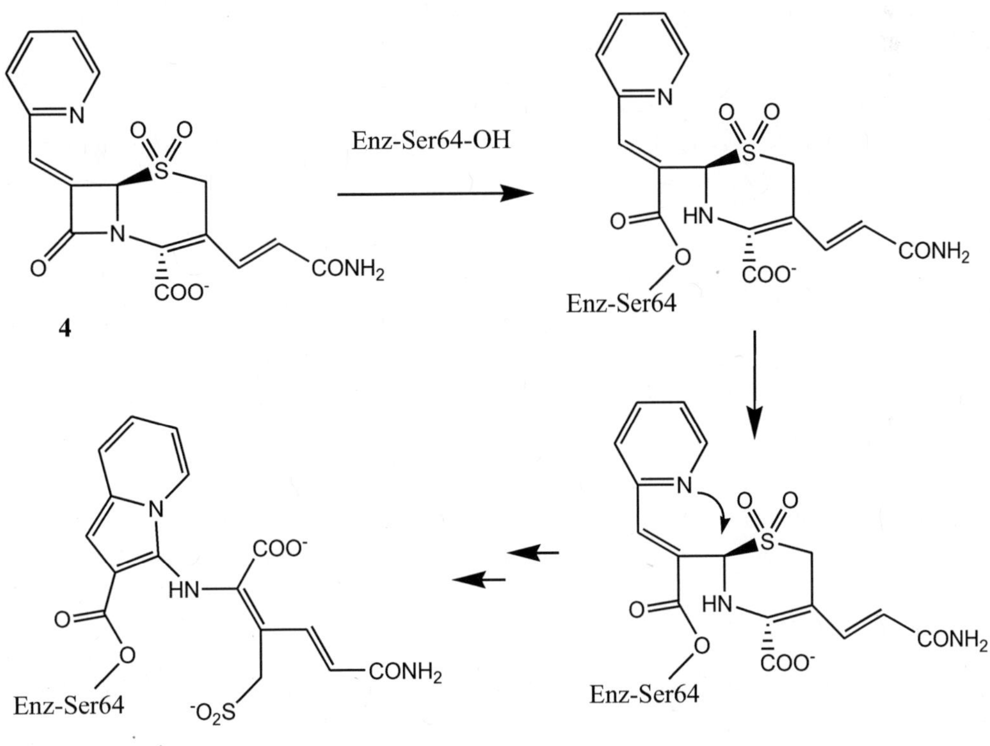

4*

Scheme 2. Proposed mechanism for the inhibition of AmpC by *4* leading to intermediate *4** after acylation by reactive serine Ser64 and intra-molecular rearrangement.

Comparison of the liganded and unliganded protein structures showed that a major movement (up to 7 Å) and refolding of part of the Ω-loop (215-224) accompanies the binding of the inhibitor. This conformational flexibility in the Ω-loop may form the basis of an extended-spectrum activity of class C β-lactamases against modern cephalosporins.

This mechanistical study was followed by a similar one using m-nitrophenyl 2-(2-aminothiazol-4-yl)-2-[(Z)-methoxyimino]acetylaminomethyl phosphonate (*5*) [25]. This phosphonate was designed to generate a transition state analog for turnover of cefotaxime.

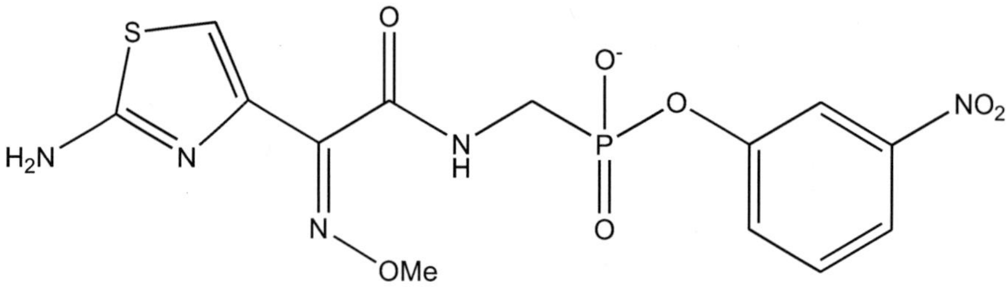

5

The crystal structures of complexes of the phosphonate with both extended-spectrum GC1 and WT *C. freundii* GN346 β-lactamases have been determined to high resolution (1.4-1.5 Å ; 1RGZ, 1RGY). In both WT GN346 and ES GC1 complexes, the catalytic Ser-64 has been phosphorylated by *5* with the departure of the *m*-nitrophenol group. The terminal P-O_1 group displaces a water molecule from the oxyanion hole formed by the backbone amide groups of Ser-64 and Ser-318/321 (using numbering for the WT GN346 and ES GC1 β-lactamases, respectively). The oxygen atom is asymmetrically hydrogen-bonded to the amides, with the stronger and more linear hydrogen bond to the NH on the B3 β-strand. The remaining P-O_2 group is exposed. It hydrogen bonds to the hydroxyl group of Tyr-150 and to a water molecule bridging to Thr-316/319 on the B3 strand. The conformations of the Ser-64-Oγ-P linkages are similar in both complexes.

However, orientation and conformation of the cefotaxime-like side chain of the phosphonate is markedly different in the two complexes. In the WT GN346 binding site, the oxymino side-chain is solvent-exposed, but in the GC1 complex it is quite buried. The serine-bound analog of the tetrahedral transition state for deacylation thus exhibits a very different binding geometry in each enzyme. In the WT β-lactamase the cefotaxime-like side chain is crowded against the Ω loop and must protrude from the binding site with its methyloxime branch exposed. In the ESBL, a mutated Ω loop adopts an alternate conformation allowing the side chain to be much more buried. During the binding and turnover of the cefotaxime substrate by this ESBL, it is proposed that ligand-protein contacts and intra-ligand contacts are considerably relieved relative to the WT, facilitating positioning and activation of the hydrolytic water molecule. The ESBL is thus able to efficiently hydrolyse third-generation cephalosporins.

Mechanism of Inhibition by Methylidene Penems

Novel penem molecules with heterocycle substitutions at position 7 *via* a methylidene linkage (*e.g. 6, 7, 8*) have been investigated at the end of the 90's for their activities and efficacy as β-lactamase inhibitors. The concentrations of these molecules that result in 50% inhibition of enzyme activity are in the low nanomolar ranges both for the TEM-1 enzyme and for AmpC. Some of them also inhibit the metalloenzyme class B β-lactamases. These inhibitors are usually more stable than imipenem against hydrolysis by dehydropeptidases. Their interest and development as potential new antibiotics is briefly discussed in the next section (point 4). These compounds also present interest from a mechanistical point of view as illustrated by a structural study of BRL 42715, C7-(N1-methyl-1,2,3-triazolylmethylene)penem, *6,* an active-site-directed inactivator of a broad range of bacterial β-lactamases, including the class C enzymes. This compound is 10 to 100 times more active than other clinically used β-lactamase inactivators such as clavulanic acid, sulbactam and tazobactam. These potent inactivating activities are reflected By the low concentrations of *6* needed to potentiate the antibacterial activity of β-lactamase-susceptible β-lactams. Although *6* has not been developed because of its chemical instability and short half-life in humans, it is still used as a reference inactivator and thus remains a useful pharmacological tool for mechanistic investigations.

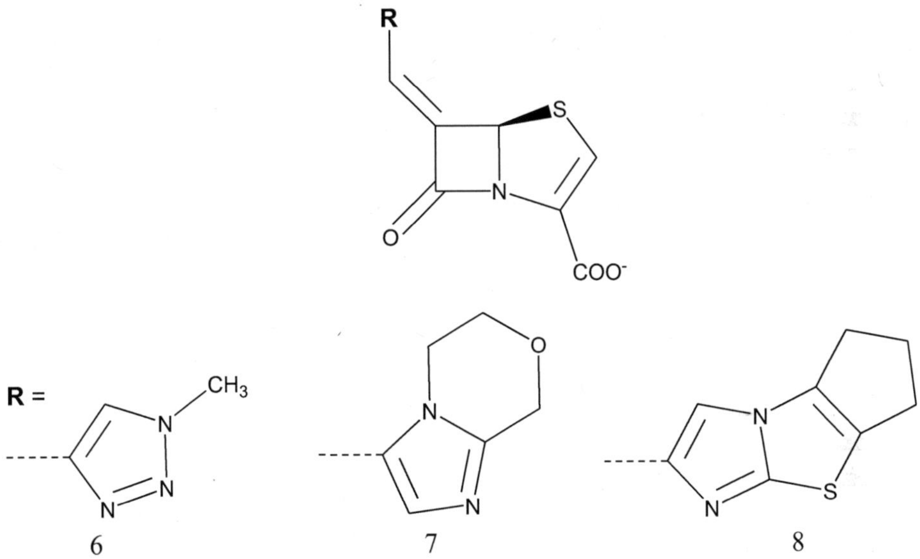

6 7 8

In particular, the mechanism of action of *6* towards class C β-lactamases was investigated and the crystal structure of its adduct with the *Enterobacter cloacae* 908R β-lactamase (1Y54) reported [26]. A stable covalent adduct, a cyclic β-aminoacrylate-enzyme complex, resulting from acylation of the active site serine by the penem followed by intramolecular rearrangement leads to the corresponding dihydrothiazepine (Figure 5).

This crystal structure confirms a mechanism that implies opening of the five-membered thiazole ring system at the C5-S bond upon alcoholysis and rearrangement via a Michael addition to form a seven-membered dihydrothiazepine ring system (*6**, Scheme 3). This is in good agreement with the spectral properties of the product that are identical to those of the dihydrothiazepine obtained after sodium hydroxide hydrolysis and with mass spectrometry results.

The interactions in and geometry of the covalent adduct of *6** with the 908R β-lactamase are quite different from those of penem analogue *7* in complex with the class C GC1 ESBL (1ONH). It is however consistent with differences observed for a transition state analogue of cefotaxime in the parental and extended-spectrum enzymes. In particular, the absolute configuration of the cyclic β-aminoacrylate-enzyme complex *6** is *S* while the other enantiomer of *7* is formed upon reaction with the GC1 β-lactamase. As already mentioned, the overall structure of the GC1 enzyme is equivalent to that of the 908R enzyme. In the GC1 structure in complex with *7*, the so-called Ω loop (189-226) is well defined. The three additional residues (213-215) after position 210 in this Ω loop in GC1 result in a more flexible binding site and allow an alternative fold ("open folding") in which Tyr224 (equivalent to Tyr221 in 908R) is displaced by ~6Å relative to its position in the free enzyme. As a result, the main difference between the two enzyme adducts is the position of this tyrosine residue. This has consequences for the binding mode of the ligand. Indeed, in 908R, the triazolyl cycle of *6* is able to stack with Tyr221, unlike the heterocyclic double ring of *7* in GC1. Moreover, the carboxylate moiety of *6* does not interact with Gln120 and Asn152 while it does so for *7* in GC1. This could explain why the absolute configuration is reversed.

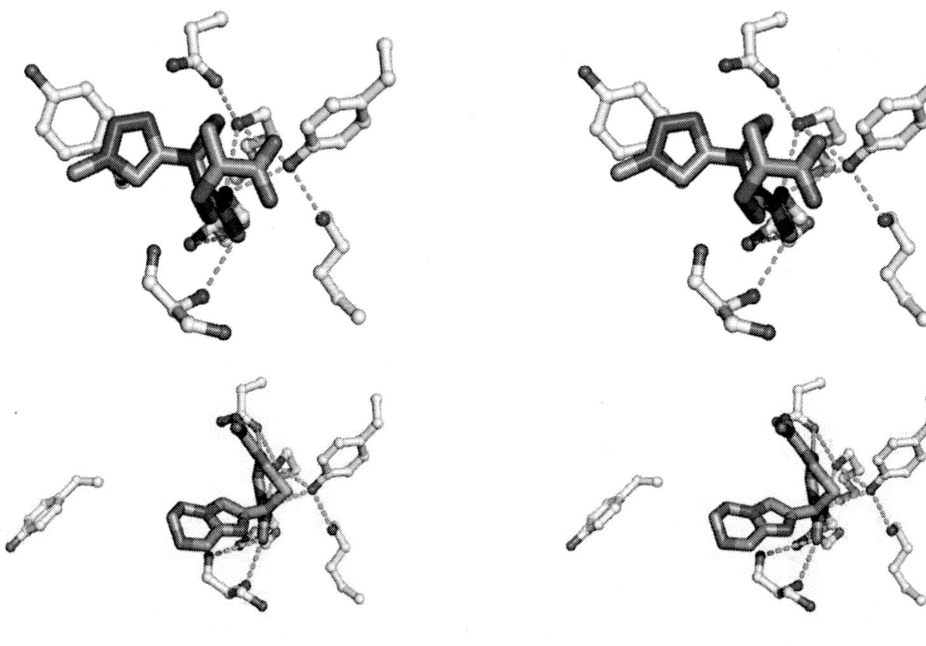

Figure 5. Comparaison of the active sites of *E. cloacae* 908R (top) and extended spectrum CG1 (bottom) class C β-lactamases after reaction with methylidene penems **6** and **7** respectively. Coordinates were taken from PDB files 1Y54 and 1ONH.

Docking and energy minimization studies were performed to further understand and quantify differences in conformation and stereochemistry between the adducts observed with the 908R and GC1 β-lactamases. In the case of *6*, the triazolyl moiety is close to Tyr221 and the thiolate group lies along the B3 β-strand. The *S* isomer of the compound is therefore strongly favoured. Indeed, the thiolate group is only able to react with one side of the double bond. In contrast, the heterocyclic ring of *7* binds along the B3 β-strand. The thiolate moiety is near Leu119 and Gln120, and is positioned in such a way that only the product with *R* configuration can be produced [8].

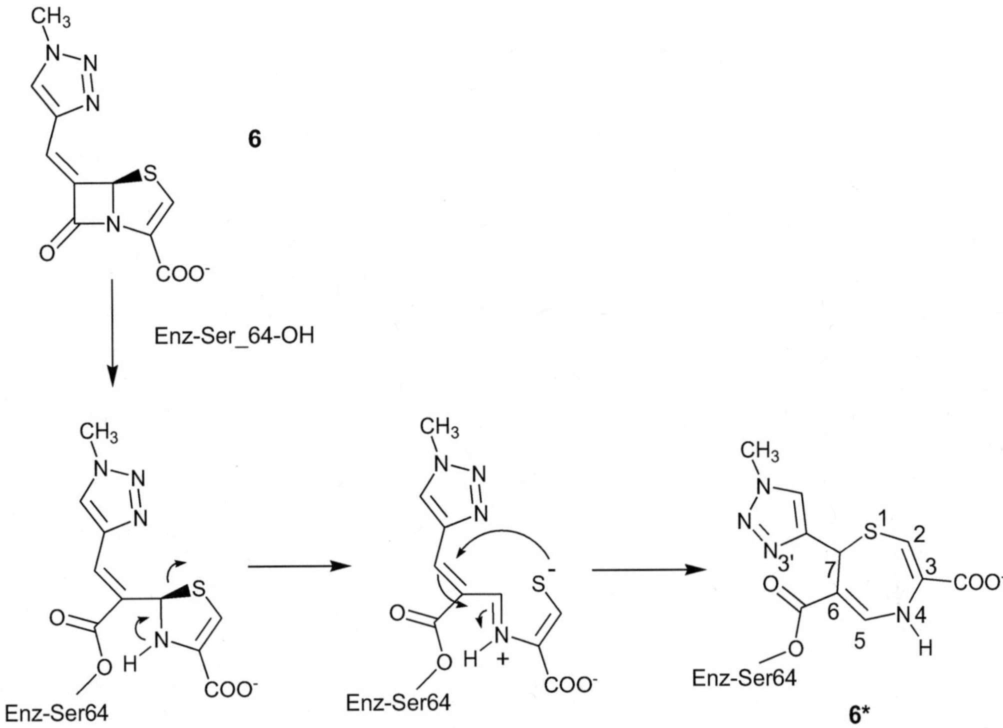

Scheme 3. Proposed mechanism for the inhibition of AmpC by *6* leading to *6** after acylation of the reactive serine Ser64 and intra-molecular rearrangement.

4. Structure-Based Drug Design: Crystal Structures of Complexes with Class C β-lactamases

Together with class A enzymes, class C β-lactamases are clinically the most commonly encountered. As already mentioned, class C enzymes such as AmpC are mainly chromosomal and are typically synthesized by Gram-negative organisms. Through the 1990s, approximately one or two new plasmid-encoded AmpC β-lactamases were discovered annually, so that this class of enzymes, both chromosomal and plasmid-born, are now present in 10-50% of the patients infected with *Citrobacter freundii, Enterobacter cloacae, Serratia marcescens*, and *Pseudomonas aeruginosa*. Complexes of class C β-lactamases with compounds belonging to different families (β–lactams, boronic acids, phosphonates, ...) have been obtained. These structures led to a better understanding of the mechanism of this class of hydrolases as just discussed (section 3) but also provided valuable starting points for the design of new selective inhibitors as illustrated in the following examples.

Methylidene Penems

As just mentioned in section 3, novel penem molecules with heterocycle substitutions at the 6 position via a methylidene linkage (*6, 7, 8*) have been investigated at the end of the 90's for their activities and efficiencies as β-lactamase inhibitors. A new β-lactamase inhibitor, a methylidene penem bearing a 5,6-dihydro-8H-imidazo[2,1-c][1,4]oxazine heterocyclic substituent at the C6 position with a Z configuration (*7*), irreversibly inhibits both class A and class C serine β-lactamases with IC_{50} values of 0.4 and 9.0 nM for TEM-1 and SHV-1 (class A), respectively, and 4.8 nM for AmpC (class C) β-lactamases. The compound also irreversibly inactivates the GC1 ESBL (IC_{50} = 6.2 nM). High-resolution crystallographic structures of a reaction intermediate of *7* with the SHV-1 β-lactamase and with the GC1 β-lactamase have been determined by X-ray diffraction to resolutions of 1.10 and 1.38 Å , respectively [27]. Cryoquenching of the reaction of *7* with each β-lactamase crystal produced a common, covalently bound intermediate. After acylation of the serine, a nucleophilic attack by the departing thiolate on the C6' atom yielded a novel seven-membered 1,4-thiazepine ring having *R* stereochemistry at the new C7 moiety (Scheme 3). The orientation of this ring in both complexes differs by a 180 degree rotation about the bond to the acylated serine. As already discussed (section 3) stereochemistry of the product of cyclization of *7* in CG1 (and SHV-1) was different from the adduct of *6* with the parent 908R *E.cloacae* enzyme [26]. On the basis of these first structures of complexes with methylidene penem *7*, design and synthesis of a series of seven tricyclic 6-methylidene penems as novel class A and C serine β-lactamase inhibitors has been described [28].

Indeed, an unexpected feature reported in the study of adducts with *7*, indicated that the 1,4-thiazepine intermediate is oriented differently in each adduct following the acylation of the active serine residue by *7*. This finding is significant in the context of designing new analogues with desired potency against both class A and C enzymes. Accordingly, on the basis of these findings, Knox *et al.* undertook the synthesis and biological evaluation of novel tricyclic methylidene penems. Seven compounds were tested against TEM-1 (class A), CcrA (class B), and AmpC (class C) β-lactamases. In comparison with tazobactam, the new inhibitors exhibited excellent activity against all these enzymes. These compounds were very potent inhibitors of TEM-1 and AmpC enzymes with IC_{50} values in the range of 1-18 and 1-3 nM, respectively. Furthermore, in comparison to tazobactam, they exhibited almost 5-100 fold greater potency against TEM-1 enzyme and were about 28,000-84,000-fold more potent against AmpC enzymes. Comparison of the potencies of these new inhibitors reveals that any substitution on the aromatic ring or any modification did not alter their potency against the AmpC enzyme significantly. *In vit*ro evaluation in cell-based assays (MICs) confirmed the effeciency of these compounds as potent broad spectrum β-lactamase inhibitors. In this context, Knox *et al.* also determined the crystallographic structures of reaction intermediates of *8* with the SHV-1 and GC1 β-lactamases. They found that the 1,4-thiazepine intermediate has distinctive hydrophobic contacts within the binding site of each β-lactamase. The absolute configuration at carbon C6' is different in both adducts. A single conformer with the *R* configuration is observed in the SHV adduct, whereas in the GC1 one the conformer with the *R* configuration exhibits only 30% occupancy and coexists with the predominant *S* conformer (70%). It is thus possible that both stereoisomers are formed initially in each β-lactamase, but the more hydrolysis resistant form predominates in the crystallographic maps. Alternatively, stereospecific reaction pathways may differ in each active site, producing one chiral form in preference to the other as discussed by Michaux *et al.* [26]. The reaction of *8* with GC1 monitored by X-ray crystallography also revealed the formation of a 1,4-thiazepine intermediate with unexpected modes of binding. The significance of hydrophobic stacking of the C6' heterocycle with Tyr105 and Tyr224 in SHV-1 and GC1, respectively, and its possible role in protecting the intermediate from hydrolytic attack were presented. At this stage, it would be very interesting to complement this study by obtaining an adduct of *8* with the 908R or P99 *E.cloacae* β-lactamase and to compare the results with the extended-spectrum enzyme. Anyway, even if pending questions remain, the reported crystallographic complexes with bicyclic and tricyclic penems supply new insights into the possibilities of inhibiting β-lactamases and further stress the benefit of those structure-based approaches in the design of new inhibitors and understanding of their mechanism(s) of inhibition.

From Boronic Acids Irreversible Inhibitors to Novel, Noncovalent Inhibitors of AmpC

The recent efforts undertaken to identify non-β-lactam-based β-lactamase inhibitors represent an excellent example of successful structure-based drug design. Starting from crystallographic structures of *E. coli* AmpC β-lactamase in complex with several aryl-boronic acids (Table 1) novel, noncovalent inhibitors of AmpC were obtained. Several types of compounds were modeled into the AmpC binding site, and a total of 37 boronic acids were ultimately tested for β-lactamase inhibition [7]. The most potent of these compounds, benzo[b]thiophene-2-boronic acid (*10*), exhibits a disociation constant of 27 nmM with *E. coli* AmpC (PDB code 1C3B). The wide range of functionalities represented by these compounds allowed the steric and chemical "mapping" of the AmpC active site in the region of the catalytic Ser64 residue, which may be useful in subsequent inhibitor discovery efforts. Also, the new boronic acid-based inhibitors were found to potentiate the activity of β-lactam antibiotics, such as amoxicillin and ceftazidime, against bacteria expressing class C β-lactamases. This suggested that boronic acid-based compounds may serve as leads for the development of therapeutic agents for the treatment of β-lactam-resistant infections.

To investigate if larger boronic acids would better interact with AmpC, structure-guided in-parallel synthesis was used to explore new inhibitors of AmpC [31]. Twenty-eight derivatives of the lead compound, 3-aminophenylboronic acid, led to an inhibitor with 80-fold better binding. Based on molecular docking results, 12 derivatives were synthesized, leading to inhibitors with Ki values of 60 nM and improved solubility. Several of these inhibitors reversed the resistance of nosocomial Gram-positive bacteria, though they showed little activity against Gram-negative bacteria.

The X-ray crystal structure of compound *11* in complex with AmpC was subsequently determined to a 2.1 Å resolution (1GA9). In the experimental structure, the inhibitor interacts with conserved residues in the active site whose role in recognition had not been previously explored. Similar recognition sites were observed in the structures of the class C β-lactamase from *Enterobacter cloacae* 908R in complex with iodo-acetamido-phenyl boronic acid (*12*) (PDB code 1S6R) [30]. The crystal structures of complexes between AmpC and other boronic acid inhibitors were further studied by computational methods and consensus binding sites were identified.

On the basis of structures of AmpC in complexes with boronic acid inhibitors and β-lactams, the programs GRID, MCSS, and X-SITE were used to predict potential binding site hot spots on AmpC.

Table 1.

compound		PDB code	Reference
1	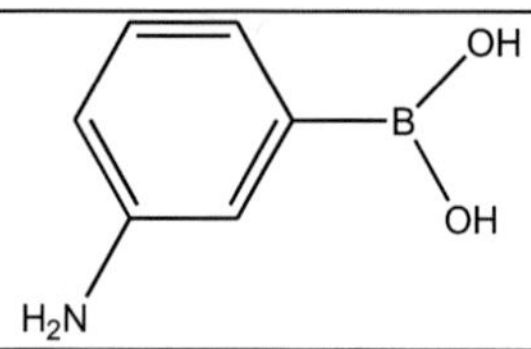	3BLS	[17]
3	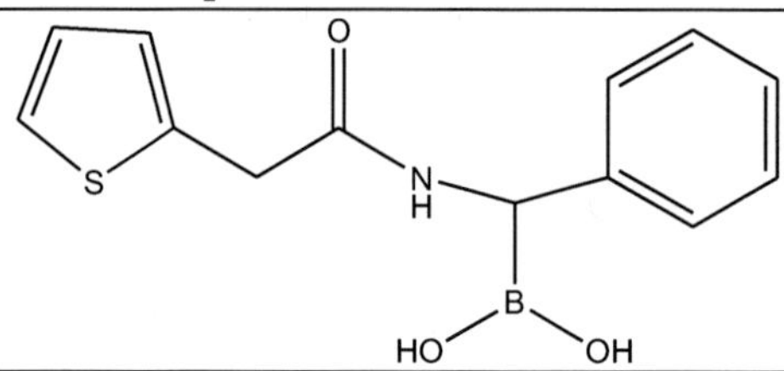	2FFY	[2]
9	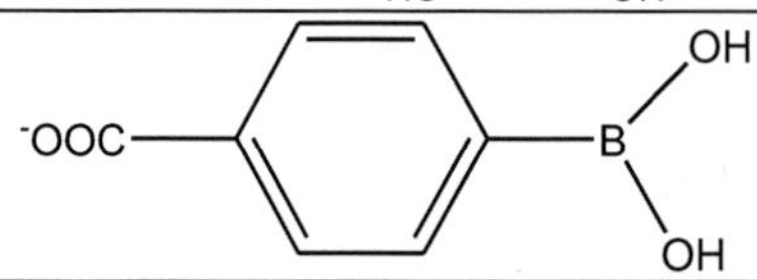	1KDW	[29]
10	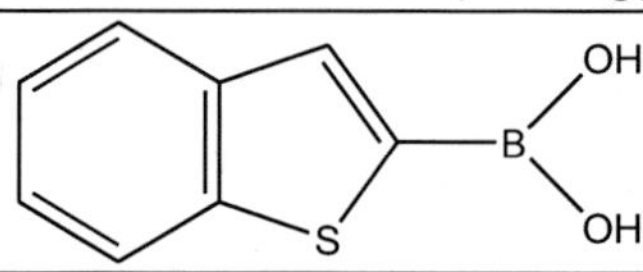	1C3B	[29]
11	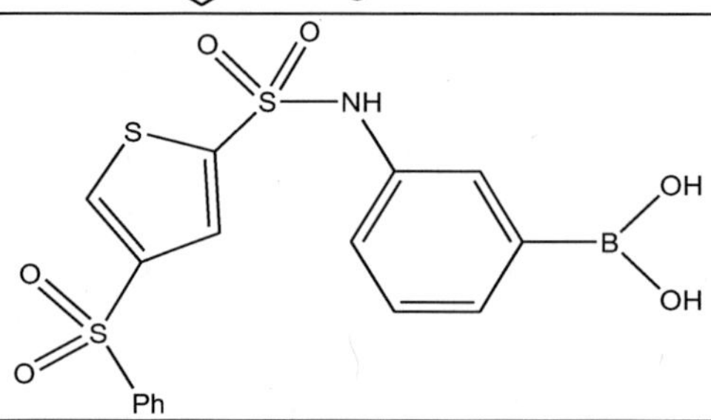	1GA9	[29]
12	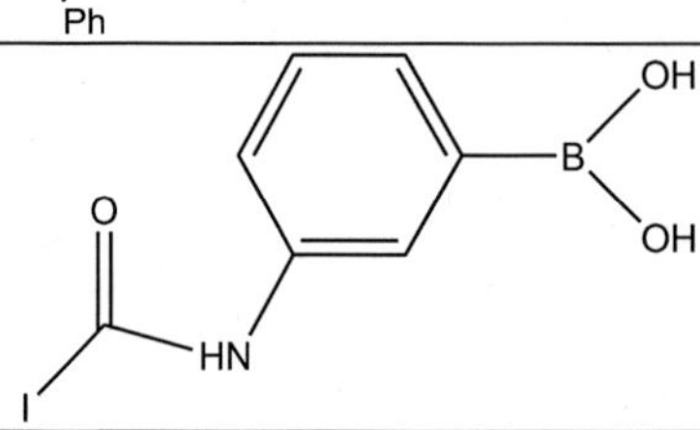	1S6R	[30]
13	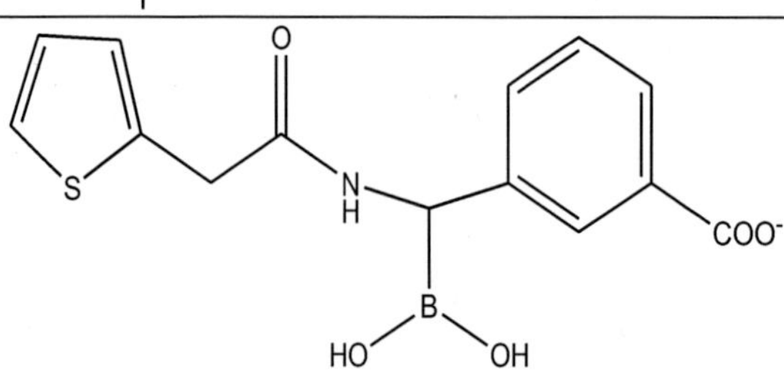	1MXO 1PI4	[8]

compound		PDB code	Reference
14	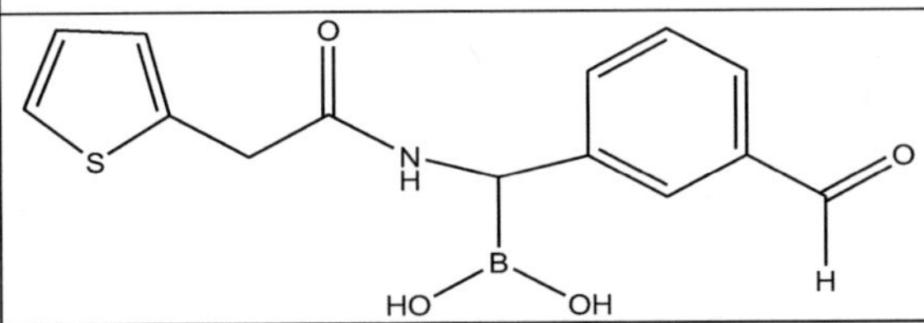	1PI5	[8]

Several consensus binding sites were identified from the crystal structures and predictions by the computational programs showed some correlation with the experimentally observed binding sites [29].

Several sites were not predicted, but novel binding sites were suggested. Taken together, a map of binding site hot spots found on AmpC, along with information on the functionality recognized by each site, was constructed. As a result of those structure based drug design approaches, new carboxyphenyl-glycylboronic acid transition-state analogue inhibitors of AmpC were designed and evaluated. The new compounds improve inhibition by over 2 orders of magnitude compared to analogous glycylboronic acids, with K_i values as low as 1 nM [8, 32]. In this context, Shoichet *et al.* designed compound *13* (Table 1) based on the β-lactam antibiotic cephalothin [8]. In the covalent adduct *13**, the benzoic acid group of *13* was intended to mimic both the dihydrothiazine ring system of the high energy intermediate of the substrate cephalothin (*14**) and the carboxylate in the active site.

Indeed β-lactams bear negatively charged groups at this position, and structural analysis of the binding determinants of AmpC suggests that there is a complementary carboxylate binding site on the enzyme [29]. Among known structures of AmpC with bound substrate, the analogous carboxylate hydrogen bonds with either Asn343 or Asn346 [19]. Thus, it was expected that the carboxylate of *13* would hydrogen bond with at least one of those residues. Surprisingly, the crystal structure of *13* in complex with AmpC revealed a hydrogen bond between the *m*-carboxylate of the benzoic acid and Asn289 [8] and not Asn343 or Asn346, as was expected and designed. To investigate the energy of this hydrogen bond, the Asn289Ala mutant was made, as was an analogue of *13* that lacked the carboxylate (compound *14*).

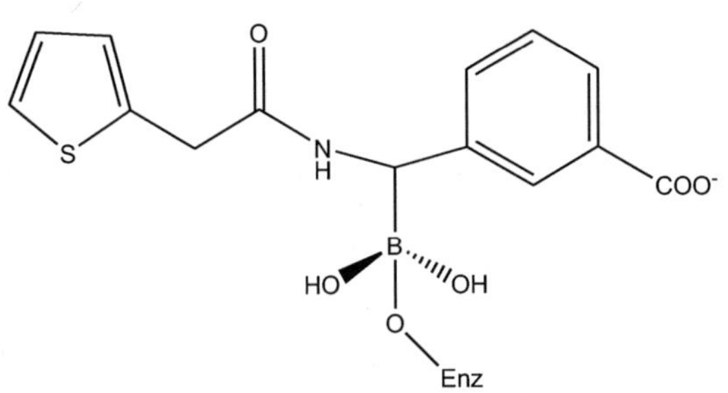

13

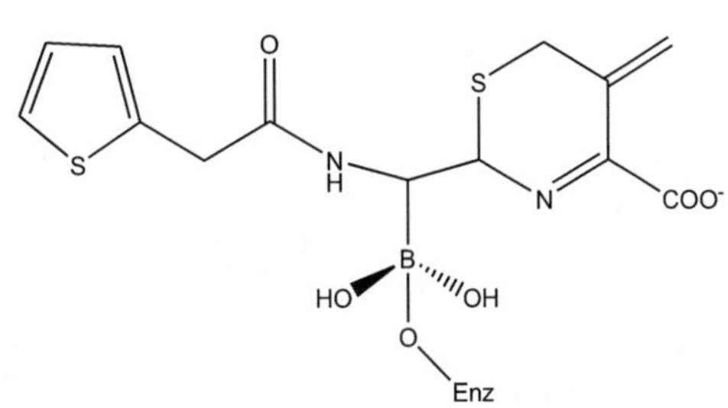

14*

The dissociation constants of the four different protein/analogue complexes indicate that the carboxylate-amide hydrogen bond contributes 1.7 kcal/mol to overall binding affinity. Synthesis of an analogue of *13* where the carboxylate was replaced with an aldehyde led to an inhibitor that lost all this hydrogen bond energy. To investigate the structural bases of these energies, X-ray crystal structures of N289A/*13* and N289A/*14* were determined to 1.49 and

1.39 Å , respectively (PDB codes 1PI4 and 1PI5). These structures suggest that no significant rearrangement occurs in the mutant compàared to the wild-type complexes with both compounds. The Leu119Ala and Leu293Ala mutant enzymes were made to investigate the interaction between a phenyl ring in *13* and these residues. Whereas deletion of the phenyl itself diminishes affinity by 5-fold, the double-mutant cycles suggest that this energy does not come from interactions with the leucines, despite the close contacts in the structure. The energies of these interactions provide key information for the design of improved inhibitors against beta-lactamases. The high magnitude of the ion-dipole interaction between Asn289 and the carboxylate of *13* is consistent with the idea that ionic interactions can provide significant net affinity in inhibitor complexes in this series of promising compounds.

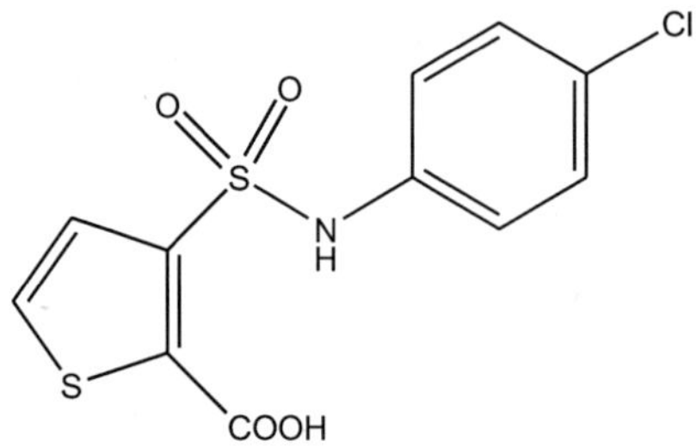

15

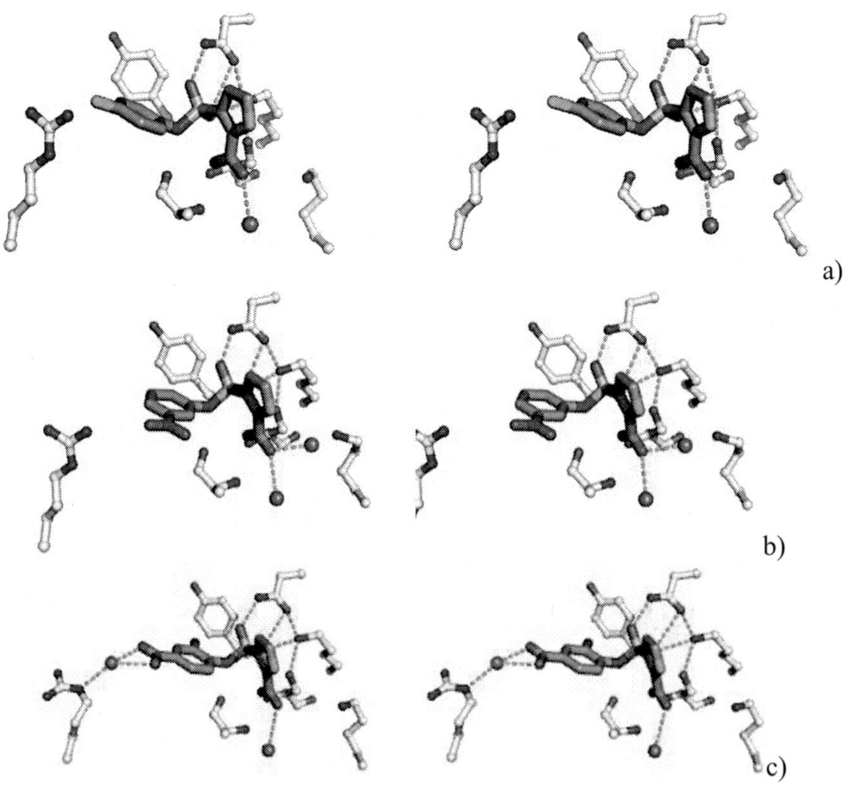

Figure 6. Comparaison of the crystal structures of complexes beteween *E. coli* AmpC and thiophene-sulfonamide derivatives designed as original non β-lactam inhibitors. Coordinates of complexes formed by AmpC and compounds *15* (a), *16* (b), and *17* (c) were obtained from PDB files 1L2S, 1XGI and 1XGJ, respectively.

The group of Shoichet extended its structure based drug design to the study of non-boronic inhibitors. A database of over 200,000 compounds was docked into the active site of the AmpC β-lactamase to identify potential inhibitors [33]. Fifty-six compounds were tested, and three had K(i) values of 650 μM or better. The best of these, 3-[(4-chloroanilino)sulfonyl]thiophene-2-carboxylic acid, *15* was a competitive noncovalent inhibitor (Ki = 26 microM), which also reversed resistance to β-lactams in bacteria expressing AmpC. The structure of AmpC in complex with this compound was determined by X-ray crystallography (1L2S) and revealed that the inhibitor interacts with key residues in sites targeted in the docking calculation. Interestingly, the experimentally determined conformation of the inhibitor closely resembles the prediction.

Lead compound *15* is dissimilar from penicillins and cephalosporins and binds to the enzymes noncovalently and reversibly, in contrast to the β-lactam substrates and boronic inhibitors. Despite these differences, the X-ray crystal structure of the AmpC/*15* complex (1L2S) revealed that *15* complements the core of the active site, interacting with key residues involved in β-lactam recognition and hydrolysis such as Ser64, Lys67, Asn152, and Tyr221 (Figure 6) [34]. Shoichet *et al.* concluded that ligand recognition by the AmpC structure was sufficiently flexible to accommodate compounds genuinely dissimilar from ß-lactams, an interesting conclusion for the design of new inhibitors.

Efforts to improve this series of inhibitors were focussed on a structure-based approach. In the structure of the AmpC/*15* complex, the inhibitor complements the core of the active site but leaves a distal region open. To take advantage of this region, Shoichet *et al.* sought derivatives that were relatively easy to synthesize and would not diminish the solubility and drug-like properties of the inhibitors. They therefore focussed on derivatives that made new interactions with polar residues in the distal part of the AmpC site, including Arg204, or which tested features of the ligands that appeared important for binding. Fourteen analogues were synthesized, the best of which bound to AmpC with a 26-fold improved affinity over *15*. To explore their biological potential, several of the new inhibitors were tested in bacterial cell cultures against clinical pathogens. Different biological effects on resistant bacteria were observed when compared to classical □β-lactam-based inhibitors such as clavulanate and cefoxitin.

To investigate the structural basis for the higher affinities found for the new series, the crystal structure of AmpC in complex with compounds *16* (1XGJ) and *17* (1XGI) were determined at 1.97 and 1.96 Å resolutions, respectively.

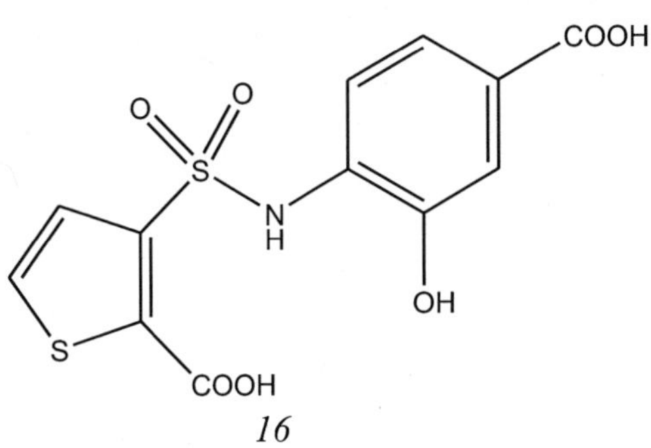

16

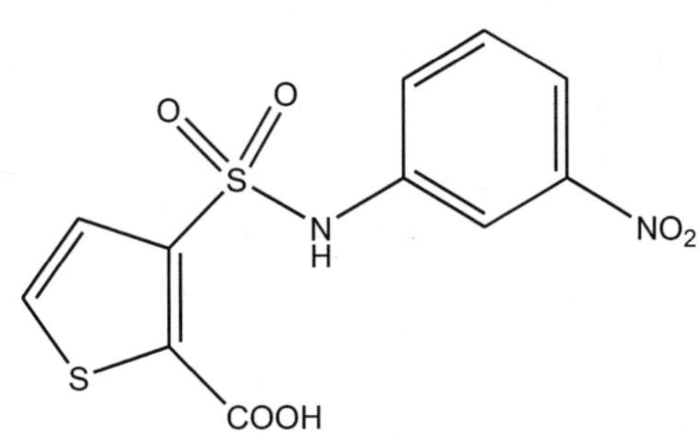

17

Overall, the complexes with *16* and *17* ressemble that of the AmpC/*15* complex, with a few interesting differences (Figure 6). Key hydrogen bond interactions are observed in the conserved parts of all three structures. The thiophene carboxylate in both *16* and *17* hydrogen bonds with both the Oγ and the main-chain NH of the hydrolytic Ser64.

As with the lead compound *15*, the thiophene ring is within van der Waals distance of residues Leu119 and Leu293, which form a hydrophobic patch on AmpC. A sulfonamide oxygen interacts with the Oγ of Ser64 and the N□ of Lys67; the other hydrogen bonds with the Nδ2 of Asn152. As in the AmpC/*15* complex, the nitrogen atom of the sulfonamide interacts with the main-chain oxygen of Ala318. Differences emerge in the 4'-substituted benzene ring, which is the point of substitution for compound *15*. The ring itself appears to form edge-to-face quadrupole interactions with Tyr221 via the introduced 2'-hydroxyl of *16*. This interaction is also present in the structure of the lead *15*, but in this case it appears to be a ring-ring or quadrupolar interaction, whereas with *16* the involvement of the hydroxyl makes it a dipole-quadrupolar interaction. The presence of the 2'-hydroxyl also displaces *16* slightly towards the “upper” part of the site, moving the ring away from Tyr221 and closer to polar residues such as Arg204 and Thr319. The 4'-carboxylate points toward the solvent accessible entry of the binding site and hydrogen bonds with an ordered water molecule, which in turn hydrogen bonds with the Nε of Arg204.

In the structure of the AmpC/*17* complex, the inhibitor is moved slightly down in the site, keeping the quadrupole edge-to-face stacking with Tyr221. The absence of the 2'-hydroxyl in this analogue excludes interaction of the phenyl ring with an ordered water and with Arg204. The loss of these polar interactions may be consistent with the diminished affinity for this inhibitor. From a molecular recognition standpoint, and as noted by the authors, it is surprising that a water-mediated hydrogen bond between the carboxylate and Arg204, in a relatively solvent-exposed region, can significantly improve affinity. Detailed conclusions as to the importance of these interactions must remain tentative pending more rigourous analyses. One of the conclusions of this promising work is that the X-ray structures of these complexes provide templates for understanding inhibition results and designing new and potentially better inhibitors.

Fragment-based Approaches

The last selected example of drug design approaches applied to β-lactamases deals with a fragment-based approach recently reported by Shoichet [35]

To investigate docking's ability to prioritize fragments, this group first docked a library of >137,000 such molecules into β-lactamase structures. Forty-eight fragments highly ranked by docking were then acquired and tested; 23 had K_i values ranging from 0.7 to 9.2 mM. X-ray crystal structures of the enzyme-bound complexes were determined for 8 of the fragments (PDB codes: 3GQZ, 3GR2, 3GRJ, 3GSG, 3GTC, 3GV9, 3GVB). For 4, the correspondence between the predicted and experimental structures was high (RMSD between 1.2 and 1.4 Å), whereas for another 2, the fidelity was lower but retained most key interactions (RMSD 2.4–2.6 Å). Two of the 8 fragments adopted very different poses in the active site owing to enzyme conformational changes. This work suggests that fragment-based methods are viable approaches in the context of structure-based drug design on drug targets like β-lactamases.

CONCLUSION

The increasing number and variety of β-lactamases produced by bacteria represent a serious threat to the clinical utility of β-lactams. It was thought that the introduction of β-lactamase inhibitors would alleviate this problem. However, alterations and mutations within β-lactamases have allowed bacteria to overcome the effects of the β-lactamase inhibitors. The currently marketed inhibitors (tazobactam, clavulanate, or sulbactam) are not active against all β-lactamases. Among the most problematic β-lactamases are the molecular class C enzymes. Bacterial genera known to produce inducible or stably derepressed AmpC β-lactamases are *Citrobacter*, *Enterobacter*, *Hafnia*, *Pseudomonas*, *Morganella*, *Providencia*, and *Serratia*. Expression of the class C enzymes by these organisms has been shown to confer resistance to the current commercial inhibitor combinations as well as extended-spectrum cephalosporins. Crystallographic structures of a large number of class C β-lactamases (alone or in complex with substrate analogues or inhibitors) have been reported and studied. As was presented in section 3, these structures produced a clear picture of different intermediates along the catalytic route of class C β-lactamases. In section 4 selected examples where structure-based drug design approches were successful in identifying and optimizing original inhibitors were presented.

REFERENCES

[1] Ambler, R. P. The structure of beta-lactamases. *Philos. Trans. R. Soc. Lond. B. Biol.* Sci., 1980. 289:321-31.

[2] Chen, Y.; G. Minasov; T. A. Roth; F. Prati; and B. K. Shoichet. The deacylation mechanism of AmpC beta-lactamase at ultrahigh resolution. *J. Am. Chem. Soc.,* 2006. 128:2970-6.

[3] Knott-Hunziker, V.; S. Petursson; G. S. Jayatilake; S. G. Waley; B. Jaurin; and T. Grundstrom. Active sites of beta-lactamases. The chromosomal beta-lactamases of Pseudomonas aeruginosa and Escherichia coli. *Biochem. J.,* 1982. 201:621-7.

[4] Matagne, A.; A. Dubus; M. Galleni; and J. M. Frere. The beta-lactamase cycle: a tale of selective pressure and bacterial ingenuity. *Nat. Prod. Rep.,* 1999. 16:1-19.

[5] Matagne, A.; J. Lamotte-Brasseur; G. Dive; J. R. Knox; and J. M. Frere. Interactions between active-site-serine beta-lactamases and compounds bearing a methoxy side chain on the alpha-face of the beta-lactam ring: kinetic and molecular modelling studies. *Biochem. J.,* 1993. 293 (Pt 3):607-11.

[6] Knott-Hunziker, V.; S. Petursson; S. G. Waley; B. Jaurin; and T. Grundstrom. The acyl-enzyme mechanism of beta-lactamase action. The evidence for class C Beta-lactamases. *Biochem. J.,* 1982. 207:315-22.

[7] Weston, G. S.; J. Blazquez; F. Baquero; and B. K. Shoichet. Structure-based enhancement of boronic acid-based inhibitors of AmpC beta-lactamase. *J. Med. Chem.,* 1998. 41:4577-86.

[8] Morandi, F.; E. Caselli; S. Morandi; P. J. Focia; J. Blazquez; B. K. Shoichet; and F. Prati. Nanomolar inhibitors of AmpC beta-lactamase. *J. Am. Chem. Soc.,* 2003. 125:685-95.

[9] Philippon, A.; G. Arlet; and G. A. Jacoby. Plasmid-determined AmpC-type beta-lactamases. *Antimicrob. Agents Chemother.,* 2002. 46:1-11.

[10] Jacoby, G. A. 2009. AmpC beta-lactamases. *Clin Microbiol Rev* 22:161-82.

[11] Nukaga, M.; S. Haruta; K. Tanimoto; K. Kogure; K. Taniguchi; M. Tamaki; and T. Sawai. Molecular evolution of a class C beta-lactamase extending its substrate specificity. *J. Biol. Chem.,* 1995. 270:5729-35.

[12] Nukaga, M.; K. Taniguchi; Y. Washio; and T. Sawai. Effect of an amino acid insertion into the omega loop region of a class C beta-lactamase on its substrate specificity. *Biochemistry,* 1998. 37:10461-8.

[13] Kim, J. Y., H. I. Jung, Y. J. An, J. H. Lee, S. J. Kim, S. H. Jeong, K. J. Lee, P. G. Suh, H. S. Lee, S. H. Lee, and S. S. Cha. 2006. Structural basis for the extended substrate spectrum of CMY-10, a plasmid-encoded class C beta-lactamase. *Mol Microbiol* 60:907-16.

[14] Lee, J. H., H. I. Jung, J. H. Jung, J. S. Park, J. B. Ahn, S. H. Jeong, B. C. Jeong, and S. H. Lee. 2004. Dissemination of transferable AmpC-type beta-lactamase (CMY-10) in a Korean hospital. *Microb Drug Resist* 10:224-30.

[15] Thomas, V.L.; A.C. McReynolds; and B.K. Shoichet. Structural Bases for Stability–Function Tradeoffs in Antibiotic Resistance *J. Mol. Biol.* 2010, 396: 47-59.

[16] Carattoli, A. 2009. Resistance plasmid families in Enterobacteriaceae. *Antimicrob Agents Chemother* 53:2227-38.

[17] Usher, K. C.; L. C. Blaszczak; G. S. Weston; B. K. Shoichet; and S. J. Remington. Three-dimensional structure of AmpC beta-lactamase from Escherichia coli bound to a transition-state analogue: possible implications for the oxyanion hypothesis and for inhibitor design. *Biochemistry,* 1998. 37:16082-92.

[18] Beadle, B. M.; and B. K. Shoichet. Structural basis for imipenem inhibition of class C beta-lactamases. *Antimicrob. Agents Chemother.,* 2002. 46:3978-80.

[19] Beadle, B. M.; I. Trehan; P. J. Focia; and B. K. Shoichet. Structural milestones in the reaction pathway of an amide hydrolase: substrate, acyl, and product complexes of cephalothin with AmpC beta-lactamase. *Structure,* 2002. 10:413-24.

[20] Kato-Toma, Y.; T. Iwashita; K. Masuda; Y. Oyama; and M. Ishiguro. pKa measurements from nuclear magnetic resonance of tyrosine-150 in class C beta-lactamase. *Biochem. J.,* 2003. 371:175-81.

[21] Chen, Y., A. McReynolds, and B. K. Shoichet. 2009. Re-examining the role of Lys67 in class C beta-lactamase catalysis. *Protein Sci* 18:662-9.

[22] Lamotte-Brasseur, J.; A. Dubus; and R. C. Wade. pK(a) calculations for class C beta-lactamases: the role of Tyr-150. *Proteins,* 2000. 40:23-8.

[23] Crichlow, G. V.; M. Nukaga; V. R. Doppalapudi; J. D. Buynak; and J. R. Knox. Inhibition of class C beta-lactamases: structure of a reaction intermediate with a cephem sulfone. *Biochemistry,* 2001. 40:6233-9.

[24] Crichlow, G. V.; A. P. Kuzin; M. Nukaga; K. Mayama; T. Sawai; and J. R. Knox. Structure of the extended-spectrum class C beta-lactamase of Enterobacter cloacae GC1, a natural mutant with a tandem tripeptide insertion. *Biochemistry,* 1999. 38:10256-61.

[25] Nukaga, M.; S. Kumar; K. Nukaga; R. F. Pratt; and J. R. Knox. Hydrolysis of third-generation cephalosporins by class C beta-lactamases. Structures of a transition state

analog of cefotoxamine in wild-type and extended spectrum enzymes. *J. Biol. Chem.,* 2004. 279:9344-52.

[26] Michaux, C.; P. Charlier; J. M. Frere; and J. Wouters. Crystal structure of BRL 42715, C6-(N1-methyl-1,2,3-triazolylmethylene)penem, in complex with Enterobacter cloacae 908R beta-lactamase: evidence for a stereoselective mechanism from docking studies. *J. Am. Chem. Soc.,* 2005. 127:3262-3.

[27] Nukaga, M.; T. Abe; A. M. Venkatesan; T. S. Mansour; R. A. Bonomo; and J. R. Knox. Inhibition of class A and class C beta-lactamases by penems: crystallographic structures of a novel 1,4-thiazepine intermediate. *Biochemistry,* 2003. 42:13152-9.

[28] Venkatesan, A. M.; Y. Gu; O. Dos Santos; T. Abe; A. Agarwal; Y. Yang; P. J. Petersen; W. J. Weiss; T. S. Mansour; M. Nukaga; A. M. Hujer; R. A. Bonomo; and J. R. Knox. Structure-activity relationship of 6-methylidene penems bearing tricyclic heterocycles as broad-spectrum beta-lactamase inhibitors: crystallographic structures show unexpected binding of 1,4-thiazepine intermediates. *J. Med. Chem.,* 2004. 47:6556-68.

[29] Powers, R. A.; and B. K. Shoichet. Structure-based approach for binding site identification on AmpC beta-lactamase. *J. Med. Chem.,* 2002. 45:3222-34.

[30] Wouters, J.; E. Fonze; M. Vermeire; J. M. Frere; and P. Charlier. Crystal structure of Enterobacter cloacae 908R class C beta-lactamase bound to iodo-acetamido-phenyl boronic acid, a transition-state analogue. *Cell. Mol. Life Sci.,* 2003. 60:1764-73.

[31] Tondi, D.; R. A. Powers; E. Caselli; M. C. Negri; J. Blazquez; M. P. Costi; and B. K. Shoichet. Structure-based design and in-parallel synthesis of inhibitors of AmpC beta-lactamase. *Chem. Biol.,* 2001. 8:593-611.

[32] Roth, T. A.; G. Minasov; S. Morandi; F. Prati; and B. K. Shoichet. Thermodynamic cycle analysis and inhibitor design against beta-lactamase. *Biochemistry*, 2003. 42:14483-91.

[33] Powers, R. A.; F. Morandi; and B. K. Shoichet. Structure-based discovery of a novel, noncovalent inhibitor of AmpC beta-lactamase. *Structure,* 2002. 10:1013-23.

[34] Tondi, D.; F. Morandi; R. Bonnet; M. P. Costi; and B. K. Shoichet. Structure-based optimization of a non-beta-lactam lead results in inhibitors that do not up-regulate beta-lactamase expression in cell culture. *J. Am. Chem. Soc.,* 2005. 127:4632-9.

[35] Teotico, D.G.; K. Babaoglu; G.J. Rocklin; R.S. Ferreira; A.M. Giannetti; and B.K. Shoichet, Docking for fragment inhibitors of AmpC beta-lactamase *Proc.Natl.Acad.Sci.USA,* 2009, 106:7455-7460.

In: Beta-Lactamases
Editor: Jean-Marie Frère, pp. 103-135

ISBN: 978-1-61324-638-2

Chapter 5

STRUCTURES OF CLASS D β- LACTAMASE

Frédéric Kerff, Eric Sauvage, Lionel Vercheval and Paulette Charlier
Centre for Protein Enginering, University of Liège,
Institute of Physics, B5, Sart Tilman, B 4000 Liège, Belgium.

ABSTRACT

Class D β-lactamases were first considered as "honorary class A members" because their sizes and general properties seemed to make them close relatives of the members of this class. However, it rapidly appeared that, besides their ability to hydrolyse oxacillin, they exhibited very specific characteristics: the absence of the conserved Glu 166 residue and, more strikingly, a carboxylated Lys residue which was shown to act as the general base in the catalytic mechanism. Interestingly, although exclusively produced by Gram-negative strains, they exhibit a close structural similarity with the exocellular domains of the BlaR penicillin receptors produced by the Gram-positive *Staphylococcus aureus* and *Bacillus licheniformis*.

INTRODUCTION

In the molecular classification [1], the class D β-lactamases were only segregated from the other serine β-lactamases in the late 1980's. The familiarly called OXA β-lactamases are typically penicillinases characterized by an efficient oxacillinase activity and not well inhibited by the inhibitors of class A enzymes such as clavulanic acid, sulbactam and tazobactam. On the basis of these considerations, the OXA β-lactamases were assigned to group 2d in the Bush-Jacoby-Meideros classification [2].

The class D enzymes are widespread and have been mostly described in Enterobacteriaceae and *Pseudomonas aeruginosa*. They are encoded by mobile genetic elements (plasmids and integrons) that promote their diffusion in Gram – hospital pathogens. However, some resident chromosome-encoded class D enzymes have been identified in environmental microbial organisms such as *Shewanella*, and some opportunistic pathogens such as *Aeromonas sp.* or *Acinetobacter sp.* One of the lastly identified enzymes, OXA-85 (FUS-1), was isolated from a human pathogenic *Fusarium nucleatum* clinical strain [3]. The number of class D β-lactamases is increasing rapidly and almost 200 OXA type enzymes have now been described (http://www.lahey.org/studies/).

In the last 25 years, the introduction of β-lactamase inhibitors and drugs which were less sensitive to hydrolysis resulted in the emergence of novel β-lactamases with new substrates profiles or decreased sensitivity to inhibitors. These β-lactamases are called extended spectrum β-lactamases (ESBLs) [4]. The group of the OXA-type enzymes also gave rise to several variants. Mutations in OXA-10 induced the emergence of enzymes with a greater cephalosporinase activity (OXA-11 [5], 14 [6; 7] and 16 [8]), some of them extending their spectrum to oxymino cephalosporins such as ceftazidime (OXA-18 [9] and -19). Another ESBL, OXA-15 [10], derives from OXA-2. Carbapenemases have also been selected. OXA-23 [11] (also called ARI-1) slowly hydrolyzes imipenem and was discovered in carbapenem resistant *Acinetobacte*r isolates.

Based on amino acid sequence alignments, the class D β-lactamases share little similarity with either class A and C enzymes. However, a stronger similarity was observed with the carboxy-terminal domain of PBPs, especially with the BlaR (and MecR) carboxy-terminal domains of the sensor-transmitters from *Bacillus subtilis* and *Staphylococcus aureus* (see chapter 17). Despite different sizes and low overall sequence identities, the three conserved motifs, also referred to as "active site elements" of the serine-active β-lactamases and PBPs were identified in the class D β-lactamases by sequence alignments. The first element includes the active serine and a strictly conserved lysine residue. In the class D β-lactamases, motif 1 corresponds to the tetrad Ser_{67}-Thr-Phe-Lys_{70}. A posttranslational modification was observed on Lys70 in one of the first OXA-10 structures [12]. The electron density was interpreted as a carboxylated lysine or Kcx70. Element 2 is a triad with the first residue being either a serine or a tyrosine residue. Considering the sequences, two motifs could possibly correspond to element 2 in class D enzymes, either Ser_{115}-Xaa-Val_{117} or Tyr_{141}-Xaa-Asn_{143}. In 2000, the determination of the first structure of a class D enzyme (OXA-10 from *P. aeruginosa*) identified Ser_{115}-Xaa-Val_{117} as the second active site motif.

Table 1. 3D structures within the PDB

Year	PDB ID	Group	Name	Protein	Source	pH	Resolution	Ref	Comments
2000	1FOF	I	OXA-10	WT	*P. aeruginosa*	6.5	2.0	[13]	Apo-enzyme
2000	1EWZ	I	OXA-10	WT	*P. aeruginosa*	8.5	2.4		Apo-enzyme
2001	1E3U	I	OXA-10	WT	*P. aeruginosa*	8.2	1.66	[12]	Apo-enzyme
2001	1E4D	I	OXA-10	WT	*P. aeruginosa*	8.3	1.8	[12]	Apo-enzyme
2001	1K4E	I	OXA-10	WT	*P. aeruginosa*	9.0	2.0		Apo-enzyme
2001	1K4F	I	OXA-10	WT	*P. aeruginosa*	9.0	1.6		Apo-enzyme
2001	1K54	I	OXA-10	WT	*P. aeruginosa*	7.5	1.7	[14]	Complex with penicillanic acid
2001	1K55	I	OXA-10	WT	*P. aeruginosa*	7.5	1.39	[14]	Apo-enzyme
2001	1K56	I	OXA-10	WT	*P. aeruginosa*	6.5	1.7	[14]	Apo-enzyme
2001	1K57	I	OXA-10	WT	*P. aeruginosa*	6.0	1.9	[14]	Apo-enzyme

Year	PDB ID	Group	Name	Protein	Source	pH	Resolution	Ref	Comments
2001	1H8Y	I	OXA-13	WT	*P. aeruginosa*	5.5	2.0	[16]	Complex with Meropenem
2001	1H8Z	I	OXA-13	WT	*P. aeruginosa*	5.5	1.8	[16]	Apo-enzyme
2002	1M6K	III	OXA-1	WT	*E. coli*	7.5	1.5	[17]	Apo-enzyme
2002	1H5X	I	OXA-13	WT	*P. aeruginosa*	5.5	1.9		Complex with Imipenem
2003	1K6R	I	OXA-10	WT	*P. aeruginosa*	7.0	2.3	[24]	Complex with Moxalactam
2003	1K6S	I	OXA-10	WT	*P. aeruginosa*	7.0	2.03		Complex with Iodophenyl boronic acid
2003	1K38	II	OXA-2	WT	*S. typhymurium*	9.0	1.5		Apo-enzyme
2006	2HP6	I	OXA-10	W154 A	*P. aeruginosa*	7.5	2.2	[25]	Apo-enzyme
2006	2HP9	I	OXA-10	W154 A	*P. aeruginosa*	6.0	2.5	[25]	Apo-enzyme
2006	2HPB	I	OXA-10	W154 A	*P. aeruginosa*	9.0	2.05	[25]	Apo-enzyme
2006	2HP5	I	OXA-10	W154 G	*P. aeruginosa*	7.0	2.7	[25]	Apo-enzyme
2007	2JC7	VI	OXA-24	WT	*A. baumanii*	6.5	2.5	[18]	Apo-enzyme
2007	2RL3	I	OXA-10	W154 H	*P. aeruginosa*	7.0	1.9	[25]	Apo-enzyme.
2009	3HBR	I	OXA-48	WT	*K. pneumoniae*	7.5	1.9	[19]	Apo-enzyme
2009	2WGI	I	OXA-10	W154 A	*P. aeruginosa*	6.0	2.85	[25]	Complex with Benzyl-penicillin
2009	3ISG	III	OXA-1	WT	*E. coli*	7.5	1.4	[26]	Complex with Doripenem
2010	3G4P	VI	OXA-24	WT	*A. baumanii*	7.5	1.97	[27]	Apo-enzyme
2010	3FV7	VI	OXA-24	WT	*A. baumanii*	7.5	2.0	[27]	Complex with SA4-44 inhibitor
2010	3FYZ	VI	OXA-24	WT	*A. baumanii*	7.5	2.1	[27]	Complex with SA4-17 inhibitor
2010	3FZC	VI	OXA-24	WT	*A. baumanii*	7.5	2.0	[27]	Complex with SA3-53 inhibitor
2010	3LCE	I	OXA-10	WT	*P. aeruginosa*	7	2.0	[28]	Complex with a cyclobutanone
2010	3IF6	II	OXA-46	WT	*P. aeruginosa*	7.5	2.4	[20]	Apo-enzyme
2010	2WGW	I	OXA-10	V117T	*P. aeruginosa*	8.0	1.8	[24]	Apo-enzyme.
2010	2WGV	I	OXA-10	V117T	*P. aeruginosa*	6.5	1.8	[24]	Cl- inhibited
2010	2WKI	I	OXA-10	K70C	*P. aeruginosa*	7.0	2.1	[24]	Apo-enzyme.
2010	2WKH	I	OXA-10	K70C	*P. aeruginosa*	7.0	1.79	[24]	Complex with Ampicillin

2010 Status

By the end of 2010, almost 200 sequences of OXA-type β-lactamases had been determined and 36 3D structures deposited into the Protein Data Bank (Table 1). After OXA-10 [12-15], the structure of OXA-13 was published in the apoform and in complex with meropenem [16] and also with imipenem. The structures of two other class D β-lactamases became available in 2002, OXA-2 from *Salmonella typhymurium* and OXA-1 from *P. aeruginosa* [17]. In 2006, one carbapenem hydrolyzing class D β-lactamase, OXA-24 from

Acinetobacter baumanii was solved [18]. More recently two other class D β-lactamases have been deposited into the Protein Data Bank: the imipenem hydrolyzing OXA-48 from *Klebsiella pneumonia* [19], and OXA-46 from *P. aeruginosa,* which exhibits a narrow substrate profile [20]. Meanwhile, the structures of several OXA-10 catalytic mutants and complexes have been solved.

On the basis of phylogenic analyses, classifications of class D β-lactamases have been established by Sanschagrin *et al.* [21] and Barlow *et al.* [22]. Danel *et al.* [23] proposed to add an additional branch to the five-branched phylogenic tree defining the groups I to V proposed by Sanschagrin. This additional group VI includes numerous more divergent proteins and contains the OXA carbapenemases. Since the number of OXA sequences has considerably increased, it appears quite difficult to make the right choice of criteria (primary and tertiary structure, kinetic behaviours, substrates preference, inhibitors sensitivity, etc.) that will give rise to a simple and clear classification. If we refer to the actualized Sanschagrin classification, the presently known 3D OXA structures belong to four groups. The OXA-10 [12-14] , OXA-13 [16] and OXA-48 [19] structures are representatives of group I. The OXA-2 and OXA-46 [20] structures belongs to group II, and OXA-1 [17] illustrates group III. The structure of the OXA-24 carbapenemase [18] belongs to the newly defined group VI.

General Fold of Class D β-lactamases

Within each subgroup, the percentage of sequence identity is quite high (between 45 and 99%) but may decrease to less than 20% between the different subgroups. The comparison of the amino acid sequences of the 7 available OXA structures illustrates well this observation and is also in good agreement with the results of the 3D structures superposition. The most distant primary structures correspond to the highest values of the root mean square deviations for the α–carbon alignments of the tertiary structures (Table 2 and Figure 1.).

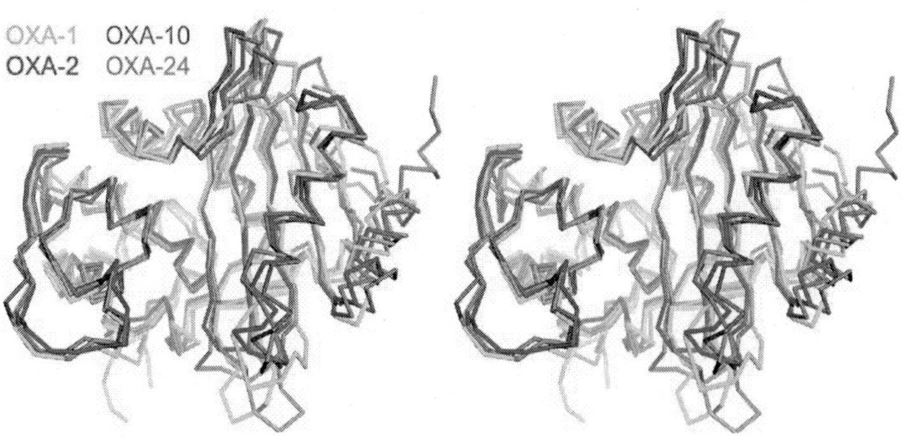

Figure 1. Conservation of the class D β-lactamase fold. Superposition of the stereo ribbon diagrams representing the different class D β-lactamases structures: group III OXA-1 (cyan), group II OXA-2 (magenta), group I OXA-10 (green) and group VI OXA-24 (orange). The active site is located in the center with the α domain on the left and the α/β domain on the right.

The class D β-lactamases are typically organized into two domains with the catalytic pocket at their interface. One α/β type domain consists of a six to seven-stranded antiparallel β-sheet covered on one face by two α-helices (corresponding to the N and C termini), and on the other face by a one-turn 3_{10}-helix. The second domain, mainly all-α, contains a hydrophobic central α–helix surrounded by four other α–helices. All those secondary structures elements are connected by several loops of variable lengths (Figure 1, Table 4).

Obviously, the class D β-lactamases enzymes share this overall common fold with the other active site serine β-lactamases (classes A and C) and in general with all serine-active penicillin-recognizing enzymes, although the distribution of secondary structure elements varies considerably in space. For example, the rms deviations in the atomic positions for the common Cα residues of OXA-10 are 2.63 Å and 2.96 Å when compared respectively with those of a typical class A (*Escherichia coli* TEM-1 [29]) and class C (*Enterobacter cloacae* 908R [30]) enzymes (Table 3). Strikingly, the rmsd values are much lower when the comparison involves the C-terminal domains of the penicillin receptors/sensors BlaR1 of *Bacillus licheniformis* [31] (1.52 Å) or MecR1 of *Staphylococcus aureus* [32] (1.31 Å), highlighting the structural homology between the C-terminal domain of the penicillin receptors/sensors and class D β-lactamases. With other penicillin-binding proteins, the rmsd values are quite high, as observed for the other serine-active β-lactamases (as an example, 2.50 Å for the *Streptomyces* K15 DD-transpeptidase [33]).

Table 2. Alignement: identity (%) and rmsd values (Å)

R.M.S.D (Å) Identity (%)	1K4F OXA-10	1H8Z OXA-13	3HBR OXA-48	1K38 OXA-2	3IF6 OXA-46	1M6K OXA-1	2JC7 OXA-24
Group I OXA-10		0.75	1.10	1.18	0.93	1.76	1.45
Group I OXA-13	96		1.26	1.28	0.96	1.84	1.60
Group I OXA-48	46	45		1.16	1.10	1.79	1.37
Group II OXA-2	31	30	39		0.74	1.98	1.63
Group II OXA-46	38	37	43	82		1.80	1.38
Group III OXA-1	23	19	21	22	30		1.71
Group VI OXA-24	34	33	32	24	31	18	

Table 3. Rmsd values (Å) obtained when comparing the structures of class D β-lactamases with those of other penicillin recognizing enzymes

R.M.S.D (Å)	1K4F OXA-10	1K38 OXA-2	1M6K OXA-1	2JC7 OXA-24
Class D OXA-10 (1K4F)		1.17	1.68	1.44
Class A TEM-1 (1XPB)	2.63	2.67	2.69	3.46
Class C 908R (1S6R)	2.96	3.06	3.06	3.12
PBP/sensor BlaR1 (1NRF)	1.52	1.30	1.47	1.31
PBP/sensor MecR1 (2IWB)	1.31	1.42	1.49	1.59
PBP K15 (1SKF)	2.50	2.49	2.64	2.79

Table 4. Sequence alignement, with α-helices represented in cyan and β-strands in magenta

Dimeric		β1 α1 β2	
I	OXA-10	-----MKTFAAYVIIACLS---------STALAGSITENTSWNKEFSAEAVNGVFVLCKS	46
I	OXA-13	-----MKTFAAYVITACLS---------STALASSITENTSWNKEFSAEAVNGVFVLCKS	46
I	OXA-48	--MRVLALSAVFLVASIIG---------MPAVAKEWQENKSWNAHFTEHKSQGVVVLWNE	49
II	OXA-2	---MAIRIFAILFSIFSLAT--------FAHAQEGTLERSDWRKFFSEFQAKGTIVVADE	49
II	OXA-46	---MAIRFFTILLSTFFLTS--------FVYAQEHVVIRSDWKKFFSDLQAEGAIVIADE	49
Monomeric			
III	OXA-1	---MKNTIHINFAIFLIIAN--------IIYSSASASTDISTVASPLFEGTEGCFLLYD	48
VI	OXA-24	MKKFILPIFSISILVSLSACSSIKTKSEDNFHISSQQHEKAIKSYFDEAQTQGVIIIKEG	60
Dimeric		β3 α2 $\beta3^{a}$ α3 $\beta3^{b}$ η3	
I	OXA-10	S--SKSCATNDLARASKEYLPASTFKIPNAIIGLETGVIKNEHQVFKWDGKPRAMKQWER	104
I	OXA-13	S--SKSCATNNLARASKEYLPASTFKIPSAIIGLETGVIKNEHQVFKWDGKPRAMKQWER	104
I	OXA-48	N--KQQGFTNNLKRANQAFLPASTFKIPNSLIALDLGVVKDEHQVFKWDGQTRDIATWNR	107
II	OXA-2	RQADRAMLVFDPVRSKKRYSPASTFKIPHTLFALDAGAVRDEFQIFRWDGVNRGFAGHNQ	108
II	OXA-46	RQAKHTLSVFDQERAAKRYSPASTFKIPHTLFALDADAVRDEFQVFRWDGVNRSFAGHNQ	108
Monomeric			
III	OXA-1	VSTNAEIAQFNKAKCATQMAPDSTFKIALSLMAFDAE-IIDQKTIFKWDKTPKGMEIWNS	107
VI	OXA-24	K--NLSTYGNALARANKEYVPASTFKMLNALIGLEN-HKATTNEIFKWDGKKRTYPMWEK	107
Dimeric		$\beta3^{c}$ α4 α5 α6 α7	
I	OXA-10	DLTLRGAIQVSAVPVFQQIAREVGEVRMQKYLKKFSYGNQNISGG------IDKFWLEGQ	158
I	OXA-13	DLSLRGAIQVSAVPVFQQIAREVGEVRMQKYLKKFSYGNQNISGG------IDKFWLEDQ	158
I	OXA-48	DHNLITAMKYSVVPVYQEFARQIGEARMSKMLHAFDYGNEDISGN------VDSFWLDGG	161
II	OXA-2	DQDLRSAMRNSTVWVYELFAKEIGDDKARRYLKKIDYGNADPSTS------NGDYWIEGS	163
II	OXA-46	DQDLRSAMRNSTVWVYELFAKDIGEDKARRYLKQIDYGNVDPSTI------KGDYWIDGN	163
Monomeric			
III	OXA-1	NHTPKTWMQFSVVWVSQEITQKIGLNKIKNYLKDFDYGNQDFSGDKERNNGLTEAWLESS	167
VI	OXA-24	DMTLGEAMALSAVPVYQELARRTGIELMQKEVKRVNFGNTNIGTQ------VDNFWLVGP	171
Dimeric		$\beta3^{d}$ α8 α9 β4 β5	
I	OXA-10	LRISAVNQVEFLESLYLNKLSASKENQLIVKEALVTEAAPE-YLVHSKTGFSGVGTESNP	217
I	OXA-13	LRISAVNQVEFLESLFLNKLSASKENQLIVKEALVTEAAPE-YLVHSKTGFSGVGTESNP	217
I	OXA-48	IRISATEQISFLRKLYHNKLHVSERSQRIVKQAMLTEANGD-YIIRAKTGYS---TRIEP	217
II	OXA-2	LAISAQEQIAFLRKLYRNELPFRVEHQRLVKDLMIVEAGRN-WILRAKTGWEG-------	215
II	OXA-46	LKISAHEQILFLRKLYRNQLPFKVEHQRLVKDLMITEAGRS-WILRAKTGWEG-------	215
Monomeric			
III	OXA-1	LKISPEEQIQFLRKIINHNLPVKNSAIENTIENMYLQDLENSTKLYGKTGAGFTANR-TL	227
VI	OXA-24	LKITPVQEVNFADDLAHNRLPFKLETQ-EVKKMLLIKEVNGS-KIYAKSGWG--MG-VTP	227
Dimeric		β6 β7 α10	
I	OXA-10	GVAWWVGWVEKETE-VYFFAFNMDIDN-ESKLPLRKSIPTKIMESEGIIGG---------	266
I	OXA-13	GVAWWVGWVEKGTE-VYFFAFNMDIDN-ENKLPLRKSIPTKIMASEGIIGG---------	266
I	OXA-48	KIGWWVGWVELDDN-VWFFAMNMDMPT-SDGLGLRQAITKEVLKQEKIIP----------	265
II	OXA-2	RMGWWVGWVEWPTG-SVFFALNIDTPNRMDDLFKREAIVRAILRSIEALPPNPAVNSDAA	274
II	OXA-46	RFGWWVGWIEWPTG-PVFFALNIDTPNRTDDLFKREAIARAILRSIDALPPN--------	266
Monomeric			
III	OXA-1	QNGWFEGFIISKSGHKYVFVSALTGNLGSNLTSSIKAKKNAIT----ILNTLNL------	276
VI	OXA-24	QVGWLTGWVEQANGKKIPFSLNLEMKE-GMSGSIRNEITYKSLENLGII-----------	274

Tertiary Structures

Group I: OXA-10, OXA-13 and OXA-48

OXA-10 and OXA-13 from *P. aeruginosa* are the most closely related class D β-lactamases of known 3D structures and exhibit a sequence identity of 96 %. They consequently also share a remarkably identical 3D structure with a rmsd value of 0.75 Å. Both proteins contain one disulfide bridge between Cys44 and Cys51 linking β2 and β3 (Figure 2). The main difference observed in the global folding is a long loop connecting helices α3 and α4, which is not defined in the OXA-13 structure from residue Trp92 to residue Trp102. Nevertheless, this loop appears ordered in the structure of OXA-13 complexed with meropenem. This observation highlights the dynamic aspect of the 3D structure and also suggests that complexation with a ligand may be necessary for loop ordering.

The active site pocket is an extended groove delineated on one side by a loop connecting helices α4 and □α5 and containing the conserved S_{115}-A-V_{117} motif 2. The other side of the groove is delineated by strand β5 with the conserved K_{205}-T-G_{207} motif 3.The bottom of the pocket is formed by the N terminus of helix α3 with the conserved S_{67}-T-F-K_{70} motif 1 (Figure 2). The entrance of the active site is defined by the so-called Ω-loop connecting helices α6 and □α8, from Asn143 to Arg160, which includes the one turn α7 helix. This Ω-loop contains Trp154 which is almost strictly conserved within all class D β-lactamases. This conserved tryptophane residue is one of the several hydrophobic residues found inside and in the vicinity of the active site of the class D β-lactamases. Trp154 with Met99, Trp102, Val117, Leu155 and Phe208 define a hydrophobic ring at the entrance of the active site cleft (Figure 3). These residues are well conserved within the class D β-lactamases catalytic pockets and their hydrophobic characteristics have obviously an impact on the substrate binding and the catalytic mechanism.

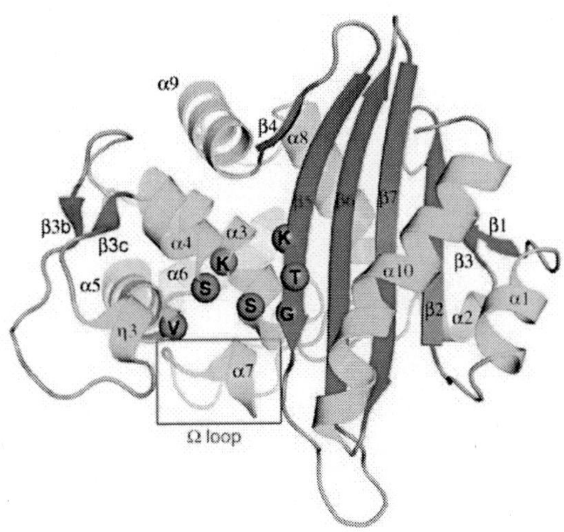

Figure 2. General fold of the class D β-lactamases. Diagram representation of the class D β-lactamase OXA-10 with labeled secondary structure elements. Helices are represented in cyan and β-strands in magenta. The characteristic Ω-loop is highlighted by a green box and the residues of the 3 conserved motifs are displayed as orange spheres.

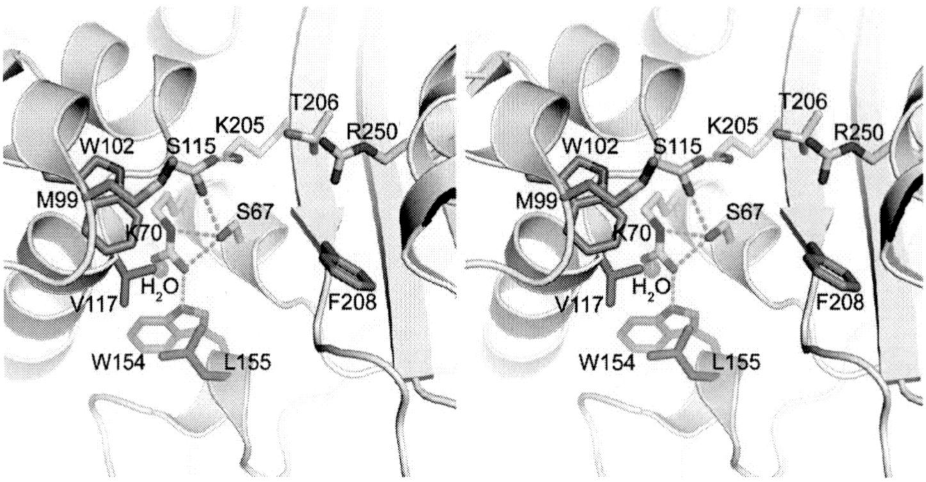

Figure 3. Active site of the class D β-lactamases. Stereo diagram representation of OXA-10 (pdb code 1K4F). Residues important for the catalytic mechanism are shown as gray sticks and hydrophobic residues as orange sticks. Lys70 is carboxylated and the two alternative conformations of Ser115 are displayed.

OXA-48 from *K. pneumonia* which displays carbapenemase activity is more distant from OXA-10 and OXA-13 (identity of 46% and 45% respectively). The overall fold is very similar to that of OXA-10 except for the β2β3 and β5β6 loops. The latter, which is 3 residues shorter and oriented differently, is critical for the carbapenemase activity [19] (see later section for details). The hydrophobic ring at the entrance of the active site is well conserved with Met99Ile as the only modification. The disulfide bridge is however not conserved.

Group II: OXA-2 and OXA-46

OXA-2 from *S. typhymurium* shares 30% of identical residues with OXA-10. Its overall structure is very similar to that of the latter with a rmsd value of 1.17 Å. The disulfide bridge is not conserved. The other main structural differences are an insertion of two residues between strands β2 and □β3 and a seven residue deletion in the loop connecting strands β5 and β6 (Figure 4).

The conformation of the catalytic site is very much like that of OXA-10, but, curiously, in this 1.5 Å resolution structure, two residues are not defined in the electron density of the Ω-loop. Moreover, the hydrophobic ring at the entrance of the active site cleft is well conserved although the residues are not the same as in OXA-10 or OXA-13, with the following replacements: Met99Phe, Trp102His, Leu155Ile and Phe208Trp.

OXA-46 and OXA-2 share 82% of identical residues. Three different conformations are observed in the asymmetric unit of the OXA-46 crystal [20]. The structure of monomer C is very close to that of OXA-2. The only differences are a modification in the backbone of the β3 strand and the absence of 8 residues at the beginning of the OXA-46 Ω-loop. This monomer is the only one with a carboxylated lysine in the active site and a conserved orientation of the 3 conserved motifs. In monomers A and B, in addition to the absence of carboxylation of the first motif lysine, the second motif is displaced and the α3α4 loop covering it is either also displaced or not observed in the structure. This type of modification of the active site is similar to that observed in OXA-10 at pH6 [14].

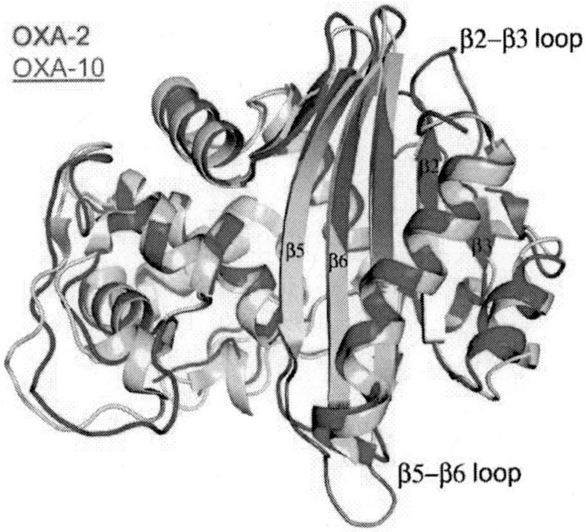

Figure 4. Diagram representation of the superposed OXA-2 (magenta) and OXA-10 (gray) β-lactamases. The major differences (loops β2-β3 and β5-β6) are highlighted in green in the OXA-10 representation.

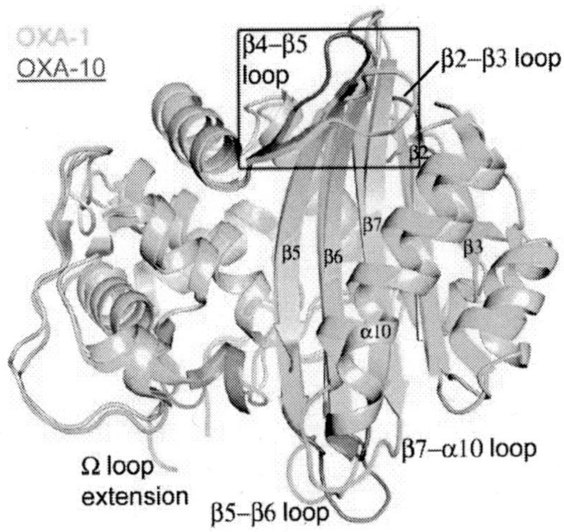

Figure 5. Diagram representation of the superposed OXA-1 (cyan) and OXA-10 (gray) class D β-lactamases. The major differences (loops β2-β3, β4-β5, β5-β6, β7-α9 and the Ω-loop) are highlighted in green in the OXA-10 representation.

Group III: OXA-1

OXA-1 from *E. coli* has a more divergent 3D structure when compared to OXA-10 and OXA-2 [17]. The sequence identity is much lower (around 20%) and the rmsd value is higher (around 1.75 Å). OXA-1 differs from the other OXA structures in some secondary structures and several loops (Figure 5). On the top of the substrate binding site, the short β4 is more twisted out of the plane of the β-sheet, toward the C-terminal helix α10, which is itself one turn shorter than in OXA-10. Differences also occur in the loop connecting β2 and □β3 which is two residues longer. The Ω-loop connecting helices α6 and □α8 and which includes the one turn helix α7 is six residues longer, and is also more disordered than in OXA-10, with electron density missing for one residue, as observed for OXA-2, although both structures were solved at 1.5 Å resolution. The loop connecting strands β5 and β6 is six residues longer than in OXA-2 and about the same length as in OXA-10, but has a different position, extending toward the Ω-loop and narrowing the entrance of the active site. These differences may determine the substrate specificities of the different OXA enzymes. On the other hand, the hydrophobic ring at the entrance of the active site cleft is well conserved with just the Phe208Ala difference.

Group VI: OXA-24

OXA-24 from *A. baumanii* has a more divergent 3D structure when compared to those of OXA-10 and OXA-2, but is more similar to that of OXA-1 (Figure 6). The sequence identities range from 18% with OXA-1 to 34% with OXA-10 and the rmsd values from 1.44 Å with OXA-10 to 1.65 Å with OXA-1.

Compared to all oxacillinase structures, OXA-24 has a one turn longer N-terminal α1 helix. As observed for OXA-1, the short β4 is twisted out of plane of the β-sheet toward the C-terminal helix α10. The well defined Ω-loop is of the same length as in the OXA-2 and OXA-10 structures and superposes well. On the other hand, the length of the loop connecting strands β5 and β6 is intermediate between that of OXA-2 and OXA-10, and is almost the same as in OXA-1. It is shifted toward the centre of the active site pocket in the direction of the Ω-loop and partially blocks the bottom part of the cleft. The hydrophobic ring at the entrance of the active site cleft is quite well conserved with two substitutions: Met99Tyr and Phe208Trp.

Quaternary Structures: Monomer versus Dimer

Dimeric Structures: OXA-10/-13/-48 and OXA-2/-46

The OXA-10, OXA-13, OXA-48, OXA-2 and OXA-46 enzymes have been described as biological dimers either from the observation of the crystal packing (two monomers are related by a two-fold symmetry axis) or by analytical ultracentrifugation. Experiments performed with the OXA-10 enzyme at different enzyme concentrations and ionic strengths showed that the dissociation constant seemed to be independent of the ionic strength and gave an estimated K_d value of 1μM [14]. Moreover, the observation of the dimer interface of the OXA-10 structure solved by Paetzel *et al.* [13] (1FOF) highlights the presence of two cobalt ions (shown as blue spheres in Figure 7A). These are coordinated by the carboxylate oxygen of Glu190 of one monomer and the imidazole nitrogen of His203 and the carboxylate oxygen of Glu227 of the second monomer. The typical octahedral coordination is completed by two water molecules. The absence of cobalt ion in the other OXA-10 reported structures may be explained by the absence of $CoCl_2$ in the crystallization buffer in contrast to the conditions used by Paetzel *et al.* (10mM $CoCl_2$). The same authors have also demonstrated that the dimer formation is promoted and, concomitantly, the enzyme activity is enhanced by addition of increasing concentrations of divalent cations. The best ones are copper, cadmium and zinc, followed by cobalt, nickel and manganese. Calcium and magnesium have no impact on the dimer formation or the enzyme activity. Since a good correlation has been found between the increase in activity and the cation mediated formation of dimer, it may be concluded that the OXA-10 enzyme is more active in the dimeric form.

In OXA-10, independently of the presence of divalent cations at the dimer interface, several residues from each monomer, mainly from helix α9 and strands β4 and β5, are involved in electrostatic (including salt bridges) and direct hydrogen bonds. The two monomers related by a two-fold symmetry axis are intimately associated with a total buried surface area of ~2500 Å², which includes mainly helices α8-α9 and strands β4-□β5 (Figure 7B). Polar interactions include residues Glu86 with Lys189 and Asn176, His87 with Tyr174, Arg104 with Glu199, Arg109 with Ala197 and His203 with Glu190 (Figure 7B). Several hydrophobic interactions complete and stabilize the dimer. Residues involved in these interactions include Leu186, Ile187, Val193, Ala 196 and Leu201 (Figure7B). These interactions are strictly conserved in OXA-13, but several variations are observed in the three other structures (OXA-48/-2/-46). However all the dimeric structures display the same orientation between molecules and the same secondary structure elements are involved in dimer stabilization.

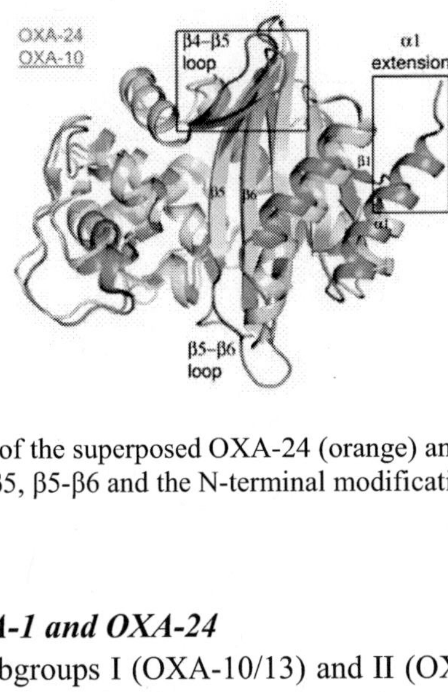

Figure 6. Diagram representation of the superposed OXA-24 (orange) and OXA-10 (gray) structures. The major differences (loops β4-β5, β5-β6 and the N-terminal modification) are highlighted in green in the OXA-10 representation.

Monomeric Structures: OXA-1 and OXA-24

Unlike the enzymes of subgroups I (OXA-10/13) and II (OXA-2), OXA-1 and OXA-24 exist exclusively as monomers over a wide range of protein concentrations (up to 200μM). The results obtained by size exclusion chromatography and analytical centrifugation are supported by the analysis of the 3D structures. Most of the residues involved in polar interactions are not conserved, in particular those belonging to α9 and β4. Consequently, β4, which is not involved in dimerization, adopts a different conformation and is bent toward the C-terminal helix α10.

Active Site Conformation and Lysine 70 Carboxylation

As observed in all serine-active β-lactamases, the active site lies at the junction between two structural domains. It also contains the three classical conserved motifs. Motif 1 contains the tetrad $S^*_{67}XXK_{70}$ where S* is the active serine situated at the N terminus of helix α3, defining the bottom of the catalytic cleft. Motif 2 is formed by the $S_{115}XV_{117}$ triad on the loop connecting helices α4 and □α5, and motif 3 lies on strand β5 with the $K_{205}T/SG_{207}$ triad. Motifs 2 and 3 delineate the two sides of the active site. The oxyanion hole consists of the main chain nitrogen atoms of Ser67 and Phe208 (in OXA-10/13, Trp208 in OXA-2/-46 and OXA-24, Ala208 in OXA-1). One particularity of the class D β-lactamases is that there is no conserved glutamic acid at the entrance of the catalytic pocket on the Ω □□loop as found in the class A β-lactamases. On the contrary, most of the residues oriented toward the inside of the pocket are mostly hydrophobic with a strictly conserved Trp154. Globally, the active sites of the class D β-lactamases are much more hydrophobic than those of the other active serine β-lactamases with at least four mostly conserved apolar residues (Met99, Val117, Trp154 and Phe208 in OXA-10).

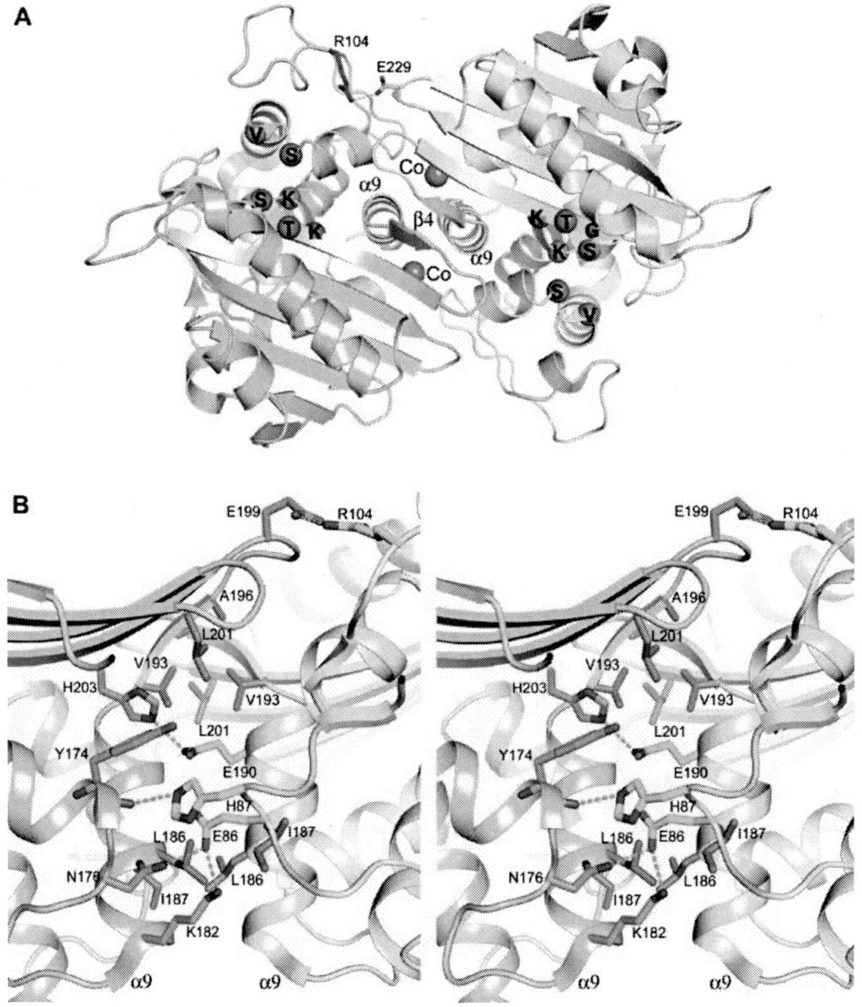

Figure 7. Dimer interface of OXA-10. *A*. Diagram representation of the OXA-10 dimer (pdb code 1FOF) with Co [2+] ions (blue spheres) at the dimer interface. The secondary structure elements which participate in the dimerisation are labeled and the residues of the 3 conserved motifs are displayed as orange spheres for orientation *B*. Stereo sticks and diagram close up of the OXA-10 dimer interface. Hydrophobic residues are shown as orange sticks and residues involved in polar interactions are cyan or pink depending on the monomer. Only one symmetry equivalent is displayed for each polar interaction.

Analysis of the different available three dimensional structures of the class D β-lactamases shows that Lys70 may be carboxylated due to a reaction between the ε-NH_2 of the lysine and carbon dioxyde [12] as shown in scheme 1.

$$\varepsilon\text{-}NH_2 + CO_2 + H_2O \rightleftharpoons \varepsilon\text{-}NH\text{-}COO^- + H_3O^+ \quad \text{(Scheme 1)}$$

Carboxylation of a lysine residue is pH dependent and is a compromise between its deprotonation, favored by high pH, and the concentration of CO_2, favored by low pH.

Several studies have demonstrated that a reaction between CO_2 and α- or ε-amino groups is not rare and that carboxylated amino groups can be in equilibrium with unprotonated amino groups and CO_2 [34]. In most of the few X-ray structures where a carboxylated lysine has been observed, the modified residue is one of the ligands of a catalytic divalent metal ion, either Mg^{2+} as in MurD [35], the ribulose biphosphate carboxylase/oxygenase from spinach [36] and dihydrofolate synthetase [37], Co^{2+} as in transcarboxylase [38] or Ni^{2+} as in urease [39]. There are only two exceptions to that with alanine racemase [40] and the class D β-lactamase where the carboxylated lysine appears to be directly involved in the catalytic activity of the enzyme. In the case of alanine racemase, an arginine residue (Arg136) which has a role in maintaining the structural integrity of the carboxylate binding site is stabilized by interactions between the side chain atoms Nη1 and Nε and the carboxylate of the carboxylated Lys129. In the case of the class D β-lactamase, the vicinity of hydrophobic residues (Val117, Trp154) is supposed to promote the deprotonation of Lys70 which appears to be fully carboxylated only above pH 7.5. This post-translational modification induces some structural modifications of the active site and is accompanied by enhancement of the enzymatic activity. Therefore the carboxylated Lys70 has been assumed to be the essential residue acting as a general base during the catalysis of the β-lactam hydrolysis [12; 13; 16].

The Active Site Geometry in the OXA-10 WT Structures

Several structures have been obtained at different pH values for the OXA-10 enzyme (Table 1).

A carboxylated Lys70 is observed in both monomers of the biological dimer above pH 7.5 (pdb codes 1K4F, 1K4E (pH 9.0), 1E4D (pH 8.3), 1K55 (pH 7.5)) [12; 14]. A mixed situation where Lys70 is carboxylated or not, depending on the monomer, is observed between pH 7 and 6 (pdb codes 1K56 (pH 6.5), 1K57 (pH 6.0)) [14]. Decarboxylation induces a reorganization of the hydrogen bonds network accompanied by a slight modification in the active site conformation. When a carboxylated Lys70 (Kcx70) is present (Figure 8), the carboxylate makes strong hydrogen bonds with Ser67-Og, Trp154-Nε and a water molecule, which is itself stabilized by a hydrogen bond with Asn73. The two hydrogen bonds between the carboxylate of Kcx70 and the indole group of Trp154 stabilize the position of the Lys70 side chain and decrease the flexibility of the Ω-loop which influences the accessibility of the active site. Residue Ser115 can occupy two positions, its hydroxyl group being hydrogen bonded with either Lys205-Nζ or Ser67-Oγ. Finally, a mostly conserved arginine residue, Arg250, may be in a suitable position for a potential salt-bridge interaction with the carboxylate group of penicillin substrates.

A non-carboxylated Lys70 is observed in the dimer of Paetzel's structure (1FOF) and in monomers C and D of Maveyraud's structures solved at pH 6.5 (1K56) and 6.0 (1K57) (Figure 8). Two distinct situations were observed. In Paetzel's structure, the absence of a carboxylate group on Lys70 does not seem to affect the geometry of the active site. The active site residues are in the same positions and all the hydrogen bonds are conserved, except that with Trp154-Nε. The hydroxyl group of Ser67 is at H-bond distance from the Lys70 side chain amino group (2.9 Å) and from the Ser115 hydroxyl group (2.9 Å /A and 3.7 Å /B). So, even in the absence of Lys70-Trp154 interaction, the conformation and orientation of Lys70 and Ser67 are the same as in the pH 9.0 structure where Lys70 is carboxylated. On the other hand, in monomers C and D of Maveyraud's structures, the hydrogen bonds network is disturbed and induces a conformational change in the active site, which appears more open. The Lys70 side chain is 4.7 Å away from the Ser67-Oγ and rather interacts with the main chain oxygen atom of Val117. The 114-116 loop, connecting helices α4 and □α5 and containing the conserved S_{115}-A-V_{117} motif 2, is moved away from the active site due to a 180° rotation of the Val114-Ser115 peptide bond. The distance between the Oγ's of Ser115 and the active Ser67 increases from 2.7 to 7 Å. This loop movement disturbs another loop connecting the short $\beta3^b$ and $\beta3^c$ strands and including the 3_{10} helix η3 which is no longer well resolved in the structure (from residue 92 to 103).

The Active Site Geometry in other OXA Structures: OXA-13, OXA48, OXA-2, OXA-46, OXA-1 and OXA-24

At pH 6.0, a non-carboxylated Lys70 prevails. The OXA-13 structure has been solved at pH 5.5, and its active site has the same conformation as that observed for monomers C and D of the OXA-10 structures at pH 6.5 and 6.0.

The structure of OXA-48 was solved at pH 7.5 and Lys70 is carboxylated in the four monomers of the asymmetric unit. The only significant difference with OXA-10 is in the β5β6 loop which adopts a different conformation and contains an arginine residue which stretches across the bottom of the active site and is related to the carbapenemase activity of this enzyme [19].

The OXA-2 structure, solved at pH 9.0, shows a carboxylated Lys70 in both monomers and an active site geometry identical to that described for the OXA-10 structures in the 7.5 - 9.0 pH range. Nevertheless, the active site of monomer B exhibits a more disturbed structure in the loop connecting the short $\beta3^b$ and $\beta3^c$ strands (residues 99-101 are missing) and curiously in the Ω-loop (residues149-151 are missing).

In the OXA-46 structure solved at pH 7.5, three monomers are present in the asymmetric unit with two different conformations of the active site [20]. In monomer C, Lys73 (equivalent to Lys70 in OXA-10) is carboxylated and the active site is very similar to that of OXA-10 with a carboxylated Lys70. In monomers A and B, Lys73 is not carboxylated and is in a similar orientation as the non-carboxylated Lys70 observed in OXA-10 at pH 6 with a displaced SXV motif (Figure 8).

The OXA-1 structure, solved at pH 7.5, exhibits the same active site geometry as OXA-10 and OXA-2, with a carboxylated Lys70. The main differences with the enzymes of subgroup I and II is the less bulky hydrophobic residue in position 208 (Ala instead of Phe or Trp) and a non-conserved Arg250 (replaced by Ile250). Also, there is no water molecule bridging the carboxylate of Kcx70 and residue 73, since Asn73 (Ser73 in OXA-13, His73 in OXA-2) is replaced by Leu73. Instead the carboxylate is stabilized by an equivalent hydrogen bond with the hydroxyl group of Ser120.

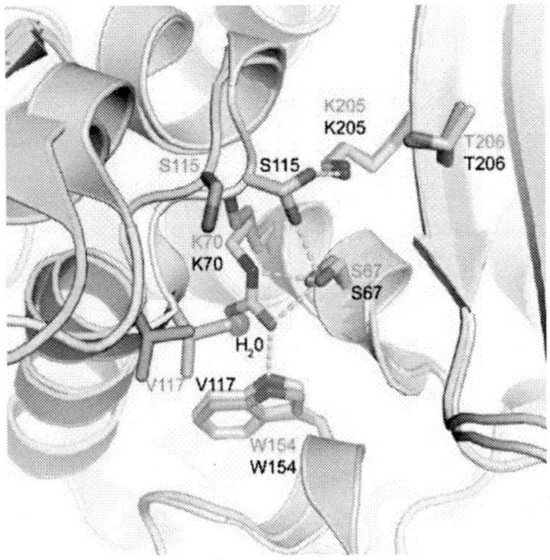

Figure 8. pH influence on the Lys70 carboxylation and the active site geometry of OXA-10. Diagram representation of the superposed OXA-10 enzymes at pH9 (pdb code 1K4F chain A, gray) and pH6 (pdb code 1K57 chain D, green). Important residues of the active site are shown as sticks and yellow dashed lines indicate the hydrogen bonds present at pH9 and absent at pH6. Note the important shift of the second conserved motif.

The OXA-24 active site is very similar to that of the OXA-10 Paetzel's structure (Table 8).Both structures were solved at pH 6.5, both active sites exhibit a non-carboxylated Lys70 and have a closed conformation with almost undisturbed hydrogen bond networks and well defined loops.

The Active Site Geometry in the OXA-10 W154A/G/H Mutants

To analyze the potential importance of residue Trp154 in the stabilization of the carboxylate group of the modified Kcx70 and/or in the carboxylation phenomenon itself, the structure of several mutants of Trp154 in the OXA-10 enzymes were solved [25].

The Trp154Ala mutant was solved at pH 7.5 and 9.0 (Figure 9). In both structures, the most striking observation is the absence of carboxylation of residue Lys70 in both monomers. This situation induces a modification in the hydrogen bonds network compared to the wild type enzyme at pH 9. It is also different from that observed in the C and D monomers of the non-carboxylated WT structure solved at pH 6.0 (pdb code 1K57) [14]. In the Trp154Ala mutant, the position of residues 114-119 and the orientation of Lys70 are well conserved in comparison with the pH 9.0 WT reference structure (pdb code 1K4F). The amino group of the side chain of Lys70 interacts with the hydroxyl group of Ser67 (2.8 Å/monomer A and 3.1 Å/monomer B) and a water molecule (3.2 Å/A and 3.1 Å/B). So, even if the interaction of the side chain nitrogen atom of Trp154 is lost, the conformation and orientation of Lys70 and Ser67 in the Trp154Ala mutant are the same as in the pH 9.0 WT structure where Lys70 is carboxylated. The hydroxyl group of Ser115 still interacts with the amino group of Lys205 (2.7 Å/A and 3.0 Å/B) but the distance between the Oγ's of Ser115 and Ser67 increases from 2.7 Å to 3.9 Å for monomer A and from 2.5 Å to 3.6 Å for monomer B. The oxyanion hole is preserved with a water molecule located at 3.2 Å and 2.7 Å of the Ser67 and Phe208 main chain nitrogen atoms respectively.

In the structure of the Trp154Gly mutant solved at pH 7.0, Lys70 is clearly not carboxylated in the four monomers and the active site undergoes more changes, especially in the Ω-loop which defines its bottom and bears the Gly154 mutation (Figure 11). The position of residues 114-119 is still well conserved and the orientation of Lys70 is also roughly the same in the four monomers and as in uncarboxylated wild type and Trp154Ala mutant structures. By contrast, the amino group of the side chain of Lys70 interacts more loosely with the hydroxyl group of Ser67. The situation is different in each monomer and the distance between Lys70-Nζ and Ser67-Oγ varies from 2.7 Å to 5.0 Å for monomers D and B respectively. The Oγ of Ser115 still interacts with the amino group of Lys205 but the distance between the Oγ's of Ser-115 and Ser-67 varies in each monomer (2.8 Å/A, 3.6 Å/B, 3.8 Å/C and 3.5 Å/D). We may conclude that the Gly154 mutation induces a very loose hydrogen bonds network which leads to a more erratic positioning of the catalytic Ser67, in addition to the non-carboxylation of Lys70.

In the structure of the Trp154His mutant, solved at pH 7, the situation in monomer B is the same as that described for both monomers of the Trp154Ala mutant, where a non-carboxylated Lys70 is observed. The amino group of the side chain of Lys70 interacts with the hydroxyl group of Ser67 (2.9 Å) and a water molecule (3.7 Å). This water molecule is at the same distance of the His154-Nε (3.6 Å) and somehow mimics the position of the carboxyl group of a carboxylated Lys70. The hydroxyl group of Ser115 interacts with the amino group of Lys205 (2.6 Å). The oxyanion hole is preserved with a water molecule located at 3.1 Å and 2.9 Å of the Ser67 and Phe208 main chain nitrogen atoms, respectively. On the other hand, in the active site of monomer A, Lys70 is modeled as partially (30%) carboxylated (Figure 9). When Lys70 is not carboxylated, the amino group of its side chain interacts with the hydroxyl group of Ser67 (3.2 Å) and a water molecule (2.7 Å). This water molecule is at the same distance of the His154-Nε (3.1 Å) and superposes well with one oxygen atom of the carboxyl group of the carboxylated form. Whether from the carboxylate or the water, this oxygen atom interacts with a water molecule (2.7 Å), itself hydrogen bonded to Asn73-Oδ1 (2.9 Å) and to the side chain nitrogen atom of His154 (3.4 Å). The second oxygen atom of the carboxyl group of the carboxylated Lys70 is hydrogen bonded to the hydroxyl group of the active Ser67 (2.6 Å) and to the side chain nitrogen atom of His154 (3.3 Å). The hydroxyl group of Ser115 interacts with the amino group of Lys205 (2.6 Å). The oxyanion hole is filled by a water molecule located at 3.2 Å and 2.7 Å of the Ser67 and Phe208 main chain nitrogen atoms, respectively.

In conclusion, the mutation of Trp154 into an alanine or a glycine is clearly responsible for the non-carboxylation of the essential Lys70 residue in each monomer. Even if the interaction with the side chain nitrogen atom of Trp154 is lost, the conformation and orientation of Lys70 are roughly conserved in both mutants. The global geometry of the active site is conserved with the exception of residues corresponding to the Ω-loop (144-158) in monomers B, C and D. The network of hydrogen bonds in the active sites of the two mutants is much less dense, especially in the Trp154Gly mutant. This may be correlated to a poor positioning of the catalytic Ser67 and thus with a much less efficient catalytic activity of both mutants. The replacement of Trp154 by a histidine residue gives an intermediary situation where Lys70 is non-carboxylated in monomer B and partially carboxylated in monomer A. In this case, one imidazole nitrogen atom of His154 interacts with one oxygen atom of the carboxyl group, whereas the second one is hydrogen bonded to the hydroxyl group of the active Ser67. This latter situation corresponds to a fully active catalytic site and explains why the Trp154His mutation has a lesser impact on the catalytic activity than the Trp154Ala/Gly mutations.

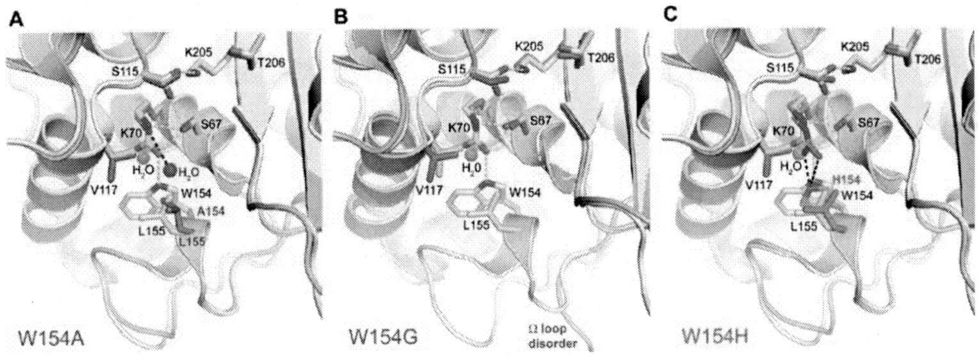

Figure 9. Importance of Trp154 for the carboxylation of Lys70. Diagram representation of the active site of the three Trp154 mutants superposed with the wild type enzyme. Important residues in the active site are shown as sticks and yellow dashed lines indicate the hydrogen bonds stabilizing the main residues of the wild type. *A*. Trp154Ala, green; *B*. Trp154Gly, blue; **C**.Trp154His, orange. The hydrogen bonds between the partially carboxylated Lys70 and His154 are shown as black dashed lines.

Importance of the Valine in the Second Conserved Motif

The second conserved motif of class D β-lactamases is significantly different from those of the rest of the penicillin recognizing enzymes. The usual SXN or YXN triads are indeed replaced by SXV that was only unambiguously identified thanks to the first structure of OXA-10. The valine of this motif is part of the hydrophobic ring surrounding the active site of class D β-lactamases (Figure 3) and is thought to influence the pKa of the ε-amino group of the first motif lysine to facilitate its carboxylation. This is confirmed by a recent study of this residue in OXA-10. Its replacement by a threonine (V117T) increased the dissociation constant of this enzyme for CO_2 by 7-8 fold [24]. The increase of mutant activity in the presence of additional sodium bicarbonate as well as its structure which displays a partially carboxylated Lys70 in conditions where the wild type enzyme is fully carboxylated further strengthen this interpretation. The structure of the OXA-10 V117T mutant did not reveal any other significant modification except a slight movement of the Thr117 side chain toward Lys70.

The class D β-lactamases and the C-terminal domains of the BlaR/MecR receptors are the most similar subfamilies within the penicillin recognizing enzyme superfamily and the major difference between them is the replacement of the SXV motif by SXT in the latter. The BlaR-CTD receptors are characterized by high acylation rates, very low deacylation rates and a non carboxylated lysine. The OXA-10 V117T mutant exhibits a reduced affinity for CO_2 and a reduced deacylation rate. However, the acylation rate of this mutant in the absence of CO_2 is orders of magnitude lower than that of BlaR-CTD. Although the V117T mutation might appear to be a step in the right direction, it is certainly not sufficient to transform OXA-10 into an efficient penicillin receptor [24].

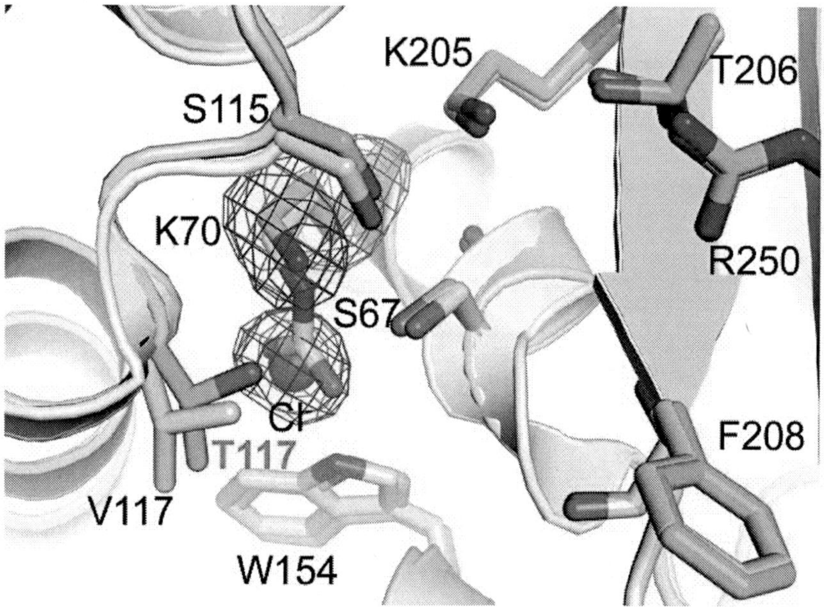

Figure 10. Inhibition by chloride ions. Structure of the V117T mutant at pH 6.5 (green) superimposed onto that of the wild-type OXA-10 (gray). The chloride ion is represented as a magenta sphere. A difference omit map calculated after removing the Lys70 side chain and chloride ion is displayed in blue.

Inhibition of Class D β-lactamase by Chloride Ions

Inhibition by chloride ions has long been recognized as a characteristic of class D β-lactamases although many of them are inhibited by chloride ions at concentrations above 20 mM [41]. The structure of the OXA-10 V117T mutant at pH 6 unambiguously identified a chloride ion in the position where the carboxylate of Lys70 is found at a higher pH value (Figure 10). The competition between chloride ions and lysine carboxylation was further confirmed [24]. The IC50 values were determined for the V117T mutant and the wild type enzyme in absence and presence of sodium bicarbonate. For the mutant, the values observed without and with 50 mM $NaHCO_3$ are 20 and 40 mM respectively. In the case of the wild-type enzyme, the IC50 value is 10 mM in the absence of sodium bicarbonate as already described by Dale *et al.* [41]. But at 50 mM $NaHCO_3$, the enzyme is not inhibited by 500 mM

Cl^-. Thus the addition of sodium bicarbonate strongly prevents the inhibition of OXA-10 by chloride ions.

STRUCTURES OF COMPLEXES BETWEEN CLASS D β-LACTAMASES AND SUBSTRATES OR INHIBITORS

Benzylpenicillin and Ampicillin

Because of the very efficient hydrolysis of β-lactam antibiotics by β-lactamases, it is not possible to trap the acylated form of the enzyme with most substrates. Structures of these complexes can only be obtained with mutants with an impaired catalytic mechanism. Two such examples are presented in this section for the OXA-10 enzyme.

The structure of the OXA-10 W154A mutant after acylation by benzylpenicillin has been determined at a 2.85Å resolution [25]. Lys70 is not carboxylated and a clear density is observed for the benzylpenicillin moiety which has acylated Ser67 in both monomers. The overall position of the antibiotic is similar to that usually observed in β-lactamase adducts (Figure 11A). The carbonyl oxygen of the former β-lactam ring is in the oxyanion hole, and the phenyl ring of the 6β side chain is oriented toward the bottom of the active site between Ala98 and Met99 on the left and Phe208 and Gly210 on the right. The thiazolidine ring is located in the upper part of the active site and two orientations differing by a 15° rotation are found in the two monomers. In monomer B, the carboxyl group attached to the thiazolidine ring interacts solely with the Arg250 side chain while in monomer A it is also involved in a hydrogen bound with the Thr206 side chain. In this conformation, Arg250 is pushed 0.5Å backwards, which is correlated with a 0.75Å displacement of the C-terminal helix and a 180° rotation of the Trp28 side chain that induces a subsequent 1.5Å shift of the N-terminal helix. The structure of the W154A mutant at pH 6 obtained in the same conditions does not display any of these changes. They are therefore clearly due to the presence of the benzylpenicillin moiety but their magnitude has to be considered with care because of the relatively low resolution.

The fact that the shifts of the N and C-terminal helices of OXA-10 in the W154A-benzylpenicillin adduct are only observed in one monomer indicates that the crystal packing influences the observed structure. This difference however emphasizes the potential plasticity of the enzyme during the hydrolysis of the substrate.

A stable acyl-enzyme adduct with ampicillin could also be visualized in the case of the OXA-10 K70C mutant [24]. This amino acid substitution has no significant effect on the overall structure of the enzyme, but the mutation induces a slight movement of the Trp-120 and Met-99 side chains in the hydrophobic core.

A slight modification is also observed in the loop between the β5-β6 strands for residues 212 to 216 as well as two conformations for the Cys-70 residue in the acyl-enzyme. The orientation of the substrate is however similar to that observed in the OXA-10 W154A-Benzylpenicillin complex [25]. The carboxylate group of the thiazolidine ring interacts with the guanidinium group of Arg-250 in monomer A. The amino group N1 of ampicillin is hydrogen bonded with the carbonyl main chain of Phe-208.

The fact that the structures of these complexes could be obtained indicates that for both mutants (W154A and K70C) the deacylation step was dramatically reduced and became rate-limiting.

Oxacephem: Moxalactam

The oxacephems are structurally very similar to cephalosporins, but the sulfur atom in position 1 is replaced by an oxygen atom. Moxalactam, one of the first 1-oxa cephalosporins, exhibited a broad spectrum of antibacterial activity against aerobic gram-negative enteric bacilli [42]. Moxalactam also possesses a methoxy group in the 7α position, a bulky side chain in the 7β position and a 1-methyl-1H-tetrazole-5 thiol side chain linked to the oxacephem ring (Figure 11B). A crystalline complex has been obtained with Moxalactam (1K6R) [24], after soaking OXA-10 crystals in a 50mM solution at pH 9. The catalytic Ser67 of both A and B monomers is acylated but, unlike the apo-form of OXA-10 solved at pH 9, Lys70 appears non-carboxylated. In monomer A, moxalactam adopts a classical position in the catalytic cleft, with the oxygen atom of the carbonyl of the lactam ring stabilized in the oxyanion hole by two hydrogen bonds with the main chain nitrogen atoms of Ser67 and Phe208.

Despite the presence of a non-carboxylated Lys70, the active site structure is the same as that of the apo-form of OXA-10 solved at pH 9. The carboxylate of the oxacephem ring makes a strong interaction with the Arg250 guanidinium group. The 7-α methoxy side chain points towards residues Met99, Trp102 and Val117. The 7-β side chain is stabilized by several interactions including H-bonds between its nitrogen atom and the main chain carbonyl oxygen of Phe208, and between its carboxylate group and the main chain carbonyl oxygens of both Leu155 and Ser209. The hydroxyphenyl group points outside the cleft and makes no interaction with active site residues. By contrast, in monomer B, moxalactam adopts a very unusual position, resulting from a rotation of almost 180° along the acyl-serine Cβ-Oγ bond, which induces quite a large modification of the active site conformation (Figure 11B). The classical interactions within the oxyanion hole are lost, as well as the salt bridge between the oxacephem ring carboxylate and Arg250.

In this case, the Arg250 guanidinium group interacts with the carboxylate of the 7-β side chain and the oxacephem ring carboxylate is hydrogen bonded to the indole Nε atom of Trp154. Moxalactam is further stabilized by two additional H-bonds between the former carbonyl oxygen of the β-lactam ring and the Nζ of Lys70, and the hydroxyl of the moxalactam side chain and the Glu244 carboxylate. The unusual position of the oxacephem ring induces a 3.5 Å shift of the S_{115}-A-V_{117} loop and consequently, may also be responsible for the perturbation of the loop connecting the short $\beta 3^b$ and $\beta 3^c$ strands. The two very dissimilar positions of moxalactam in monomers A and B is another illustration of the high plasticity of the OXA-10 active site. Kinetic experiment showed that the activity of OXA-10 against moxalactam is strongly increased in the presence of sodium bicarbonate and that the apparent dissociation constant for CO_2 is highly dependent on the substrate used to measure it. This result also indicates that moxalactam induces a decarboxylation of Lys70 which partly explains the poor activity of OXA-10 on this substrate [24].

β-Hydroxyisopropylpenicillanate

6-hydroxylalkylpenicillanates have been described as active serine β-lactamases inhibitors [43]. These compounds act by impairing the approach of the hydrolytic water molecule that attacks the acyl-enzyme intermediate. In the case of class A β-lactamases, the attack occurs on the α-face and thus α-hydroxylalkylpenicillanates are good inhibitors [44; 45]. In contrast, for class C β-lactamases, β-hydroxylalkylpenicillanates are good inhibitors since water attacks via the β-face. Curiously, for the class D β-lactamase OXA-10, both types of compounds are potent inhibitors with similar K_i values in the micromolar range. However, NMR studies have shown that the α-compound, unlike the β-compound, interacts more strongly with the carboxylate of the modified Lys70 [15]. A crystalline complex (1K54) has been obtained only with the β-hydroxyisopropylpenicillanate, after soaking in a 100mM solution at pH 8.5 [15]. Soaking of OXA-10 crystals with the α-hydroxylalkylpenicillanate induces the disruption of the crystal lattice, suggesting that the binding of the α-hydroxyisopropylpenicillanate induces a large conformational change in the acyl-enzyme. By contrast with the apo-form of OXA-10 at the same pH, the two biological dimers in the crystal asymmetric unit exhibits a mixed situation with two acylated monomers (A and B), and two monomers (C and D) exhibiting an empty active site. The active site conformation of monomers C and D resembles that described in the low pH structures. A non-carboxylated Lys70 is observed and quite a large conformational change occurs when compared to the carboxylated apo form.

The Lys70 side chain is 4.6 Å away from the Ser67-Og and rather interacts with the main chain oxygen atom of Val117. The 114-116 loop, connecting helices α4 and □α5 and containing the conserved S_{115}-A-V_{117} motif 2, has moved out of the active site and Ser115 is distant by more than 7 Å from the active Ser67. This loop movement may be correlated with the poor definition of the loop connecting the short $\beta3^b$ and $\beta\beta3^c$ strands (residues 92 to 103).

On the other hand, the catalytic Ser67 of the A and B monomers is acylated and Lys70 appears carboxylated (Figure 11C). As observed in most structures of acyl-enzyme adducts of class A and class C β-lactamases, the oxygen atom of the carbonyl of the opened lactam ring is stabilized in the oxyanion hole by two hydrogen bonds with the main chain nitrogen atoms of Ser67 and Phe208. By contrast, the well conserved interaction between the carboxylate of the thiazolidine ring with the conserved Arg250 is lost and the carboxylate is oriented toward the outside of the active site and interacts loosely with water molecules. The hydroxyl group of the β-hydroxyisopropyl side chain makes two hydrogen bonds with the main chain nitrogen atoms of Ser67 and on oxygen atom of the Lys70 carboxylate, which is in very good agreement with the NMR results.

The position of the hydroxyl group may be compared to that of a hydrolytic water molecule in the acyl-enzymes of class A β-lactamases. Thus the β-hydroxyisopropylpenicillanate is an irreversible inactivator because its β-hydroxyisopropyl side chain prevents the income of the water molecule that would hydrolyze the acyl-enzyme species.

Carbapenems: Meropenem, Imipenem and Doripenem

Carbapenems, originally developed from thienamycin, are β-lactam antibiotics with a broad spectrum of antibacterial activity [46]. They are highly resistant to β-lactamases. The carbapenems are structurally very similar to the penicillins, but the sulfur atom in position 1 has been replaced by a carbon atom. They also possess a large substituent in the 6α position, a hydroxyethyl group as in meropenem [47], imipenem [48] and doripenem [49].

Crystalline complexes have been obtained with meropenem (1H8Y) and imipenem (1H5X), after soaking crystals of OXA-13 in a 10mM solution at pH 5.5 [16]. As described for the apo-form of the OXA-13 enzyme solved at the same pH, Lys70 appears non-carboxylated in both monomers of the two acyl-enzyme structures. Meropenem and imipenem are deeply anchored in the catalytic cleft and covalently bound to Ser67. Unlike most class A and class C carbapenem adducts, the oxygen atom of the carbonyl of the opened lactam ring is stabilized in the oxyanion hole by two hydrogen bonds with the main chain nitrogen atoms of Ser67 and Phe208. The carboxylate and the hydroxyethyl moieties lie in two cavities facing each other in the catalytic pocket. The carboxylate is tightly fixed by a strong salt bridge with the Arg250 side chain.

Despite these similarities in the binding of both compounds, the two structures show two different conformations of the active site. In the Meropenem complex, the hydroxyl group of Ser67 is more or less at H-bond distance of the ε-amine group of Lys70 and the hydroxyl group of Ser115, highlighting a more constrained form of the catalytic site, as described for the other complex structures where Lys70 is carboxylated (Figure 11D).

By contrast, in the imipenem complex, the distance between Ser67-Oγ and Ser115-Oγ increases to about 7 Å. This defines a more open active site with a shifted S_{115}-A-V_{117} loop and an undefined loop connecting the short $\beta3^b$ and $\beta3^c$ strands, as observed for the apo-form of OXA-13. The difference between the two complexes may be correlated to the position of the C2 carbapenem side chain lying at the entrance of the site (an extended aminomethylideneamino ethylsulfanyl side chain for imipenem and a much bulkier dimethylcarbamoyl pyrrolidin-2-sulfanyl side chain for meropenem). Whereas the meropenem side chain has no obvious interaction with the loops at the active site entrance, the long sulphur-containing side chain of imipenem interacts with the loop connecting β7 and the C-terminal α10, by two hydrogen bonds with residue Glu244.

The third complex between a class D β-lactamase and a carbapenem available was obtained with OXA-1 and doripenem [26] at pH7.5. In this case, the acyl carbonyl oxygen in the OXA-1-doripenem complex is found in the oxyanion hole (Figure 11E). The side chain of Lys70 is fully carboxylated and forms a hydrogen bond with the alcohol of the 6R-hydroxyethyl moiety. Unlike the situation in the meropenem and imipenem complexes, the carboxylate of the doripenem pyrrolidine ring is within binding distance from the Lys212 ε-amino and the Thr213 and Ser115 hydroxyl groups. In the OXA-13 complexes, this carboxylate interacts with Arg244 which has no equivalent in OXA-1. These novel sets of interactions with the carboxylate results in a major shift of the carbapenem pyrrolidine ring compared to the position of the same ring in the meropenem-OXA-13 adduct. Interestingly, no water molecules were observed in the vicinity of the acyl linkage, providing an explanation for the inhibition of OXA-1 by carbapenems.

Iodo-Acetamidophenylboronic Acid

Boronic acid inhibitors are highly effective deacylation transition state analogues for class A and class C β-lactamase [50-55]. The tetrahedral geometry adopted by the boronic acid, the putative concentration of its negative charge on the boronic acid oxygens, and the different side-chain functionalities displayed by several designed boronic inhibitors make this type of compounds the best inhibitors designed for serine β-lactamases.

A crystalline complex (1K6S) has been obtained with the iodo-acetamidophenylboronic acid, after soaking OXA-10 crystals in a 1mM solution at pH 9. In both A and B monomers of the biological dimer an electron density is observed connecting the hydroxyl group of Ser67 to the boron atom of the inhibitor (Figure 11F).

The geometry around the boron is tetrahedral, as expected. The boronic acid O1 is within good hydrogen-bonding distance of the backbone nitrogens of Ser67 and Phe208. The boronic acid O2 hydrogen bonds with hydroxyl group of Ser115. Lys70 appears to be carboxylated and the carboxylate makes the classical hydrogen bonds with Ser67-Oγ, Trp154-Nε and a water molecule, which is itself stabilized by a hydrogen bond with Asn73. The side chain of the inhibitor is further stabilized by different interactions. The phenyl ring of the iodo-acetamidophenyl side chain is trapped between the side chains of Trp102, Val 117 and Phe208.

The proximity of the Met99 side chain leads to a further interaction where the iodo-acetamide reacts with the thioether group of Met99 and concomitantly releases its iodine atom. This unusual carboxymethylation reaction is quite evident in the light of the difference electron density maps and has been confirmed by mass spectrometry experiments.

Penicillanic Acid Sulfones

Sulbactam and tazobactam are two penicillanic acid sulfones utilized clinically in combination with other antibiotics because of their efficient inhibition of class A β-lactamases.

Their efficiency against class D oxacillinases is unfortunately very low. In the search of potent inhibitors of class D enzymes, several more active substituted penicillanic acid sulfones have recently been tested [27; 56] and three structures of complexes between the carbapenemase OXA-24 and 6-alkylidene-2′-substituted penicillanic acid sulfones are now available in the Protein Data Bank (3FV7, 3FYZ and 3FZC). These compounds display IC50 values ranging from 10 to 338 nM and can serve as leads in the search of new efficient class D inhibitors [27].

These structures reveal the formation of stable bicyclic aromatic intermediates with their carbonyl oxygen in the oxyanion hole (Scheme 2). The network of interactions between OXA-24 and the compounds are very similar and only the 3FYZ structure ((2S,3R)-2-[(7-aminocarbonyl-2-methanoyl-indolizin-3-yl)amino]-4-aminocarbonyloxy-3-methyl-3- sulfino-butanoic acid) will be described (Figure 11G).

The sulfinate makes a salt-bridge with the guanidium group of Arg261 and hydrogen bonds with hydroxyl groups of Ser128 and Ser259. The carboxylate group which is usually found at the position of the sulfinate anion in other antibiotic-β-lactamase complexes is instead oriented in the opposite direction and interacts with the hydroxyl group of Tyr212. The two adjacent cycles resulting from the molecular reorganization are found at the centre of the characteristic hydrophobic ring formed by Trp115, Val130, Leu168 and Met223 at the entrance of the active site. The remaining contacts with the enzyme are specific for each inhibitor. In the described structure an additional hydrogen bond is observed between the hydroxyl group of Thr111 and the carbamoyl group of the compound.

According to the authors, the poor deacylation observed for these compounds could result from the interference between the bicyclic aromatic ring and the attacking water molecule. Therefore 6-alkylidene-2′-substituted penicillin sulfones seem to be effective mechanism-based inactivators of class D β-lactamases.

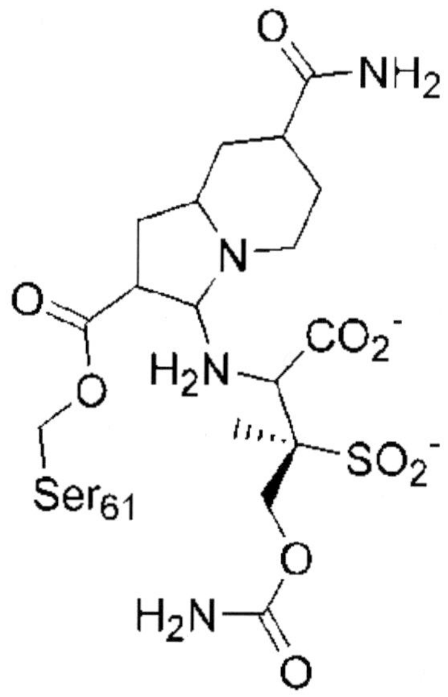

Scheme 2.

Cyclobutanones

The inhibition of OXA-10 by cyclobutanones has recently been published. The structure of the enzyme-inhibitor complex could be obtained at 2 Å resolution with the best compound (1S,3S,4S,5S)-7,7-dichloro-3-methoxy-2-thiabicyclo[3.2.0]heptan-6-one-4-carboxylic acid) [28]. This cyclobutanone covalently binds to the active serine and forms a serine hemiketal in two of the four monomers present in the asymmetric unit (Figure 11H). The first motif lysine (Lys70) is carboxylated in both monomers.

In this structure, the inhibitor is in the *exo* envelope with the carboxylate in an equatorial orientation interacting with Arg250, Thr206 and Ser115, and the methoxy group in an axial orientation. The oxygen atom bound to the β-lactam ring is in the oxyanion hole and the chlorine atoms are not well defined probably because of x-ray radiation damage.

The IC_{50} value of this molecule is only 156 μM. This is however the first type of reversible inhibitor to show some inhibition of the four classes of β-lactamases.

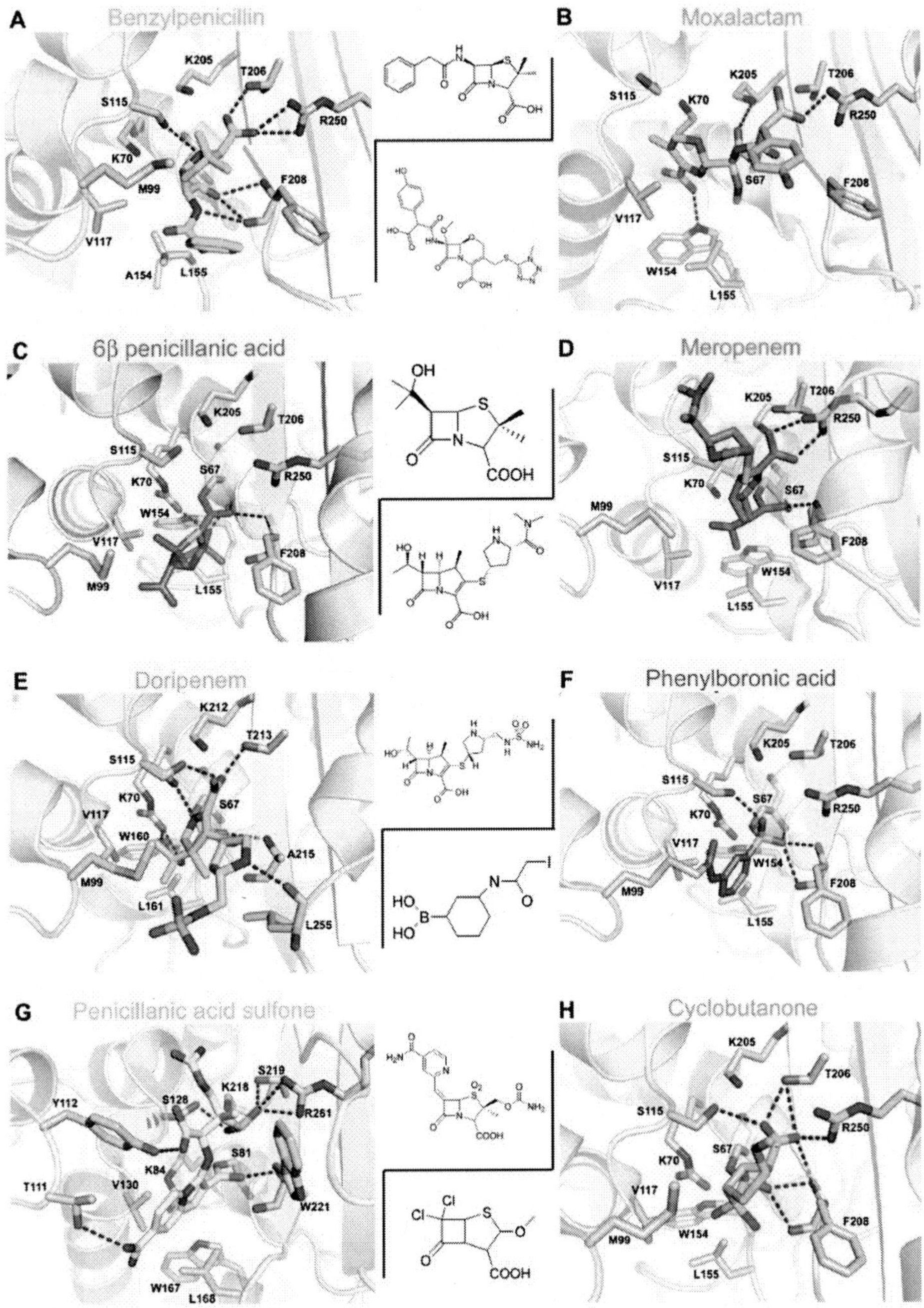

Figure 11. Class D β-lactamases complexes. Diagram representation of the active site of the different class D β-lactamases complexes. The important residues and the compounds of interest are shown as sticks. Polar interactions with the enzymes are represented by dashed lines. *A*. OXA-10 with benzylpencillin (yellow); *B*. OXA-10 with moxalactam, monomer A (magenta); *C*. OXA-10 with 6β - (1-hydroxy-1-methylethyl) penicillanic acid (blue); *D*. OXA-13 with meropenem (orange); *E*. OXA-1 with doripenem (light green); *F*. OXA-10 with 4-iodo-acetamido phenylboronic acid (green); *G*. OXA-24 with penicillanic acid sulfone (cyan); *H*. OXA-10 with cyclobutanone (light blue).

THE PROPOSED CATALYTIC MECHANISM

The active serine Ser67 must be activated in order to attack the β-lactam carbonyl group to form the acyl-enzyme intermediate. The activation may be performed by Lys73 either directly [13; 16] or indirectly when Lys73 is carboxylated [12]. When Lys73 is carboxylated, the carboxylate may activate the deprotonation of Ser70 and participate in the proton transfer to the nitrogen atom of the β-lactam ring, directly or with the help of the conserved Ser115. In the case of a non-carboxylated Lys70, the ε-amino group is supposed to activate Ser67. Ser115, Lys70 and also Lys205 were also suggested to be involved in the proton transfer to the nitrogen atom of the β-lactam ring.

The hydrophobic environment of the active site may be responsible for lowering the pKa of Lys70 and thus facilitating the reaction and/or favoring the carboxylation of the side chain.

Hydrolysis of the acyl-enzyme intermediate should be performed by an activated water molecule. When Lys70 is carboxylated there is no water molecule in a suitable position for attacking the acyl-enzyme. Anyway, it is possible for an incoming water molecule to approach the carboxylate group and be in an adequate position for the attack. On the other hand, when Lys70 is non-carboxylated, a water molecule is hydrogen bonded to the Lys70 ε-amino group and the nitrogen Nε of the Trp154 indole group and thus at the right position for the nucleophilic attack.

Individual rate constants have recently been determined for the wild type OXA-10 and its V117T, K70A and K70C mutants [24]. These results clearly show the importance of the carboxylation for the activity of this class of enzymes and suggest that the decarboxylation of Lys70 impairs both the acylation and deacylation steps but that the latter is more affected. This behavior can be compared with that of the E166Q mutant of the TEM class A β-lactamase for which the deacylation was also more affected than the acylation.

EXTENDED SPECTRUM ENZYMES AND CARBAPENEMASES IN THE LIGHT OF THE 3D STRUCTURES

Extented-spectrum Enzymes

The term extended-spectrum was first applied in 1987 to variants of the class A TEM and SHV enzymes that were able to hydrolyze oxymino-cephalosporins [57] and were inhibited by clavulanic acid [58]. More recently, it appeared that point mutations in the class D β-lactamases could also significantly increase the activity on these substrates. OXA-type ß-lactamases are characterized by their high hydrolytic activity against oxacillin and cloxacillin and are poorly inhibited by clavulanic acid. Variants in groups I (OXA-10), II (OXA-2), III (OXA-1) and IV (OXA-18) have been reported.

Variants in group I exhibit one or two point mutations: OXA-11 [5] (Asn143Ser, Gly159Asp), OXA-14 [6; 7] (Gly157Asp), OXA-16 [8] (Ala124Thr, Gly157Asp) and OXA-17 [59] (Asn73Ser). These mutations confer a wider hydrolytic activity, against ceftriaxone, cefotaxime and ceftazidime. In the light of the 3D structures of OXA-10 and-13, it is quite difficult to rationalize the activity profiles of the variants. Residue 73 follows the essential Lys70 and defines the bottom of the catalytic cleft. Residue 157 follows Trp154 and is part of the Ω-loop, defining the entrance of the active site. Residues Ser73 and Asp157 are also found in the OXA-13 enzyme, which has no extended-spectrum behavior. Residues 124 and 143 lie respectively on α5 and on the loop connecting α6 and α7, and are quite distant from the active site.

Variants in group II exhibit only one mutation: OXA-15 [10] (Asp145Gly) and OXA-32 [60] (Leu159Ile). OXA-15 had a greater cephalosporinase activity, with increased relative hydrolysis rates for cephaloridine and cephalothin and, most importantly, for ceftazidime. Cefotaxime and carbapenems are poorly hydrolyzed. OXA-32 confers a high resistance to ceftazidime and a low-level resistance to cefotaxime. Both mutations occur at the entrance of the catalytic cleft. Residue 145 is on the loop connecting α6 and α7, with its side chain pointing out into the solvent. Residue 159 is part of the Ω-loop and its side chain points into the cavity. However a Leu/Ile mutation is usually considered as conservative and does not have a dramatic influence. Nevertheless, the mutations in group II also occur at the entrance of the substrate binding groove defined by structural elements that have been shown to possess a potential flexibility.

OXA-30 [61] is a variant of OXA-1 from group III. Oxa-30 was found in a clinical isolate of *Escherichia coli and was associated to resistance to* cefotaxime, ceftazidime, and cefepime. There was some controversy about the sequence corresponding to the name OXA-1 or OXA-30, as noticed by Boyd et al. [62]. So, it was quite difficult to highlight the point mutation that could be associated to the extended-spectrum behavior.

OXA-18 [9] from group IV confers high resistance to extended-spectrum cephalosporins. It has a broad substrate profile and hydrolyzes amoxicillin, ticarcillin, cephalothin, ceftazidime, cefotaxime, and aztreonam, but not imipenem or cephamycins. Its activity is strongly inhibited by clavulanic acid. OXA-45 [63] shares 65 % identity with OXA-18 and is very similar to OXA-18 in its substrate profile and inhibition by clavulanic acid. No three dimensional structure is known for a group IV enzyme but a homology model could be obtained using the OXA-1 structure (1M6K) as template. In the model, it appears that the loops defining the entrance of the active site are longer with insertions of 6 residues in the Ω-loop and of 3 residues in the β5-β6 β-turn respectively□ Both insertions may contribute additional interactions with the substrate.

Carbapenemases

The number of class D βlactamases with carbapenemase activity has greatly increased over the last years. More than 45 class D carbapenemases have been identified [64]. They are more frequently found in *Acinetobacter baumannii* and they mostly belong to group VI with OXA-24 as template structure. They hydrolyze imipenem and meropenem weakly, do not hydrolyze extended-spectrum cephalosporins and aztreonam and are not inhibited by clavulanic acid. Based on sequence similarities, nine subgroups have been defined in group VI. The identity between members of each subgroup is very high (> 90%) with sequence divergence of 1 to 20 amino acids. Members of subgroups 1, 2 and 3 are the most abundant. The three subgroups are defined on the basis of the sequence of OXA-23 [11], -24 [65] and -51 [66] respectively. The OXA-24 structure shows that the the carbapenem substrate specificity is determined by a hydrophobic barrier defined by the Tyr112, Trp221 and Met223 side chains. Tyr112 is equivalent to Met99, and Trp221 is equivalent to Phe208 in OXA-10. With the conserved Val117 and Leu155, they define the hydrophobic ring at the entrance of the active site. With the much bulkier Tyr112 and Trp221 side chains, plus the presence of Met223, access to the catalytic pocket is more hindered and the environment more hydrophobic. This allows favorable interactions with carbapenems. These residues are strictly conserved among the carbapenemases, with the exception of Tyr112 which is sometimes replaced by phenylalanine. The importance of these residues was further confirmed by mutagenesis studies.

OXA-48 [67] shares 46 % identity with OXA-10, and has a poor hydrolytic efficiency against expanded-spectrum cephalosporins but hydrolyzes imipenem at a very high rate, and thus contributes to carbapenem resistance. The 3D structure of OXA-48 is very similar to that of OXA-10 and indicates that the hydrolysis of carbapenems depends of very small changes in the active site area [19]. This structure also highlights the lack of similarity between the active site groove of OXA-48 and OXA-24. In the case of OXA-48, the authors identified the β5β6 loop and its surroundings as a key region for the carbapenemase activity. Residues in this area would modify the acyl-enzyme conformation to subsequently allow the attack of the acylated serine by the deacylating water molecule [19].

CONCLUSIONS

The class D β-lactamases also known as the OXA-type β-lactamases represent an emerging and rapidly growing molecular class that includes enzymes with narrow to broad substrate profiles, some of which exhibiting carbapenemase activity.

Almost 200 sequences are known but only a few 3D structures have been determined. As all serine-active β-lactamases, the OXA-enzymes share a typical two domains structure with the catalytic pocket at their interface. However, it is known from structural, biochemical and kinetic analysis that their activity is related to the prevalence of monomeric or dimeric species and depends critically on an unusual carboxylated lysine as the general base for both acylation and deacylation steps of catalysis. A comparison of the structures of the OXA-10 enzyme at different pH values highlights two active site geometries, only one being functional with a carboxylated lysine residue located in close proximity to the hydroxyl group of the active serine. The class D β-lactamases have evolved a distinct catalytic mechanism when compared to other β-lactamases. They illustrate the fact that disparate classes of β-lactamases have divised different mechanisms to carry out the same reaction.

REFERENCES

[1] Ambler, R. P. (1980). The structure of beta-lactamases. *Philos Trans R Soc Lond B Biol Sci* 289, 321-31.

[2] Bush, K., Jacoby, G. A. & Medeiros, A. A. (1995). A functional classification scheme for beta-lactamases and its correlation with molecular structure. *Antimicrob Agents Chemother* 39, 1211-33.

[3] Voha, C., Docquier, J. D., Rossolini, G. M. & Fosse, T. (2006). Genetic and biochemical characterization of FUS-1 (OXA-85), a narrow-spectrum class D beta-

lactamase from Fusobacterium nucleatum subsp. polymorphum. *Antimicrob Agents Chemother* 50, 2673-9.

[4] Livermore, D. M. (2008). Defining an extended-spectrum beta-lactamase. *Clin Microbiol Infect* 14 Suppl 1, 3-10.

[5] Hall, L. M., Livermore, D. M., Gur, D., Akova, M. & Akalin, H. E. (1993). OXA-11, an extended-spectrum variant of OXA-10 (PSE-2) beta-lactamase from Pseudomonas aeruginosa. *Antimicrob Agents Chemother* 37, 1637-44.

[6] Yan, J. J., Tsai, S. H., Chuang, C. L. & Wu, J. J. (2006). OXA-type beta-lactamases among extended-spectrum cephalosporin-resistant Pseudomonas aeruginosa isolates in a university hospital in southern Taiwan. *J Microbiol Immunol Infect* 39, 130-4.

[7] Danel, F., Hall, L. M., Gur, D. & Livermore, D. M. (1995). OXA-14, another extended-spectrum variant of OXA-10 (PSE-2) beta-lactamase from Pseudomonas aeruginosa. *Antimicrob Agents Chemother* 39, 1881-4.

[8] Danel, F., Hall, L. M., Gur, D. & Livermore, D. M. (1998). OXA-16, a further extended-spectrum variant of OXA-10 beta-lactamase, from two Pseudomonas aeruginosa isolates. *Antimicrob Agents Chemother* 42, 3117-22.

[9] Philippon, L. N., Naas, T., Bouthors, A. T., Barakett, V. & Nordmann, P. (1997). OXA-18, a class D clavulanic acid-inhibited extended-spectrum beta-lactamase from Pseudomonas aeruginosa. *Antimicrob Agents Chemother* 41, 2188-95.

[10] Danel, F., Hall, L. M., Gur, D. & Livermore, D. M. (1997). OXA-15, an extended-spectrum variant of OXA-2 beta-lactamase, isolated from a Pseudomonas aeruginosa strain. *Antimicrob Agents Chemother* 41, 785-90.

[11] Donald, H. M., Scaife, W., Amyes, S. G. & Young, H. K. (2000). Sequence analysis of ARI-1, a novel OXA beta-lactamase, responsible for imipenem resistance in Acinetobacter baumannii 6B92. *Antimicrob Agents Chemother* 44, 196-9.

[12] Maveyraud, L., Golemi, D., Kotra, L. P., Tranier, S., Vakulenko, S., Mobashery, S. & Samama, J. P. (2000). Insights into class D beta-lactamases are revealed by the crystal structure of the OXA10 enzyme from Pseudomonas aeruginosa. *Structure* 8, 1289-98.

[13] Paetzel, M., Danel, F., de Castro, L., Mosimann, S. C., Page, M. G. & Strynadka, N. C. (2000). Crystal structure of the class D beta-lactamase OXA-10. *Nat Struct Biol* 7, 918-25.

[14] Golemi, D., Maveyraud, L., Vakulenko, S., Samama, J. P. & Mobashery, S. (2001). Critical involvement of a carbamylated lysine in catalytic function of class D beta-lactamases. *Proc Natl Acad Sci U S A* 98, 14280-5.

[15] Maveyraud, L., Golemi-Kotra, D., Ishiwata, A., Meroueh, O., Mobashery, S. & Samama, J. P. (2002). High-resolution X-ray structure of an acyl-enzyme species for the class D OXA-10 beta-lactamase. *J Am Chem Soc* 124, 2461-5.

[16] Pernot, L., Frenois, F., Rybkine, T., L'Hermite, G., Petrella, S., Delettre, J., Jarlier, V., Collatz, E. & Sougakoff, W. (2001). Crystal structures of the class D beta-lactamase OXA-13 in the native form and in complex with meropenem. *J Mol Biol* 310, 859-74.

[17] Sun, T., Nukaga, M., Mayama, K., Braswell, E. H. & Knox, J. R. (2003). Comparison of beta-lactamases of classes A and D: 1.5-A crystallographic structure of the class D OXA-1 oxacillinase. *Protein Sci* 12, 82-91.

[18] Santillana, E., Beceiro, A., Bou, G. & Romero, A. (2007). Crystal structure of the carbapenemase OXA-24 reveals insights into the mechanism of carbapenem hydrolysis. *Proc Natl Acad Sci U S A* 104, 5354-9.

[19] Docquier, J. D., Calderone, V., De Luca, F., Benvenuti, M., Giuliani, F., Bellucci, L., Tafi, A., Nordmann, P., Botta, M., Rossolini, G. M. & Mangani, S. (2009). Crystal structure of the OXA-48 beta-lactamase reveals mechanistic diversity among class D carbapenemases. *Chem Biol* 16, 540-7.

[20] Docquier, J. D., Benvenuti, M., Calderone, V., Giuliani, F., Kapetis, D., De Luca, F., Rossolini, G. M. & Mangani, S. (2010). Crystal structure of the narrow-spectrum OXA-46 class D beta-lactamase: relationship between active-site lysine carbamylation and inhibition by polycarboxylates. *Antimicrob Agents Chemother* 54, 2167-74.

[21] Sanschagrin, F., Couture, F. & Levesque, R. C. (1995). Primary structure of OXA-3 and phylogeny of oxacillin-hydrolyzing class D beta-lactamases. *Antimicrob Agents Chemother* 39, 887-93.

[22] Barlow, M. & Hall, B. G. (2002). Phylogenetic analysis shows that the OXA beta-lactamase genes have been on plasmids for millions of years. *J Mol Evol* 55, 314-21.

[23] Danel, F., Page, M. G. & Paetzel, M. . (2007). Class D beta-Lactamases. In *Enzyme-Mediated Resistance to Antibiotics: Mechanisms, Dissemination and Prospects for Inhibition.*, pp. 163-194.

[24] Vercheval, L., Bauvois, C., di Paolo, A., Borel, F., Ferrer, J. L., Sauvage, E., Matagne, A., Frere, J. M., Charlier, P., Galleni, M. & Kerff, F. (2010). Three factors that modulate the activity of class D beta-lactamases and interfere with the post-translational carboxylation of Lys-70. *Biochem J* 432, 495-504.

[25] Baurin, S., Vercheval, L., Bouillenne, F., Falzone, C., Brans, A., Jacquamet, L., Ferrer, J. L., Sauvage, E., Dehareng, D., Frere, J. M., Charlier, P., Galleni, M. & Kerff, F. (2009). Critical role of tryptophan 154 for the activity and stability of class D beta-lactamases. *Biochemistry* 48, 11252-63.

[26] Schneider, K. D., Karpen, M. E., Bonomo, R. A., Leonard, D. A. & Powers, R. A. (2009). The 1.4 A crystal structure of the class D beta-lactamase OXA-1 complexed with doripenem. *Biochemistry* 48, 11840-7.

[27] Bou, G., Santillana, E., Sheri, A., Beceiro, A., Sampson, J. M., Kalp, M., Bethel, C. R., Distler, A. M., Drawz, S. M., Pagadala, S. R., van den Akker, F., Bonomo, R. A., Romero, A. & Buynak, J. D. (2010). Design, synthesis, and crystal structures of 6-alkylidene-2'-substituted penicillanic acid sulfones as potent inhibitors of Acinetobacter baumannii OXA-24 carbapenemase. *J Am Chem Soc* 132, 13320-31.

[28] Johnson, J. W., Gretes, M., Goodfellow, V. J., Marrone, L., Heynen, M. L., Strynadka, N. C. & Dmitrienko, G. I. (2010). Cyclobutanone analogues of beta-lactams revisited: insights into conformational requirements for inhibition of serine- and metallo-beta-lactamases. *J Am Chem Soc* 132, 2558-60.

[29] Fonze, E., Charlier, P., To'th, Y., Vermeire, M., Raquet, X., Dubus, A. & Frere, J. M. (1995). TEM1 beta-lactamase structure solved by molecular replacement and refined structure of the S235A mutant. *Acta Crystallogr D Biol Crystallogr* 51, 682-94.

[30] Wouters, J., Fonze, E., Vermeire, M., Frere, J. M. & Charlier, P. (2003). Crystal structure of Enterobacter cloacae 908R class C beta-lactamase bound to iodo-acetamido-phenyl boronic acid, a transition-state analogue. *Cell Mol Life Sci* 60, 1764-73.

[31] Kerff, F., Charlier, P., Colombo, M. L., Sauvage, E., Brans, A., Frere, J. M., Joris, B. & Fonze, E. (2003). Crystal structure of the sensor domain of the BlaR penicillin receptor from Bacillus licheniformis. *Biochemistry* 42, 12835-43.

[32] Marrero, A., Mallorqui-Fernandez, G., Guevara, T., Garcia-Castellanos, R. & Gomis-Ruth, F. X. (2006). Unbound and acylated structures of the MecR1 extracellular antibiotic-sensor domain provide insights into the signal-transduction system that triggers methicillin resistance. *J Mol Biol* 361, 506-21.
[33] Fonze, E., Vermeire, M., Nguyen-Disteche, M., Brasseur, R. & Charlier, P. (1999). The crystal structure of a penicilloyl-serine transferase of intermediate penicillin sensitivity. The DD-transpeptidase of streptomyces K15. *J Biol Chem* 274, 21853-60.
[34] Gros, G., Forster, R. E. & Lin, L. (1976). The carbamate reaction of glycylglycine, plasma, and tissue extracts evaluated by a pH stopped flow apparatus. *J Biol Chem* 251, 4398-407.
[35] Bertrand, J. A., Auger, G., Martin, L., Fanchon, E., Blanot, D., Le Beller, D., van Heijenoort, J. & Dideberg, O. (1999). Determination of the MurD mechanism through crystallographic analysis of enzyme complexes. *J Mol Biol* 289, 579-90.
[36] Lundqvist, T. & Schneider, G. (1991). Crystal structure of the ternary complex of ribulose-1,5-bisphosphate carboxylase, Mg(II), and activator CO2 at 2.3-A resolution. *Biochemistry* 30, 904-8.
[37] Mathieu, M., Debousker, G., Vincent, S., Viviani, F., Bamas-Jacques, N. & Mikol, V. (2005). Escherichia coli FolC structure reveals an unexpected dihydrofolate binding site providing an attractive target for anti-microbial therapy. *J Biol Chem* 280, 18916-22.
[38] Hall, P. R., Zheng, R., Antony, L., Pusztai-Carey, M., Carey, P. R. & Yee, V. C. (2004). Transcarboxylase 5S structures: assembly and catalytic mechanism of a multienzyme complex subunit. *EMBO J* 23, 3621-31.
[39] Jabri, E., Carr, M. B., Hausinger, R. P. & Karplus, P. A. (1995). The crystal structure of urease from Klebsiella aerogenes. *Science* 268, 998-1004.
[40] Morollo, A. A., Petsko, G. A. & Ringe, D. (1999). Structure of a Michaelis complex analogue: propionate binds in the substrate carboxylate site of alanine racemase. *Biochemistry* 38, 3293-301.
[41] Dale, J. W. (1971). Characterization of the -lactamase specified by the resistance factor R-1818 in E. coli K12 and other Gram-negative bacteria. *Biochem J* 123, 501-5.
[42] DeMaria, A., Alvarez, S., Klein, J. O. & McCabe, W. R. (1980). In vitro studies of moxalactam (LY127935), a new beta-lactam antibiotic with significant activity against gram-negative bacteria. *Infection* 8 Suppl 3, S 261-7.
[43] Golemi D, M. L., Ishiwata A, Tranier S, Miyashita K, Nagase T, Massova I, Mourey L, Samama JP, Mobashery S. (2000). 6-(hydroxyalkyl)penicillanates as probes for mechanisms of beta-lactamases. *J Antibiot* 53, 1022-1027.
[44] Maveyraud, L., Massova, I., Birck, C., Miyashita, K., Samama, J.P., Mobashery, S. (1996). Crystal Structure of 6Alpha-Hydroxymethylpenicillanate Complexed to the Tem-1 Beta-Lactamase from Escherichia Coli: Evidence on the Mechanism of Action of a Novel Inhibitor Designed by a Computer-Aided Process *J.Am.Chem.Soc.* 118, 7435
[45] Mourey, L., Swaren, P., Miyashita, K., Bulychev, A., Mobashery, S., Samama, J.P. (1998). Inhibition of the Nmc-A B-Lactamase by a Penicillanic Acid Derivative, and the Structural Bases for the Increase in Substrate Profile of This Antibiotic Resistance Enzyme *J.Am.Chem.Soc.* 120, 9382

[46] MH, R. I. (1981). The semi-synthetic thienamycin derivative MK0787 and its properties with respect to a range of beta-lactamases from clinically relevant bacterial species. *J Antimicrob Chemother* 7, 279-85.

[47] Edwards, J. R., Turner, P. J., Wannop, C., Withnell, E. S., Grindey, A. J. & Nairn, K. (1989). In vitro antibacterial activity of SM-7338, a carbapenem antibiotic with stability to dehydropeptidase I. *Antimicrob Agents Chemother* 33, 215-22.

[48] Kropp, H., Sundelof, J. G., Kahan, J. S., Kahan, F. M. & Birnbaum, J. (1980). MK0787 (N-formimidoyl thienamycin): evaluation of in vitro and in vivo activities. *Antimicrob Agents Chemother* 17, 993-1000.

[49] Pillar, C. M., Torres, M. K., Brown, N. P., Shah, D. & Sahm, D. F. (2008). In vitro activity of doripenem, a carbapenem for the treatment of challenging infections caused by gram-negative bacteria, against recent clinical isolates from the United States. *Antimicrob Agents Chemother* 52, 4388-99.

[50] Usher, K. C., Blaszczak, L. C., Weston, G. S., Shoichet, B. K. & Remington, S. J. (1998). Three-dimensional structure of AmpC beta-lactamase from Escherichia coli bound to a transition-state analogue: possible implications for the oxyanion hypothesis and for inhibitor design. *Biochemistry* 37, 16082-92.

[51] Ness, S., Martin, R., Kindler, A. M., Paetzel, M., Gold, M., Jensen, S. E., Jones, J. B. & Strynadka, N. C. (2000). Structure-based design guides the improved efficacy of deacylation transition state analogue inhibitors of TEM-1 beta-Lactamase. *Biochemistry* 39, 5312-21.

[52] Tondi, D., Powers, R. A., Caselli, E., Negri, M. C., Blazquez, J., Costi, M. P. & Shoichet, B. K. (2001). Structure-based design and in-parallel synthesis of inhibitors of AmpC beta-lactamase. *Chem Biol* 8, 593-611.

[53] Wang, X., Minasov, G., Blazquez, J., Caselli, E., Prati, F. & Shoichet, B. K. (2003). Recognition and resistance in TEM beta-lactamase. *Biochemistry* 42, 8434-44.

[54] Venturelli, A., Tondi, D., Cancian, L., Morandi, F., Cannazza, G., Segatore, B., Prati, F., Amicosante, G., Shoichet, B. K. & Costi, M. P. (2007). Optimizing cell permeation of an antibiotic resistance inhibitor for improved efficacy. *J Med Chem* 50, 5644-54.

[55] Morandi, S., Morandi, F., Caselli, E., Shoichet, B. K. & Prati, F. (2008). Structure-based optimization of cephalothin-analogue boronic acids as beta-lactamase inhibitors. *Bioorg Med Chem* 16, 1195-205.

[56] Drawz, S. M., Bethel, C. R., Doppalapudi, V. R., Sheri, A., Pagadala, S. R., Hujer, A. M., Skalweit, M. J., Anderson, V. E., Chen, S. G., Buynak, J. D. & Bonomo, R. A. (2010). Penicillin sulfone inhibitors of class D beta-lactamases. *Antimicrob Agents Chemother* 54, 1414-24.

[57] Jarlier, V., Nicolas, M. H., Fournier, G. & Philippon, A. (1988). Extended broad-spectrum beta-lactamases conferring transferable resistance to newer beta-lactam agents in Enterobacteriaceae: hospital prevalence and susceptibility patterns. *Rev Infect Dis* 10, 867-78.

[58] Bush, K. (1989). Classification of beta-lactamases: groups 2c, 2d, 2e, 3, and 4. *Antimicrob Agents Chemother* 33, 271-6.

[59] Danel, F., Hall, L. M., Duke, B., Gur, D. & Livermore, D. M. (1999). OXA-17, a further extended-spectrum variant of OXA-10 beta-lactamase, isolated from Pseudomonas aeruginosa. *Antimicrob Agents Chemother* 43, 1362-6.

[60] Poirel, L., Gerome, P., De Champs, C., Stephanazzi, J., Naas, T. & Nordmann, P. (2002). Integron-located oxa-32 gene cassette encoding an extended-spectrum variant of OXA-2 beta-lactamase from Pseudomonas aeruginosa. *Antimicrob Agents Chemother* 46, 566-9.

[61] Siu, L. K., Lo, J. Y., Yuen, K. Y., Chau, P. Y., Ng, M. H. & Ho, P. L. (2000). beta-lactamases in Shigella flexneri isolates from Hong Kong and Shanghai and a novel OXA-1-like beta-lactamase, OXA-30. *Antimicrob Agents Chemother* 44, 2034-8.

[62] Boyd, D. A. & Mulvey, M. R. (2006). OXA-1 is OXA-30 is OXA-1. *J Antimicrob Chemother* 58, 224-5.

[63] Toleman, M. A., Rolston, K., Jones, R. N. & Walsh, T. R. (2003). Molecular and biochemical characterization of OXA-45, an extended-spectrum class 2d' beta-lactamase in Pseudomonas aeruginosa. *Antimicrob Agents Chemother* 47, 2859-63.

[64] Queenan, A. M. & Bush, K. (2007). Carbapenemases: the versatile beta-lactamases. *Clin Microbiol Rev* 20, 440-58.

[65] Bou, G., Oliver, A. & Martinez-Beltran, J. (2000). OXA-24, a novel class D beta-lactamase with carbapenemase activity in an Acinetobacter baumannii clinical strain. *Antimicrob Agents Chemother* 44, 1556-61.

[66] Brown, S., Young, H. K. & Amyes, S. G. (2005). Characterisation of OXA-51, a novel class D carbapenemase found in genetically unrelated clinical strains of Acinetobacter baumannii from Argentina. *Clin Microbiol Infect* 11, 15-23.

[67] Poirel, L., Heritier, C., Tolun, V. & Nordmann, P. (2004). Emergence of oxacillinase-mediated resistance to imipenem in Klebsiella pneumoniae. *Antimicrob Agents Chemother* 48, 15-22.

PART 2: KINETICS, MECHANISMS, INHIBITORS AND INACTIVATION

In: Beta-Lactamases
Editor: Jean-Marie Frère, pp. 139-160
ISBN: 978-1-61324-638-2

Chapter 6

KINETICS OF β-LACTAMASES IN THEORY AND IN PRACTICE

Jean-Marie Frère
Centre for Protein Engineering, Institute of Chemistry, B6, University of Lièg, Sart-Timan, B4000 Liège, Belgium.

ABSTRACT

Beta-lactamases are probably the enzymes for which the largest quantity of kinetic parameters have been accumulated due to the large numbers of both available substrates and enzymes. Unfortunately, many of these parameters are not very reliable because the experimental conditions were rather poor. For instance, low K_m values are quite difficult to determine directly since the variation of the monitored property (e.g. absorbance or proton release) can become very small.

In this chapter, we summarize the main features of β-lactamase kinetic analysis. If, in practice, the k_{cat} and K_m values are sufficient to correlate β-lactamase activity and MIC values (see chapter 16), the active-site serine and metallo-β-lactamases are analysed separately since they follow distinct kinetic pathways.

1. ACTIVE-SERINE β-LACTAMASES

In the simple cases, the three-step, linear model 1 [1] which involves the formation of an acylenzyme (see chapter 7) applies

$$E + S \rightleftharpoons ES \rightarrow ES^* \rightarrow E + P \qquad \text{(model 1)}$$

where S is the substrate, E the enzyme, ES a non-covalent complex, ES* the acylenzyme and P the reaction product, penicilloic acid in the case of penicillin. k_1, k_2 and k_3 are first-order rate constants while k_{+1} is a second-order one.

1.1. Steady-state Equations

1.1.1. k_{cat} and K_m

The steady-state equations are

$$k_{cat} = k_2k_3/(k_2 + k_3) \quad (1)$$

and $K_m = k_3K'/(k_2 + k_3)$ (2)

where $K' = (k_{-1} + k_2)/k_{+1}$ (3)

The k_{cat}/K_m ratio is equal to k_2/K'. It is easy to deduce that if $k_{-1} >> k_2$, $K' = k_{-1}/k_{+1} = K_S$, the dissociation constant of ES.

β-lactamases are often very active enzymes with k_{cat} values up to 2000 s^{-1} for their best substrates. In a few cases, the k_{cat}/K_m ratio is so high (close to 10^8 M^{-1} s^{-1}) that the reaction rate might become diffusion-limited. In consequence, it can be strongly misleading to consider, as do many authors that there is "no activity" if, under standard conditions, the rate of hydrolysis of a "poor" substrate is below 1% of that of the best one. It has indeed been shown that the MIC of the superproducing 908R strain of *Enterobacter cloacae* for cefotaxime was 660 μM (*vs* 0.5 for a non-producing construct) despite the fact that the k_{cat} value was 2000-fold lower than that of benzylpenicillin and 60 000-fold lower than that of cephaloridine [2].

In practice, the k_{cat} and K_m values for a set of representative substrates supply a full characterisation of the enzyme activity spectrum and sometimes allow to rationalise the MIC values for ß-lactamase producing Gram-negative bacterial strains [3, 4].

The k_{cat} and K_m values can be derived from initial rate measurements or from complete time-courses. With the latter, care must be taken to make sure that no enzyme inactivation and no product inhibition occur. Although product inhibition is rather exceptional, substrate-induced inactivation is observed with many class D enzymes (see below). Product inhibition can easily be detected by directly estimating the effect of product addition on the initial rate or by adding a new aliquot of substrate to the sample after a first hydrolysis reaction is completed. Moreover, in the case of product inhibition, initial rates and complete time-courses yield different values for the kinetic parameters.

The analysis of complete time-courses rests on a careful determination of a stable final value of the measured parameter (most often, absorbance). Any side-reaction which might modify this value thus interferes with the analysis. For example, when the hydrolysis of cephalexin is monitored by UV absorbance measurements, a slow degradation of the product results in a further decrease of the absorbance [5]. In all cases, it is careful to confirm the k_{cat} and K_m values derived from complete time-course analyses by a few initial rate measurements.

When the K_m values are very low, it becomes impossible to determine them directly because, for instance, the absorbance variation becomes too small. A low K_m value can then be estimated as a K_i by using the tested compound as a competitive inhibitor for a good substrate for which the kinetic parameters are known [6] and whose rate of hydrolysis remains constant during the experiment.

In a few cases, inhibition by excess substrate is observed.

1.1.2. Concentration of Acylenzyme at the Steady-state

It is clear that the four rate constants cannot be deduced from the sole values of k_{cat} and K_m and it is not possible to decide which of the three steps is rate-limiting on the basis of steady-state rate measurements. However, determining the acylenzyme concentration at the steady-state allows to derive the k_2/k_3 ratio. Indeed, at the steady-state

$$(ES^*)/E_0 = k_f/(k_3 + k_f) \quad (4)$$

where $k_f = k_2S_0/(K' + S_0)$ (5)

At substrate concentrations much larger than K', the acylenzyme concentration is thus

$$k_2E_0(/k_2 + k_3) \quad (6)$$

Measuring the acylenzyme concentration can most easily be performed with a radioactive substrate by quenching the reaction for instance with strong acid [7] . But even in this case if a good substrate is studied, the reaction proceeds so rapidly that the substrate concentration significantly decreases when manual mixing is followed by manual quenching. A quenched-flow apparatus must thus be used. With penicillins, Waley and his coworkers also devised a quenched-flow method where quantification of the acylenzyme was based on the penamaldate reaction [7]. This allowed them to measure the k_2 and k_3 values of some compounds with several enzymes (unfortunately, the penamaldate test does not work with cephalosporins). In these experiments, it was quite difficult to determine k_2/k_3 ratios higher than 4 because (ES*) was then close to E_0 and experimental errors strongly influenced the result. Mass spectrometric methods allow a more accurate determination of the $(ES^*)/E_0$ ratio since they visualise both ES* and E and maybe ES* + ES under non-denaturing conditions. Similarly, very low k_2/k_3 ratios are difficult to determine accurately. In this case, the acylenzyme does not accumulate, $k_{cat} = k_2$ and k_3 becomes impossible to estimate. However, the reaction is usually quenched by denaturing the enzyme so that the mass spectra allow to quantify (ES*) and (E) + (ES). Mass spectrometry has been used extensively to characterize the adducts formed with inactivators and poor substrates [8-12] but to our knowledge, a combination of quenched flow and mass spectrometry has only been used recently [59] to determine the $(ES^*)/E_0$ ratio with nitrocefin and the OXA-10 class D β-lactamase. A difficulty of the mass spectrometry method is that deacylation might occur during the recording of the spectra.

The conclusion that k_3 is much larger than k_2 must rest on experimental results obtained with (S) ~ K_m or (S) > K_m. Indeed, when K' is very large and even if $k_2 > k_3$, k_f can remain much smaller than k_3 with the practically accessible substrate concentrations. In this case, no detectable intermediate is formed, be it ES or ES*. By contrast, if no acylenzyme is detected when (S) ~ K_m, one can be certain that the only intermediate which accumulates is ES and that $k_2 >> k_3$.

It has been suggested that the K' value of a relatively poor substrate could be determined by utilising it as a competitive inhibitor *versus* a good (reporter) substrate over short periods of time, i.e. before the k_2-step becomes influential [13]. If K' is thus estimated, one can derive the individual k_2 and k_3 values with the help of eq 1 and 2. With manual mixing methods, this is only possible if k_2 is quite low, less than 0.01 s^{-1}. In most of the examples proposed by the authors, this condition was not fulfilled and the measured K' value was in fact the K_m of the poor substrate. When higher k_2 values prevail, the utilisation of stopped-flow equipment becomes necessary.

1.1.3. Influence of Viscosity and Estimation of k_{+1} and k_{-1}

When the value of k_2 is relatively low, (e.g. < 100 s^{-1}) it can safely be assumed that it is smaller than k_{-1} (although exceptions can be encountered) and that K' = K_S and only this latter value can be determined. In this case, measuring k_{-1} and k_{+1} would require that the non-covalent ES complex exhibit specific easily measurable properties, a situation which has never been encountered.

By contrast, if the k_2 and k_{-1} values are similar or if k_2 is larger than k_{-1}, the reaction becomes diffusion-limited. In this case, the viscosity of the solvent influences K_m because k_{+1} and k_{-1} decrease with increasing solvent viscosity, while k_2 and k_3 should not. The first thing to verify is thus that the viscosity does not modify k_{cat}. If this condition is fulfilled, k_{+1} and k_{-1} can be derived. Indeed,

$$K_m/k_{cat} = \eta_r/k_{+1o} + k_{-1o}/k_{+1o}\, k_2 \qquad (7)$$

where η_r is the relative viscosity and k_{+1o} and k_{-1o} the values of the parameters in the absence of viscogen. The value of k_{+1o} can thus be obtained from the slope of the line K_m/k_{cat} *vs* η_r.

This method, together with the estimation of the concentration of acylenzyme at the steady-state, has been successfully applied by Waley and his coworkers to determine all the constants involved in the interaction between benzylpenicillin and several enzymes [14].

Table 1 shows that the *Bacillus cereus* and TEM enzymes closely approach "catalytic perfection".

Table 1. Rate constants for the hydrolysis of benzylpenicillin at 20°C as determined by Christensen et al. [14]

Rate constants	*B. cereus* β-lactamase I	*S. aureus* PC I	TEM-1
k_{+1} ($\mu M^{-1}s^{-1}$)	41 ± 3	22 ± 2	123 ± 10
k_{-1} (s^{-1})	2300 ± 700	200 ± 30	1200 ± 1500
k_2 (s^{-1})	4100 ± 200	173 ± 10	2800 ± 300
k_3 (s^{-1})	3600 ± 200	96 ± 10	1500 ± 200
k_{cat} (s^{-1})	1920 ± 60	62 ± 3	980 ± 100
k_{cat}/K_m ($\mu M^{-1}s^{1}$)	26 ± 4	10.3 ± 1	24 ± 5
K_m (μM)	73 ± 5	6 ± 1	42 ± 5
$K_S = k_{-1}/k_{+1}$(μM)	57 ± 14	9 ± 2	96 ± 9
$K' = \frac{k_{-1} + k_2}{k_{+1}}$ (μM)	160 ± 30	17 ± 3	120 ± 24

The k_2/k_3 ratios were determined on the basis of (ES*) at the steady-state and the $k_{+1}$and k_{-1} values from experiments performed in the presence of glycerol as a viscogen.

1.2. Transient State Kinetics

If the initial concentration of S (S_0) is much larger than that of the enzymes (E_0), the formation of acylenzyme usually follows pseudo-first order kinetics (as long as k_{-1} is $>>k_2$) and is characterised by eq. 8 an 9

$$\frac{\varepsilon - \varepsilon_{ss}}{E_{0} - \varepsilon_{ss}} = e^{-k_a t} \qquad (8)$$

where $\varepsilon = (E) + (ES)$ [i.e. $\varepsilon = E_0 - (ES^*)$], ε_{ss} the value of ε at the steady-state, k_a the pseudo-first order rate constant and t the time of contact.

$$(ES^*)/(ES^*)_{SS} = 1 - e^{-k_a t} \qquad (9)$$

where $(ES^*)_{SS}$ is the concentration of acylenzyme at the steady-state.

The value of k_a depends upon S_0 according to equations 10 and 11

$$k_a = k_3 + k_2 S_0 / (K' + S_0) \qquad (10)$$

or

$$k_a = k_3 + k_f \qquad (11)$$

The curve k_a vs S_0 is a rectangular hyperbola which intercepts the ordinate at $k_a = k_3$ (for $S_0 = 0$) and with a horizontal asymptote $k_a = k_3 + k_2$ (for $S_0 >> K'$).

If S_0 is $<< K'$,

$$k_a = k_3 + k_2 S_0 / K' \qquad (12)$$

from which k_3 can be obtained as the intercept of the line k_a *vs* S_0 with the ordinate. For larger S_0 values, eq 10 applies and k_2 and K are obtained by fitting $k_a - k_3$ *vs* S_0 (eq. 10). Eq. 10 can also be linearized as

$$S_0/(k_a - k_3) = (K' + S_0)/k_2 \qquad (13)$$

A 2-step procedure can thus be utilised to obtain the three constants:

- firstly, plot k_a *vs* S_0 at substrate concentrations where k_a varies linearly with S_0 and deduce k_3 (intercept) and k_2/K' (slope) according to eq. 12
- secondly, measure k_a at higher S_0 concentrations where the dependence is not linear and fit $k_a - k_3$ to $k_2S_0/(K'+S_0)$ or use the linearized form (eq. 13).

It is also possible to directly fit all the data to eq. 5 but this yields a very inaccurate k_3 value if $k_3 << k_2$ and it is usually better to use the two-step procedure described above.

If the excess of substrate is eliminated, the recovery of free enzyme also follows first order kinetics

$$(ES^*)/(ES^*)_0 = e^{-k_3 t} \quad (14)$$

where $(ES^*)_0$ is the concentration of acylenzyme when the excess of S has ben completely eliminated.

With many substrates (cephalosporins, carbapenems), the opening of the β-lactam ring results in a decrease of absorbance (at 260 nm for many cephalosporins and 300 nm for carbapenems). This decrease already takes place when the acylenzyme is formed. The accumulation of acylenzyme and product formation yield "burst" type kinetics (Figure 1) where a rapid decrease of the absorbance is followed by a slower, linear one when the steady-state is reached (assuming that the initial concentration of substrate is sufficiently high). The k_a value can be deduced from the shape of the pre-steady state curve according to eq. 15

$$\frac{v - v_{ss}}{v_0 - v_{ss}} = e^{-k_a t} \quad (15)$$

where v, v_0 and v_{ss} are the rates of absorbance variation at times t, 0 and the steady-state respectively. Extrapolation of the linear steady-state part of the curve to t = 0 defines the size of the "burst" and yields an absorbance variation (π) corresponding to the concentration of enzyme if $k_3 << k_f$ and to lower values in other cases (eq. 16)

$$\frac{\pi}{\Delta\varepsilon} = E_0\left(\frac{k_f}{k_f + k_3}\right)^2 \quad (16)$$

where Δε is the variation of the molar extinction coefficient of the substrate upon opening of the β-lactam ring.

With penicillins and monobactams, the Δε value is much lower and takes place in the far UV (230 nm or below) so that direct measurements become much more difficult and less accurate. Under these conditions, it is easier to monitor the accumulation of acylenzyme with the help of a reporter substrate with which the steady-state is established within the mixing dead-time. The rate of reporter substrate utilisation decreases as the acylenzyme of the tested compound accumulates until the steady-state is reached with both substrates. The k_a value can be determined with the help of eq. 15 where the rates now correspond to the utilisation of the reporter substrate. However, the reporter substrate immobilises some enzyme along its own pathway, so that the observed k_a value is somewhat lower than the real one

$$(k_a)_{obs} = \frac{k_a K_{mR}}{K_{mR} + (R)} \quad (17)$$

where (R) and K_{mR} are respectively the concentration and the K_m of the reporter substrate. The preferred reporter substrates are those whose hydrolysis results in absorbance variations in the visible part of the spectrum, such as nitrocefin or CENTA (Figure 1b). This avoids interferences between the absorbance variations due to the hydrolysis of the tested compound (in the UV) and that of the reporter substrate. Note however that the size of the burst π' no longer obeys equation 16 and can be orders of magnitude larger than E_0.

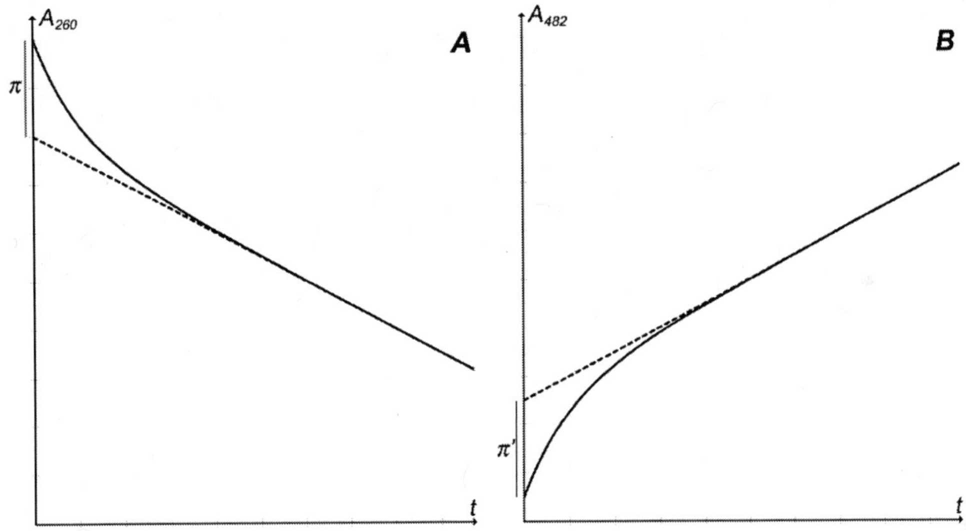

Figure 1. A) Burst-type kinetics resulting from the formation of acylenzyme and product release when a cephalosporin is hydrolysed by a □-lactamase with k2 >> k3. B) Burst-type kinetics in the presence of nitrocefin as a reporter substrate. The value of □' does not obey eq. 16 and is often orders of magnitude larger than E0. The non-zero value of A482 is due to the absorbance of intact nitrocefin.

1.2.1. k_2 Is Smaller than k_3

At substrate saturation, no or little acylenzyme accumulates. In the first case, as mentioned above, k_3 cannot be measured, but $k_{cat} = k_2$ and $K_m = K'$. In the second case, a small proportion of acylenzyme is formed and its accumulation can possibly be monitored. However, when the initial concentration of substrate is decreased, k_f also decreases and less and less acylenzyme is formed, which means that the amplitude of the observed phenomenon also decreases (the size of the burst becomes very small, see eq. 16) and this is also true when a reporter substrate is utilised. In consequence, the measured k_f values can become very inaccurate.

1.2.2. k_2 Is Larger than k_3

At the steady-state, a large proportion of the enzyme accumulates as acylenzyme and the size of the burst is close to the enzyme concentration, which allows accurate measurements of the k_a values.

Some interactions between class C enzymes and poor substrates supply very good examples of low k_3 values [6, 15]. The *Enterobacter cloacae* P99 and 908R enzymes exhibit

k_3 values in the range of 0.2-10*10^{-3} s^{-1} with sustrates such as aztreonam, imipenem, oxacillin and carbenicillin. The k_2 values are all above 150 s^{-1} with the sole exception of carbenicillin (3.3 s^{-1}). In all cases, the k_2/K' values are quite high (15 to 30000 $mM^{-1}s^{-1}$) and with all compounds but carbenicillin, the acylation reaction is so fast at high concentrations that the individual k_2 and K' values cannot be determined even by stopped-flow methods since the k_a value remains proportional to S_0 (eq. 12) throughout the range of accessible substrate concentrations [16].

Since the accumulating acylenzyme is expectedly inactive *vs* good substrates, the reporter substrate method has been successfully utilized to measure the k_2/K' values of several poor substrates of class C β-lactamases using nitrocefin or cephaloridine as reporter substrates. An interesting additional information resulting from these measurements is due to the fact that k_2/K' = k_{cat}/K_m. For the poor substrates, k_{cat} was measured as V/E_0 directly at substrate saturation (which is easy since the K_m values are in the range of 1 nM – 1 µM) and K_m was measured as a K_i in a simple competition experiment with a good substrate. The independent measurement of k_2/K' thus allows to verify the k_{cat}/K_m value and supplies a rough estimate of the purity of the enzyme since knowledge of the enzyme concentration is not needed to determine k_2/K' while it directly reflects on k_{cat}/K_m (since k_{cat} = V/E_0) [6, 15].

1.2.3. Uncertainty

The hydrolysis of ceftazidime by the class A TEM-1 enzyme supplies an example where the relative values of k_2 and k_3 cannot even be approximated. The k_{cat} and K_m values are about 0.3 s^{-1} and 4 mM respectively and the second order acylation rate constant, k_2/K' is thus 70 $M^{-1}s^{-1}$. However, K' is higher than the largest substrate concentrations which can be practically achieved so that no acylenzyme has ever been detected ($k_f = k_z/K' << k_3$). It is thus not possible to tell whether k_{cat} represents k_2, k_3 or a combination of both [17]. As explained above, the fact that no acylenzyme can be detected does not mean that k_3 is larger than k_2. But it can be safely concluded that within the range of physiologically significant ceftazidime concentrations, acylation remains rate-limiting and no acylenzyme accumulates. In contrast, the values of the three constants have been determined for the interaction between the same enzyme and cefoxitin [18]. Both k_2 (10*10^{-3} s^{-1}) and k_3 (2*10^{-3} s^{-1}) are low and K' is high (10mM) so that the acylenzyme can be visualised, but only at high cefoxitin concentrations (> 5mM). Below 1 mM cefoxitin, acylation thus becomes rate-limiting since k_2(S)/K' < $10^{-3}s^{-1}$ and k_3 = 2*10^{-3} s^{-1}.

One can be tempted to conclude that, again within the range of physiologically meaningful concentrations, poor substrates are characterised by a slow deacylation rate with class C and a slow acylation rate with class A, but this generalisation should be considered with a lot of care.

1.2.4. Value of k_3 by the Reporter Substrate Method

When k_2 is >> k_3, k_3 is obtained as the k_{cat} value. It can also be determined with the help of a reporter substrate. A sample of enzyme is incubated with a concentration of poor substrate such that the activity *vs* the reporter substrate is close to zero. The sample is then diluted so that k_f becomes negligible when compared to k_3 and the reporter substrate is added. The disappearance of ES* results in an increase of the hydrolysis rate of the reporter substrate. When (ES*) = 0, the hydrolysis of the reporter substrate becomes linear and k_3 can be derived from equation 18

$$\frac{v_f - v}{v_f - v_0} = e^{-k_3 t} \qquad (18)$$

where v_0, v and v_f are respectively the hydrolysis rates of the reporter substrate at time zero (i.e. just after dilution), t and after complete reactivation. Care must be taken to utilise a sufficient concentration of reporter substrate so that its own hydrolysis does not influence v_f.

1.3. Activity *vs* Linear Substrates

ß-lactamases hydrolyse linear esters, thiolesters and even amides (Figure 2) with highly variable efficiencies [19, 20].

I $C_6H_5 - CO - NH - CH(CH_3) - CO - O - CH(C_6H_5) - COO^-$

II $C_6H_5 - CO - NH - CH(CH_2 - C_6H_5) - CO - S - CH_2 - COO^-$

III $C_6H_5 - CO - NH - CH(CH_3) - CO - S - CH_2 - COO^-$

IV $C_6H_5 - CO - NH - CH_2 - CO - NH - CH(CH_3) - COO^-$

Figure 2. Structures of some linear substrates. The asymmetric carbons are usually D but, surprisingly, when the central amino acid is Phe (compound II) some beta-lactamases prefer the L isomer whereas the D isomer of Ala is preferred for compound III [55]. The hydrolysis of the peptide IV is orders of magnitude slower than that of the esters I and thiolesters II et III [20].

Some esters and thiolesters are hydrolysed relatively well, mainly by the class C enzymes. Peptides are very poor substrates, but in a few cases the enzyme rate enhancement factor, as defined by Page (see chapter 21) is far from negligible, due to the high intrinsic stability of peptides [20].

1.4. Branched Pathways

1.4.1. In the Presence of an Acceptor

The class C β-lactamases can transfer the penicilloyl moiety of penicillins onto methanol. When the acylenzyme accumulates, addition of methanol (the acceptor) increases both the k_{cat} and the K_m values by identical factors so that the k_{cat}/K_m ratio remains unchanged [21, 22]. When no acylenzyme accumulates, no increase of the reaction rate is observed. In both cases, the acylenzyme partitions between the hydrolysis and transfer pathways and the ratio between the transfer and hydrolysis products increases with the acceptor concentration. In a simple branched pathway where the two competing reactions are

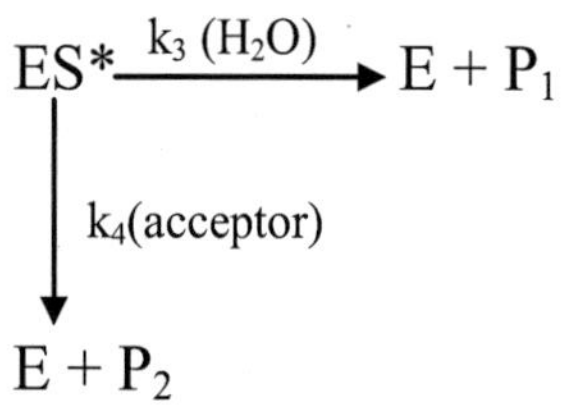

the P_2/P_1 ratio is proportional to the acceptor concentration [23]. With the class C enzymes and ester or thiolester substrates, transfer reactions onto a suitable acceptor have also been observed [24]. In most cases, the rate of substrate disappearance is not increased by the acceptor, since no acylenzyme accumulates. However, with the *Enterobacter cloacae* P99 enzyme, the acylenzyme accumulates in the presence of some linear substrates and in consequence, addition of the acceptor increases the reaction rate [25].

1.4.2. Cephalosporins: Expulsion of the C_3' Leaving Group

Many cephalosporins contain a good leaving group (for instance O-CO-CH_3, O-CO-NH_2) on C_3', i.e. the C atom on C_3 of the dihydrothiazine cycle. When the fused ß-lactam ring is opened, this group is expelled at a rate characteristic of the exact structure of the cephalosporin molecule [26, 27]. This expulsion occurs at the levels of both the product and the acylenzyme (and the expulsion rate also depends upon the studied enzyme) leading to the formation of an exomethylene group (Figure 3). The rearranged acylenzyme ES** undergoes hydrolysis more slowly than ES* and this decreases the k_{cat} value when compared to the simple linear pathway.

The reaction pathway is described by the following scheme where LG is the leaving group.

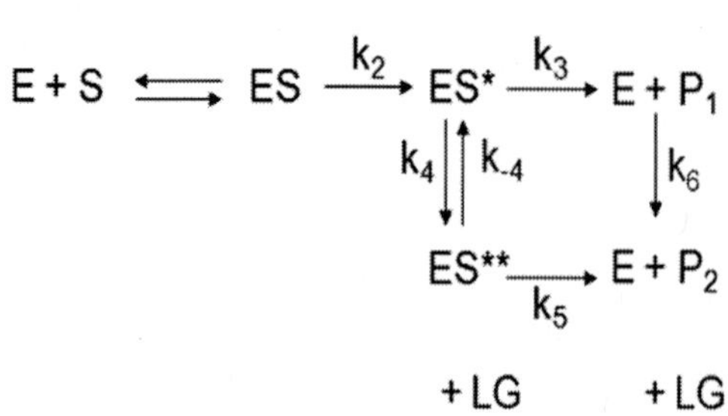

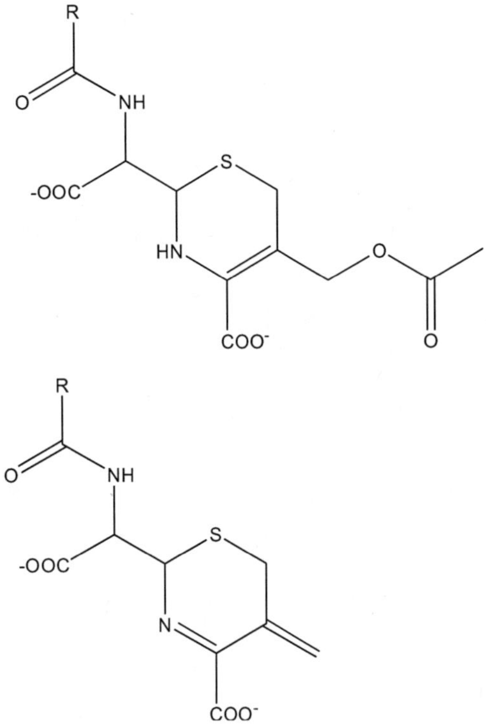

Figure 3. Structures of an opened cephalosporin before and after the expulsion of the C'$_3$ leaving group.

It is sometimes possible to determine the k_{cat} and K_m values of the first phase, before accumulation of ES** becomes significant. These kinetic parameters can then be compared to those determined at the steady-state. An analysis of the model shows that at the steady-state, the k_{cat} and K_m values are decreased by the same factor so that k_{cat}/K_m is not affected and remains equal to k_2/K'.

With β-lactamases exhibiting very high k_3 values, the cephalosporin can be completely hydrolysed before ES** significantly accumulates and the kinetic parameters determined under both initial rate or complete time-course conditions are identical. However, when complete time-courses are recorded with poorer substrates or with very low enzyme concentrations, a fraction of acylenzyme equal to $k_4/(k_3 + k_4)$ undergoes rearrangement at each turn-over and this induces a progressive decrease of the reaction rate characterised as "substrate-induced inactivation". If the reaction is not completed before a significant amount of ES** has accumulated, the k_{cat} and K_m values thus obtained will be lower than those derived from initial rate experiments, while the k_{cat}/K_m ratio remains the same. Using a constant initial substrate concentration, k_{cat} and K_m determined on the basis of complete time-courses decrease progressively with decreasing total enzyme concentrations. When possible, the occurrence of this phenomenon can also be demonstrated by using an initial substrate concentration well above the K_m value (determined on the basis of initial rates) and showing that the time-course is not zero-order but that the rate decreases with time faster than predicted on the basis of substrate depletion. This yields hydrolysis curves similar to that shown in figure 1 but π does not obey eq. 16. $\pi/\Delta\varepsilon$ can be many times the enzyme concentration and increases with the k_3/k_4 ratio. The interactions between β-lactamases and cephalosporins have been analysed in depth by Pratt and Faraci [26, 27]. Note that when k_2 is

<< k_3, neither ES* nor ES** accumulate and the kinetic behaviour cannot be distinguished from that of the simple linear pathway.

1.4.3. Simple Suicide Substrates of ß-lactamases

Suicide substrates of β-lactamases (ß-halogenopenicillanates, penicillanic acid sulphones, clavulanic acid) also rearrange at the acylenzyme level [28, 29]. Clavulanic acid and ß-lactamases often react according to very complex pathways which are briefly examined below. The situation with the other compounds is simpler. The acylenzyme irreversibly rearranges into a stable ES** adduct.

$$
\begin{array}{ccccccc}
E + S & \rightleftharpoons & ES & \rightarrow & ES^* & \xrightarrow{k_3} & E + P \\
 & & & & k_4 \downarrow & & \\
 & & & & ES^{**} & &
\end{array}
$$

The final enzyme activity depends upon the S_0/E_0 ratio. If this ratio is > than or equal to $(k_3 + k_4)/k_4$, the enzyme is completely inactivated. When S_0/E_0 is smaller than $(k_3 + k_4)/k_4$, complete hydrolysis of the suicide substrate occurs before the enzyme is totally inactivated so that, at the end of the reaction, a residual activity is observed.

$$(E)/E_0 = 1 - k_4S_0/(k_3 + k_4)\,E_0 \qquad (19)$$

The enzyme can thus be "titrated" with the substrate and the k_3/k_4 ratio derived from such a titration. Clearly, the simplest situation prevails when $k_3 < k_4$, where the enzyme is completely inactivated for $S_0 \approx E_0$. Detailed studies of various examples have been performed by Knowles, Waley, Frère and their co-workers [29-31].

1.4.4. Clavulanic Acid

At least two rearranged acylenzymes are formed upon the interaction between clavulanic acid and the TEM β-lactamase first studied in depth by Knowles and his co-workers [29, 32]. It involves 3 different adducts: the acylenzyme ES*, a second adduct ES** whose rearrangement is slowly reversible and a third adduct, ES_i which is irreversibly modified.

$$
\begin{array}{ccccccc}
 & & & & ES^{**} & & \\
 & & & & \upharpoonleft\downharpoonright & & \\
E + S & \rightleftharpoons & ES & \rightarrow & ES^* & \rightarrow & E + P \\
 & & & & \downarrow & & \\
 & & & & ES_i \text{ [irreversibly rearranged adduct]} & &
\end{array}
$$

The characteristic of this 3-branch pathway is that, at relatively low S_0/E_0 ratios (< 200 in the case of the TEM-clavulanic acid interaction), a transient inactivation is observed. When all the substrate is hydrolysed, a part of the activity is recovered due to the relatively slow reversal of the ES* $\rightleftharpoons$ ES** step. Note that a slow decay of ES**, regenerating the free enzyme is also possible and cannot be kinetically distinguished from the simple reversal of ES* ⇌ ES**. If the S_0/E_0 ratio is sufficient, inactivation is complete (all the enzyme is immobilised as ES_i) and no recovery occurs. Brown, Aplin and Schofield [9] have performed a very elegant analysis of this interaction by mass spectrometry and proposed an even more complex pathway where one of the final complexes involves the bridging between Ser 130 and the active Ser70.

Not all β-lactamases react with clavulanic acid according to such a complex scheme. Although the 3-branch pathway appears to be valid for the *Actinomadura* R39 class A enzyme, that for the inactivation of the *Streptomyces albus* G β-lactamase, another class A enzyme, seems to obey the more simple scheme described in section 1.4.3. Class C enzymes also appear to follow this more simple scheme, but the efficieny of the inactivation is very low, due to a low k_2/K' value [34]. In consequence, clavulanic acid is of no clinical utility against strains producing the class C enzymes.

1.5. Inactivation due to an Enzyme Conformation Change

1.5.1. Class A Enzymes

Citri and his coworkers [35] showed that the hydrolysis of so-called type A penicillin substrates (methicillin, oxacillin and cloxacillin) by the *Bacillus cereus* I (BcI) class A β-lactamase was accompanied by a reversible substrate-induced inactivation. It was later noticed that k_{cat} and K_m decreased by similar factors so that the k_{cat}/K_m ratio was not affected [36]. This could again be explained by a rearrangement at the level of the acylenzyme. In contrast to what has been described above for cephalosporins, a chemical rearrangement of the penicilloyl moiety can be safely excluded and many results suggest a conformation change of the enzyme itself, induced by the large, sterically hindered side-chain of the substrates. However, from a kinetic point of view, the equations are the same as those derived for the situation described in section 1.4.2 with $k_5 = 0$ if ES** is unable to undergo hydrolysis but, to our knowledge, it was never determined whether the modified acylenzyme retained some catalytic competence or not. With these type A substrates, k_6 is also zero.

1.5.2. Class D β-lactamases

Several class D β-lactamases also exhibit substrate-induced inactivation (or biphasic burst kinetics) with a significant number of substrates. In the case of the OXA-2 enzyme, substrates exhibiting a normal, Henri-Michaelis behaviour are even the exceptions [37]. But in contrast to all the cases discussed above, the k_{cat}/K_m ratio is smaller in the steady-state phase than in the early phase of the reaction. This shows that the phenomenon cannot be explained by an event taking place only at the acylenzyme level. Class D enzymes exhibit two specific properties. Many have a tendency to dimerise [38] and the active site Lys side-chain found three residues after the active Serine can be reversibly carboxylated [39, 40], the carboxylated form being more active than the unmodified one. Both properties can explain the peculiar substrate-induced inactivation if one assumes that binding of the substrate can modify the monomer/dimer equilibrium or the equililibrium between the free and the carboxlayted Lys side-chain. Indeed, in some, but not all cases, the substrate-induced inactivaction disappeared in the presence of saturating concentrations of bicarbonate which were assumed to maintain the Lys in a completely carboxylated form [39]. This suggests that the interaction with the substrate somewhat destabilises the carbamyl group and displaces the equilibrium in favour of a free Lys side-chain, but to explain the variation of the k_{cat}/K_m ratio, one must assume that this destabilisation already occurs at the level of the non-covalent ES complex. Similarly, a monomer/dimer equilibrium should also be perturbed at the level of the ES complex. However, apart from the the k_{cat}/K_m decrease, no indication of a modification at the level of the non covalent ES complex has ever been obtained. Moreover, when the presence of bicarbonate does not eliminate the burst, the phenomenon remains unexplained. Finally, in at least one case (Ledent and Frère, unpublished data), the pseudo-first order inactivation rate constant even decreases when the substrate concentration is increased, a result which is not accounted for by any of the simple substrate-induced inactivation models.

2. The Class B Metallo-Beta-Lactamases

These enzymes (see also chapter 3) contain one or two Zn^{++} ions in their active site. They do not form covalent adducts with their substrates, at least not during a "normal" turn-over. An intermediate exhibiting very special properties has been characterised only with the chromogenic substrate, nitrocefin [41] and some B1 and B3 enzymes. Another unexpected non-covalent intermediate has been highlighted by X-ray crystallography when a large concentration of biapenem was added to crystals of the B2 *Aeromonas hydrophila* enzyme before flash-freezing [42].

2.1. Nitrocefin

With several enzymes, the hydrolysis of nitrocefin appears to follow the mechanism described by the linear model 1. Here however, ES* is a non-covalent complex where the ß-lactam amide bond has been opened, but the nitrogen atom remains to be protonated so that it bears a negative charge [43].

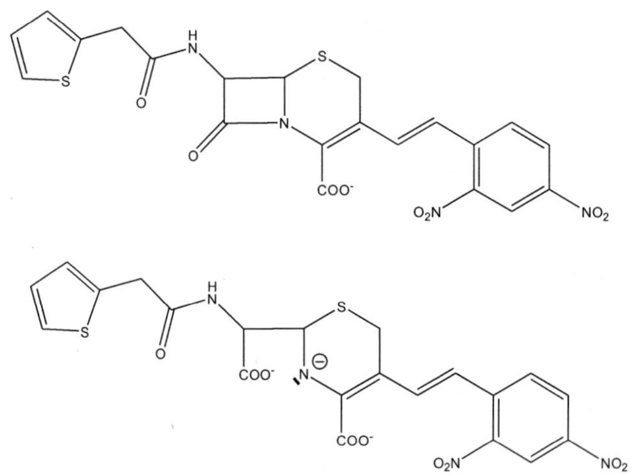

Figure 4. Structures of nitrocefin and of the intermediate which absorbs at 665 nm.

This intermediate has been detected thanks to its spectral properties: it exhibits an absorption maximum at 665 nm. At the steady-state and at substrate saturation, it accumulates in very different proportions according to the enzyme [43-45]: from nearly 100 % with the *Stenotrophomonas maltophilia* enzyme to less than 3% with that from *Bacillus cereus* (BcII). It is probably stabilised by the possibility of electron delocalisation over the unique C3 dinitrostyryl side-chain of nitrocefin (Figure 4).

2.2. Other Substrates

The other substrates do not contain a side-chain which might stabilise a negative charge on the leaving nitrogen atom upon opening of the ß-lactam amide bond. In consequence, it seems likely that protonation of the nitrogen is concommitant with the opening of the C-N bond. No intermediate but the classical ES complex has thus been detected with all other substrates, including the most clinically relevant carbapenems so that the kinetic data can be analysed on the basis of the simplest pathway, i.e. $E + S \rightleftharpoons ES \rightarrow E + P$ (with the exception of B2 enzymes, see the next paragraph)

2.3. The *Aeromonas hydrophila* CphA Enzyme and Biapenem

In contrast to nearly all the other β-lactamases, B2 enzymes exhibit a very narrow activity spectrum and only hydrolyse carbapenems at a significant rate. Indeed, the few class A carbapenemases and the B1 and B3 metallo-β-lactamases hydrolyse a vast number of β-lactams in addition to the carbapenems. The specificity of the B2 enzymes is well explained by the shape of the active site which is very well defined (see chapter 3).

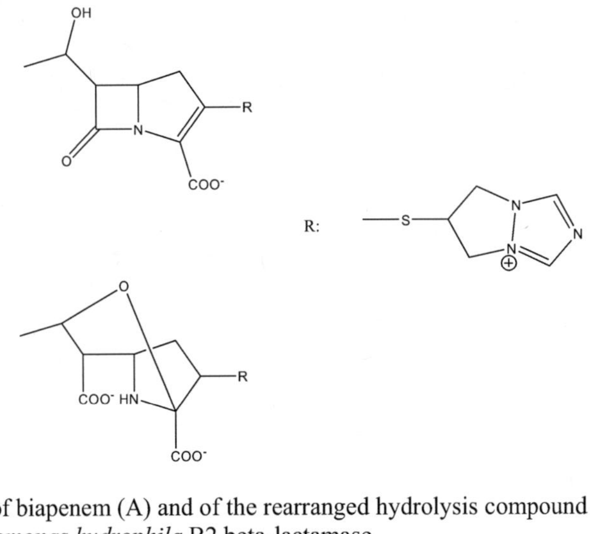

Figure 5. Structures of biapenem (A) and of the rearranged hydrolysis compound (B) observed in the active site of the *Aeromonas hydrophila* B2 beta-lactamase.

By soaking crystals of the *Aeromonas hydrophila* enzyme in a large concentration of biapenem, Garau et al [42] obtained a complex in which the β-lactam amide bond was hydrolysed but where a new 6-membered ring was formed by an internal molecular rearrangement resulting in a direct bonding of C3 to the oxygen atom of the hydroxyethyl group on C6 (Figure 5). Such a molecule is very unlikely to exist in solution and is stabilised by specific interactions with the enzyme. Inhibition by the product is also observed with a K_i of about 300 μM and this might be due to the formation of the same intermediate. It remains to be firmly confirmed that this complex is an on-pathway, catalytically relevant intermediate and not a dead-end EP complex.

2.4. One or Two Zn^{++} Ions? (See also Chapter 3)

Although it is clear that subclass B1 enzymes exhibit an optimal activity in the di-Zn^{++} form, the level of residual activity of the mono-Zn^{++} species has not been clearly established and might change according to the enzyme. With the BcII enzyme the binding of the second Zn^{++} ion was first found to be accompanied by a marginal increase of activity [46, 47] but recent results indicate a cooperative binding of the two ions [48]. By contrast, binding of a second Zn^{++} ion to the mono-Zn^{++} form of VIM-4 increases the activity between 2 and 10-fold, depending on the substrate, with the exception of Biapenem for which no modification is observed [60]. It has not been possible to prepare the mono-Zn^{++} form of IMP-1 without first denaturing the protein [49]. Interestingly, after removal of the two Zn^{++} ions from the active site of the enzyme, the IMP-1 apoenzyme becomes unable, upon refolding, to bind more than one Zn^{++} and the kinetic properties of the new "mono" form are quite different from those of the native "di-Zn^{++}" enzyme [49]. The structural basis of this phenomenon remains to be explained.

The subclass B2 enzyme from *Aeromonas hydrophila* is optimally active with one Zn^{++} [50] and is even inhibited when binding a second Zn^{++} ion. Finally, the class B3 enzymes tightly bind two Zn^{++} ions per monomer. The *Stenotrophomonas maltophilia* enzyme is a tetramer [51], which exhibits weak negative cooperativity with Nitrocefin [52]. To our knowledge, no mono-Zn^{++} form of a class B3 enzyme has ever been obtained.

Wommer et al [53] have obtained results indicating an increase of affinity for the first Zn^{++} ion in the presence of the substrate and suggested that the apo form might prevail in the absence of substrates and that formation of the active, mononuclear form would occur in the presence of the latter at picomolar concentrations of free Zn^{++}. Heinz and Adolph [54] further discussed these results in the light of the possibly variable metal ion content under physiological conditions. This underlines the fact that the situation might be different when the β-lactamase producing bacteria are grown under laboratory conditions where free Zn^{++} is usually easily available or in the bloodstream or infectious foci where a strong competition is expected with other Zn^{++}-binding molecules.

3. A Summary of Kinetic Parameters

Class A β-lactamases were first considered as penicillinases, class C as cephalosporinases, class D as oxacillinases and class B as broad spectrum enzymes which could hydrolyse everything but monobactams. Tables 2 and 3 show that the situation is not so simple. Class C enzymes hydrolyse several cephalosporins with high k_{cat} values but they also hydrolyse a limited number of penicillins with lower, but still significant k_{cat} values. With the same enzyme, second and third generation cephalosporins usually exhibit very low k_{cat} values, but in many clinical isolates, this is compensated by the enormous quantities of enzymes which are produced.

Table 2. Some examples of k_{cat} values (s^{-1})

		BP	Ampi	Oxa	CR	CTX	CTZ	IMI
A	*S. aureus* PC1	130	240	> 7	0.07	ND	ND	ND
	B. licheniformis	2200	1500	10	630	7	15	ND
	TEM-1	1600	1050	150	1500	9	0.3	0.04
	TEM-7	40	18	10	26	1.5	9	ND
	Sme-1	28	220	ND	1800	ND	ND	340
	TEM-19	62	36	5.3	ND	66	26	ND
B	*B. fragilis* CfiA	190	190	ND	42	98	ND	200
	S. maltophiliae ULA511	280	175	285	28	66	ND	65
C	*E. cloacae* P 99	15	0.7	0.006	700	0.015	0.012	0.003
	P. aeruginosa 18 SH	76	4.4	0.04	110	0.15	ND	0.03
D	OXA-10 (PSE-2)	120	220	300	70	ND	ND	ND

BP = benzylpenicillin, Ampi = ampicillin, Oxa = oxacillin, CR = cephaloridin, CTX = cefotaxime, CTZ = ceftazidime, IMI = imipenem. The values for OXA-10 were obtained in the presence of 50 mM $NaHCO_3$ [56]. Other results are from ref. 57.

ND: Not determined.

Table 3. Some examples of k_{cat}/K_m values (mM^{-1} s^{-1})

		BP	Ampi	Oxa	CR	CTX	CTZ	IMI
A	*S. aureus*	40000	20000	1.3	100	8	ND	ND
	B. licheniformis	29000	11000	1200	5000	2500	33	ND
	TEM-1	84000	33000	20000	2200	1.5	0.07	1.5
	TEM-7	13000	1100	9500	300	15	9	ND
	Sme-1	1400	730	ND	1500	5000	ND	1600
	TEM-19	1000	6100	5300	ND	230	5	ND
B	*B. fragilis* CfiA	4800	1600	ND	7000	3600	ND	740
	S. maltophiliae ULA511	1400	4400	11000	93	2600	ND	730
C	*E. cloacae* P 99	23000	1800	7000	10000	1500	2.5	60
	P. aeruginosa 18 SH	45000	9000	1600	5000	750	ND	88
D	OXA-10 (PSE-2)	6000	6200	3000	180	ND	ND	ND

BP = benzylpenicillin, Ampi = ampicillin, Oxa = oxacillin, CR = cephaloridin, CTX = cefotaxime, CTZ = ceftazidime, IMI = imipenem. The values for OXA-10 were obtained in the presence of 50 mM $NaHCO_3$ [56]. Other results are from ref. 57.

ND: Not determined.

The fact that class A enzymes were first considered as penicillinases mainly rests on the properties of the *Staphylococcus aureus* PCI enzyme which did only act on the older penicillins, benzypenicillin, penicillin V and ampicillin. However, as already mentioned by Matagne et al [5], nearly all other class A enzymes exhibit much broader activity spectra, including many early cephalosporins. This diversity was later highlighted by the emergence of TEM mutants (chapters 4 and 14] which could efficiently hydrolyse second and third generation cephalosporins (but, interestingly, not carbapenems) and of the class A carbapenemases.

If class D enzymes do indeed show an elevated activity *vs* oxacillin and other isoxazolylpenicillins, they significantly hydrolyse a wide variety of β-lactams and, in this family, extended-spectrum mutants have also been selected. Finally, and as mentioned above, whereas B1 and B3 metallo-β-lactamases hydrolyse a large number of compounds, large variations in the activity spectra are observed and carbapenems are not systematically the best substrates. In turn, the B2 enzymes exhibit the most specific properties with an activity restricted to carbapenems. It is however interesting to note that, by introduction of only two mutations, the activity spectrum of the CphA enzyme was significantly broadened, even if the activity *vs* imipenem was decreased about 30 fold [58].

REFERENCES

[1] Frère, J.M. (1995) Beta-lactamases and bacterial resistance to antibiotics. *Mol Microbiol* 16, 385-95.

[2] Lakaye, B., Dubus, A., Lepage, S., Groslambert, S., Frère, J.M. (1999) When drug inactivation renders the target irrelevant to antibiotic resistance: a case story with beta-lactams. *Mol Microbiol* 31, 89-101.

[3] Frère, J.M. (1989) Quantitative relationship between sensitivity to beta-lactam antibiotics and beta-lactamase production in gram-negative bacteria--I. Steady-state treatment. *Biochem Pharmacol* 38, 1415-26.

[4] Nikaido, H., Normark, S. (1987) Sensitivity of *Escherichia coli* to various beta-lactams is determined by the interplay of outer membrane permeability and degradation by periplasmic beta-lactamases: a quantitative predictive treatment. *Mol Microbiol* 1, 29-36.

[5] Matagne, A., Misselyn-Bauduin, A.M., Joris, B., Erpicum, T., Granier, B., Frère, J.M. (1990) The diversity of the catalytic properties of class A beta-lactamases. *Biochem J* 265, 131-46.

[6] Galleni, M., Amicosante, G., Frère, J.M. (1988) A survey of the kinetic parameters of class C beta-lactamases. Cephalosporins and other beta-lactam compounds. *Biochem J* 255, 123-9.

[7] Martin, M.T., Waley, S.G. (1988) Kinetic characterization of the acyl-enzyme mechanism for beta-lactamase I. *Biochem J* 254, 923-5.

[8] Aplin, R.T., Baldwin, J.E., Schofield, C.J., Waley, S.G. (1990) Use of electrospray mass spectrometry to directly observe an acyl enzyme intermediate in beta-lactamase catalysis. *FEBS Lett* 277, 212-4.

[9] Brown, R.P., Aplin, R.T., Schofield, C.J. (1996) Inhibition of TEM-2 beta-lactamase from *Escherichia coli* by clavulanic acid: observation of intermediates by electrospray ionization mass spectrometry. *Biochemistry* 35, 12421-32.

[10] Pagan-Rodriguez, D., Zhou, X., Simmons, R., Bethel, C.R., Hujer, A.M., Helfand, M.S., Jin, Z., Guo, B., Anderson, V.E., Ng, L.M., Bonomo, R.A. (2004) Tazobactam inactivation of SHV-1 and the inhibitor-resistant Ser130 -->Gly SHV-1 beta-lactamase: insights into the mechanism of inhibition. *J Biol Chem* 279, 19494-501.

[11] Drawz, S.M., Bethel, C.R., Doppalapudi, V.R., Sheri, A., Pagadala, S.R., Hujer, A.M., Skalweit, M.J., Anderson, V.E., Chen, S.G., Buynak, J.D., Bonomo, R.A. (2010) Penicillin sulfone inhibitors of class D beta-lactamases. *Antimicrob Agents Chemother* 54, 1414-24.

[12] Bethel, C.R., Distler, A.M., Ruszczycky, M.W., Carey, M.P., Carey, P.R., Hujer, A.M., Taracila, M., Helfand, M.S., Thomson, J.M., Kalp, M., Anderson, V.E., Leonard, D.A., Hujer, K.M., Abe, T., Venkatesan, A.M., Mansour, T.S., Bonomo, R.A. (2008) Inhibition of OXA-1 beta-lactamase by penems. *Antimicrob Agents Chemother* 52, 3135-43.

[13] Vakulenko, S.B., Taibi-Tronche, P., Toth, M., Massova, I., Lerner, S.A., Mobashery, S. (1999) Effects on substrate profile by mutational substitutions at positions 164 and 179 of the class A TEM(pUC19) beta-lactamase from Escherichia coli. *J Biol Chem 274*, 23052-60.

[14] Christensen, H., Martin, M.T., Waley, S.G. (1990) Beta-lactamases as fully efficient enzymes. Determination of all the rate constants in the acyl-enzyme mechanism. *Biochem J* 266, 853-61.

[15] Galleni, M., Frère, J.M. (1988) A survey of the kinetic parameters of class C beta-lactamases. Penicillins. *Biochem J* 255, 119-22.

[16] Monnaie, D., Virden, R., Frère, J.M. (1992) A rapid-kinetic study of the class C beta-lactamase of *Enterobacter cloacae* 908R. *FEBS Lett* 306, 108-12.

[17] Raquet, X., Lamotte-Brasseur, J., Fonze, E., Goussard, S., Courvalin, P., Frère, J.M. (1994) TEM beta-lactamase mutants hydrolysing third-generation cephalosporins. A kinetic and molecular modelling analysis. *J Mol Biol* 244, 625-39.

[18] Fisher, J., Belasco, J.G., Khosla, S., Knowles, J.R. (1980) beta-Lactamase proceeds via an acyl-enzyme intermediate. Interaction of the *Escherichia coli* RTEM enzyme with cefoxitin. *Biochemistry* 19, 2895-901.

[19] Pratt, R.F., Govardhan, C.P. (1984) beta-Lactamase-catalyzed hydrolysis of acyclic depsipeptides and acyl transfer to specific amino acid acceptors. Proc *Natl Acad Sci U* S A 81, 1302-6.

[20] Rhazi, N., Galleni, M., Page, M.I., Frère, J.M. (1999) Peptidase activity of beta-lactamases. *Biochem J* 341 (Pt 2), 409-13.

[21] Knott-Hunziker, V., Petursson, S., Waley, S.G., Jaurin, B., Grundstrom, T. (1982) The acyl-enzyme mechanism of beta-lactamase action. The evidence for class C Beta-lactamases. *Biochem J* 207, 315-22.

[22] Jamin, M., Adam, M., Damblon, C., Christiaens, L., Frère, J.M. (1991) Accumulation of acyl-enzyme in DD-peptidase-catalysed reactions with analogues of peptide substrates. *Biochem J* 280 (Pt 2), 499-506.

[23] Frère, J.M. (1973) Enzymic mechanisms involving concomitant transfer and hydrolysis reactions. *Biochem J* 135, 469-81.

[24] Pazhanisamy, S., Govardhan, C.P., Pratt, R.F. (1989) Beta-lactamase-catalyzed aminolysis of depsipeptides: amine specificity and steady-state kinetics. *Biochemistry* 28, 6863-70.

[25] Xu, Y., Soto, G., Hirsch, K.R., Pratt, R.F. (1996) Kinetics and mechanism of the hydrolysis of depsipeptides catalyzed by the beta-lactamase of *Enterobacter cloacae* P99. *Biochemistry* 35, 3595-603.
[26] Faraci, W.S., Pratt, R.F. (1986) Mechanism of inhibition of RTEM-2 beta-lactamase by cephamycins: relative importance of the 7 alpha-methoxy group and the 3' leaving group. *Biochemistry* 25, 2934-41.
[27] Faraci, W.S., Pratt, R.F. (1985) Mechanism of inhibition of the PC1 beta-lactamase of *Staphylococcus aureus* by cephalosporins: importance of the 3'-leaving group. *Biochemistry* 24, 903-10.
[28] Cohen, S. A., Pratt, R.F. (1980) Inactivation of *Bacillus cereus* beta-lactamase I by 6 beta-bromopencillanic acid: mechanism. *Biochemistry* 19, 3996-4003.
[29] Fisher, J., Charnas, R.L., Knowles, J.R. (1978) Kinetic studies on the inactivation of *Escherichia coli* RTEM beta-lactamase by clavulanic acid. *Biochemistry* 17, 2180-4.
[30] Knott-Hunziker, V., Orlek, B.S., Sammes, P.G., Waley, S.G. (1979) 6 beta-Bromopenicillanic acid inactivates beta-lactamase I. *Biochem J* 177, 365-7.
[31] Frère, J.M., Dormans, C., Duyckaerts, C., De Graeve, J. (1982) Interaction of beta-iodopenicillanate with the beta-lactamases of *Streptomyces albus* G and *Actinomadura* R39. *Biochem J* 207, 437-44.
[32] Charnas, R.L., Fisher, J., Knowles, J.R. (1978) Chemical studies on the inactivation of *Escherichia coli* RTEM beta-lactamase by clavulanic acid. *Biochemistry* 17, 2185-9.
[33] Frère, J.M., Dormans, C., Lenzini, V.M., Duyckaerts, C. (1982) Interaction of clavulanate with the beta-lactamases of *Streptomyces albus* G and *Actinomadura* R39. *Biochem J* 207, 429-36.
[34] Monnaie, D., Frère, J.M. (1993) Interaction of clavulanate with class C beta-lactamases. *FEBS Lett* 334, 269-71.
[35] Citri, N., Samuni, A., Zyk, N. (1976) Acquisition of substrate-specific parameters during the catalytic reaction of penicillinase. *Proc Natl Acad Sci U S A* 73, 1048-52.
[36] Frère, J.M. (1981) Interaction between serine beta-lactamases and class A substrates: a kinetic analysis and a reaction pathway hypothesis. *Biochem Pharmacol* 30, 549-52.
[37] Ledent, P., Raquet, X., Joris, B., Van Beeumen, J., Frère, J.M. (1993) A comparative study of class-D beta-lactamases. *Biochem J* 292 (Pt 2), 555-62.
[38] Danel, F., Paetzel, M., Strynadka, N.C., Page, M.G. (2001) Effect of divalent metal cations on the dimerization of OXA-10 and -14 class D beta-lactamases from *Pseudomonas aeruginosa*. *Biochemistry* 40, 9412-20.
[39] Golemi, D., Maveyraud, L., Vakulenko, S., Samama J.P., Mobashery, S. (2001) Critical involvement of a carbamylated lysine in catalytic function of class D beta-lactamases. *Proc Natl Acad Sci U S* A 98, 14280-5.
[40] Paetzel, M., Danel, F., de Castro, L., Mosimann, S.C., Page, M.G., Strynadka, N.C. (2000) Crystal structure of the class D beta-lactamase OXA-10. *Nat Struct Biol* 7, 918-25.
[41] Herzberg, O., Fitzgerald, P.M.D. (2004) *Metallo beta-lactamases,* vol. 3. John Wiley & Sons, Chichester.
[42] Garau, G., Bebrone, C., Anne, C., Galleni, M., Frère, J.M., Dideberg, O. (2005) A metallo-β-lactamase enzyme in action: crystal structure of the monozinc carbapenemase CphA and its complex with biapenem. *J Mol Biol* 345, 785-95.

[43] Wang, Z., Fast, W., Benkovic, S.J. (1999) On the mechanism of the metallo-beta-lactamase from *Bacteroides fragilis*. *Biochemistry* 38, 10013-23.

[44] Fast, W., Wang, Z., Benkovic, S.J. (2001) Familial mutations and zinc stoichiometry determine the rate-limiting step of nitrocefin hydrolysis by metallo-beta-lactamase from *Bacteroides fragilis*. *Biochemistry* 40, 1640-50.

[45] Moali, C., Anne, C., Lamotte-Brasseur, J., Groslambert, S., Devreese, B., Van Beeumen, J., Galleni, M., Frère, J.M. (2003) Analysis of the importance of the metallo-beta-lactamase active site loop in substrate binding and catalysis. *Chem Biol* 10, 319-29.

[46] Paul-Soto, R., Bauer, R., Frère, J.M., Galleni, M., Meyer-Klaucke, W., Nolting, H., Rossolini, G.M., de Seny, D., Hernandez-Valladares, M., Zeppezauer, M., Adolph, H.W. (1999) Mono- and binuclear Zn^{2+}-β-lactamase. Role of the conserved cysteine in the catalytic mechanism. *J Biol Chem* 274, 13242-9.

[47] de Seny, D., Heinz, U., Wommer, S., Kiefer, M., Meyer-Klaucke, W., Galleni, M., Frère, J.M., Bauer, R., Adolph, H.W. (2001) Metal ion binding and coordination geometry for wild type and mutants of metallo-β-lactamase from *Bacillus cereus* 569/H/9 (BcII). A combined thermodynamic, kinetic and spectroscopic approach. *J Biol Chem* 276, 45065-78.

[48] Jacquin, O., Balbeur, D., Damblon, C., Marchot, P., De Pauw, E., Roberts, G.C., Frère, J.M., Matagne, A. (2009) Positively cooperative binding of zinc ions to *Bacillus cereus* 569/H/9 beta-lactamase II suggests that the binuclear enzyme is the only relevant form for catalysis. *J Mol Biol,* 392, 1278-91.

[49] Siemann, S., Brewer, D., Clarke, A.J., Dmitrienko, G.I., Lajoie, G., Viswanatha, T. (2002) IMP-1 metallo-beta-lactamase: effect of chelators and assessment of metal requirement by electrospray mass spectrometry. *Biochim Biophys Acta* 1571, 190-200.

[50] Hernandez Valladares, M., Felici, A., Weber, G., Adolph, H.W., Zeppezauer, M., Rossolini, G.M., Amicosante, G., Frère, J.M., Galleni, M. (1997) Zn(II) dependence of the *Aeromonas hydrophila* AE036 metallo-beta-lactamase activity and stability. *Biochemistry* 36, 11534-41.

[51] Spencer, J., Clarke, A.R., Walsh, T.R. (2001) Novel mechanism of hydrolysis of therapeutic beta-lactams by *Stenotrophomonas maltophilia* L1 metallo-beta-lactamase. *J Biol Chem* 276, 33638-44.

[52] Simm, A. M., Higgins, C.S., Carenbauer, A.L., Crowder, M.W., Bateson, J.H., Bennett, P.M., Clarke, A.R., Halford, S.E., Walsh, T.R. (2002) Characterization of monomeric L1 metallo-beta -lactamase and the role of the N-terminal extension in negative cooperativity and antibiotic hydrolysis. *J Biol Chem* 277, 24744-52.

[53] Wommer, S., Rival, S., Heinz, U., Galleni, M., Frère, J.M., Franceschini, N., Amicosante, G., Rasmussen, B., Bauer, R., Adolph, H.W. (2002) Substrate-activated Zinc binding of metallo-β-lactamases. *J Biol Chem* 277, 24142-7.

[54] Heinz, U., Adolf, H.W. (2004) Metallo-beta-lactamases: two binding sites for one catalytic metal ion? *Cell Mol Life Sci* 61, 2827-39.

[55] Damblon, C., Zhao, G.H., Jamin, M., Ledent, P., Dubus, A., Vanhove, M., Raquet, X., Christiaens, L., Frère, J.M. (1995) Breakdown of the stereospecificity of DD-peptidases and β-lactamases with thiolester substrates. *Biochem J* 309, 431-436.

[56] Baurin, S., Vercheval, L., Bouillenne, F, Falzone, C., Brans, A., Jacquamet, L., Ferrer, J.L., Sauvage, E., Dehareng, D., Frère, J.M., Charlier, P., Galleni, M., Kerff, F. (2009).

[57] Critical role of tryptophan 154 for the activity and stability of class D beta-lactamase. *Biochemistry,* 48, 11252-63.

[58] Matagne, A., Dubus, A., Galleni, M., Frère, J.M. (1999) The beta-lactamase cycle: a tale of selective pressure and bacterial ingenuity. *Nat Prod Rep* 16, 1-19.

[59] Bebrone, C., Anne, C., De Vriendt, K., Devreese, B., Rossolini, G.M., Van Beeumen, J., Frère, J.M., Galleni, M. (2005) Dramatic broadening of the substrate profile of the *Aeromonas hydrophila* CphA metallo-beta -lactamase by site-directed mutagenesis. *J Biol Chem,* 280, 28195-202.

[60] Vercheval L., Bauvois C., di Paolo A., Borel F., Ferrer JL., Sauvage E., Matagne A., Frère JM., Charlier P., Galleni M. and Kerff F. (2010) Three factors that modulate the activity of class D beta-lactamases and interfere with the post-translational carboxylation of Lys 70. *Biochem J,* 432, 495-504.

[61] Lassaux P., Traoré DA., Loisel E., favier A., Docquier JD., Sohier JS., Laurent C., Bebrone C., Frère JM., Ferrer JL. and Galleni M. (2011) *Biochemical and structural characterization of the subclass B1 metallo-beta-lactamase VIM-4.* Antimicrob Agents Chemother, in press (epub ahead of print 21149620).

In: Beta-Lactamases
Editor: Jean-Marie Frère, pp. 161-198
ISBN: 978-1-61324-638-2

Chapter 7

THE MECHANISMS OF CATALYSIS BY β-LACTAMASES

Anthony L. Fink[1] and Michael I. Page[2]
[1] Department of Chemistry and Biochemistry,
University of California, Santa Cruz, California 95064, U.S.A.
[2] Department of Chemical and Biological Sciences,
University of Huddersfield, Huddersfield HD1 3DH, UK.

ABSTRACT

In this chapter, the catalytic mechanisms of β-lactamases are discussed in detail. After an introduction on the reactivity of compounds containing the the β-lactam ring, each class of β-lactamases is reviewed and analysed in terms of essential residues and active site chemistry. Special attention is given to class B enzymes (metallo-β-lactamases or MBLs) for which many questions remain presently unanswered and which seem to exhibit specific mechanisms depending on the subclass.

ß-Lactamases catalyse the hydrolysis of the ß-lactam of penicillins *(1)* and cephalosporins *(2)* to give the ring opened and bacterially inert ß-amino acid (Scheme 1). [1] Four membered ß-lactams occur relatively rarely in nature, therefore it is not surprising that the biological activity of these compounds should be attributed to the chemical reactivity of the ß-lactam ring. Shortly after the introduction of penicillin to the medical world it was suggested that the antibiotic's activity was due to the inherent strain of the four-membered ring or to reduced amide resonance. [2] The non-planar butterfly shape of the penicillin molecule *(3)* could reduce amide resonance and thus increase the susceptibility of the carbonyl group to nucleophilic attack, compared with planar amides. However, there is little evidence to suggest that the kinetic reactivity of ß-lactams in penicillins and cephalosporins is due to an unusually strained or an amide-resonance inhibited ß-lactam. [3] The rate of alkaline hydrolysis of the simple ß-lactam *(4)* is only 3-fold greater than that of *(5)*.

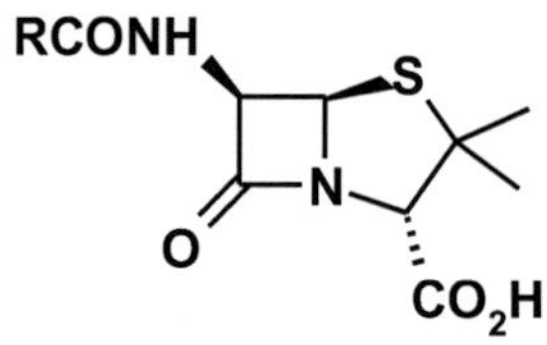

(1)

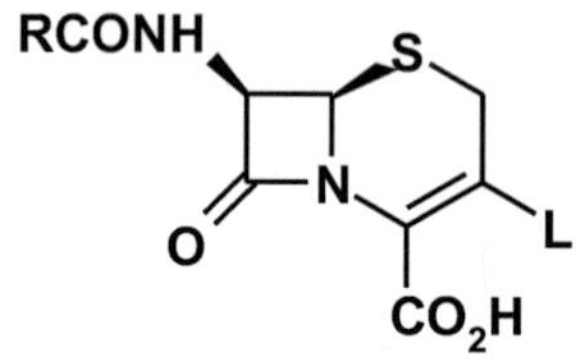

(2)

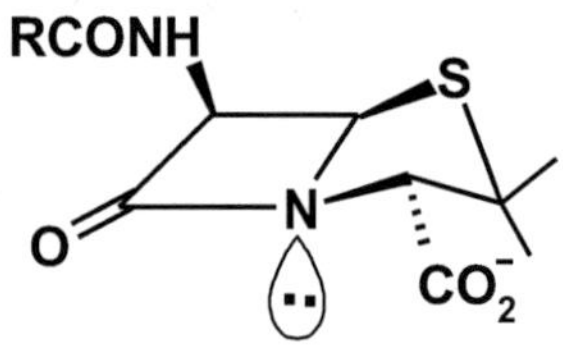

(3)

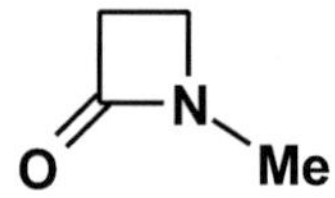

(4)

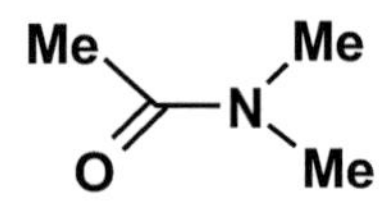

(5)

Nucleophilic substitution at the carbonyl of ß-lactams is an acyl transfer process involving covalent bond formation between the carbonyl carbon and the nucleophile and C-N bond fission of the ß-lactam (Scheme 2). In these types of reactions the mechanism involves, at least, a two-step process. [2,3] Covalent bond formation to the incoming nucleophile occurs before C-N bond fission resulting in the reversible formation of a tetrahedral intermediate. Contrary to expectations, opening the four-membered ring is not a facile process. [4] In many of these nucleophilic substitution reactions the rate limiting step is often not the first addition step but a subsequent one which may sometimes even be ring opening itself. [1]

Those reactions which involve the attack of a neutral nucleophile which has an ionisable hydrogen (Scheme 2) invariably require general base catalysis to remove the proton. [2,3] The requirement for proton removal is paramount - and in extreme cases only the reaction of the anionic nucleophile is observed (Scheme 2). For example, there is no pH independent reaction of water with penicillin, and alcohols react only through their anions. [5] The importance of general base catalysis is a reflection of the fact that contrary to expectations penicillins are not powerful acylating agents. Similarly, C-N bond fission requires protonation of the amine nitrogen and when this is the rate limiting step general acid catalysis is observed. [2,3] It has been suggested that ring opening does not occur by stretching of the C-N bond but rather by a rotational motion. [4,6] This minimises strain effects and maximises favourable orbital interactions. The unusual mode of C-N bond fission could have interesting consequences in the enzyme catalysed hydrolysis of ß-lactams for the geometrical relationship of the proton donor in the protein and the amine leaving group.

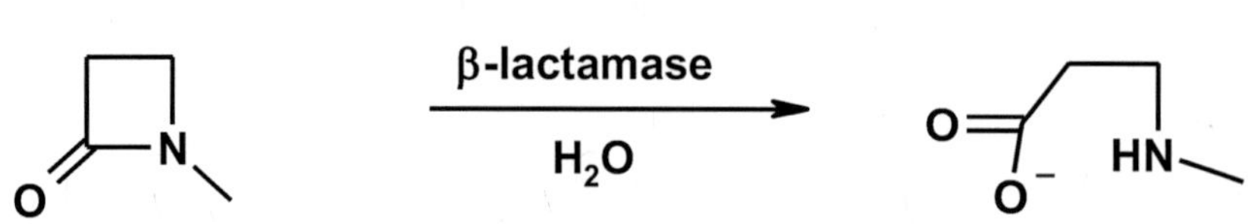

Scheme 1.

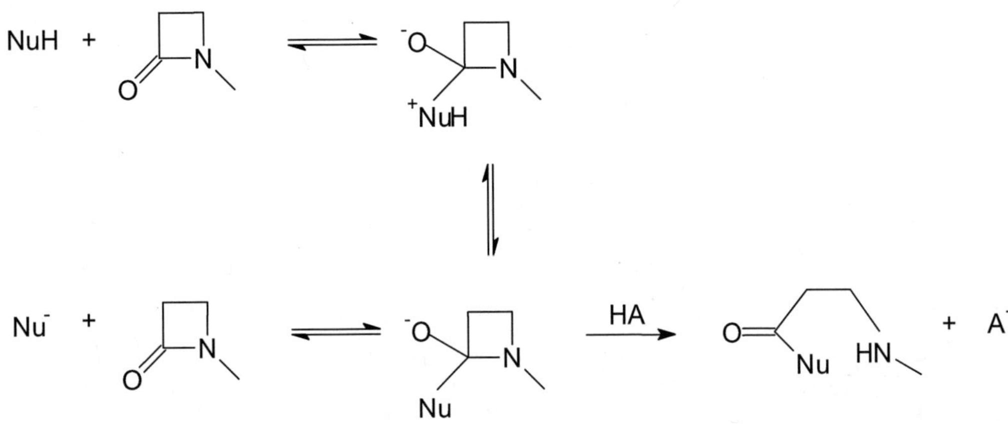

Scheme 2.

Another interesting difference between nucleophilic substitution in penicillins and peptides/amides is the preferred direction of attack and the geometry of the initially formed tetrahedral intermediate. It is usually assumed that nucleophilic attack on the carbonyl carbon of a planar peptide will generate a tetrahedral intermediate with the lone pair on nitrogen *anti* to the incoming nucleophile, *(6)*, *(7)*. Conversely, nucleophilic attack on ß-lactams occurs from the least hindered α-face (*exo*) so that the ß-lactam nitrogen lone pair is *syn* to the incoming nucleophile in the tetrahedral intermediate *(8)*. [7] This has obvious consequences for the placement of catalytic groups - particularly the general acid donating a proton to the departing amine of the ß-lactam or metal-ions coordinating to this atom. [6]

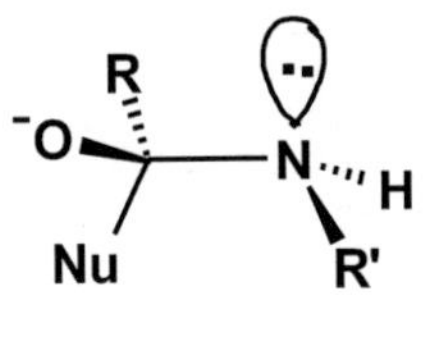

(6)

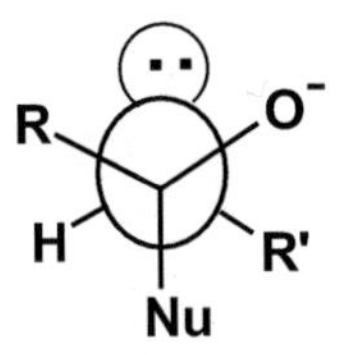

(7)

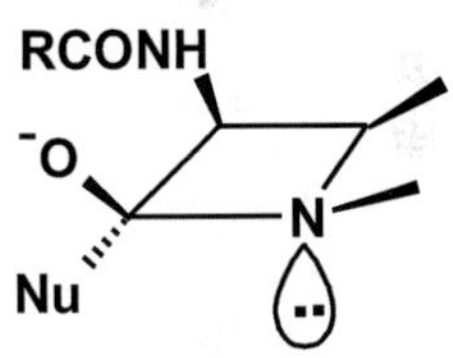

(8)

ß-LACTAMASES

The main mechanistic division of ß-lactamases is into serine enzymes and zinc enzymes. [1] The former have an active site serine residue and the catalytic mechanism involves the formation of an acyl-enzyme intermediate. The metallo enzymes appear to involve only non-covalently bound intermediates. On the basis of their amino acid sequences, the serine ß-lactamases are sub-divided into three classes - A, C and D whereas the class B ß-lactamases consist of the zinc enzymes. [1]

SERINE ß-LACTAMASES

There is very strong evidence for the formation of an acyl enzyme intermediate (Scheme 3) - including electro-spray mass spectrometry [8], infra-red measurements, [9] trapping experiments, [10] the determination of the rate constants for their formation and breakdown [11] and even an X-ray crystal structure. [12] The two experimental kinetic parameters associated with simple systems exhibiting Henri-Michaelis-Menten behaviour are the second-order rate constant k_{cat}/K_m determined below saturation and the first order rate constant k_{cat} determined above saturation. The *interpretation* of these experimental macroscopic rate constants as they vary with parameters such as pH, substituents in the substrate and amino acid replacement in the enzyme depends upon the model assumed for the reaction pathway. The overall reaction pathway for the serine ß-lactamases is given in eq. *(1)*.

$$E + S \underset{k_{-1}}{\overset{k_1}{\rightleftharpoons}} ES \underset{k_{-2}}{\overset{k_2}{\rightleftharpoons}} EA \xrightarrow{k_3} E + P \qquad (1)$$

It is generally assumed that formation of the usually relatively unstable acylenzyme (EA) is irreversible ie $k_{-2} = 0$ and the second-order rate constant k_{cat}/K_m, determined below saturation conditions, is then given by eq. *(2)* and hence always reflects the rate of acylation *irrespective* of the value of k_2 and k_3.

$$\frac{\text{Rate}}{(E)(S)} = \frac{k_{cat}}{K_m} = \frac{k_1 k_2}{k_{-1} + k_2} \qquad (2)$$

Under saturation conditions the rate constant, k_{cat}, is independent of substrate concentration and is given by eq. 3. and so simplifies to the rate of acylation k_2, if $k_3 >> k_2$ and the enzyme is essentially present as the enzyme-substrate complex. Conversely, k_{cat} reflects the rate of deacylation, k_3, when $k_2 >> k_3$ and the enzyme is present mainly as the acyl enzyme.

$$k_{cat} = \frac{k_2 k_3}{k_2 + k_3} \qquad (3)$$

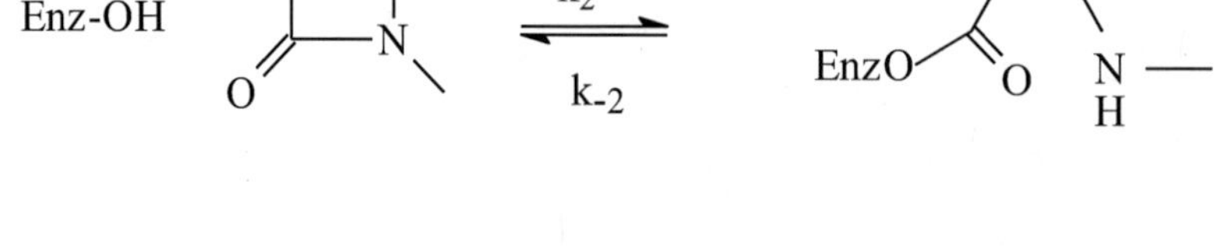

Scheme 3.

However, if formation of the acylenzyme is *reversible* then k_{cat}/K_m and k_{cat} are given by eq. *(4)* and *(5)*, respectively. For serine proteases, k_{-2} represents the "intermolecular" aminolysis of the

$$\frac{k_{cat}}{K_m} = \frac{k_1 k_2 k_3}{k_3(k_{-1} + k_2) + k_{-1}k_{-2}} \tag{4}$$

$$k_{cat} = \frac{k_2\, k_3}{k_2 + k_{-2} + k_3} \tag{5}$$

acylenzyme to regenerate the peptide substrate.

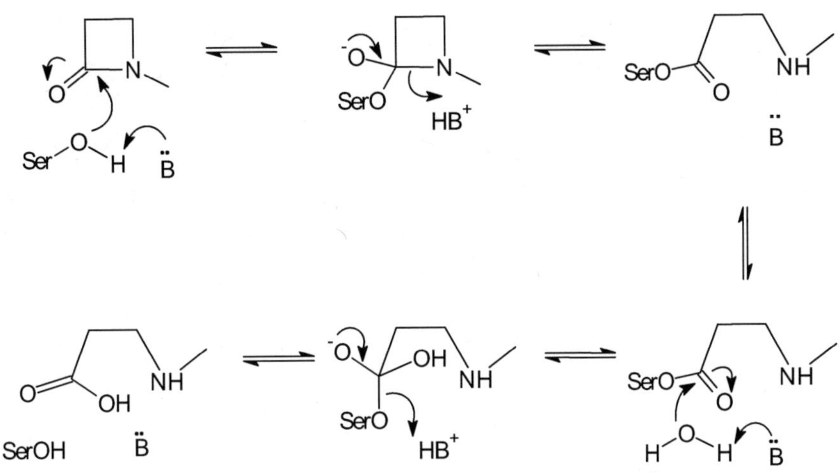

Scheme 4.

For the hydrolysis of ß-lactams catalysed by ß-lactamases the k_{-2} step corresponds to "intramolecular" aminolysis to regenerate the ß-lactam substrate (Scheme 3). Although the formation of a strained four-membered ß-lactam is likely to have a significant activation energy it is not totally inconceivable that this intramolecular step has an entropic advantage [13] and is competitive with intermolecular hydrolysis of the acylenzyme. The interpretation of the macroscopic rate constant k_{cat}/K_m may then become ambiguous because it is made of all microscopic rate constants up to and including the first effectively irreversible step.

Formation of the acylenzyme intermediate requires at least two proton transfers: proton removal from the attacking serine and proton donation to the departing ß-lactam amine (Scheme 4). Despite the availability of a number of X-ray crystal structures of several class A and class C ß-lactamases and many site directed mutagenesis studies the indentity of the catalytic groups involved in these proton transfer steps remain elusive. In principle, the general base which accepts a proton from the nucleophilic serine is not necessarily the same residue which then acts as a general acid to donate a proton to the departing ß-lactam nitrogen. Conceptually, the latter process is is neater and our prejudices are reinforced by the juxtaposition of the nitrogen lone pair, the serine and the general base-acid being on the same α (*exo*) face of the substrate in the tetrahedral intermediate *(8)*. Similarly, the hydrolysis of the acylenzyme requires two proton transfer steps - from water and to the departing serine - and only our preference for simplicity require these to be one and the same residue *and* the same as that involved in the acylation step (Scheme 4).

There are therefore at least four transition states corresponding to formation and breakdown of the tetrahedral intermediate to give initially the acyl enzyme followed by its hydrolysis (Scheme 4). The concept of a unique "transition state analogue" as an inhibitor for an enzyme catalysing a reaction through many steps becomes ambiguous - there could be many analogues corresponding to different transition states. The evolutionary perfect enzyme reduces the transition state energies for all steps in the conversion of reactants to products without "over-stabilising" any intermediates. The maximum second order rate constant for an enzyme catalysed reaction is that corresponding to diffusion control (*ca.* 10^8 M^{-1} s^{-1}) and some ß-lactamases appear to be near this limit and are "perfect catalysts". [14] Of course, for enzymes such as some of the Penicillin Binding proteins (PBPs) which only catalyse the acylation step and the rate of the deacylation step k_3 (*eq. 1*) is very low, the value of k_{cat}/K_m is still large and the meaning of 'catalytic efficiency' can be misleading.

There has been some speculation about the role of the carboxylic acid residue in β-lactam antibiotics and that it may be involved in some sort of 'substrate assisted catalysis', especially as esterification of the C3 carboxyl in penicillins greatly decreases antibacterial activity and susceptibility towards ß-lactamases. In principle, this residue, in its undissociated form, could act as a general acid catalyst providing the proton required for C-N bond fission and opening of the ring (Scheme 4). Alternatively, in its ionised form, it could act as a general base facilitating attack by the active site serine or hydrolysis of the acyl-enzyme intermediate (Scheme4). There is no evidence from the non-enzyme catalysed hydrolysis that the carboxylic acid group is involved with any neighbouring group participation as the rates of reaction of benzylpenicilin and its methyl ester are similar to that for the methyl ester at low pH and the ester is actually more reactive towards alkaline hydrolysis. Cephalosporin lactones are actually better substrates for *B.cereus* serine β-lactamase than are cephalosporins with a free carboxyl group, so the main role of the carboxylate group is in molecular recognition and binding.

ß-Lactamases have been shown to exhibit elements of 'promiscuity' in that some of the serine enzymes react not only with acyl centres but also sulfyl and phosphyl groups. Nucleophilic substitution at acyl centres generally proceeds through the formation of an unstable tetrahedral intermediate (TI). [15] The reaction pathway involves a change in geometry as the carbonyl carbon is converted from three to four coordination. Furthermore, it is usually assumed that there is some preferential direction of nucleophilic attack such that the incoming nucleophile approaches at approximately the tetrahedral angle to the carbonyl group. By contrast, the associative mechanism for phosphoryl and sulfonyl group transfer involves a pentacoordinate intermediate with trigonal bipyramidal geometry. An initially four coordinate and tetrahedral phosphorus or sulphur centre is converted to a five coordinate one and, in general, it is assumed that the preferential pathway involves the nucleophile taking up the apical position and the leaving group departing from an apical position of the trigonal bipyramidal intermediate (TBPI). [16] Given the preferential geometrical requirements for incoming nucleophiles and departing leaving groups, there is probably a favoured relative positioning of the general base and general acid catalysts (Scheme 5).

It is often assumed, but with little actual supporting evidence, that enzymes catalyse reactions by an exquisite positioning of the catalytic groups. This is obviously not the case for ß-lactamases which efficiently catalyse the hydrolysis of both penicillins and cephalosporins where there are significant geometrical differences, for example, between the ß-lactam carbonyl and nitrogen and the C-3 carboxylic acid in penicillins and that at C4 in cephalosporins. Also, if rigid positioning of catalytic groups was so important then it is doubtful if an enzyme with a primary function as a catalyst for acyl transfer could be an effective catalyst for phosphyl transfer because of the geometrical differences in the displacement mechanisms (Scheme 5). The class C P99 ß-lactamase is, in fact, extremely efficient in enhancing the rate of phosphoryl [17] and sulfonyl [18] transfer and the pH dependence of the kinetic parameters indicate that similar catalytic machinery is used for both acyl and phosphoryl sulfonyl transfer.

The fundamental requirement for the acylation reaction of the serine β-lactamases is a general base to activate the active site Ser hydroxyl and a general acid to protonate the β-lactam nitrogen. Similarly, deacylation requires activation of the deacylating water molecule by a base catalyst and protonation of the departing Ser O-γ by an acid catalyst (Scheme 4). Over the years, many candidates for these catalytic groups have been proposed; although there is now a consensus on those involved in deacylation for the Class A β-lactamases there is still some controversy about the critical catalysts for acylation for the Class A β-lactamases, and for the general acid/general base catalysts for the Class C enzymes.

All serine hydrolase β-lactamases have three conserved sequence motifs that are located in similar positions in the active site [19]. The first is an SXXK motif, including the active-site serine nucleophile. The second motif, (S/Y)XN, shows some variation between β-lactamase classes: the class A enzymes usually are SXN, whereas the class C enzymes are YXN and the class D enzymes SXV. The third motif is a K(T/S)G triad: the class A and class D enzymes each have an additional conserved potential general base in the active site: this is Glu 166 in the class A enzymes, and a carbamoyl group on the conserved lysine of the SXXK motif in the Class D enzymes [20].

Class A β-lactamases

The catalytic mechanism of the class A β-lactamases remains controversial, as it has been for two decades, in spite of many investigations and, perhaps, due to the surfeit of potential catalytic residues [21]. It is well established that these enzymes catalyze the hydrolysis of the β-lactam bond via a covalent acyl-enzyme [22]. Although there is agreement that Ser70 acts as the nucleophile in the acylation reaction, the role of other active site residues is less well understood. Glu166 is generally accepted to be critical in deacylation, probably acting as a general base, but its role in acylation is unclear. The role of the conserved Lys73 remains contentious. In the structure of the free enzyme Lys73 forms hydrogen bonds to Ser70, Ser130, Asn132, and Glu166. In addition there is a conserved water hydrogen bonded to Ser70, Glu166, and Asn170. It is thought that the evolution of Glu166 into the active site of the serine Penicillin Binding Proteins (PBPs) led to the origin of the hydrolytic capacity of the class A β-lactamases [23]. The lysine equivalent to Lys73 in the PBPs appears to be required for acyl-enzyme formation in these enzymes, so it is reasonable to assume that the Lys also plays a critical role in the β-lactamase acylation mechanism as well. Proposed catalysts in the acylation reaction for the Class A β-lactamases include Glu166 and Lys73 as general bases and Ser130, Lys73, Glu166 as general acids, and S130/K73/water involving a proton shuttle mechanism. In deacylation the proposed catalysts are Glu166 or Lys73 acting as general acid/base catalysts or a proton shuttle involving S130/K73/water. [21, 24-34]

Mechanistically it is reasonable to expect that acylation must involve general base activation of Ser 70. Unlike the serine proteases, which catalyze a very similar reaction to β-lactamase, there is no clear-cut acid/base catalyst in β-lactamase equivalent to His57 in the serine proteases. In the serine proteases His57 initially activates the serine O-γ in the nucleophilic attack on the peptide carbonyl carbon by proton abstraction that is concerted with C-O bond formation. It then also supplies the proton to the leaving nitrogen. In the β-lactamases there are several significantly different factors involved in the acylation reaction: the substrate amide is in the 4-membered lactam ring which is more strained than a typical peptide bond, in bicyclic β-lactams the lone pair on nitrogen remains pro-S and *syn* to the serine 0-γ following nucleophilic attack and formation of the tetrahedral intermediate in contrast to the *anti* – arrangement that occurs with serine proteases *(6)* and the attacking serine/substrate carbonyl are in the positive electrostatic field of the dipole of helix-2. These factors would be expected to make the acylation reaction significantly more facile in the β-lactamases. However, there is considerable evidence that indicates that nucleophilic attack by a protonated nucleophile, such as Ser 70, would definitely require the assistance of (general) base catalysis and that C-N bond cleavage requires general acid catalysis [35].

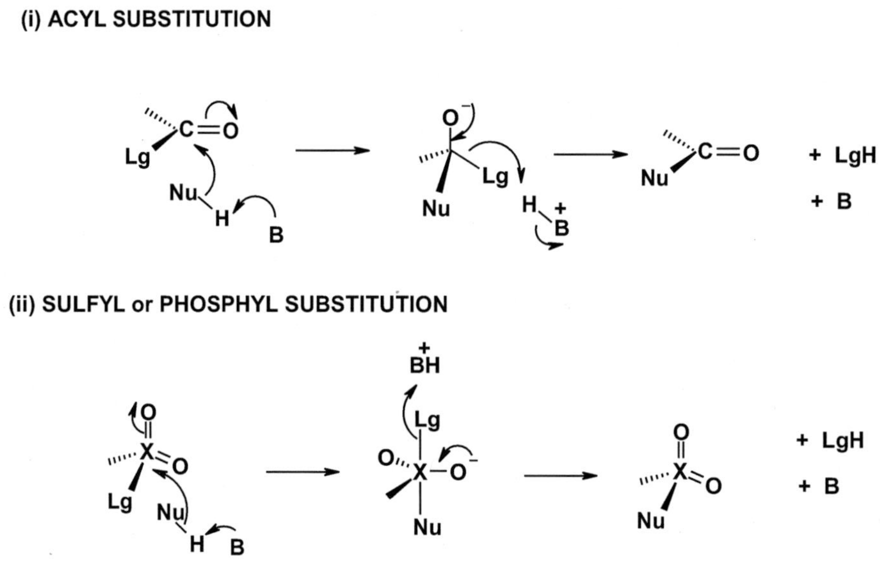

Scheme 5.

General Acid/general Base Catalysts in Acylation

Two main mechanisms for the catalysis in acylation have been proposed, which differ in whether Glu166 or Lys73 acts as the general base to activate Ser70 for nucleophilic attack. Most experimental studies, and some theoretical calculations e. g. [36], suggest a critical role for Glu166 as a general base in acylation. However, the fact that mutation of Glu166 (e. g. to Ala) still results in formation of acyl-enzyme indicates that Glu166 is not absolutely essential to acylation [29, 34, 37, 38], although the rate is significantly decreased [29] (it should be noted that even removal of the nucleophilic Ser70 still leads to an enzyme with significant rate enhancement over the non-catalyzed reaction [39]).

Any reasonable mechanistic proposal for the catalytic mechanism of the class A β-lactamases, must account for the conserved nature of Lys73 in all penicillin-sensitive enzymes, which suggests a role either in acylation and/or binding. This does not preclude a role in deacylation as well, and in fact, data for the K73A mutant suggest that Lys73 may play a substantial role in deacylation [31]. A particular point of debate is whether the side-chain of Lys73 acts as a general base or general acid catalyst or even both. Proposed mechanisms in which it functions as a general base (by analogy to His-57 in the serine proteases) [34] require that Lys73 be in a non-protonated state at neutral pH, and thus have a very low pK for an ammonium group. It has been suggested that such a dramatic perturbation of the pK is possible due to local electrostatic effects [34].

A major problem has been the difficulty in ascertaining the pK of the side chain of Lys 73. This situation is further compounded by the possibility of significantly different pK's in the free enzyme and the various enzyme-substrate complexes.

Typically class A β-lactamases show a bell-shaped pH profile for k_{cat}/K_m, with $pK_1 \sim 5.0$ and $pK_2 \sim 8.5$ at 30°C. This kinetic parameter reflects the pK's of the essential ionizable catalytic groups in the free enzyme. Mutation of Lys73 to alanine in *B. licheniformis* β-lactamase resulted in two very unexpected findings: the values of k_{cat}/K_m for penicillins at the pH optima for the K73A mutant were only decreased by 100-200-fold over the values for the wild-type β-lactamase, and plots of k_{cat}/K_m for K73A β-lactamase were bell-shaped, as with the wild-type, although pK_1 was shifted from a value of ~5.0 in the wild-type to ~8.0 in the K73A mutant, and a small increase in pK_2 was observed [31]. The increase in pK_1 is consistent with the removal of the positive ammonium group of the lysine from the proximity of Glu166, to which the acid limb was ascribed (the solvent accessibility of Glu166 (and Lys73) is very low, consequently with the removal of the neighboring ammonium group from Lys73 in the K73A mutant, the environment around the Glu166 carboxyl will become significantly more non-polar, consistent with a substantial increase in its pK). As far as the proton transfer steps are concerned, making the general base, required for serine attack, a stronger base (increased pK) should *increase* activity; whereas making the general acid, required for C-N bond cleavage, a weaker acid (increased pK) should *decrease* activity. The lack of change in pK_2 indicates that pK_2 must be due to some group other than Lys73 - perhaps Lys234, since clearly the ionization reflects a group with a high pK. The K234A mutant showed similar catalytic properties to the K73A mutant [26]. Although the pK_2 for k_{cat}/K_m for the K234A mutant is similar to that for the K73A mutant, pK_1 is not significantly affected by the mutation at Lys234, consistent with the effect of K73A on pK_1 being due to the loss of the proximal Lys73 on Glu166. The value of k_{cat}/K_m at the pH optimum for both Lys to Ala mutants was similar, about a 200-fold decrease from that of the wild-type with penicillin substrates. Thus the alkaline limb for the wild-type enzyme reflects the ionization of either or both Lys73 and Lys234 or it represents the ionization of another group. In the former case we are presented with two options: in the wild-type enzyme either both lysine residues participate in catalysis or only one does, but it can be replaced by the other. In contrast to either single Lys to Ala mutant, the double mutant, K73A/K234A, possessed very little catalytic activity: k_{cat}/K_m at the pH optimum was down a million-fold for penicillin substrates relative to the wild-type value [31]. Since plots of k_{cat}/K_m vs. pH indicate only a single ionizing group is responsible for pK_2, it would seem that either lysine can replace the other if one is missing. The heat of ionization of the group responsible for pK_2 in the wild-type enzyme is large (10 kcal/mol) indicative of a lysine residue [40]. Data for the K234H mutant in *S. albus* G β-lactamase also suggest that pK_2 may reflect the pK of residue 234 [41]. Thus it is probable that pK_2 in the K73A mutant reflects the ionization of Lys234. This residue makes several key hydrogen bonds in the active site, including one to Ser130, and is highly conserved in the serine β-lactamases. Replacement of Lys234 by Arg in the TEM β-lactamase has little affect on the reaction with penicillin substrates [42]. There are two structural features that Lys73 and Lys234 have in common that are relevant to the catalytic mechanism: they both are hydrogen-bonded to Ser130, and they both have positively charged side-chains.

Formation of the tetrahedral intermediate increases the basicity of the β-lactam nitrogen. However, although an amine, it is not strongly basic as evidenced by the relatively low pK_a of about 5 for the N-protonated species of penicilloic acid. Facilitation of C-N bond cleavage by N-protonation would therefore require a reasonably strong general acid. Protonation by Ser130 or Lys234/73 does not readily fit this requirement. Maximal catalytic activity would be provided by a general acid of pK_a about 7. However, the two lysines could, in principle, provide a positively charged electrostatic environment around the region of the substrate undergoing reaction, and both could shuttle a proton to Ser130 which acts as the proton donor to nitrogen. It is very likely that the position normally occupied by the ammonium group of Lys73 will be occupied by a water molecule in the K73A mutant. Such a water molecule, although a weaker acid, could substitute for the lysine side-chain if it acted as a proton donor/acceptor, whereas if its main role were electrostatic, then the substituting water would be unable to undertake that role.

Mutations at Glu166 have been shown to result in a greater affect on deacylation than acylation, but the rate of acylation is also decreased in Glu166 mutants [24, 27, 29, 37, 43, 44]. The simplest mechanism would involve Glu166 acting as a general base, possibly via the bridging water as a transient proton acceptor, in formation of the tetrahedral transition state/intermediate. The partitioning of this species is determined by the relative rates of C-O and C-N cleavage, the latter leading to the acyl-enzyme, the former returning to the ES complex. Partitioning toward products will be facilitated by proton donation to the lactam nitrogen by either the now undissociated carboxylic acid of Glu166 or a proton shuttle involving Ser130 and Lys73.

Among the strongest evidence that Glu166 is the general base in acylation is that from the ultrahigh resolution structure of TEM-1 β-lactamase complexed with a boronic acid transition-state analog [32]. This complex mimics the acylation tetrahedral transition state/intermediate, and shows that Glu166 is clearly protonated. This supports the hypothesis that Glu166 is the general base in the acylation half of the reaction cycle, activating Ser70 for nucleophilic attack on the β-lactam via the catalytic water. This structure also suggests that Ser130 is the general acid in the acylation reaction, activating the leaving group. Similarly a structure of the SHV-2 β-lactamase at 0.91 Å resolution is consistent with Glu166 acting as the general base through the active site water molecule, and with Lys73 being protonated in the free enzyme [33].

The role of Lys73, which hydrogen bonds with both Ser70 and Ser130, may be to activate both residues electrostatically. Lys73 may also transfer a proton to replace the one Ser130 has donated in activating the lactam nitrogen. Such a proton transfer might be essential for an efficient deprotonation of Glu166 for deacylation. Lys73 is optimally positioned to function in such a proton transfer. Following tetrahedral intermediate formation, nitrogen protonation must occur for scission of the β-lactam C-N bond ring to give the acyl-enzyme. While the closest proton source for this event is the Ser130 hydroxyl, it is not known whether Lys234 or Lys73 acts as the ultimate proton source. Mutagenesis studies are consistent with the involvement of Ser130 in this proton transfer [45, 46].

Mobashery and colleagues recently addressed the question of the protonation state of Lys 73 in TEM-1 β-lactamase using both experimental and theoretical calculations [47]. The results indicated that the pK of Lys-73 is decreased to 8.0–8.5 (previous experiments and calculations suggested that Lys73 has a pK >10 in both free enzyme and the ES complex [48, 49]). The suppression of the Lys73 pK was attributed to several factors: the hydrophobic environment around the side chain amine, the nearly complete lack of solvent accessibility of this lysine, and the proximity of Lys234 (which has a normal pK) [47]. Regardless of these differences, it is reasonably clear that Lys73 is protonated in the free enzyme; however, the possibility that binding of the substrate perturbs the pK even further downward, so that Lys73 acts as a general base, cannot be completely excluded.

Although, as noted, most experimental data support a mechanism in which Glu166 acts as the general base and Lys73 as a general acid in the acylation reaction, some computational studies suggest that Lys73 may act as the general base. It is possible that β-lactamase is an unusual, perhaps unique, enzyme in having sufficient flexibility in the active site to allow multiple modes of catalysis, depending on the nature of the substrate and mutations in the enzyme. Thus, it is possible that there are alternate pathways for acylation of similar energy, one involving Glu166, the other Lys73, as the general base [49, 50]. The existence of alternative conformations of Lys73 in the structure of Toho-1 β-lactamase allowing potential proton transfer between Lys73 and Glu166 or Ser130 is consistent with the possibility of multiple pathways involving this residue [49]. Theoretical support for multiple catalytic modes with class A β-lactamase comes from recent QM/MM analyses of three proposed mechanisms for acylation (Glu166 as general base, Lys73 as general base, and a general acid pathway) [50]. The results showed a concerted Lys73 general base pathway competing favorably with Glu166 acting as the general base.

Based on the preponderance of experimental data at this time we suggest that the most likely mechanism for acylation involves Glu166 as a general base, activating the active site serine via a conserved water molecule, and Lys73, as part of a proton shuttle involving Ser130, acting as a general acid to protonate the leaving lactam N. Perturbations of these and other active site residues due to mutations can lead to significant changes in the mechanism of acylation.

General Acid/general Base Catalysts in Deacylation

In contrast to the situation with the acylation reaction of the class A β-lactamases, there is general agreement on the mechanism of deacylation, namely that Glu166 acts as a general base to activate the active site water molecule to attack the acyl-enzyme. Crystallographic studies have shown that the side-chain of Glu166 in the acyl-enzyme is well-positioned to act as a general base in activating the deacylating water molecule for its attack on the acyl-group [29, 34, 38, 51-54]. Mutagenesis experiments on Glu166 have generally shown smaller effects on acylation and major effects on deacylation rates [24, 37, 43, 44, 55, 56]. Typically, these mutant enzymes undergo slow acylation and yield stable or relatively stable acylenzymes. The fact that deacylation is rate-limiting in the reaction of penicillins with K73A β-lactamase most likely reflects the fact that the role of Lys73 is more critical in deacylation than in acylation [31]. From the TEM β-lactamase acyl-enzyme structure [34] Lys73 is within hydrogen bonding distance of Ser70 O-γ, and could thus directly donate a proton during collapse of the tetrahedral transition state in deacylation (Lys234 is too far away to do this). However, Ser130 is also within H-bonding distance of the Ser70 O-γ, so a proton shuttle could also operate in deacylation, in which case Lys234 might be able to substitute for the missing Lys73 in the K73A mutant, since Lys234 is situated about the same distance from Ser130 O-γ as Lys73.

ENDOR studies show a water molecule in the active site of the acyl-enzyme that was not seen in the deacylation-impaired Glu166Asn mutant [57]. On the basis of the X-ray structure of the enzyme, the ENDOR distance constraints placed this water molecule within hydrogen-bonding distance to the carboxylate side chain of Glu166 as expected if the latter were involved as the general base catalyst for hydrolytic breakdown of the acyl-enzyme intermediate [57].

An additional complication with the Class A β-lactamases is that of branched-pathway kinetics, which are observed with both certain substrates and certain mutations. The data suggest that in any enzyme/substrate combination in which the deacylation reaction is sufficiently slow, the acyl-enzyme may isomerize into an alternate more stable conformation (EA*) [31, 40, 58, 59]. Thus the observation of the branched pathway in the case of K73A β-lactamase is consistent with rate-limiting deacylation [31].

We thus conclude that the deacylation half of the catalytic reaction involves essentially the reverse process of acylation, with Glu166 activating the hydrolytic water for attack on the acyl bond and general acid catalysis by the proton shuttle involving Lys73 and Ser130 supplying the proton for the Ser leaving group.

Class C β-lactamases

Even more than in the case of the class A β-lactamases, the identification of the residue responsible for activating the nucleophilic Ser64 in the acylation reaction of the class C β-lactamases has proven elusive. Presumably a general base is required to activate the serine residue and abstract a proton to facilitate formation of the tetrahedral intermediate, and a general acid is then required to protonate the departing lactam nitrogen in the collapse of the tetrahedral intermediate to the acyl-enzyme. Potential catalytic groups for these roles are those of Tyr150 and Lys67. However, in spite of several high resolution structures of the free enzymes, and complexes with inhibitors or transition state analogs [60-74], and several studies with mutations at the active site residues [75-81], there is no consensus on the identity of the general base catalyst for the acylation reaction. Other residues important for catalysis, based on mutagenesis experiments, are Asn152, Lys315, Thr316, and Gly317 [29]. Crystallographic structures indicate that all these residues are involved in a hydrogen bonding network, and it has been suggested that the class C enzymes do not have a single residue that acts as the catalytic general base. Instead, the catalytic activity has been proposed to arise from an active site hydrogen-bonding network, possibly with assistance from the ß-lactam substrate [78]. For example, Asn152, Lys67, and Tyr150 also form a hydrogen-bonding network, and mutation of any of these residues reduces the activity of the enzyme by several orders of magnitude [75-77, 82].

General acid/general base catalysts in acylation: Based on the initial crystal structure of a class C β-lactamase, Tyr150 was proposed to act as a general base [67]. However, subsequent investigations, both experimental and theoretical, have cast some doubt on this proposal, which requires the tyrosine to be in the ionized, phenoxide, form, which means a significantly lower pK than normal for the tyrosine. In addition to results with Tyr150 mutants that suggest this residue is not essential for catalysis, a lowered pK does not appear to be in accord with ^{13}C NMR titration experiments, which indicated that Tyr150 is protonated throughout the pH range of enzyme stability [83].

A second suggested acylation mechanism involves Lys67 as the general base [76]. This also requires a substantially perturbed pK for the lysine to be in its basic form at neutral pH. The proton it abstracts from Ser64 is proposed to be shuttled via Tyr150 to the lactam nitrogen. Mutations at Lys67 cause a major decrease in catalytic activity, with the exception of Lys67Arg, which is significantly active at higher pHs [82, 84]. It has been proposed that in the Lys67Arg mutant the anionic form of Tyr150 is favored, and acts as the general base catalyst [85]. Mutations of Asn152 lead to a large drop in activity which is consistent with Lys67 playing an important role in catalysis, since Asn152 interacts directly with Lys67, but not with Tyr150. Recent molecular dynamics simulations support the assignment of Lys67 as the general base catalyst for the acylation step [85].

Based simply on inspection of high-resolution structures of class C β-lactamases, both Lys67 and Tyr150 are in suitable positions to abstract a proton from Ser64 or the hydrolytic water molecule and both side chains are also part of a hydrogen-bond network: in fact an equilibrium between the ionized and neutral forms of Lys67 and Tyr150 has been suggested for the ES complex, based on both deuterium kinetic effects [86] on the catalysis and calculations [85]. A critical role for a H-bonding proton shuttle would be consistent with the results of kinetic isotope effects on catalysis [87].

Thus, we suggest that acylation in the class C β-lactamases involves coordinated proton transfers, in which Tyr150 acts as the initial general base catalyst to activate the hydroxyl of Ser64, and that Lys67 is a critical component of the proton shuttle system.

General Acid/general Base Catalysts in Deacylation

A viable mechanism for deacylation must take into account for the difference in activity between the class C β-lactamases and the PBPs, since the latter do not catalyze deacylation. All of the functionalities that have been proposed to participate in the hydrolysis by the class C β-lactamases are also present in the PBPs and occupy similar positions in the respective active sites [88], although most PBPs have a serine in the second conserved element rather than tyrosine Both Tyr150 and Lys67 have been proposed to act as a general base in the deacylation reaction, among various proposals. As has been proposed for acylation, it therefore seems possible that instead of a single residue acting as a general base, there is a charge relay system, involving several of the residues Gln120, Asn152, Lys67, Tyr150, and Lys315 and possibly even the substrate. Alternatively, it has been suggested that Tyr150 in the anionic state may activate the hydrolytic water molecule that attacks the acyl-enzyme intermediate in deacylation [85, 89]. The problem with this latter interpretation is to explain why the similar PBPs do not undergo deacylation. Based on a study using ab initio-based mixed quantum mechanics/molecular mechanics techniques the faster rate of deacylation in the β-lactamase is attributed to a more favorable electrostatic environment around Tyr150, due the ability of the β-lactamase to covalently bind the substrate without concurrent elimination of hydrogen bonds to Tyr150 [89].

Based on these investigations, we propose that Tyr150 acts as the general base in deacylation, with concerted transfer of a proton from Tyr150 to Lys67; the proton shuttle in both acylation and deacylation accounts for the observed protonated state of Tyr150.

Class D β-lactamases

The class D β-lactamases have been far less investigated than other members of the β-lactamase family. Recent crystallographic structures show that in spite of limited sequence homology, there is substantial 3D structural similarity with the class A β-lactamases. Of particular note is the fact that the class D enzymes appear to have evolved a distinct catalytic mechanism for β-lactam hydrolysis, which involves carboxylation of the active site lysine (Lys70; using the numbering system of the OXA-10 β-lactamase) to form a carbamate [20]. Several lines of evidence indicate that this carboxylation is necessary for catalytic activity [90, 91]. For example, the fully decarboxylated form is inactive, and the degree of carboxylation correlates with the degree of catalytic activity [91], NMR experiments demonstrated the presence of a carboxylated lysine in the active enzyme and conversion of Lys70 to Ala leads to total inactivation [20]. The formation of the carboxylated lysine is readily reversible. Several observations indicate that full carboxylation of the lysine only occurs at $pH > 6$ [91]. Another unique aspect of the class D β-lactamases is that they typically exhibit biphasic kinetics [91, 92], which is now thought to be due to decreased levels of Lys70 carboxylation, leading to decreased specific activity of the enzyme. The ionized form of the carboxylate group is stabilized by a hydrogen bond between the carboxylate and the side-chain amide of the conserved Trp154.

Based on the available data at this time the most likely mechanism for the class D β-lactamases is as follows. The carboxylated active site lysine acts as the general base to activate Ser67 for nucleophilic attack on the substrate carbonyl carbon [90, 91, 93]. Cleavage of the β-lactam bond is facilitated by general acid catalysis by Ser115 to form the acyl-enzyme. Deacylation proceeds via general base catalysis of water by the carboxylated lysine. This mechanism is supported by the observation that mutation of Ser115 (Ser104) to Pro in the OXA-57 β-lactamase leads to complete loss of activity [93]. As with the class A and class C β-lactamases it is likely that proton shuttles are important in the catalytic process.

METALLO B-LACTAMASES

Class B β-lactamases or metallo-β-lactamases (MBLs) require one or two zinc ions to catalyse the hydrolysis of β-lactams. They have no sequence or structural similarity to the serine β-lactamases and exhibit a broad spectrum substrate profile, hydrolysing penicillins, cephalosporins, carbapenems and even some mechanism based inhibitors of class A β-lactamases [94]. The first metallo-β-lactamase to be discovered was produced by an innocuous strain of *Bacillus cereus*, but in the last 20 years, MBL-mediated resistance has appeared in several pathogenic strains and is being rapidly spread by horizontal transfer, involving both plasmid and integron-borne genetic elements [95]. MBLs represent a huge potential clinical threat to β-lactam antibiotic therapy as presently there is no clinically useful inhibitor for this class of β-lactamases.

MBLs can be divided in three subclasses: subclass B1, B2 and B3 according to their aminoacid sequences, substrate profile and metal ion requirement [96]. Subclass B1 is the largest and contains three well-studied β-lactamases: BcII from *Bacillus cereus,* CcrA from *Bacteroides fragilis*, IMP-1 from *Pseudomonas aeruginosa* and BlaB from *Cryseobacterium meningosepticum*. These enzymes efficiently catalyse the hydrolysis of a wide range of substrates, including penicillins, cephalosporins and carbapenems. BcII hydrolyses penicillins at significantly higher rates than cephalosporins and carbapenems, although CcrA does not show this preference, both enzymes exhibit lower K_m values for cephalosporins [97].

The most common representatives of subclass B2 are CphA from *Aeromonas hydrophila* and ImiS from *Aeromonas veronii*, which preferentially hydrolyse carbapenems, e.g., imipenem and meropenem, and have poor activity against penicillins and cephalosporins [98]. Finally, subclass B3 contains the only known tetrameric zinc β -lactamase, the L1 enzyme from *Stenotrophomonas maltophilia,* and the monomeric FEZ-1 from *Legionella gormanii*. Both enzymes hydrolyze a wide spectrum of β-lactam antibiotics [99] with L1 having higher catalytic rate constants (k_{cat}) for penicillins compared to FEZ-1 which has higher k_{cat} values for cephalosporins. Generally, the K_m values are smaller for L1 than for FEZ-1. However, mutation of a methionine residue, important for the subunit interaction in L1, gives a monomer, with significantly higher K_m values, and except for nitrocefin, smaller k_{cat} values compared with the wild type tetramer [100].

The number of zinc ions required for MBL activity has been a matter of controversy and the different crystal structures reported has added to the confusion. The structures of several MBLs have been determined by X-ray diffraction and all show a similar αββα fold, which was first seen with MBLs, but since recognised in other enzymes, such as: glyoxalase II, aryl sulfatase, cAMP phosphodiesterase, which became members of the MBL fold superfamily [101] . The active site of MBLs is situated at the bottom of a wide shallow groove between two β-sheets and has two potential zinc ion binding sites at the active site often referred to as sites 1 and 2 [102-104].

The zinc ligands in the two sites are not the same and are not fully conserved between the different MBLs. Table 1 shows the enzyme residues involved in zinc coordination in the two binding sites in the subclasses B1, B2 and B3 [101] .

In the subclass B1 such as the *Bacillus cereus* enzyme BcII the zinc in site 1 (the histidine site or His_3 site) is tetra-coordinated by the imidazoles of three histidine residues (116, 118, and 196) and a water molecule, Wat_1. In site 2 (or Cys site) the metal is penta-coordinated by histidine 263, aspartic acid 120, cysteine 221 and one water molecule; the fifth ligand at site 2 is carbonate or water, often referred to as the apical water, or Wat_2 [103, 104] but is missing in structures with inhibitors bound [105]. The two metal ions are relatively close to each other, but the distance between them varies from 3.4 to 4.4 Å in different structures of the BcII and CcrA enzymes. Several structures of the CcrA enzyme show a bridging water ligand between the two metals which is thought to exist as a hydroxide ion. In a structure of BcII containing two zinc ions determined at pH 7.5 there is also a similar bridging water molecule but in structures of this enzyme at lower pH this solvent molecule is strongly associated to the zinc in site 1 [106].

Although all MBLs have two conserved zinc binding sites, they have different metal ion affinities for the two binding sites and different metal ion requirements for catalysis. Initial studies indicated that the BcII enzyme from *Bacillus cereus* has very different dissociation constants for the two metal binding sites. Although the first crystal structure, obtained at low pH [104] shows only one zinc ion bound to the histidine site, equilibrium dialysis studies showed two binding events, with a dissociation constant of 0.3μM for the first zinc ion (K_{mono}) and 3 μM for the second zinc ion (K_{bi}) [107]. Later metal binding studies by fluorescence spectroscopy using a chromophoric chelator reported a K_{mono} of 0.62nM and a K_{bi} of 1.5 μM [108]. As the conditions of the experiments were similar for the two studies, the large discrepancy in K_{mono} may be due to the fact that different strains of BcII enzyme from *Bacillus cereus* were used. The dissociation constants K_{mono} and K_{bi} have also been determined in presence of substrates from the steady state rates of imipenem hydrolysis in presence of EDTA, as a metal ion buffer, at different zinc concentrations: K_{mono} was found to decrease significantly, from nM to pM, in presence of substrate, while K_{bi} decreased only by two fold [109]. This led to the suggestion that the monozinc enzyme is responsible for the catalytic activity in physiological conditions, where the concentration of the free Zn^{2+} is in the pM region. However, a recent extensive study of BcIIusing circular dichroism, competitive chelation, mass spectrometry and NMR concluded that the dizinc form is the only relevant species for catalysis [139]. Similarly, ITC studies show only one binding event with two zinc ions bound to BcII, except below pH 5.6 where only one zinc binds to the enzyme (H. Motara, R. Brown and M. I. Page in preparation). It now appears that there is no significant difference between BcII and the homologous class B1, CcrA from *Bacteroides fragilis* which binds both zinc ions very tightly [97] although in CcrA a cysteine replaces the positively charged Arg 121 residue found in BcII. However, replacing Arg 121 in BcII by a Cys, showed no increase in the affinity for the second zinc ion [110].

Early kinetic studies of CcrA led to the proposal that both the mono and the dinuclear forms of the enzyme were catalytically active, with slightly different activities, at physiological pH [111]. However, later studies showed that only the dinuclear species was active and that the previously observed "monozinc" CcrA was a mixture of the dizinc and the apo (metal free) enzyme [112].

Class B2 metallo-β-lactamases were shown to be catalytically active with one bound zinc ion, the binding of the second zinc ion non-competitively inhibiting the enzyme, with a K_i of 50μM [113]. The dissociation constant of the first zinc ion was found to be 7.0 and 1.2pM in the absence and presence of substrate (imipenem), respectively [109]. Subclass B3 MBLs , L1 and FEZ-1 bind both metal ions tightly, the dissociation constants are in the nM region, and are fully catalytically active in binuclear form [114, 115].

Table 1. The zinc ligands in class B β-lactamases

Subclass	Zn1 ligands			Zn2 ligands		
B1	His116	His118	His196	Asp120	Cys221	His263
B2	(Asn116)	His118	His196	Asp120	Cys221	His263
B3	His/Gln116	His118	His196	Asp120	His121	His263

Although corresponding to His 116 in both other subclasses, it was recently shown that the B2 Asn 116 did not participate in the binding of the inhibitory Zn^{++} ion in the Zn1 site [144].

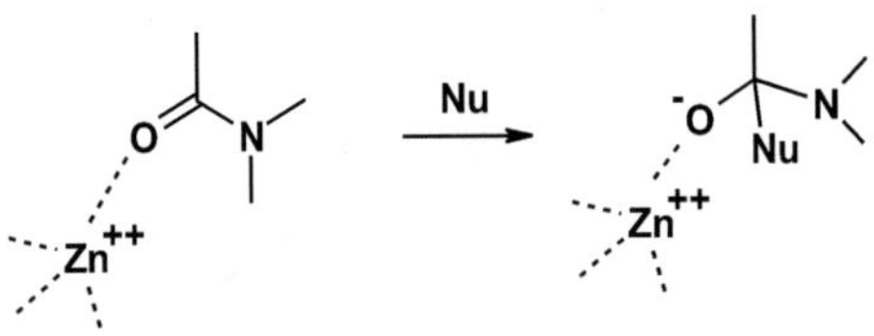

Scheme 6

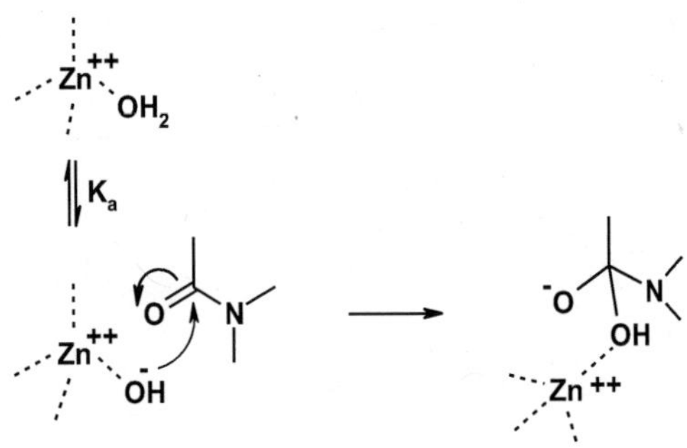

Scheme 7

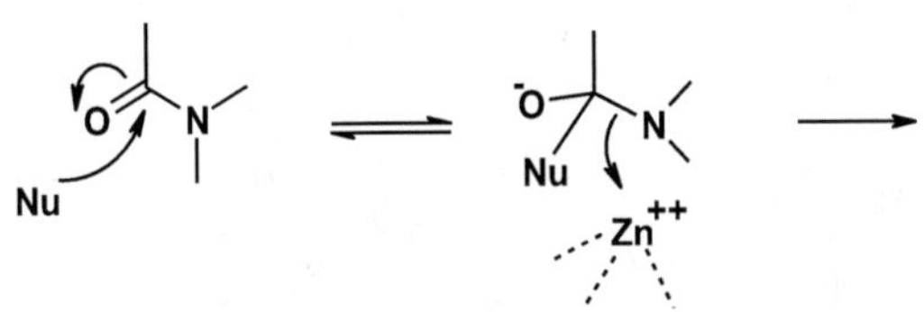

Scheme 8

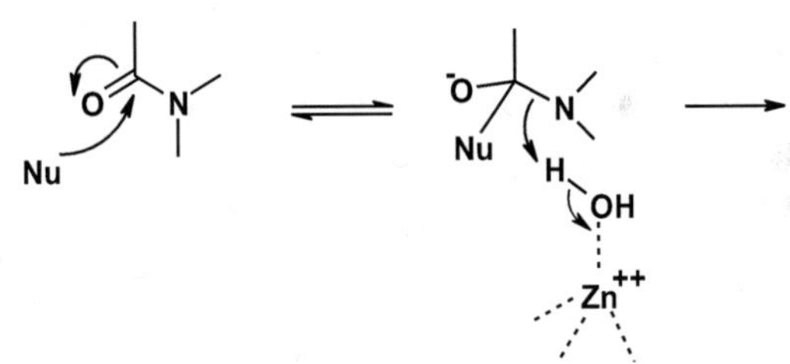

Scheme 9.

The Catalytic Role of Zinc

There are many potential mechanistic roles for the metal-ion in metallo-proteases and they may well vary from enzyme to enzyme [15, 116, 117]. The role of zinc in catalysis is related to its ability to participate in tight but readily exchangeable ligand binding and its exceptional flexibility of its coordination number and geometry. In addition, zinc shows no redox properties and this facilitates its evolution in living systems without the risk of oxidative damage. Finally, its intermediate hard-soft behaviour allows it to bind a variety of atoms, as seen, for example, in the second binding site of class B1 MBLs which involve nitrogen, oxygen and sulphur as ligands. The Lewis acidity, flexible geometry and co-ordination number and the lack of redox properties make zinc an ideal metal cofactor for many enzymes. The small energy difference between 4, 5 or 6 coordination geometries and the rapid exchange of the kinetically labile zinc-bound water molecule is an important feature in all zinc hydrolases, including MBLs.

It is commonly suggested that the metal-ion acts as a Lewis acid by co-ordination to the peptide carbonyl oxygen giving a more electron deficient carbonyl carbon which facilitates nucleophilic attack. The metal-ion thus stabilises the negative charge developed on the carbonyl oxygen of the tetrahedral intermediate anion (Scheme 6). Many metallo-proteases have a water molecule directly co-ordinated to the metal-ion which may act as the nucleophile to attack the carbonyl carbon. The role of the metal-ion is to lower the pK_a of the co-ordinated water so that the concentration of metal-bound hydroxide ion, albeit different, is increased relative to bulk solvent hydroxide ion at neutral pH and is a better nucleophile than water (Scheme 7). Although C-N bond fission is the most energetically difficult process in peptide hydrolysis, little attention is normally given to the mechanism of the breakdown of the tetrahedral intermediate. Breakdown of the tetrahedral intermediate could be facilitated by direct co-ordination of the departing amine nitrogen to the metal-ion (Scheme 8). This is the mechanism adopted for the zinc catalysed hydrolysis of penicillin. Alternatively, a metal-bound water could act as a general acid catalyst protonating the amine nitrogen leaving group to facilitate C-N bond fission [118] (Scheme 9). Despite intense mechanistic studies, the detailed roles of the metal-ion in metallo-proteases remain controversial and distinguishing between the relative importance of the possible roles for zinc is complex.

The effective positive charge on the zinc ion depends on the number and nature of its ligands. Coordination of a ligand, with an ionisable hydrogen, to zinc lowers its pK_a and the ionised ligand obviously is better at neutralising the positive charge density on the metal. The pK_a of water bound to zinc in aqueous solution is 9.5 but is there evolutionary pressure to lower the pKa of the zinc bound water in an enzyme? This could be achieved, for example, by replacing say a negatively charged carboxylate ligand by a neutral histidine. Changing zinc-bound histidine for aspartate in carbonic anhydrase increases the pK_a of the zinc-bound water from 6.8 to ≥ 9.6 [119]. Does a higher or lower pK_a metal-bound water lead to a faster reaction and more efficient catalysis? A low pK_a implies a more electron deficient metal-ion centre which would give a better Lewis acid to stabilise the negative charge developed on the oxyanion of the tetrahedral intermediate. Similarly, a high pK_a for the metal-bound water implies a weaker Lewis acid and so the zinc ion will be less efficient at stabilising the tetrahedral intermediate. Conversely, the lower the pK_a of metal-bound water the more "tightly bound" and stabilised is the resulting hydroxide-ion which, although it becomes the dominant species even at low pH, a low pK_a corresponds to a more weakly nucleophilic hydroxide-ion. For example, if the pK_a of the zinc bound water is about 5 then the nucleophilicity of the metal bound hydroxide-ion is only similar to that of a carboxylate anion. If a major role of the metal-ion is to provide a better nucleophile than water then the net effect depends on the relative importance of concentration and the dependence of the rate upon nucleophilicity. If the pK_a is "too high" metal-coordinated water will be the dominant species over the desired pH range but deprotonation will give a more nucleophilic metal-bound hydroxide. How reactivity changes with changing pK_a and pH will depend on the susceptibility of the rate of reaction to the basicity of the nucleophile - the hydroxide-ion bound to the metal – as indicated by the Bronsted β_{nuc} value.

It has been suggested that general base catalysis by a protein residue to remove the proton from the zinc bound water contributes to catalysis. A problem with the commonly accepted mechanism for metallo-proteases of rate-limiting deprotonation of zinc-bound water concerted with nucleophilic attack is the pK_a of this co-ordinated water molecule. Coordination to zinc(II) lowers the pK_a of water to 5-9 depending on the number and type of other ligands. The concentration of zinc-bound hydroxide ion is therefore quite high over the normal pH range studied. There is little or no catalytic advantage in having a general base remove a proton! Even if the pK_a of the zinc(II)-bound water is about 9 then 10% of the species already exists in the fully deprotonated form at pH 8 and 1% at pH 7. Presumably, the deprotonated form is a much better nucleophile than the species which is only partially deprotonated. There would be no catalytic advantage of the general base catalysed mechanism.

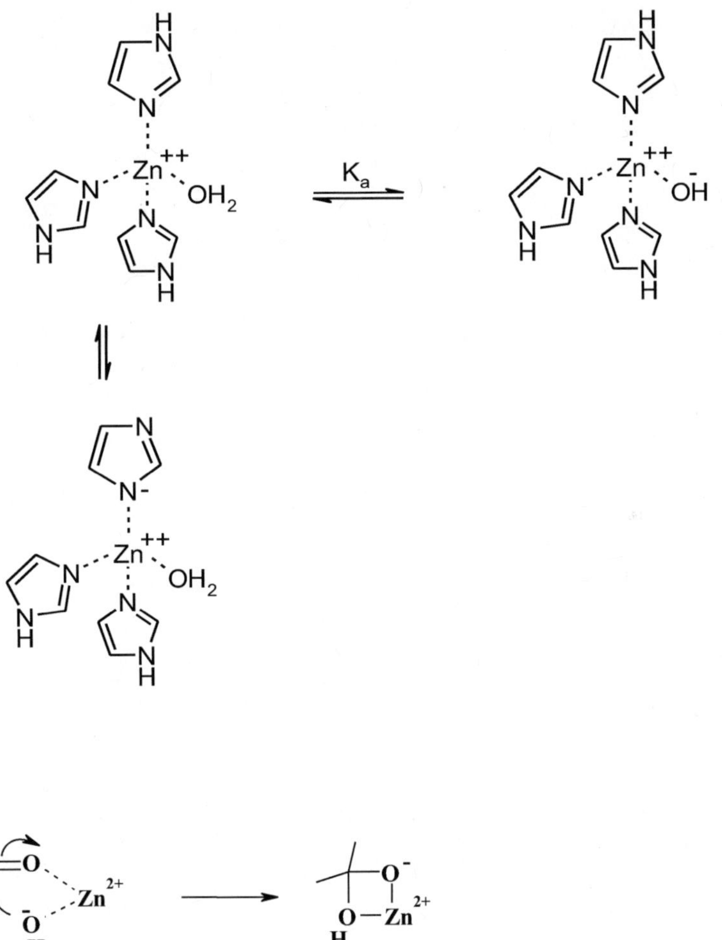

Scheme 10.

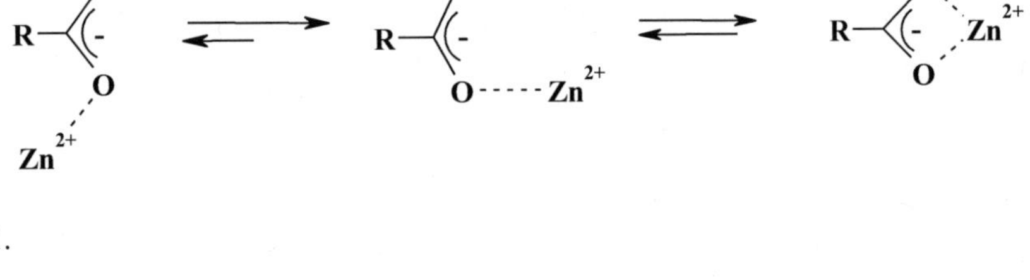

Scheme 11.

The zinc of *B. cereus* ß-lactamase is co-ordinated to three protein ligands in the zinc1 site - His 86, His 88 and His 149 and a water molecule [104]. In aqueous solution, zinc is coordinated to six water molecules and the pK_a of the zinc bound water is 9.5. The reduced coordination number of four in the zinc1 site of β-lactamase reduces the pKa to less than 6. In principle one or more of the imidazole residues could be ionised, the pKa for proton loss from imidazole is similar to that of water, about 14. However, NMR evidence suggests that the imidazole of the bound histidines are neutral and the most likely group to ionise is therefore the bound water (Scheme 10).

The simplest way to modify the pK_a of the zinc bound water is to change the ligands or the metal-ion. If the activity of the resulting enzyme is *inversely* proportional to the pK_a of the zinc-bound water ligand this would be compatible with the zinc coordinating to the carbonyl oxygen and stabilisation of the negative charge developed on this oxygen following nucleophilic attack. The lower pK_a of zinc-bound water indicates a more electrophilic zinc which is better at stabilising negative charge giving rise to a better catalyst. Stabilisation of the intermediate anion must be more important that the nucleophilicity of the zinc hydroxide. Conversely, if activity increases with basicity, *i.e.* with increasing pK_a then this could indicate a greater role for nucleophilicty of the zinc hydroxide ion compared with the Lewis acid role of the metal ion stabilising the negative charge development on the carbonyl oxygen.

The *dual* role of zinc in metalloproteases, *i.e.*, acting as a Lewis acid in polarising the carbonyl bond and as source of nucleophilic hydroxide, requires formally the formation of a strained four membered ring (Scheme 11). However, this coordination may occur in the bidentate complexes of zinc with carboxylate anions. Monodentate complexes between zinc and a carboxylate anion exist predominantly as the *syn* stereoisomer, due to the more basic *syn* lone pair and favourable opposition of the C=O and O-Zn dipoles. The coordination adopted for the *syn* conformation is similar to a bidentate one which is favoured by electrostatic interactions [120].

One final point to consider for the hydrolysis of β-lactams by metallo-β-lactamases is that if the zinc bound hydroxide-ion is the nucleophile then it is *consumed* during turnover. Consequently, regeneration of the catalyst requires the coordination of a new water molecule to the active site zinc. For example, if product release occurs through displacement by water or is introduced in a separate step following product release, *deprotonation* must occur to generate a catalytically active species (Scheme 12). It is conceivable that one of these steps may become kinetically significant under some conditions, particularly if the water is used both as a nucleophile and as a bridging ligand in binuclear enzymes..

Mechanisms

Although the metal ion requirement in MBLs catalysed hydrolysis of β-lactam antibiotics is still a matter of debate, catalytic mechanisms have been proposed for both the mono and dinuclear enzymes. The first catalytic mechanism suggested for BcII [104] (Scheme 13) was based on its crystal structure at a resolution of 2.5Å, where only one zinc ion was bound to the enzyme, in the histidine site. The main features of this mechanism resemble those often proposed for zinc peptidases such as carboxypeptidase A and thermolysin. Following substrate binding, the zinc bound water molecule, deprotonated by the Asp120 residue, attacks the carbonyl centre with the formation of a negatively charged tetrahedral intermediate, which is stabilized by its interactions with the metal ion. The Asp120 residue donates a proton to the nitrogen and C-N bond cleavage occurs, followed by product dissociation from the enzyme active site. The main disadvantage of this model is simply the unnecessary step of removing a proton from the zinc bound water as it is already ionised, although this may be required during turnover.

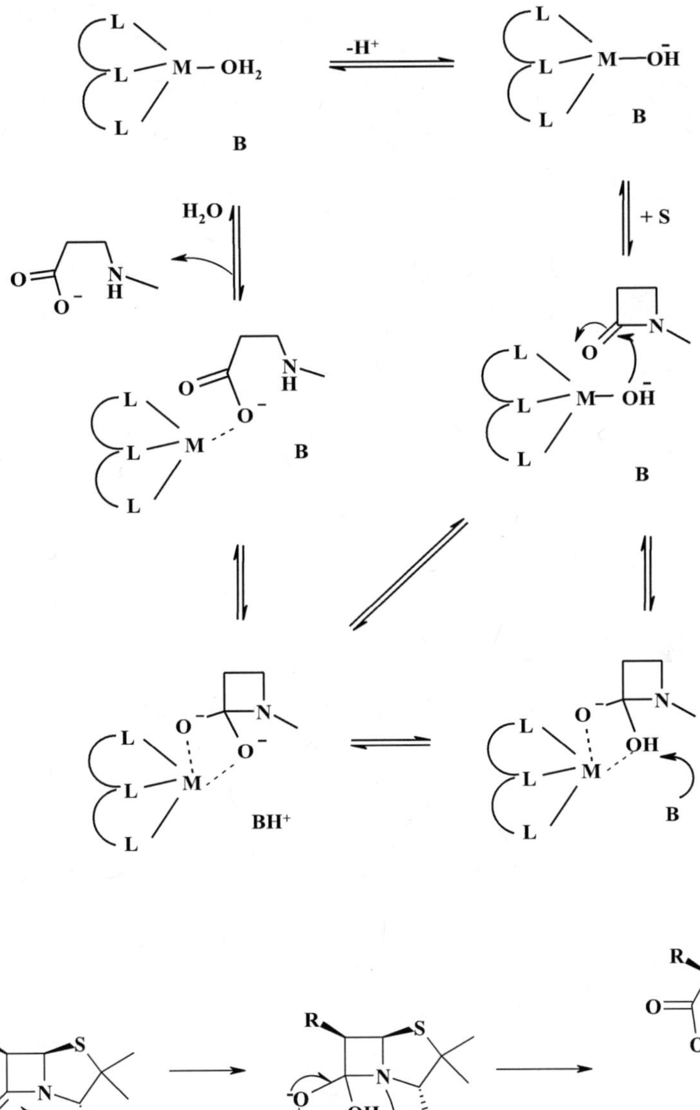

Scheme 12.

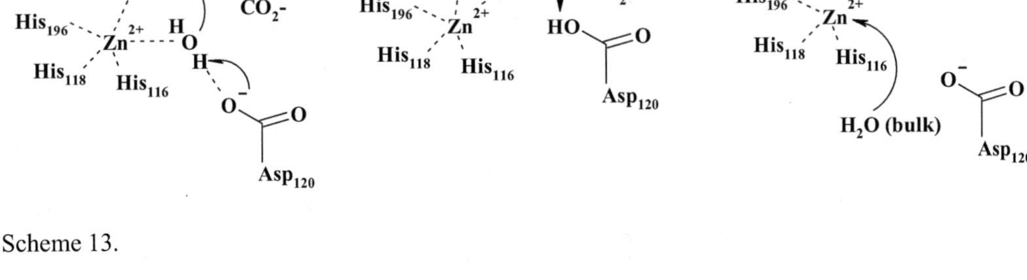

Scheme 13.

The pH-rate profile for the BcII catalysed hydrolysis of benzylpenicillin and cephaloridine [121] was taken to indicate that the zinc ion bound water has a low pK_a of <5 and is therefore fully ionized at neutral pH. Nucleophilic attack by the metal-bound hydroxide ion on the carbonyl followed by a proton abstraction from the Asp120 gives a dianionic tetrahedral intermediate (Scheme 14). It was suggested that the same aspartate residue functions as proton donor to facilitate C-N bond fission, and either step k_2 or k_3 could be rate limiting. A di-anionic intermediate assists β-lactam ring opening and generates a carboxylate anion rather than the undissociated acid.

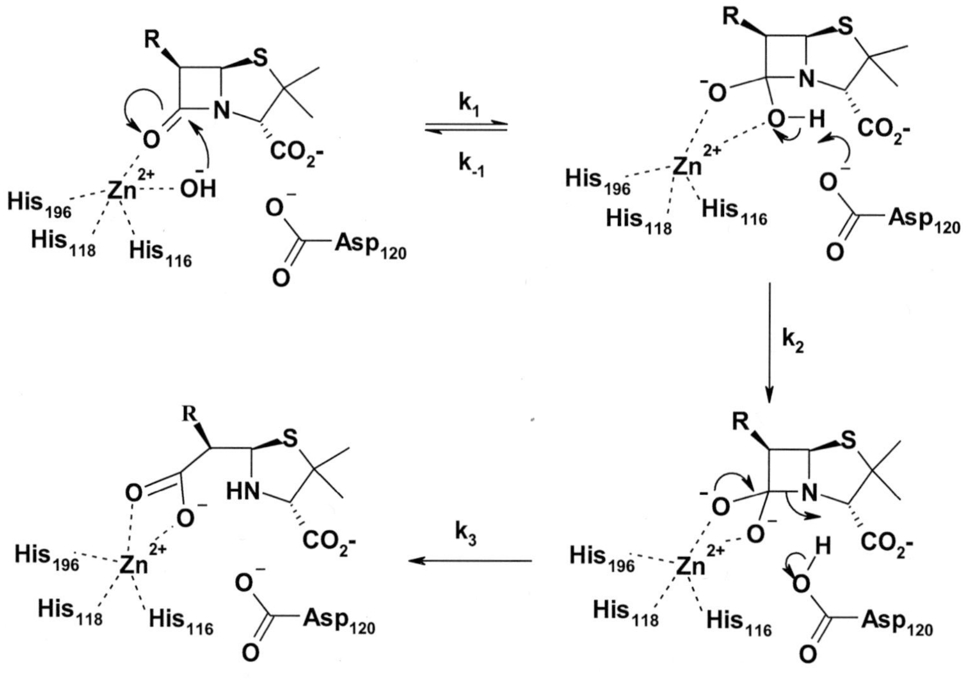

Scheme 14.

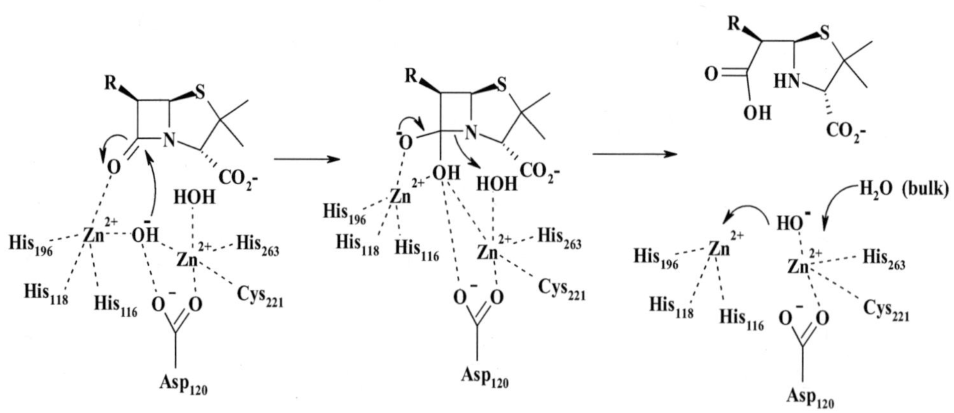

Scheme 15.

Based on the crystal structure of CcrA MBL from *Bacteroides fragilis* and on models of β-lactam substrates bound in the enzyme active site, a mechanism of hydrolysis for the dizinc enzyme has been suggested to be analogous to mechanisms proposed for other binuclear metallo-hydrolases [122]. In this mechanism, (Scheme 15), the bridging hydroxide ion is responsible for the nucleophilic attack which results in a negatively charged intermediate stabilized by the oxyanion hole. The apical water molecule bound to zinc is optimally positioned to donate a proton to the leaving nitrogen, and the newly formed hydroxide ion moves to occupy the vacated Wat1 site, followed by product dissociation from the enzyme active site. The original proposal involved zinc coordination to the β-lactam nitrogen, but this is unlikely because of its relatively low electron density due to amide resonance. However, nitrogen binding to zinc is more likely once the tetrahedral intermediate is formed because of its increased basicity.

With the unusual β-lactam but chromophoric nitrocefin as substrate for CcrA, rapid scanning and single-wavelength stopped-flow studies revealed the accumulation, during turnover, of an enzyme bound intermediate with an intense absorbance at 665nm [123, 124]. This was postulated to be due to the generation of a species in which the leaving nitrogen atom is not protonated during the cleavage of the C-N bond and remains negatively charged *(9)*. The anion is stabilized by extensive conjugation with the dinitrostyryl substituent in the C_3 position and by the zinc ion in the second active site and so is an atypical leaving group. Most amine anions are very unstable and are unlikely to be expelled without N-protonation.

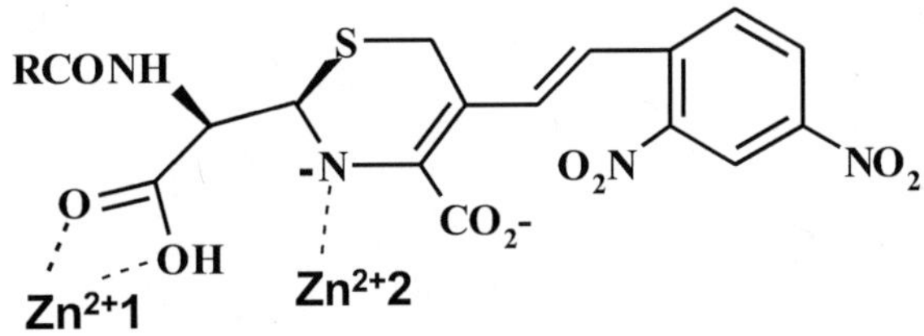

(9)

A similar intermediate has been observed during the class B3 L1 enzyme catalysed hydrolysis of nitrocefin, and, as in the case of CcrA, its breakdown was rate-limiting [125]. However, based on tryptophan fluorescence stop-flow studies, the mechanism of nitrocefin hydrolysis by binuclear metallo-β-lactamases may be atypical and cleavage of the β-lactam amide bond is the rate determining step for the breakdown of the majority of β-lactam substrates by the L1 enzyme [126].

In contrast to these results, no accumulation of an anionic intermediate could be detected in reactions catalysed by BcII and the rate of substrate depletion and product formation are similar [110]. However, for the BcII catalysed hydrolysis under cryoenzymological conditions, a slightly red shifted intermediate (440nm) was observed, which was suggested to be a nitrocefin-like tetrahedral intermediate [127].

An unusual mechanism has been proposed for subclass B2 metallo-β-lactamase CphA from *Aeromonas hydrophila* based on the crystal structures of the free enzyme and of a reaction intermediate between the enzyme and the substrate meropenem [128] (Scheme 16). Nucleophilic attack is performed by a water molecule *not* coordinated to zinc, but is activated by general base catalysis by His118, followed by C-N bond fission which occurs *prior* to nitrogen protonation. This proposal uses zinc as a Lewis acid to facilitate C-N bond fission and stabilize the leaving group – a mechanism proposed much earlier based on model studies [118]. Most mechanisms use the zinc ion to act as a Lewis acid to stabilize oxyanion formation from the β-lactam carbonyl oxygen. Rearrangement of the proposed intermediate amine anion, whose negatively charged nitrogen is stabilized by an interaction with the zinc ion, gives the intermediate shown (Scheme 16), which is observed in the crystal structure, although this rearrangement may not occur in the active site of the enzyme. A solvent molecule replacing the water used in the nucleophilic attack protonates the nitrogen and releases the final product.

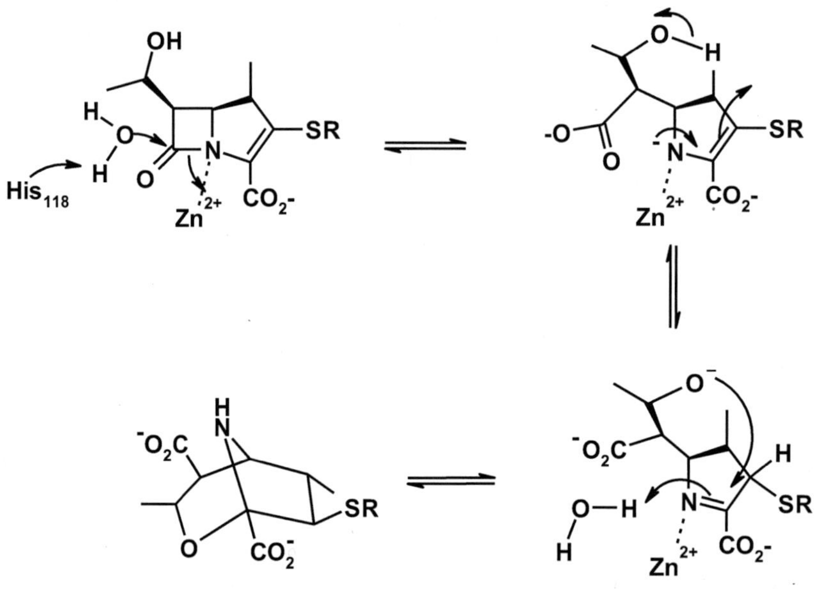

Scheme 16.

Mutation Studies with MBLs

Many mutant enzymes from the three subclasses have been produced and characterized. In general, the residues substituted have been the metal ligands, those providing the framework for orienting the metal ligands or those thought to interact with the substrates.

For BcII, replacing the histidine residues in the first binding site with serine, resulted in decrease of catalytic activity, although the affinity for zinc was not significantly altered [126]. The decrease in activity was mainly due to an increase in K_m which was rationalized on the basis that, in absence of a His ligand, there is a higher degree of flexibility of the substrate inside the catalytic site.

The replacement of the Cys221 residue with an alanine or a serine in the active sites of BcII [126], CcrA [127] and IMP-1 [128] gives a drastic decrease in the rate of hydrolysis at low zinc concentrations, but the mutants are "reactivated" by addition of excess zinc, at least in case of BcII and IMP1. It was proposed that the presence of Cys has a crucial importance for the catalytic activity of the monozinc enzyme but not for the dinuclear species [126].

Substituting the second zinc ligand Asp120 in BcII with Asn, reduces the catalytic activity by more than 100 fold but does not impair the binding of the second zinc ion [126]. Although these results support the hypothesis that Asp120 plays an important role in catalysis, the significant catalytic activity of the mutants is inconsistent with Asp playing the role of sole general base [101]. The same amino-acid substitution in subclass B3 MBL enzyme L1, led to an average 10 fold decrease in catalytic activity in the binuclear mutant [129]. It is interesting that, for the hydrolysis of nitrocefin, this mutant shows an unusually large kinetic solvent isotope effect (KSIE) of 5.36 compared with the wild type (KSIE = 2.08) and other aspartate mutants: Asp120Cys and Asp120Ser (KSIE<2). It is concluded that the Asp120 residue is essential for the catalytic activity in the orientation of the bridging group of the L1 enzyme in the protonation step rather than in the nucleophilic attack as previously suggested [130].

The mutation of the second shell ligand, Arg121Cys, in BcII gives an enzyme with a lower affinity for the second metal ion compared to the wild type [110]. Moreover, the electronic spectrum of the dicobalt Arg121Cys BcII mutant is identical to that of the wild type enzyme, indicating that the mutation has not altered the metal coordination sphere. The catalytic activity of the mutant is only two fold lower than that of the wild type for both the mono and dinuclear forms of the enzyme. The Cys104Arg mutation in CcrA led to a binuclear enzyme, which unlike the wild type, was able to loose one bound metal ion in presence of a chelator, and had significantly decreased activity [112]. Pre-steady state kinetics suggested a change in the rate limiting step in the hydrolysis of nitrocefin, from nitrogen protonation in the wild type, to C-N bond fision in the mutant, which showed a catalytic rate constant similar to that of the dizinc BcII enzyme. The monozinc form of the mutant manifested a similar value of k_{cat} with the monozinc BcII, for nitrocefin hydrolysis, with a further decrease of the rate of C-N bond fission compared with the dizinc mutant, but with a similar protonation rate. Based on these observations it was concluded that both the replacement of the arginine residue and the introduction of the second metal ion are an evolutionary tool for accelerating C-N bond fission in MBL catalysed hydrolysis of β-lactams. However, an important role in lowering the energy barrier for breaking the C-N bond must come from other enzyme structural rearrangements, since the reciprocal, i.e. the sole replacement of the arginine by a cysteine residue and the insertion of a second zinc ion in BcII, only marginally, i.e., two fold, improves the catalytic activity [110].

Another interesting mutagenesis study has been carried out on class B2 CphA enzyme from *Aeromonas hydrophila*, which is a special case of MBL, both regarding its narrow substrate profile and the metal ion requirement [131]. The Asn116 residue in site one was mutated to a histidine in an attempt to create a B1 type first binding site; this mutation changed the properties of the CphA enzyme toward a broader substrate profile (characteristic to subclass B1 and which is even more marked if the N220G mutation is added to the N116H one), as both penicillins and cephalosporins were significantly better substrates. Moreover, the activity of the Asn116His mutant increased with increasing metal concentration for the hydrolysis of benzylpenicillin and cephaloridine, as opposed to imipenem, where addition of zinc to both the mutant and the wild type lead to non-competitive inhibition. The Cys221Ser and Cys221Ala mutants were seriously impaired in their ability to bind the first, catalytic zinc ion and were nearly completely inactive, indicating a major role for Cys221 in binding of the catalytic metal ion.

METAL ION SUBSTITUTION

The exchange of the spectroscopically silent zinc in zinc enzymes with probes, such as cobalt, copper and cadmium, enables the study of the metal interactions in the enzyme active site with its ligands, substrates and inhibitors of the metallo-enzyme, using techniques such as electronic spectroscopy, NMR, EPR and PAC (Perturbed Angular Correlation) spectroscopy. The zinc of metallo β-lactamases can be exchanged with cadmium, cobalt and manganese to give catalytically active enzymes. The use of a combination of NMR and PAC spectroscopy to study cadmium binding to *B. cereus* MBL has revealed a rapid intramolecular exchange of the metal between the two sites in the mono-cadmium enzyme and negative cooperativity in metal binding [132]. The enzyme inhibitor (R)-thiomandelate induces a very strong positive cooperativity for binding the second cadmium cation [133].

The metal environment of cobalt substituted metallo β-lactamases has been studied by UV-VIS spectroscopy, NMR and ESR. The UV-VIS spectra of cobalt substituted class B1 metallo-β-lactamases CcrA from *Bacteroidis fragilis* [97, 134] and BcII from *Bacillus cereus,* [134, 135] show similar features with an intense S to Co(II) ligand to metal charge transfer transition (LMCT) band at 340 and 348nm and four characteristic d-d transition bands between 500 and 650 nm. The spectrum of Co-BcII in the visible region is also very similar to that of Co(II) carbonic anhydrase at alkaline pH [136], suggesting that the contributions from the 3 His metal centre dominate this part of the spectrum of Co substituted class B1 MBLs.

On titration of monozinc BcII with Co(II), the spectrum shows a LMCT band but without the four ligand field bands which suggests that zinc binds to the 3His site while cobalt preferentially occupies the Cys site in the heterosubstituted CoZn-enzyme [134]. A similar discrimination between metals in occupying the two binding sites in BcII was noted for the CdZnenzyme, where zinc also binds in the 3His site and Cd to the Cys site [132].

Studies of cobalt binding to apo-BcII, using HEPES as a buffer and no added sodium chloride, showed that increasing the ratio Co(II)/enzyme above one resulted in a decrease of the charge transfer band at 344nm and the appearance of an additional charge transfer band at 383nm. This has been explained by the binding of a second cobalt ion to the monocobalt BcII species and the existence of two different Co(II) LMCT (Ligand Metal Charge Transfer) band positions, one for the mononuclear (344 nm) and one for the binuclear enzyme (383nm); this latter band disappears upon addition of sodium chloride, accompanied by an increase in absorbance at 344nm, which led to the suggestion that chloride hinders the binding of a second cobalt ion [108].

Low temperature EPR spectra of class B1, B2 and B3 metallo-β-lactamases indicate, respectively, the presence of high-spin, but not coupled, Co(II) ions in a rhombically distorted pentacoordination sphere in Co substituted BcII [135], a penta/hexacoordination environment in Co substituted CcrA enzyme [97] and a tetracoordination environment for the first cobalt ion and penta/hexacoordination sphere for the second cobalt ion in Co substituted class B2 ImiS [137]. For this latter enzyme, it has been shown by EPR that the second cobalt ion is magnetically isolated, suggesting a distance of more than 7 Å from the first bound cobalt ion. Furthermore, NMR titration of the monocobalt ImiS enzyme with Co(II) indicated that the second cobalt ion is not bound to histidine, as the newly observed resonances are not solvent exchangeable; these findings were taken to indicate that the Zn1 site has no catalytic or metal binding role in ImiS, that Zn2 site binds the metal ion that is required for catalysis and a remote, lower affinity, metal binding site is responsible for the noncompetitive inhibition of the enzyme [129]. From the mechanistic point of view, it has been shown by time dependent UV-VIS spectroscopy and fluorescence quenching, using nitrocefin as a substrate, that Co substituted L1, class B3 metallo-β-lactamase, probably utilizes a reaction mechanism similar to that of the native zinc enzyme [126, 138]. Rapid-freeze-quench (RFQ) EPR shows that a short-lived intermediate is a metal-bound species, and the role of the metal ion in catalysis is similar for nitrocefin, cephalothin, meropenem and benzylpenicillin.

The kinetics and mechanism of hydrolysis of substrates catalysed by BcII substituted with various metal-ions have been investigated to determine the role of the active site metal-ion. [140] The pH and metal-ion dependence of k_{cat} and k_{cat}/K_m for the cobalt substituted BcII catalysed hydrolysis of cefoxitin, cephaloridine and cephalexin indicate that an enzyme residue of apparent pK_a 6.3 is required in its deprotonated form for metal-ion binding and catalysis. The k_{cat}/K_m for cefoxitin and cephalexin hydrolysis with cadmium substituted BcII is dependent on two ionising groups on the enzyme: one of pK_{a1} 8.7, required in its deprotonated form, and the other of pK_{a2} 9.3, required in its protonated form for activity. The identity of these residues was determined from the pH-dependence of the competitive inhibition constant, K_i, of the CdBcII by L-captopril which showed that the pK_{a1} of 8.7 corresponds to the cadmium bound water. For the manganese substituted BcII, the pH dependence of k_{cat}/K_m for the hydrolysis of β-lactam antibiotics similarly indicated the importance of two catalytic groups: one of pK_{a1} 8.5 which needs to be deprotonated and the other of pK_{a2} 9.4 which needs to be protonated for catalysis; the pK_{a1} was assigned to the manganese bound water. [140] Interestingly, the metal substituted enzymes have similar or higher catalytic activities compared with the native zinc enzyme, albeit at pHs above 7 and, for the Co enzyme, at all pHs (Figure 1). With cefoxitin, a very poor substrate for ZnBcII, both k_{cat} and k_{cat}/K_m increase with increasing pK_a of the metal bound water, in the order Zn<Co<Mn<Cd. A higher pK_a for the metal-bound water for cadmium and manganese BcII leads to more reactive enzymes than the native zinc BcII, suggesting that the role of the metal-ion is predominantly to provide the nucleophilic hydroxide, rather than to act as a Lewis acid to polarise the carbonyl group and stabilise the oxyanion tetrahedral intermediate. [140]

The bridging water is an important ligand helping to bind the second metal-ion along with the protein's amino acid residues in site 2 but it is *consumed* during the catalytic cycle of hydrolysis, so it is possible that the 'loss' of this ligand could result in the second metal-ion dissociating from the protein during turnover. In fact, the kinetics of the hydrolysis of benzylpenicillin catalysed by the cobalt substituted β-lactamase from BcII are biphasic with an initial burst of product formation followed by a steady-state rate of hydrolysis. [141] This was interpreted as being due to a branched kinetic pathway with two enzyme intermediate species, ES^1 and ES^2, which have different metal/enzyme stoichiometries. ES^1 is a di-cobalt enzyme intermediate and is catalytic, but is slowly loses one bound cobalt ion during turnover via the branching route, to give the mono-nuclear and inactive enzyme intermediate ES^2. The dependence of enzyme activity on pH and metal-ion concentration indicates that only the di-cobalt enzyme is catalytically active. A mono-cobalt enzyme species is formed during the catalytic cycle, which is virtually inactive and requires the association of another cobalt ion for turnover. The di-cobalt enzyme intermediate is responsible for the direct catalytic route, which is pH-independent between 5.5 and 9.5, but also slowly loses one bound cobalt ion via the branching route to give the mono-cobalt inactive enzyme intermediate. This inactivation pathway of metal-ion dissociation occurs by both an acid catalysed and a pH-independent reaction, which is dependent on the presence of an enzyme residue of pK_a 8.9 in its protonated form and shows a large kinetic solvent isotope effect (H_2O/D_2O) of 5.2, indicative of a rate limiting proton transfer. [141] The pseudo first-order rate constant to regenerate the di-cobalt β-lactamase from the mono-cobalt enzyme intermediate has a first-order dependence on cobalt ion concentration. This unusual behaviour is attributed to an intrinsic property of metallo hydrolytic enzymes that depend on a metal bound water both as a ligand for the second metal-ion and as the nucleophile which is consumed during hydrolysis of the substrate and so has to be replaced to maintain the catalytic cycle. [141,142,143]

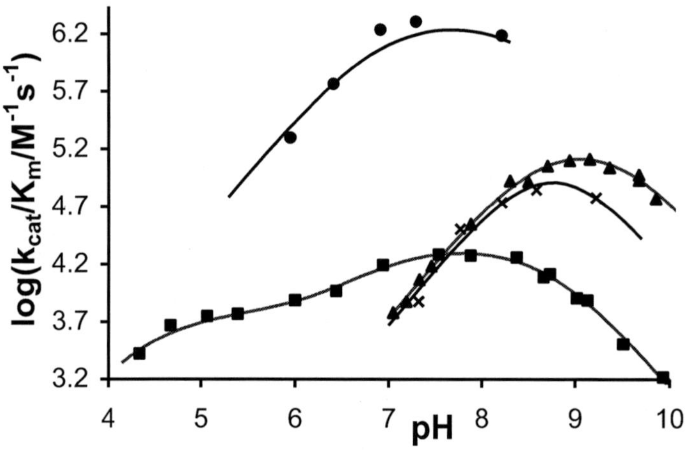

Figure 1. Plot of log k_{cat}/K_m for (■) ZnBcII, (●) CoBcII, (▲) CdBcII, (x) MnBcII catalysed hydrolysis of cephalexin.

There remain questions regarding the necessity of the binuclear metal centre itself: (i) what metallo-enzyme species is responsible for the physiological activity, the binuclear one or the mononuclear one, or are they interchangeable under certain conditions, and if so, (ii) are these changes relevant to any other functions, apart from catalysis, such as regulation of activity? It does seem intrinsically odd in the binuclear enzyme that the bridging hydroxide-ion is used as a nucleophile because of its presumed weak basicity. Hydroxide-ion has three lone pairs, two of which are taken up by metal-ion coordination and one of which is required for nucleophilic attack and covalent bond formation. If this bridging hydroxide-ion is used as the attacking nucleophile is it facilitated by general base catalysis so that concerted proton removal generates a di-anionic tetrahedral intermediate?

REFERENCES

[1] Waley, S.G., (1992), *The Chemistry of β-Lactams*, Ed. Page, M.I., Blackie, Glasgow, pp 198-226.

[2] Page, M.I., (1992), Structure-Activity Relationships: Chemical. *The Chemistry of β-Lactams*, Ed. Page, M.I., Blackie, Glasgow, pp 79-100.

[3] Page, M.I., (1987), The Mechanisms of Reactions of β-Lactam Antibiotics. *Adv. Phys. Org. Chem.*, 1987, 23, 165.

[4] Webster, P., Ghosez, L. and Page, M.I., (1990), The Hydrolysis of Azetidinyl Amidinium Salts Part 2. Substituent Effects, Buffer Catalysis and the Reaction Mechanism. *J. Chem. Soc. Perkin Trans 2*, 805.

[5] Davis, A.M., Proctor, P. and Page, M.I., (1991), The Alcohol Catalysed Hydrolysis of Benzylpenicillin. *J. Chem. Soc. Perkin Trans 2,* 1213.

[6] Page, M.I., (1991), The Energetics of Intramolecular Reactions and Enzyme Catalysis. *Phil. Trans. R. Soc. London Ser B*, 332, 149.

[7] Martin, A.F., Morris, J.J. and Page, M.I., (1979), Intramolecular General Acid Catalysis in the Aminolysis of Benzylpenicillin. A Preferred Direction of Nucleophilic Attack. *J.Chem. Soc. Chem. Commun.*, 298; Gensmantel, N.P. and Page, M.I., (1979), The Aminolysis of Penicillin Derivatives. Rate Constants for the Formation and Breakdown of the Tetrahedral Addition Intermediate. *J. Chem. Soc. Perkin Trans. 2*, 137; Morris, J.J. and Page, M.I., (1980), Intra- and Inter-molecular Catalysis in the Aminolysis of Benzylpenicillin. *J. Chem. Soc. Perkin Trans. 2*, 212; Llinas-Marti, A., Vilanova, B., Frau, J., Munoz, F., Donoso, J. and Page, M.I., (1998), The Chemical Reactivity of Pencilllins and Cephalosporins. Intramolecular Involvement of the Acyl-Amido Side Chain. *J.Org.Chem.* 63, 9052.

[8] Brown, R.P.A., Aplin, R.T. and Schofield, C.J., Inhibition of TEM-2 β-Lactamase from *Escherichia coli* by Clavulanic Acid: Observation of Intermediates by Electrospray Ionization Mass Spectrometry, *Biochemistry,* 1996, 35, 12421.

[9] Fisher, J., Belasco, J.G., Khosla, S. and Knowles, J.R., β.-Lactamase proceeds via an acyl-enzyme intermediate. Interaction of the Escherichia coli RTEM enzyme with cefoxitin, *Biochemistry*, 1980, 19, 2895.

[10] Page, M.I., Vilanova, B. and Layland, N.J., (1995), The pH Dependence of and Kinetic Solvent Isotope Effects on the Methanolysis and Hydrolysis of β-Lactams Catalysed by Class C β-Lactamase. *J. Amer. Chem. Soc.* 117, 12092.

[11] Christensen, H., Martin, M.T. and Waley, S.G., β-Lactamases as fully efficient enzymes. Determination of all the rate constants in the acyl-enzyme mechanism *Biochem. J.*, 1990, 266, 853.

[12] Strynadka, N.C.J., Adachi, H., Jensen, S.E., Johns, K. Sielecki, A., Betzel, C., Sutoh, K. and James, M.N.G., *Nature*, 1992, 359, 700.

[13] Page, M.I., (1973), The Energetics of Neighbouring Group Participation. *Chem. Soc. Revs.*, 2, 295; Page, M.I. and Jencks, W.P., (1987), In Defence of Entrophy and Strain as Explanations for the Rate Enhancement shown in Intramolecular Reactions. *Gazz. Chim. Ital.*, 117, 455.

[14] J J. Burbaum, R T. Raines, W. J Albery, J R. Knowles (1989) Evolutionary optimization of the catalytic effectiveness of an enzyme, *Biochemistry;*; 28; 9293.

[15] Page, M.I. and Williams, A. in "Organic and Bioorganic Mechanisms" Longman 1997

[16] Thatcher, G.R.J. and Kluger, R., *Adv. Phys. Org. Chem.* (1989) 25, 99.

[17] Page, M.I., Laws, A.P., Slater, M.J. and Stone, J.R., (1995), Reactivity of β-Lactams and Phosphonamidates and Reactions with β-Lactamase. *Pure Appl.* Chem. 67, 711; Laws, A.P., Page, M.I. and Slater, M.J., (1993), The Mechanism of Reactions Catalysed by the Serine β-Lactamases. *Bioorg Med. Chem. Lett.*, 3, 2317.

[18] Page, M.I., Hinchliffe, P.S., Wood, J.M. and Laws, A.P., (2003), Novel Mechanism of Inhibiting β-Lactamases by Sulfonylation using β-Sultams. *Bioorg Med. Chem. Letts.,*

13, 4489; Ahmed, N., Harding, C.P., Hinchliffe, P.S., Tsang, W.Y., Wood, J.M., Laws, A.P. and Page, M.I., (2005), Different Transition State Structures for the Reactions of β-Lactams and Analogous β-Sultams with Serine β-Lactamases. *J. Amer. Chem. Soc.* 127, 17556.

[19] Joris, B., Ghuysen, J.-M., Dive, G., Renard, A., Dideberg, O., Charlier, P. and Frere, J.-M. (1988). The active-site-serine penicillin-recognizing enzymes as members of the Streptomyces R61 DD-peptidase family. *Biochem.J.*, 250, 313-324.

[20] Maveyraud, L., Golemi, D., Kotra, L.P., Tranier, S., Vakulenko, S., Mobashery, S. and Samama, J.P. (2000). Insights into class D beta-lactamases are revealed by the crystal structure of the OXA10 enzyme from Pseudomonas aeruginosa. *Structure.*, 8, 1289-1298.

[21] Galleni, M., Lamotte-Brasseur, J., Raquet, X., Dubus, A., Monnaie, D., Knox, J.R. and Frere, J.M. (1995). The enigmatic catalytic mechanism of active-site serine beta-lactamases. *Biochem.Pharmacol.*, 49, 1171-1178.

[22] Matagne, A., Lamottebrasseur, J. and Frere, J.M. (1998). Catalytic properties of class A ß-lactamases: efficiency and diversity. *Biochemical Journal*, 330, 581-598.

[23] Samraoui, B., Sutton, B.J., Todd, R.J., Artymiuk, P.J., Waley, S.G. and Phillips, D.C. (1986). Tertiary structural similarity between a class A beta-lactamase and a penicillin-sensitive D-alanyl carboxypeptidase-transpeptidase. *Nature*, 320, 378-380.

[24] Adachi, H., Ohta, T. and Matsuzawa, H. (1991). Site-directed mutants, at position 166, of RTEM-1 beta-lactamase that form a stable acyl-enzyme intermediate with penicillin. *J.Biol.Chem*, 266, 3186-3191.

[25] Atanasov, B.P., Mustafi, D. and Makinen, M.W. (2000). Protonation of the beta-lactam nitrogen is the trigger event in the catalytic action of class A beta-lactamases. *Proc.Natl.Acad.Sci.U.S.A*, 97, 3160-3165.

[26] Ellerby, L.M., Escobar, W.A., Fink, A.L., Mitchinson, C. and Wells, J.A. (1990). The role of lysine-234 in beta-lactamase catalysis probed by site-directed mutagenesis. *Biochemistry*, 29, 5797-5806.

[27] Escobar, Tan, Lewis and Fink. (1994). Site-directed mutagenesis of glutamate-166 in beta-lactamase leads to a branched path mechanism. *Biochemistry*, 33, 7619-26.

[28] Hata, M., Fujii, Y., Ishii, M., Hoshino, T. and Tsuda, M. (2000). Catalytic mechanism of class A beta-lactamase. I. The role of Glu166 and Serl30 in the deacylation reaction. *Chem Pharm.Bull.* (Tokyo), 48, 447-453.

[29] Knox, J.R., Moews, P.C., Escobar, W.A. and Fink, A.L. (1993). A catalytically-impaired class-a beta-lactamase - 2 angstrom crystal structure and kinetics of the bacillus-licheniformis E166A mutant. *Protein Engineering*, 6, 11-18.

[30] Lamotte-Brasseur, J., Dive, G., Dideberg, O., Charlier, P., Frere, J.M. and Ghuysen, J.M. (1991). Mechanism of acyl transfer by the class A serine beta-lactamase of Streptomyces albus G. *Biochem.J.*, 279 (Pt 1), 213-221.

[31] Lietz, E.J., Truher, H., Kahn, D., Hokenson, M.J. and Fink, A.L. (2000). Lysine-73 is involved in the acylation and deacylation of beta-lactamase. *Biochemistry,* 39, 4971-4981.

[32] Minasov, G., Wang, X. and Shoichet, B.K. (2002). An ultrahigh resolution structure of TEM-1 beta-lactamase suggests a role for Glu166 as the general base in acylation. *J.Am.Chem Soc.*, 124, 5333-5340.

[33] Nukaga, M., Mayama, K., Hujer, A.M., Bonomo, R.A. and Knox, J.R. (2003). Ultrahigh resolution structure of a class A beta-lactamase: on the mechanism and specificity of the extended-spectrum SHV-2 enzyme. *J.Mol.Biol.*, 328, 289-301.

[34] Strynadka, Adachi, H., Jensen, Johns, K., Sielecki, A., Betzel, C., Sutoh, K. and James (1992). Molecular structure of the acyl-enzyme intermediate in beta-lactam hydrolysis at 1.7 A resolution. *Nature*, 359, 700-5.

[35] Page, M.I. and and Laws, A.P. (1998). The mechanism of catalysis and the inhibition of beta-lactamases. *Chem.Commun.*, 1609-1617.

[36] Massova, I. and Kollman, P.A. (2002). pKa, MM, and QM studies of mechanisms of beta-lactamases and penicillin-binding proteins: acylation step. *J.Comput.Chem*, 23, 1559-1576.

[37] Escobar,W.A., Tan, A.K. and Fink, A.L. (1991). Site-directed mutagenesis of beta-lactamase leading to accumulation of a catalytic intermediate. *Biochemistry*, 30, 10783-10787.

[38] Fonze, E., Vanhove, M., Dive, G., Sauvage, E., Frere, J.M. and Charlier, P. (2002). Crystal structures of the Bacillus licheniformis BS3 class A beta-lactamase and of the acyl-enzyme adduct formed with cefoxitin. *Biochemistry*, 41, 1877-1885.

[39] Hokenson, M.J., Cope, G.A., Lewis, E.R., Oberg, K.A. and Fink, A.L. (2000). Enzyme-induced strain/distortion in the ground-state ES complex in beta-lactamase catalysis revealed by FTIR. *Biochemistry,* 39, 6538-6545.

[40] Cartwright, S.J., Tan, A.K. and Fink, A.L. (1989). Trapping the acyl-enzyme intermediate in beta-lactamase I catalysis. *Biochem J*, 263, 905-12.

[41] Brannigan, J., Matagne, A., Jacob, F., Damblon, C., Joris, B., Klein, D., Spratt, B.G. and Frere, J.M. (1991). The mutation lys234his yields a class a beta-lactamase with a novel ph-dependence. *Biochemical Journal*, 278 (Pt 3), 673-8.

[42] Lenfant, F., Labia, R. and Masson, J.M. (1991). Replacement of lysine 234 affects transition state stabilization in the active site of beta-lactamase tem1. *Journal Of Biological Chemistry*, 266, 17187-94.

[43] Delaire, M., Lenfant, F., Labia, R. and Masson, J.M. (1991). Site-directed mutagenesis on tem-1 beta-lactamase: role of glu166 in catalysis and substrate binding. *Protein Engineering,* 4, 805-10.

[44] Gibson, R.M., Christensen, H. and Waley, S.G. (1990). Site-directed mutagenesis of beta-lactamase-i - single and double mutants of glu-166 and lys-73. *Biochemical Journal*, 272, 613-619.

[45] Jacob, F., Joris, B., Lepage, S., Dusart, J. and Frere, J.M. (1990). Role of the conserved amino acids of the 'sdn' loop (ser130, asp131 and asn132) in a class a beta-lactamase studied by site-directed mutagenesis. *Biochemical Journal*, 271, 399-406.

[46] Thomas, V.L., Golemi-Kotra, D., Kim, C., Vakulenko, S.B., Mobashery, S. and Shoichet, B.K. (2005). Structural consequences of the inhibitor-resistant Ser130Gly substitution in TEM beta-lactamase. *Biochemistry*, 44, 9330-9338.

[47] Golemi-Kotra, D., Meroueh, S.O., Kim, C., Vakulenko, S.B., Bulychev, A., Stemmler, A.J., Stemmler, T.L. and Mobashery, S. (2004). The importance of a critical protonation state and the fate of the catalytic steps in class A beta-lactamases and penicillin-binding proteins. *J.Biol.Chem*, 279, 34665-34673.

[48] Damblon, C., Raquet, X., Lian, L.Y., Lamotte-Brasseur, J., Fonze, E., Charlier, P., Roberts, G.C. and Frere, J.M. (1996). The catalytic mechanism of beta-lactamases:

NMR titration of an active-site lysine residue of the TEM-1 enzyme. *Proc.Natl.Acad.Sci.U.S.A*, 93, 1747-1752; Lamotte-Brasseur,J., Lounnas,V., Raquet,X. and Wade,R.C. (1999) pKa calculations for class A beta-lactamases: influence of substrate binding. *Protein Sci.*, 8, 404-409; Raquet,X., Lounnas,V., Lamotte-Brasseur,J., Frère,J.-M. and Wade,R.C. (1997) pK_a Calculations for Class A ß-Lactamases: Methodological and Mechanistic Implications. *Biophys.J.*, 73, 2416-2426.

[49] Ibuka, A.S., Ishii,Y., Galleni, M., Ishiguro, M., Yamaguchi, K., Frere, J.M., Matsuzawa, H. and Sakai, H. (2003). Crystal structure of extended-spectrum beta-lactamase Toho-1: insights into the molecular mechanism for catalytic reaction and substrate specificity expansion. *Biochemistry*, 42, 10634-10643.

[50] Meroueh, S.O., Fisher, J.F., Schlegel, H.B. and Mobashery, S. (2005). Ab initio QM/MM study of class A beta-lactamase acylation: dual participation of Glu166 and Lys73 in a concerted base promotion of Ser70. *J.Am.Chem Soc.*, 127, 15397-15407.

[51] Chen, C.C. and Herzberg, O. (1999). Relocation of the catalytic carboxylate group in class A beta-lactamase: the structure and function of the mutant enzyme Glu166-->Gln:Asn170-->Asp. *Protein Eng*, 12, 573-579.

[52] Chen, C.C. and Herzberg, O. (2001). Structures of the acyl-enzyme complexes of the Staphylococcus aureus beta-lactamase mutant Glu166Asp:Asn170Gln with benzylpenicillin and cephaloridine. *Biochemistry*, 40, 2351-2358.

[53] Duval, V., Swinnen, M., Lepage, S., Brans, A., Granier, B., Franssen, C., Frere, J.M. and Joris, B. (2003). The kinetic properties of the carboxy terminal domain of the Bacillus licheniformis 749/I BlaR penicillin-receptor shed a new light on the derepression of beta-lactamase synthesis. *Mol.Microbiol.*, 48, 1553-1564.

[54] Wang, X., Minasov, G. and Shoichet, B.K. (2002). Noncovalent interaction energies in covalent complexes: TEM-1 beta-lactamase and beta-lactams. *Proteins*, 47, 86-96.

[55] Guillaume, G., Vanhove, M., Lamotte-Brasseur, J., Ledent, P., Jamin, M., Joris, B. and Frere, J.M. (1997). Site-directed mutagenesis of glutamate 166 in two beta-lactamases. Kinetic and molecular modeling studies. *J.Biol.Chem.*, 272, 5438-5444.

[56] Leung, Robinson, Aplin and Waley. (1994). Site-directed mutagenesis of beta-lactamase i: role of glu-166. *Biochemical Journal*, 299 (Pt 3), 671-8.

[57] Mustafi, D., Sosa-Peinado, A. and Makinen, M.W. (2001). ENDOR structural characterization of a catalytically competent acylenzyme reaction intermediate of wild-type TEM-1 beta-lactamase confirms glutamate-166 as the base catalyst. *Biochemistry*, 40, 2397-2409.

[58] Escobar, W.A., Tan, A.K., Lewis, E.R. and Fink, A.L. (1994). Site-directed mutagenesis of glutamate-166 in beta-lactamase leads to a branched path mechanism. *Biochemistry*, 33, 7619-7626.

[59] Fink, A.L., Behner, K.M. and Tan, A.K. (1987). Kinetic and structural characterization of reversibly inactivated beta-lactamase. *Biochemistry*, 26, 4248-58.

[60] Beadle, B.M. and Shoichet, B.K. (2002). Structural bases of stability-function tradeoffs in enzymes. *J.Mol.Biol.*, 321, 285-296.

[61] Beadle,B.M. and Shoichet, B.K. (2002). Structural basis for imipenem inhibition of class C beta-lactamases. *Antimicrob.Agents Chemother.*, 46, 3978-3980.

[62] Crichlow, G.V., Kuzin, A.P., Nukaga, M., Mayama, K., Sawai, T. and Knox, J.R. (1999). Structure of the extended-spectrum class C beta-lactamase of Enterobacter

cloacae GC1, a natural mutant with a tandem tripeptide insertion. *Biochemistry*, 38, 10256-10261.

[63] Crichlow, G.V., Nukaga, M., Doppalapudi,V.R., Buynak, J.D. and Knox, J.R. (2001). Inhibition of class C beta-lactamases: structure of a reaction intermediate with a cephem sulfone. *Biochemistry*, 40, 6233-6239.

[64] Lobkovsky, E., Billings, Moews, Rahil, J., Pratt and Knox (1994). Crystallographic structure of a phosphonate derivative of the enterobacter cloacae p99 cephalosporinase: mechanistic interpretation of a beta-lactamase transition-state analog. *Biochemistry*, 33, 6762-72.

[65] Lobkovsky, E., Moews, Liu, H., Zhao, H., Frere and Knox. (1993). Evolution of an enzyme activity: crystallographic structure at 2-a resolution of cephalosporinase from the ampc gene of enterobacter cloacae p99 and comparison with a class a penicillinase. *Proceedings Of The National Academy Of Sciences Of The United States Of America*, 90, 11257-61.

[66] Nukaga, M., Mayama, K., Crichlow, G.V. and Knox, J.R. (2002). Structure of an extended-spectrum class A beta-lactamase from Proteus vulgaris K1. *J.Mol.Biol.*, 317, 109-117.

[67] Oefner, C., D'Arcy, A., Daly, J.J., Gubernator, K., Charnas, R.L., Heinze, I., Hubschwerlen, C. and Winkler, F.K. (1990). Refined crystal structure of beta-lactamase from Citrobacter freundii indicates a mechanism for beta-lactam hydrolysis. *Nature,* 343, 284-288.

[68] Powers, R.A., Caselli, E., Focia, P.J., Prati, F. and Shoichet, B.K. (2001). Structures of ceftazidime and its transition-state analogue in complex with AmpC beta-lactamase: implications for resistance mutations and inhibitor design. *Biochemistry,* 40, 9207-9214.

[69] Powers, R.A., Morandi, F. and Shoichet, B.K. (2002). Structure-based discovery of a novel, noncovalent inhibitor of AmpC beta-lactamase. *Structure.*, 10, 1013-1023.

[70] Powers, R.A. and Shoichet, B.K. (2002) Structure-based approach for binding site identification on AmpC beta-lactamase. *J.Med.Chem*, 45, 3222-3234.

[71] Tondi, D., Morandi, F., Bonnet, R., Costi, M.P. and Shoichet, B.K. (2005) Structure-based optimization of a non-beta-lactam lead results in inhibitors that do not up-regulate beta-lactamase expression in cell culture. *J.Am.Chem Soc.*, 127, 4632-4639.

[72] Tondi, D., Powers, R.A., Caselli, E., Negri, M.C., Blazquez, J., Costi, M.P. and Shoichet, B.K. (2001) Structure-based design and in-parallel synthesis of inhibitors of AmpC beta-lactamase. *Chem Biol.*, 8, 593-611.

[73] Trehan, I., Beadle, B.M. and Shoichet, B.K. (2001) Inhibition of AmpC beta-lactamase through a destabilizing interaction in the active site. *Biochemistry*, 40, 7992-7999.

[74] Trehan, I., Morandi, F., Blaszczak, L.C. and Shoichet, B.K. (2002) Using steric hindrance to design new inhibitors of class C beta-lactamases. *Chem Biol.*, 9, 971-980.

[75] Dubus, A., Ledent, P., Lamotte-Brasseur, J. and Frere, J.M. (1996) The roles of residues Tyr150, Glu272, and His314 in class C beta-lactamases. *Proteins*, 25, 473-485.

[76] Dubus, A., Normark, S., Kania, M. and Page, M.G. (1994) The role of tyrosine 150 in catalysis of beta-lactam hydrolysis by ampc beta-lactamase from escherichia coli investigated by site-directed mutagenesis. *Biochemistry*, 33, 8577-86.

[77] Dubus, A., Normark, S., Kania, M. and Page, M.G. (1995) Role of asparagine 152 in catalysis of beta-lactam hydrolysis by Escherichia coli AmpC beta-lactamase studied by site-directed mutagenesis. *Biochemistry*, 34, 7757-7764.

[78] Goldberg, S.D., Iannuccilli,W., Nguyen,T., Ju, J. and Cornish,V.W. (2003) Identification of residues critical for catalysis in a class C beta-lactamase by combinatorial scanning mutagenesis. *Protein Sci.*, 12, 1633-1645.

[79] Jacobs, C., Dubus, A., Monnaie, D., Normark, S. and Frere. (1992) Mutation of serine residue 318 in the class c beta-lactamase of enterobacter cloacae 908r. Fems *Microbiology Letters*, 71, 95-100.

[80] Vakulenko, S.B., Golemi, D., Geryk, B., Suvorov, M., Knox, J.R., Mobashery, S. and Lerner, S.A. (2002) Mutational replacement of Leu-293 in the class C Enterobacter cloacae P99 beta-lactamase confers increased MIC of cefepime. *Antimicrob.Agents Chemother.*, 46, 1966-1970.

[81] Zawadzke, L.E., Chen, C.C., Banerjee, S., Li, Z., Wasch, S., Kapadia,G., Moult, J. and Herzberg, O. (1996) Elimination of the hydrolytic water molecule in a class a beta-lactamase mutant: crystal structure and kinetics. *Biochemistry*, 35, 16475-82.

[82] Monnaie, D., Dubus, A. and Frere. (1994) The role of lysine-67 in a class c beta-lactamase is mainly electrostatic. *Biochemical Journal*, 302 (Pt 1), 1-4.

[83] Kato-Toma,Y., Iwashita, T., Masuda, K., Oyama,Y. and Ishiguro, M. (2003) pKa measurements from nuclear magnetic resonance of tyrosine-150 in class C beta-lactamase. *Biochem.J.*, 371, 175-181.

[84] Tsukamoto, K., Tachibana, K., Yamazaki, N., Ishii,Y., Ujiie, K., Nishida, N. and Sawai,T. (1990) Role of lysine-67 in the active site of class C beta-lactamase from Citrobacter freundii GN346. *Eur.J.Biochem.*, 188, 15-22.

[85] Diaz, N., Suarez, D. and Sordo, T.L. (2006) Molecular Dynamics Simulations of Class C beta-Lactamase from Citrobacter freundii: Insights into the Base Catalyst for Acylation. *Biochemistry*, 45, 439-451.

[86] Adediran, S.A., Deraniyagala, S.A., Xu,Y. and Pratt, R.F. (1996) Beta-secondary and solvent deuterium kinetic isotope effects on beta- lactamase catalysis. *Biochemistry*, 35, 3604-3613.

[87] Adediran, S.A. and Pratt, R.F. (1999) Beta-secondary and solvent deuterium kinetic isotope effects on catalysis by the Streptomyces R61 DD-peptidase: comparisons with a structurally similar class C beta-lactamase. *Biochemistry*, 38, 1469-1477.

[88] Knox, J.R., Moews, P.C. and Frere, J.M. (1996) Molecular evolution of bacterial beta-lactam resistance. *Chemistry & Biology*, 3, 937-947.

[89] Gherman, B.F., Goldberg, S.D., Cornish,V.W. and Friesner, R.A. (2004) Mixed quantum mechanical/molecular mechanical (QM/MM) study of the deacylation reaction in a penicillin binding protein (PBP) versus in a class C beta-lactamase. *J.Am.Chem Soc.*, 126, 7652-7664.

[90] Maveyraud, L., Golemi-Kotra, D., Ishiwata, A., Meroueh, O., Mobashery, S. and Samama, J.P. (2002) High-resolution X-ray structure of an acyl-enzyme species for the class D OXA-10 beta-lactamase. *J.Am.Chem Soc.*, 124, 2461-2465.

[91] Golemi, D., Maveyraud, L., Vakulenko, S., Samama, J.P. and Mobashery, S. (2001) Critical involvement of a carbamylated lysine in catalytic function of class D beta-lactamases. *Proc.Natl.Acad.Sci.U.S.A*, 98, 14280-14285.

[92] Ledent, P. and Frere, J.M. (1993) Substrate-induced inactivation of the OXA2 beta-lactamase. *Biochem J*, 295, 871-8.

[93] Keith, K.E., Oyston, P.C., Crossett, B., Fairweather, N.F., Titball, R.W., Walsh, T.R. and Brown, K.A. (2005) Functional characterization of OXA-57, a class D beta-lactamase from Burkholderia pseudomallei. *Antimicrob.Agents Chemother.*, 49, 1639-1641.

[94] Frere, J. M., Beta-lactamases and bacterial resistance to antibiotics, *Molecular Microbiology*, 1995, *16(3)*, 385-395 .

[95] Payne, D. J., Metallo-beta-lactamases-a new therapeutic challenge, *J. Med. Microbiol,* 1993,. *39*, 93-99.

[96] Galleni, M., Lamotte-Brasseur, J., Rossolini, G. M., Spencer, J., Dideberg, O. and Frère, J. M., Standard numbering scheme for class B β-lactamases, *Antimicrob. Agents Chemother.,* 2001, *45*, 660-663.

[97] Crowder, M. W., Wang, Z., Franklin, S. L., Zovinka, E. P., and Benkovic, S. J., Characterization of the metal-binding sites of the β-lactamase from *Bacteroides fragilis, Biochemistry,* 1996, *35*, 12126-12132.

[98] Felici, A., Amicosante, G., Oratore, A., Strom, R., Ledent, P., Joris, B., Fanuel, L., and Frere, J. M., An overview of the kinetic parameters of class B β-lactamases, *Biochem. J.*, 1993, *291*, 151-155.

[99] Felici, A., Amicosante, G., Kinetic analysis of extension of substrate specificity with *Xanthomonas maltophilia*, *Aeromonas hydrophila*, and *Bacillus cereus* metallo-β-lactamases, *Antimicrob. Agents & Chemother.*, 1995, *39*, 192-199.

[100] Simm, A. M., Higgins, C. S., Carenbauer, A. L., Crowder, M. W., Bateson, J. H., Bennett, P. M., Clarke, A. R., Halford, S. E., and Walsh, T. R., Characterization of monomeric L1 metallo-β-lactamase and the role of the N-terminal extension in negative cooperativity and antibiotic hydrolysis, *J. Biol. Chem.*, 2002, *277*, 24744-24752.

[101] Herzberg, O., and Fitzgerald, P., Metallo-β-lactamases, *Handbook of Metalloproteins, vol.3*, eds. Messerschmidt, A., Bode, W., Cygler, M., John Wiley & Sons, Ltd, Chichester, 2004, 217-234 .

[102] Fitzgerald, P. M., Wu, J. K., and Toney, J. H., Unanticipated inhibition of the metallo-β-lactamase from *Bacteroides fragilis* by 4-morpholineethanesulfonic acid (MES): a crystallographic study at 1.85-A resolution, *Biochemistry,* 1998, *37*, 6791-6800 .

[103] Concha, N. O., Rasmussen, B. A., Bush, K., and Herzberg, O., Crystal structure of the cadmium- and mercury-substituted metallo-β-lactamase from Bacteroides fragilis, *Protein Sci.,* 1997, *6*, 2671-2676 .

[104] Carfi, A., Duee, E., Galleni, M., Frère, J. M. and Dideberg, O., 1.85 A resolution structure of the zinc (II) β-lactamase from *Bacillus cereus*, *Acta Crystallogr. Biol. Crystallogr.* 1998, *D54*, 313-323.

[105] Carfi, A., Pares, S., Duee, E., Galleni, M., Duez, C., Frère, J. M., and Dideberg, O., The 3-D structure of a zinc metallo-β-lactamase from Bacillus cereus reveals a new type of protein fold, *EMBO J.,* 1995, *14*, 4914-4921.

[106] Concha, N. O., Janson, C. A., Rowling, P., Pearson, S., Cheever, C. A., Clarke, B. P., Lewis, C., Galleni, M., Frère, J. M., Payne, D. J., Bateson, J. H. and Abdel-Meguid, S. S., Crystal structure of the IMP-1 metallo β-lactamase from Pseudomonas aeruginosa and its complex with a mercaptocarboxylate inhibitor: binding determinants of a potent, broad-spectrum inhibitor, *Biochemistry,* 2000, *39*, 4288-4298.

[107] Fabiane, S. M., Sohi, M. K., Wan, T., Payne, D. J., Bateson, J. H., Mitchell, T. and Sutton, B. J., Crystal structure of the zinc-dependent β-lactamase from Bacillus cereus at 1.9 A resolution: binuclear active site with features of a mononuclear enzyme, *Biochemistry,* 1998, *37*, 12404-12411.

[108] Paul-Soto, R., Bauer, R., Frère, J. M., Galleni, M., Meyer-Klaucke, W., Nolting, H., Rossolini, G. M., de Seny, D., Hernandez-Valladares, M., Zeppezauer, M. and Adolph, H. W., Mono- and binuclear Zn^{2+}-β-lactamase. Role of the conserved cysteine in the catalytic mechanism, *J. Biol. Chem.*, 1999, *274*, 13242-13249.

[109] De Seny, D, Heinz, U., Wommer, S., Kiefer, M., Meyer-Klaucke, W.,Galleni, M., Frere, J. M., Bauer, R., and Adolph, H. W., Metal ion binding and coordination geometry for wild type and mutants of metallo-β-lactamase from *Bacillus cereus 569/H/9* (BcII); a combined thermodynamic, kinetic and spectroscopic approach, *J. Biol. Chem.,* 2001, *276 (48)*, 45065-45078.

[110] Wommer, S., Rival, S., Heinz, U., Galleni, M., Frere, J. M., Franceschini, N., Amicosante, G., Rasmussen, B., Bauer, R., and Adolph, H. W., Substrate activated zinc binding of metallo-β-lactamases; physiological importance of the mononuclear enzymes, *J. Biol. Chem.,* 2002, *277 (27)*, 24142-24147.

[111] Rasia, R. M., and Vila, A. J., Exploring the role and the binding affinity of a second zinc equivalent in *B. cereus* metallo-β-lactamase, *Biochemistry,* 2002, *41*, 1853-1860.

[112] Rasia, R. M., and Vila, A. J., Mechanistic study of the hydrolysis of nitrocefin mediated by *B. cereus* metallo-β-lactamase, *ARKIVOC*, 2003, *x*, 507-516.

[113] Paul-Soto, R., Hernadez-Valladares, M., Galleni, M., Bauer, R., Zeppezauer, M., Frère, J. M. and Adolph, H. W., Mono- and binuclear Zn^{2+}-β-lactamase from Bacteroides fragilis: catalytic and structural roles of the zinc ions, *FEBS Lett.,* 1998, *438*, 137-140 .

[114] Fast, W., Wang, Z., and Benkovic, S. J., Familial mutations and zinc stoichiometry determine the rate-limiting step of nitrocefin hydrolysis by metallo-β-lactamase from *Bacteroides fragilis*, *Biochemistry,* 2001, *40*, 1640-1650.

[115] Hernandez Valladares, M., Felici, A., Weber, G., Adolph, H. W., Zeppezauer, M., Rossolini, G. M., Amicosante, G., Frère, J. M. and Galleni, M., Zn(II) dependence of the Aeromonas hydrophila AE036 metallo-β-lactamase activity and stability, *Biochemistry,* 1997, *36*, 11534-11541.

[116] Crowder, M. W., Walsh, T. R., Banovic, L., Pettit, M. and Spencer, J., Overexpression, purification, and characterization of the cloned metallo-β-lactamase (L1) from *Stenotrophomonas maltophilia, Antimicrob. Agents Chemother.* 1998, *42*, 921-926.

[117] Mercuri, P. S., Bouillenne, F., Boschi, L., Lammote-Brasseur, J., Amicosante, G., Devreese, B., Van Beeumen, J., Frere, J. M., Rossolini, G. M., and Galleni, M., Biochemical characterization of the FEZ-1 metallo-β-lactamase of *Legionella gormanii* ATCC 33297^T produced in *Escherichia coli*, *Antimicrob. Agents & Chemother.*, 2001, *45*, 1254-1262.

[118] Weston, J., Mode of action of bi- and trinuclear zinc hydrolases and their synthetic analogues. *Chem. Rev*. 2005, *105*, 2151-2174.

[119] Parkin, G., Synthetic analogues relevant to the structure and function of zinc enzymes, *Chem. Rev*. 2004, *104*, 699-767.

[120] Gensmantel, N.P, Proctor, P. and Page, M.I., Metal-ion Catalysed Hydrolysis of some β-Lactam Antibiotics, *J. Chem. Soc. Perkin II*, 1980, 1725-1732.

[121] Kiefer, L.L. and Fierke, C.A., Functional characterisation of human carbonic anhydrase II variants with altered zinc binding sites, *Biochemistry*, *33,* 15233.

[122] Christianson, D. W., and Lipscomb, W. N., Comparison of carboxypeptidase A and thermolysin: inhibition by phosphonamidates, *J. Am. Chem. Soc.*, 1988, *110*, 5560-5565.

[123] Bounaga, S., Laws, A. P., Galleni, M., and Page, M. I., The mechanism of catalysis and the inhibition of the Bacillus cereus zinc-dependent β-lactamase, *Biochem J.,* 1998, *331*, 703-711.

[124] Concha, N. O., Rasmussen, B. A., Bush, K. and Herzberg, O. Crystal structure of the wide-spectrum binuclear zinc β-lactamase from Bacteroides fragilis, *Structure*, 1996, *4*, 823-836.

[125] Wang, Z., Fast, W., and Benkovic, Direct observation of an enzyme-bound Intermediate in the catalytic cycle of the metallo-β-lactamase from *Bacteroides fragilis*, *J. Am Chem. Soc.*, 1998, *120*, 10788-10789.

[126] Wang, Z., Fast, W. and Benkovic, S. J., On the mechanism of the Bacteroides fragilis metallo-β-lactamase, *Biochemistry,* 1999, *38*, 10013-10023.

[127] McManus-Munoz, S., and Crowder, M. W., Kinetic mechanism of metallo-β-lactamase from *Stenotrophomonas maltophilia, Biochemistry,* 1999, *38*, 1547-1553.

[128] Spencer, J., Clarke, A. R., and Walsh, T. R., Novel mechanism of hydrolysis of therapeutic β-lactamas by *Stenotrophomonas maltophilia* L1 metallo-β-lactamase, *J. Biol. Chem.*, 2001, *276*, 33638-33644.

[129] De Seny, D., Prosperi-Meys, C., Bebrone, C., Rossolini, G. M., Page, M. I., Noel, P., Frere, J. M., and Galleni, M., Mutational analysis of the two zinc-binding sites of the *Bacillus cereus 569/H/9* metallo-β-lactamase, *Biochem. J.*, 2002, *363*, 687-696.

[130] Bicknell, R., and Waley, S. G., Cryoenzymology of *Bacillus cereus* β-lactamase II, *Biochemistry,* 1985, *24*, 6876-6887.

[131] Yang, Y., Keeney, D., Tang, X., Canfield, N. and Rasmussen, B. A., Kinetic properties and metal content of the metallo-β -lactamase CcrA harboring selective amino acid substitutions, *J. Biol. Chem.,* 1999, *274*, 15706-15711.

[132] Haruta, S., Yamaguchi, H., Yamamoto, E. T., Eriguchi, Y., Nukaga, M., O'Hara, K. and Sawai, T., Functional analysis of the active site of a metallo-β-lactamase proliferating in Japan, *Antimicrob. Agents Chemother.*, 2000, *44*, 2304-2309.

[133] Garrity, J. D., Bennet, B., and Crowder, M. W., Direct evidence that the reaction intermediate of metallo-β-lactamase L1 is metal bound, *Biochemistry,* 2005, *44*, 1078-1087.

[134] Ullah, J. H., Walsh, T. R., Taylor, I. A., Emery, D. C., Verma, C. S., Gamblin, S. J. and Spencer, J., The crystal structure of the L1 metallo-β-lactamase from Stenotrophomonas maltophilia at 1.7 A resolution, *J. Mol. Biol.,* 1998, *284*, 125-136.

[135] Vanhove, M., Zakhem, M., Devreese, B., Franceschini, N., Anne, C., Bebrone, C., Amicosante, G., Rossolini, G. M., Van Beeumen, J., Frere, J. M., and Galleni, M., Role of Cys221 and Asn116 in the zinc-binding sites of the Aeromonas hydrophila metallo-β-lactamase, *Cell. Mol. Life Sci.*, 2003, *60*, 1-9.

[136] Hemmingsen, L., Damblon, C., Antony, J., Jensen, M., Adolph, H.W., Wommer, S., Roberts, G.C.K., and Bauer, R., Dynamics of mononuclear cadmium beta-lactamase revealed by the combination of NMR and PAC spectroscopy, *J. Amer. Chem. Soc.,* 2001, *123*, 10329-10335.

[137] Damblon, C., Jensen, M., Ababou, A., Barsukov, I., Papamicael, C., Schofield, C. J., Olsen, L., Bauer, R., and Roberts, G. C., The inhibitor thiomandelic acid binds to both metal ions in metallo-beta-lactamase and induces positive cooperativity in metal binding, *J. Biol. Chem.*, 2003, *31*, 29240-29251.

[138] Orellano, E. G., Girardini, J. E., Cricco, J. A., Ceccarelli, E. A. and Vila, A. J., Spectroscopic characterization of a binuclear metal site in Bacillus cereus β-lactamase II, *Biochemistry,* 1998, *37*, 10173-10180 .

[139] Wang, Z., and Benkovic, S. J., Purification, characterization, and kinetic studies of a soluble *Bacteroides fragilis* metallo-β-lactamase that provides multiple antibiotic resistance, *J. Biol. Chem.,* 1998, *273 (35)*, 22402-22408 .

[140] Myers, J. L., and Shaw, R. W., Production, purification and spectral properties of metal-dependent β-lactamase from *Bacillus cereus, Biochim. Biophys. Acta,* 1989, *995,* 264-272 .

[141] Bertini, L., Luchinat, C., Cobalt(II) as a probe of the structure and function of carbonic anhydrase, *Acc. Chem. Res.,* 1983, *16,* 272.

[142] Crawford, P. A., Yang, K. W., Sharma, N., Bennett, B., and Crowder, M. W., Spectroscopic studies on cobalt(II) – substituted metallo-β-lactamase ImiS from *Aeromonas veronii* bv. *sobria, Biochemistry,* 2005, *44*, 5168-5176.

[143] Crowder, M. W., and Walsh, T. R., Structure and function of metallo-β-lactamases, *Recent Res. Devel. Antimicrob. Agents & Chemother.*, 1999, *3*, 105-132.

[144] Crawford, P. A., Sharma, N., Chandrasekar, S., Sigdel, T., Walsh, T. R., Spencer, J., and Crowder, M. W., Over-expression, purification, and characterization of metallo-β-lactamase ImiS from *Aeromonas veronii* bv. *Sobria*, *Protein Expression Purif.*, 2004, *36*, 272-279.

[145] Jacquin,O., Balbeur, D., Damblon, C., Marchot, P.,Pauw, E. De, Roberts, G.C.K., Frere, J-M. And Matagne, A. Positively cooperative binding of zinc ions to *Bacillus cereus* 569/H/9 β-lactamase II suggests that the binuclear enzyme is the only relevant form for catalysis *J. Mol. Biol*., in press.

[146] Badarau, A. and Page, M.I., The Variation of Catalytic Efficiency of *Bacillus cereus* Metallo-β-lactamase with Different Active Site Metal Ions, *Biochemistry,* 2006, *45,* 10654 .

[147] Badarau, A. and Page, M.I., Enzyme Deactivation Due to Metal-Ion Dissociation during Turnover of the Cobalt-β-Lactamase Catalyzed Hydrolysis of β-Lactams *Biochemistry,* 2006, *45,* 11012.

[148] Badarau, A., Damblon, C. and Page, M.I. The Activity of the Dinuclear Cobalt-β-Lactamase from *Bacillus cereus* in Catalysing the Hydrolysis of β-Lactams, *Biochem. J.*, 2007, *401,* 197-203.

[149] Badarau, A. and Page, M.I., Loss of enzyme activity during turnover of the *Bacillus cereus* β-lactamase catalysed hydrolysis of β-lactams due to loss of zinc ion, *J. Biol. Inorg. Chem.,* 13, 919-928 (2008).

[150] Bebrone C., Delbrück H., Kupper M.B., Philipp Schlömer P., Willmann C., Frère J-M., Fischer R., Galleni M, and Hoffmann K.M.V(2009) The Structure of the Dizinc Subclass B2 Metallo-β-Lactamase CphA Reveals that the Second Inhibitory Zinc Ion Binds in the Histidine Site. *Antimicrobial Ag. Chemother*. 53, 4464-4471.

In: Beta-Lactamases
Editor: Jean-Marie Frère, pp. 199-216

ISBN: 978-1-61324-638-2

Chapter 8

INSIGHTS INTO THE MECHANISMS OF β-LACTAMASES FROM NUCLEAR MAGNETIC RESONANCE STUDIES

Christian Damblon and Gordon Roberts
Henry Wellcome Laboratories for Structural Biology
Department of Biochemistry, University of Leicester,
Leicester LE1 9HN, UK.

ABSTRACT

Nuclear Magnetic resonance (NMR) provides a substantial amount of structural and dynamic information short of a full protein structure. This technique has been successfully utilised to study the role of various residues in the catalytic mechanisms of the four classes of β-lactamases. In the active-site serine enzymes, NMR has been useful in elucidating the nature of the residue which acts as a general base in activating the active serine and for the metallo- β-lactamases, the focus has been on studies of the active-site metal ions environment.

INTRODUCTION

Nuclear magnetic resonance spectroscopy (NMR) is second only to X-ray crystallography in the detailed structural information it can provide on proteins and their complexes. With the rapid advance in NMR techniques over the last 15 years, it is now possible to use NMR to obtain high-resolution structures of proteins in solution, albeit with a limitation in the size of protein whose complete structure can be determined – currently, for most purposes, about 45 kDa. As yet, NMR has been little used to determine the structures of β-lactamases, although the structures of the class B metallo-β-lactamase from *Bacillus cereus* and of its complex with an inhibitor have been determined in this way [1]. One of the major advantages of NMR is that, since NMR allows one to observe individual atoms within a complex molecular structure such as a protein it provides a substantial amount of structural and dynamic information short of a full structure determination. It is this kind of experiment that NMR has, to date, been most fruitfully applied to the study of β-lactamases, in studying enzyme-ligand interactions, metal-protein interactions and protein folding and determining ionisation constants.

To illustrate these contributions of NMR to the study of β-lactamases we will consider each of the different classes of β-lactamase in turn. For the serine β-lactamases, classes A, C & D, NMR has been particularly useful in addressing mechanistic questions – notably the nature of the residue which acts as a general acid/base in activating the active site serine. For the metallo-β-lactamases, class B, the focus has been on studies of the active site metal ions and on studies of the binding of candidate inhibitors.

Class A β-Lactamases

Because of the continuous antibiotic resistance problems linked to class A β-lactamases, studies to improve understanding of the catalytic properties of these enzymes and the search for inhibitors or inactivators continue to be important subjects of research.

Although class A β-lactamases are mechanistically closely analogous to the serine proteases such as chymotrypsin in which a serine sidechain is the nucleophile which attacks the amide bond of the substrate, in the β-lactamases the base which activates the serine is not a histidine as it is in chymotrypsin. Two different hypotheses have been proposed for the identity of this general base: glutamate 166 [2, 3] or lysine 73 [4] would be the general base. The second hypothesis would imply an unusually low pK_a for the ε-amino group of Lys73, postulated to be due to a strongly positive environment in the active site. These two possibilities can thus be distinguished by determining the ionization state of the corresponding residues, particularly Lys73. Other than in exceptionally high-resolution structures, X-ray crystallography cannot generally determine directly the charge state of ionisable groups in proteins, and NMR is the method of choice. Titration of an ionisable group is commonly accompanied by large chemical shift changes which have been used to determine pK_a values of individual groups within enzyme active sites since the earliest days of protein NMR, and this is an area where NMR has made major contributions to understanding catalytic mechanisms of enzymes. The main difficulty in this approach has always been the unambiguous identification of the signals corresponding to the amino acid side chain of interest. Specific isotope labelling and/or mutagenesis have commonly been used to help in the assignment procedure.

In the case of the TEM-1 β-lactamase, two independent studies have used selective ^{13}C-labelling of the lysine ε-CH_2 to determine the pK values of the lysine side chains by NMR [5, 6] . The fact that the ε-CH_2 is immediately adjacent to the amino group makes this methylene group an excellent probe to monitor the ionization state of the amino group. The selectively labelled methylene carbon can be indirectly detected by its directly bonded protons and the pK can be determined by the variation of the chemical shift of both protons and carbon as a function of pH. The difficulty, then, lies in the identification of the signals coming from Lys 73 among those from a total of 11 lysine residues in the enzyme. Two different strategies have been used for this, and since the two reports come to different conclusions they will be described separately.

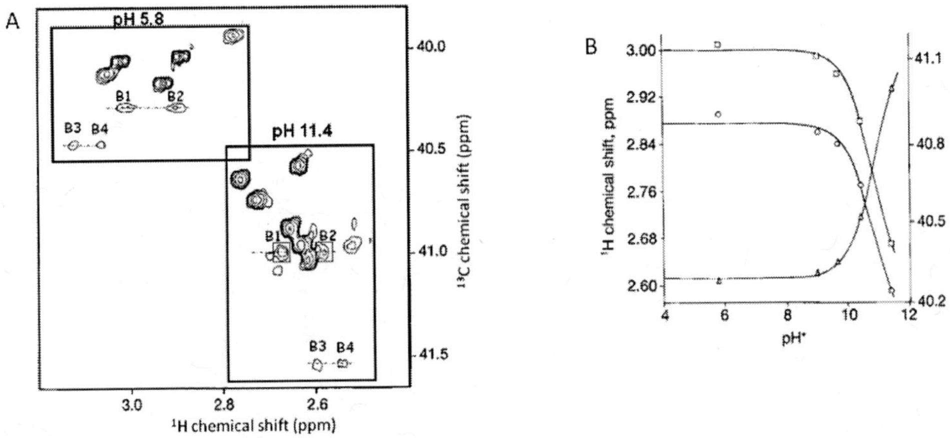

Figure 1. pH titration of the TEM β-lactamase Lys73 by NMR. (A) ^{1}H–^{13}C HSQC spectra of [^{13}C]lysine-labeled β-lactamase at pH* 5.8 and pH* 11.4. (B) pH* dependence of the ^{1}H and ^{13}C chemical shifts of peaks B1 and B2 (arising from Lys73). □ and ○, ^{1}H chemical shifts for B1 and B2, respectively; Δ, ^{13}C chemical shifts for B1 and B2. The notation pH* indicates a direct meter reading, not corrected for the deuterium isotope effect of 0.4 unit on the glass electrode. For nitrogen bases the isotope effect on the ionization equilibrium is opposite to that on the glass electrode, so that a pK_a estimated from meter readings in $^{2}H_2O$ is close to that in H_2O. From ref 5, reprinted by permission of PNAS.

Figure 1A shows the ^{1}H-^{13}C correlation experiment on selectively [ε-$^{13}CH_2$] lysine-labelled enzyme reported by Damblon *et al.* [5]. In general, the two prochiral protons of the ε-CH_2 group of a lysine residue in a protein are predicted to have different chemical shifts; however, since they are located relatively far from the chiral centres and most lysine residues in proteins are on the surface and show highly dynamic behaviour, significant shift non-equivalence is not generally observed. Thus in Figure 1A, the majority of the lysine residues give rise to single cross peaks, representing a correlation between the ε-^{13}C and both of the two methylene protons. However, for two residues a clear non-equivalence of the two ε-CH_2 methylene protons is seen (labelled B1-B2 and B3-B4). This kind of non-equivalence will only be observed if the rotation within the side chain is restricted by interactions with other residues of the protein. A similar behaviour has been described in studies of a domain of DNA polymerase λ, where chemical shift non-equivalence was observed for 3 out of 10 lysine residues and could readily be explained by the environment of these residues in the protein [7]. In the TEM-1 β-lactamase, Damblon *et al.* concluded from the crystal structure that only the two lysine residues in the active site, Lys73 and Lys234, were likely to be motionally restricted so as to lead to this non-equivalence. They then carried out an HMQC-NOESY experiment, which allowed the observation of nuclear Overhauser effects (NOEs) between the ε-CH_2 protons and other protons in the protein. A cross-section at the ^{13}C frequency of peaks B1 and B2 revealed an NOE to a proton with a resonance frequency of 7.65 ppm, characteristic of an aromatic residue. Thus, one of the protons of this ε-CH_2 is within 5Å of an aromatic ring proton. The crystal structure of the enzyme shows that the only situation in which an aromatic ring is close enough to a lysine ε-CH_2 to give rise to an NOE signal is Lys73/Phe 72 where the internuclear distance is 2.35Å. On this basis peaks B1-B2

(Figure 1) were assigned to the ε-CH_2 of Lys73. It is clear that this assignment depends upon the proposition that the solution and crystal structure are the same, or at least closely similar; the next closest lysine ε-CH_2 □ aromatic proton distance is 5.3Å (for Lys34/Phe60), and for the assignment to be incorrect the crystal-solution difference would have to be such as to interchange the relative magnitudes of these distances. The pH-dependence of the ^{1}H and ^{13}C chemical shifts of peaks B1-B2 are shown in Figure 1B. It is clear that only incomplete titration curves are obtained, limited by the stability of the enzyme at high pH; since the determination of pK values depends critically on the availability of accurate values for the chemical shifts at the two extremes of the titration curves, the data do not permit the determination of the pK of Lys73. However, they do demonstrate that the pK of Lys73 must be above 10, and hence that this residue cannot be the base responsible for activating the catalytic serine [5].

The second study [6] used mutagenesis to assign the ε-CH_2 resonances, specifically a mutant in which Lys73 was replaced by alanine, the principle being that the signal in the ^{1}H-^{13}C spectrum of [ε-$^{13}CH_2$]lysine-labelled enzyme which is present in the wild-type but missing in the mutant indicates which signal corresponds to the Lys73. A comparison of the spectra of the wild-type and mutant enzymes at pH 6.0 revealed no very obvious difference, presumably due to the considerable overlap of peaks at this pH. At pH 9.3, however, the authors identified a low intensity peak in the wild-type spectrum which was not present in that of the mutant. However, this was not the only difference between these two spectra. This highlights the central problem with the use of mutants for resonance assignments: the substitution can produce changes in structure and/or chemical shifts which can lead to ambiguities. The authors fitted the pH-dependence of the chemical shifts of the low intensity peak to a pK of ~8.5. However, as noted above the pK value obtained depends critically on having accurate values for the chemical shifts at the two extremes, and good high-pH values are clearly not available in this case. Moreover, the chemical shift variations between the fully protonated and the free amino group are significantly lower than expected, for both proton (0.06 vs 0.4 ppm) and carbon (0.1 vs 0.8-2.0 ppm). Finally, Gao et al [7] showed that the experimental data in both studies could be interpreted on the basis of pK values of 10.2-11.0 for the lysine amino group(s) under consideration.

These two studies demonstrate the value of NMR in determining ionization constants, but emphasise the care needed in resonance assignment and in the interpretation of the experimental data.

Lysine 234 is highly conserved among class A β-lactamases and is the second lysine residue pointing into the active site; it has been assumed to participate in binding of the substrate through the formation of a salt-bridge with the free carboxylate on C-3 of penicillins or C-4 of cephalosporins [8]. In the *Streptomyces albus G* β-lactamase this lysine has been replaced by a histidine residue in order to establish whether this replacement would confer a characteristic pH dependence for the catalytic parameters of the mutant enzyme. Since the wild-type enzyme contains no histidine residues, the pK of the sole histidine residue in the mutant enzyme was easily determined by NMR using the characteristic resonance of the imidazole ring C2 proton, yielding a pK of 6.44 ± 0.13, in very reasonable agreement with the value of 6.15 ± 0.02 determined from the pH-dependence of the k_{cat} and k_{cat}/K_m values [9].

As yet, only limited NMR studies of ligand binding to class A β-lactamases have been reported. In 1994, it was shown that during the very slow hydrolysis of cefoxitin by the

Bacillus licheniformis enzyme, a conformational change occurred in the acyl-enzyme complex [10]. ^{1}H NMR showed that both methyl resonances around 0 ppm and aromatic resonances were perturbed during turnover (Figure 2). These are linked, since the appearance of methyl resonances around 0 ppm is commonly a consequence of the ring current shift caused by proximity to aromatic residues. Since it was established that the shifts observed were not caused by a direct interaction of the affected methyl groups with the substrate, a conformational change within the enzyme was invoked. Further NMR observation suggested that the methyl/aromatic pairs were unlikely to be localized within the active site of the enzyme. The structure of a complex between cefoxitin and another class A β-lactamase of *B. licheniformis* differing only by seven point mutations has been later determined by X-Ray crystallography and this has confirmed the conformational change qualitatively indicated by NMR [11].

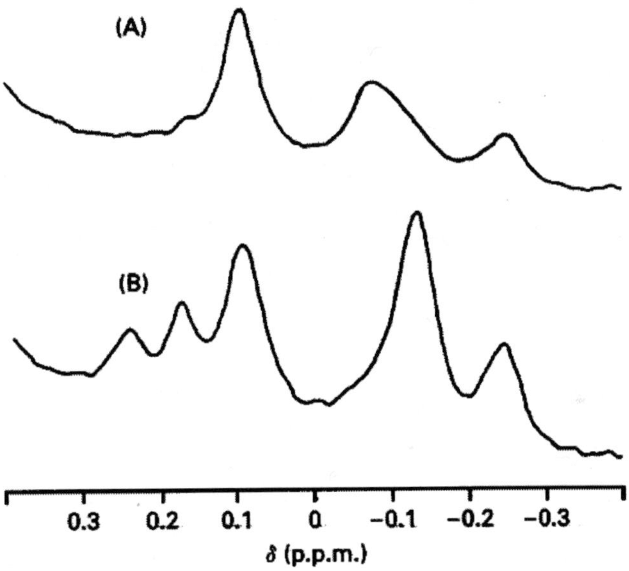

Figure 2. Substrate-induced conformational changes in the Bacillus licheniformis class A □-lactamase The spectra were recorded 15-20 min after mixing of the reagents at 303 K and pH* 7.0. The enzyme was 0.85 mM in both cases. (A) Free enzyme (B) enzyme in the presence of 1.2 mM cefoxitin. From ref [10], reprinted by permission of the Biochemical Society.

In the case of the TEM-1 β-lactamase, backbone ^{1}HN, ^{15}N ^{13}CO, ^{13}Cα and ^{13}Cβ assignments are now available [12], opening the door for more in-depth NMR studies of the enzyme, no doubt including complexes with inhibitors or substrate analogues. Another promising area is studies of the folding of the protein – with the unique ability of NMR to provide structural information at the level of individual residues, it has proved extremely valuable in the study of folding pathways. Indeed, one of the earliest NMR studies of a class A β-lactamase involved studies of the folding of the *Staphylococcus aureus* enzyme, whose folding had been extensively studied by optical method. The NMR study of histidine C2 protons at different concentrations of denaturant allowed Thomas *et al.* to demonstrate the existence of a stable folding intermediate (H) in slow conformational exchange with the native state (N) [13]. This was followed by the use of NMR to show that the molten globule had lost most of the tertiary interactions [14]. With the availability of extensive resonance assignments, this kind of experiment can now be extended to detailed studies of the folding pathway and the structure of any intermediate(s).

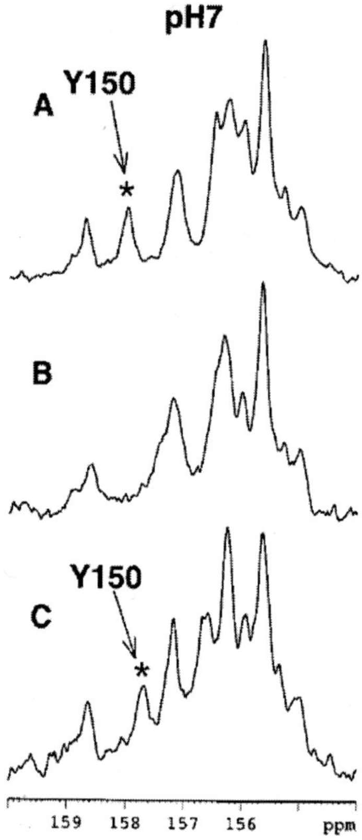

Figure 3. Comparison of wild-type (A), Y150C (B) and K67C (C) C. freundii class C β-lactamase 13C-NMR spectra at pH 7. The peak for Tyr150 is indicated by an asterisks. Reprinted from ref 15 by permission of the Biochemical Society.

Classes C and D

In class C β-lactamases as well, the question of the identification of the general acid/base involved in the enzyme reaction has arisen; one suggestion has been Tyr150, which would require that it be deprotonated at physiological pH. Kato-Toma *et al.* labeled the 15 tyrosine residues of *C. freundii* GN346 β-lactamase using [*phenol*-4-^{13}C]tyrosine and carried out 1D ^{13}C NMR experiments [15], using a mutant (Y150C) to assign the signal of the tyrosine 150. In this study the comparison between the spectra of the wild-type mutant enzymes reveals an intense and well resolved resonance which is missing in the spectrum of the mutant, allowing a clear assignment of Tyr150 (Figure 3). At pH 7.0, the chemical shift for the signal of Tyr150 was 158 ppm and this remained unchanged until pH 11; beyond pH 11, the protein started to denature and at pH 12, all tyrosine residues have a chemical shift of 166 ppm, corresponding to the unprotonated form. The crystal structure indicates strong hydrogen bonds between Tyr150 and two lysine ε-ammonium groups. A mutant (K67C) in which one of these nearby positive charges next to tyrosine 150 has been removed shows only a minor change (0.2 ppm) in the Tyr150 chemical shift, suggesting that the primary determinant of the tyrosine chemical shift in the enzyme is the charge state of the tyrosine hydroxyl. The chemical shift for the free tyrosine in solution is 156 ppm for the protonated form and 166 ppm for the unprotonated form, in good agreement with the measured values in the β-lactamase. The authors thus concluded that Tyr150 is protonated in the active site of the enzyme, and hence cannot be the general base/acid required in the catalytic mechanism.

The class D β-lactamases are peculiar among the serine β-lactamases, since they depend critically on a carboxylated lysine as the basic residue for both the acylation and deacylation steps of catalysis. Since the formation of carboxylated lysine is reversible, the active site lysine was modified by $NaH^{13}CO_3$ at pH 7.5 which led to the observation of a characteristic ^{13}C NMR signal at 164 ppm corresponding to a carbamate. By decreasing the pH to 4.5, the free lysine was restored and the ^{13}C signal was lost in parallel with the catalytic activity of the enzyme. These results demonstrate that carboxylation of lysine is indispensable for the activity of the class D β-lactamase [16].

It has been found that phosphonate monoesters inhibit class A & C serine β-lactamases forming a relatively stable tetrahedral covalent adduct that appears to be a good transition state analogue for these enzymes. ^{31}P NMR was among the techniques which provided evidence for the structure of this adduct [17]. The transition state for the phosphonylation reaction is presumably a pentacoordinated phosphorus. It would be interesting to design a molecule that would react with these enzymes to generate a stable pentacoordinate species. Vanadate complexes have been employed as inhibitors of phosphoryl transfer enzymes and structural studies show that the inhibitory complexes contain pentacoordinated vanadium [18]. Bell *et al.* have investigated the possibility of using a hydroxamic acid, [carbonyl-^{13}C]4-nitrobenzo-hydroxamic acid, complexed to vanadate (*1*) to inhibit class C β-lactamase [19].

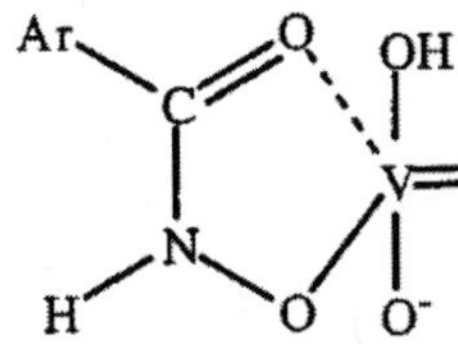

(1)

When the hydroxamic acid was mixed with vanadate and the *Enterobacter cloacae* β-lactamase of P99, three resonances corresponding to carbonyl carbons were observed. The resonance at 174.04 ppm was assigned to the Enzyme-bound Vanadium Hydroxamate complex (EVH) (Figure 4A).

The assignment was confirmed by an experiment in which an irreversible covalent inhibitor was added to the NMR tube containing the sample described above. The ^{13}C NMR spectrum of the reaction mixture after addition of the covalent inhibitor showed that free hydroxamic acid (H) and vanadate hydroxamate (VH) were present, but that the resonance at 174.04 ppm has disappeared (Figure 4B). This observation confirms the assignment of this peak to an enzyme-bound species and supports the idea that VH binds to the active site of the enzyme. The vanadium isotope ^{51}V can be observed directly by NMR; the vanadium chemical shift was almost identical (497 ppm) for the inhibitor in solution or bound to the enzyme showing clearly that the vanadium must be in very similar environment in both cases. Therefore, the difference in ^{13}C chemical shift between the vanadate hydroxamate (VH) in solution (166.3 ppm) and bound to the enzyme (EVH) (174.0 ppm) is not caused by an increased interaction between the vanadium and the carbonyl but, most probably, by a polarization of the inhibitor carbonyl by a group on the enzyme. Based on these observations and on molecular modeling, the authors proposed an open structure for the inhibitor bound to the enzyme (**3**), similar to phosphonate-inhibited complexes (**2**), where E is the enzyme.

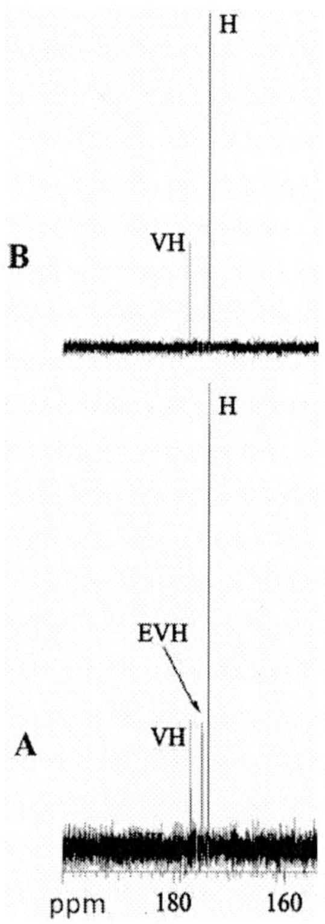

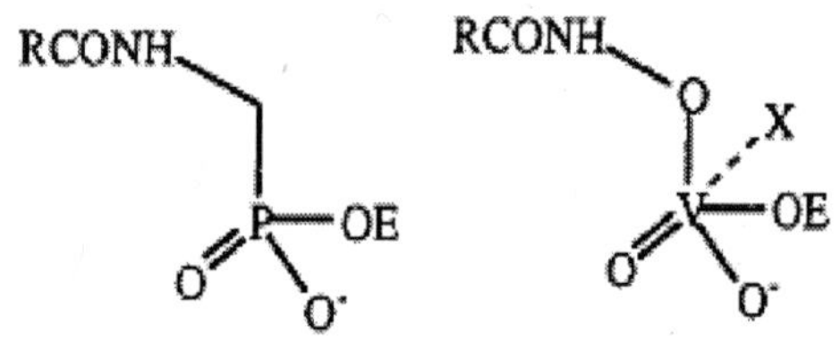

(**2**)

Figure 4. Interaction between the *Enterobacter cloacae* P99 β-lactamase and 1:1 complexes of vanadate with hydroxamic acids. (A) ^{13}C NMR spectrum of a mixture of vanadate (V, 1 mM), 4-nitrobenzo[^{13C}]hydroxamic acid (H, 5 mM), and the P99 β-lactamase (E, 0.25 mM). Calculated concentrations of H, VH, and EVH at equilibrium were 3.9, 0.71, and 0.25 mM, respectively. (B) The same mixture after the addition of 1 mM *m*-nitrophenyl phenylacetamidomethylphos-phonate as described in the text. Reprinted from ref 19 by permission of Biochemistry (ACS).

CLASS B METALLO-β-LACTAMASES

Since resistance arising from the action of the metallo-β-lactamases (MBLs) represents a new threat to the usage of β-lactams as antibiotics, there is considerable current interest in the mechanism of these enzymes and in the development of inhibitors of their action.

Inhibitor Binding

Although structures of several MBLs have been solved by X-ray crystallography, some inhibitor complexes have stubbornly resisted all attempts to crystallise them. NMR is the most powerful tool for the study of protein-ligand interactions in solution and to date has been used to study the interaction of two MBLs with inhibitors. The simplest NMR experiment for the study of inhibitor binding is the so-called ‘chemical shift mapping experiment’, which attempts to identify the inhibitor binding site by identifying those residues whose chemical shift is affected by addition of the inhibitor. This requires the assignment of at least resonances of the backbone amide groups of the protein. For proteins larger than 20k Da, ^{13}C and ^{15}N double labelling and triple resonance NMR experiments are used to perform the backbone NH assignment [1, 20, 21].

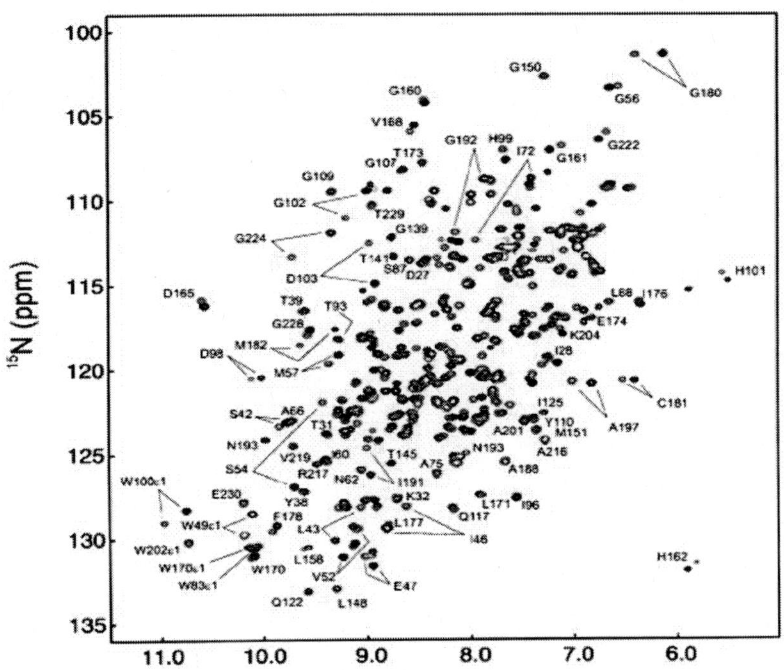

Figure 5. Interaction between the *Bacteroides fragilis* metallo-β-lactamase and a tight-binding inhibitor. Superposition of the 750 MHz ^{15}N NHSQC spectrum of free metallo-β-lactamase (red) and the same protein bound to the inhibitor SB225666 (4, black). In both samples the protein was 0.9 mM in 10 mM HEPES buffer (pH 7.0) containing 10 mM $ZnCl_2$ and 0.01% NaN3. An aliquot of DMSO was added to the free metallo-β-lactamase sample to allow for accurate comparison with the inhibitor-bound form, since the inhibitor was added to the sample as a DMSO solution. The temperature was 295 K. Reprinted from ref 25 by permission of Biochemistry (ACS).

In the case of MBLs, the amino acid side chains involved in metal binding are of particular importance; like other sidechains, their resonances are also best assigned by triple-resonance experiments, but metal removal or metal substitution have also been used to assign these signals [20, 22].

Once the backbone NH resonances are assigned, the perturbation induced by inhibitor binding can be observed figuby means of simple and quick ^{1}H-^{15}N HSQC NMR experiments (Figure 5). If the inhibitor binds relatively weakly, so that its exchange on and off the enzyme is fast, the signals of residues affected by binding will shift progressively as the inhibitor concentration is increased. For more tightly binding inhibitors, where exchange is slow, progressive addition of the inhibitor will lead to a decrease in intensity of the spectrum of the free enzyme and an increase in intensity of the spectrum of the complex. In the first, fast exchange, case, the magnitude of the change in chemical shift of individual residues on binding can readily be determined. In the slow exchange case this can only be done precisely if resonance assignments are available for the complex as well as for the free enzyme. This is often not the case, but valuable information can still be obtained by using the so-called 'minimum chemical shift' approach in which the chemical shift difference from a given peak in the free protein to the closest peak in the complex is calculated; while this method can of course lead to underestimation of chemical shift changes, particularly in crowded regions of the spectrum, it has proved useful and reliable. Visualising the perturbed aminoacids on the X-ray structure of the free enzyme leads to a map of the inhibitor interaction [23-25]. These two studies involving two different MBLs (from *Bacteroides fragilis* and *Bacillus cereus*) and two different inhibitors (<u>4</u> and <u>5</u>) both showed that, in addition to residues in the active site, residues in a loop relatively distant from the active site were also significantly perturbed by inhibitor binding.

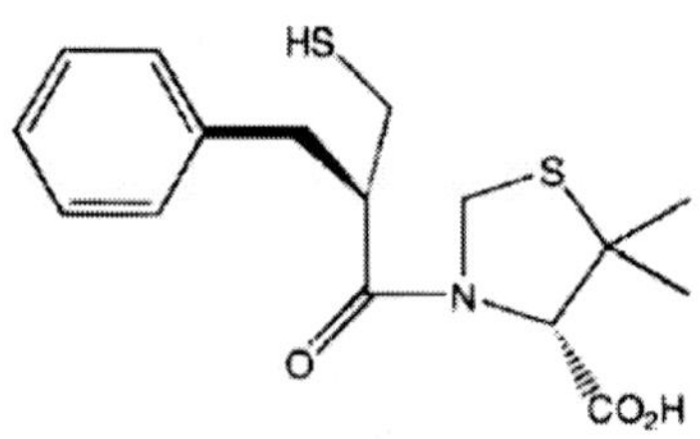

(<u>4</u>)

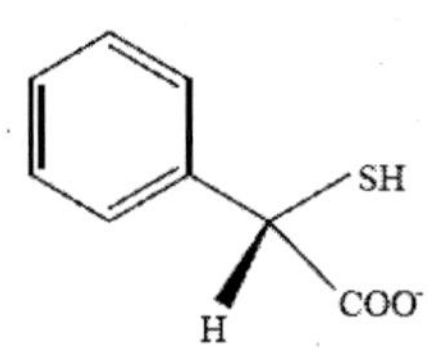

(<u>5</u>)

The dynamic properties of this loop in the *B. fragilis* enzyme, free and bound to the inhibitor, have been studied by NMR [26] and it has been proposed that the mobility of this loop in the free enzyme could explain the broad substrate specificity of the enzyme by allowing it to accommodate both small and bulky substrates.

To further characterise this loop, mutations of the apical tryptophan within the loop have been performed and the mutant enzyme studied by NMR; this work has indicated that the hydrophobic nature of the tryptophan was necessary for catalytic efficiency [27].

Although a hydrophobic interaction between the loop and the relatively large inhibitors was invoked to explain the perturbation of the loop, it seems likely that this is not the whole story, since small inhibitors such as thiomandelate [24] and even mercaptoacetate [28] also affect the chemical shifts of residues in the loop.

Studies of the Metal Ion and its Ligands

At least one divalent metal cation is absolutely required for catalytic activity of MBLs, although the exact role of the second metal in the different MBLs remains uncertain. NMR has been extensively used to study the metals in metallo-β-lactamases, both directly and through studies of the aminoacid side-chains which bind them. Indeed the first reported NMR experiment on β-lactamases [29] involved the use of ^{1}H NMR and a combination of zinc titration and deuterium exchange on the apo and holo-enzymes to infer that in the *B. cereus* MBL four imidazoles have to be involved in zinc binding, three in the binding of the first zinc and one for the second zinc.

Although the native metal, zinc, is not detectable by high resolution NMR, its presence introduces perturbation of the nearby protons and nitrogen atoms which can be detected and studied by NMR. When the *B. cereus* MBL (BcII) apoenzyme was more recently titrated with Zn^{++} ions, the 4 imidazole rings involved in metal binding exhibited identical behaviours, i.e. the perturbations due to Zn^{++} binding occurred concomitantly. Accordingly, mass spectrometric analysis indicated that, during titration, the mono-Zn^{++} species never represented more than 15 % of the total enzyme. In conclusion, all these results indicate a cooperative binding of both Zn^{++} ions [30].

In other approaches, metal substitution, which is tolerated well by many MBLs, can be used to introduce very powerful NMR 'probes' into the active site of the enzyme.

Cobalt (II) is paramagnetic, and therefore has dramatic effects on the resonances of all the nearby protons which undergo shifts of as much as 220 ppm together with marked broadening. Galdes and Baldwin [31] demonstrated that cobalt and zinc were likely to be bound by the same histidine residues in BcII. They also showed that the dissociation constant for the first cobalt was higher than for the first zinc, while the converse was true for the second metal binding. Orellano and coworkers took advantage of this observation to study the cobalt/zinc hybrid of the *B. cereus* MBL [32]. When the spectrum of the di-cobalt enzyme is compared with that of the cobalt/zinc hybrid enzyme, the missing resonance in the hybrid can immediately be assigned to the high affinity metal binding site that is the three histidine site. Upon $^{2}H_2O$ exchange, four signals in the di-cobalt enzyme (CDEK in Figure 6) disappear which indicates that the protons are exchangeable and most probably arise from the imidazole NH of the four cobalt-bound histidines. Only one of these resonances is observed in the hybrid, indicating that one histidine is present in the low affinity binding site. Resonances AB are characteristic of the β protons of the metal bound cysteine and their observation in both species indicate that the cysteine is in the low affinity binding site.

The same approach has been used to characterise BcII mutants in which additional potential metal ligands have been introduced into the low affinity metal binding site [33] . The NMR results confirmed that the cobalt was still bound in the low affinity metal binding site containing the cysteine residue and suggested that in the R121H mutant the "new" histidine was replacing histidine 263 as metal ligand in the low affinity metal binding site, while in the R121E mutant, there was no histidine binding the metal in the low affinity site.

Similar paramagnetic NMR experiments, coupled with deuterium exchange, have been used to study metal binding in ImiS, a sub class B2 MBL which is active with only one metal ion bound. These experiments indicate that in this enzyme metal binding involves one histidine, one cysteine and an additional residue proposed to be an aspartate (on the basis of the chemical shift of its hyperfine shifted resonance) [34].

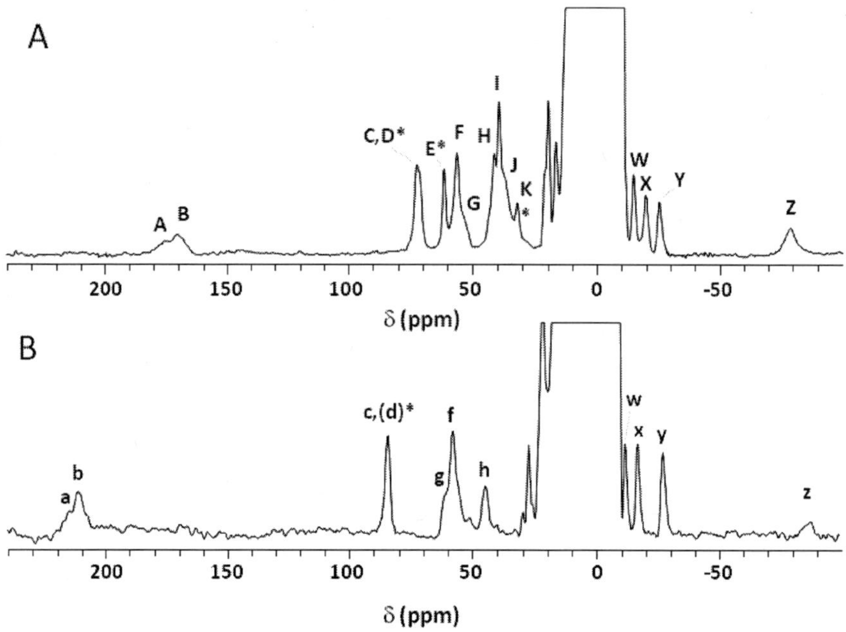

Figure 6. Utilisation of Cobalt to probe the metal binding site of the *B. cereus* metallo-β-lactamase (BcII). ^{1}H NMR spectra of (A) CoCo-βcII, and (B) ZnCo- βcII. Signals marked with an asterisk are absent when the spectra are recorded in D_2O. The spectra were recorded in H_2O at 200 MHz and 298 K in 20 mM sodium succinate/1 M NaCl at pH 6.0. Reprinted from ref 32 by permission of Biochemistry (ACS).

An attractive route to the study of the metal centres in MBLs is to substitute the zinc ions with cadmium, a substitution which is often well tolerated with relatively small differences in kinetic parameters compared to the zinc enzymes. Cadmium isotopes 111 and 113 have a nuclear spin I=1/2, which allows high resolution NMR and hence provides a very direct way of obtaining information about the metal and its environment.

The sensitivity of cadmium for NMR detection is similar to that for ^{13}C. It has a very large chemical shift range (~800 ppm) making it very sensitive to even small changes in its environment. Because of this large chemical shift range, the chemical shift anisotropy (CSA) can in some cases lead to rather broad lines but nevertheless ^{113}Cd NMR has been successfully used to study MBLs.

For example, in the case of BcII, separate signals can be observed for cadmium in the two binding sites at 142 and 262 ppm (Figure 7A) [35]. From the characteristic dependence of cadmium chemical shifts on the nature of the ligands [36] a tentative assignment of these signals can be made, the higher field (142 ppm) signal to a cadmium ligated by nitrogens (histidines) and the lower field (262 ppm) signal to a cadmium with a sulphur (cysteine).

An additional advantage in using a metal with a nuclear spin I=1/2 is the existence of resolvable scalar (through-bond) couplings between the metal nucleus and nuclei (^{1}H, ^{13}C,

^{15}N) of its ligands, and hence the ability to transfer magnetization to and from these nuclei, particularly ^{1}H, leading to indirect detection of the cadmium with increased sensitivity.

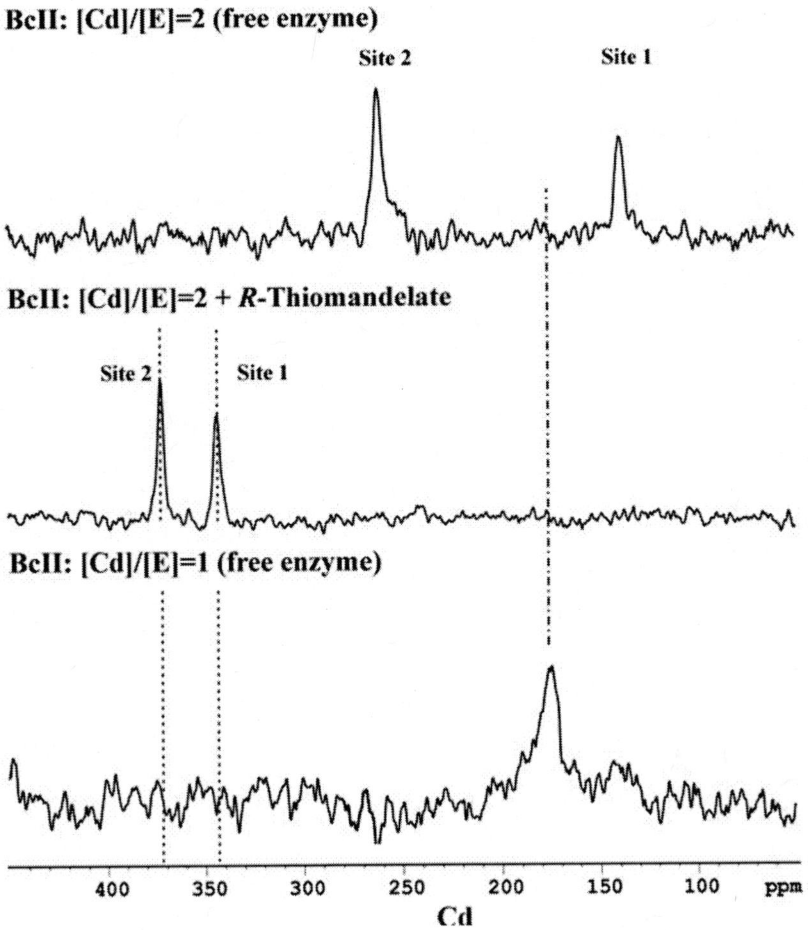

Figure 7. ^{113}Cd spectra of dicadmium BcII and of monocadmium BcII with and without one equivalent of *R*-thiomandelate (5). This figure was originally published in the Journal of Biological Chemistry. Damblon, C. *et al.* . The inhibitor thiomandelic acid binds to both metal ions in metallo-beta-lactamase and induces positive cooperativity in metal binding. *Journal of Biological Chemistry*. 2003; 278:29240-51.(ref 35) © the American Society for Biochemistry and Molecular Biology.

In the case of MBLs, two scalar couplings have been successfully used: one-bond ^{15}N-^{113}Cd ($^{1}J_{N\text{-}Cd}$) and three-bonds ^{1}H-C-N-^{113}Cd and ^{1}H-C-S-^{113}Cd ($^{3}J_{H\text{-}Cd}$). They can be observed in a single 2D NMR experiment [35]. The coupling constant for the direct interaction between the imidazole nitrogen and the cadmium is rather large (100-200Hz) and can be observed as a splitting of the nitrogen-proton correlation in the nitrogen frequency domain [22]. Three bonds H-Cd scalar coupling constants range from ~2 to 50 Hz for imidazole C2 and C4 protons and cysteine β protons and are also readily observable in the spectra. The observation of these scalar couplings have been used to make unambiguous assignments of the two ^{113}Cd signals [35].

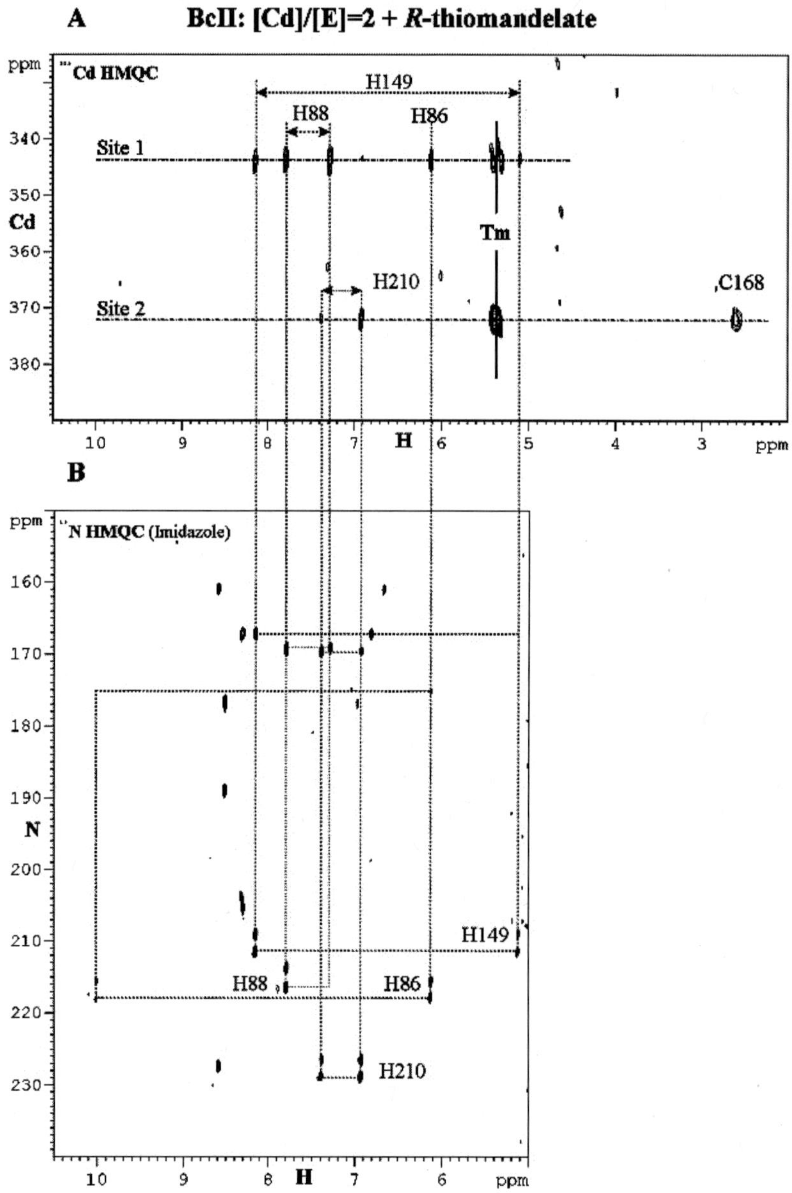

Figure 8. NMR spectra of dicadmium BcII in the presence of one equivalent of *R*-thiomandelate (5). A) $^{1}H,^{113}Cd$ HMQC spectrum. B) $^{1}H,^{15}N$ HMQC spectrum of the imidazole resonances, allowing assignment of the imidazole crosspeaks in A. This figure was originally published in the Journal of Biological Chemistry. Damblon, C. *et al.* . The inhibitor thiomandelic acid binds to both metal ions in metallo-beta-lactamase and induces positive cooperativity in metal binding. *Journal of Biological Chemistry*. 2003; 278:29240-51. (rerf 35) © the American Society for Biochemistry and Molecular Biology.

Similar couplings have provided precise information on the mode of binding of the inhibitor thiomandelic acid (5). When this inhibitor, which has a thiol group, binds to the enzyme both ^{113}Cd resonances are shifted substantially downfield (Figure 7), consistent with the idea that both cadmiums have gained an additional sulfur in their coordination sphere –

that is, that the thiol of the inhibitor binds to both metals. This was unambiguously confirmed by the observation of scalar coupling between the α-proton of the inhibitor and *both* cadmium nuclei. (Figure 8) Detailed analysis showed that the inhibitor thiol was bound simultaneously to both metals [35].

The mono-cadmium species displays only one broad ^{113}Cd signal at 176 ppm (Figure 7). This could be interpreted as indicating that the single cadmium binds in the three-histidine site (with a somewhat altered chemical shift from the di-cadmium enzyme). However, PAC spectroscopy, another technique allowing the direct observation of cadmium, gives signals corresponding to cadmium in two different environments. These two observations can be reconciled by postulating that a single cadmium is exchanging between the two sites at a rate of $\sim 10^6$ s^{-1}, in fast exchange for the NMR experiment but slow exchange for the PAC experiment [37]. By contrast, similar experiments show that when thiomandelate (**5**) binds to the mono-cadmium enzyme it converts it to a mixture of di-cadmium enzyme and apo-enzyme, the cooperativity in metal binding presumably being induced by the simultaneous binding of the inhibitor to both metals noted above [35].

CONCLUSION

As we have shown in this review, many studies on all classes of β-lactamases which involve NMR spectroscopy have been performed and have provided valuable information on different structural and catalytic aspects such as pK determinations, metal binding, inhibitor binding, conformational changes and folding. Because of its ability to observe individual nuclei within a molecule, NMR spectroscopy applied to proteins in solution allows structural and dynamic information to be gained which complement X-ray spectroscopy.

Because of the constant improvement in sensitivity and in resolution, it is likely that in a near future, NMR will provide as well the full determination of 3D structure for some β-lactamases and their complexes with inhibitors that have resisted crystallisation.

REFERENCES

[1] Karsisiotis, A. I., Damblon, C., and Roberts, G. C. (2006) Solution structures of the 25 kDa BcII Metallo-β-Lactamase, from *Bacillus cereus* and the BcII – R-thiomandelic acid enzyme-Inhibitor complex, *Unpublished data.*

[2] Lamotte-Brasseur, J., Dive, G., Dideberg, O., Charlier, P., Frere, J. M., and Ghuysen, J. M. (1991) Mechanism of acyl transfer by the class A serine beta-lactamase of Streptomyces albus G, *Biochem J 279 (Pt 1)*, 213-221.

[3] Gibson, R. M., Christensen, H., and Waley, S. G. (1990) Site-directed mutagenesis of beta-lactamase I. Single and double mutants of Glu-166 and Lys-73, *Biochem J 272*, 613-619.

[4] Strynadka, N. C., Adachi, H., Jensen, S. E., Johns, K., Sielecki, A., Betzel, C., Sutoh, K., and James, M. N. (1992) Molecular structure of the acyl-enzyme intermediate in beta-lactam hydrolysis at 1.7 A resolution, *Nature 359*, 700-705.

[5] Damblon, C., Raquet, X., Lian, L. Y., Lamotte-Brasseur, J., Fonze, E., Charlier, P., Roberts, G. C., and Frere, J. M. (1996) The catalytic mechanism of beta-lactamases:

NMR titration of an active-site lysine residue of the TEM-1 enzyme, *Proc Natl Acad Sci U S A 93*, 1747-1752.

[6] Golemi-Kotra, D., Meroueh, S. O., Kim, C., Vakulenko, S. B., Bulychev, A., Stemmler, A. J., Stemmler, T. L., and Mobashery, S. (2004) The importance of a critical protonation state and the fate of the catalytic steps in class A beta-lactamases and penicillin-binding proteins, *J Biol Chem 279*, 34665-34673.

[7] Gao, G., Prasad, R., Lodwig, S. N., Unkefer, C. J., Beard, W. A., Wilson, S. H., and London, R. E. (2006) Determination of lysine pK values using [5-13C]lysine: application to the lyase domain of DNA Pol beta, *J Am Chem Soc 128*, 8104-8105.

[8] Herzberg, O., and Moult, J. (1987) Bacterial resistance to beta-lactam antibiotics: crystal structure of beta-lactamase from Staphylococcus aureus PC1 at 2.5 A resolution, *Science 236*, 694-701.

[9] Brannigan, J., Matagne, A., Jacob, F., Damblon, C., Joris, B., Klein, D., Spratt, B. G., and Frere, J. M. (1991) The mutation Lys234His yields a class A beta-lactamase with a novel pH-dependence, *Biochem J 278 (Pt 3)*, 673-678.

[10] Jamin, M., Damblon, C., Bauduin-Misselyn, A. M., Durant, F., Roberts, G. C., Charlier, P., Llabres, G., and Frere, J. M. (1994) Direct n.m.r. evidence for substrate-induced conformational changes in a beta-lactamase, *Biochem J 301 (Pt 1)*, 199-203.

[11] Fonze, E., Vanhove, M., Dive, G., Sauvage, E., Frere, J. M., and Charlier, P. (2002) Crystal structures of the Bacillus licheniformis BS3 class A beta-lactamase and of the acyl-enzyme adduct formed with cefoxitin, *Biochemistry 41*, 1877-1885.

[12] Savard, P. Y., Sosa-Peinado, A., Levesque, R. C., Makinen, M. W., and Gagne, S. M. (2004) 1H, 13C and 15N backbone resonance assignments for TEM-1, a 28.9 kDa beta-lactamase from E. coli, *J Biomol NMR 29*, 433-434.

[13] Thomas, R. M., Feeney, J., Nicholson, R. B., Pain, R. H., and Roberts, G. C. (1983) Identification by n.m.r. spectroscopy of a stable intermediate structure in the unfolding of staphylococcal beta-lactamase, *Biochem J 215*, 525-529.

[14] Ptitsyn, O. B., Pain, R. H., Semisotnov, G. V., Zerovnik, E., and Razgulyaev, O. I. (1990) Evidence for a molten globule state as a general intermediate in protein folding, *FEBS Lett 262*, 20-24.

[15] Kato-Toma, Y., Iwashita, T., Masuda, K., Oyama, Y., and Ishiguro, M. (2003) pKa measurements from nuclear magnetic resonance of tyrosine-150 in class C beta-lactamase, *Biochem J 371*, 175-181.

[16] Golemi, D., Maveyraud, L., Vakulenko, S., Samama, J. P., and Mobashery, S. (2001) Critical involvement of a carbamylated lysine in catalytic function of class D beta-lactamases, *Proc Natl Acad Sci U S A 98*, 14280-14285.

[17] Rahil, J., and Pratt, R. F. (1992) Mechanism of inhibition of the class C beta-lactamase of Enterobacter cloacae P99 by phosphonate monoesters, *Biochemistry 31*, 5869-5878.

[18] Holtz, K. M., Stec, B., and Kantrowitz, E. R. (1999) A model of the transition state in the alkaline phosphatase reaction, *J Biol Chem 274*, 8351-8354.

[19] Bell, J. H., and Pratt, R. F. (2002) Mechanism of inhibition of the beta-lactamase of Enterobacter cloacae P99 by 1:1 complexes of vanadate with hydroxamic acids, *Biochemistry 41*, 4329-4338.

[20] Scrofani, S. D., Wright, P. E., and Dyson, H. J. (1998) The identification of metal-binding ligand residues in metalloproteins using nuclear magnetic resonance spectroscopy, *Protein Sci 7*, 2476-2479.

[21] Scrofani, S. D., Wright, P. E., and Dyson, H. J. (1998) 1H, 13C and 15N NMR backbone assignments of 25.5 kDa metallo-beta-lactamase from Bacteroides fragilis, *J Biomol NMR 12*, 201-202.

[22] Damblon, C., Prosperi, C., Lian, L. Y., Barsukov, I., Soto, R. P., Galleni, M., Frere, J. M., and Roberts, G. C. K. (1999) H-1-N-15 HMQC for the identification of metal-bound histidines in Cd-113-substituted Bacillus cereus zinc beta-lactamase, *J. Am. Chem. Soc. 121*, 11575-11576.

[23] Lian, L. Y., and Roberts, G. C. (1993) Effects of chemical exchange on NMR spectra., In *NMR of Biological Macromolecules* (Roberts, G. C., Ed.), pp 153-182, Oxford University Press.

[24] Mollard, C., Moali, C., Papamicael, C., Damblon, C., Vessilier, S., Amicosante, G., Schofield, C. J., Galleni, M., Frere, J. M., and Roberts, G. C. (2001) Thiomandelic acid, a broad spectrum inhibitor of zinc beta-lactamases: kinetic and spectroscopic studies, *J Biol Chem 276*, 45015-45023.

[25] Scrofani, S. D., Chung, J., Huntley, J. J., Benkovic, S. J., Wright, P. E., and Dyson, H. J. (1999) NMR characterization of the metallo-beta-lactamase from Bacteroides fragilis and its interaction with a tight-binding inhibitor: role of an active-site loop, *Biochemistry 38*, 14507-14514.

[26] Huntley, J. J., Scrofani, S. D., Osborne, M. J., Wright, P. E., and Dyson, H. J. (2000) Dynamics of the metallo-beta-lactamase from Bacteroides fragilis in the presence and absence of a tight-binding inhibitor, *Biochemistry 39*, 13356-13364.

[27] Huntley, J. J., Fast, W., Benkovic, S. J., Wright, P. E., and Dyson, H. J. (2003) Role of a solvent-exposed tryptophan in the recognition and binding of antibiotic substrates for a metallo-beta-lactamase, *Protein Sci 12*, 1368-1375.

[28] Kumar, S., Roberts, G. C., and Damblon, C. (2002) Study by NMR of the inhibition of B. cereus MBL by various mercaptocaboxylic compounds.

[29] Baldwin, G. S., Galdes, A., Hill, H. A., Smith, B. E., Waley, S. G., and Abraham, E. P. (1978) Histidine residues of zinc ligands in beta-lactamase II, *Biochem J 175*, 441-447.

[30] Jacquin, O., Balbeur, D., Damblon, C., Marchot, P., De Pauw, E., Roberts, G. C., Frere, J. M., and Matagne, A. (2009) Positively cooperative binding of zinc ions to Bacillus cereus 569/H/9 beta-lactamase II suggests that the binuclear enzyme is the only relevant form for catalysis, *J Mol Biol 392*, 1278-1291.

[31] Galdes, A., Hill, H. A., Baldwin, G. S., Waley, S. G., and Abraham, E. P. (1980) The 1H nuclear-magnetic-resonance spectroscopy of cobalt(II)-beta-lactamase II, *Biochem J 187*, 789-795.

[32] Orellano, E. G., Girardini, J. E., Cricco, J. A., Ceccarelli, E. A., and Vila, A. J. (1998) Spectroscopic characterization of a binuclear metal site in Bacillus cereus beta-lactamase II, *Biochemistry 37*, 10173-10180.

[33] Rasia, R. M., Ceolin, M., and Vila, A. J. (2003) Grafting a new metal ligand in the cocatalytic site of B. cereus metallo-beta-lactamase: structural flexibility without loss of activity, *Protein Sci 12*, 1538-1546.

[34] Crawford, P. A., Yang, K. W., Sharma, N., Bennett, B., and Crowder, M. W. (2005) Spectroscopic studies on cobalt(II)-substituted metallo-beta-lactamase ImiS from Aeromonas veronii bv. sobria, *Biochemistry 44*, 5168-5176.

[35] Damblon, C., Jensen, M., Ababou, A., Barsukov, I., Papamicael, C., Schofield, C. J., Olsen, L., Bauer, R., and Roberts, G. C. (2003) The inhibitor thiomandelic acid binds to

both metal ions in metallo-beta-lactamase and induces positive cooperativity in metal binding, *J Biol Chem 278*, 29240-29251.

[36] Coleman, J. E. (1993) Cadmium-113 nuclear magnetic resonance applied to metalloproteins, *Methods Enzymol 227*, 16-43.

[37] Hemmingsen, L., Damblon, C., Antony, J., Jensen, M., Adolph, H. W., Wommer, S., Roberts, G. C., and Bauer, R. (2001) Dynamics of mononuclear cadmium beta-lactamase revealed by the combination of NMR and PAC spectroscopy, *J Am Chem Soc 123*, 10329-10335.

In: Beta-Lactamases
Editor: Jean-Marie Frère, pp. 217-258

ISBN: 978-1-61324-638-2

Chapter 9

β-LACTAMS AS INHIBITORS OF β-LACTAMASES

John D. Buynak*
Department of Chemistry, Southern Methodist University,
Dallas TX 75275-0314, U.S.A.

ABSTRACT

The co-administration of a β-lactam antibiotic together with a β-lactamase inhibitor represents an efficacious treatment for infections caused by many β-lactamase producing strains. Most known inhibitors of β-lactamase, including the current commercial products, clavulanate, sulbactam, and tazobactam, are themselves β-lactams. These molecules are mechanism-based inhibitors which, subsequent to acylation of the active site serine, undergo a series of transformations leading to formation of a hydrolytically stabilized acyl-enzyme. Not surprisingly few display inhibition of the metallo-β-lactamases. This chapter discusses β-lactams which function as β-lactamase inhibitors from a mechanistic perspective. A review of inhibitor-resistance is also presented.

INTRODUCTION

The co-administration of a β-lactam antibiotic and a β-lactamase inhibitor represents a clinically proven strategy for overcoming β-lactamase-mediated resistance [1-10]. This section will focus on inhibitors of β-lactamases that possess a β-lactam ring, a category that includes the majority of such compounds. Most are mechanism-based irreversible inactivators. Due to a combination of exacting kinetic studies and structural elucidations of key intermediates by X-ray crystallography and electrospray ionization mass spectrometry (ESIMS), the respective mechanisms of inactivation are now reasonably well established. Therefore, mechanism will be emphasized, with the hope that such an analysis will have value in the design of future inhibitors.

* Email: jbuynak@smu.edu

Structural and Mechanistic Considerations

Obviously, those β-lactams that inhibit β-lactamases must structurally resemble the β-lactam antibiotic substrates themselves and, in fact, must react as substrates during acylation of the active site serine. This resemblance is probably an advantageous feature, since mutational resistance to the initial recognition of such inhibitors might also be expected to reduce recognition of the antibiotic substrate [11]. To illustrate structural similarities of both substrates and inhibitors, a simplified schematic of the class A TEM-1 β-lactamase active site, with a generic substrate, is presented in Figure 1.

Recognition Elements in the Serine β-lactamase Active Site Include

Positively charged residues (in the case of TEM-1, Lys234 and Arg244) insure recognition of the anionic carboxylate of the substrate or inhibitor. In some β-lactamases, one or more polar serine or threonine side-chain O-H bonds are proximally located to assist stabilization of the carboxylate. As further verification of the positive nature of the enzyme active site, many of the (non-ligated) PDB structures of β-lactamases have inorganic sulfate located at the precise position of the eventual antibiotic carboxylate.

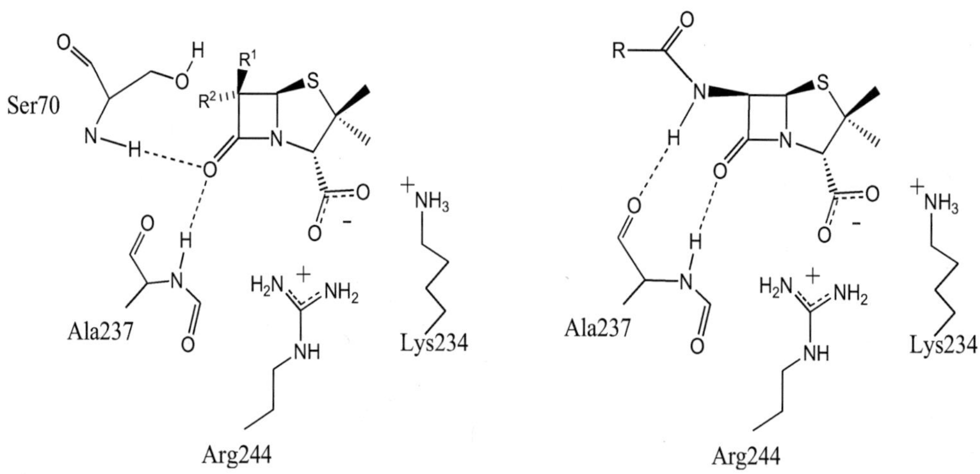

Figure 1. Simplified visualization of key interactions in the binding site of TEM-1 β-Lactamase.

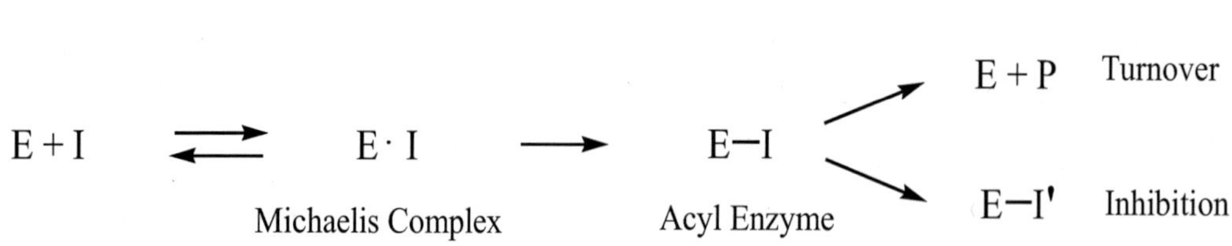

Scheme 1. Simplified mechanism for inhibitors of the serine β-lactamases.

An oxyanion hole is placed approximately 6 - 7 Å away (to the 'west') of the lysine to bind the β-lactam carbonyl oxygen to stabilize the negative charge of the eventual anionic tetrahedral intermediate (during formation and also during hydrolysis of the acyl-enzyme).

In the case of the antibiotic substrates, an additional hydrogen bond is formed from the N-H of the (in case of penicillin, C6β) side chain acylamino group to a backbone carbonyl oxygen. This recognition element is usually absent in the case of the β-lactamase inhibitors, which have atypical substituents at C6.

The importance of these recognition features is evident in the following: a) With very few exceptions, all β-lactam-containing inhibitors of β-lactamase have an appropriately placed negative charge (usually a carboxylate). b) All such inhibitors have an activated β-lactam carbonyl group, positioned at an appropriate distance [12] from this negatively charged group, to acylate the active site serine.

Mechanistically, the serine (i.e., Ambler classes A, C, and D) β-lactamases operate via a multistep process involving the formation and hydrolytic destruction of an intermediate acyl-enzyme. Most β-lactam inhibitors of the serine β-lactamases function through the formation of a hydrolytically stabilized ester. Depending on the specific inhibitor, this acyl-enzyme may be stabilized electronically (i.e., by favorable resonance interactions that improve the stability of the acyl-enzyme ester bond to the serine), stabilized by covalently bonding to a second nucleophilic residue in the active site, or stabilized due to its position in the active site (e.g., the carbonyl of the bound inhibitor may be removed from the oxyanion hole and/or the hydrolytic water). This last type of stabilization may result from movement of the inhibitor or from a conformational change in the enzyme itself. β-Lactam-containing compounds that inactivate serine β-lactamases are typically also substrates of the enzyme, with partitioning of the two processes (turnover and inhibition) at the acyl enzyme stage. For any particular inhibitor, the substrate/inhibitor ratio can range from below 10 to nearly 10,000. A simplified scheme is shown below. As will be discussed, there is typically more than one inactivated form of the enzyme, including both irreversibly and also transiently inhibited intermediates.

With a few exceptions, known β-lactam-containing inhibitors fall into a small number of distinct structural classes. These classes can be distinguished by two factors: a) the substituent α to the β-lactam carbonyl group, and b) the nature of the fused (five or six-membered) ring, particularly including whether the second ring is capable of fragmentation (with consequent opening) subsequent to acylation of the β-lactamase. β-Lactam structural motifs include unsubstituted compounds, **1**; hydroxyalkyl compounds, **2**; alkylidene compounds **3**; and compounds which have a leaving group α to the β-lactam carbonyl, **4**.

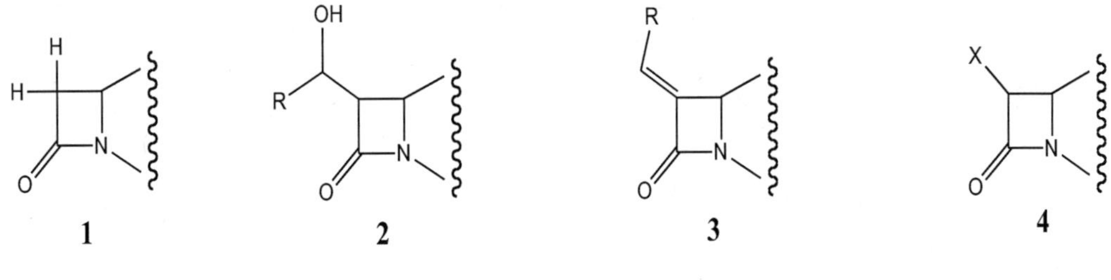

1 **2** **3** **4**

Figure 2. Common structural motifs of inhibitors of the serine-β-lactamases.

As would be anticipated from Scheme 1, the potency of such inhibitors is determined by the acylating ability of the molecules (formation of E-I) and also by their ability to produce stabilized acyl-enzymes (formation of E-I′). The structural motifs of Figure 2 are typically fused to a second ring, the exact structure of which may enhance the acylating ability of the β-lactam and is also important to the inhibitory mechanism, in particular to the chain of events occurring after acylation of the serine hydroxyl.

Regarding the acylating ability, it should be noted that the strain inherent in the geometrically enforced 90° bond angles of a four-membered β-lactam ring, particularly when fused to an adjacent five-membered ring, results in a significant deviation from planarity for the bicyclic β-lactam nitrogen (which would therefore prefer to adapt the 109° bond angles of sp^3 hybridization rather than the 120° bond angles of a planar sp^2 hybridized nitrogen). This deviation, in turn, decreases the ability of the lone pair of electrons on nitrogen to interact with the pi electrons of the β-lactam carbonyl group, thus making the O=C-N linkage less amide-like and activating the CO bond toward nucleophilic addition. Thus, those bicyclic structures which are also most reactive toward non-specific nucleophiles (i.e., the best acylating agents) tend to be molecules with such a fused four-five membered ring, including the penams, clavams, penems, and carbapenems. The fused four-six membered ring systems of the cephems are not inherently as reactive toward nucleophilic attack at the β-lactam carbonyl, but the endocyclic Δ-3,4 double bond does enhance acylating capability as do electron-withdrawing groups at C3, which interact with the lone pair of electrons on the nitrogen via resonance through the intervening double bond. β-Lactamases preferentially hydrolyze molecules which are themselves more reactive acylating agents rather than the less reactive amide linkages.

As regards the transformation to a hydrolytically stabilized acyl enzyme (i.e. formation of E-I′), it should be noted that when an appropriate heteroatom (or a heteroatom in the appropriate oxidation state) is a part of the ring, a second fragmentation may ensue (subsequent to acylation of the serine) as shown in Scheme 2. This fragmentation initially results in a relatively energetic imine intermediate. This ring-opening process also ‘linearizes’ the system, thus generating a more conformationally unrestricted acyl-enzyme. Since most organic chemical processes have relatively stringent geometric requirements, this conformationally mobile linear system has significantly more mechanistic options for structural transformation than does the initial cyclic intermediate, and intramolecular transformations can lead to a yet more stabilized acyl-enzyme. This conformational freedom can also facilitate numerous new interactions with surrounding residues, potentially leading to reaction with a second active site nucleophile.

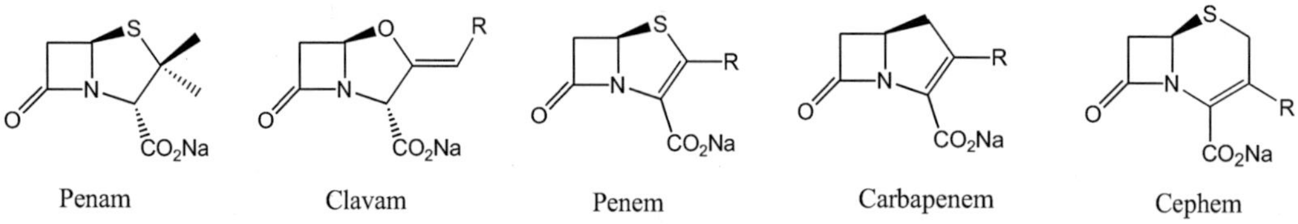

Figure 3. Common antibiotic/β-lactamase inhibitor scaffolds (C6/C7 substituents not shown).

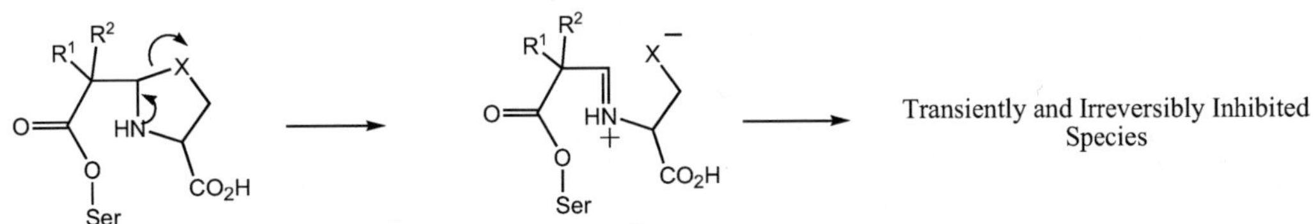

5

Scheme 2. Fragmentation of the second (fused) ring.

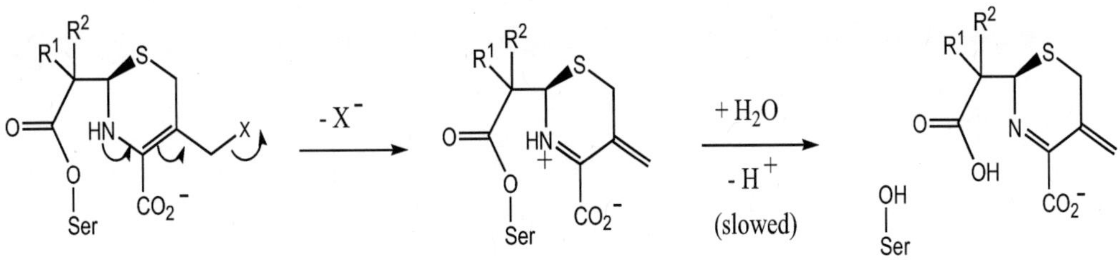

13

Scheme 3. Fragmentation of 3'-Substituted Cephalosporins, Leading to Turnover.

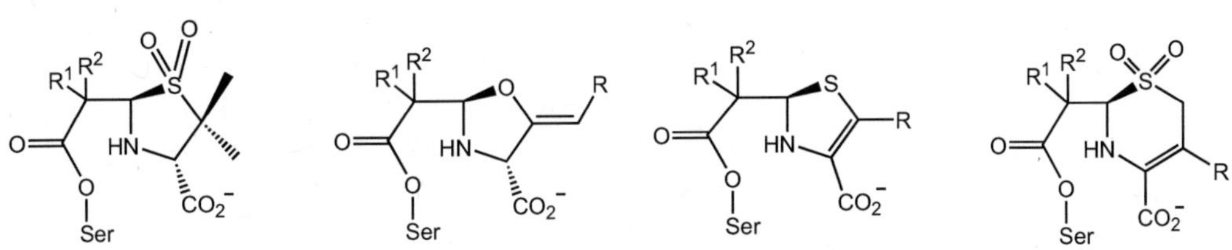

6 7 8 9

Figure 4. Ring systems that open readily subsequent to formation of the acyl-enzyme..

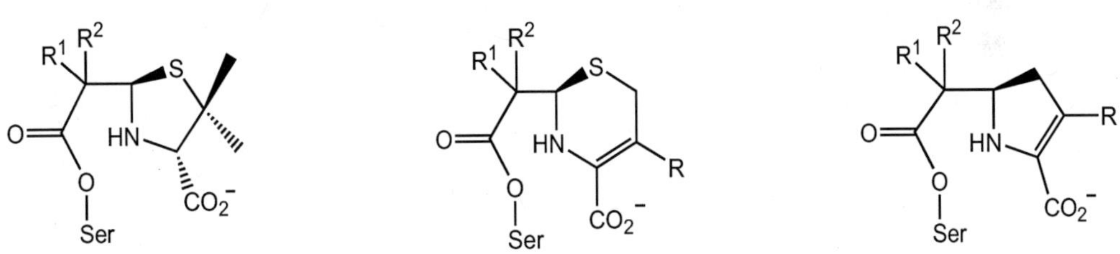

10 11 12

Figure 5. Ring systems that are less likely to undergo fragmentation subsequent to formation of the acyl-enzyme.

Five and six-membered rings that are known to fragment, with consequent opening of the second ring, include the dioxothiazolidine ring system (6) of sulbactam and tazobactam, the (methylene)oxazolidine ring system (7) of clavulanic acid, the dihydrothiazole (8) of the penems, and the dihydrothiazine 1,1-dioxide (9) of the cephalosporin sulfones. These are shown in Figure 4 (shown at the acyl-enzyme stage, and in all cases without the C6 substituent). Rings which are less likely to open (Figure 5) are the (non-oxidized) thiazolidine (10) of the penicillins and the dihydrothiazine (11) of the cephalosporins, and, of course, the dihydropyrrole (12) of the carbapenems, which does not possess a second heteroatom as a potential leaving group in the five-membered ring. As will be shown below, the penems probably represent a borderline case, with some inhibitor-derived acyl enzymes undergoing fragmentation, while others likely remaining intact.

Notice that the ring-opening fragmentation described in Scheme 2 is distinct from the exocyclic eliminative fragmentation of the traditional cephalosporins, which have a leaving group at the C3′ position. This latter fragmentation, shown in Scheme 3, does form an imine, but does not open the second ring, and does not increase conformation flexibility of the resultant intermediate acyl enzyme. Such compounds were studied extensively by the Pratt group [13-16] and found to result in transient inhibition, potentially due to a conformational shift of the enzyme, promoted by the resultant imine. This effect is particularly significant in cases where there is steric hindrance at C7 to slow deacylation, which may sometimes occur faster than elimination of the C3′ group.

STRUCTURAL MOTIF 1: α-UNSUBSTITUTED-β-LACTAMS

Commercial β-Lactamase Inhibitors

Current commercial inhibitors include clavulanic acid, sulbactam and tazobactam, shown below (note the historic non-IUPAC numbering). These inhibitors all fall into the α-unsubstituted-β-lactam class, and all possess a fused ring capable of additional fragmentation. Thus it is not surprising that they have similar inhibitory mechanisms. Although several commercial inhibitors are known to bind one or more of the primary antibiotic targets, the penicillin binding proteins (PBPs, which are believed to be the evolutionary precursors to the β-lactamases [17-19]) these inhibitors have inadequate independent antibacterial activity and are thus administered in the antibiotic/inhibitor combinations: Augmentin™ (amoxicillin/clavulanic acid) [20, 21], Timentin™ (ticarcillin/clavulanic acid) [22], Unasyn™ (ampicillin/sulbactam) [23], and Zosyn™ (piperacillin/tazobactam) [24]. In particular, these commercial combinations are effective against susceptible organisms expressing Ambler Class A enzymes [25], which remain the most commonly encountered. However, due to the continued increase in the number of infections caused by the class C β-lactamases, and also due to the emergence of inhibitor-resistant class A β-lactamases, the search for more effective inhibitors continues.

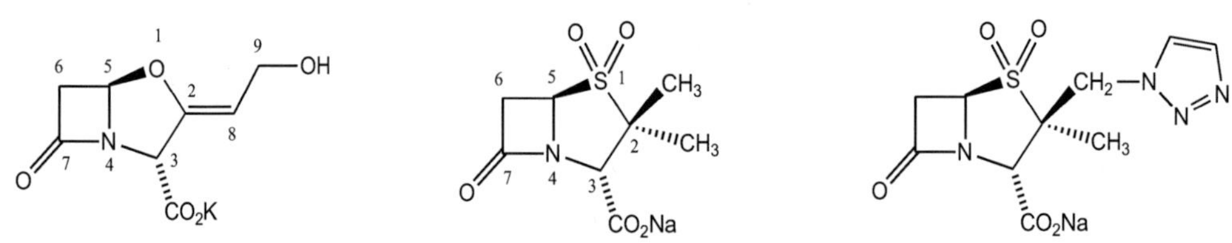

Potassium Clavulanate Sulbactam Sodium Salt Tazobactam Sodium Salt

14 **15** **16**

Figure 6. Current Commercial β-Lactamase Inhibitors.

Clavulanic Acid

The amoxicillin/clavulanate combination represents a broad-spectrum antibacterial, that has been in use for more than 20 years and continues to be widely utilized. In the first thorough mechanistic studies of clavulanic acid with the class A *Eschericia coli* RTEM β-lactamase, the Knowles group [26] found that clavulanic acid, in addition to undergoing turnover, formed both a transiently-inhibited and also three irreversibly-inhibited species, identified by isoelectric focusing. The transient intermediate formed within a few minutes, and had a half-life of less than 30 min. On the basis of its absorbance at 280 nm, they assigned an enamine structure **22**. This intermediate is electronically stabilized toward hydrolysis of the acyl-enzyme through resonance interaction of the lone pair of electrons on nitrogen, through the pi electrons of the carbon-carbon double bond, with the pi electrons of the carbonyl group of the intermediate ester. This enamine (or 'vinylogous urethane') first forms in the *cis*-configuration (21), which is rapidly isomerized to a more stable *trans* form.

This group [27] showed that the C9-hydroxyl group of clavulanate was not necessary for activity and, using radioactively labeled 9-deoxyclavulanic acid, demonstrated that one of the three irreversibly bound species had lost a large portion of the antibiotic skeleton, retaining only the β-lactam carbons. By contrast, the irreversibly inhibited species form more slowly, over the course of an hour or more, depending on clavulanate concentration. Initially, the irreversibly inhibited species was proposed to have a second covalent attachment to the enzyme, potentially due to a trans-imination reaction with a nearby Lys234. Imtiaz *et al.* later proposed [28], on the basis of molecular dynamics simulations performed on the structure of a similar class A β-lactamase from *Bacillus licheniformis* 749/C, that the second attachment was probably due to Ser130, as shown below. This group also suggested an important role of Arg244 in protonating the exocyclic enol ether double bond, through an intervening water molecule, and thus activating the oxazolidine oxygen for departure, and cleavage of the five-membered ring. As supportive evidence, they note that the Arg244 → Ser mutant is resistant to clavulanic acid.

Initial crystallographic studies were performed, on the class A PC1 β-lactamase from *Staphylococcus aureus*, by Chen and Herzberg [29], who observed two acyl-enzyme species attached to the catalytic Ser70: a *cis*-enamine, and a decarboxylated *trans*-enamine. (Note that loss of the antibiotic carboxylate by decarboxylation is chemically facilitated due to its being a β-keto acid.) No intermediate with two covalent bonds to the enzyme was observed. ESIMS studies were performed by the Schofield group [30] who conducted a study of the reaction of clavulanate with the TEM-2 β-lactamase from *E. coli*, and observed the appearance of four intermediates corresponding to mass increments of 52, 70, 88, and 155 Da. More recent crystallographic studies by van den Akker on the deacylation deficient E166A variant of the SHV-1 β-lactamase have also seen a decarboxylated trans-enamine bound to Ser70 [31]. More recent ESIMS studies by the Bonomo group [32] on the inhibitor resistant S130G SHV-1 β-lactamase have seen, in addition to the intermediates observed by Schofield, two new species at mass increments of 113 and 174 that have yet to be assigned. A consensus scheme for the inhibition due to clavulanate is shown below. The departure of the oxygen leaving group (and consequent cleavage of the oxazolidine ring) is undoubtedly assisted by the electrons on the (newly formed) amine nitrogen. However, the fragmentation of the oxazolidine ring is probably not concurrent with cleavage of the β-lactam ring, since, in such a single-step process, the geometry of the liberated nitrogen electrons would not be immediately appropriate for a resonance interaction with the p-orbital of a newly generated C5 carbocation. The 113 intermediate observed by Bonomo [32] may come from the aldehyde **19** (mass 70) in addition to transfer of the clavulanate CO_2 to lysine (mass 44 – 1). Knowles and others have shown that the hydroxymethyl group is not essential for activity, and considerable modification of this position has been performed, leading to some highly active compounds.

Sulbactam

The Knowles group also studied sulbactam [33]. In preliminary work, they noted that a number of penicillin sulfones, including the sulfones of several antibiotic penicillins, inactivate β-lactamase, with a partitioning of the initial acyl enzyme between turnover and inhibition. Antibiotics which were poor substrates (as sulfides) tended to produce better inhibitors as sulfones. Again they found a partitioning between hydrolysis, the formation of a transiently stabilized intermediate, and the formation of an irreversibly inactivated species. This inactivated species again showed loss of radioactivity incorporated into one of the gem-dimethyl groups, indicating substantial fragmentation, leading to loss of the thiazolidine portion of the inhibitor. The inactivated species appeared identical, by isoelectric focusing, with one of the inactivated species derived from incubation of the enzyme with clavulanate. They therefore proposed a nearly identical inhibitory mechanism for sulbactam, using the sulfone sulfur as a (sulfinate) leaving group in place of the enol ether oxygen of clavulanate as shown. Once again, the Knowles group proposed that the irreversibly stabilized species was due to dual attachment of the inhibitor to both the serine and to an active site lysine, and, once again, Mobashery suggested that the second residue was Ser130, or possibly, Lys234. The consensus is shown in Scheme 5.

Tazobactam is structurally related to sulbactam, the sole difference being the addition of a triazole group on the C2β-methyl group. This structural modification improves activity against several β-lactamases (IC_{50} 2 to 100 fold better) and microorganisms (MIC values 4 to 64 fold lower). The partition ratios, against representative class A β-lactamases for clavulanate, sulbactam and tazobactam are 160, 10,000, and 125, respectively. Tazobactam also has modest activity against some class C β-lactamases, but not enough to be clinically useful. Crystallographic [34, 35] and ESIMS [36, 37] studies support a mechanism closely related to that of sulbactam and clavulanic acid. Knox has proposed that the role of the triazole functionality, especially including the reason for tazobactam's improved activity, may be due to a steric clash of the triazole in the bound intermediate, which restricts movement of the tazobactam-derived segment of the complex to a region closer to Ser130 [34]. Based on the proximity (3.9 Å) of the triazole to C6 in the published van den Akker acyl-enzyme crystal structure (PDB code 1RCJ), Buynak has speculated that the triazole may promote tautomerization of the initially formed imine to the enamine (β-aminoacrylate) via intramolecular abstraction of the C6 proton [38]. There appears to be little conformational restriction of the triazole in the respective linearized intermediates of either of the acyl-enzyme structures.

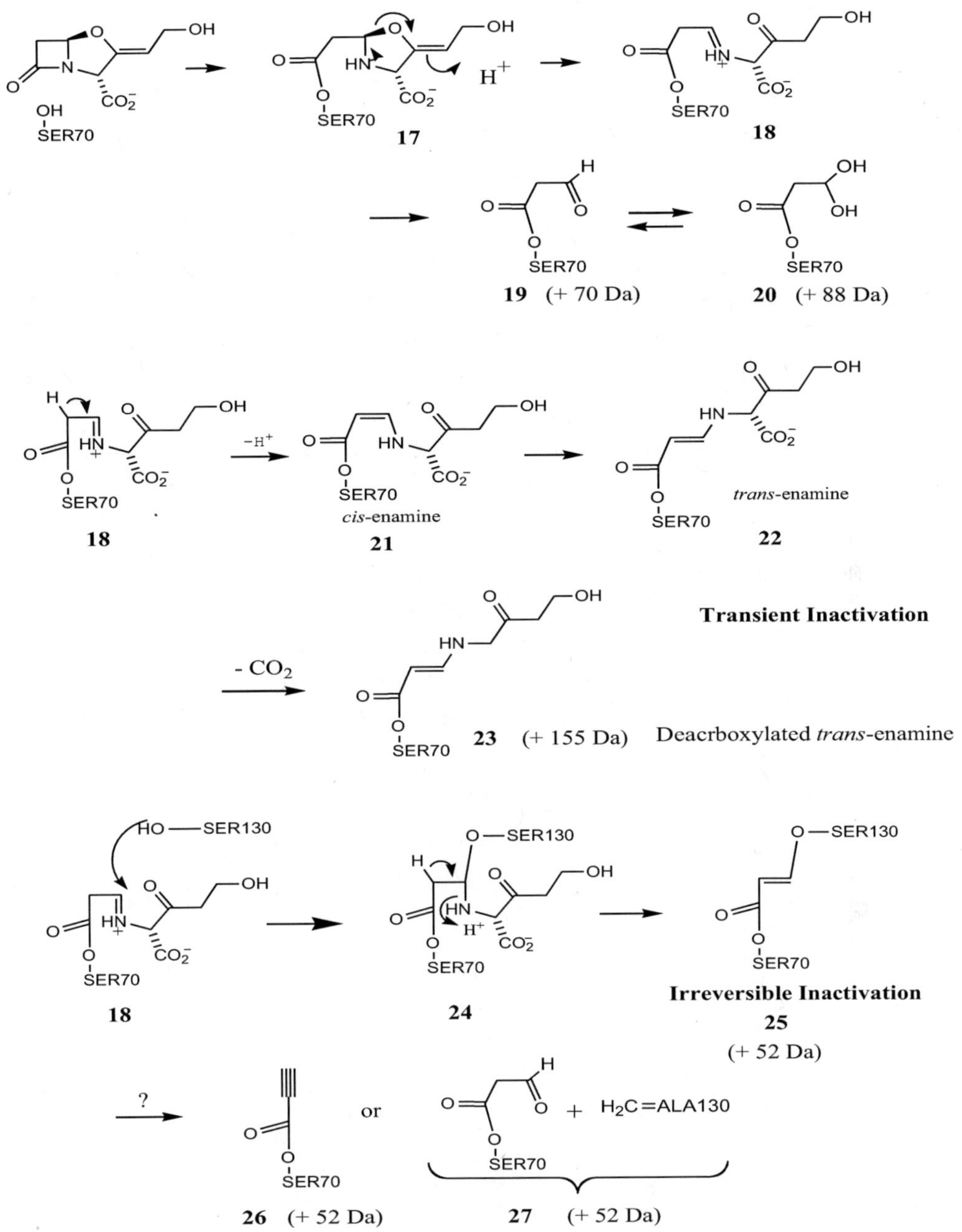

Scheme 4. Mechanism for Inhibition of Class A β-Lactamases by Clavulanic Acid.

Other 6-(Unsubstituted)penicillin Sulfones

A number of groups have continued to report improvements in β-lactamase inhibitory activity by substitution at (penicillin) C2′. Ro 48-1220 (33), displays excellent activity against both class A and class C β-lactamases [39] and synergy with representative cephalosporins against class A and class C producing microorganisms [40]. Buynak *et al.*

[41] have recently reported a C3 homologue of sulbactam with improved activity against class C β-lactamases.

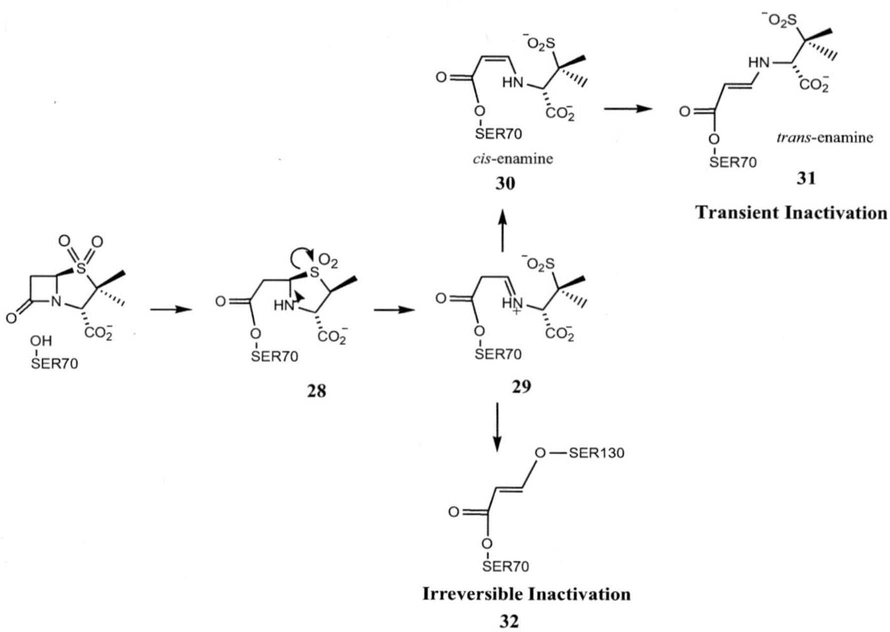

Scheme 5. Mechanism for the Inactivation of Class A β-Lactamases by Sulbactam.

Figure 7. Other 6-(Unsubstituted)penicillin Sulfones.

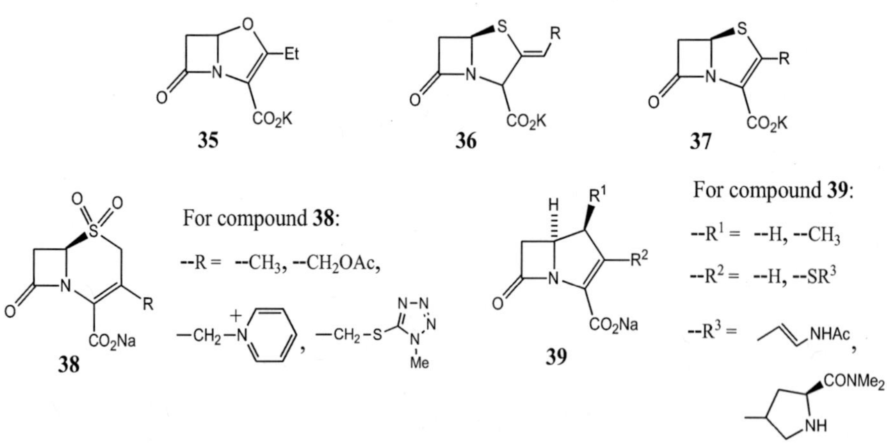

Figure 8. Structures of Other α-Unsubstituted-β-lactams.

Other α-Unsubstituted-β-lactams

The racemic oxapenem **35** was prepared and shown to be a potent inhibitor of isolated β-lactamases from both Gram-positive and Gram-negative species, particularly of the class C cephalosporinases [42]. 6-Unsubstituted-2-(exomethylidene)penams, such as **36** have been prepared and reported to be less active than the corresponding oxygen analogs [43, 44]. 6-Unsubstituted penems, **37**, are also known, some of which display antibiotic activity [45, 46] and also modest β-lactamase inhibitory activity [47]. Cephalosporin analogs of sulbactam, **38**, are less potent than the corresponding penams [48]. C6-unsubstituted carbapenems, **39**, are, depending on the C2 substituent, known to be antibacterial agents [49, 50], but separate studies of their abilities to inhibit β-lactamases are unavailable. These compounds have variable antimicrobial activity against β-lactamase-producing strains [51].

STRUCTURAL MOTIF 2: α-(α'-HYDROXYALKYL)- β-LACTAMS

6-(Hydroxyalkyl)carbapenems

While most β-lactamase inhibitors lack significant antibacterial activity, the carbapenems, having a C6 hydroxyethyl side chain, are recognized as both broad-spectrum antibiotics and also as either very poor substrates or as competitive inhibitors of many serine β-lactamases. Commercial carbapenems include imipenem (**40**) /cilastatin (Primaxin™), panipenem (**41**), biapenem (**42**), meropenem (**43**) (Merrem™), and ertapenem (**44**) (Invanz™). The cilastatin is included in the imipenem mixture as an inhibitor of the renal metabolic enzyme, dehydropeptidase (DHP-1), whereas the C1β-methyl group of later carbapenems renders the molecules poorer substrates of DHP.

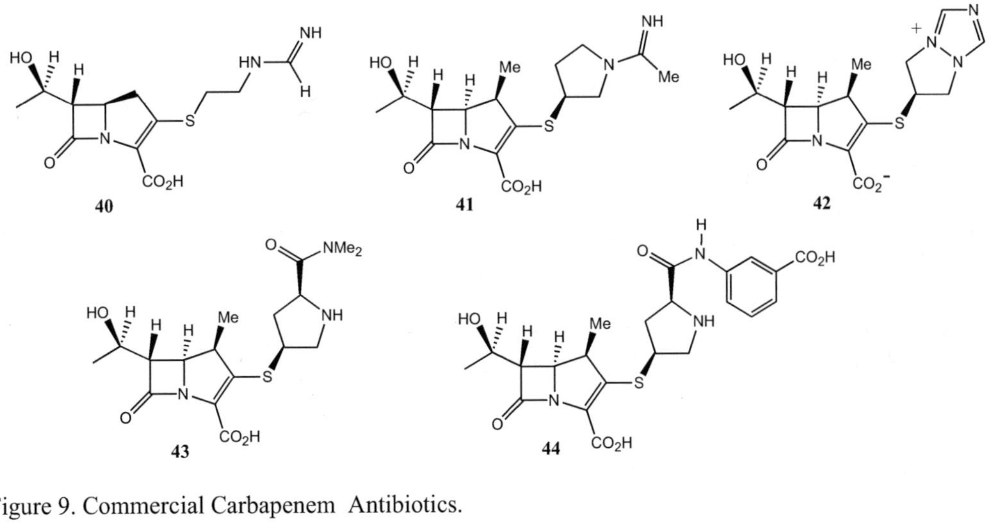

Figure 9. Commercial Carbapenem Antibiotics.

Early patents by Beecham researchers describe C6-(hydroxyethyl)-*O*-sulfonic acids (note: older literature uses an alternate numbering system in parentheses) isolated from

Streptomyces olivaceus (subsequently known as the olivanic acids) with potent β-lactamase inhibitory activity [52, 53]. Among the compounds isolated, which included both alcohols and *O*-sulfonic acids, of both 6*S* and 6*R* stereochemistry, the best overall β-lactamase inhibitor, based on synergy with amoxicillin and comparison to amoxicillin/clavulanate was the *O*-sulfonic acid MM 4550 (**45**) [54]. (A comparison of a related alcohol and corresponding *O*-sulfonic acid as inhibitors of the class A TEM β-lactamase, verified this comparison in the 6*R*, 8*S* series [55].) This sulfonic acid has nanomolar activity against both class A and class C β-lactamases. The C2 side chain sulfoxide (MM 4550, 45) is somewhat better against class A enzymes, whereas the corresponding sulfide, MM 13902, is more active against class C cephalosporinases [56]. Merck reported the isolation and structural elucidation of structurally related compounds, (in their case, as the alcohols) and identified thienamycin, **46**, on the basis of its combination antibacterial and β-lactamase stability [57, 58]. Researchers at Fujisawa, Takeda, Sankyo, and other companies followed with a number of carbapenems, including the C6 ethyl deriviative PS-5, **47**. Like most carbapenems, PS-5 has both excellent antimicrobial activity and also inhibitory activity against β-lactamases, thus indicating that the hydroxyl group of the C6 side chain may not be essential for inhibition [59].

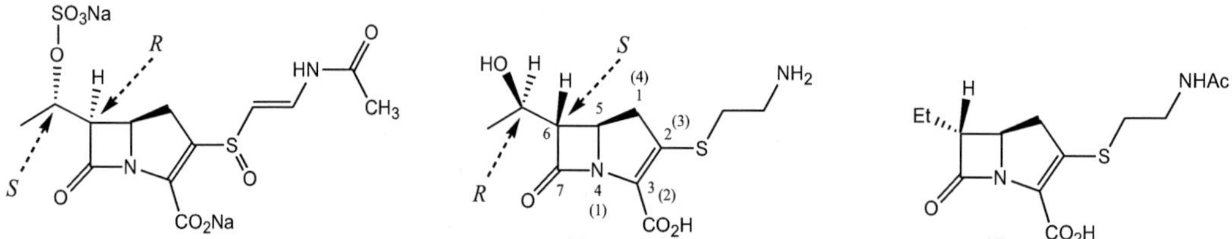

45 **46** **47**

Figure 10. Early carbapenems.

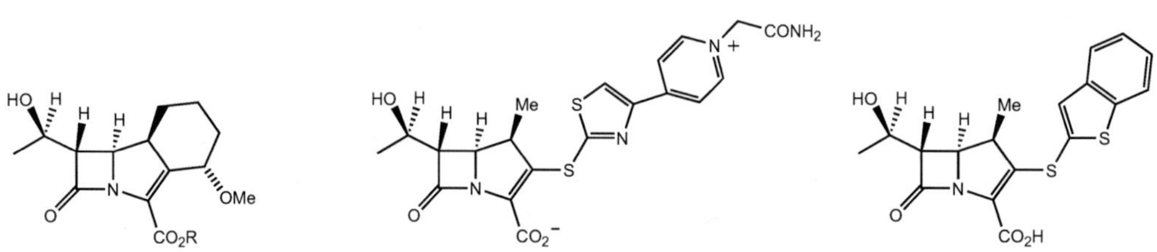

48: R = Na
49: R = —CH-O—

50 **51**

Figure 11. Recent Carbapenems.

The O-sulfonic acids are good antibacterials and β-lactamase inhibitors [60] but have less stability *in vivo* than the corresponding alcohols [61]. Among the 6-(hydroxyalkyl)carbapenems, optimal antimicrobial activity was attained from the 6*S*, 8*R* stereochemistry of thienamycin [61]. Based on data published on the corresponding C2 side chain dehydro-derivative, the 6*S*, 8*R* isomer has MIC's ten to thirty times lower than the corresponding 6*S*, 8*S* diastereomer. This discovery was commercialized as the formamidine, imipenem.

The carbapenems are regarded as the β-lactam class with the broadest spectrum of antibacterial activity. These antibiotics are also inhibitors (or very slow substrates [62]) of most serine β-lactamases. However, carbapenems are substrates of class B metallo-β-lactamases [63, 64] and also substrates of some serine β-lactamases [65, 66].

All carbapenems of current commercial interest retain the C6α-hydroxyethyl side chain of thienamycin. Recent developments include the addition of a C1β-methyl group to protect against renal dehydropeptidase 1 and modifications of the C2 side chain, including tricyclic carbapenems (**48** and **49**) [67], carbapenems having improved activity against MRSA [68, 69] (e.g., **50**, through insertion of a suitable lipophilic group at C2) and also orally available carbapenem prodrugs (e.g., **49**) [70]. The β-lactamase inhibitory activity of this group appears to be a function of the substituents at C2 and at C6, with the newer derivatives (having more elaborate C2 side chains) being better β-lactamase inhibitors (or at least more stable to the enzymes) than the original discoveries/products. Although MIC values are available for these newer carbapenems, little has been published on their independent ability to inactivate β-lactamases. A recent series (e.g., **51**) displays nanomolar activity against some class B metallo-β-lactamases notably including IMP-1 [71].

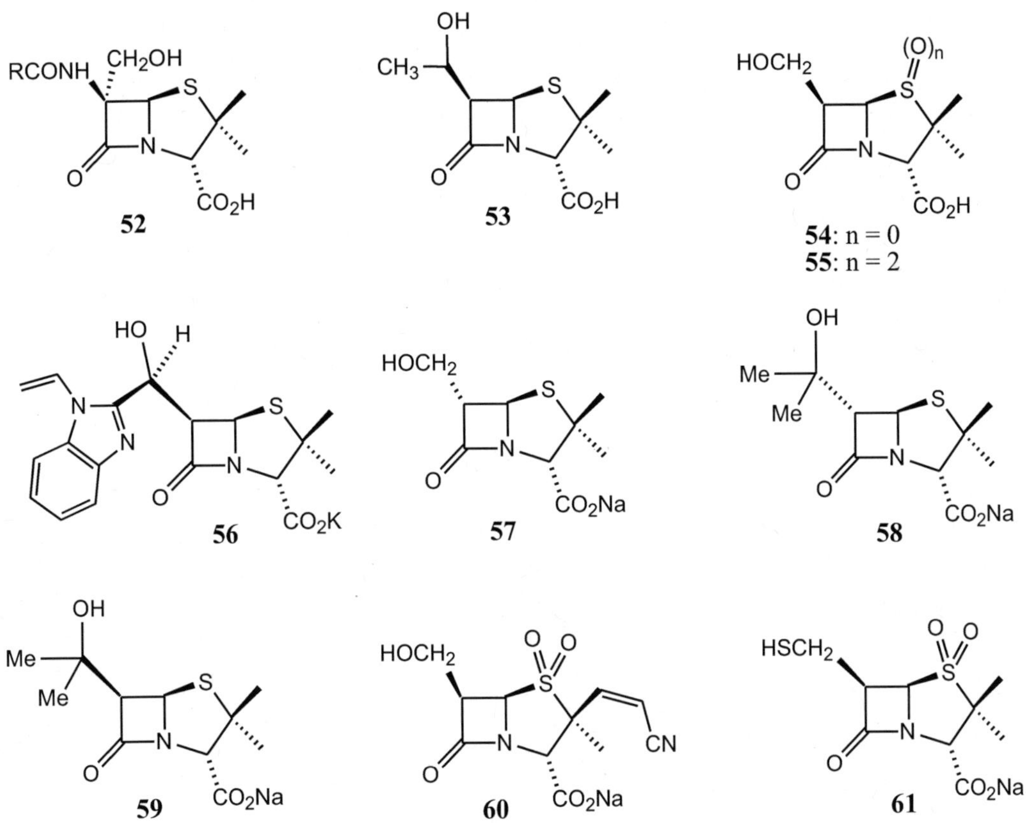

Figure 12. 6-(Hydroxyalkyl)penicillinates and Corresponding Sulfones.

As regards the precise mechanism of β-lactamase inactivation by the carbapenems, several kinetic investigations [55, 62, 72-75] and three crystallographic studies, including complexes of imipenem with TEM-1 (class A) and AmpC (class C) [77], and a complex of meropenem with OXA-13 (class D) [78], have been published. In both imipenem structures, the reason for inhibition appears to be that the carbonyl oxygen of the resultant acyl-enzyme has been displaced from the oxyanion hole. In the case of the meropenem structure with the class D oxacillinase, the carbonyl oxygen is in the oxyanion hole, but a conformational change of the enzyme has displaced the hydrolytic water to a distance of nearly 5 Å from the carbonyl carbon. The C6α hydroxyalkyl side chain may play a role in the positional movements that occur subsequent to enzyme acylation.

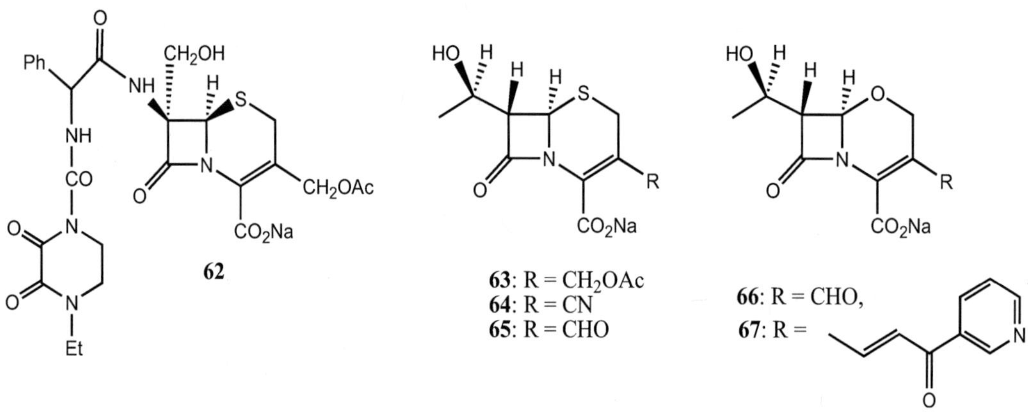

62

63: R = CH_2OAc
64: R = CN
65: R = CHO

66: R = CHO,
67: R =

Figure 13. C7-(Hydroxyalkyl)cephalosporins and oxacephalosporins.

6-(Hydroxyalkyl)penicillinates

Early researchers discovered that C6 hydroxyalkyl derivativates, such as **52**, were stabilized toward β-lactamase, with the loss of some antibacterial activity [79, 80]. DiNinno *et al.,* working at Merck, first prepared simple 6-(hydroxyethyl)penicillinates **53** (lacking the acylamino group; the hydroxyethy group being analogous to the carbapenem side chain) and showed these compounds to be of modest antibacterial activity [81]. Subsequently, a number of companies simultaneously filed patents describing the generation of C6 hydroxyalkyl penicillinates, such as sulfide **54**, and sulfone **55**, and claimed their usefulness as inhibitors of β-lactamase [82, 83]. The Pfizer group attempted to further develop benzimidazole-substituted (hydroxymethyl)penicillinates, such as **56**, which display both antibacterial activity (against Gram-positive and a few Gram-negative strains) as well as β-lactamase inhibitory activity [84]. Recently, Mobashery and Samama described and provided structures of the class A TEM-1 β-lactamase from *E. coli* inhibited with the C6α -hydroxymethylpenam **57** [85, 86], the NMC-A class A carbapenemase inhibited with the C6α-[(1'-hydroxy-1'-methyl)ethyl]penam **58** [87] (the corresponding hydroxymethyl analog was a substrate) and the the class D Oxa-10 β-lactamase inhibited with the C6α-[(1'-hydroxy-1'-methyl)ethyl]penam **59** [88]. The Mobashery group has been able to utilize the hydroxyalkylpenicillinates as probes to determine the direction of addition of water to the acyl-enzyme [89]. Wyeth researchers evaluated the epimeric C6-(hydroxymethyl) penicillinates, both as sulfides and as sulfones, presenting data that the best inhibitors in this series were the sulfones (e.g., C6β-(hydroxymethyl)penicillin sulfone **55** exhibited an IC_{50}

value of 8 nM against TEM-1) [90]. They also investigated the potential for the inhibitory activity to be modulated through manipulation of the C2 substituent, with their most active molecule being the acrylonitrile **60** [91]. Recently, Buynak has prepared the corresponding mercaptomethyl analogs **61** and has demonstrated their activity as dual inhibitors of both serine-β-lactamases and also certain class B metallo-β-lactamases [92].

7-(Hydroxyalkyl)cephalosporins

As was the case with the corresponding penicillins, the 7-(hydroxymethyl)cephems, **62**, with a C7 acylamino group are known to possess antibiotic activity (particularly against Gram-negative strains) and β-lactamase stability [93-95]. Monosubstituted C7-hydroxyethyl analogs, **63**, were first prepared at Merck [81]. Later, the corresponding 7-(hydroxyethyl)cephalosporins, **64** and **65**, were reported as inhibitors of β-lactamase, with particularly good activity against class C cephalosporinases [96, 97]. The corresponding oxacephems, **66** and **67**, also inhibit β-lactamase with preferential inactivation of the class C enzymes [98].

6-(Hydroxyalkyl)clavams and 6-(Hydroxyalkyl)oxapenems

Simple 6-(hydroxyethyl)clavams were prepared (as diastereomers) and shown to be potent β-lactamase inhibitors, with the 6β analogs **68** in excess of ten times more potent than the corresponding 6α analogs **69** against both representative class A and class C β-lactamases (69 is active against class C cephalosporinases, but inactive against penicillinases) [99, 100]. Similarly, 6-(hydroxyethyl)oxapenems **70** have been prepared and shown to be potent cephalosporinase inhibitors, but only weak inhibitors of the class A penicillinases [101].

The C2-isopropyl compound **70** also had independent antibacterial activity.

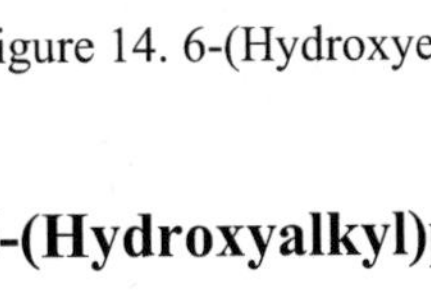

68
(BRL 23547)

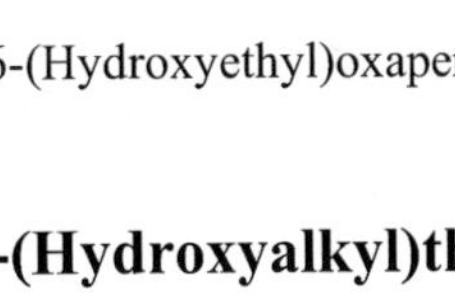

69
(BRL 23546)

70

Figure 14. 6-(Hydroxyethyl)clavams and 6-(Hydroxyethyl)oxapenems.

6-(Hydroxyalkyl)penems and 6-(Hydroxyalkyl)thiaclavams

The penems were designed to be a cross between the fused 4- and 5-membered-ring system of the penicillins and the cephalosporins, which have an endocyclic double bond conjugated to the nitrogen [102]. In terms of structure-activity relationships, penems most closely parallel carbapenems. Further reinforcement of this comparison, is evident in the identical 6*S*, 8*R* stereochemistry of the hydroxyethyl side chain, which, as in the carbapenems, was optimized for antibacterial activity [103]. Penems possess both antimicrobial activity and β-lactamase stability. However, several studies also document their β-lactamase inhibitory activity, which seems to be specific for class C cephalosporinases [104-106]. The precise inhibitory mechanism is unknown. By contrast, the hydroxyethyl thiaclavulanate derivative **75** was prepared and found to be selective for class A and class D β-lactamases [107]. Faropenem is scheduled (2006) to become the first oral penem available outside Japan.

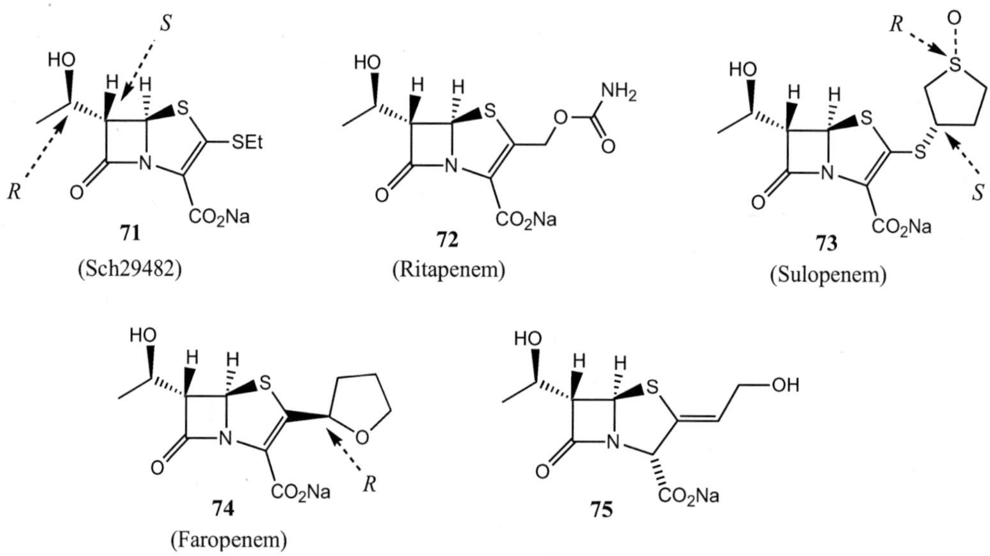

Figure 15. 6-(Hydroxyalkyl)penems and 6-(Hydroxyalkyl)thiaclavams.

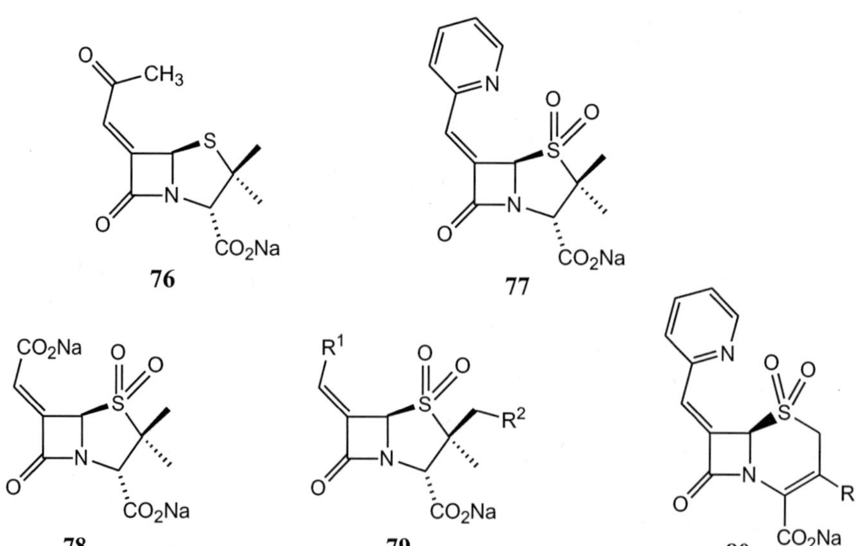

Figure 16. 6-(Alkylidene)penicillinates and Sulfones.

STRUCTURAL MOTIF 3: α-(ALKYLIDENE)- β-LACTAMS

6-(Alkylidene)penams

The first structure of this motif to be prepared was Ro 15-1903, 6-(acetylmethylene)penicillanic acid, **76**. This compound was found to be a nanomolar to low micromolar inhibitor against several class A and class C β-lactamases, typically 10 to 100 fold more potent than clavulanic acid [108, 109]. Based on kinetic and spectroscopic data, a mechanism of inhibition was proposed and shown below [110]. Note that the stabilized acyl-enzyme, **81**, is a trans-enamine, although the nitrogen is part of an aromatic pyrrole. The conjugation of the nitrogen lone pairs and also the overlap of the aromatic ring with the carbonyl group should stabilize the acyl-enzyme. This mechanism has not been verified crystallographically.

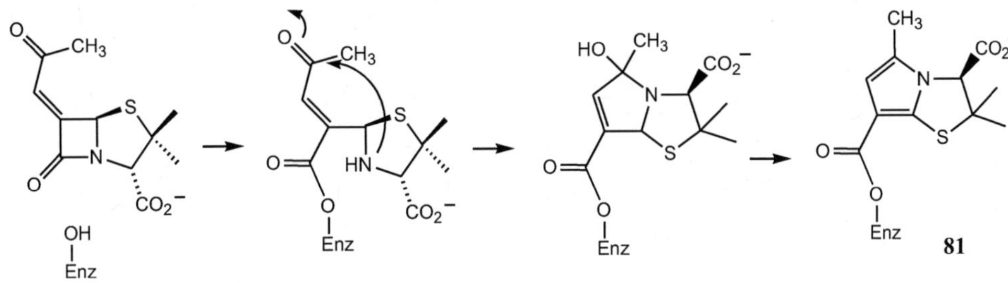

Scheme 6. Proposed Inhibitory Mechanism of Ro 15-1903.

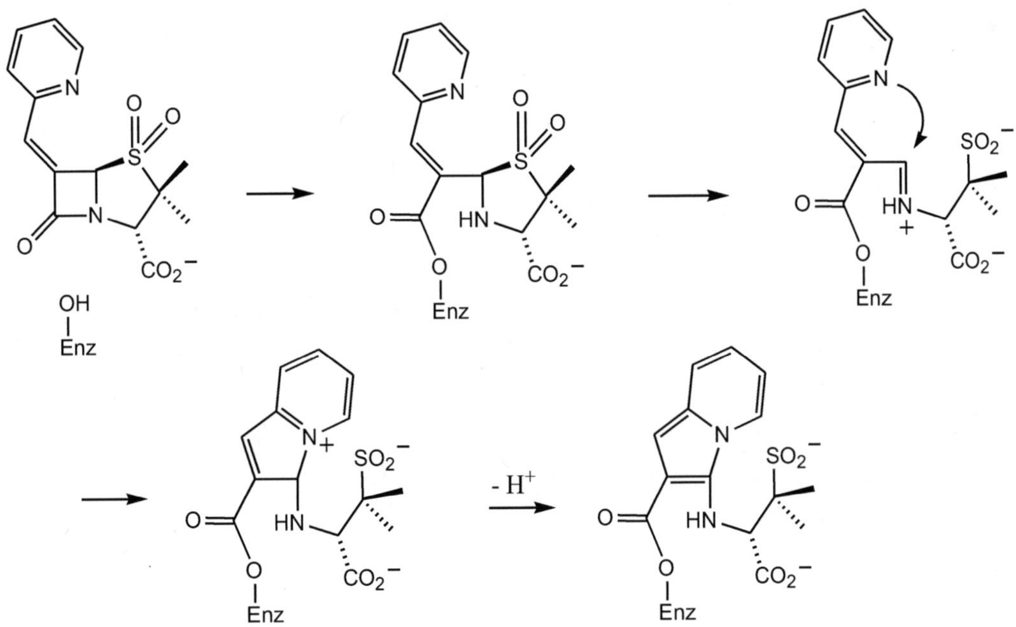

83

Scheme 7. Inhibitory Mechanism of 6-(Pyridylmethylidene)penicillin Sulfone.

The next major development within this substructure was the Pfizer discovery that 6-(heterocyclylmethylene)penicillinic acid sulfones (e.g., **77**) were excellent β-lactamase inhibitors [111].

This series was optimized and a mechanism proposed (Scheme 7) [112], based on analogous chemical reactions. The Buynak group also noted that 6-(carboxymethylidene)penicillin sulfones, **78**, were inhibitors of β-lactamase, in some cases superior to Ro 15-1903 [113]. Buynak [114] subsequently showed that inhibitory activity of a number of 6-(alkylidene)penam sulfones could be modulated through manipulation of the 2′ substituent, resulting in potent β-lactamase inhibitors of both class A and class C β-lactamases and also reported the first siderophore-linked β-lactamase inhibitor.

7-(Alkylidene)cephems

The corresponding 7-(alkylidene)cephalosporin sulfones **80** were first prepared by Buynak [115]. This compound class was systematically developed into nanomolar inhibitors of both class A and C β-lactamases, or, depending on the C3 substituent, highly selective class C inhibitors. Those selective inhibitors have been utilized diagnostically [116]. A crystal structure of an extended spectrum class C β-lactamase inhibited with one of these molecules was recently obtained, and identified stabilized acyl-enzyme **83**, verifying a mechanism (Scheme 8) analogous to that proposed by Chen in the case of the analogous penicillinates as shown in Scheme 8 [117].

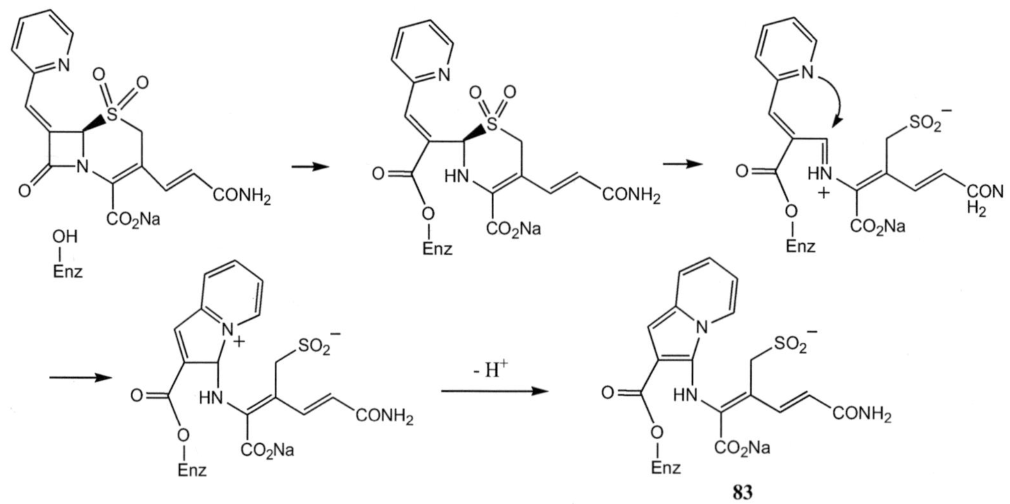

Scheme 8. Inhibitory Mechanism of 6*Z*-(Pyridylmethylidene)cephalosporin Sulfone.

6-(Vinylidene)penams and 7-(Vinylidene)cephems

Buynak has prepared the C6 and C7 allenic analogs of the penicillins (e.g., **84**) and cephalsoporins (e.g., **85**, **86**), respectively, and shown that the sulfones are modest inhibitors of class C β-lactamases [118, 119]. Based on a six-fold slowing of the rate of inactivation with deuterium labeling of the allene, they provided evidence for the inhibitory mechanism shown in Scheme 9.

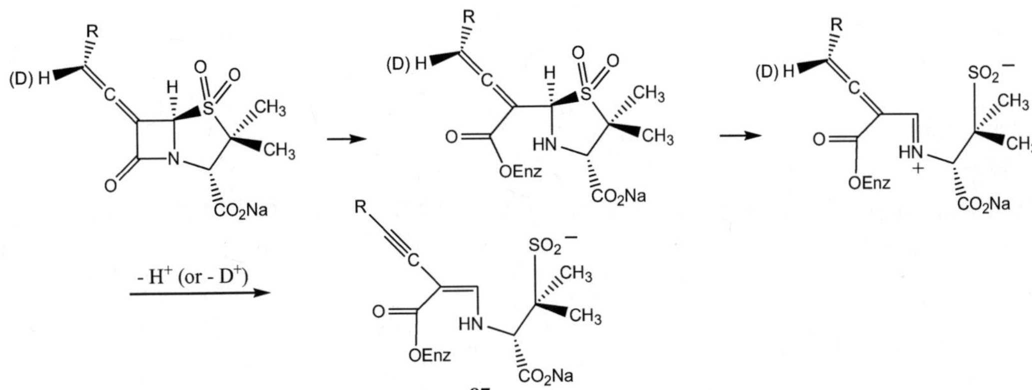

87

Scheme 9. Proposed Inhibitory Mechanism of the 6-(Vinylidene)penicillin Sulfones.

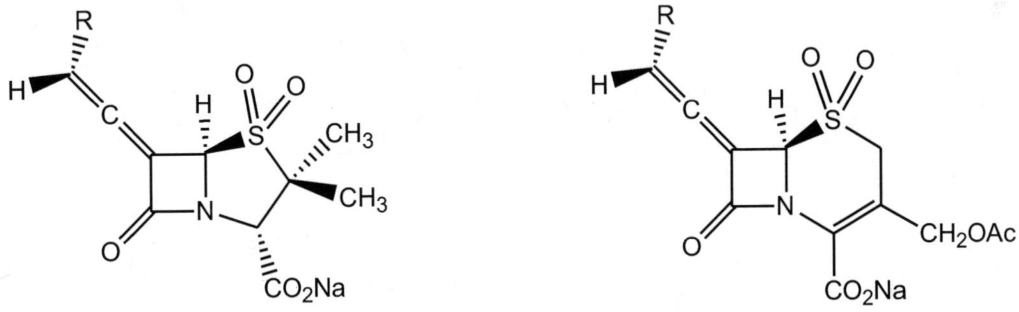

84: R = *t*-Bu

85: R = H
86: R = *t*-Bu

Figure 17. 6-(Vinylidene)penicillin Sulfones and 7-(Vinylidene)cephalosporin Sulfones.

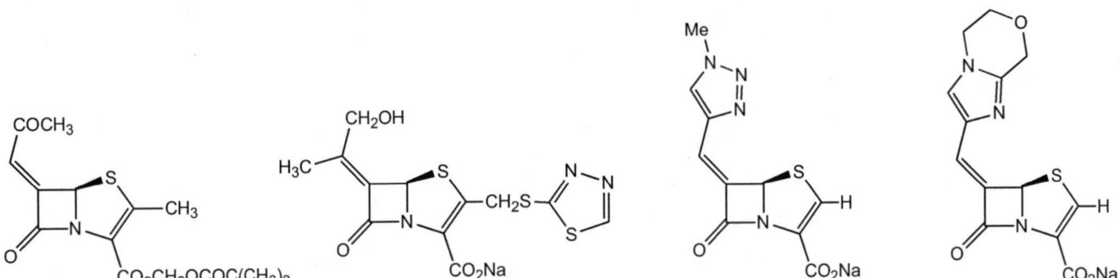

88 **89** **90** **91**

Figure 18. Structures of 6-(Alkylidene)penems

6-(Alkylidene)penems

The first 6-(alkylidene)penem to be disclosed was **88**, the penem analog of 6-(acetylmethylidene)penicillinate, Ro 15-1903 [120]. This penem, synthesized as the pivaloyloxymethyl ester, was reported to be a 'potent inhibitor of a number of isolated β-lactamases' with no data provided. A subsequent disclosure of the penem analog of the carbapenem antibiotic asparenomycin, **89**, which displayed potent independent *in vitro* antimicrobial activity including MICs in the range of less than 1 μg/mL against a number of strains [121, 122]. In a series of articles [123-127], the SmithKline group systematically optimized the β-lactamase inhibitory activity of 6-alkylidenepenems, resulting in the production of the extremely potent molecule, BRL 42715, **90**. This compound displays IC_{50} values in the nanomolar range against classes A, C, and D β-lactamases and displays synergy against β-lactamase-producing strains when combined with a wide range of β-lactam antibiotics. More recently, this compound has undergone further optimization, resulting in the attachment of bi- and tricyclic heterocycles to the exocyclic double bond (e.g., **91**) [128, 129].

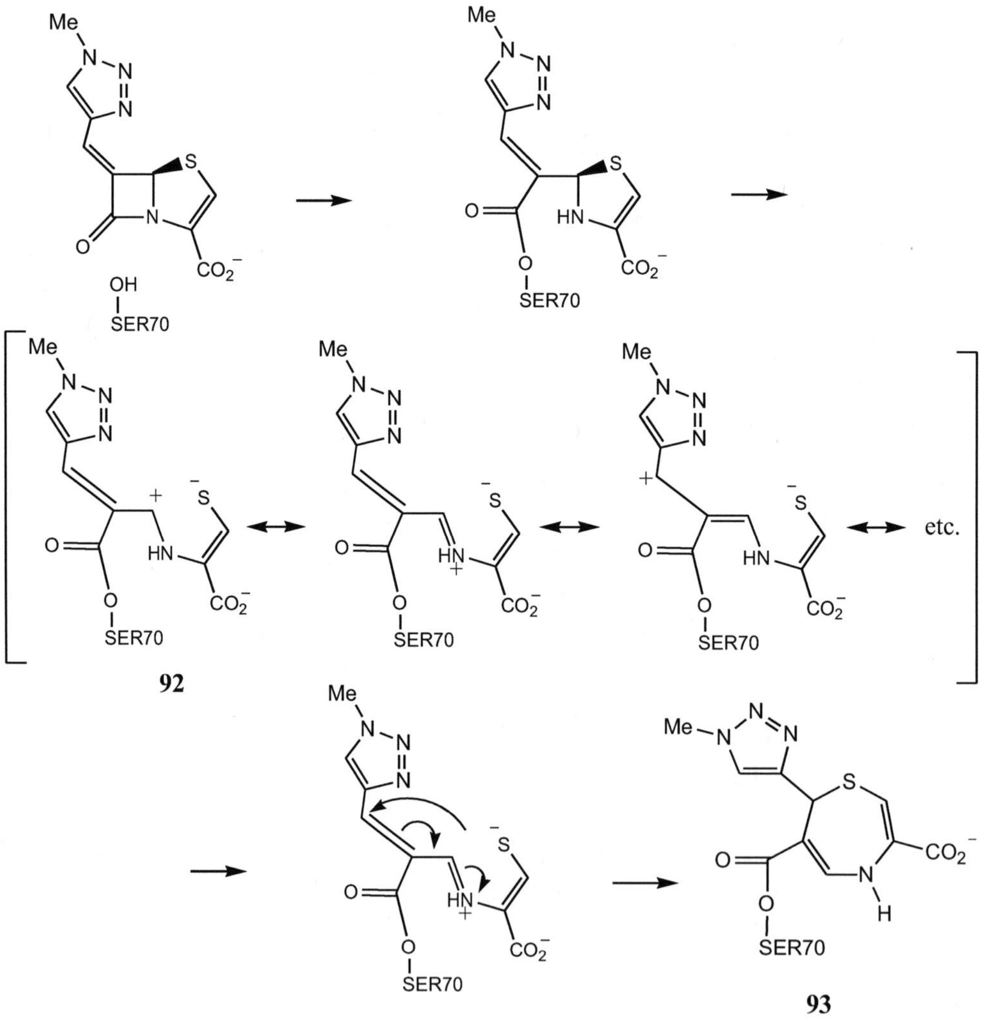

Scheme 10. Proposed Mechanism of Inhibition by BRL 42715 (**90**).

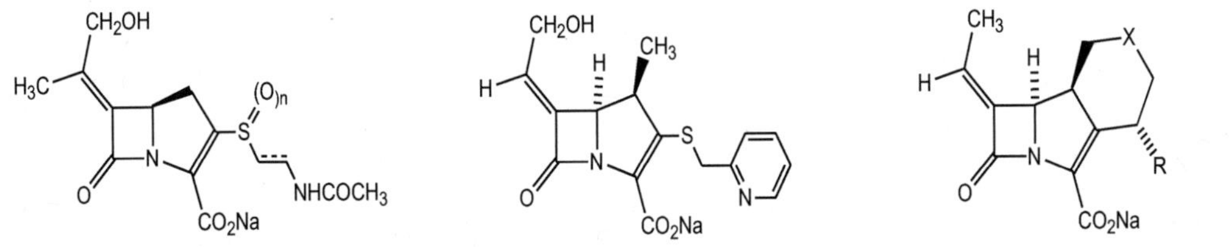

94: A: n = 1, double bond in C2 side chain
95: B: n = 1, single bond in C2 side chain
96: C: n = 0, double bond in C2 side chain

97

98: X = CH_2, S; R = H
99: X = CH_2, R = OCH_3

Figure 19. Structures of 6-(Alkylidene)carbapenems.

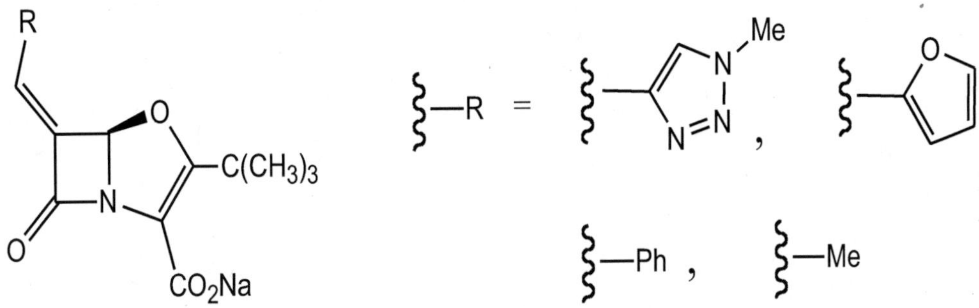

100

Figure 20. 6-(Alkylidene)-1-oxapenems.

The mechanism of action of the 6-(alkylidene)penems has been investigated through chemical [130], kinetic [131, 132], modeling [133], ESIMS [134], and crystallographic [135-137] studies, and is shown in Scheme 10. Subsequent to acylation of the active site serine, a fragmentation of the dihydrothiazole ring occurs, followed by conjugate addition of the liberated thiolate anion to the C6 double bond to generate an enamine (vinylogous urethane) motif in a seven-membered dihydro[1,4]thiazepine structure **93** (observed crystallographically). As was observed in the transiently stabilized intermediates for clavulanate, sulbactam, and tazobactam, this intermediate is chemically stabilized due to the ability of the nitrogen lone pair to interact with the ester (acyl-enzyme) carbonyl group. It is likely that departure of sulfur (and consequent ring opening of the dihydrothiazole) is assisted by the additional resonance-stabilization provided to the resultant (C5) carbocation (92) by the C6 double bond and appended heterocycle (in addition, of course, to the adjacent nitrogen, as shown.)

6-(Alkylidene)carbapenems

The first isolated 6-(alkylidene)carbapenems were the asparenomycins A, B, and C (**94** – **96**), which have minor structural variations at the C2 side chain. These compounds display both antibiotic and β-lactamase inhibitory activity, with asparenomycin A and C displaying the best antibacterial activity [138] and asparenomycin B displaying the best β-lactamase inhibitory activity [139]. These compounds are potent, time-dependent β-lactamase inactivators, with activity better than clavulanate against class A enzymes, and also excellent activity against the class C β-lactamases (approx. 100 times more potent than clavulanate against the latter). An attempt to further develop this structural motif, by removal of one of the alkylidene methyl groups and addition of a C1 methyl (to stabilize against renal dehydropeptidase), resulted in **97**, which displayed reduced β-lactamase inhibitory activity [140]. Two recent studies have described the synthesis and activity of **95** [141] and also a kinetic study of compounds **99** [142], all of which appear to possess dual class A and class C inhibitory activity. The precise mechanism of β-lactamase inhibition of the 6-(alkylidene)carbapenems has not been elucidated.

6-(Alkylidene)-1-oxapenems

6-(Alkylidene)-1-oxapenems, **100**, above, have been synthesized and described as β-lactamase inhibitors, with no data provided [143]. The electron-donating C2 *tert*-butyl group stabilizes the bicyclic system toward hydrolysis.

STRUCTURAL MOTIF 4: α-HALO-β-LACTAMS

Compounds containing a halogen α to the β-lactam carbonyl represent some of the earliest reported inhibitors. 6β-bromopenicillanic acid (**101**) and the corresponding iodide were found to be effective against class A β-lactamases, but significantly less active against the class C enzymes [144-148]. A proposed inhibitory mechanism is shown in Scheme 11.

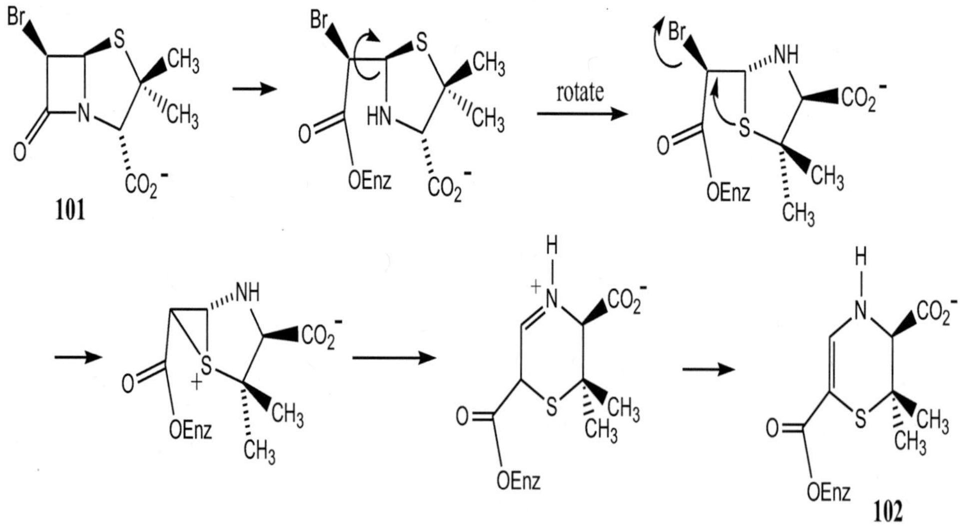

Scheme 11. Proposed inhibitory mechanism of 6β-bromopenicillanic acid.

The 6-(spiroepoxy)penicillinates, such as **103**, also have β-lactamase inhibitory activity and likely function through a related mechansism, as shown in Scheme 12.

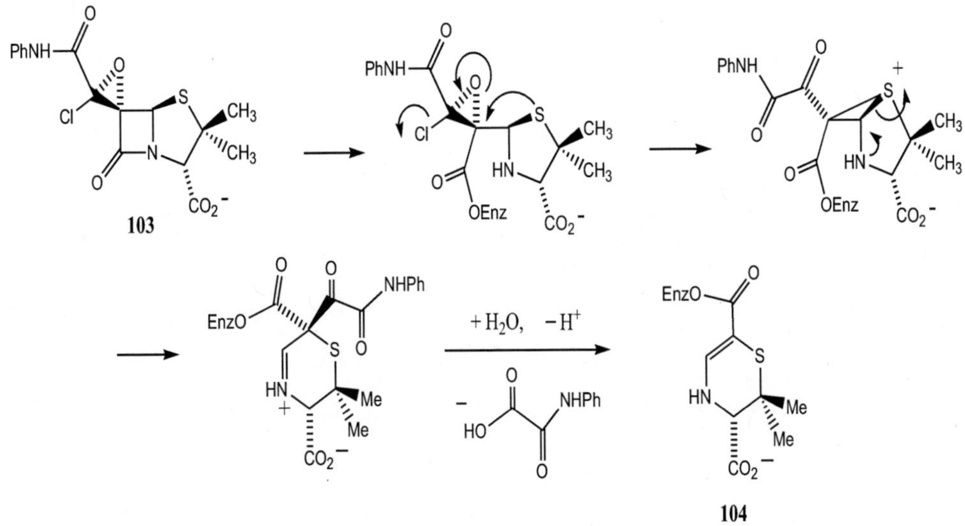

Scheme 12. Proposed Inhibitory Mechanism of 6-(Spiroepoxy)penicillinates.

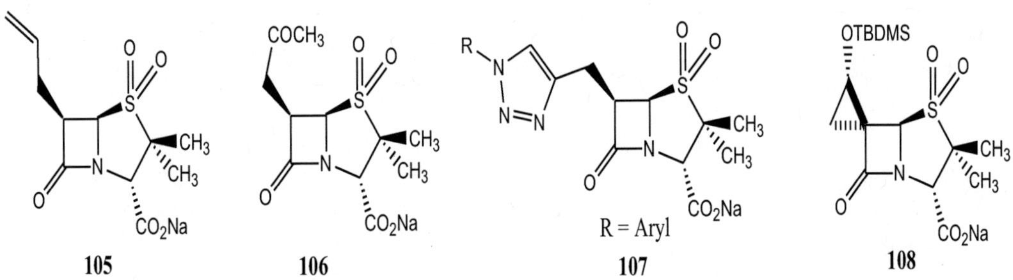

Figure 21. 6-Alkylpenicillin Sulfones.

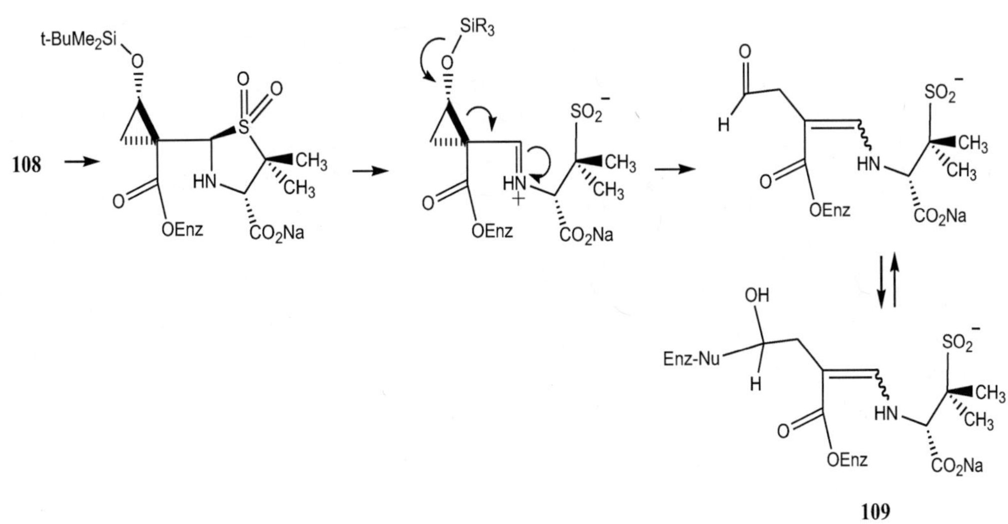

Scheme 13. Proposed mechanism of β-lactamase inactivation by **108**.

MISCELLANEOUS β-LACTAM-CONTAINING INHIBITORS

6-Alkylpenicillin Sulfones

That C6-substituted penicillin sulfones (apparently only with the 6β configuration shown) could be effective inhibitors of β-lactamase was first reported by Hanessian, who observed the inhibitory activity of **105**, but did not provide details of the inhibition [149]. A later report noted similar activity for **106**, which was derived from reduction of double bond and oxidation of the sulfur of the structurally related, previously known inhibitor, Ro15-1903 [150]. Compound **106** was shown to possess inhibitory activity between that of clavulanate and sulbactam. This general class of compounds was further developed by Wyeth, into the triazoles **107**, which now have low micromolar inhibitory activity against both class A and class C β-lactamases [151]. The Wyeth group has also developed the spirocyclopropyl compounds **108**, which have sub-micromolar activity against both class A and class C β-lactamases [152]. A proposed mechanism of inhibition is shown in Scheme 13.

N-Sulfonyl- and *N*-Sulfonyloxy-β-lactams

Monocylic, *N*-sulfonated-β-lactams, **110**, were first reported as antibacterials in a 1979 Takeda patent [153], and such compounds were the subjects of articles from Takeda and Squibb (where the name 'monobactam' was suggested), both published in 1981 [154, 155]. Squibb developed this series into the commercial product, aztreonam, **111**. Aztreonam is highly active against Gram-negative aerobic bacteria, but relatively inactive against Gram-positive bacteria or Gram-negative anaerobes [156, 157].

It has high affinity for PBP 3. While initially discovered monobactams had poor affinities for class A β-lactamases, and were micromolar inhibitors of class C β-lactamases, aztreonam was described as forming a covalent complex with the class C P99 enzyme [158] and has been characterized as 'an efficient transient inactivator' of class C β-lactamases [159]. More recently, Syn 2190, **112**, a monobactam which is devoid of antibacterial activity, but highly efficient as a selective inhibitor of class C β-lactamases, has been described [160-163].

A structurally related series of compounds are the monosulfactams, represented by tigemonam (**113**). These compounds are synthetic in origin and were designed, synthesized, and evaluated at Squibb [164, 165].

Structure-activity relationships demonstrate the importance of the substituent at C4 to β-lactamase stability and inhibition [166]. Tigemonam has a spectrum of activity similar to that of aztreonam, but is orally bioavailable. This monosulfactam is resistant to most β-lactamases, and is reported to be an inhibitor of the class C enzymes, such as *Enterobacter cloacae*, P99 [167]. Another series of *N*-sulfonyloxy-β-lactams (e.g., **114**) was developed as inactivators of class A β-lactamases by Miller and Mobashery [168, 169].

Recently, new series of bridged monobactams [170, 171] and bridged sulfactams [172] have been developed as β-lactamase inhibitors by Hoffmann-La Roche. Monobactam Ro 48-1256, **115**, had no antibacterial activity, but acted as a selective inhibitor of class C β-lactamases and displayed synergy with ceftazidime and piperacillin against strains derepressed for the AmpC enzyme [173].

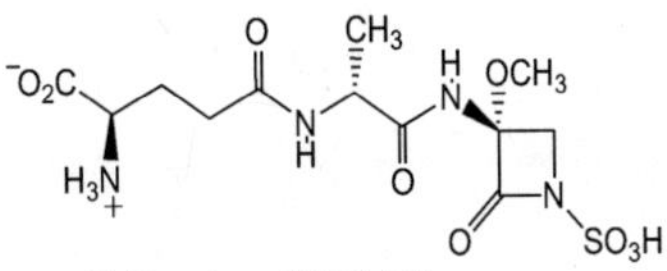
Sulfazecin or SQ 26,445
110

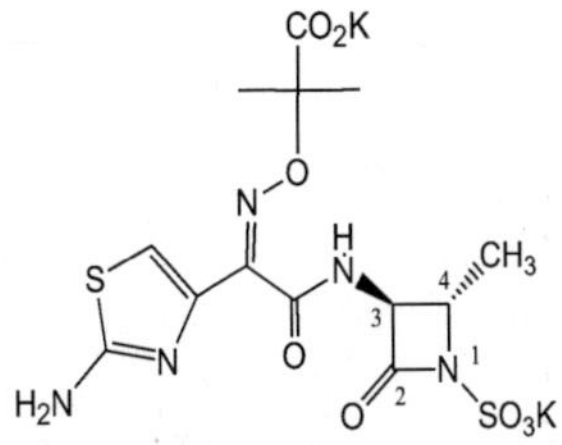
Aztreonam
111

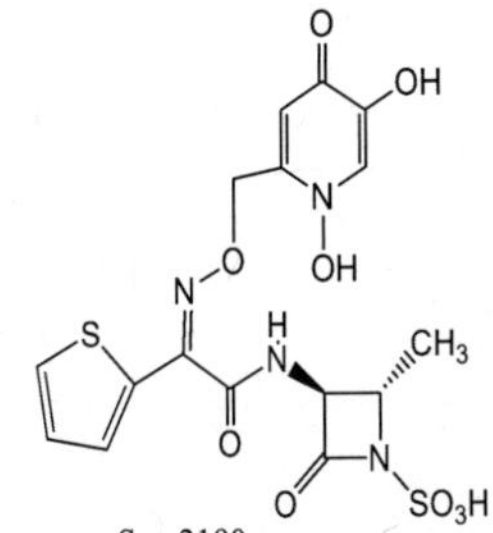
Syn 2190
112

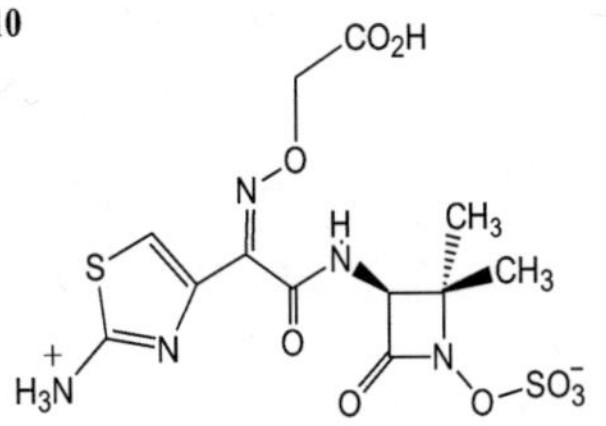
Tigemonam
113

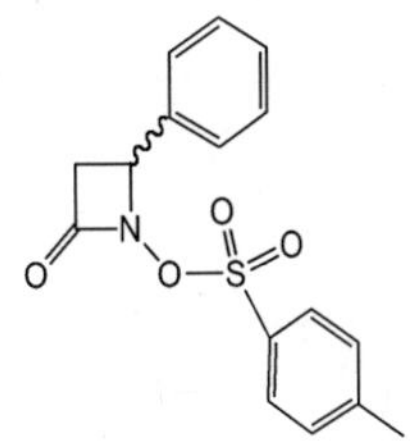
114

Figure 21. Monocyclic *N*-Sulfated-β-lactams (Monobactams) and *N*-Sulfonyloxy-β-lactams (Monosulfactams).

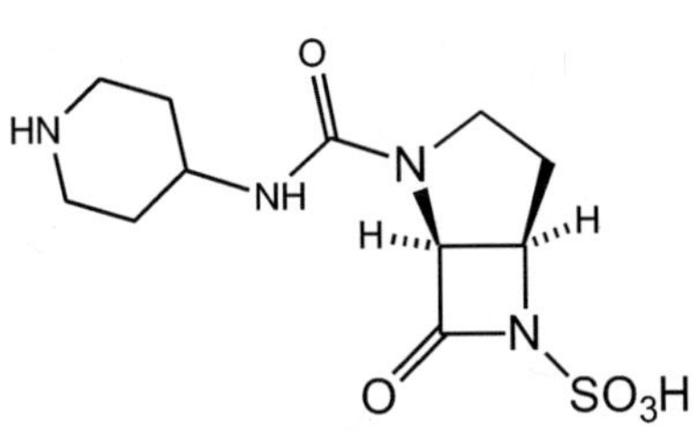
Ro 48-1256
115

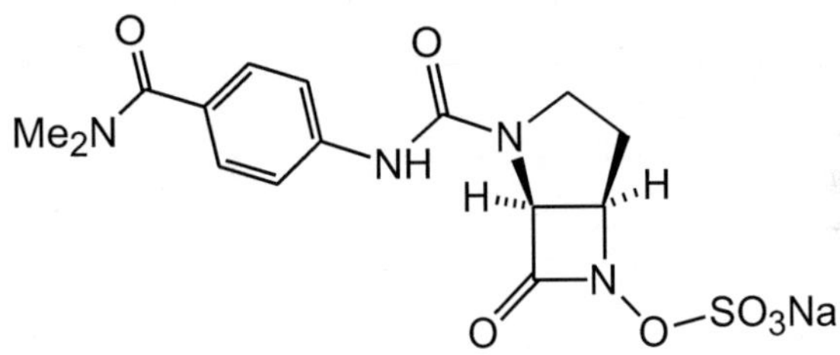
116

Figure 22. Bridged Monobactams and Bridged Monosulfactams.

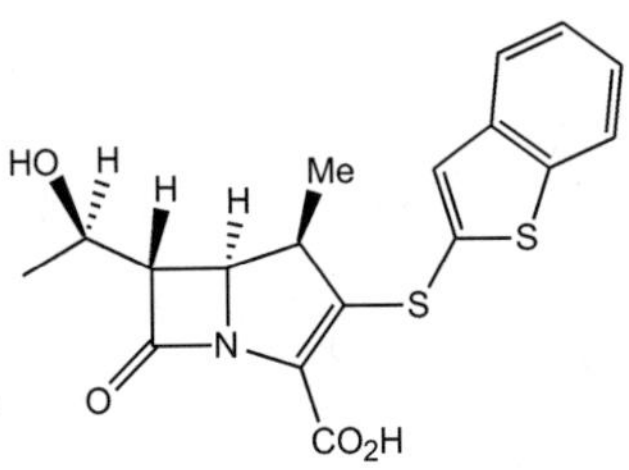
51

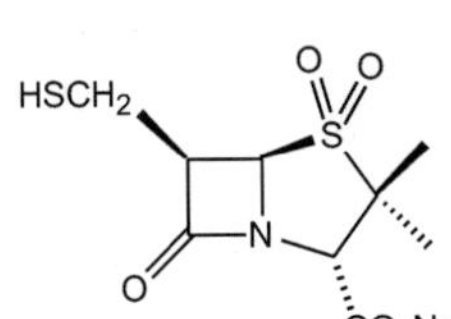
61

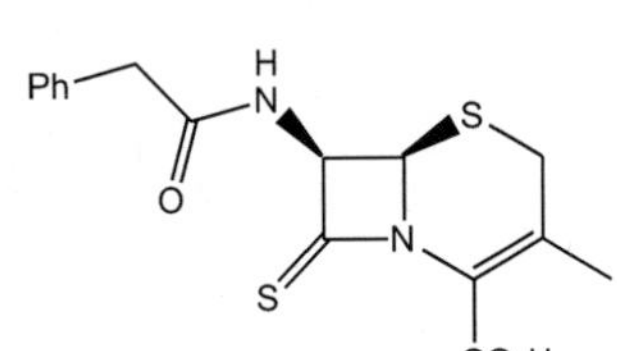
117

Figure 23. β-Lactams Investigated as Inhibitors of the Class B Metallo-β-lactamases.

Mechanistically, these compounds were designed to block access of the active site water to the intermediate acyl-enzyme [170]. Previously, the crystal structure of aztreonam bound to the class C β-lactamase from *Citrobacter freundii* [174] indicated that, subsequent to formation of the acyl-enzyme, a 70° rotation about the C3-C4 bond had to occur in order for the hydrolytic water to have access to the ester linkage. (The C4α-methyl group of aztreonam is believed to interfere with the requisite rotation through a steric interaction with the aryl ring of Tyr150.) By structurally prohibiting this rotation through construction of a bridge between C3 and C4 (i.e. the five-membered ring), they were able to stabilize the acyl-enzyme toward attack of the hydrolytic water and inhibit the enzyme. The corresponding bridged sulfactams **116**, were also potent inhibitors, of both class C as well as class A β-lactamases. Analysis of the inhibition indicated that the inhibition by the bridged sulfactams was due to a high initial affinity for the enzymes and a high rate of acylation. However, the deacylation rate of the sulfactams was also high, rendering the overall synergy with ceftriaxone relatively weak, relative to the corresponding monobactams.

Metallo-β-lactamase Inhibitors

Relatively few β-lactams have been reported as inhibitors of the class B metallo-β-lactamases. As the mechanism of these metalloenzymes does not involve formation of an acyl-enzyme, as was the case with the serine β-lactamases, the inhibitors of the metallo-β-lactamases (at least in the β-lactam series of inhibitors) has centered around finding an appropriate zinc ligand to stabilize the antibiotic in the active site. The aforementioned carbapenem **51** was reported a sub-micromolar inhibitor of IMP-1 and other metallo-β-lactamases [71] and aforementioned thiol **61** has recently been reported as an inhibitor of the L1 and BCII metallo-β-lactamases [92]. Thiono-β-lactams, **117**, first reported in 1975 [175], were initially shown to have reduced antimicrobial activity relative to the natural oxo analogs. Pratt subsequently explored these compounds as potential inhibitors of the class B metallo-β-lactamase from *Bacillus cereus* and found that these molecules were poor substrates instead of inhibitors [176]. However, by varying the active site metal, he was able to provide initial evidence for an interaction between the metal and the thiocarbonyl (or carbonyl) group. Similar results were obtained in a recent study [177]. Page has provided evidence that thiolates **119**, produced during normal turnover of cephalosporins, are micromolar-level inhibitors of the BCII metallo-β-lactamase [178].

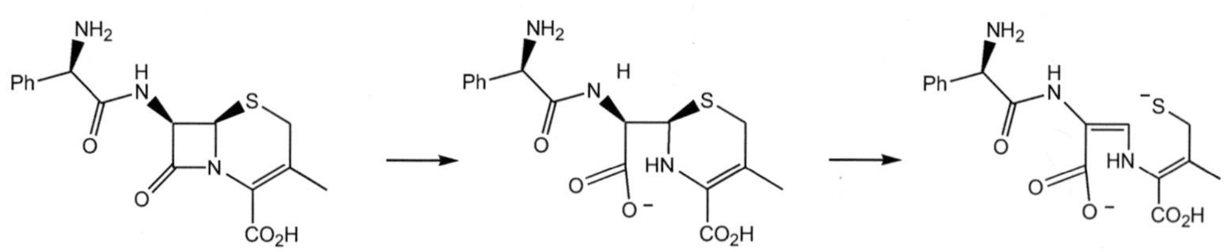

118 **119**

Scheme 14. Production of Thiolate Upon Hydrolysis of Cephalexin.

INHIBITOR RESISTANCE

Given the frequency of clinical administration of antibiotic/inhibitor combinations, as well as the high reproduction rate and mutational frequency of bacteria, it is not surprising that inhibitor resistance has developed. The term 'inhibitor-resistant' usually refers to resistance to amoxicillin/clavulanate and does not necessarily imply resistance to other inhibitors. Even the (mechanistically) related penicillin sulfones are observed to have resistance profiles somewhat different from that of clavulanate [179] and no data are currently available on the effect of non-commercial β-lactamase inhibitors on 'inhibitor-resistant' enzymes. Such resistance may occur through hyperproduction of unmodified β-lactamase [180-182], by modification of outer membrane proteins [183], as well as through production of a mutant form of the β-lactamase which is more resilient toward the commercial inhibitors [184-187]. It is this last form that will be discussed.

Since commercial inhibitors target the clinically prevalent, plasmid-mediated class A β-lactamases, it is logical that the inhibitor-resistant enzymes are also of class A, specifically the inhibitor resistant TEM β-lactamases (IRT's), as well as resistant SHV β-lactamases, close structural relatives of TEM. The frequency of IRT production among resistant strains is a matter of current study. A 1993 French study examined nearly 3000 *E. coli* isolates and found the amoxicillin-clavulanate resistance rate to be 25.0% (hospital) and 10.0% (community). Of these resistant strains, 27.5% (hospital) and 45.0% (community) showed an unusual β-lactam resistance pattern, suggestive of IRT production. Thus, they estimate that the overall IRT production rate, among *E. coli* to be around 4.9%, indicating the IRT producing strains may be beginning to make an important contribution [188]. A recent U.S. study found the overall incidence of amoxicillin/clavulanate resistance was 3 to 5%, but although IRT-production was detected in some of these resistant strains, the precise frequency of such IRT production was not determined [189]. A study of clinical *E. coli* strains from a Spanish hospital showed that about 7% had reduced susceptibility to amoxicillin/clavulanate, with the major cause (75% of those resistant strains) being probable overproduction of class C β-lactamase, with only approximately 5% of the amoxicillin/clavulanate-resistant strains producing IRT's (i.e. 0.35% of the total resistant strains) [190]. The overall IRT incidence may be higher than recognized, since inhibitor-resistant strains cannot be detected by routine susceptibility tests and are identified by iso-electric points and kinetic investigation of the produced β-lactamases, as well as by molecular biology techniques [191, 192]. Obviously, this can lead to discrepancies in determining their current clinical significance.

Several studies of selected inhibitor-resistant enzymes were recently completed and some of these are briefly summarized. In the class A TEM β-lactamases, active site mutations leading to inhibitor resistance include single amino acid substitutions at position 69 (M69I, M69L, M69V), at position 130 (S130G), at position 244 (R244S, R244C, R244H, and R244G), at position 275 (R275L), and at position 276 (N276D). Given the structural similarity between the inhibitors and the substrates, the key mechanistic question involves determining how the enzyme is able to maintain its ability to hydrolyze the antibiotic while reducing the efficacy of inhibitors. Indeed, in most mutations, some attenuation of catalytic efficacy is observed.

The mutations of Met69 are commonly observed in inhibitor-resistant enzymes and are poorly understood mechanistically. The three mutations lead to enzymes with an approximate tenfold increase in the apparent K_i and also an approximate tenfold decrease in the k_{inact} toward all three commercial inhibitors (and consequent hundred-fold degradation of the inhibitory efficacy, k_{inact}/K_i) and with a comparatively less dramatic effects on either the k_{cat} or K_m of the antibiotics [193, 194]. This side chain lies directly behind the oxyanion hole. Recent hypotheses include hydrophobicity and steric constraints [193] a distortion in the conformation of Ser70 [195], and a difference in the dynamic motions of the residues in the vicinity of the active site [196]. It is interesting that replacement of the methionine by less sterically demanding residues (i.e. M69A and M69G) results in an enzyme *more* susceptible toward inhibitors [197].

Ser130 is currently believed to play key roles in recognition of the C3 (C4) carboxylate of the penams and cephems respectively, providing a proton to the departing nitrogen during initial acylation of Ser70, as well as serving as the second nucleophile in irreversible inactivation (attack of the intermediate imine, resulting in double attachment to the enzyme) of the β-lactamase by commercial inhibitors, clavulanate and the penicillin sulfones. Particularly in view of this last factor, it might be anticipated that an S130G mutant would be entirely resistant to inhibitors. The fact that commercial inhibitors still have (albeit diminished) efficacy [198] against such a mutant underscores the importance of the transiently-stabilized intermediates (see Schemes 4 and 5) on the in vitro inactivation of the enzyme (particularly considering the 20 min replication cycle of the bacterium). A recent structure of an S130G mutant of SHV-1, complexed with tazobactam as the cis-vinylogous urethane further emphasizes this fact [199]. In a study of this SHV-1 mutant, it was found that the mutation increased the K_m for the substrate ampicillin (approx. 2 fold) and the apparent K_i values for the inhibitors (approx. 40 - 300 fold) [200]. The k_{cat} values for ampicillin was decreased (approx. 10 fold), while k_{inact} of the inhibitors was relatively unchanged. As a measure of efficiency, the S130G k_{cat}/K_m value for ampicillin was 3.6% that of the wild type, while the k_{inact}/K_i values for clavulanate and tazobactam, were 0.24% and 1% that of the wild type, respectively. The partition ratios for the two inhibitors were approximately the same for the wild type and the S130G mutant. Thus the enzyme loses efficiency in the hydrolysis of ampicillin, to enjoy a larger enhancement in inhibitor resistance. In both cases, the dominating factor is the increase in the (apparent) dissociation constants for the preacylation complex of either the inhibitor or the antibiotic. Similar results were obtained in a study of a TEM 1 S130G mutant [201]. In both SHV and TEM S130G mutants, a water molecule is believed to be taking the place of Ser130 in the donation of a proton to the departing nitrogen.

The IRT data obtained thus far seem to indicate that the inhibitors, with their smaller structures and two point recognition (i.e. C3 (C4) carboxylate and β-lactam carbonyl oxygen, via the oxyanion hole) are more susceptible to mutations affecting these two critical points than are the β-lactam antibiotics, which have additional hydrogen bonding recognition at the C6 (C7) acylamino side chain.

CONCLUSION AND ANALYSIS

β-Lactam antibiotic/β-lactamase inhibitor combinations are one of the success stories of the anti-infectives area. Thus far, these drugs have been reasonably successful in circumventing the bacterial evolutionary drive toward resistance. Hopefully, this section of the review can provide a starting point for the design of new, more effective β-lactam-derived inhibitors. To be certain, there are still several important questions to be addressed. The most important of these remains the identification of a suitable inhibitor which can simultaneously inactivate both class A and class C serine β-lactamases. Other issues involve the creation of inhibitors of the class B metallo-β-lactamases, and molecules that are more effective against inhibitor-resistant β-lactamases. I personally believe that we have only scratched the structural surface of potential inhibitory molecules. I also believe that the design and synthesis of new inhibitors, particularly when coupled with crystallographic and ESIMS studies, illuminates important features of evolution, molecular recognition, and enzyme mechanism that may have broader significance in the understanding and inhibition of other classes of enzymes.

ACKNOWLEDGMENT

J.D.B. acknowledges the support of the Robert A. Welch Foundation and the Centers for Disease Control during the writing of this review.

REFERENCES

[1] Maiti, S. N., Phillips, O. A., Micetich, R. G., Livermore, D. M. (1998) β-Lactamase inhibitors: agents to overcome bacterial resistance. *Curr. Med. Chem.* 5, 441–456.

[2] Miller, L. A., Ratnam, K., Payne, D. J. (2001) β-lactamase-inhibitor combinations in the 21st century: current agents and new developments. *Curr. Opin. Pharm.* 1, 451–458.

[3] Micetich, R. G., Salama, S. M., Maiti, S. N., Reddy, A. V. N., Singh, R. (2002) β-Lactamases and their inhibitors: an update. *Curr. Med. Chem. Anti Infect. Agents* 1, 193–213.

[4] Sandanayaka, V. P., Prashad, A. S. (2002) Resistance to β-lactam antibiotics: Structure and mechanism based design of β-lactamase inhibitors. *Curr. Med. Chem.* 9, 1145–1165.

[5] Page, M. G. P. (2000) ββ-Lactamase inhibitors. Drug Resist. Updates. 3, 109–125.

[6] Theuretzbacher, U. (2004) β-Lactamases and β-lactamase inhibitors *Chemother. J.* 13, 206-217.

[7] Georgopapadakou, N. H. (2004) β-Lactamase inhibitors: evolving compounds for evolving resistance targets. *Expert Opin. Investig. Drugs* 13, 1307-1318.

[8] Bush, K. (2002) The impact of β-lactamases on the development of novel antimicrobial agents. Curr. Opin. Investig. *Drugs* 3, 1284-1290.

[9] Coleman, K. (1995) An update on β-lactamases and β-lactamase inhibitors. *Expert Opin. Investig. Drugs* 4, 693-704.

[10] Pratt, R. F. (1992*) β-Lactamase: inhibition. In Chemistry of* β*-Lactams*, Page, M. I. Ed. Springer, New York, pp 229-271.

[11] Varghese, J. N., Smith, P. W., Sollis, S. L., Blick, T. J., Sahasrabudhe, A., McKimm-Breschkin, J. L., Colman, P. M. (1998) Drug design against a shifting target: a

structural basis for resistance to inhibitors in a variant of influenza virus neuraminidase. *Structure* 6, 735-46.

[12] Cohen, N. C. (1983) β-Lactam antibiotics: geometrical requirements for antibacterial activities. *J. Med. Chem.* 26, 259-264.

[13] Faraci, W. S., Pratt, R. F. (1985) Mechanism of inhibition of the PC1 β-lactamase of Staphylococcus aureus by cephalosporins: importance of the 3'-leaving group. *Biochemistry* 24, 903-910.

[14] Faraci, W. S., Pratt, R. F. (1986) Mechanism of inhibition of RTEM-2 β-lactamase by cephamycins: relative importance of the 7α-methoxy group and the 3' leaving group. *Biochemistry* 25, 2934-2941.

[15] Pratt, R. F., Faraci, W. S. (1986) Direct observation by proton NMR of cephalosporoate intermediates in aqueous solution during the hydrazinolysis and β-lactamase-catalyzed hydrolysis of cephalosporins with 3' leaving groups: kinetics and equilibria of the 3' elimination reaction. *J. Am. Chem. Soc.* 108, 5328-5333.

[16] Mazzella, L. J., Pratt, R. F. (1989) Effect of the 3'-leaving group on turnover of cephem antibiotics by a class C β-lactamase. *Biochem. J.* 259, 255-260.

[17] Massova, I., Mobashery, S. (1998) Kinship and diversification of bacterial penicillin-binding proteins and β-lactamases. Antimicrob. *Agents Chemother.* 42, 1-17.

[18] Knox, J. R., Moews, P. C., Frère, J.-M. (1996) Molecular evolution of bacterial β-lactam resistance. *Chem. Biol.* 3, 937-947.

[19] Pratt, R. F. (2002) Functional evolution of the serine β-lactamase active site. *J. Chem. Soc., Perkin Trans.* 2, 851-861.

[20] White, A. R., Kaye, C., Poupard, J., Pypstra, R., Woodnutt, G., Wynne, B. (2004) Augmentin (amoxicillin/clavulanate) in the treatment of community-acquired respiratory tract infection: a review of the continuing development of an innovative antimicrobial agent. *J. Antimicrob. Chemother.* 53 (S1), i3-i20.

[21] Stein, G. E., Gurwith, M. J. (1984) Amoxicillin-potassium clavulanate, a β-lactamase-resistant antibiotic combination.*Clin. Pharm.* 3, 591-9.

[22] Reed, M. D. (1998) Rational prescribing of extended-spectrum penicillin β-lactamase inhibitor combinations: focus on ticarcillin/clavulanic acid. *Annal. Pharmacother.* 32, S17-S21.

[23] Lode, H. (2001) Role of sultamicillin and ampicillin/sulbactam in the treatment of upper and lower bacterial respiratory tract infections. *Int. J. Antimicrob. Agents* 18, 199-209.

[24] For a review, see: Bryson, H. M., Brogden, R. N. (1994) Piperacillin/tazobactam. A review of its antibacterial activity, pharmacokinetic properties and therapeutic potential. *Drugs* 47, 506-535.

[25] Bush, K., Mobashery, S. (1998) How β-lactamases have driven pharmaceutical drug discovery. From mechanistic knowledge to clinical circumvention. *Adv. Exp. Med. Biol.* 456, 71–98.

[26] Charnas, R. L., Fisher, J., Knowles, J. R. (1978) Chemical studies on the inactivation of Escherichia coli RTEM β-lactamase by clavulanic acid. *Biochemistry* 17, 2185-2189.

[27] Charnas, R. L., Knowles, J. R. (1981) Inactivation of RTEM β-lactamase from Escherichia coli by clavulanic acid and 9-deoxyclavulanic acid. *Biochemistry* 20, 3214-3219.

[28] Imtiaz, U., Billings, E., Knox, J. R., Manavathu, E. K., Lerner, S. A., Mobashery, S. (1993) Inactivation of class A β-lactamases by clavulanic acid: the role of arginine-244 in a proposed nonconcerted sequence of events. *J. Am. Chem. Soc.* 115, 4435-4442.

[29] Chen, C. C. H., Herzberg, O. (1992) Inhibition of β-lactamase by clavulanate. Trapped intermediates in cryocrystallographic studies. *Mol. Biol.* 224, 1103-1113.

[30] Brown, R. P. A., Aplin, R. T., Schofield, C. J. (1996) Inhibition of TEM-2 β-lactamase from Escherichia coli by clavulanic acid: observation of intermediates by electrospray ionization mass spectrometry. *Biochemistry* 35, 12421-12432.

[31] van den Akker, F., personal communication.

[32] Sulton, D.; Pagan-Roderiguez, D.; Zhou, X.; Liu, Y.; Hujer, A.; Bethel, C. R.; Helfand, M. S.; Thomson, J. M.; Anderson, V. E.; Buynak, J. D.; Ng, L. M.; Bonomo, R. A. (2005) Clavulanic acid inactivation of SHV-1 and the inhibitor resistant Ser130Gly SHV-1 b-lactamase: Insights into the mechanism of inhibition. *J. Biol. Chem.* (in press).

[33] Fisher, J., Charnas, R. L., Bradley, S. M., Knowles, J. R. (1981) Inactivation of the RTEM β-lactamase from Escherichia coli. Interaction of penam sulfones with enzyme. *Biochemistry* 20, 2726-2731.

[34] Kuzin, A. P., Nukaga, M., Nukaga, Y., Hujer, A., Bonomo, R. A., Knox, J. R. (2001) Inhibition of the SHV-1 β-lactamase by sulfones: crystallographic observation of two reaction intermediates with tazobactam. *Biochemistry* 40, 1861-1866.

[35] Padayatti, P. S., Helfand, M. S., Totir, M. A., Carey, M. P., Hujer, A. M., Carey, P. R., Bonomo, R. A., van den Akker, F. (2004) Tazobactam forms a stoichiometric trans-enamine intermediate in the E166A variant of SHV-1 β-lactamase: 1.63 Å crystal structure. *Biochemistry* 43, 843-848.

[36] Yang, Y., Janota, K., Tabei, K., Huang, N., Siegel, M. M., Lin, Y-I, Rasmussen, B. A., Shlaes, D. M. (2000) Mechanism of inhibition of the class A -lactamases PC1 and TEM-1 by tazobactam. Observation of reaction products by electrospray ionization mass spectrometry. *J. Biol. Chem.* 275, 26674-26682.

[37] Pagan-Rodriguez, D., Zhou, X., Simmons, R., Bethel, C. R., Hujer, A. M., Helfand, M. S., Jin, Z., Guo, B., Anderson, V. E., Ng, L. M., Bonomo, R. A. (2004) Tazobactam inactivation of SHV-1 and the inhibitor-resistant Ser130 □ Gly SHV-1 β-lactamase: Insight into the mechanism of inhibition.*J. Biol. Chem.* 279, 19494-19501.

[38] Buynak, J. D. *9th β-Lactamase Meeting, Leonessa,* Italy 2005.

[39] Richter, H. G. F., Angehrn, P., Hubschwerlen, C., Kania, M., Page, M. G. P., Specklin, J.-L., Winkler, F. K. (1996) Design, synthesis, and evaluation of 2β-alkenyl penam sulfone acids as inhibitors of β-lactamases. *J. Med. Chem.* 39, 3712-3722.

[40] Tzouvelekis, L., Gazouli, M., Prinarakis,, E. E., Tzelepi, E., Legakis, N. J. (1997) Comparative evaluation of the inhibitory activities of the novel penicillanic acid sulfone Ro 48-1220 against β-lactamases that belong to groups 1, 2b, and 2be. Antimicrob. *Agents Chemother.* 41, 475-477.

[41] Buynak, J. D., Gadhachanda, V. R., Vogeti, L., Zhang, H., Chen, H. (2005) Synthesis and Evaluation of 3-(Carboxymethylidene)- and 3-(Carboxymethyl)penicillinates as Inhibitors of β-Lactamase. *J. Org. Chem.* 70, 4510-4513.

[42] Cherry, P. C., Newall, C. E., Watson, N. S. (1978) Preparation of the 7-oxo-4-oxa-1-azabicyclo[3.2.0]hept-2-ene system and the reversible cleavage of its oxazoline ring. *J. Chem. Soc., Chem. Commun.* 469-470.

[43] Cherry, P. C., Evans, D. N., Newall, C. E., Watson, N. S. (1980) Conversion of clavulanic acid into thiadeoxa nuclear analogs. *Tetrahedron Lett.* 21, 561-564.

[44] Douglas, J. L., Martel, A., Caron, G., Menard, M. (1984) Nuclear analogs of β-lactam antibiotics. XX. Synthesis and x-ray structure determination of exocyclic penems: sulfur analogs of clavulanic acid. *Can. J. Chem.* 62, 2282-2286.

[45] Oida, S., Yoshida, A., Hayashi, T., Takeda, N., Nishimura, T., Ohki, E. (1980) Synthesis of penems and their antibacterial activities. *J. Antibiot.* 33, 107-109.

[46] Franceschi, G., Foglio, M., Arcamone, F., Sanfliippo, A., Schioppacassi, G. (1980) Antibacterial activity of novel broad-spectrum "(5R)-penem" derivatives. *J. Antibiot.* 33, 453-454.

[47] Tanaka, H., Kameyama, Y., Sumida, S., Torii, S. (1993) 2-Substituted penems, new candidates for cephalosporinase inhibitors. *Bioorg. Med. Chem. Lett.* 3, 2253-2254.

[48] De Angelis, F., Attorrese, G., Cavicchio, G., Ciampa, S., Di Tullio, A., Fattori, D., Nicoletti, R., Domenici, E. (2001) Synthesis and preliminary biological evaluation of 3'-substituted cephem sulfones as potential β-lactamase inhibitors. *Eur. J. Org. Chem.* 3075-3081.

[49] Sunagawa, M., Matsumura, H., Sasaki, a., Yamaga, H., Kitamura, Y., Sumita, Y., Nouda, H. (1997) Structure-activity relationships of 1β-methyl-carbapenems to antimicrobial activity: effect of C-6 substituent. *J. Antibiot.* 50, 621-627.

[50] Parker, W. L., Rathnum, M. L., Wells, J. S., Trejo, W. H., Pincipe, P. A., Sykes, R. B. (1982) SQ 27,860, a simple carbapenem produced by species of Serratia and Erwinia.*J. Antibiot.* 35, 653-660.

[51] Basker, M. J., Bateson, J. H., Baxter, A. J. G., Ponsford, R. J., Roberts, P. M., Southgate, R., Smale, T. C., Smith, J. (1981) Synthesis of 6-unsubstituted olivanic acid analogs and their antibacterial activities. *J. Antibiot.* 34, 1224-1226.

[52] Butterworth, D., Rolinson, G. N. Antibacterial biosynthetic material. (1972) *German patent DE 2146400* 19720323.

[53] Brown, A. G., Corbett, D. F., Eglington, A. J., Howarth, T. T. (1977) Structures of olivanic acid derivatives MM 4550 and MM 13902; two new, fused β-lactams isolated from Streptomyces olivaceus. *J. Chem. Soc., Chem. Commun.* 523-525.

[54] Basker, M. J., Boon, R. J., Hunter, P. A. (1980) Comparative antibacterial properties in vitro of seven olivanic acid derivatives: MM 4550, MM 13902, MM 17880, MM 22380, MM 22381, MM 22382 and MM 22383. *J. Antibiot.* 33, 878-884.

[55] Charnas, R. L., Knowles, J. R. (1981) Inhibition of the RTEM β-lactamase from Escherichia coli. Interaction of the enzyme with derivatives of olivanic acid. *Biochemistry* 20, 2732-2737.

[56] Okonogi, K., Harada, S., Shinagawa, S., Imada, A., Kuno, M. (1982) β-Lactamase inhibitory activities and synergistic effects of 5,6-cis-carbapenem antibiotics. *J. Antibiot.* 35, 963-971.

[57] Kahan, J. S., Kahan, F. M., Stapley, E. O., Goegelman, R. T., Hernandez, S. (1976) *Antibiotics from Streptomyces species.* US patent 3950357 19760413.

[58] Albers-Schonberg, G., Arison, B. H., Hensens, O. D., Hirshfield, J., Hoogsteen, K., Kaczka, E. A., Rhodes, R. E., Kahan, J. S., Kahan, F. M., Ratcliffe, R. W., Walton, E., Ruswinkle, L. J., Morin, R. B., Christensen, B. G. (1978) Structure and absolute configuration of thienamycin. *J. Am. Chem. Soc.* 100, 6491-6499.

[59] Sunagawa, M., Matsumura, H., Sasaki, A., Yamaga, H., Kitamura, Y., Sumita, Y., Nouda, H. (1997) Structure-activity relationships of 1β-methyl-carbapenems to antimicrobial activity: effect of C-6 substituent. *J. Antibiot.* 50, 621-627.

[60] Hood, J. D., Box, S. J., Verrall, M. S. (1979) Olivanic acids, a family of β-lactam antibiotics with β-lactamase inhibitory properties produced by Streptomyces species. II. Isolation and characterization of the olivanic acids MM 4550, MM 13902 and MM 17880 from Streptomyces olivaceus. *J. Antibiot.* 32, 295-304.

[61] Basker, M. J., Boon, R. J., Box, S. J., Prestige, E. A., Smith, G. M., Spear, S. R. (1983) Properties of MM 13902 and other carbapenem antibiotics in vitro and in vivo *J. Antibiot.* 36, 416-422.

[62] Monks, J., Waley, S. G. (1988) Imipenem as substrate and inhibitor of β-lactamases. *Biochem. J.* 253, 323-328.

[63] Livermore, D. M., Neil, W. (2000) Carbapenemases: A problem in waiting? *Curr. Opin. Microbiol.* 3, 489-495.

[64] Rice, L. B., Bonomo, R. A. (2000) β-Lactamases: Which ones are clinically important? *Drug Resist.* Updates 3, 178-189.

[65] Rasmussen, B. A., Bush, K. (1997) Carbapenem-hydrolyzing β-lactamases. Antimicrob. *Agents Chemother.* 41, 223-232.

[66] Yigit, H., Queenan, A. M., Rasheed, J. K., Biddle, J. W., Domenech-Sanchez, A., Alberti, S., Bush, K., Tenover, F. C. (2003) Carbapenem-resistant strain of Klebsiella oxytoca harboring carbapenem-hydrolyzing β-lactamase KPC-2. Antimicrob. *Agents Chemother.* 47, 3881-3889.

[67] Copar, A.; Prevec, T.; Anzic, B.; Mesar, T.; Selic, L.; Vilar, M.; Solmajer, T. (2002) Design, synthesis and bioactivity evaluation of tribactam β lactamase inhibitors. *Bioorg. Med. Chem. Lett.* 12, 971-975.

[68] Nagano, R., Shibata, K., Adachi, Y., Imamura, H., Hashizume, T., Morishima, H. (2000) In vitro activities of novel trans-3,5-disubstituted pyrrolidinylthio-1β-methylcarbapenems with potent activities against methicillin-resistant Staphylococcus aureus and Pseudomonas aeruginosa. Antimicrob. *Agents Chemother.* 44, 489-495.

[69] Sumita, Y., Nouda, H., Kanazawa, K., Fukasawa, M. (1995) Antimicrobial activity of SM-17466, a novel carbapenem antibiotic with potent activity against methicillin-resistant Staphylococcus aureus. *Antimicrob. Agents Chemother.* 39, 910-916.

[70] Fukuoka, T., Ohya, S., Utsui, Y., Domon, H., Takenouchi, T., Koga, T., Masuda, N., Kawada, H., Kakuta, M., Kubota, M., Ishii, C., Ishii, C., Sakagawa, E., Harasaki, T., Hirasawa, A., Abe, T., Yasuda, H., Iwata, M., Kuwahara, S., (1997) In vitro and in vivo antibacterial activities of CS-834, a novel oral carbapenem. Antimicrob. *Agents Chemother.* 41, 2652-2663.

[71] Nagano, R., Adachi, Y., Imamura, H., Yamada, K., Hashizume, T., Morishima, H. (1999) Carbapenem derivatives as potential inhibitors of various β-lactamases, including class B metallo-β-lactamases. *Antimicrob. Agents Chemother.* 43, 2497-2503.

[72] Matagne, A., Ghuyesen, M., Frère, J. M. (1993) Interactions between active-site-serine β-lactamases and mechanism-based inactivators: a kinetic study and an overview. *Biochem. J.* 295, 705-711.

[73] Galleni, M., Amicosante, G., Frère, J. M. (1988) A survey of the kinetic parameters of class C β-lactamases. Cephalosporins and other β-lactam compounds. *Biochem. J.* 255, 123-129.

[74] Babini, G. S., Yuan, M., Livermore, D. M. (1998) Interactions of β-lactamases with sanfetrinem (GV 104326) compared to those with imipenem and with oral β-lactams. Antimicrob. *Agents Chemother.* 42, 1168-1175.

[75] Hashizume, T., Yamaguchi A., Hirata, T., Sawai, T. (1984) Kinetic studies on the inhibition of Proteus vulgaris β-lactamase by imipenem. Antimicrob. *Agents Chemother.* 25, 149-151.

[76] Maveyraud, L., Mourey, L., Kotra, L. P., Pedelacq, J.-D., Guillet, V., Mobashery, S., Samama, J.-P. (1998) Structural basis for clinical longevity of carbapenem antibiotics in the face of challenge by the common class A β-lactamases from the antibiotic-resistant bacteria. *J. Am. Chem. Soc.* 120, 9748-9752.

[77] Beadle, B. M., Shoichet, B. K. (2002) Structural basis for imipenem inhibition of class C β-lactamases. *Antimicrob. Agents Chemother.* 46, 3978-3980.

[78] Pernot, L., Frenois, F., Rybkine, T., L'Hermite, G., Petrella, S., Delettre, J., Jarlier, V., Collatz, E., Sougakoff, W. (2001) Crystal structures of the class D β-lactamase OXA-13 in the native form and in complex with meropenem. *J. Mol. Biol.* 310, 859-874.

[79] Dixon, R. A., Edmondson, R. A., Hardy, K. D., Milner, P. H. (1984) The synthesis and antibacterial activity of some β-lactamase stable 6α-(hydroxymethyl)penicillins. *J. Antibiot.* 37, 1729-1731.

[80] Reiner, R., Zeller, P. (1968) Substitution of 6-aminopenicillanic acid at the 6th carbon atom. *Helv. Chim. Acta.* 51, 1905-1918.

[81] DiNinno, F., Beattie, T. R., Christensen, B. G. (1977) Aldol condensations of regiospecific penicillinate and cephalosporanate enolates. Hydroxyethylation at C-6 and C-7. *J. Org. Chem.* 42, 2960-2965.

[82] Kellogg, M. S. *Derivatives of 6β-hydroxyalkylpenicillanic acids as β-lactamase inhibitors.* (Oct 22, 1979) U.S. Patent 4,287,181.

[83] Schneider, P., Scartazzini, R. *Penam-dioxide compounds, processes for their manufacture and their use.* (May 15, 1981) British Patent GB 2076812A.

[84] Chen, Y. L., Hedberg, K., Guarino, K., Retsema, J. A., Anderson, M., Manousos, M., Barrett, J. (1991) (6R,8S)-(2-Benzimidazolyl)hydroxymethylpenicillanic acids as potent antibacterial agents and β-lactamase inhibitors. *J. Antibiot.* 44, 870-884.

[85] Miyashita, K., Massova, I., Taibi, P., Mobashery, S. (1995) design, synthesis, and evaluation of a potent mechanism-based inhibitor for the TEM β-lactamase with implications for the Enzyme Mechanism. *J. Am. Chem. Soc.* 117, 11055-11059.

[86] Maveyraud, L., Massova, I., Birck, C., Miyashita, K., Samama, J.-P., Mobashery, S. (1996) Crystal structure of 6α-(hydroxymethyl)penicillanate complexed to the TEM-1 β-lactamase from Escherichia coli: Evidence on the mechanism of action of a novel inhibitor designed by a computer-aided process. *J. Am. Chem. Soc.* 118, 7435-7440.

[87] Mourey, L., Miyashita, K., Swaren, P., Bulychev, A., Samama, J.-P., Mobashery, S. (1998) Inhibition of the NMC-A β-lactamase by a penicillanic acid derivative and the structural bases for the increase in substrate profile of this antibiotic resistance enzyme. *J. Am. Chem. Soc.* 120, 9382-9383.

[88] Maveyraud, L., Golemi-Kotra, D., Ishiwata, A., Meroueh, O., Mobashery, S., Samama, J.-P. (2002) High-resolution X-ray structure of an acyl-enzyme species for the class D OXA-10 β-lactamase. *J. Am. Chem. Soc.* 124, 2461-2464.

[89] Golemi, D., Maveyraud, L., Ishiwata, A., Tranier, S., Miyashita, K., Nagase, T., Massova, I., Mourey, L., Samama, J.-P., Mobashery, S. (2000) 6-(Hydroxyalkyl)penicillanates as probes for mechanisms of β-lactamases. *J. Antibiot.* 53, 1022-1027.

[90] Bitha, P., Li, Z., Francisco, G. D., Rasmussen, B. A., Lin, Y.-I. (1999) 6-(1-Hydroxyalkyl)penam sulfone derivatives as inhibitors of class A and class C β-lactamases I. *Bioorg. Med. Chem. Lett.* 9, 991-996.

[91] Bitha, P., Li, Z., Francisco, G. D., Yang, Y., Petersen, P. J., Lenoy, E., Lin, Y.-I. (1999) 6-(1-Hydroxyalkyl)penam sulfone derivatives as inhibitors of class A and class C β-lactamases II. *Bioorg. Med. Chem. Lett.* 9, 997-1002.

[92] Buynak, J. D., Chen, H., Vogeti, L., Gadhachanda, V. R., Buchanan, C. A., Palzkill, T., Shaw, R. W., Spencer, J., Walsh, T. R. (2004) Penicillin-derived inhibitors that simultaneously target both metallo- and serine-β-lactamases. *Bioorg. Med. Chem. Lett.* 14, 1299-1304.

[93] Johnston, D. B. R., Schmitt, S. M., Firestone, R. A., Christensen, B. G. (1972) Substituted penicillins and cephalosporins. V. 6(7)-Substituted alkyl derivatives. *Tetrahedron Lett.* 13, 4917-4920.

[94] Rasmusson, G. H., Reynolds, G. F., Arth, G. E. (1973) 6-Substituted penicillin derivatives. *VI. Tetrahedron Lett.* 14, 145-148.

[95] Dixon, R. A., Hardy, K. D., Kaura, A. C., Milner, P. H., Taylor, A. W. (1984) The synthesis and antibacterial activities of novel 7α substituted cephalosporins. *J. Antibiot.* 37, 1732-1735.

[96] Nishimura, S., Yasuda, N., Sasaki, H., Matsumoto, Y., Kamimura, T., Sakane, K., Takaya, T. (1989) Synthesis and β-lactamase inhibitory activity of 7α-hydroxyethylcephem derivatives. *J. Antibiot.* 42, 159-162.

[97] Nishimura, S., Yasuda, N., Sasaki, H., Matsumoto, Y., Kamimura, T., Sakane, K., Takaya, T. (1990) Synthesis and β-lactamase inhibitory activity of 7α-hydroxyethyl cephem sulfone and sulfoxide derivatives. *J. Antibiot.* 43, 114-117.

[98] Nishimura, A., Sasaki, H., Yasuda, N., Matsumoto, Y., Kamimura, T., Sakane, K., Takaya, T. (1989) Synthesis and biological activity of 7α-hydroxyethyl-1-oxacephem derivatives. *J. Antibiot.* 42, 1124-1132.

[99] Reading, C., Cole, M. (1986) Structure-activity relationships amongst β-lactamase inhibitors. *J. Enzyme Inhibition* 1, 83-104.

[100] Howarth, T. T., Luk, K. (1978) Derivatives of clavulanic acid substituted in the six position. *German patent DE* 2822001 19781130.

[101] Murakami, M., Aoki, T., Matsuura, M., Nagata, W. (1990) Synthesis and biological properties of 1-oxapenems. *J. Antibiot.* 43, 1441-1449.

[102] Ernest, I., Gosteli, J., Greengrass, C. W., Holick, W., Pfaendler, H. R., Woodward, R. B. (1978) The penems, a new class of β-lactam antibiotics: 6-acylaminopenem-3-carboxylic acids. *J. Am. Chem. Soc.* 100, 8214-8222.

[103] Ganguly, A. K., Girijavallabhan, V. M., McCombie, S., Pinto, P., Rizvi, R., Jeffrey, P. D., Lin, S. (1982) Synthesis of Sch 29482 - a novel penem antibiotic. *J. Antimicrob. Chemother.* 9 (Suppl. C), 1-5.

[104] Barry, A. L., Jones, R. N., Wilson, H. W., Badal, R. E., Thornsberry, C. (1982) Sch 29482, a new oral penem: comparative in vitro activity, β-lactamase stability and inhibition. *J. Antimicrob. Chemother.* 9 (Suppl. C), 97-112.

[105] Pechere, J. C., Letarte, R., Guay, R., Asselin, C., Morin, C. (1982) Sch 29482: stability and inhibitory potency towards β-lactamases from gram-negative bacteria. J. Antimicrob. *Chemother. 9 (Suppl. C),* 123-132.

[106] Inoue, E., Mitsuhashi, S. (1994) In vitro antibacterial activity and β-lactamase stability of SY5555, a new oral penem antibiotic. *Antimicrob. Agents. Chemother.* 38, 1974-1979.

[107] Cabri, W., Bedeschi, A., Perrone, E., Visentin, G., Jabes, D., Franceschi, G. (1993) Synthesis and β-lactamase inhibitory activity of 6-hydroxyethyl thiaclavulanic acid. *Bioorg. Med. Chem. Lett.* 3, 2255-2258.

[108] Arisawa, M., Then, R. L. (1982) 6-Acetylmethylenepenicillanic acid (Ro 15-1903), a potent β-lactamase inhibitor. I. Inhibition of chromosomally and R-factor-mediated β-lactamases. *J. Antibiot.* 35, 1578-1583.

[109] Angehrn, P., Arisawa, M. (1982) 6-Acetylmethylenepenicillanic acid (Ro 15-1903), a potent β-lactamase inhibitor. II. Antibacterial properties. *J. Antibiot.* 35, 1584-1589.

[110] Arisawa, M., Adam, S. (1983) Mechanism of inactivation of TEM-1 β-lactamase by 6-acetylmethylenepenicillanic acid. *Biochem. J.* 211, 447-454.

[111] Chen, Y. L., Chang, C.-W., Hedberg, K., Guarino, K., Welch, W. M., Kiessling, L. (1987) Structure-activity relationships of 6-(heterocyclyl)-methylene penam sulfones: a new class of β-lactamase inhibitors. *J. Antibiot.* 40, 803-822.

[112] Chen, Y. L., Chang, C.-W., Hedberg, K. (1986) Synthesis of a potent β-lactamase inhibitor - 1,1-dioxo-6-(2-pyridyl)methylenepenicillanic acid and its reaction with sodium methoxide. *Tetrahedron Lett.* 27, 3449-3452.

[113] Buynak, J. D., Geng, B., Bachmann, B., Hua, L. (1995) The synthesis and lactamase inhibitory activity of 6-(carboxymethylene)penicillins and 7-(carboxymethylene)cephalosporins. *Bioorg. Med. Chem. Lett.* 5, 1513-1518.

[114] Buynak, J. D., Rao, A. S., Doppalapudi, V. R., Adam, G., Petersen, P. J., Nidamarthy, S. D. (1999) The synthesis and evaluation of 6-alkylidene-2'β-substituted penam sulfones as β-lactamase inhibitors. *Bioorg. Med. Chem. Lett.* 9, 1997-2002.

[115] Buynak, J. D., Wu, K., Bachmann, B., Khasnis, D., Hua, L., Nguyen, H. K., Carver, C. L. (1995) Synthesis and biological activity of 7-alkylidenecephems. *J. Med. Chem.* 38, 1022-1034.

[116] Black, J. A., Thomson, K. S., Pitout, J. D. D. (2004) Use of β-lactamase inhibitors in disk tests to detect plasmid-mediated AmpC β-lactamases. J. *Clin. Microbiol.* 42, 2203-2206.

[117] Crichlow, G. V., Nukaga, M., Doppalapudi, V. R., Buynak, J. D., Knox, J. R. (2001) Inhibition of class C β-lactamases: Structure of a reaction intermediate with a cephem sulfone. *Biochemistry* 40, 6233-6239.

[118] Buynak, J. D., Borate, H. B., Lamb, G. W., Khasnis, D. D., Husting, C., Isom, H., Siriwardane, U. (1993) Synthesis of 6-vinylidenepenams. *J. Org. Chem.* 58, 1325-1335.

[119] Buynak, J. D., Khasnis, D., Bachmann, B., Wu, K., Lamb, G. (1994) Synthesis and mechanistic evaluation of 7-vinylidenecephem sulfones as β-lactamase inhibitors. *J. Am. Chem. Soc.* 116, 10955-10965.

[120] Adam, S. (1984) Synthesis of methylene (R)-6-acetonylidene-3-methyl-7-oxo-4-thia-1-azabicyclo[3.2.0]hept-2-ene-carboxylate pivalate. *Heterocycles* 22, 1509-1512.

[121] Imuta, M., Uyeo, S., Nakano, M., Yoshida, T. (1985) Carbapenem and penem antibiotics. III. Synthesis and antibacterial activity of 2-functionalized-methyl penems related to asparenomycins. *Chem. Pharm. Bull.* 33, 4371-4381.

[122] Ona, H., Uyeo, S., Fukao, T., Doi, M., Yoshida, T. (1985) Carbapenem and penem antibiotics. I. Total synthesis and antibacterial activity of dl-asparenomycins A, B and C and related carbapenem antibiotics. *Chem. Pharm. Bull.* 33, 4382-4394.

[123] Basker, M. J., Osborne, N. F. (1990) 6-(Substituted methylene)penems, potent broad spectrum inhibitors of bacterial β-lactamase. I. Racemic 6-ethylidenepenems. *J. Antibiot.* 43, 70-75.

[124] Broom, N. J. P., Coleman, K., Hunter, P. A., Osborne, N. F. (1990) 6-(Substituted methylene)penems, potent broad spectrum inhibitors of bacterial β-lactamase. II. Racemic furyl and thienyl derivatives. *J. Antibiot.* 43, 76-82.

[125] Bennett, I., Broom, N. J. P., Bruton, G., Calvert, S., Clarke, B. P., Coleman, K., Edmondson, R., Edwards, P., Jones, D., Osborne, N. F., Walker, G. (1991) 6-(Substituted methylene)penems, potent broad spectrum inhibitors of bacterial β-lactamase. III. Structure-activity relationships of the 5-membered heterocyclic derivatives. *J. Antibiot.* 44, 331-337.

[126] Bennett, I., Broom, N. J. P., Coleman, K., Coulton, S., Edwards, P. D., Francois, I., Griffin, D. R. J., Osborne, N. F., Woodall, P. M. (1991) 6-(Substituted methylene)penems, potent broad spectrum inhibitors of bacterial β-lactamase. IV. Kidney stability, serum binding and additional biological evaluation of racemic derivatives. *J. Antibiot.* 44, 338-343.

[127] Bennett, I. S., Brooks, G., Broom, N. J. P., Calvert, S. H., Coleman, K., Francois, I. (1991) 6-(Substituted methylene)penems, potent broad-spectrum inhibitors of bacterial β-lactamase. V. Chiral 1,2,3-triazolyl derivatives. *J. Antibiot.* 44, 969-978.

[128] Weiss, W. J., Petersen, P. J., Murphy, T. M., Tardio, L., Yang, Y., Bradford, P. A., Venkatesan, A. M., Abe, T., Isoda, T., Mihira, A., Ushirogochi, H., Takasake, T., Projan, S., O'Connell, J., Mansour, T. S. (2004) In vitro and in vivo activities of novel 6-methylidene penems as β-lactamase inhibitors. Antimicrob. *Agents Chemother.* 48, 4589-4596.

[129] Venkatesan, A. M., Agarwal, A., Abe, T., Ushirogochi, H., Yamamura, I., Kumagai, T., Petersen, P. J., Weiss, W. J., Lenoy, E., Yang, Y., Shlas, D. M., Ryan, J. L., Mansour, T. S. (2004) Novel imidazole substituted 6-methylidene-penems as broad-spectrum β-lactamase inhibitors. *Bioorg. Med. Chem. Lett.* 12, 5807-5817.

[130] Broom, N. J. P., Farmer, T. H., Osborne, N. F., Tyler, J. W. (1992) Studies on the mechanism of action of (5R)-(Z)-6-(1-methyl-1,2,3-triazol-4-ylmethylene)penem-3-carboxylic acid (BRL 42715), a potent inhibitor of bacterial β-lactamase. *J. Chem. Soc., Chem. Commun.,* 1663-1664.

[131] Farmer, T. H., Page, J. W. J., Payne, D. J., Knowles, D. J. C. (1994) Kinetic and physical studies of β-lactamase inhibition by a novel penem, BRL 42715. *Biochem. J.* 303, 825-830.

[132] Mantagne, A., Ledent, P., Monnaie, D., Felici, A., Jamin, M., Raquet, X., Galleni, M., Klein, D., Francois, I., Frère, J.-M. (1995) Kinetic study of interaction between BRL 427i5, β-lactamases, and D-alanyl-D-alanine peptidases. Antimicrob. *Agents Chemother.* 39, 227-231.

[133] Bulychev, A., Massova, I., Lerner, S. A., Mobashery, S. (1995) Penem BRL 42715: An Effective Inactivator for β-Lactamases. *J. Am. Chem. Soc.* 117, 4797-4801.

[134] Tabei, K., Feng, X., Venkatesan, A. M., Abe, T., Hideki, U., Mansour, T. S., Siegel, M. M. (2004) Mechanism of inactivation of β-lactamases by novel 6-methylidene penems elucidated using electrospray ionization mass spectrometry. *J. Med. Chem.* 47, 3674-3688.

[135] Nukaga, M., Abe, T., Venkatesan, A. M., Mansour, T. S., Bonomo, R. A., Knox, J. R. (2003) Inhibition of class A and class C β-lactamases by penems: Crystallographic structures of a novel 1,4-thiazepine intermediate. *Biochemistry* 42, 13152-13159.

[136] Venkatesan, A. M., Gu, Y., Santos, O. D., Abe, T., Agarwal, A., Yang, Y., Petersen, P. J., Weiss, W. J., Mansour, T. S., Nukaga, M., Hujer, A. M., Bonomo, R. A., Knox, J. R. (2004) Structure-activity relationship of 6-methylidene penems bearing tricyclic heterocycles as broad-spectrum β-lactamase inhibitors: Crystallographic structures show unexpected binding of 1,4-thiazepine intermediates *J. Med. Chem.* 47, 6556-6568.

[137] Michaux, C., Charlier, P., Frère, J.-M., Wouters, J. (2005) Crystal structure of BRL 42715, C6-(N1-methyl-1,2,3-triazolylmethylene)penem, in complex with Enterobacter cloacae 908R β-lactamase: Evidence for a stereoselective mechanism from docking studies. *J. Am. Chem. Soc.* 127, 3262-3263.

[138] Kimura, Y., Motokawa, K., Nagata, H., Kameda, Y., Matsuura, S., Mayama, M., Yoshida, T. (1982) Asparenomycins A, B and C, new carbapenem antibiotics. IV. Antibacterial activity. *J. Antibiot.* 35, 32-38.

[139] Murakami, K., Doi, M., Yoshida, T. (1982) Asparenomycins A, B and C, new carbapenem antibiotics. V. Inhibition of β-lactamases.*J. Antibiot.* 35, 39-45.

[140] Bouthillier, G., Mastalerz, H., Menard, M., Fung-Tomc, J., Gradelski, E. (1992) The synthesis, antibacterial, and β-lactamase inhibitory activity of a novel asparenomycin analog. *J. Antibiot.* 45, 240-245.

[141] Copar, A., Prevec, T., Anzic, B., Mesar, T., Selic, L., Vilar, M., Solmajer, T. (2002) Design, synthesis and bioactivity evaluation of tribactam β lactamase inhibitors. *Bioorg. Med. Chem. Lett.* 12, 971-975.

[142] [1]Vilar, M., Galleni, M., Solmajer, T., Turk, B., Frère, J. M., Matagne, A. (2001) Kinetic study of two novel enantiomeric tricyclic β-lactams which efficiently inactivate class C β-lactamases. Antimicrob. *Agents Chemother.* 45, 2215-2223.

[143] Wild, H., Hartwig, W. (1992) First synthetic access to 6-(methylene)oxapenems: a new class of β-lactamase inhibitors. *Synthesis,* 1099-1103.

[144] Pratt, R. F., Loosemore, M. J. (1978) 6-β-Bromopenicillanic acid, a potent β-lactamase inhibitor. *Proc. Natl. Acad. Sci. USA* 75, 4145-4149.

[145] Knott-Hunziker, V., Orlek, B. S., Sammes, P. G., Waley, S. G. (1979) Kinetics of inactivation of β-lactamase I by 6β-bromopenicillanic acid. *Biochem. J.* 177, 365-367.

[146] von Daehne, W. (1980) 6β-Halopenicillanic acids, a group of β-lactamase inhibitors. *J. Antibiot.* 33, 451-452.

[147] Pratt, R. F., Cahn, D. J. (1988) The mechanism of rearrangement of 6-halopenicilloates to 2,3-dihydro-1,4-thiazines. *J. Am. Chem. Soc.* 110, 5096-5104.

[148] De Meester, F., Matagne, A., Dive, G., Frère, J. M. (1989) Unexpected influence of ionic strength on branched-pathway interactions between β-lactamases and β-halogenopen-icillanates. *Biochem. J.* 257, 245-249.

[149] Hanessian, S., Alpegiani, M. (1989) Highly stereoselective free radical C-6-allylation of penams. Synthesis of a novel β-lactamase inhibitor. *Tetrahedron* 45, 941-950.

[150] Adam, S., Then, R., Angehrn, P. (1993) (6R)-6-(substituted methyl)penicillanic acid sulfones: new potent beta-lactamase inhibitors. *J. Antibiot.* 46, 641-646.

[151] Sandanayaka, V. P., Feigelson, G. B., Prashad, A. S., Yang, Y., Petersen, P. J. (2001) Allyl and propargyl substituted penam sulfones as versatile intermediates toward the syntheses of new β-lactamase inhibitors. *Bioorg. Med. Chem. Lett.* 11, 997-1000.

[152] Sandanayaka, V. P., Prashad, A. S., Yang, Y., Williamson, R. T., Lin, Y. I., Mansour, T. S. (2003) Spirocyclopropyl β-lactams as mechanism-based inhibitors of serine β-lactamases. Synthesis by rhodium-catalyzed cyclopropanation of 6-diazopenicillanate sulfone. *J. Med. Chem.* 46, 2569-2571.

[153] Imada, A., Kitano, K., Asai, M. (1979) *Antibiotic G-6320*. German patent DE 2855949 19790705.

[154] Imada, A., Kitano, K., Kintaka, K., Muroi, M., Asai, M. (1981) Sulfazecin and isosulfazecin, novel β-lactam antibiotics of bacterial origin. *Nature* 289, 590-591.

[155] Sykes, R. B., Cimarusti, C. M., Bonner, D. P., Bush, K., Floyd, D. M., Georgopapadakou, N. H., Koster, W. H., Liu, W. C., Parker, W. L., Principe, P. A., Rathnum M. L., Slusarchyk, W. A., Trejo, W. H., Wells, J. S. (1981) Monocyclic β-lactam antibiotics produced by bacteria. *Nature* 291, 489-491.

[156] Georgopapadakou, N. H., Smith, S. A., Sykes, R. B. (1982) Mode of action of azthreonam. Antimicrob. *Agents Chemother.* 21, 950-956.

[157] Jacobus, N. V., Ferreira, M. C., Barza, M. (1982) In vitro activity of azthreonam, a monobactam antibiotic. Antimicrob. *Agents Chemother.* 22, 832-838.

[158] Bush, K., Freudenberger, J. S., Sykes, R. B. (1982) Interaction of azthreonam and related monobactams with ββ-lactamases from gram-negative bacteria. *Antimicrob. Agents Chemother.* 22, 414-420.

[159] Galleni, M., Amicosante, G., Frère, J.-M. (1988) A survey of the kinetic parameters of class C β-lactamases. Cephalosporins and other β-lactam compounds. *Biochem. J.* 255, 123-129.

[160] Maiti, S. N., Setti, E. L., Phillips, O. A., Reddy, A. V. N., Micetich, R. G., Singh, R., Higashitani, F., Kunugita, C., Nishida, K., Uji, T. (1998) Preparation of 2-oso-1-azetidine sulfonic acid derivatives as potent β-lactamase inhibitors. *World Patent WO* 9847895 A1 19981029.

[161] Nishida, K., Kunugita, C., Uji, T., Higashitani, F., Hyodo, A., Unemi, N., Maiti, S. N., Phillips, O. A., Spevak, P., Atchison, K. P., Salama, S. M., Atwal, H., Micetich, R. G. (1999) In vitro and in vivo activities of Syn2190, a novel β-lactamase inhibitor. Antimicrob. *Agents Chemother.* 43, 1895-1900.

[162] Babini, G. S., Livermore, D. M. (2000) Effect of conalbumin on the activity of Syn 2190, a 1,5 dihydroxy-4-pyridon monobactam inhibitor of AmpC β-lactamases.*J. Antimicrob. Chemother.* 45, 105-109.

[163] Danes, C., Navia, M. M., Ruiz, J., Marco, F., Jurado, A., Jimenez de Anta, M. T., Vila, J. (2002) Distribution of β-lactamases in Acinetobacter baumannii clinical isolates and the effect of Syn 2190 (AmpC inhibitor) on the MICs of different β-lactam antibiotics. *J. Antimicrob. Chemother.* 50, 261-264.

[164] Gordon, E. M., Ondetti, M. A. (1982) *O-Sulfated* β-*lactam hydroxamic acids. European patent* EP 51381 A1 19820512.

[165] Gordon, E. M., Ondetti, M. A., Pluscec, J., Cimarusti, C. M., Bonner, D. P., Sykes, R. B. (1982) O-Sulfated β-lactam hydroxamic acids (monosulfactams). Novel monocyclic β-lactam antibiotics of synthetic origin. *J. Am Chem. Soc.* 104, 6053-6060.

[166] Yoshida, C., Hori, T., Momonoi, K., Nagumo, K., Nakano, J., Kitani, T., Fukuoka, Y., Saikawa, I. (1985) Studies on monocyclic β-lactam antibiotics. II. Synthesis and antibacterial activity of 3-acylamino-2-azetidinone-1-oxysulfonic acids. *J. Antibiot.* 38, 1536-1549.

[167] Chin, N.-X., Neu, H. C. (1988) Tigemonam, an oral monobactam. *Antimicrob. Agents Chemother. 32*, 84-91.

[168] Teng, M., Miller, M. J., Nicas, T., Grissom-Arnold, J., Cooper, R. D. G. (1993) β-Lactamase inhibitors derived from N-tosyloxy-β-lactams. *Bioorg. Med. Chem.* 1, 151-154.

[169] Bulychev, A., O'Brien, M. E., Massova, I., Teng, M., Gibson, T. A., Miller, M. J., Mobashery, S. (1995) Potent mechanism-based inhibition of the TEM-1 β-lactamase by novel N-sulfonyloxy β-lactams. *J. Am. Chem. Soc.* 117, 5938-5943.

[170] Charnas, R., Gubernator, K., Heinze, I., Hubschwerlen, C. (1992) *Preparation of* β-*lactams as* β-*lactamase inhibitors.* European patent EP 508234 A2 19921014.

[171] Heinze-Krauss, I., Angehrn, P., Charnas, R. L., Gubernator, K., Gutknecht, E.-M., Hubschwerlen, C., Kania, M., Oefner, C., Page, M. G. P., Sogabe, S., Specklin, J.-L., Winkler, F. (1998) Structure-based design of β-lactamase inhibitors. 1. Synthesis and evaluation of bridged monobactams. *J. Med. Chem.* 41, 3961-3971.

[172] Hubschwerlen, C., Angehrn, P., Gubernator, K., Page, M. G. P., Specklin, J. –L. (1998) Structure-based design of β-lactamase inhibitors. 2. Synthesis and evaluation of bridged sulfactams and oxamazins. *J. Med. Chem.* 41, 3972-3975.

[173] Livermore, D. M., Chen, H. Y. (1997) Potentiation of β-lactams against Pseudomonas aeruginosa strains by Ro 48-1256, a bridged monobactam inhibitor of AmpC β-lactamases. *J. Antimicrob. Chemother.* 40, 335-343.

[174] Oefner, C., D'Arcy, A., Daly, J. J., Gubernator, K., Charnas, R. L., Heinze, I., Hubschwerlen, C., Winkler, F. K. (1990) Refined crystal structure of β-lactamase from Citrobacter freundii indicates a mechanism for beta-lactam hydrolysis. *Nature* 343, 284-288.

[175] Wojtkowski, P. W., Dolfini, J. E., Kocy, O., Cimarusti, C. M. (1975) β-Thionolactam analogs of cephalosporins and penicillins. *J. Am. Chem. Soc.* 97, 5628-5630.

[176] Murphy, B. P., Pratt, R. F. (1989) A thiono-β-lactam substrate for the β-lactamase II of Bacillus cereus. Evidence for direct interaction between the essential metal ion and substrate. *Biochem. J.* 258, 765-768.

[177] Tsang, W. Y., Dhanda, A., Schofield, C. J., Frére, J.-M., Galeni, M., Page, M. I. (2004) The inhibition of metallo-β-lactamase by thioxo-cephalosporin derivatives. *Bioorg. Med. Chem. Lett.* 14, 1737-1739.

[178] Badarau, A., Llinas, A., Laws, A. P., Damblon, C., Page, M. I. (2005) Inhibitors of metallo-β-lactamase Generated from β-Lactam Antibiotics. *Biochemistry* 44, 8578-8589.

[179] Bonomo, R. A., Rudin, S. A., Shlaes, D. M. (1997) Tazobactam is a potent inactivator of selected inhibitor-resistant class A β-lactamases. *FEBS Microbiol. Lett.* 148, 59-62.

[180] Martinez, J. L., Vicente, M. F., Delgado-Iribarren, A., Perez, Diaz, J. C., Baquoro, F. (1989) Small plasmids are involved in amoxicillin-clavulanate resistance in Escherichia coli. Antimicrob. *Agents Chemother.* 33, 595.

[181] Wu, P.-J., Shannon, K., Phillips, I. (1994) Effect of hyperproduction of TEM-1 β-lactamase on in vitro susceptibility of Escherichia coli to β-lactam antibiotics. Antimicrob. *Agents Chemother.* 38, 494-498.

[182] Wu, P.-J., Shannon, K., Phillips, I. J. (1995) Mechanisms of hyperproduction of TEM-1 βββ-lactamase by clinical isolates of Escherichia coli. *J. Antimicrob. Chemother.* 36, 927-939.

[183] Reguera, J. A., Baquero, F., Perez-Diaz, J. C., Martinez, J. L. (1991) Factors determining resistance to β-lactam combined with β-lactamase inhibitors in Escherichia coli. *J. Antimicrob. Chemother.* 27, 569-575.

[184] Bonomo, R. A., Rice, L. B. (1999) Inhibitor resistant class A beta-lactamases. *Front. Biosci.* 4, e34-e41.

[185] Nicolas-Chanoine, M. H. (1997) Inhibitor-resistant β-lactamases. *J. Antimicrob. Chemother.* 40, 1-3.

[186] Chaïbi, E. B., Sirot, D., Paul, G., Labia, R. (1999) Inhibitor-resistant TEM β-lactamases: phenotypic, genetic and biochemical characteristics. *J. Antimicrob. Chemother.* 43, 447-458.

[187] Yang, Y., Rasmussen, B. A., Shlaes, D. M. (1999) Class A ββ-lactamases-enzyme-inhibitor interactions and resistance. *Pharmacol. Ther.* 83, 141-151.

[188] Henquell, C., Sirot, D., Chanal, C., De Champs, C., Chatron, P., Lafeuille, B., Texier, P., Sirot, J., Cluzel, R. (1994) Frequency of inhibitor-resistant TEM β-lactamases in Escherichia coli isolates from urinary tract infections in France. *J. Antimicrob. Chemother.* 34, 707-714.

[189] Kaye, K. S., Gold, H. S., Schwaber, M. J., Venkataraman, L., Qi, Y., De Girolami, P. C., Samore, M. H., Anderson, G., Rasheed, J. K., Tenover, F. C. (2004) Variety of β-lactamases produced by amoxicillin-clavulanate-resistant Escherichia coli isolated in the northeastern United States. *Antimicrob. Agents Chemother*. 48, 1520-1525.

[190] Miro, E., Navarro, F., Mirelis, B., Sabate, M., Rivera, A., Coll, P., Prats, G. (2002) Prevalence of clinical isolates of Escherichia coli producing inhibitor-resistant β-lactamases at a university hospital in Barcelona, Spain, over a 3-year period. Antimicrob. *Agents Chemother.* 46, 3991-3994.

[191] Chaïbi, E. B., Farzaneh, S., Morand, A., Péduzzi, J., Barthélémy, M., Sirot, D., Labia, R. (1996) *J. Antimicrob. Chemother.* 37, 190-1.

[192] Chardon, H., Farzaneh, S., Labia, R., Jarlier, V., Nicolas, M. H., Paul, G., Poyart, C., Sirot, D., Sirot, J. (1995) Analysis of β-lactamases produced by cephalothin-susceptible Escherichia coli clinical isolates resistant to co-amoxiclav and ticarcillin-clavulanic acid. *J. Antimicrob. Chemother.* 36, 267-269.

[193] Lin, S., Thomas, M., Shlaes, D. M., Rudin, S. D., Knox, J. R., Anderson, V., Bonomo, R. A. (1998) Kinetic analysis of an inhibitor-resistant variant of the OHIO-1 β-lactamase, an SHV-family class A enzyme. *Biochem. J.* 333, 395-400.

[194] Chaïbi, E. B., Péduzzi, J., Farzaneh, S., Barthélémy, M., Sirot, D., Labia, R. (1998) Clinical inhibitor-resistant mutants of the β-lactamase TEM-1 at amino-acid position 69. Kinetic analysis and molecular modeling. *Biochim. Biophys. Acta* 1382, 38-46.

[195] Wang, X., Minasov, G., Shoichet, B. K. (2002) The structural bases of antibiotic resistance in the clinically derived mutant β-lactamases TEM-30, TEM-32, and TEM-34. *J. Biol. Chem.* 277, 32149-32156.

[196] Meroueh, S. O., Roblin, P., Golemi, D., Maveyraud, L., Vakulenko, S. B., Zhang, Y., Samama, J.-P., Mobashery, S. (2002) Molecular dynamics at the root of expansion of function in the M69L inhibitor-resistant TEM β-lactamase from Escherichia coli. *J. Am. Chem. Soc.* 124, 9422-9430.

[197] Madec, S., Blin, C., Krishnamoorthy, R., Picard, B., Chaïbi, E. B., Fouchereau-Peron, M., Labia, R. (2002) Substitution of Met-69 by Ala or Gly in TEM-1 β-lactamase confer an increased susceptibility to clavulanic acid and other inhibitors. *FEMS Microbiol. Lett.* 211, 13-16.

[198] Pagan-Rodriguez, D., Zhou, X., Simmons, R., Bethel, C. R., Hujer, A. M., Helfand, M. S., Jin, Z., Guo, B., Anderson, V. E., Ng, L. M., Bonomo, R. A. (2004) Tazobactam inactivation of SHV-1 and the inhibitor-resistant Ser130 □ Gly SHV-1 β-lactamase: Insight into the mechanism of inhibition. *J. Biol. Chem.* 279, 19494-19501.

[199] Sun, T., Bethel, C. R., Bonomo, R. A., Knox, J. R. (2004) Inhibitor-Resistant Class A β-Lactamases: Consequences of the Ser130-to-Gly mutation seen in apo and tazobactam structures of the SHV-1 variant. *Biochemistry* 43, 14111-14117.

[200] Helfand, M. S., Bethel, C. R., Hujer, A. M., Hujer, K. M., Anderson, V. E., Bonomo, R. A. (2003) Understanding resistance to β-lactams and β-lactamase inhibitors in the SHV β-lactamase: lessons from the mutagenesis of SER-130. *J. Biol. Chem.* 278, 52724-52729.

[201] Thomas, V. L., Golemi-Kotra, D., Kim, C., Vakulenko, S. B., Mobashery, S., Shoichet, B. K. (2005) Structural consequences of the inhibitor-resistant Ser130Gly substitution in TEM β-lactamase. *Biochemistry* 44, 9330-9338.

In: Beta-Lactamases
Editor: Jean-Marie Frère, pp. 259-292

ISBN: 978-1-61324-638-2

Chapter 10

β-LACTAMASE INHIBITORS: NON-β-LACTAMS

Rex F. Pratt
Department of Chemistry, Wesleyan University,
Middletown, CT, U.S.A.

ABSTRACT

Besides the β-lactam containing inactivators, a variety of non-β-lactam containing compounds have been described as inactivators or inhibitors of β-lactamases. It is not surprising that very different compounds were found to be active against metallo-β-lactamases on the one hand and active-site serine enzymes on the other. Several of them were isolated from natural sources while many others are due to the imagination of chemists.

The best-known β-lactamase inhibitors, as described in Chapter 9, are themselves β-lactams. This includes the three produced commercially and employed in clinical practice, clavulanic acid, sulbactam, and tazobactam. These have been effective for a number of years but suffer from their susceptibility, real and potential, to mutant β-lactamases that are selected for by their use. Extended spectrum β-lactamases have, in fact, emerged that are insensitive to the classical inhibitors [1,2]. It is important, therefore, to maintain an armamentarium of diverse β-lactamase inhibitors in order to preserve the effectiveness of β-lactams as antibiotics for as long as possible. As part of this effort, it is important to keep watch for and, indeed, specifically search for, β-lactamase inhibitors that are not β-lactams. This review of such inhibitors is concentrated on developments over the past 15 years.

It is necessary first to subdivide the discussion into two sections. As described in detail in chapters 2-5 and 7 of this volume, β-lactamases can be divided mechanistically into two groups, those of classes, A, C, and D, the serine β-lactamases, and the distinctly different class B metallo-β-lactamases. Enzymes of the former group share active site structural and mechanistic features and thus, for them, the concept of a common inhibitor is logical. The metallo-enzymes, however, catalyze β-lactam hydrolysis in a completely different fashion in a completely differently sculpted active site and it is therefore more difficult to imagine or design an inhibitor that would be effective against both groups. The discussion below will therefore consider the groups separately.

The design of β-lactamase inhibitors, as opposed to their discovery by random or somewhat less than random screening, and although, from all precedent, the latter procedure is at least as likely to be successful as the former, has, in general, taken two overlapping paths, those of substrate analogues and of transition state analogues. The former may, in principle be cyclic compounds other than β-lactams or acyclic. They may inhibit by covalent interaction with the active site, in the fashion of a substrate, or purely by non-covalent means. To a considerable degree, the design of acyclic inhibitory substrates has been linked to the issue of β-lactamase evolution. It is now generally accepted that serine β-lactamases are derived from ancestral bacterial DD-peptidases. This was first suggested on the basis of the structural similarity between the substrates of these enzymes, β-lactams in the former case, and D-alanyl-D-alanine C-terminal peptides in the latter [3]. This proposal has been solidly supported in recent years by protein structural analogies, particularly at the actives sites of these enzymes [4]. Consequently, one branch of the search for acyclic β-lactamase inhibitors began with the concept of D-alanyl-D-alanine analogs [5].

Transition state analogues may closely resemble substrates, but need not, provided that the center of their activity closely resembles the transition state of β-lactam hydrolysis with respect to molecular shape and charge distribution, i.e. they must resemble *1* or *2*, the transition states of acylation, or *3* or *4*, the transition states for deacylation, respectively; such molecules would therefore, necessarily, be covalent inhibitors.

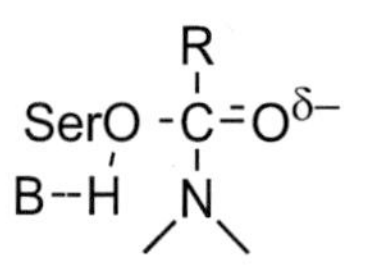

1

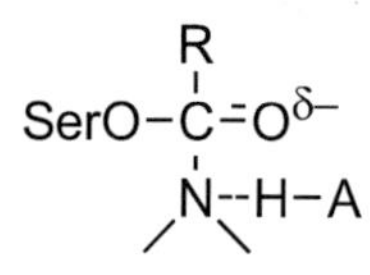

2

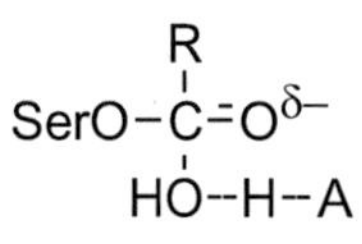

3

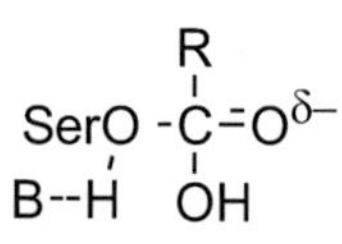

4

SERINE β-LACTAMASES

(a) Non-covalent Inhibitors

Non-covalent binding of small molecules to β-lactamase active sites is usually not strong ($K_i \geq$ 1mM] so there are few potent examples of this class. The screening of chemical libraries, however, led to a series of rhodanines *5* that are effective against class A and class C β-lactamases at the micromolar level [6]. Similar compounds were subsequently found to have DD-peptidase inhibitory and antibiotic activity [7]; the inhibitory activity was apparently non-competitive, so the rhodanines may not bind at the active sites of β-lactam recognizing enzymes. High throughput screening has also produced micromolar phthalimide inhibitors [8].

The testing of natural products revealed the inhibitory properties of anacardic acids *6* [9] and various compounds from tea extracts [10]. The anacardic acids appear to have major activity against class C β-lactamases [11] although synergistic effects with methicillin against MRSA have been reported [12]; the inhibition observed may be non-competitive [11]. There may well be some similarity between the mode of action of these compounds and that of izumenolide and panosialin [13,14]. The tea extracts have a number of biological activities, where the most relevant here is the time-dependent inhibition of the *S. aureus* β-lactamase by epigallocatechin gallate *7* [15]. This compound appears less effective against other β-lactamases [16]. Epigallocatechin and its analogues also have some direct antibacterial activity [10] and a number of gallates have been prepared to investigate this further [17].

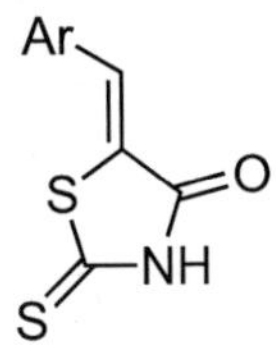

5

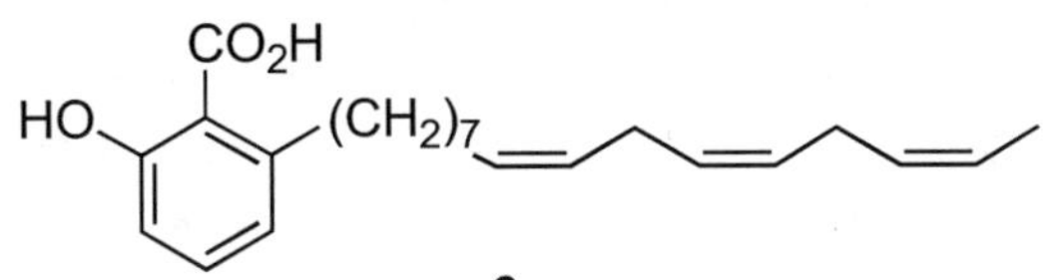

6

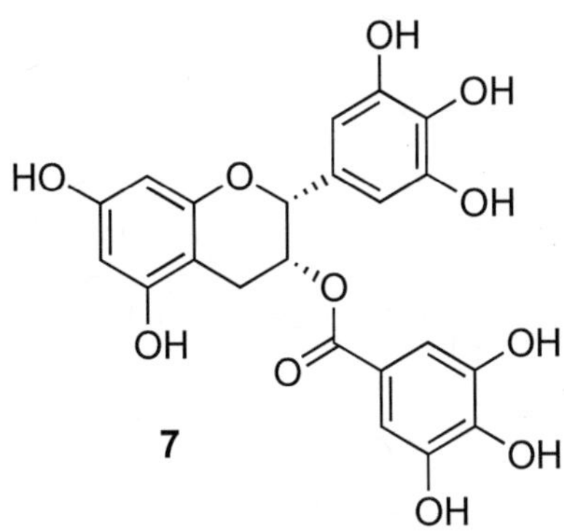

7

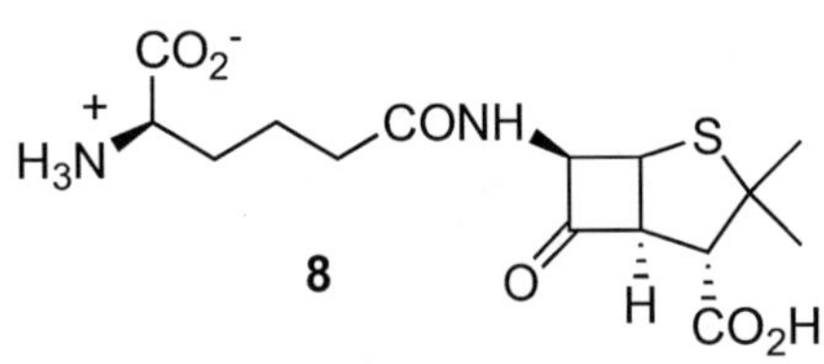

8

An interesting recent advance in the design and synthesis of non-covalent substrate analogues of the β-lactam-recognizing enzymes was that of the deaza-β-lactam *8*, a penicillin N analogue, by Baldwin and coworkers [18]. Although other deaza-β-lactams have been previously described, none until now has included an amido side chain. No biochemical studies of *8* with β-lactamases, however, have yet been reported. More recently, 6-halodeazapenicillins have been shown to be micromolar inhibitors of all classes of β-lactamases, including class B, the metalloenzymes. A crystal structure of one of these compounds bound to the class D OXA-10 β-lactamase revealed covalent attachment to the active site serine in a hemiketal linkage [19].

Structure-based design has been employed to obtain inhibitors of class C β-lactamases. Shoichet and coworkers have computationally screened a commercial collection of 200,000 compounds to pinpoint *9* as a 26 μM inhibitor of the Amp C β-lactamase [20]. A crystal structure of the enzyme-inhibitor complex revealed the inhibitor at the active site, oriented essentially as predicted by the computational docking procedure. The carboxylate occupies the general region of the oxyanion hole, hydrogen-bonded to Ser 64 and Ala 318, the sulfonamide oxygens are hydrogen-bonded to the Ser 64 hydroxyl group, the Lys 67 ammonium group and the Asn 152 side chain NH_2, and the aromatic rings interact with Leu 119, Leu 293, and Tyr 221. Further analysis of the crystal structure led to *10*, which had a K_i of 1 μM against the AmpC β-lactamase [21]. This compound was also shown to act synergistically with ceftazidime against organisms that expressed the AmpC β-lactamase and to not itself induce β-lactamase expression. Fragment-based screening has also been employed to find micromolar inhibitors [22].

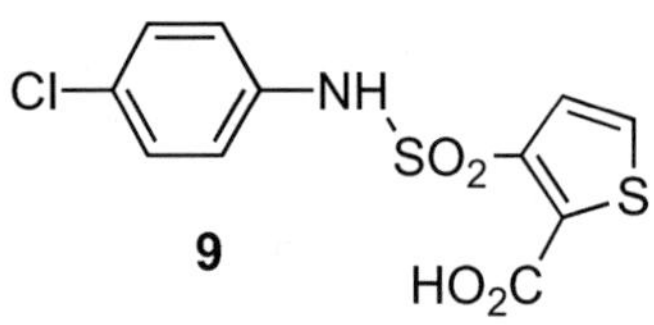

9

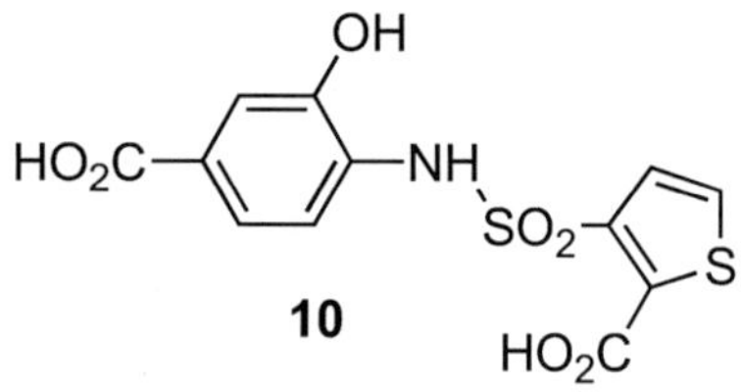

10

The tightest-binding non-covalent inhibitors of β-lactamases yet discovered are the β-lactamase inhibitory proteins BLIP and BLIP II [23, 24]. These appear to be primarily specific for class A β-lactamases and particularly for TEM variants: K_i values for TEM-1, for example, are sub-nanomolar. A full discussion of the structure and activity of these inhibitors is found in chapter 11 of this volume. For the purposes of the present chapter, it needs only be noted that BLIP makes a large number of contacts with TEM-1 in the inhibited complex, particularly including those between a specific type II turn, at the surface of BLIP, which protrudes into the β-lactamase active site, and certain active site residues. The carboxylate group of Asp 49 of BLIP hydrogen bonds to Ser 130, Lys 234, Ser 235, and Arg 244 of TEM-1, much as does the carboxylate of a bound penicillin, and the phenyl ring of Phe 142 stacks with Tyr 105 as does the thiazolidine ring of benzylpenicillin [25]. Mutation of these residues of BLIP leads to significant losses of inhibitory activity [26]. BLIP II does not occupy the active site but binds elsewhere on the β-lactamase from which position it blocks substrate access to the active site [24].

Given that BLIP binds very tightly to TEM-1, an important question was whether a small molecule analogue of BLIP could be constructed. Palzkill and coworkers have investigated this issue [27]. They prepared a peptide containing residues Ala 46-Tyr 51 of the inhibitory β-turn from BLIP and containing a cysteine residue at each end so that a cyclic (less mobile) peptide could also be obtained by oxidation. A K_i value of only 0.6 mM, however, was achieved against TEM-1. Apparently a much larger part of BLIP must be incorporated in order to obtain tight binding. It would be interesting to see if appropriate incorporation of another aromatic substituent would be of any substantial assistance. The same group has also screened small peptides for TEM-1 inhibition by phage display [28]. After optimization of a lead, they produced the peptide RRGHYY that inhibited with a K_i of 0.14 mM. Interestingly, this same peptide inhibited the class C P99 β-lactamase with the same K_i value. It may be possible, therefore, to discover small peptidomimetics with broad inhibitory powers.

The common theme, therefore, in non-covalent inhibitors is the stereo-specific combination of (one or more) (hetero) aromatic rings and a carboxylate group. This of course, imitates the specificity elements of good β-lactam substrates. A general survey of progress in this area suggests, however that it may not be possible to achieve broad specificity with purely non-covalent inhibitors. It may be possible to design an excellent inhibitor for a particular β-lactamase but this, in itself, is probably not sufficient for practical applications. BLIP, for example, can be modified to enhance its activity for particular β-lactamases [29], and specific inhibitory antibodies can be obtained for any β-lactamase of choice [30], but simple mutations of the β-lactamases can avoid the trap [26]. For this reason, the most generally effective β-lactamase inhibitors will continue to be molecules that act by covalent modification of the active site. This is demonstrated by the very effective mechanism-based inhibitors derived from β-lactams (see Chapter 9). Other scaffolds, however, are also possible, as described below.

(b) Covalent Inhibitors

(i) Substrate Analogues

Under this heading, a number of acylating agents, as are β-lactams themselves, will be considered. Early research showed that simple γ-and δ-lactams were insufficiently chemically reactive to act as substrates or covalent inhibitors of β-lactam-recognizing enzymes. On suitable activation by a variety of mechanisms, however, a number of potential leads were identified. The early examples of this were nicely reviewed by Jungheim and Ternansky in 1992 [31]. They identified a number of activated γ-lactam systems, γ-lactams themselves, *11*, pyrazolidinones, *12*, isoxazolidinones, for example lactivicin *13*, and imidazolidinones, *14*. In these structures, the R groups include classical β-lactam side chains. Many compounds of this type were found to react covalently with DD-peptidases and β-lactamases. Although particularly effective β-lactamase inhibitors were not discovered, the compounds themselves remain as potential platforms for the design of mechanism-based inhibitors. Interest in the lactivicins, for example, has been recently renewed [32].

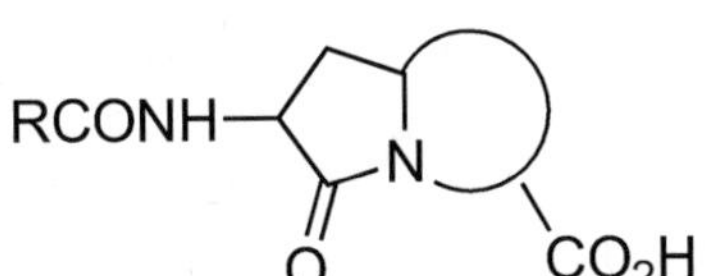

11

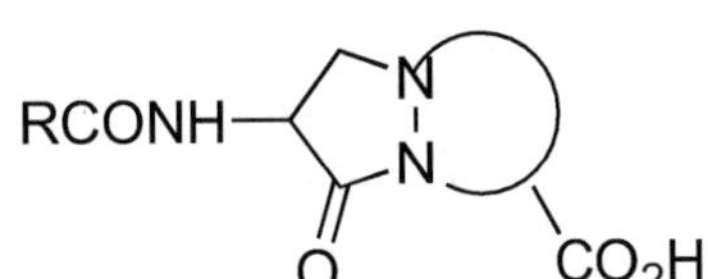

12

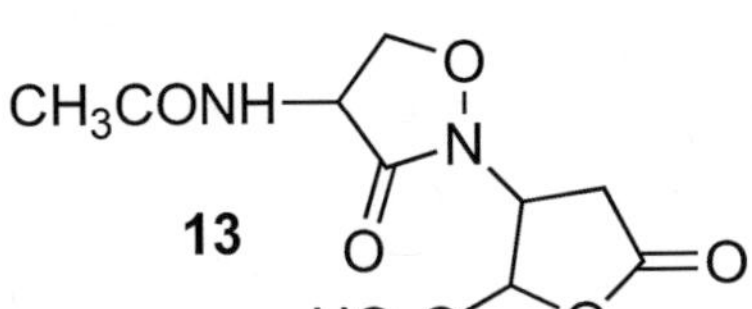

13

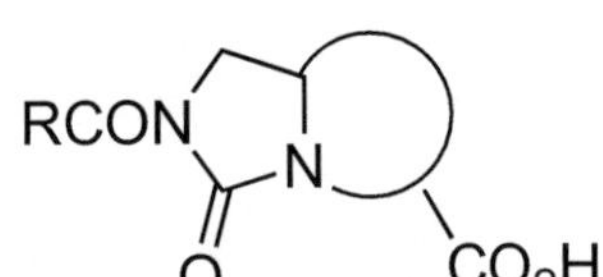

14

There have also been some interesting developments in bridged γ-lactams in more recent years. Hadfield et al., for example, have described the synthesis of several derivatives of *15* [33].

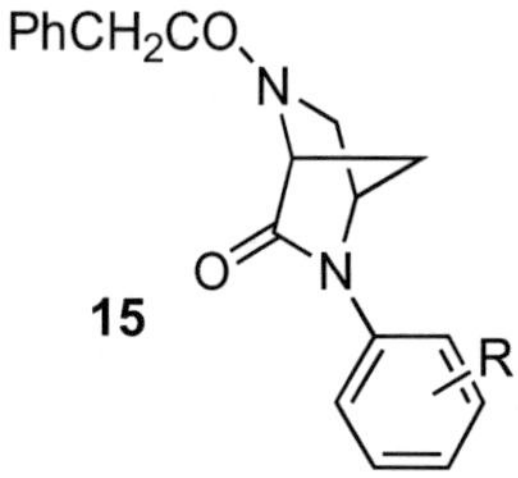

15

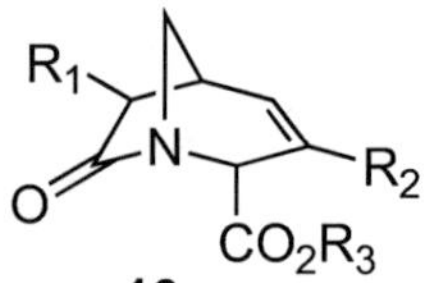

16

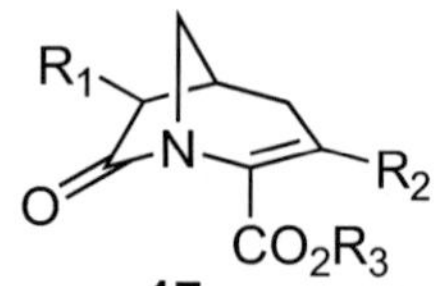

17

One of these (*15*; R = p-NO_2) inhibited the class A *B. cereus* β-lactamase and the class C P99 enzyme with K_i values of 90 μM and 71 μM, respectively. No indication is given, however, as to whether covalent interaction was involved. Aszodi et al. have described the synthesis and some biological properties of *16* and *17* [34]. These compounds have been structurally characterized and thus shown to have a highly pyramidal bridgehead nitrogen atom (the sum of angles around nitrogen is 322°) and a carbonyl oxygen to carboxyl carbon distance of 3.30 Å [within the distance criterion of Cohen [35] for an active antibiotic]. They reacted with typical DD-peptidases, and, at micromolar concentrations, were time-dependent inhibitors of the TEM-1 and P99 β-lactamases.

Finally, and perhaps most exciting, is the disclosure of *18* (NXL104) by Bonnefoy et al. [36]. This molecule has no antibiotic activity but essentially irreversibly inactivates both TEM-1 and P99 β-lactamases at nanomolar concentrations. The aim of the project was to find an inhibitor of class C β-lactamases that would overcome resistance to third generation cephalosporins such as ceftazidime. Combinations of *18* with ceftazidime were indeed more effective than ceftzidime alone against several AmpC-producing bacteria and against *Enterobacteriaceae* carrying ESBL (except metallo-β-lactamases) [37]. Its activity against KPC carbapenemases has also been demonstrated [38, 39]. The kinetics of reaction of *18* with the TEM and P99 β-lactamases have now been described [40]. It had, in general, lower IC_{50} values against a range of Class A and class C β-lactamases than the classical mechanism-based β-lactamase inhibitors.

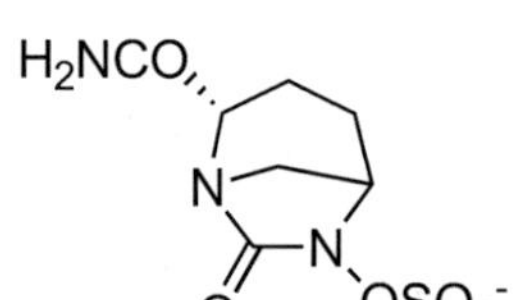

18

It is clear from these examples that inhibitory complexes, most likely in the form of hydrolytically inert acyl-enzymes, can be achieved with cyclic platforms other than β-lactams. Further research in this direction may well be profitable.

Acyclic substrates of β-lactamases may also lead, directly or indirectly, to inhibitors. As noted above, the first of these were modeled after D-alanyl-D-alanine peptides following the presumed evolutionary relationship between the enzymes. Thus, depsipeptides of general structure *19,* where considerable variation of the structure is possible, were found to be substrates of all classes of β-lactamases, [5, 41, 42, 43]. They are also substrates of DD-peptidases [44] and are thus very useful for comparisons of DD-peptidase and β-lactamase active sites [45-47], and have been used to investigate transition state structures by means of kinetic isotope effects [48, 49].

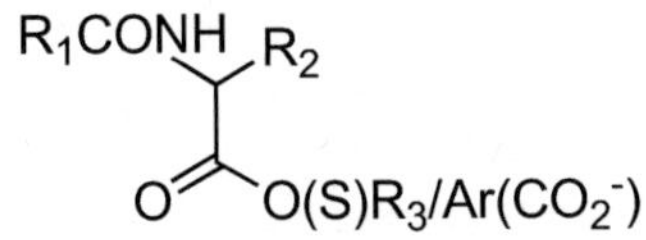

19

To date, class C β-lactamases seem most reactive with *19*, but little has been done to optimize structures for any particular β-lactamase or class of β-lactamases. The D-methyl group at R_2, favored by DD-peptidases, is less acceptable to the P99 β-lactamase, and may have been selected against in β-lactamase evolution [47, 50]. Some investigation of leaving groups has been made with the P99 enzyme [41, 42, 49, 51, 52], leading to the order of reactivity indicated below:

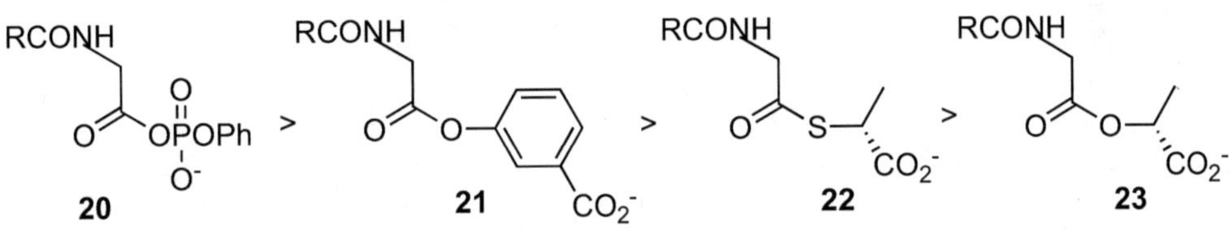

R = $PhCH_2$

Molecular modeling suggested that the negative charges on these substrates interact with the same group of active site residues as do the carboxylates of β-lactams, e.g. the phosphate oxyanions of *20* with the side chains of Lys 315 and Tyr 150 [53], and the carboxylates of *21* - *23* with Lys 315, Thr 316 and Tyr 150 [44, 49]. All evidence from the above-cited references suggests that serine β-lactamases catalyze hydrolysis of *19* by the same double displacement mechanism with an acyl-enzyme intermediate as they employ with β-lactams. With good leaving groups such as in *20* and *21*, deacylation of the P99 enzyme is rate-determining, while acylation is the slow step with the poorer leaving groups of *22* and *23* [41, 42, 49, 51]. Class C enzymes [41, 42, 54], at least one class A enzyme [*S. aureus* PCI, but, notably, not TEM-2 [41]], and the class D OXA-1 enzyme [55]. The P99 β-lactamase also catalyzes acyl-enzyme aminolyis by D-amino acids in a reaction that is effectively the reverse of the acylation step of the transpeptidation reaction of DD-peptidases [56 - 58]. It is interesting to note that a retro-amide side chain, such as in *24*, also, in general, produces substrates of β-lactamases [43, 59].

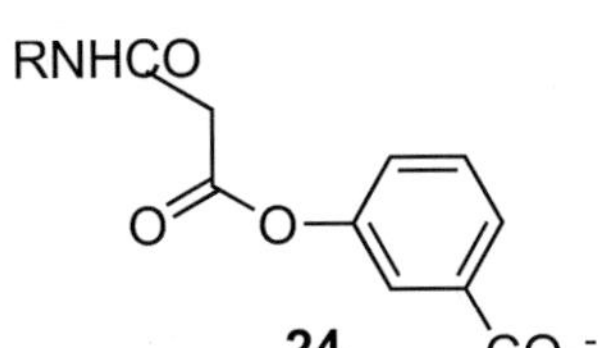

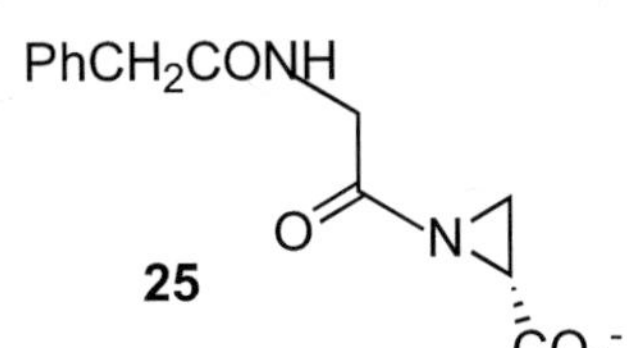

Molecular modeling suggested that the retro-amide carbonyl oxygen is in a position to hydrogen bond to the side chain ammonium ion of Lys 67. In a normal substrate, of course, the amide hydrogen bonds to a backbone carbonyl of the β-2 strand and to the Asn side chain NH of the S(Y)XN box. The acyl-aziridine *25* represents the unusual situation of a good amide substrate of β-lactamases [60]. It is likely that this situation stems from the presence of a pyramidal nitrogen in the scissile bond. This leads to its much closer resemblance to the shape of a bicyclic β-lactam than present in a normal amide with its planar nitrogen; depsipeptides *19* can assume a similar conformation to that of *25* by rotation around the carbonyl carbon-leaving group oxygen bond.

Attempts to produce mechanism-based inhibitors on the basis of *19* have not been successful to date. For example, the potential quinone methide generators, *26*, were substrates of serine β-lactamases but had only modest irreversible inactivating power [61]. They had also little effect on the class B β-lactamase II of *B. cereus.*

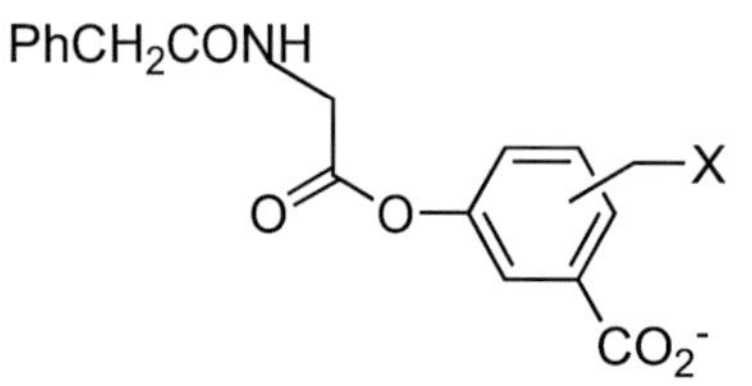

26: X = OAc, OCOAr,Cl, Br, Me_2S^+

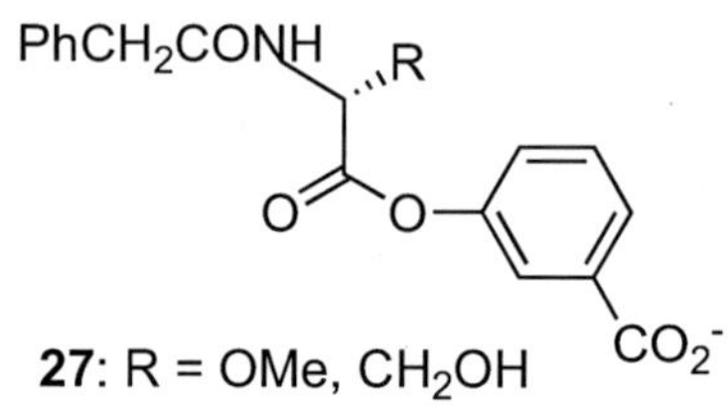

27: R = OMe, CH_2OH

Replacement of the benzyl side chain of *21* with one taken from a third generation cephalosporin did not seem to impede depsipeptide hydrolysis [62], nor did addition of substituents to the glycyl α-carbon as in *27* [63]. On the other hand, these depsipeptides led directly to the phosphonate and acyl phosphate inhibitors (see below).

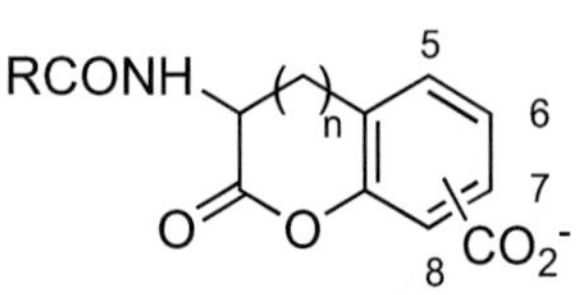

28a: n = 0, 7-CO_2^-
28b: n = 0, 8-CO_2^-
28c: n = 1, 7-CO_2^-
28d: n = 1, 8-CO_2^-

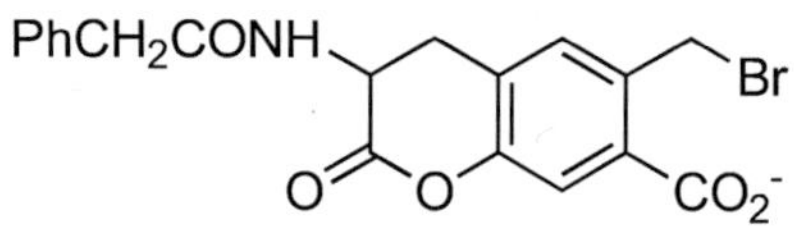

29

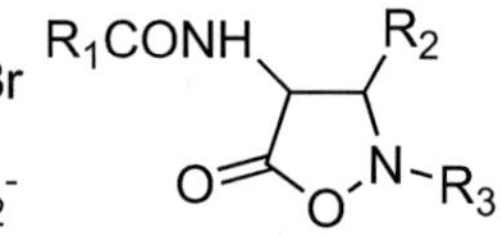

30

Following the lead of β-lactams, it seemed reasonable to assess the activity of cyclic variants of *19*. Thus, the lactones *28* were prepared [64-66]. It seemed possible that, as with β-lactams, the tethered leaving group might impede deacylation, thereby leading to inhibition of the enzyme. The compounds *28a* - *28d*, however, turned out to be substrates of class A and class C β-lactamases of comparable reactivity to that of *19*; the k_{cat} values were smaller, but not sufficiently so for *28* to become transiently inactivating substrates. Incorporation of a tethered quinone methide generator, as in *29*, also did not lead to a potent inhibitor [64].

It may still be possible to design better inhibitors based on the scaffold of *28* – substituents that block movement of the leaving group after acylation might be suggested by inspection of structural models – but the modest stability of *28* in aqueous solution would probably remain an issue. Another lactone system *30*, has also recently been reported to contain β-lactamase and DD-peptidases inhibitors, although kinetic data has not yet been provided [67]. Classes A, B and D but not C β-lactamases were inhibited.

(ii) Transition State Analogues

As noted above, transition state analogue inhibitors for acyl transfer reactions such as β-lactam hydrolysis will have transition states resembling *1 – 4*, i.e. tetrahedral anions, probably bound, in the case of serine β-lactamases, to the active site serine. The precedent of serine proteases has indicated several possible sources of such inhibitors [68]. Scheme 1 shows common variants.

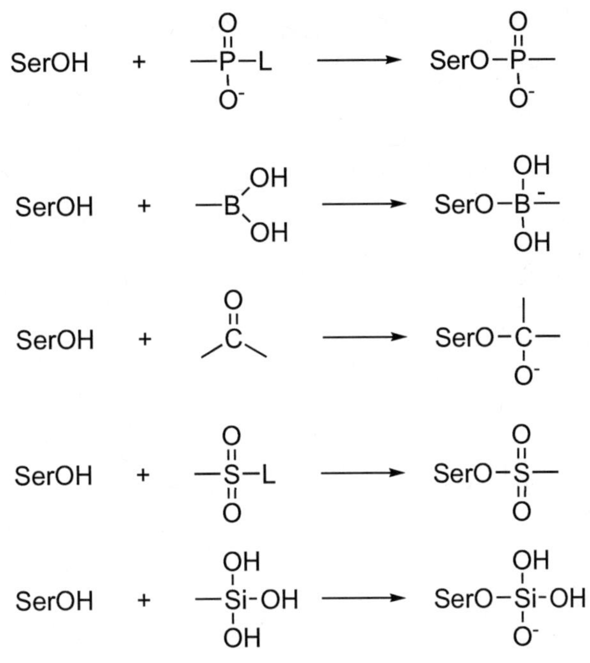

Scheme 1.

1. Phosph(on)ates

It has long been known that neutral phosphyl reagents are strong inhibitors of serine proteases [68]. Simple reagents of this type, such as di-isopropyl phosphoryl fluoride did not, however, have significant effect on β-lactamases [69], thus delaying their identification as serine enzymes. When it first became clear that the β-lactamase active site contained considerable positive charge [70, 71], phosphonate monoester monoanions were examined. Thus, molecules such as *31* [L represents a leaving group, e.g. O(S)R, O(S)Ar, OCOR(Ar), F] were found to be β-lactamase inhibitors, particularly, with respect to these first generation structures, of class C β-lactamases [72-75]. Class A [76] and class D [77] appeared less affected and class B not at all. Kinetics studies indicated that these molecules were irreversible inactivators and that one equivalent, with respect to enzyme, of L was released in the inactivation reaction [72-74]. X-ray crystal structures of the inhibited complexes show phosphylation of the active site serine hydroxyl group [78-80]. These crystal structures have been interpreted as analogues of the acyl transfer transition states for depsipeptide and β-lactam hydrolysis. This interpretation was supported by an appropriate thermodynamic test [81] although the general applicability of this method has been questioned [82].

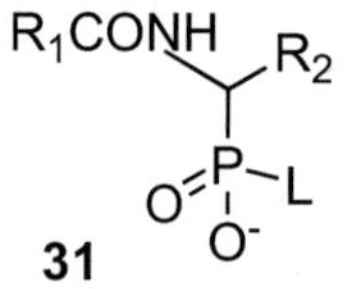

31

Structure-activity studies of *31* established that the specificity of these inhibitors could be expanded to other serine β-lactamases by modification of R and L [83]. In particular, the effectiveness of *31* usually increases as L becomes a better leaving group [73]. In general, only leaving groups that do not require acid catalysis to depart [O(S)Ar, OCOR(Ar), F] yield good inhibitors. This has been rationalized on the basis of the different transition states geometries of acyl transfer (tetrahedral) and phosphyl transfer (trigonal bipyramidal) transition states. A general acid catalyst at the β-lactamase active site that is optimally placed to donate a proton to the leaving group in an acyl transfer reaction, such as β-lactam hydrolysis, would not be optimally placed for phosphyl transfer [73]. This point comes across clearly when phosphonamidates (*31*, L = NHR/Ar) are considered. Compounds such as *32* and *33* are relatively poor inhibitors at neutral pH [84, 85]. Careful analysis of pH profiles for the inhibition reaction makes clear that the major reason for this is the dearth of protons [84]. Neutral phosphonate esters of *33* are, as expected from the discussion above, even less effective as inhibitors [85, 86]. Further development of *31* has recently been described [87, 88].

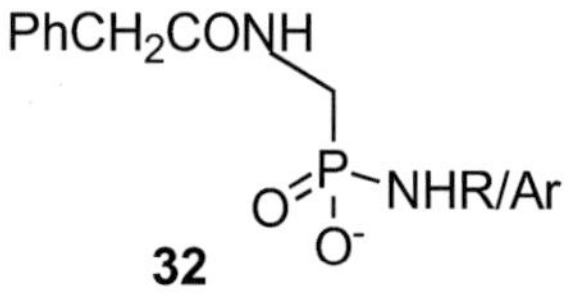

32

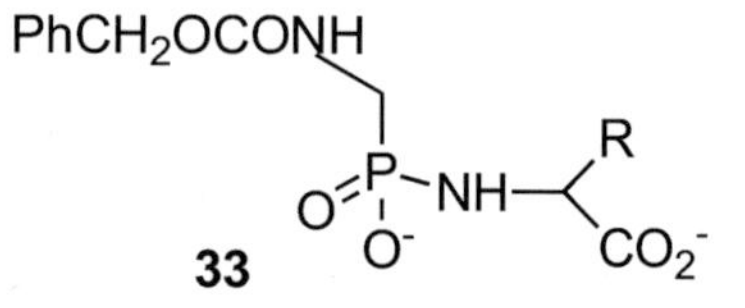

33

The search for better leaving groups for *31* led eventually to acyloxy, OCOR/Ar. Compounds *34,* for example, were found to be intriguing new inhibitors, and particularly, with simple substituents, of class C β-lactamases, yielding inert complexes that slowly (2 x 10^{-2} s^{-1}) reverted to free enzyme [53, 89]. Cyclic analogues of these, such as *35*, were also inhibitors with interesting properties [90-92]. Kinetics and spectroscopic evidence suggested that the acyclic compounds *34* generally inhibited class A and class C β-lactamases by acylation rather than phosphylation [53, 89]. The phosph(on)ate leaving group probably interacts in a catalytically favorable fashion with the active site [53].

Subsequently, however, the bicyclic analogues *36* were shown to inactivate the P99 β-lactamase by phosphylation. Slow reversal of inactivation occurred in these cases by recyclization of the inhibitor at the active site with reformation of *36*, rather than by hydrolysis of the phosphyl-enzyme species [91].

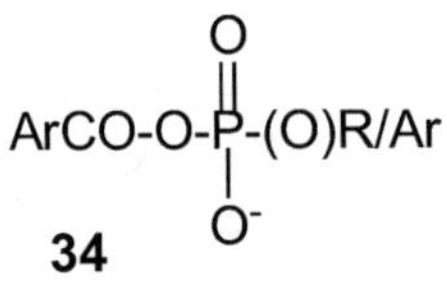

34

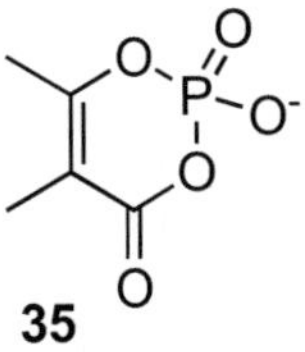

35

An increase in the flexibility of the inhibitors, as in *37*, led to irreversible inhibition of the P99 enzyme; the evidence in this case, however, pointed to enzyme acylation [92]. Interestingly, crystal structures showed that *36* (X=O) and *37* (X=O) inhibit the *Streptomyces* R61 DD-peptidase (similar in structure to the P99 β-lactamase) by phosphorylation. In these covalent complexes, the covalently-bound inhibitor had rearranged into a position where recyclization was difficult and where the carboxylate group strongly interacted with Tyr 159, Arg 285, and Thr 301 [93]. The influence of substituents on the covalent nature of the inhibition by these compounds is nicely shown by comparison of *38* and *39* [51]. Compound *38* is a β-lactamase substrate, particularly of the P99 enzyme, where its k_{cat}/K_m value of 1.0 x10^6 $s^{-1}M^{-1}$ makes it the best acyclic substrate yet discovered for this β-lactamase, and shows again the beneficial influence of the phosphate leaving group [52]. On the other hand, *39* is a good irreversible inhibitor (k_i =5.9 x10^4 $s^{-1}M^{-1}$) of the P99 enzyme, presumably by active site phosphonylation, and showing the beneficial effect of the benzoate leaving group.

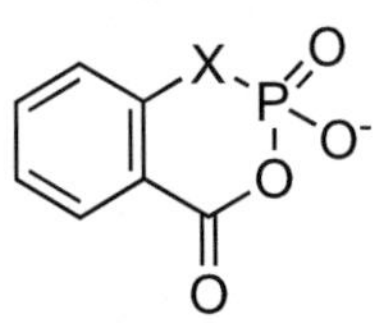

36; X = O, CH_2

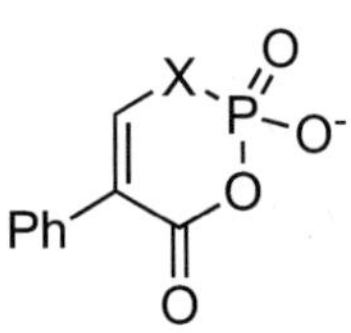

37: X = O, CH_2

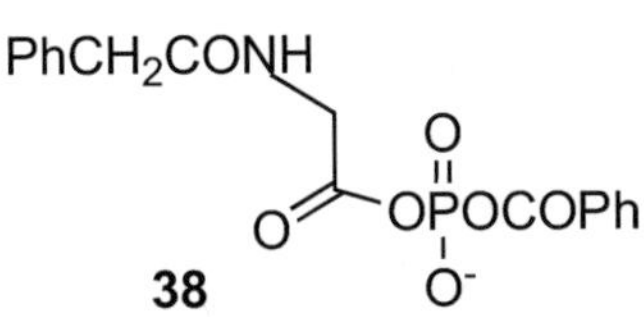

38

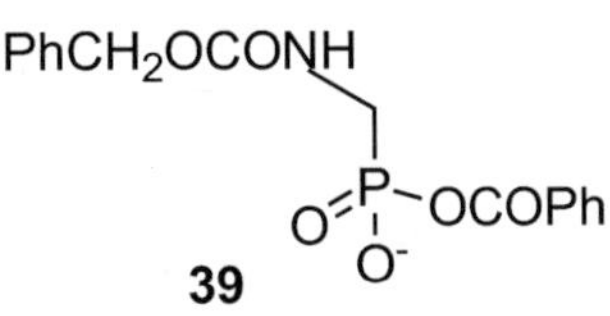

39

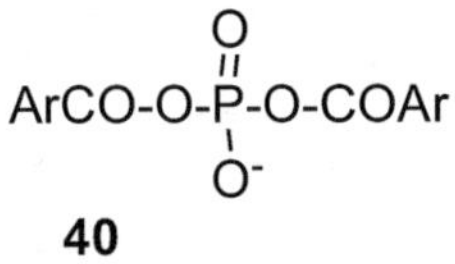

40

The diacylphosphates *40* are more effective than *34* against class A and Class D β-lactamases [55, 77, 89] and are also powerful inhibitors of class C enzymes [94, 95]. No significant attempt has yet been made, however, to optimize the inhibitory properties of the acyl phosph(on)ates.

The rather extraordinary reactivity of anionic phosphyl derivatives at the β-lactamase active site [73, 96] indicates that this environment must be able to strongly stabilize not only the tetrahedral anionic transition state of normal acyl transfer reactions but also the anionic penta-coordinated transition states of phosphyl transfer reactions. This suggests that anionic penta-coordinated species should also be inhibitors. This was shown to be true by an investigation of a series of 1:1 complexes of hydroxamic acids and vanadate, *41*, which are submicromolar inhibitors of the P99 β-lactamase [97, 98]. Strikingly, however, they seem to have no effect on the TEM enzyme. On the other hand, these complexes were able to inhibit serine proteases [97]. ^{51}V NMR spectra of a complex of the P99 β-lactamase with vanadate and pnitrobenzohydroxamic acid (K_i = 0.48 μM) indicated the presence of 5- or 6-coordinated vanadium [98]. A crystal structure of such a complex may give further information on how to effectively distribute anionic heteroatoms at the class C β-lactamase active site to achieve inhibition. Such a structure has not yet been obtained for a β-lactamase but one has been for the serine protease chymotrypsin. This displays hexa-coordinated vanadium and shows how the complex is able to orient itself to place the aryl group into the S1 subsite [99]. Vanadate complexes of catechols have been shown to inhibit typical examples of all classes of serine β-lactamases at micromolar levels [100].

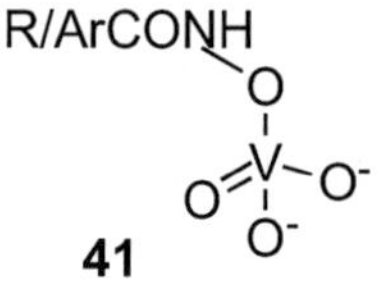

41

Consideration of the results from *19* and *41* suggested that diacylhydrazines and diacylhydroxylamines might be reactive with β-lactamases. Although this was not true with the former [101], examples of the latter, $ArCH_2OCONHOCO_2Ar'$, were found to be time-dependent irreversible inhibitors of serine β-lactamases [102]. An X-ray crystal structure of the complex of one such compound with the class C AmpC β-lactamase showed that inhibition had resulted in the cross-linking of Ser 64 and Lys 315 of the active site, by formation of a carbamate and loss of the two acyl groups [102]. More detained structure-activity and mechanistic studies of these inhibitors have now been reported [103].

2. Boronates

Borate itself and boronates have long been recognized as inhibitors of class A and class C β-lactamases, just as they are of serine proteases [104-106]. Waley and coworkers showed that the incorporation of classical β-lactam side chains into boronic acids, such as in *42*, could readily achieve submicromolar inhibitors [107]. This theme was also taken up by a HoffmanLaRoche group [108].

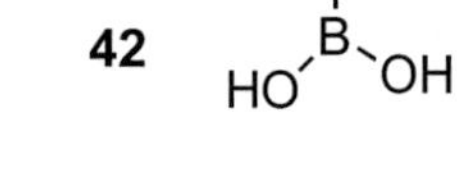

42

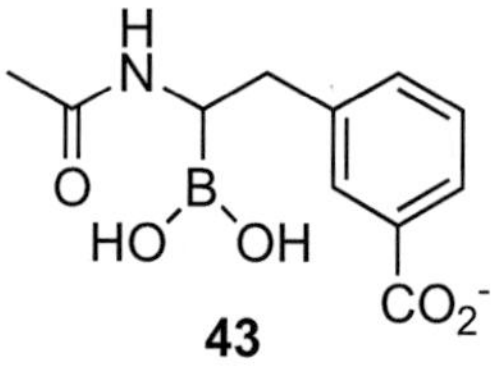

43

In 1995, Martin and Jones showed how further structural considerations, aided by the crystal structure of the TEM-1 enzyme, led to *43* as a 0.1 μM inhibitor of that enzyme [109]. A subsequent crystal structure of the complex [110] showed the boronate covalently bound to the active site serine O_γ, the amido side chain bound in the same way as β-lactam side chains, and the carboxylate group hydrogen bonded to Arg 244, Ser 130 and Ser 235 in much the same way as expected of a β-lactam carboxylate.

The geometry around boron was tetrahedral, as in *44*, with one boronate oxygen hydrogen-bonded into the oxyanion hole of the active site, and the other to Glu 166 and Asn 170. The structure was thus interpreted as an analogue of a tetrahedral transition state for deacylation,

3, where the second boronate oxygen would be the analogue of the oxygen of the deacylating water molecule. Further design yielded *45* and *46* as genuine nanomolar inhibitors of the TEM-1 β-lactamase. A crystal structure of the complex of *46* with the enzyme showed the additional phenolic hydroxyl group hydrogen-bonded to Ser 130 O_γ, as predicted from modeling [111]. From the above studies, it is clear that boronates inhibit β-lactamases reversibly, but covalently, according to Scheme 2. The overall negative charge on the boronate is, of course, largely localized on the boronate oxygen atoms.

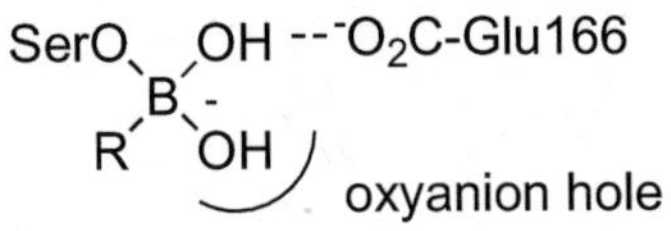

44

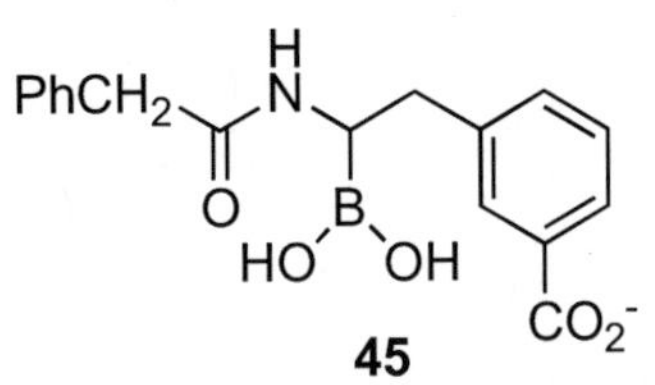

45

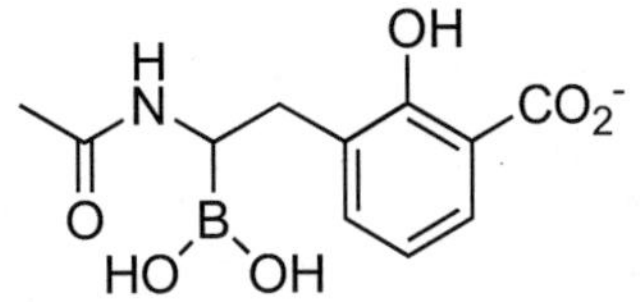

46

Structure-based computational approaches to inhibitor design have been employed to an even greater degree by Shoichet and coworkers using the β-lactamase /boronate system as a case study. A series of aryl boronic acids, for example, was investigated. The crystal structure of the complex between the class C AmpC β-lactamase and m-aminobenzeneboronic acid showed the boronate bound to the active site serine, but with the aryl substituent "upside down" in the active site, i.e. in the leaving group site rather than in the acyl donor site [112]. Computational docking experiments, based on this crystal structure, identified a variety of aryl boronates that were predicted to be, and were found to be, better inhibitors of the AmpC β-lactamase than m-aminobenzeneboronic acid [113]. The best of these, in this initial study, was benzothiophene-2-boronic acid, *47,* with a K_i of 27 nM. A crystal structure of the complex of *47* with the AmpC enzyme [114]

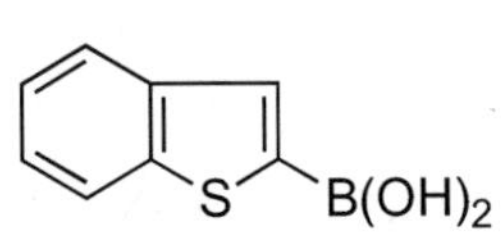

47

showed the expected covalent complex but with the aryl group in the acyl donor site, interacting with the side chain amide group of Asn 161 and the aromatic ring of Tyr 221. This structure could also be interpreted in terms of an acyl transfer transition state, one involved in the hydrolysis of an aroyl donor substrate. One such substrate would, of course be the acyl phosphate *34*; free energies of activation for reaction of the P99 β-lactamase with substituted aroyl phosphates correlated linearly with free energies of formation of aryl boronate complexes with this enzyme [53]. The activity of boronates against class D β-lactamases has also been demonstrated [115].

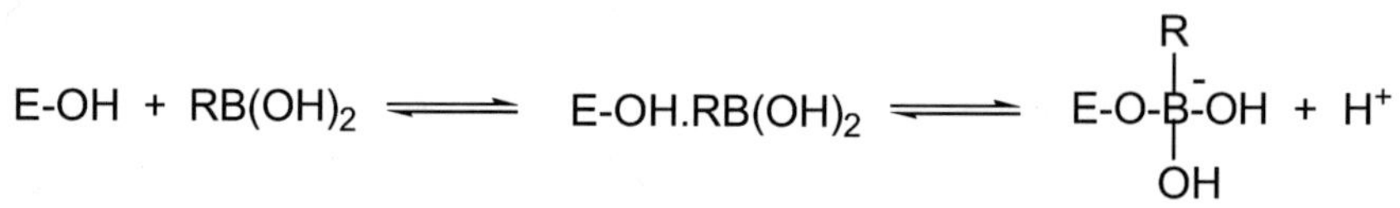

Scheme 2.

Structure-guided parallel synthesis of boronic acid inhibitors, based on the m-aminobenzeneboronic acid platform led to *48*, a 60 nM inhibitor of the AmpC β-lactamase, which also inhibited the class A TEM-1 enzyme. A crystal structure of the complex with the former enzyme showed interaction of the thiophene ring with Asn 161 and Tyr 221, and hydrophobic interactions between the terminal phenyl ring and leucines 119 and 293 [116]. Further design along these lines has been described [117] and arylazobenzeneboronic acids have been screened [118].

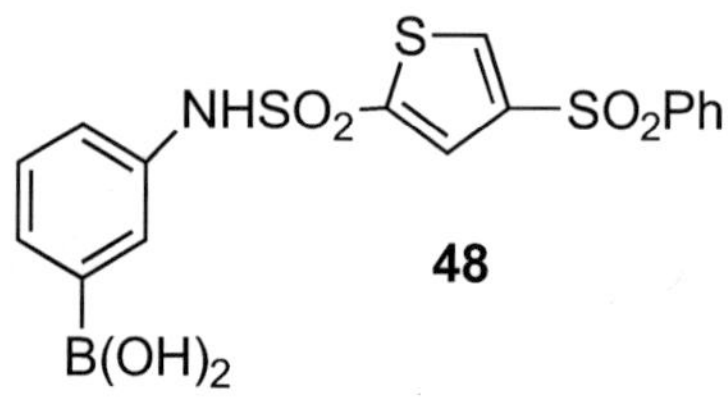

48

The incorporation of traditional β-lactam side chains, as in *49*, which contains the ceftazidime side chain, also led Shoichet et al. to powerful boronate inhibitors [119,120]. These were improved further by incorporation of carboxyphenyl groups, as in *50*, where, with R = thienylmethyl, a 1 nM inhibitor of the AmpC β-lactamase was generated [121].

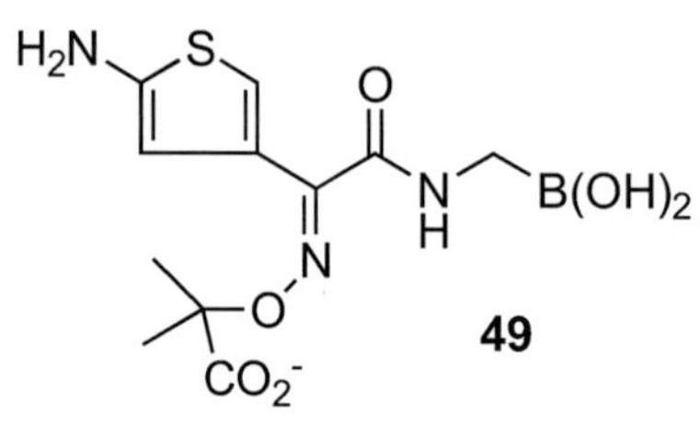

49

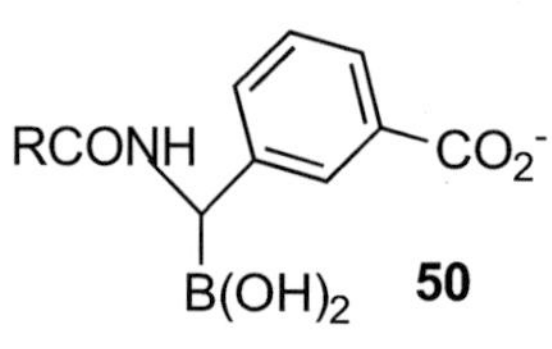

50

A crystal structure of the inhibited enzyme showed the phenyl ring in hydrophobic contact with Leu 119 and Leu 293, and the carboxylate hydrogen-bonding with the amido side chain of Asn 289. As Jones and coworkers had previously demonstrated [109, 110], these compounds were also powerful inhibitors of the TEM β-lactamase [122]. A crystal structure of the TEM complex showed the phenyl ring in contact with Tyr 105 and the carboxylate within hydrogen-bonding distance of Arg 244, Ser 235, and an ordered water molecule. Here too, the carboxylate position resembles those of bound β-lactams. Further design in this direction has been described recently [123, 124].

By a combination of computational and structural approaches, Shoichet et al. have thus nicely demonstrated how "hot spots" for inhibitor recognition can be discovered [125]. The effectiveness of specific boronates against inhibitor-resistant and extended spectrum class A β-lactamases has also been assessed [126, 127]. To date, only one boron-containing compound has been employed pharmaceutically (the anticancer drug Velcade). The reason for this is not entirely obvious. Boron in general, is not strongly toxic [128], although there are some specific compounds that are dangerous. It is possible that boronic acid derivatives may play a more dominant part in future pharmaceuticals than they have in the past [129, 130] and a β-lactamase inhibitor may well be next across the "boron barrier". Boronates have also been shown to inhibit certain DD-peptidases and thus may have antibiotic activity [131-133].

3. Carbonyl Compounds

A few aldehydes and electrophilic ketones have been designed and tested as β-lactamase inhibitors but the results to date have been unimpressive. The simple carbonyls *51-53* were ineffective as inhibitors of the P99 β-lactamase, for example, with K_i values

$PhCH_2CONHCH_2CHO$ **51**

$PhCH_2CONHCH_2COCO_2^-$ **52**

$PhCH_2CONHCH_2COCO_2Me$ **53**

well above 1 mM [134]. It is interesting to note that this result was predicted by a computational method that identified the electrostatic interactions between MOX^- (*54*) and the Lys 67 and Lys 315 cations as the dominant favorable interactions between the

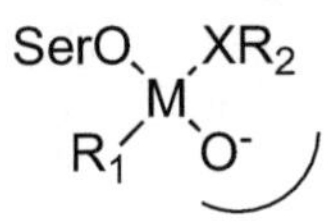

54

active site and simple tetrahedral transition state analogues [134]. A series of heterocyclic ketones *55* were also only modestly effective β-lactamase inhibitors [135]. Analysis of models of the tetrahedral adducts that would be formed, suggested that the heterocyclic rings of *55* would probably not interact with β-lactamase active site residues as favorably as those of analogous compounds with the active sites of serine proteases [136]. The favorable interaction of suitably placed phosph(on)ate groups with the β-lactamase active site [53, 89] suggested that ketophosph(on)ates of general structure *56* might hold promise.

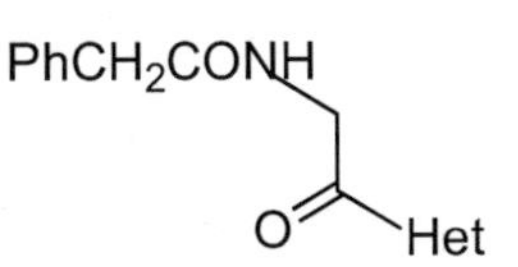

55: Het = 2-benzoxazolyl, 2-thiazolyl, 2-imidazolyl, 2-(4-carboxy)thiazolyl, 5-tetrazolyl

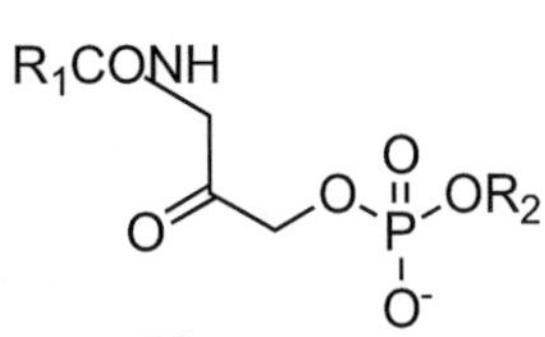

56

This was found to be true and revealed micromolar inhibitors of class C and D β-lactamases [137]. A series of comparably effective ketophosphonates of structure $ArCOCH_2PO_2^-(O)R/Ar$ have also been reported [138]. Both groups of these compounds, however, probably inhibit non-covalently. Note, however, the recent results with 6-halodeazapenicillins mentioned above. [18, 19]. The active site lysine residues of β-lactamases are susceptible to modification by certain reactive, but non-specific, aldehydes [139] but not by specific penicillin and cephalosporin aldehydes [140, 141].

4. Sulfonyl Compounds

Although β-lactam-mimetic β-sultams *57* have been known for some time, there was no evidence of activity as β-lactamase inhibitors among these compounds until recently when Page and coworkers introduced a new, more reactive structural variant, the N-acyl-β-sultams *58* [142].

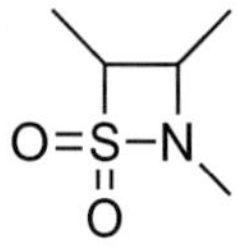

57

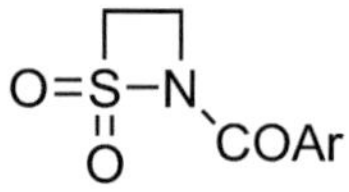

58

Acylic sulfonamides are very unreactive but can be strongly activated by incorporation into a four-membered ring and by an electron-withdrawing N-acyl substituent. Although, in a fashion resembling the acyl phosph(on)ates [53, 89], there are potentially two electrophilic centers in *58*, experiments to date suggest that nucleophiles, especially oxyanions, including the β-lactamase active site serine, react essentially exclusively at the sulfonyl center [142, 143]. The ring-opening reaction is thought to involve a penta-coordinated sulfur transition state/intermediate [144]. N-Acyl-β-sultams, unlike acyl phosph(on)ates, are also significant inhibitors of serine proteases [145]. The greater specificity of the latter compounds for β-lactamases probably results from the negative charge on the phosph(on)ates which has little affinity for most serine protease active sites.

Some unusual sulfonyloxadiazole inhibitors have been discovered by high throughput screening [8]. These compounds appear to covalently modify the active site serine residue of the AmpC β-lactamase by its nucleophilic attack on the oxadiazole ring, with a sulfinate as the leaving group.

5. Silanediols

Silanediols represent another opportunity to achieve an anionic tetrahedral transition state analogue at the active site of serine hydrolases (*59*). Inhibition of several proteases, for example, has been achieved by these compounds [146,147]. Their only tested compound, *60*, against β-lactamases, however proved inactive [134]. The calculations referred to above [134], suggested that the silanoate *59* would not interact strongly enough with positive charges at the class C β-lactamase active site to achieve significant inhibition.

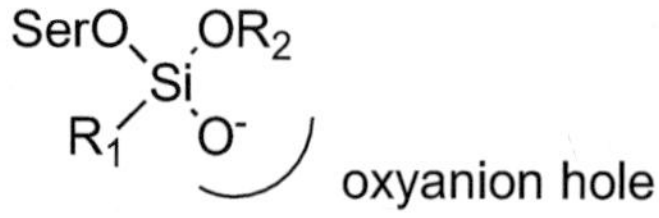

59

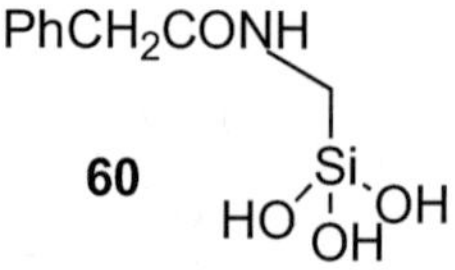

60

Despite the considerable effort, described above, devoted to the search for non-β-lactam inhibitors, there is no example yet fully developed for clinical application. Nonetheless, there are clearly a number of promising indications that, in the future, could lead to practical β-lactam synergists.

METALLO-β-LACTAMASES

The class B β-lactamases are zinc metallo-hydrolases. There are three (or perhaps only two [148]) subclasses, and crystal structures of representative enzymes of all types have been obtained; these are reviewed in detail in Chapter 3 of this volume. Although it is clear that at least one of the (two possible) zinc ions at the active site participates directly in catalysis, details of the mechanism of hydrolysis remain not universally accepted - see the discussion of mechanism in Chapter 7. Reflecting to some degree, perhaps, the absence of an established mechanism, no tight-binding transition-state analogue inhibitors have yet been described for the metallo-β-lactamases. Phosphon(amid)ates, for example, which are well-defined transition state analogue inhibitors of metallo-proteases, have little activity against class B β-lactamases [72, 149].

The thiophosphonate *61* was also inactive against *B. cereus* β-lactamase II (J. Rahil and R.F. Pratt, unpublished). Moderate inhibitory activity has been observed with some specific trifluoromethyl alcohols and ketones [150], but to date their mechanisms of action has not been examined.

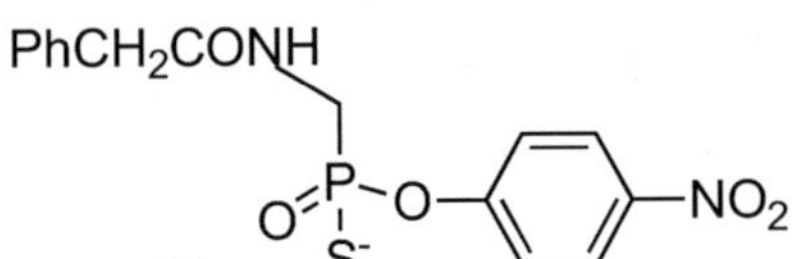

61

Design of metallo-β-lactamase inhibitors has therefore largely concentrated on molecules that directly bind to the active site zinc ion(s). Such molecules include thiols, carboxylates and dicarboxylates, hydroxamates, hydrazones, and various heterocycles. Care must be taken, of course, with such inhibitors since they may act simply by external chelation of zinc [151, 152]. All of the crystal structures published to date, however, of complexes of these inhibitors with class B β-lactamases do show the inhibitors bound at the active site and acting as a ligand of one or both of the zinc ions. Other polar functionality on the inhibitors tends to interact with polar residues at the active site and most structures show interactions, hydrophobic in particular, with the flexible loop or flap that lies adjacent to the active site.

Thiol carboxylates, including D-captropril, *62*, for example, inhibit the dizinc IMP-1, BlaB and VIM-2 β-lactamases, by displacement of the zinc-bridging hydroxyl group by thiolate [153-155]. This result was also observed with various dansylamido-thiols that were developed as fluorescent probes of the metallo-β-lactamase active site [156]. The succinate *63* inhibits by displacement of the bridge by one carboxylate oxygen while an oxygen of the second carboxylate displaces the water molecule from the coordination sphere of Zn2 of the IMP-1 enzyme [157]. There seems considerable interest in other dicarboxylate inhibitors, such as maleates [158] and phthalates [159] as general metallo-β-lactamase inhibitors.

Mercaptophosphonates also appear promising as broad-spectrum metallo-β-lactamase inhibitors [160].

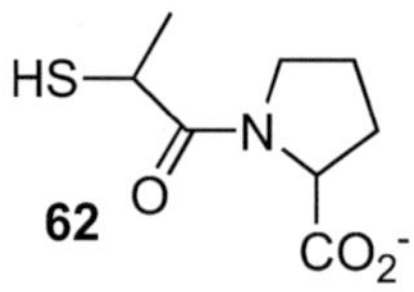

62

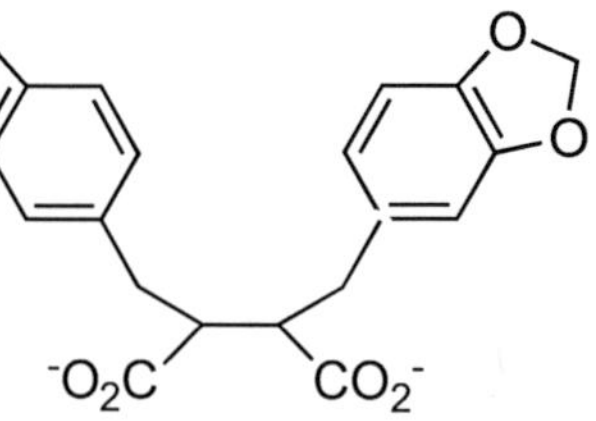

63

The carboxylate of *64* displaced the water molecule from Zn2 of the CfiA metallo-β-lactamase [161] while a tetrazole (anion) nitrogen of *65* had the same effect on a CcrA enzyme [162]. Pyridine carboxylates chelate the active zinc of the CphA β-lactamase [163] while a complex of the products formed on reaction between biapenem and this enzyme has the erstwhile β-lactam nitrogen and the thiazoline carboxylate directly bound to the active site zinc [164].

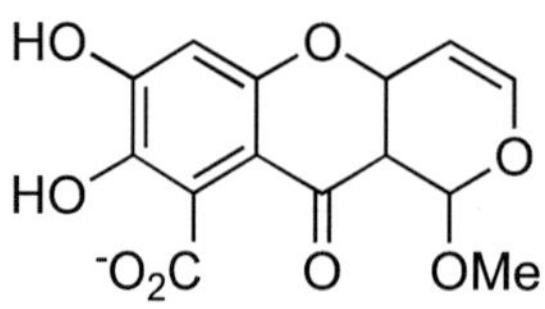

64

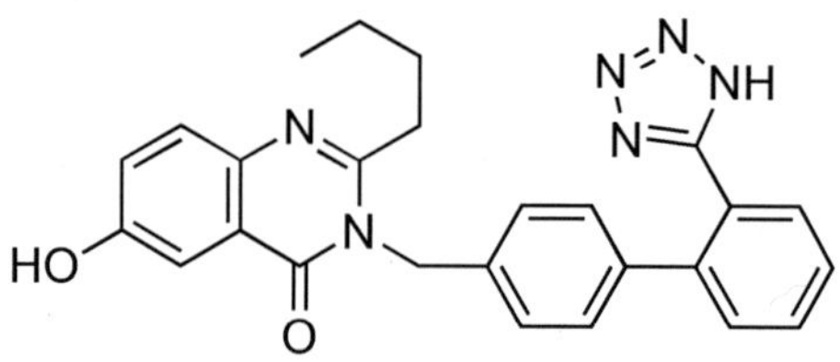

65

It seems likely that other strong inhibitors that are also potential zinc ligands bind in the fashion of those described above. Although neutral thiols such as *66* are micromolar inhibitors of certain class B β-lactamases [165, 166], those incorporating carboxylate groups as well seem generally more effective. The inhibition produced by neutral thiols, for example of the IMP-1 and *B. cereus* II enzymes, may be time dependent while that of thiol carboxylates is generally fast and reversible [167]. Even simple molecules such as mercaptoacetic and mercaptopropionic acids can be submicromolar inhibitors [168]. Cysteine peptides may also have significant inhibitory activity [169] and homocysteine peptides may be more effective under some circumstances [170]. Dynamic combinatorial libraries of thiols have been constructed to promote lead compound discovery [171]. Thiomandelic acid, another simple thiol carboxylate, appears to have a broader spectrum among class B enzymes than most of these compounds [172].

Bateson and coworkers of SmithKlineBeecham have patented a wide range of β-thiopropionoyl amino acid derivatives such as *67* (see references 173 and 174, for example), many of which are submicromolar inhibitors of a number of the class B enzymes. NMR studies of the binding of thiomandelic acid and *67* to relevant enzymes [172,175,176] support displacement of the zinc-bridging hydroxyl group by these thiols and their significant interaction with the active site loop. Motion of the loop and its relevance to inhibitor specificity have been explored by molecular dynamics simulations [177].

A variety of thiolesters, such as *68*, also possess significant metallo-β-lactamase inhibitory activity [178-181]. There may be more than one mode of inhibition by these compounds since they are also substrates of the β-lactamases [178] and thus the thiol leaving group, which may itself be an inhibitor, must be released to some degree during the interaction. In several cases, however, the thiolester was more effective as an inhibitor than the thiol and thus the thiolester itself must be an inhibitor; the converse, however has also been observed [181].

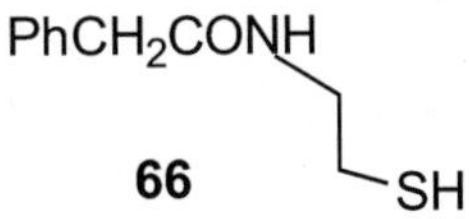

66

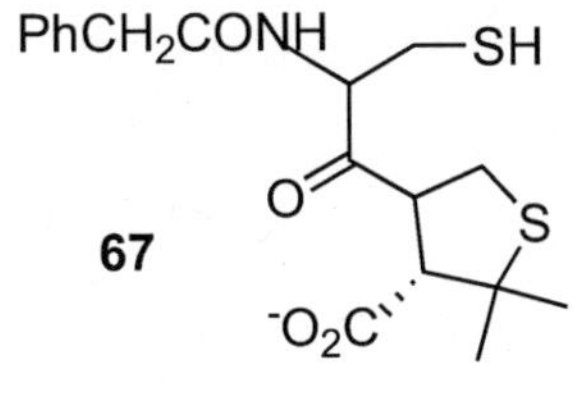

67

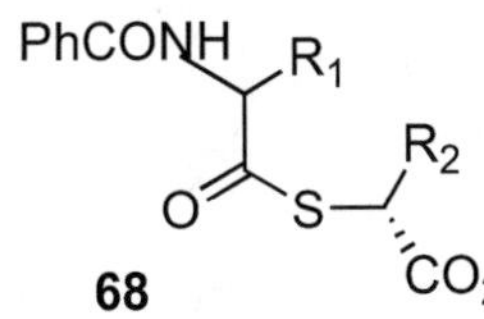

68

Inhibition by intact thiolesters appears to be fast, reversible and competitive [179]. In such cases, one might expect that the analogous oxo-esters would be equally effective, but this does not seem to have been examined. Thiols, if released by thiolester hydrolysis [178], could presumably inhibit by direct interaction with the active site zinc, as described above. There is also evidence, from one study [178], that in some cases inhibition by thiolesters may involve covalent modification of the enzyme, specifically by formation of a disulfide linkage between the thiol leaving group and the active site zinc cysteine ligand. The mechanism of this oxidative interaction does not seem to have been elucidated at this time. In other cases, however, covalent modification was not observed [179].

Recently, Kurosaki et al. designed a covalent IMP-1 inactivator based on the premise of a combination of a thiol to target the active site zinc ions and an electrophilic center to trap a nucleophilic residue adjacent to the active site [182]. The compounds *69* and *70* were thus prepared and both were found to irreversibly inactivate the IMP-1 β-lactamase. A crystal structure of the complex with *70* showed that the thiol has displaced the zinc-bridging hydroxyl group and the pentafluorophenyl ester had acylated Lys 224. This residue is conserved through most enzymes of the B1 subclass. This study seems to represent the first example of a class B β-lactamase inhibitor that covalently modifies a specific active site residue.

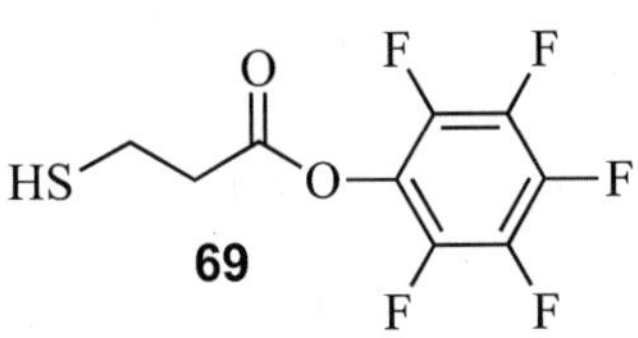

69

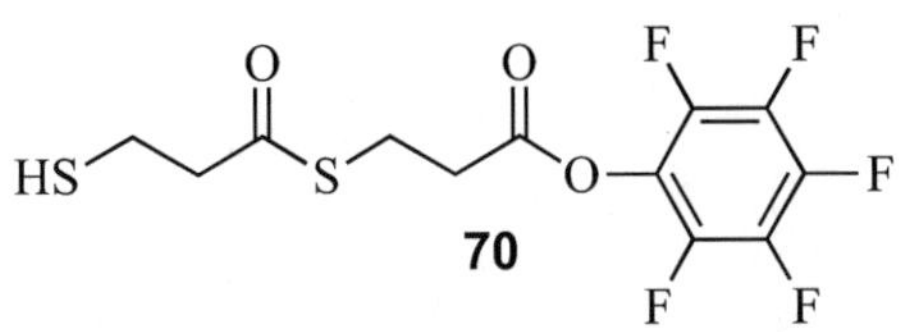

70

Other inhibitors described in the literature but not yet developed beyond intitial communication include hydroxamates, *71* [183, 184], N-arylsulfonylhydrazones, *72* [185], arylsulfonylamido-1,2,3-triazoles [186], bulgecin [187], and various β-lactam reaction products [188-190]. These may all inhibit by forming complexes with the active site zinc ions. There have also been reports of screening for Class B β-lactamase inhibitors. The screening of natural products led to the discovery of phenazine inhibitors, although the evidence suggests that these act by external zinc chelation [152]. The NCI chemical diversity set yielded some novel succinates and the phenol *73* [191], and peptide libraries yielded CysValHisSerProAsnArgGlnCys as a micromolar inhibitor of several class B β-lactamases [192]. These are also intriguing recent reports of RNA and DNA aptamer based inhibitors [193-195].

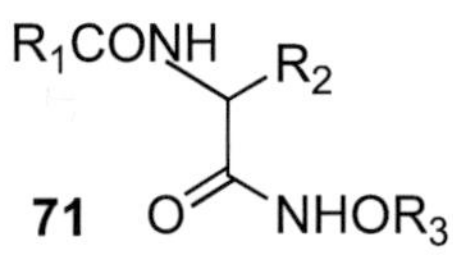

71

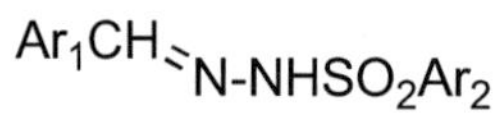

72

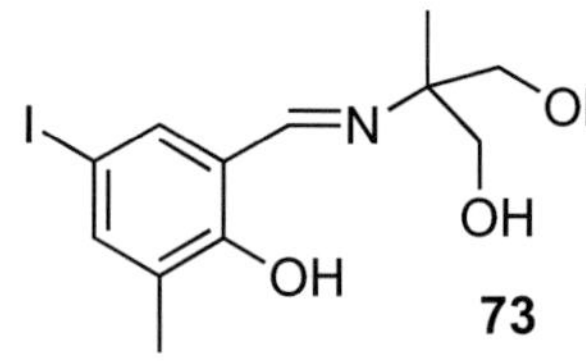

73

Thus, to date, no broadly specific and potent class B β-lactamase inhibitors have been discovered, or none of sufficient quality to be developed for medical application.

There has been, however, over the last decade, an impressively large amount of basic research on these enzymes on which to base future inhibitor design. It does seem that reasonably broad inhibitors may yet be discovered, by application of mass spectral screening and structure-based design, for example [196, 197].

Although the zinc ligands appear to hold the most promise at present, their non-specific metal-binding effects would have to be carefully evaluated in any future development. It may be that β-lactam based inhibitors will yet prevail (see chapter 9).

OVERVIEW

The prospects for new β-lactamase inhibitors to assist established β-lactams against bacterial infections appear to be good. Certainly, it seems at present that, in the immediate and middle distant future, the bolstering of β-lactams with inhibitors may be a better bet than waiting for new targets to be developed [198, 199]. At present, however, despite threats from extended spectrum and (classical) inhibitor-resistant β-lactamases, most major pharmaceutical companies appear to be resting on their laurels (clavulanic acid, sulbactam, and tazobactam).

It is generally believed that this attitude derives from the lower return on investment from antibacterials than from a variety of chronic diseases [200, 201], and thus one might anticipate that the early development of new antibacterials, including β-lactamase inhibitors, will come from smaller companies.

As described in detail above, there are a number of both β-lactam (see chapter 9) and non-β-lactam systems that are promising leads. Development of these leads will probably involve mainly the now-traditional combinations of medicinal chemistry and computational methods. The threat posed by the class B metallo-β-lactamases is probably less ready to be met at present. More screening may be necessary to produce novel leads for these enzymes, but it still may be, as for the serine β-lactamases of classes A, C and D, that transition state analogues and mechanism-based inhibitors will be more effective in the long run and thus these approaches need more serious attention at this time.

ACKNOWLEDGMENTS

The author of this chapter is grateful to the National Institutes of Health, U.S.A. and Wesleyan University for support of his research.

REFERENCES

[1] Chaibi, E.B., Sirot, D., Paul, G., Labia, R. (1999) Inhibitor-resistant TEM β-lactamases: phenotypic, genetic and biochemical characteristics. *J. Antimicrob. Chemother.* 43, 447-458.

[2] Georgopapadakou, N. (2004) □-Lactamase inhibitors: evolving compounds for evolving resistance targets. *Expert Opin. Invest. Drugs* 13, 1307-1318.

[3] Tipper, D.J., Strominger, J.L. (1965) Mechanism of action of penicillins: a proposal based on their structural similarity to acyl-D-alanyl-D-alanine. *Proc. Natl. Acad. Sci.* USA 54, 1133-1141.

[4] Knox, J.R., Moews, P.C., Frère J.-M. (1996) Molecular evolution of bacterial β-lactam resistance. *Chem. & Biol.* 3, 937-947.

[5] Pratt, R.F., Govardhan, C.P. (1984) β-Lactamase-catalyzed hydrolysis of acyclic depsipeptides and acyl transfer to specific amino-acid acceptors. *Proc. Natl. Acad. Sci. USA* 81, 1302-1306.

[6] Grant, E.B., Guiadeen, D., Baum, E.Z., Foleno, B.D., Jin, H., Montenegro, D.A., Nelson, E.A., Bush, K., Hlasta D.J. (2000) Mechanism of action of penicillins: a proposal based on their structural similarity to acyl-D-alanyl-D-alanine. *Bioorg. Med. Chem. Lett. 10*, 2179-2182.

[7] Zervosen, A., Lu, W.-P., Chen, Z., White, R.E., Demuth, T.P., Jr., Frère, J.-M. (2004) Interactions between penicillin-binding proteins (PBPs) and two novel classes of PBP inhibitors, arylalkylidene rhodanines and arylalkylidene iminothiazolidin-4-ones. *Antimicrob. Agents Chemother.* 48, 961-969.

[8] Babaoglu, K., Simeonov, A., Irwin, J. J., Nelson, M. E., Feng, B., Thomas, C. J., Cancian, L., Costi, M. P., Maltby, D. A., Jadhav, A., et al. (2008) Comprehensive mechanistic analysis of hits from high-throughput and docking screens against β-lactamase. *J. Med. Chem.* 51, 2502-2511.

[9] Coates, N.J., Gilpin, M.L., Gwynn, M.N., Lewis, D.E., Milner, P.H., Spear, S.R., Tyler, J.W. (1994) SB-202742, a novel β-lactamase inhibitor isolated from *Spondias mombin. J. Nat. Prod.* 5, 654-657.

[10] Hamilton-Miller, J.M.T., Shah, S. (2000) Activity of the tea component epicatechin gallate and analogues against methicillin-resistant *Staphylococcus aureus. J. Antimicrob. Chemother.* 46, 852-853.

[11] Hird, N.W., Milner, P.H. (1994) Synthesis and β-lactamase inhibition of anacardic acids and their analogs. *Bioorg. Med. Chem. Lett.* 12, 1423-1428.

[12] Muroi, H., Nihei, K., Tsujimoto, K., Kubo, I. (2004) Synergistic effects of anacardic acids and methicillin against methicillin resistant *Staphylococcus aureus*. *Bioorg. Med. Chem.* 12, 583-587.

[13] Bush, K., Bonner, D.P., Sykes, R.B. (1980) Izumenolide - a novel β-lactamase inhibitor produced by *Micromonospora*. II. Biological properties. *J. Antibiot.* 33, 1262-1269.

[14] Bush, K., Freudenberger, J., Sykes, R.B. (1980) Inhibition of *Escherichia coli* TEM-2 β-lactamase by the sulfated compounds izumenolide, panosialin and sodium dodecyl sulfate. *J. Antibiot.* 33, 1560-1562.

[15] Zhao, W.-H., Hu, Z.-Q., Hara, Y., Shimamura, T. (2002) Inhibition of penicillinase by epigallocatechin gallate resulting in restoration of antibacterial activity of penicillin against penicillinase-producing *Staphylococcus aureus*. *Antimicrob. Agents Chemother.* 46, 2266-2268.

[16] Zhao, W.-H., Asano, N., Hu, Z.-Q., Shimamura, T. (2003) Restoration of antibacterial activity of β-lactams by epigallocatechin gallate against β-lactamase-producing species depending on location of β-lactamase. *J. Pharm. Pharmacol.* 55, 735-740.

[17] Kubo, I., Fujita, K., Nihei, K. (2003) Molecular design of multifunctional antibacterial agents against methicillin resistant *Staphylococcus aureus*. *Bioorg. Med. Chem.* 11, 4255-4262.

[18] Ferguson, A.C., Adlington, R.M., Martyres, D.H., Rutledge, P.J., Cowley, A., Baldwin, J.E. (2003) Total synthesis of a novel 2-thiabicyclo[3.2.0]heptan-6-one analogue of penicillin N. *Tetrahedron* 59, 8233-8243.

[19] Johnson, J.R., Gretes, M., Goodfellow, V.J., Marrone, L., Heynen, M.L., Strynadka, N.C.J., Dmitrienko, G.I. (2010) Cyclobutanone analogues of β-lactams revisited: Insights into conformational requirements for inhibition of serine and metallo-β-lactamases. *J. Am. Chem. Soc.* 132, 2558-2560.

[20] Powers, R.A., Morandi, F., Shoichet, B.K. (2002) Structure-based discovery of a novel, non-covalent inhibitor of AmpC β-lactamase. *Structure* 10, 1013-1023.

[21] Tondi, D., Morandi, F., Bonnet, R., Costi, M.P., Shoichet, B.K. (2005) Structure-based optimization of a non-β-lactam lead results in inhibitors that do not up-regulate β-lactamase expression in cell culture. *J. Amer. Chem. Soc.* 127, 4632-4639.

[22] Teotico, D. G., Babaoglu, K., Rocklin, G. J., Ferreira, R. S., Giannetti, A. M., Shoichet, B. K. (2009) Docking for fragment inhibitors of AmpC β-lactamase. *Proc. Natl. Acad. Sci.* U.S.A. 106, 7455-7460.

[23] Strynadka, N.C.J., Jensen, S.E., Alzari, P.M., James M.N.G. (1996) A potent new mode of β-lactamase inhibition revealed by the 1.7 A X-ray crystallographic structure of the TEM-1-BLIP complex. *Nat. Struct. Biol.* 3, 290-297.

[24] Lim, D., Park, H.U., DeCastro, L., Kang, S.G., Lee, H.S., Jensen, S., Lee, K.J., Strynadka, N.C.J. (2001) Crystal structure and kinetic analysis of β-lactamase inhibitor protein-II in complex with TEM-1 β-lactamase. *Nat. Struct. Biol.* 8, 848-852.

[25] Strynadka, N.C.J., Adachi, H., Jensen, S.E., Johns, K., Sielecki, A., Betzel, C., Sutoh, K., James, M.N.G. (1992) Molecular structure of the acyl-enzyme intermediate in β-lactam hydrolysis at 1.7 ANG. resolution. *Nature* 359, 700-705.

[26] Petrosino, J., Rudgers, G., Gilbert, H., Palzkill, T. (1999) Contributions of aspartate 49 and phenylalanine 142 residues of a tight binding inhibitory protein of β-lactamases. *J. Biol. Chem.* 274, 2394-2400.

[27] Rudgers, G.W., Huang, W., Palzkill, T. (2001) Binding properties of a peptide derived from β-lactamase inhibitory protein. *Antimicrob. Agents Chemother.* 45, 3279-3286.

[28] Huang, W., Beharry, Z., Zhang, Z., Palzkill, T. (2003) A broad-spectrum peptide inhibitor of β-lactamase identified using phage display and peptide arrays. *Protein Eng.* 11, 853-860.

[29] Zhang, Z., Palzkill, T. (2003) Determinants of binding affinity and specificity for the interaction of TEM-1 and SME-1 β-lactamase with β-lactamase inhibitory protein. *J. Biol. Chem.* 278, 45706-45712.

[30] Conrath, K.E., Lauwereys, M., Galleni, M., Matagne, A., Frère, J.-M., Kinne, J., Wyns, L., Muyldermans, S. (2001) β-Lactamase inhibitors derived from single-domain antibody fragments elicited in the *Camelidae. Antimicrob. Agents Chemother.* 45, 2807-2812.

[31] Jungheim, L.N., Ternansky, R.J. (1992) Non-β-lactam mimics of β-lactam antibiotics. in The Chemistry of Beta-Lactams, Page, M.I., Ed., Chapman and Hall, London, Ch 9.

[32] Brown, T., Jr., Charlier, P., Herman, R., Schofield, C.J., Sauvage, E. (2010) Structural basis for the interaction of lactivicins with serine β-lactamases. *J. Med. Chem.* 53, 5890-5894.

[33] Hadfield, P.S., Galt, R.H.B., Sawyer, Y., Layland, N.J., Page, M.I. (1997) An unexpected Mitsunobu reaction. A direct route to the 2,5-diazabicyclo[2.2.1]heptan-3-one skeleton as a γ-lactam mimic of β-lactam antibiotics. *J. Chem. Soc. Perkin* 1 503-509.

[34] Aszodi, J., Rowlands, D.A., Mauvais, P., Collette, P., Bonnefoy, A., Lampilas, M. (2004) Design and synthesis of bridged γ-lactams as analogues of β-lactam antibiotics. *Bioorg. Med. Chem. Lett.* 14, 2489-2492.

[35] Cohen, N.C. (1983) β-Lactam antibiotics: geometrical requirements for antibacterial activities. *J. Med. Chem.* 26, 259-264.

[36] Bonnefoy, A., Dupuis-Hamelin, C., Steier, V., Delachaume, C., Seys, C., Stachyra, T., Fairley, M., Guitton, M., Lampilas, M. (2004) In vitro activity of AVE1330A, an innovative broad-spectrum non-β-lactam β-lactamase inhibitor. *J. Antimicrob. Chemother.* 54, 410-417.

[37] Livermore, D. M., Mushtaq, S., Warner, M., Mioussec, C., Woodford, N. (2008) NXL 104 combinations versus *Enterobacteriaceae* with CTX-M extended-spectrum β-lactamases and carbapenemases. *J. Antimicrob. Chemother.* 62, 1053-1056.

[38] Endimiani, A., Choudhary, Y., Bonomo, R.A. (2009) In vitro activity of NXL104 in combination with β-lactams against *Klebsiella pneumoniae* isolates producing KPC carbapenamases. *Antimicrob. Agents Chemother.* 53, 3599-3601.

[39] Stachyra, T., Levasseur, P., Pechereau, M.-C., Girard, A.-M., Claudon, M., Miossec, C., Black, M.T. (2009) In vitro activity of the β-lactamase inhibitor NXL104 against KPC-2 carbapenemase and *Enterobacteriaciae* expressing KPC carbapenemases. *J. Antimicrob. Chemother.* 64, 326-329.

[40] Stachyra, T., Péchereau, M.-C., Bruneau, J.-M., Claudon, M., Frère J.-M., Miossec, C., Coleman, K., Black, M.T. (2010) Mechanistic studies of the inactivation of TEM-1 and P99 by NXL104, a novel non-β-lactam β-lactamase inhibitor. *Antimicrob. Agents Chemother.* 54, 5132-5138.

[41] Govardhan, C.P., Pratt, R.F. (1987) Kinetics and mechanism of the serine β-lactamase catalyzed hydrolysis of depsipeptides. *Biochemistry* 26, 3385-3395.

[42] Xu, Y., Soto, G., Hirsch, K.R., Pratt, R.F. (1996) Kinetics and mechanism of the hydrolysis of depsipeptides catalyzed by the β-lactamase of *Enterobacter cloacae* P99. *Biochemistry* 35, 3595-3603.

[43] Adediran, S.A., Cabaret, D., Lohier, J.-F., Wakselman, M., Pratt, R.F. (2010) Substituted aryl malonamates as new serine β-lactamase substrates: Structure-activity studies. *Bioorg. Med. Chem.* 18, 282-291.

[44] Adam, M., Damblon, C., Plaitin, B., Christiaens, L., Frère J.-M. (1990) Chromogenic depsipeptide substrates for β-lactamases and penicillin-sensitive DD-peptidases. *Biochem. J.* 270, 525-529.

[45] Xu, Y., Soto, G., Adachi, H., van der Linden, M.P.G., Keck, W., Pratt, R.F. (1994) Relative specificities of a series of β-lactam-recognizing enzymes towards the side-chains of penicillins and of acyclic thioldepsipeptides *Biochem. J.* 302, 851-856.

[46] Xu, Y., Pratt, R.F. (1994) β-Lactam-recognizing enzymes exhibit different structural specificity in acyclic amide and ester substrates: a starting point in β-lactamase evolution? *Bioorg. Med. Chem. Lett.* 4, 2291-2296.

[47] Bernstein, N.J., Pratt, R.F. (1999) On the importance of a methyl group in β-lactamase evolution: Free energy profiles and molecular modeling. *Biochemistry* 32, 10499-10510.

[48] Adediran, S.A., Deraniyagala, S.A., Xu, Y., Pratt, R.F. (1996) β-Secondary and solvent deuterium kinetic isotope effects on β-lactamase catalysis. *Biochemistry* 35, 3604-3613.

[49] Adediran, S.A., Pratt, R.F. (1999) β-Secondary and solvent deuterium kinetic isotope effects on catalysis by the *Streptomyces* R61 DD-peptidase: Comparisons with a structurally similar class C β-lactamase. *Biochemistry* 38, 1469-1477.

[50] Adediran, S.A., Zhang, Z., Nukaga, M., Palzkill, T., Pratt, R.F. (2005) The D-methyl group in β-lactamase evolution: Evidence from the Y221G and GC1 mutants of the class C β-lactamase of *Enterobacter cloacae* P99. *Biochemistry* 44, 7543-7552.

[51] Morrison, M.J., Li, N., Pratt, R.F. (2001) Inverse acyl phosph(on)ates: Substrates or inhibitors of β-lactam-recognizing enzymes? *Bioorg. Chem.* 29, 271-281.

[52] Ahn, Y-M., Pratt, R.F. (2004) Kinetic and structural consequences of the leaving group in substrates of a class C β-lactamase. *Bioorg. Med. Chem.* 12, 1537-1542.

[53] Kaur, K., Pratt, R.F. (2001) Mechanism of reaction of acyl phosph(on)ates with the β-lactamase of *Enterobacter cloacae* P99. *Biochemistry* 40, 4610-4621.

[54] Knott-Hunziker, V., Petursson, S., Waley, S.G., Jaurin, B., Grundstrom, T. (1982) The acyl-enzyme mechanism of β-lactamase action. The evidence for class C β-lactamases. *Biochem. J.* 207, 315-322.

[55] Majumdar, S., Adediran, S.A., Nukaga, M., Pratt, R.F. (2005) Inhibition of class D β-lactamases by diaroyl phosphates. *Biochemistry* 44, 16121-16129.

[56] Pazhanisamy, S., Govardhan, C.P., Pratt, R.F. (1989) β-Lactamase-catalyzed aminolysis of depsipeptides: amine specificity and steady-state kinetics. *Biochemistry* 28, 6863-6870.

[57] Pazhanisamy, S., Pratt, R.F. (1989) β-Lactamase-catalyzed aminolysis of depsipeptides: proof of the nonexistence of a specific D-phenylalanine/enzyme complex by double-label isotope trapping. *Biochemistry* 28, 6870-6875.

[58] Pazhanisamy, S., Pratt, R.F. (1989) β-Lactamase-catalyzed aminolysis of depsipeptides: peptide inhibition and a new kinetic mechanism. *Biochemistry* 28, 6875-6882.

[59] Cabaret, D., Adediran, S.A., Pratt, R.F., Wakselman, M. (2003) New substrates for β-lactam-recognizing enzymes: Aryl malonamates. *Biochemistry* 42, 6719-6725.

[60] Murphy, B.P., Pratt, R.F. (1991) N-(Phenylacetyl)glycyl-D-aziridine-2-carboxylate, an acylic amide substrate of β-lactamases: importance of the shape of the substrate in β-lactamase evolution. *Biochemistry* 30, 3640-3649.

[61] Cabaret, D., Liu, J., Wakselman, M., Pratt, R.F., Xu, Y. (1994) Functionalized depsipeptides, substrates and inhibitors of β-lactamases and DD-peptidases. *Bioorg. Med. Chem.* 2, 757-771.

[62] Nukaga, M., Kumar, S., Nukaga, K., Pratt, R.F., Knox, J.R. (2004) Hydrolysis of Third-generation cephalosporins by class C β-lactamases: structures of a transition state analog of cefotaxime in wild-type and extended spectrum enzymes. *J. Biol. Chem.* 270, 9344-9352.

[63] Adediran, S.A., Cabaret, D., Flavell, R.R., Sammons, J.A., Wakselman, M., Pratt, R.F. (2006) Synthesis and β-lactamase reactivity of α-substituted phenaceturates. *Bioorg. Med. Chem.* 14, 7023-7033.

[64] Cabaret, D., Adediran, S.A., Garcia Gonzalez, M.J., Pratt R.F., Wakselman M. (1999) Synthesis and reactivity with β-lactamases of "penicillin-like" cyclic depsipeptides. *J. Org. Chem.* 64, 713-720.

[65] Adediran, S.A., Cabaret, D., Pratt, R.F., Wakselman, M. (1999) A "cephalosporin-like" cyclic depsipeptide: synthesis and reaction with β-lactam-recognizing enzymes. *Bioorg. Med. Chem. Lett.* 9, 341-346.

[66] Adediran, S.A., Cabaret, D., Drouillat, B., Pratt, R.F., Wakselman, M. (2001) The synthesis and evaluation of benzofuranones as β-lactamase substrates. *Bioorg. Med. Chem.* 9, 1175-1183.

[67] Cao, X., Igbal, A., Patel, A., Gretz, P., Huang, G., Crowder, M., Day, R.A. (2003) 3-Alkoxy-5-isoxazolidinones mimic β-lactams. *Biochem. Biophys. Res. Commun.* 311, 267-271.

[68] Kraut, J. (1977) Serine proteases: structure and mechanism of catalysis. *Annu. Rev. Biochem.* 46, 333-358.

[69] Pain, R.H., Virden, R. (1979) *The Structural and Conformational Basis of β-Lactamase Activity. in Beta-Lactamases,* Hamilton-Miller, J.M.T., Smith, J.T., Eds. Academic Press, Ch. 7.

[70] Herzberg, O., Moult, J. (1987) Bacterial resistance to β-lactam antibiotics: crystal structure of β-lactamase from *Staphylococcus aureus* PC1 at 2.5 ANG. resolution. *Science,* 236, 694-701.

[71] Knap, A., Pratt, R.F. (1991) Inactivation of the RTEM-1 cysteine β-lactamase by iodoacetate. The nature of active-site functional groups and comparisons with the native enzyme. *Biochem. J.* 273, 85-91.

[72] Pratt, R.F. (1989) Inhibition of a class-C β-lactamase by a specific phosphonate monoester. *Science* 246, 917-919.

[73] Rahil, J., Pratt, R.F. (1992) Mechanism of inhibition of the class C β-lactamase of *Enterobacter cloacae* P99 by phosphonate monoesters. *Biochemistry* 31, 5869-5878.

[74] Rahil, J., Pratt, R.F. (1993) Structure-activity relationships in the inhibition of serine β-lactamases by phosphonic acid derivatives. *Biochem. J.* 296, 389-393.

[75] Li, N., Rahil, J., Wright, M., Pratt, R.F. (1997) Structure-activity studies of the inhibition of serine β-lactamases by phosphonate monoesters. *Bioorg. Med. Chem.* 5, 1783-1788.

[76] Rahil, J., Pratt, R.F. (1991) Phosphonate monoester inhibitors of class A β-lactamases. *Biochem. J.* 275, 793-795.

[77] Adediran, S.A., Nukaga, M., Baurin, S., Frère, J.-M., Pratt, R.F. (2005) Inhibition of class D β-lactamases by acyl phosphates and phosphonates. *Antimicrob. Agents Chemother.* 49, 4410-4412.

[78] Chen, C.C.H., Rahil, J., Pratt, R.F., Herzberg, O. (1993) Structure of a phosphonate-inhibited β-lactamase. An analog of the tetrahedral transition state/intermediate of β-lactam hydrolysis. *J. Mol. Biol.* 234, 165-178.

[79] Lobkovsky, E., Billings, E.M., Moews, P.C., Rahil, J., Pratt, R.F., Knox, J.R. (1994) Crystallographic structure of a phosphonate derivative of the *Enterobacter cloacae* P99 cephalosporinase: Mechanistic interpretation of a β-lactamase transition-state analog. *Biochemistry* 33, 6762-6772.

[80] Maveyraud, L., Pratt, R.F., Samama, J.-P. (1998) Crystal structure of an acylation transition-state analog of the TEM-1 β-lactamase. Mechanistic implications for class A β-lactamases. *Biochemistry* 37, 2622-2628.

[81] Rahil, J., Pratt, R.F. (1994) Characterization of covalently bound enzyme-inhibitors as transition-state analogs by protein stability measurements - Phosphonate monoester inhibitors of a β-lactamase. *Biochemistry* 33, 116-125.

[82] Nagarajan, R., Pratt, R.F. (2004) Thermodynamic evaluation of a covalently bonded transition state analogue inhibitor: Inhibition of β-lactamases by phosphonates. *Biochemistry* 43, 9664-9673.

[83] Besterman, J.M., Delorme, D., Rahil, J. (2001) Sulfonamidomethyl phosphonate inhibitors of β-lactamase. Patent WO 2000-US18344, *Chem. Abst.* 134, 95480.

[84] Rahil, J., Pratt, R.F. (1993) Kinetics and mechanism of β-lactamase inhibition by phosphonamidates: The quest for a proton. *Biochemistry* 32, 10763-10772.

[85] Bateson, J.H., Gasson, B.C., Khushi, T., Neale, J.E., Payne, D.J., Tolson, D.A., Walker, G. (1994) The synthesis and serine β-lactamase inhibitory activity of some phosphonamidate analog of dipeptides. *Bioorg. Med. Chem. Lett.* 4, 1667-1672.

[86] Laws, A.P., Page, M.I., Slater, M.J. (1993) The mechanism of reactions catalyzed by the serine β-lactamases. *Bioorg. Med. Chem. Lett.* 11, 2317-2322.

[87] Di Ninno, F., Hammond, M. L., Dykstra, K., Kim, S., Tan, Q., Young, K., Hermes, J. D., Chen, H., Raeppel, S., Mannion, M., Zhou, N. Z., et al. (2007) Sulfonylamidomethylphosphonates as inhibitors of beta-lactamases for decreasing of bacterial resistance to antibiotics. PCT Int. Appl. WO 2007139729.

[88] Di Ninno, F., Hammond, M. L., Dykstra, K., Kim, S., Tan, Q., Young, K., Hermes, J. D., Chen, H., Raeppel, S., Mannion, M., et al. (2008) *(Hetero)arylsulfonamidomethylphosphonate derivatives as β-lactamase inhibitors and their preparation, pharmaceutical compositions and use in the treatment of bacterial infection.* PCT Int. Appl. WO 2008073142.

[89] Li, N., Pratt, R.F. (1998) Inhibition of serine β-lactamases by acyl phosph(on)ates: A new source of inert acyl [and phosph(on)yl] enzymes. *J. Am. Chem. Soc.* 120, 4264-4268.

[90] Pratt, R.F., Hammar, N.J. (1998) Salicyloyl cyclic phosphate, a "penicillin-like" inhibitor of β-lactamases. *J. Am. Chem. Soc.* 120, 3004-3006.

[91] Kaur, K., Lan, M.J.K., Pratt, R.F. (2001) Mechanism of inhibition of the class C β-lactamase of *Enterobacter cloacae* P99 by cyclic acyl phosph(on)ates: Rescue by return. *J. Am. Chem. Soc.* 123, 10436-10443.

[92] Kaur, K., Adediran, S.A., Lan, M.J.K., Pratt, R.F. (2003) Inhibition of β-lactamases by monocyclic acyl phosph(on)ates. *Biochemistry* 42, 1529-1536.

[93] Silvaggi, N., Kaur, K., Adediran, S.A., Pratt, R.F., Kelly, J.A. (2004) Toward better antibiotics: Crystallographic studies of a novel class of DD-peptidase/β-lactamase inhibitors. *Biochemistry* 43, 7046-7053.

[94] Majumdar, S., Pratt, R.F. (2009) Inhibition of class A and class C β-lactamases by diaroyl phosphates. *Biochemistry* 48, 8285-8292.

[95] Majumdar, S., Pratt, R.F. (2009) Intramolecular cooperativity in the reaction of diacyl phosphates with serine β-lactamases. *Biochemistry* 48, 8293-8298.

[96] Slater, M.J., Laws, A.P., Page, M.I. (2001) The relative catalytic efficiency of beta-lactamase catalyzed acyl and phosphyl transfer. *Bioorg. Chem.* 29, 77-95.

[97] Bell, J.H., Curley, K., Pratt, R.F. (2000) Inhibition of serine amidohydrolases by complexes of vanadate with hydroxamic acids. *Biochem. Biophys. Res. Commun.* 274, 732-735.

[98] Bell, J.H., Pratt, R.F. (2002) Mechanism of inhibition of the β-lactamase of *Enterobacter cloacae* P99 by 1:1 complexes of vanadate with hydroxamic acids. *Biochemistry,* 41, 4329-4338.

[99] Moulin, A., Bell, J.H., Pratt, R.F., Ringe, D. (2007) Inhibition of chymotrypsin by a complex of ortho-vanadate and benzohydroxamic acid: structure of the inert complex and its mechanistic interpretation. *Biochemistry* 46, 5982-5990.

[100] Adediran, S. A., Pratt, R. F. (2008) Inhibition of serine β-lactamses by vanadate-catechol complexes. *Biochemistry* 47, 9467-9474.

[101] Cabaret, D., Garcia Gonzalez, M., Wakselman, M., Adediran, S.A., Pratt, R.F. (2001) Synthesis, hydrolysis, and evaluation of 3-acylamino-3,4-dihydro-2-oxo-2*H*-1,3-benzoxazinecarboxylic acids and linear azadepsipeptides as potential substrates/inhibitors of β-lactam-recognizing enzymes. *Eur. J. Org. Chem.* 141-149.

[102] Wyrembak, P.N., Babaoglu, K., Pelto, R.B., Shoichet, B.K. and Pratt, R.F. (2007) O-Aryloxycarbonyl hydroxamates: new β-lactamase inhibitors that cross-link the active site. *J. Am. Chem. Soc.* 129, 9548-9549.

[103] Pelto, R. B., Pratt, R. F. (2008) Kinetics and mechanism of inhibition of a serine β-lactamase by O-aryloxycarbonyl hydroxamates. *Biochemistry* 47, 12037-12046.

[104] Dobozy, O., Mile, I., Ferencz, I., Csanyi, V. (1971) Effect of electrolytes on the activity and iodine sensitivity of penicillinase from *B. cereus*. *Acta Biochem. Biophys. Acad. Sci. Hung.* 6, 97-105.

[105] Kiener, P.A., Waley, S.G. (1978) Reversible inhibitors of penicillinases. *Biochem. J.* 169, 197-204.

[106] Beesley, T., Gascoyne, N., Knott-Hunziker, V., Petursson, S., Waley, S.G., Jaurin, B., Grundstrom, T. (1983) The inhibition of class C β-lactamases by boronic acids. *Biochem. J.* 209, 229-233.

[107] Crompton, I.E., Cuthbert, B.K., Lowe, G., Waley, S.G. (1988) β-Lactamase inhibitors. The inhibition of serine β-lactamases by specific boronic acids. *Biochem. J.* 251, 453-459.

[108] Then, R.L., Hubschwerlen, C., Charnas, R. (1991) Antagonism between some novel boronic acids and cephalosporins in inducible strains of *Enterobacter cloacae.* in *50 Years Of Penicillin Application,* Kleinkauf, H., von Dohren, H., Eds., Public Ltd, Czech Republic, pp 383-387.

[109] Martin, R., Jones, J.B. (1995) Rational design and synthesis of a highly effective transition state analog inhibitor of the RTEM-1 β-lactamase. *Tetrahedron Lett.* 46, 8399-8402.

[110] Strynadka, N.C.J., Martin, R., Jensen, S.E., Gold, M., Jones, J.B. (1996) Structure-based design of a potent transition state analogue for TEM-1 β-lactamase. *Nat. Struct. Biol.* 3, 688-695.

[111] Ness, S., Martin, R., Kindler, A.M., Paetzel, M., Gold, M., Jensen, S.E., Jones, J.B., Strynadka, N.C.J. (2000) Structure-based design guides the improved efficacy of deacylation transition state analogue inhibitors of TEM-1 β-lactamase. *Biochemistry* 39, 5312-5321.

[112] Usher, K.C., Blaszczak, L.C., Weston, G.S., Shoichet, B.K., Remington, S.J. (1998) Three-dimensional structure of AmpC β-lactamase from *Escherichia coli* bound to a transition-state analogue: Possible implications for the oxyanion hypothesis and for inhibitor design. *Biochemistry* 37, 16082-16092.

[113] Weston, G.S., Blazquez, J., Baquero, F., Shoichet, B.K., (1998) Structure-based enhancement of boronic acid-based inhibitors of AmpC β-lactamase. *J. Med. Chem.* 41, 4577-4586.

[114] Powers, R.A., Blazquez, J., Weston, G.S., Morosini, M.-I., Baquero, F., Shoichet, B.K. (1999) The complexed structure and antimicrobial activity of a non-β-lactam inhibitor of AmpC β-lactamase. *Protein Sci.* 8 2330-2337.

[115] Tan, Q., Ogawa, A.M., Painter, R.E., Park, Y.-W., Young, K., Di Ninno, F.P. (2010) 4,7-Dichlorobenzothien-2-yl sulfonylaminomethyl boronic acid: first boronic-acid-derived β-lactamase inhibitor with class A, C, and D activity. *Bioorg. Med. Chem. Lett. 20, 2622-2624.*

[116] Tondi, D., Powers, R.A., Caselli, E., Negri, M.-C., Blazquez, J., Costi, M.P., Shoichet, B.K. (2001) Structure-based design and in-parallel synthesis of inhibitors of AmpC β-lactamase. *Chem. Biol.* 8, 593-610.

[117] Tondi, D., Calo, S., Shoichet, B.K., Costi, M.P. (2010) Structural study of phenyl boronic acid derivatives as ampC β-lactamase inhibitors. *Bioorg. Med. Chem. Lett.* 20, 3416-3419.

[118] Buzzoni, V., Blazquez, J., Ferrari, S., Calo, S., Venturelli, A., Costi, M.P. (2004) Aza-boronic acids as non-β-lactam inhibitors of AmpC β-lactamase. *Bioorg. Med Chem. Lett.* 14, 3979-3983.

[119] Powers, R.A., Caselli, E., Focia, P.J., Prati, F., Shoichet, B.K. 2001 Structures of ceftazidime and its transition-state analogue in complex with AmpC β-lactamase: Implications for resistance mutations and inhibitor design. *Biochemistry* 40, 9207-9214.

[120] Caselli, E., Powers, R.A., Blaszczak, L.C., Wu, C.Y.E., Prati, F., Shoichet, B.K. (2001) Energetic, structural, and antimicrobial analyses of β-lactam side chain recognition by β-lactamases. *Chem. Biol.* 8, 17-31.

[121] Morandi, F., Caselli, E., Morandi, S., Focia, P.J., Blaquez, J., Shoichet, B.K., Prati, F. (2003) Nanomolar inhibitors of AmpC β-lactamase. *J. Am. Chem. Soc.* 125, 685-695.

[122] Wang, X., Minasev, G., Blazquez, J., Caselli, E., Prati, F., Shoichet, B.K. (2003) Recognition and resistance in TEM β-lactamase. *Biochemistry* 42, 8434-8444.

[123] Morandi, S., Morandi, F., Caselli, E., Shoichet, B. K., Prati, F. (2008) Structure-based optimization of cephalothin-analogue boronic acids as β-lactamase inhibitors. *Bioorg. Med. Chem.* 16, 1195-1205.

[124] Drawz, S.M., Babic, M., Bethel, C.R., Taracila, M., Distler, A.-M., Ori, C., Caselli, E., Prati, F., Bonomo, R.A. (2010) Inhibition of the class C β-lactamase from *Acinetobacter* spp.: Insights into effective inhibitor design. *Biochemistry* 49, 329-340.

[125] Powers, R.A., Shoichet, B.K. (2002) Structure-based approach for binding site identification on Amp C β-lactamase. *J. Med. Chem.* 45, 3222-3234.

[126] Thomson, J.M., Distler, A.M., Prati, F., Bonomo, R.A. (2006) Probing active site chemistry in SHV β-lactamase variants at Ambler position 244: Understanding unique properties of inhibitor resistance. *J. Biol. Chem.* 281, 26734-26744.

[127] Thomson, J.M., Prati, F., Bethel, C.R., Bonomo, R.A. (2007) Use of novel boronic acid transition state inhibitors to probe substrate affinity in SHV-type extended-spectrum β-lactamases. *Antimicrob. Agents Chemother.* 51, 1577-1579.

[128] Hubbard, S.A. (1998) Comparative toxicology of borates. *Biol. Trace. Elem. Res.* 66, 343-357.

[129] Groziak, M.P. (2001) Boron therapeutics on the horizon. *Am. J. Therapeut.* 8, 321-328.

[130] Hunter, P. (2009) Not boring at all. Boron is the new carbon in the quest for novel drug candidates. *EMBO reports* 10, 125-128.

[131] Pechenov, A., Stefanova, M.E., Nicholas, R.A., Peddi, S., Gutheil, W.G. (2003) Potential transition state analogue inhibitors for the penicillin-binding proteins. *Biochemistry* 42, 579-588.

[132] Inglis, S. R., Zervosen, A., Woon, E. C.Y., Gerards, T., Teller, N., Fischer, D. S., Luxen, A., Schofield, C. J. (2009) Synthesis and evolution of 3-(dihydroxy-boryl) benzoic acids as DD-carboxypeptidase R39 inhibitors, *J. Med. Chem.* 52, 6097-6106.

[133] Dzhekieva, L., Rocaboy, M., Kerff, F., Charlier, P., Sauvage, E., Pratt, R.F. (2010) Crystal structure of a complex between the *Actinomadura* R39 DD-peptidase and a peptidoglycan-mimetic boronate inhibitor: Interpretation of a transition state analogue in terms of catalytic mechanism. *Biochemistry* 49, 6411-6419.

[134] Curley, K., Pratt, R.F. (1997) Effectiveness of tetrahedral adducts as transition-state analogs and inhibitors of the class C β-lactamase of *Enterobacter cloacae* P99. *J. Am. Chem. Soc.* 119, 1529-1538.

[135] Kumar, S., Pearson, A.L., Pratt, R.F. (2001) Design, synthesis, and evaluation of α-ketoheterocycles as class C β-lactamase inhibitors. *Bioorg. Med. Chem.* 9, 2035-2044.

[136] Edwards, P.D., Meyer, E.F., Jr., Vijayalakshmi, J., Tuthill, P.A., Andisik, D.A., Gomes, B., Strimpler, A. (1992) Design, synthesis, and kinetic evaluation of a unique class of elastase inhibitors, the peptidyl α-ketobenzoxazoles, and the x-ray crystal structure of the covalent complex between porcine pancreatic elastase and Ac-Ala-Pro-Val-2-benzoxazole. *J. Am. Chem. Soc.* 114, 1854-1863.

[137] Perumal, S., Pratt, R.F. (2006) Synthesis and evaluation of ketophosph(on)ates as β-lactamase inhibitors. *J. Org. Chem.* 71, 4778-4785.

[138] Perumal, S. K., Adediran, S. A., Pratt, R. F. (2008) β-Ketophosphonates as β-lactamase inhibitors: intramolecular cooperativity between the hydrophobic subsites of a class D β-lactamase. *Bioorg. Med. Chem.* 16, 6987-6994.

[139] Grace, M.E., Schenkein, D.P., Pratt, R.F. 1987 Kinetics and mechanism of inactivation of the RTEM-2 β-lactamase by phenylpropynal - Identification of the characteristic chromophore. *J. Biol. Chem.* 262, 16778-16785.

[140] Laws, A.P., Layland, N.J., Procter, D.G., Page, M.I. (1993) The roles of the carboxy group in β-lactam antibiotics and lysine 234 in β-lactamase-I. *J. Chem. Soc. Perkin Trans. 2* 17-21.

[141] Layland, N.J., Laws, A.P., Vilanova, B., Page, M.I. (1995) Penicillin 3-aldehyde is a good substrate and not an inhibitor of β-lactamase A and β-lactamase C. *J. Chem. Soc. Perkin Trans. 2* 869-870.

[142] Page, M.I., Hinchliffe, P.S., Wood, J.M., Harding, L.P., Laws, A.P. (2003) Novel mechanism of inhibiting β-lactamases by sulfonylation using β-sultams. *Bioorg. Med. Chem. Lett.* 13, 4489-4492.

[143] Wood, J.M., Hinchliffe, P.S., Laws, A.P., Page, M.I. (2002) Reactivity and the mechanisms of reactions of β-sultams with nucleophiles. *J. Chem. Soc. Perkins Trans* 2 938-946.

[144] Page, M.I. (2004) β-sultams - Mechanism of reactions and use as inhibitors of serine proteases. *Acc. Chem. Res.* 37, 297-303.

[145] Beardsell, M., Hinchliffe, P.S., Wood, J.M., Wilmouth, R.C., Schofield, C., Page, M.I. (2001) β-Sultams - A novel class of serine protease inhibitors. *Chem. Commun.* 497-498.

[146] Sieburth, S.M., Mutahi, A.M. (1998) Synthesis of silanol-based peptide analogs and their use to inhibit protease enzymes. Patent WO9802578; *Chem Abst* 128, 154218.

[147] Chen, C.A., Sieburth, S.M., Glekas, A., Hewitt, G.W., Trainer, G.L., Erickson-Viitanen, S., Garber, S.S., Cordova, B., Jeffry, S., Klabe, R.M. (2001) Drug design with a new transition state analog of the hydrated carbonyl: silicon-based inhibitors of the HIV protease. *Chem. Biol.* 8, 1161-1166.

[148] Hall, B.G., Salipante, S.J., Barlow, M. (2003) The metallo-β-lactamases fall into two distinct phylogenetic groups. *J. Mol. Evol.* 57, 249-254.

[149] Payne, D.J., Bateson, J.H., Tolson, D., Gasson, B., Khushi, T., Ledent, P., Frère, J-M. (1996) Phosphonamidate analogues of dipeptides with carboxypeptidase A and β-lactamase-inhibitory activity: Elucidation of the mechanism of β-lactamase inhibition by electrospray mass spectrometry. *Biochem. J.* 314, 457-461.

[150] Walter, M.A., Felici, A., Galleni, M., Soto, R.P., Adlington, R.M., Baldwin, J.E., Frère, J.-M., Gololobov, M., Schofield, C.J. (1996) Trifluoromethyl alcohol and ketone inhibitors of metallo-β-lactamases. *Bioorg. Med. Chem. Lett.* 6, 2455-2458.

[151] Van Hove, F., Vanwetswinkel, S., Marchand-Brynaert, J., Fastrez, J. (1995) Synthesis and rearrangement of potential zinc □-lactamase inhibitors. *Tetrahedron Lett.* 36, 9313-9316.

[152] Gilpin, M.L., Folston, M., Payne, D., Cramp, R., Hood, I. (1995) Isolation and structure determination of two novel phenazines from a Streptomyces with inhibitory activity against metallo-enzymes, including metallo-β-lactamase. *J. Antibiot.* 48, 1081-1085.

[153] Concha, N.O., Janson, C.A., Rowling, P., Pearson, S., Cheever, C.A., Clarke, B.P., Lewis, C., Galleni, M., Frère, J.-M., Payne, D.J., Bateson, J.H., Abdel-Meguid, S.S. (2000) Crystal structure of the IMP-1 metallo β-lactamase from *Pseudomonas aeruginosa* and its complex with a mercaptocarboxylate inhibitor: Binding determinants of a potent, broad-spectrum inhibitor. *Biochemistry* 39, 4288-4298.

[154] Garcia-Saez, I., Hopkins, J., Papamicael, C., Franceschini, N., Amicosante, G., Rossolini, G.M., Galleni, M., Frère, J.-M., Dideberg, O. (2003) The 1.5-angstrom structure *of Chryseobacterium meningosepticum* zinc β-lactamase in complex with the inhibitor, D-captopril. *J. Biol. Chem.* 278, 23868-23873.

[155] Yamaguchi, Y., Jin, W., Matsunaga, K., Ikemizu, S., Yamagata, Y., Wachino, J. –I., Shibata, N., Arakawa, Y., Kurosaki, H. (2007) Crystallographic investigation of the inhibition mode of a VIM-2 metallo-β-lactamase from *Pseudomonas aeruginosa* by a mercaptocarboxylate inhibitor. *J. Med. Chem.* 50, 6647-6653.

[156] Kurosaki, H., Yamaguchi, Y. Yasuzawa, H., Jin, W., Yamagata, Y., Arakawa, Y. (2006) Probing, inhibition, and crystallographic characterization of metallo-β-lactamase (IMP-1) with fluorescent agents containing dansyl and thiol groups. *ChemMedChem* 1, 969-972.

[157] Toney, J.H., Hammond, G.G., Fitzgerald, P.M.D., Sharma, N., Balkovec, J.M., Rouen, G.P., Olson, S.H., Hammond, M.L., Greenlee, M.L., Gao, Y.-D. (2001) Succinic acids as potent inhibitors of plasmid-borne IMP-1 metallo-β-lactamase. *J. Biol. Chem.* 276, 31913-31918.

[158] Chikauchi, K., Ida, M., Abe, T., Hiraiwa, Y., Morinaka, A., Kudo, T. (2008) Preparation of maleic acid derivatives as metallo-β-lactamase inhibitors. U.S. *Pat. Appl. Publ.* US 2008090825.

[159] Hiraiwa, Y., Ida, M., Kudo, T., Morinaka, A., Chikauchi, K., Mori, K. (2008) Metallo-β-lactamase inhibitor. *PCT Int. Appl.* WO 2008016007.

[160] Lassaux, P., Hamel, M., Gulea, M., Delbruck, H., Meruri, P.S., Horsfall, L., Dehareng, D., Kupper, M., Frère, J.-M., Hoffman, K., Galleni, M., Bebrone, C. (2010) Mercaptophosphonate compounds as broad spectrum inhibitors of the metallo-β-lactamases. *J. Med. Chem.* 53, 4862-4876.

[161] Payne, D.J., Hueso-Rodriguez, J.A., Boyd, H., Concha, N.O., Janson, C.A., Gilpin, M., Bateson, J.H., Cheever, C., Niconovich, N.L, Pearson, S., Rittenhouse, S., Tew, D., Diez, E., Perez, P., de la Fuente, J., Rees, M., Rivero-Sagredo, A., (2002) Identification of a series of tricyclic natural products as potent broad-spectrum inhibitors of metallo-β-lactamases. *Antimicrob Agents Chemother*. 46, 1880-1886.

[162] Toney, J.H., Fitzgerald, P.M.D., Grover-Sharma, N., Olson, S.H., May, W.J., Sundelof, J.G., Vanderwall, D.E., Cleary, K.A., Grant, S.K., Wu, J.K., Kozarich, J.W., Pompliano, D.L., Hammond, G.G. (1998) Antibiotic sensitization using biphenyl

tetrazoles as potent inhibitors of *Bacteroides fragilis* metallo-β-lactamase. *Chem. Biol.* 5, 185-196.

[163] Horsfall, L.E., Garau, G., Lienhard, B.M.R, Dideberg, O., Schofield, C., Frère, J.-M., Galleni, M. (2007) Competitive inhibitors of the CphA metallo-β-lactamase from *Aeromonas hydrophila. Antimicrob. Agents Chemother.* 51, 2136-2142.

[164] Garau, G., Bebrone, C., Anne, C., Galleni, M., Frère, J.-M., Dideberg, O. (2005) A metallo-β-lactamase enzyme in action: Crystal structures of the monozinc carbapenemase CphA and its complex with biapenem. *J. Mol. Biol.* 345, 785-795.

[165] Bounaga, S., Laws, A.P., Galleni, M., Page, M.I. (1998) The mechanism of catalysis and the inhibition of the *Bacillus cereus* zinc-dependent β-lactamase. *Biochem. J.* 331, 703-711.

[166] Yang, K.W., Crowder, M.W. (1999) Inhibition studies on the metallo-β-lactamase from *Stenotrophomonas maltophilia. Arch. Biochem. Biophys.* 368, 1-6.

[167] Siemann, S., Clarke, A.J., Viswanatha, T., Dmitrienko, G.I. (2003) Thiols as classical and slow-binding inhibitors of IMP-1 and other binuclear metallo-β-lactamases. *Biochemistry* 42, 1673-1683.

[168] Goto, M., Takahashi, T., Yamashita, F., Koreeda, A., Mori, H., Ohta, M., Arakawa, Y. (1997) Inhibition of the metallo-β-lactamase produced from *Serratia marcescens* by thiol compounds. *Biol. Pharm. Bull.* 20, 1136-1140.

[169] Bounaga, S., Galleni, M., Laws, A.P., Page, M.I. (2001) Cysteinyl peptide inhibitors of *Bacillus cereus* zinc β-lactamase. *Bioorg. Med. Chem.* 9, 503-510.

[170] Sun, Q., Law, A., Crowder, M.W., Geyson, H.M. (2006) Homo-cysteinyl peptide inhibitors of the L1 metallo-β-lactamase, and SAR as determined by combinatorial library synthesis. *Bioorg. Med. Chem. Lett.* 16, 5169-5175.

[171] Lienhard, B.M.R., Hueting, R., Lassaux, P., Galleni, M., Frère, J.-M., Schofield, C. J. (2008) Dynamic combinatorial mass spectrometry leads to metallo-β-lactamase inhibitors. *J. Med. Chem.* 51, 684-688.

[172] Mollard, C., Moali, C., Papamicael, C., Damblon, C., Vessiler, S., Amicosante, G., Schofield, C.J., Galleni, M., Frère, J.-M., Roberts, G.C.K. (2001) Thiomandelic acid, a broad spectrum inhibitor of zinc β-lactamases - Kinetic and spectroscopic studies. *J. Biol. Chem.* 276, 45015-45023.

[173] Bateson, J.H., Witty, D.R., Gasson, B.C., Best, D.J., Payne, D.J. (1997) β-thiopropionyl-amino acid derivatives and their use as β-lactamase inhibitors. Patent WO 9730027, *Chem. Abst.* 127, 220985.

[174] Gilpin, M.L., Payne, D.J., Bateson, J.H. (1997) Pyrrolidine and thiazole derivatives with antibacterial and metallo-beta-lactamase inhibitory properties. Patent WO 9710225, *Chem. Abst.* 126, 277475.

[175] Scrofani, S.D.B., Chung, J., Huntley, J.J.A., Benkovic, S.J., Wright, P.E., Dyson, H.J. (1999) NMR characterization of the metallo-β-lactamase from *Bacteroides fragilis* and its interaction with a tight-binding inhibitor: Role of an active-site loop. *Biochemistry* 38, 14507-14514.

[176] Huntley, J.J.A., Scrofani, S.D.B., Osbourne, M.J., Wright, P.E., Dyson, H.J. (2000) Dynamics of the metallo-β-lactamase from *Bacteroides fragilis* in the presence and absence of a tight-binding inhibitor. *Biochemistry* 39, 13356-13364.

[177] Salsbury, F.R., Jr., Crowder, M.W., Kingsmore, S.F., Huntley, J.J.A. (2009) Molecular dynamics simulations of the metallo-β-lactamase from *Bacteroides fragilis* in the presence and absence of a tight-binding inhibitor. *J. Mol. Model.* 15, 133-145.

[178] Payne, D.J., Bateson, J.H., Gasson, B.C., Proctor, D., Khushi, T., Farmer, T.H., Tolson, D.A., Bell, D., Skett, P.W., Marshall, A.C., Reid, R., Ghosez, L, Combret, Y., Marchand-Brynaert, J. (1997) Inhibition of metallo-β-lactamases by a series of mercaptoacetic acid thiol ester derivatives. *Antimicrob. Agents Chemother.* 41, 135-140.

[179] Payne, D.J., Bateson, J.H., Gasson, B.C., Khushi, T., Proctor, D., Pearson, S.C., Reid, R. (1997) Inhibition of metallo-β-lactamases by a series of thiol ester derivatives of mercaptophenylacetic acid. *FEMS Microbiol. Lett.* 157, 171-175.

[180] Hammond, G.G., Huber, J.L., Greenlee, M.L., Laub, J.B., Young, K., Silver, L.L., Balkovec, J.M., Pryor, K.D., Wu, J.K., Leiting, B., Pompliano, D.L., Toney, J.H. (1999) Inhibition of IMP-1 metallo-β-lactamase and sensitization of IMP-1-producing bacteria by thioester derivatives. *FEMS Microbiol. Lett.* 179, 289-296.

[181] Greenlee, M.L., Laub, J.B., Balkovec, J.M., Hammond, M.L., Hammond, G.G., Pompliano, D., Epstein-Toney, J.H. (1999) Synthesis and SAR of thioester and thiol inhibitors of IMP-1 metallo-β-lactamase. *Bioorg. Med. Chem. Lett.* 9, 2549-2554.

[182] Kurosaki, H., Yamaguchi, Y., Higashi, T., Soga, K., Matsueda, H., Yumoto, H., Misumi, S., Yamagata, Y., Arakawa, Y., Goto, M. (2005) Irreversible inhibition of metallo-β-lactamase (IMP-1) by 3-(3-mercaptopropionylsulfanyl)-propionic acid pentafluorophenyl ester. *Angew. Chem. Int. Ed.* 44, 3861-3864.

[183] Walter, M.W., Valladares, M.H., Adlington, R.M., Amicosante, G., Baldwin, J.E., Frère, J.-M., Galleni, M., Rossolini, G.M., Schofield, C.J. (1999) Hydroxamate inhibitors of *Aeromonas hydrophila* AE036 metallo-β-lactamase *Bioorg. Chem.* 27, 35-40.

[184] Lienhard, B.M.R., Horsfall, L.E., Galleni, M., Frère, J.-M., Schofield, C.J. (2007) Inhibitors of the FEZ-1 metallo-β-lactamase. *Bioorg. Med. Chem. Lett.* 17, 964-968.

[185] Sieman, S., Evanoff, D.P., Marrone, L., Clarke, A.J., Viswanatha, T., Dmitrienko, G.I. (2002) N-arylsulfonyl hydrazones as inhibitors of IMP-1 metallo-β-lactamase. *Antimicrob. Agents Chemother.* 46, 2450-2457.

[186] Weide, T., Saldhana, S.A., Minond, D., Spicer, T.P., Fotsing, J.R., Spaargaren, M., Frère, J.-M., Bebrone, C., Sharpless, K.B., Hodder, P.S., Fokin, V.V. (2010) NH-1,2,3-Triazole inhibitors of the VIM-2 metallo-β-lactamase. *ACS Med. Chem. Lett.* 1, 150-154.

[187] Simm, A.M., Loveridge, E.J., Crosby, J., Avison, M.B., Walsh, T.R., Bennett, P.M. (2005) Bulgecin A: a novel inhibitor of binuclear metallo-β-lactamases. *Biochem. J.* 387, 585-590.

[188] Navarro, P.G., Osso, B.Q., Ortiz, R.G., de las Parvas, P.J.M., Puentedura, M.I.M., Gonzalez, M.C.C. (2004) Inhibition of β-lactamase II of *Bacillus cereus* by penamaldic derivatives of penicillins. *Antimicrob. Agents Chemother.* 48, 1058-1060.

[189] Tsang, W.T., Dhanda, A., Schofield, C.J., Frère, J.-M., Galleni, M., Page, M.I. (2004) The inhibition of metallo-β-lactamase by thioxo-cephalosporin derivatives. *Bioorg. Med. Chem. Lett.* 14, 1737-1739.

[190] Badarau, A., Llinas, A., Laws, A.P., Damblon, C., Page, M.I. (2005) Inhibitors of metallo-β-lactamase generated from β-lactam antibiotics. *Biochemistry* 44, 8578-8589.

[191] Moloughney, J.G., Thomas, J.D., Toney, J.H. (2005) Novel IMP-1 metallo-β-lactamase inhibitors can reverse meropenem resistance in *Escherichia coli* expressing IMP-1. *FEMS Microbiol. Lett.* 243, 65-71.

[192] Sanschagrin, F., Levesque, R.C. (2005) A specific peptide inhibitor of the class B metallo-β-lactamase L-1 from *Stenotrophomonas maltophilia* identified using phage display. *J. Antimicrob. Chemother* 55, 252-255.

[193] Shaw, R.W., Kim, K.M. (2006) RNA inhibitors of bacterial metallo-β-lactamases and their use with β-lactam antibiotics to kill bacteria. Patent WO 2006135431, *Chem. Abst.* 146, 77060.

[194] Shaw, R.W., Cottenoir, M. (2007) Inhibition of metallo-β-lactamase by double-stranded DNA aptamers and killing β-lactamase-producing bacteria. Patent WO 2007025016, *Chem. Abst.* 146, 268941.

[195] Kim, S.-K., Sims, C.L., Wozniak, S.E., Drude, S.H., Whitsun, D., Shaw, R.W. (2009) Antibiotic resistance in bacteria: novel metalloenzyme inhibitors. *Chem. Biol. Drug Design* 4, 343-348.

[196] Lienard, B. M. R., Garau, G., Horsfall, L., Karsisiotis, A. I., Damblon, C., Lassaux, P., Papamicael, C., Roberts, G. C. K., Galleni, M., Dideberg, O., Frère, J.-M., Schofield, C.J. Structural basis for the broad-spectrum inhibition of metallo-β-lactamases by thiols. (2008) *Org. Biomol. Chem.* 6, 2282-2294.

[197] Nauton, L., Kahn, R., Garau, G., Hernandez, J. F., Dideberg, O. (2008) Structural insights into the design of inhibitors for the L1 metallo-β-lactamase from *Stenotrophomonas maltophilia. J. Mol. Biol.* 375, 257-269.

[198] Pucci, M.J., Dougherty, T.J., Barrett, J.F. (2000) Targets, targets everywhere – so where are the new antibacterial drugs? *Curr. Opin. Anti-Infect. Invest. Drugs* 2, 123-124.

[199] Payne, D. J., Gwynn, M. N., Holmes, D. J., Pompliano, D. L. (2007) Drugs for bad bugs: confronting the challenges of antibacterial discovery. *Nature Rev. Drug. Discov.* 6, 29-40.

[200] Georgopapadakou, N. (2004) Superbugs and Superdrugs: a focus on antibacterials. *Expert Opin. Emerg. Drugs* 9, 191-195.

[201] Projan, S.J., Bradford, P. A. (2007) Late stage antibacterial drugs in the clinical pipeline. *Curr. Opin. Microbiol.* 10, 441-446.

In: Beta-Lactamases
Editor: Jean-Marie Frère, pp. 293-314

ISBN: 978-1-61324-638-2

Chapter 11

ß-Lactamase Inhibitory Proteins

Michael Gretes[1], Susan Jensen[2], and Natalie Strynadka[1]

[1] Department of Biochemistry and Molecular Biology,
University of British Columbia, Vancouver, British Columbia, V6T 1Z4 Canada.
[2] Department of Biological Sciences
University of Alberta, Edmonton, Alberta, Canada T6G 2E9.

Abstract

β-lactamase inhibitory proteins (BLIPs) are extracellular proteins produced by species of the soil bacteria *Streptomyces*. Three have been characterized and shown *in vitro* to selectively inhibit primarily class A β-lactamases. BLIP-0 and BLIP-I are homologous, 18 kDa proteins that are highly similar in function and with respect to their two-domain, tandemly repeated structure. BLIP-0 competitively inhibits TEM-1 β-lactamase using a large binding interface and loops that form specific contacts with key enzyme active site residues. BLIP-II has a molecular mass of 33 kDa and possesses a distinct fold and inhibitory mechanism, involving no direct contact with TEM-1 active site residues. Both BLIP inhibition mechanisms are reversible and show 1:1 stoichiometry. In addition to their structure and mechanism, the potential roles for BLIPs in nature and their implications for ß-lactamase inibitor design are discussed.

Introduction

The ß-lactamases are the most prevalent factor in bacterial resistance to penicillin and related antibiotics. ß-lactamases are widely observed to confer resistance to ß-lactams, including clinically indispensable broad-spectrum antibiotics as well as to coadministered ß-lactamase inhibitors. This has manifestly serious and potentially devastating implications for the treatment of bacterial infections. Extended spectrum and inhibitor resistant derivatives of the widespread and clinically important ß-lactamase TEM-1 [1] exploded in number and ubiquity during the 1990s. Meanwhile, prospects improved for novel means to restore clinical effectiveness to the ß-lactams: potent, natural protein inhibitors of ß-lactamases, termed ß-lactamase inhibitory proteins (BLIPs) were discovered.

What follows is an effort to catalogue and describe the few known and characterized BLIPs in terms of their functions and molecular structures, to explore their possible clinical impact, and shed some light on their natural physiological roles.

BLIP Discovery, Genetic Regulation, and Molecular Biology

The first BLIP was identified from *Streptomyces clavuligerus* during culture filtrate assays for the ß-lactamase inhibitor clavulanic acid. In addition to clavulanic acid and related clavam metabolites, that bacterium is known to produce penicillin and cephamycin antibiotics [2]. Unexpectedly, measured levels of clavulanic acid could account for only 1% of the total ß-lactamase inhibition reported by spectrophotometry. Heating to 98ºC or treating with methanol abolished this additional inhibition, suggesting the unidentified inhibitory factor to be a protein. Once ß-lactamase-specific protease activity was ruled out, investigators isolated an 18 kDa protein, named ß-lactamase inhibitory protein (BLIP-0)[3], from the growth media by ammonium sulfate precipitation. Purification to homogeneity allowed BLIP-0 to be characterized; N-terminal amino acid sequencing made possible a reverse genetic approach to cloning its corresponding gene (bli_{scl}). The bli_{scl} gene (603 bp) encodes a 201 residue protein precursor including a 36 residue leader peptide that is cleaved upon export from the cell (Figure 1). Its sequence contains two homologous, 76 residue tandemly repeated segments, suggesting that it has evolved by gene duplication and fusion. This mature form of BLIP-0 has a calculated molecular mass of 17,532 Da. Gel filtration chromatography shows it to be monomeric [2]. The amino acid composition of mature BLIP-0 includes 16.4% acidic and basic residues and 49% hydrophobic residues and also includes four cysteine residues, which form disulfide bridges in the mature protein.

The bli_{scl} gene is expressed as a monocistronic transcript from a promoter located 47 bp upstream from the start codon [3]. Expression from the bli_{scl} promoter (monitored using a reporter gene) is much more active in *S. clavuligerus* than in the heterologous host *Streptomyces lividans*, indicating the involvement of a species-specific transcriptional regulator. Mutants of *S. clavuligerus* defective in CcaR, the transcriptional activator that acts as a master regulator of the cephamycin-clavulanic acid pathway, also show greatly impaired production of BLIP-0 [4]. This makes CcaR a strong candidate for the regulator responsible for these species-specific expression effects and suggests that BLIP-0 production is tied to that of the ß-lactam metabolites. Furthermore, within *S. clavuligerus* both bli_{scl} expression and clavam metabolite production (except for clavulanic acid) show evidence of nutritional regulation: BLIP-0 is not produced during growth on a defined starch asparagine-based medium, yet it is produced on a complex trypticase soy starch-based medium; the clavam metabolites likewise are not produced on starch asparagine medium and are produced only on some complex media, although the pattern of expression is not identical. Despite these apparent relationships to ß-lactam antibiotic production, however, the bli_{scl} gene is not physically linked to the cephamycin-clavulanic acid gene cluster on the chromosome of *S.*

[3] Since it was the first such protein recognized, it was simply denoted "BLIP" and is known only as such in the literature. However, to distiguish specific references to this inaugural BLIP from BLIPs in general, in this chapter the *S. clavuligeris* BLIP will be referred to as BLIP-0.

clavuligerus. Similarly, no linkage is evident between bli_{scl} and gene clusters involved in the production of the other clavam metabolites by *S. clavuligerus* [5].

Proteinaceous ß-lactamase inhibitors have also been isolated from *Streptomyces exfoliatus*, a non-ß-lactam producing species. BLIP activity was detected in *S. exfoliatus* by a spectrophotometric ß-lactamase inhibition assay (similar to that used for *S. clavuligerus*) and two distinct BLIPs (named BLIP-I and BLIP-II) were isolated and characterized [6-8]. Oligonucleotide probes based upon the N-terminal amino acid sequences of the proteins then allowed the genes encoding each of these BLIPs to be cloned and sequenced. The BLIP-I gene, *bliA*, is 558 bp in length encoding a pro-protein of 186 residues (Figure 1). Removal of the 29 residue signal peptide generates a 157 amino acid mature protein comprised of 21.6% acidic and basic residues and 49% hydrophobic residues. BLIP-I shows 38% identity and 53% similarity to BLIP-0 (Figure 2), indicating common evolutionary origin [9] as well as structural and functional similarity, discussed in detail below.

***Streptomyces clavuligerus* BLIP-0: accession P35804**

MRTVGIGAGVRRLGRAVVMAAAVGGLVLGSAGASNAAGVMTGAKFTQIQF 50
GMTRQQVLDIAGAENCETGGS**F**GDSI**H**CRGHAAG**D**YYA**Y**ATFGFTSAAAD 100
AKVDSKSQE**K**LLAPSAPTLTLAKFNQVTVGMTRAQVLATVGQGSCTT**W**SE 150
YYPAYPSTAGVTLSLSCFDVDGYSSTG**F**YRGSA**H**L**W**FTDGVLQGK**R**Q**W**DL 200
V

***Streptomyces exfoliatus* BLIP-I: accession AAF74027**

MVKARIARTGISLVAAAAAVVATTSSAQANSGFSAEKYEQIQFGMTFDEV 50
WEIGGGEAACDTGGVIGDSILCFTESGDYAPYGGFSFTDEGELWSKRNEY 100
LYKAKTPSVKLSHYNRTALGMTEAQLWAAVPKDSCVSQGESYPNWPAKTG 150
FEEKYYCAAATGLFPPSASFHLTDGVLTYRYQRSLT

***Streptomyces exfoliatus* BLIP-II: accession AAC35427**

MPNGIVRLHRFGAALVMSGLLSALSAAPGVAAEPSSVSVAATSVVAWGGN 50
NDWGEATVPAEAQSGVDAIAGGYFHGLALKGGKVLGWGANLNGQLTMPAA 100
TQSGVDAIAAGNYHSLALKDGEVIAWGGNEDGQTTVPAEARSGVDAIAAG 150
AWASYALKDGKVIAWGDDSDGQTTVPAEAQSGVTALDGGVYTALAVKNGG 200
VIAWGDNYFGQTTVPAEAQSGVDDVAGGIFHSLALKDGKVIAWGDNRYKQ 250
TTVPTEALSGVSAIASGEWYSLALKNGKVIAWGSSRTAPSSVQSGVSSIE 300
AGPNAAYALKG

Figure 1. Amino acid sequences of all known BLIPs. Signal peptide sequences are underlined, and NCBI accession numbers are given. For BLIP-0, residues underlined and in boldface type have been identified by mutagenic studies as critical for binding TEM-1.

BLIP-II is distinct from the BLIPs described thus far. Encoded by the 936 bp *bliB* gene, the mature BLIP-II protein is composed of 280 amino acids with a predicted molecular mass of 33,336 Da following processing to remove a 31 residue signal peptide. The mature protein contains 16.4% acidic and basic residues and 62.8% hydrophobic residues, and lacks the cysteine residues that typify the other BLIPs. BLIP-II shows no sequence similarity to BLIP-I or BLIP-0, and comparison with other proteins in the public databases yields only unknown proteins or those associated with chromosome condensation in eukaryotes. The behaviour of BLIP-II upon size exclusion chromatography indicates it also exists as a monomer, but the protein does not show the two-domain structure with tandem duplication seen in BLIP-0. Instead, it features seven tandem repeats of a shorter ~40 amino acid domain.

```
C30-D49    1 ------------------------------CETGGSFGDSIHCRGHAAGD--------------------
A46-Y51    1 ----------------------------------------------AAGDYY------------------
H41-Y50    1 -----------------------------------------HCRGHAAGDY-------------------
P1         1 ------------------------------------F--TIHCSVTAAGDYYCVHGANGTSF--------
BLIP-0     1 AGVMTGAKFTQIQFGMTRQQVLDIAG-AENCETGGSFGDSIHCRGHAAGDYYAYATFGFTSAAADAKVDS
BLIP-I     1 NSGFSAEKYEQIQFGMTFDEVWEIGGGEAACDTGGVIGDSILC-FTESGDYAPYGGFSFTDEGE---LWS

C30-D49      ----------------------------------------------------------------------
A46-Y51      ----------------------------------------------------------------------
H41-Y50      ----------------------------------------------------------------------
P1           ----------------------------------------------------------------------
BLIP-0    70 KSQEKLLAPSAPTLTLAKFNQVTVGMTRAQVLATVGQGSCTTWSEYYPAYPSTAGVTLSLSCFDVDGYSS
BLIP-I    67 KRNEYLYKAKTPSVKLSHYNRTALGMTEAQLWAAVPKDSCVSQGESYPNWPAKTGFEEKYYCAAATG---

C30-D49      --------------------------
A46-Y51      --------------------------
H41-Y50      --------------------------
P1           --------------------------
BLIP-0   140 TGFYRGSAHLWFTDGVLQGKRQWDLV
BLIP-I   134 --LFPPSASFHLTDGVLTYRYQRSLT
```

Figure 2. Amino acid alignment of BLIP-0 and BLIP-I. Also included is the alignment of the various derivative peptides of BLIP-0: C30-D49, A46-Y51, H41-Y50, and P1.

There is possibly one other documented BLIP. Three studies from the 1970s note the discovery of the high molecular mass ß-lactamase inhibitor KA-107 from *Streptomyces gedanensis* [10-12]. While KA-107 appears likely to be a protein (its production is inducible over time, its size that of a large macromolecule, and it is precipitated by ~50% ammonium sulfate), this is neither conclusively shown nor explicitly stated; in any case, no related genes or gene products were subsequently isolated.

Of note is the likely BLIP-0 and BLIP-I homologue ß-lactamase-inhibitory-protein-like protein (BLP). Of unknown biological significance, BLP was discovered in *S. clavuligerus* by DNA sequence analysis of the cephamycin-clavulanic acid biosynthesis gene cluster [13, 14]. It is predicted to be 154 amino acid in length, processed from a 182 residue precursor polypeptide and secreted. Despite its apparent homology to BLIP-0 (32% identity) and BLIP-I (42% identity), it has no detectable ß-lactamase inhibitory activity.

BLIP Structure and Function

Considering the small number of BLIPs reported to date, there is an abundance of relevant structural and functional information available. The most common traits in the literature are: (1) all BLIPs described thus far invariantly show 1:1, reversible, non-covalent inhibition of ß-lactamases; (2) TEM-1 is the ß-lactamase known to be most strongly inhibited by all BLIPs.

BLIP-0

The first BLIP to be characterized, BLIP-0 from *S. clavuligerus*, was also the first to yield up its high-resolution crystal structure to the x-ray beam [15]. Soon afterward, the structure of BLIP-0 in action—in an inhibitory complex with TEM-1 ß-lactamase—was solved [16]. This complex has become widely used as a model system for the study of protein-protein interactions [17-32]. Far more detail is therefore available regarding the binding and inhibition mechanism of BLIP-0 than of any other BLIP.

In its crystalline form, BLIP-0 adopts an overall flat, mildly concave conformation, with dual helix-loop-helix motifs packed behind an extended 8-stranded ß-sheet to form a slightly concave face (Figure 3a) [15]. This topology strongly resembles the TATA-box binding protein [33, 34]; in functional terms, BLIP-0 features a similar large, saddle-like concave surface tailored for binding large, convex surfaces of DNA or protein molecules.

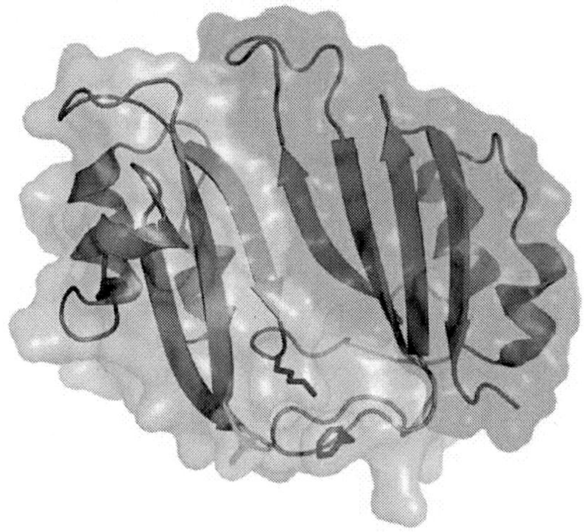

(A)

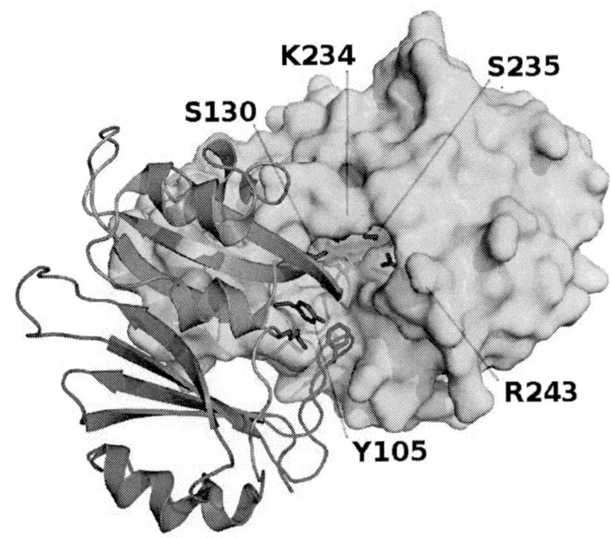

(B)

Figure 3. (Continued).

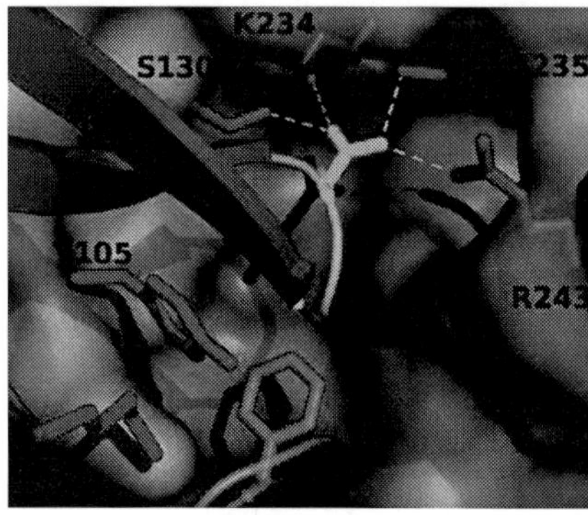

(C)

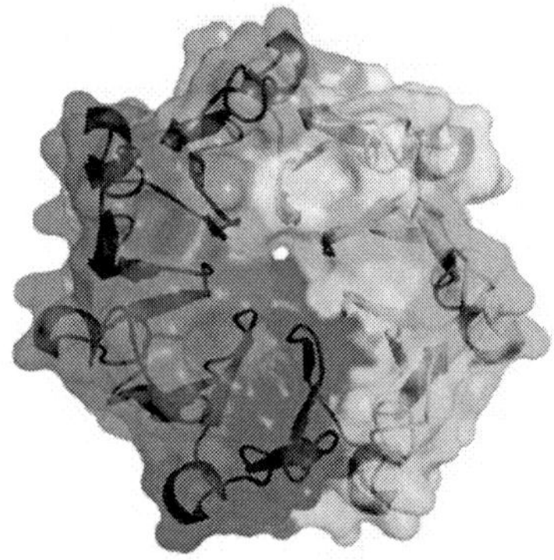

(D)

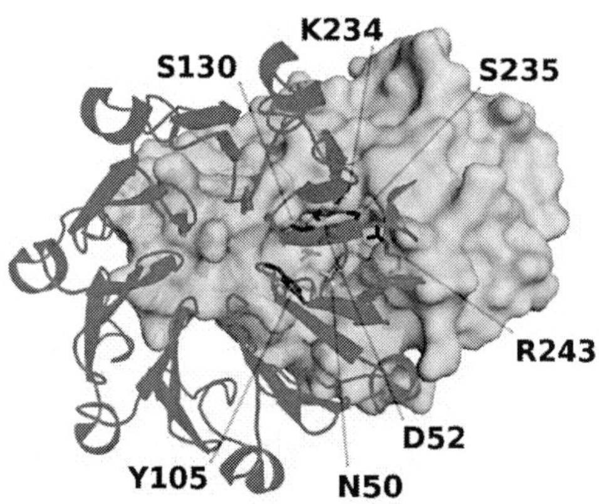

(E)

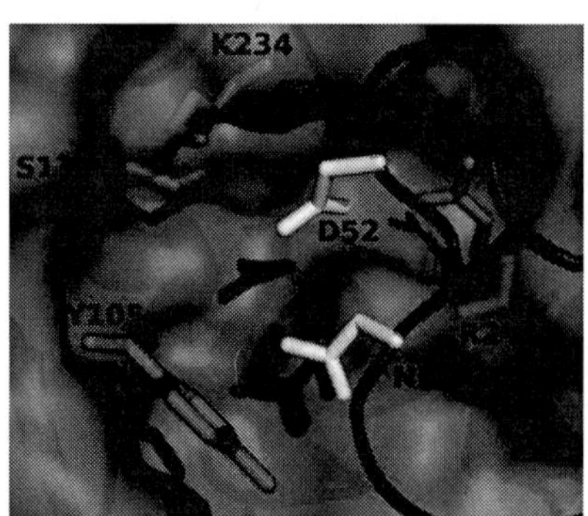

(F)

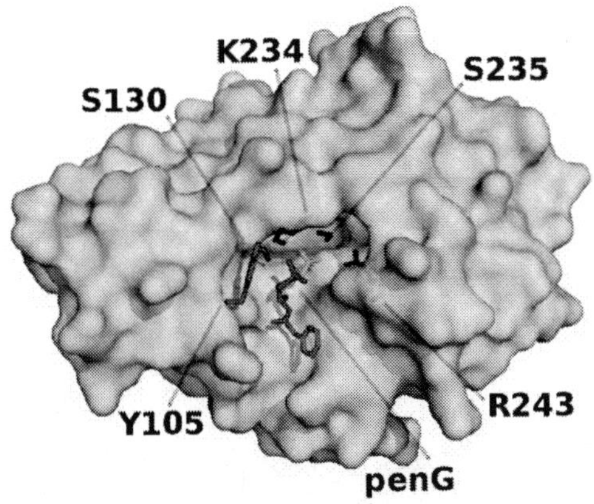

(G)

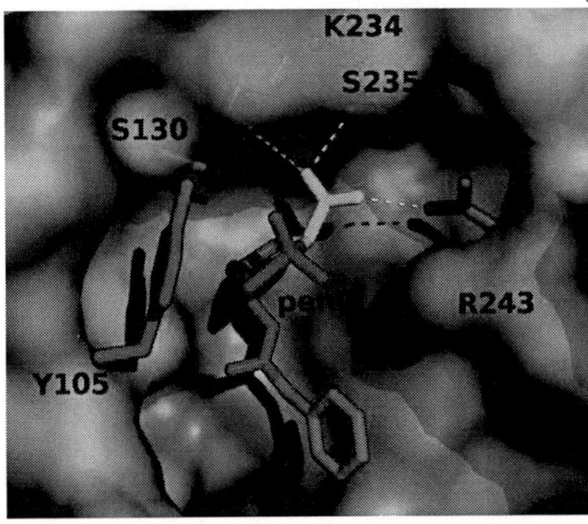

(H)

Figure 3. Three-dimensional ribbon cartoons (showing α-helices and ß-strands) and van der Waals molecular surface depictions of BLIP-0, BLIP-II, and TEM-1 ß-lactamase, drawn using PyMOL 0.99beta31 (DeLano Scientific LLC); *(a) BLIP-0* drawn as a cartoon inside its molecular surface, viewed facing its concave binding surface and showing its two-domain structure along with key residues. Domain 1 is drawn in blue (also showing a stick model of side chain K74), ß-hairpin A46-Y51 (showing D49) in green, domain 2 in pink, and extended ß-loop (showing F142) in purple; *(b) BLIP-0* (coloured as above) in its inhibited complex with TEM-1 ß-lactamase, drawn in yellow as a van der Waals surface. Shown in red as side-chain stick models are TEM-1 active site residues S130, K234, S235, R243, and Y105 (c) close view of BLIP-0 loops making contacts in the TEM-1 active site. BLIP-0 D49 (green) forms hydrogen and ionic bonds (dashed white lines) with TEM-1 S130, K234, S235, and R243 (red), and BLIP-0 F142 (purple) and K74 (blue) interacts with TEM-1 Y105 (red); *(d) BLIP-II* drawn as a cartoon inside its molecular surface, viewed facing its concave binding surface and showing each of its seven tandemly-repeated domains in a different colour; *(e)* BLIP-II (drawn in blue, with side chains N50 and D52 drawn as white stick models) in its inhibited complex with the TEM-1 ß-lactamase (drawn and coloured as above); *(f)* close view of BLIP-II residues N50 and D52 (white) inserted into the TEM-1 active site. While no direct contacts are made, the active site is occupied, thus precluding substrate binding. Notice that D52 is too distant from TEM-1 active site residues and at an incorrect angle for the formation of hydrogen bonds (contrasted with (c), above); *(g)* TEM-1 ß-lactamase (coloured as above) with bound substrate penicillin-G (penG). Comparing with (b) and (c), above, the penicillin-G carboxylate functional group is drawn in green (mimicked by BLIP-0 D49) and its phenyl group is drawn in purple (mimicked by BLIP-0 F142); *(h)* close view of TEM-1 ß-lactamase active site with bound penicillin-G, its carboxylate forming hydrogen bonds (dashed white lines) with TEM-1 K234, S235 and R243, and its 5-membered penem ring stacking with TEM-1 Y105.

Inhibition arises from three features of the BLIP-0•ß-lactamase interaction. First, the concavity and vastness of the binding surface (2636 $Å^2$ are buried at the BLIP-0•TEM-1 intermolecular interface [16]) gives rise to tight binding: inhibitory constants (K_i) were measured from 0.11–0.6 nM (Table 1). Second, a layer of water molecules at the interface mediates numerous hydrogen bond contacts [16], making these contacts flexible and adaptable; it has been proposed that this 'water adapter' (which allows tolerance of mutations in either BLIP-0 or TEM-1) is largely responsible for the ability of BLIP-0 to bind and inhibit a wide selection of ß-lactamase enzymes [17]. Lastly, two loops insert into the active site pocket of TEM-1, making numerous favourable contacts there (Figure 3b) [27]. Most notably, BLIP-0 residue D49 forms hydrogen bonds with four critical TEM-1 active site residues, which are ordinarily bound to the ß-lactam substrate: S130, K234, S235, and R243 (Figures 3c, h) [16]. Also, TEM-1 Y105, otherwise involved in aromatic ring stacking with a bound ß-lactam thiazolidine ring, is rotated 90 degrees in the complex and stacked between BLIP-0 F142 and K74.

Of 165 residues in the processed form of BLIP-0, 152 comprise two 76-residue tandemly repeated segments. One helix-loop-helix motif and four ß-strands are formed by each tandem repeat, comprising the bulk of the molecule and lending it a degree of internal symmetry. In addition, two disulfide bonds, C30-C42 and C109-C131, respectively join ß-strand 1 with ß-strand 2 in domain 1 and ß-strand 5 with ß-strand 6 in domain 2. Although such domain symmetry is far from geometrically perfect, the analogous domains show an elegant form of tandem functionality. In the inhibitory complex, the ß-hairpin A46-Y51 from BLIP-0 domain 1 and the analogous extended ß-loop from domain 2 insert into the TEM-1 active site and mimic the carboxylate (D49) and benzyl (F142) groups, respectively, of bound penicillin-G (compare Figures 3b and 3c with Figures 3g and 3h) [24]. Also, while the co-crystal structure of BLIP-0 in complex with TEM-1 shows relatively little variation from the individual, unbound structures, upon binding TEM-1 the two analogous BLIP-0 domains rotate 5.3° toward each other on the concave face like a grasping hand [16].

Table 1. Inhibition Constants (K_i) for BLIPs with Various Class A ß-Lactamases

BLIP	β-lactamase	β-lactam	K_i (nM)	Reference
BLIP-0	*Proteus vulgaris* 1028	nitrocefin	0.018	15
BLIP-0	*E. coli* TEM-3	nitrocefin	0.3	15
BLIP-0	*E. coli* TEM-1	cephaloridine	0.11	24
BLIP-0	*E. coli* TEM-1	cephalosporin C	0.5	17
BLIP-0	*E. coli* TEM-1	nitrocefin	0.6	16
BLIP-0	*Klebsiella oxytoca* 1082	nitrocefin	1.0	15
BLIP-0	*Serratia marcescens* SME-1	cephalosporin C	2.4	17
BLIP-0	*Bacillus anthracis* Bla1	nitrocefin	2.5	18
BLIP-0	*Pseudomonas aeruginosa* PSE-1	nitrocefin	3	15
BLIP-0	*B. cereus* ATCC1-27348	nitrocefin	1×10^3	15
BLIP-0	*Klebsiella pneumoniae* SHV-1	cephalosporin C	1.1×10^3	18
BLIP-0	*S. aureus* 853	nitrocefin	3×10^3	15
BLIP-0	*B. licheniformis* 749/C	nitrocefin	3×10^3	15
BLIP-0	*Streptomyces albus* G	nitrocefin	3×10^3	15
BLIP-0	*Actinomadura* R39	nitrocefin	11×10^3	15

BLIP	β-lactamase	β-lactam	K_i (nM)	Reference
BLIP-0	*S. aureus* 157	nitrocefin	activity increased	15
BLIP-0	*Lysobacter enzymogenes*	nitrocefin	activity diminished	15
BLIP-0	*B. cereus* NCIB 8933	nitrocefin	no effect	15
BLIP-0	*Kluyvera Georgiana* CTX-M-14	nitrocefin	810	38
BLIP-0	*S. aureus* PC1	nitrocefin	350	38
BLIP-I	*E. coli* TEM-1	cephaloridine	0.047	8
BLIP-I	Bactopenase	penicillin G	0.062	6
BLIP-I	*B. subtilis* Type I	cephaloridine	70×10^3	8
BLIP-I	*B. subtilis* Type II	cephaloridine	70×10^3	8
BLIP-I	*E. cloacae* Type III	cephaloridine	$>10^6$	8
BLIP-I	*E. cloacae* Type IV	cephaloridine	$>10^6$	8
BLIP-I	*B. cereus* penicillinase	penicillin G	no effect	6
BLIP-I	*E. cloacae* cephalosporinase	penicillin G	no effect	6
BLIP-II	*E. coli* TEM-1	penicillin G	0.0272	39
BLIP-II	*E. coli* TEM-1	nitrocefin	0.0025	38
BLIP-II	*Serratia marcescens* SME-1	nitrocefin	0.0084	38
BLIP-II	*Bacillus anthracis* Bla1	nitrocefin	<0.025	38
BLIP-II	*Klebsiella pneumoniae* SHV-1	nitrocefin	0.012	38
BLIP-II	Bactopenase	penicillin G	274	6
BLIP-II	*B. cereus* penicillinase	penicillin G	no effect	6
BLIP-II	*E. cloacae* cephalosporinase	penicillin G	no effect	6
BLIP-II	*Kluyvera Georgiana* CTX-M-14	nitrocefin	0.0091	38
BLIP-II	*S. aureus* PC1	nitrocefin	<0.016	38

Despite its considerable degree of inhibited-enzyme specificity compared with a more general small-molecule inhibitor such as clavulanic acid, BLIP-0 is capable of inhibiting a wide range of ß-lactamases with a nM or sub-nM K_i [15]. To the extent that BLIP-0 functionally mimics a ß-lactam, it binds penicillin-binding protein *Enterococcus faecalis* PBP 5 weakly with a K_i of 12 μM but otherwise does not inhibit two of the best-studied exocellular cell wall biosynthetic enzymes produced by actinomycetes (*Streptomyces* R61 or *Actinomadura* R39 exocellular DD-carboxypeptidases/transpeptidases).

Table 1 summarizes several studies that have reported BLIP inhibition of various ß-lactamases versus a given ß-lactam. For BLIP-0, the BLIP for which the most comprehensive data exists, it was found that all Class C enzymes (cephalosporinases) are either unaffected or stimulated by the presence of BLIP [15]. Likewise, no Class B ß-lactamases (metallo-ß-lactamases) included in this extensive, initial study were inhibited. However, all Class A enzymes (with the exception of *Staphylococcus aureus* 157, which was stimulated and *Bacillus cereus* NCIB 8933 which was unaffected as well as the *Salmonella typhimurium* class D OXA-2), were strongly inhibited with μM to nM and sub-nM K_i.

Why is there such wide variation in the extent of BLIP-0 inhibition of select ß-lactamases, when the majority of these share a highly similar fold? The most likely explanation is that specific BLIP-0 surface features either accommodate or clash with corresponding features on the inhibited enzyme [17, 18]. From an extensive study comparing the effects of dozens of BLIP-0 binding surface mutations on its ability to inhibit two different ß-lactamases (TEM-1 and SME-1) it was first shown that an aromatic patch on each BLIP-0 domain is largely responsible for specificity [17]. A subsequent study, again employing alanine-scanning mutagenesis, identified distinct but overlapping functional epitopes of BLIP-0 binding to four different ß-lactamases from various bacterial species: TEM-1 (*Escherichia coli*), SHV-1 (*Klebsiella pneumoniae*), SME-1 (*Serratia marcescens*) and Bla1 (*Bacillus anthracis*) [18]. Of primary interest here is the identification of residues that explain why BLIP inhibits SHV-1 [35] far more weakly ($K_i = 1.1$ μM) than it does TEM-1 ($K_i = 0.5$ nM), despite the fact that the two ß-lactamases have 68% amino acid identity and themselves hydrolyze a similar spectrum of substrates. The authors also report key residues that permit BLIP to strongly inhibit the divergent ß-lactamases SME-1 and Bla1, which have no more than 30% identity with TEM-1. More recently, five clusters of surface residues that act in a modular, independent, and additive fashion to bind TEM-1 have been identified [29]. Their interrelationships as well as their implications for the study and design of protein-protein interfaces are being actively pursued.

Additional features of BLIP-0 that affect its binding affinity for TEM-1 have been identified beyond the binding interface [19, 21]. These act via 'long-range' charge-charge interactions; certain charged residues exert electrostatic attractive forces on others to promote the properly oriented association of otherwise randomly-diffusing proteins. Since such charge properties are relatively independent of protein conformation, they may help to explain the ability of BLIP-0 to selectively inhibit structurally similar ß-lactamases.

A recent noteworthy structural and biochemical study [36] shows that BLIP-0 binds KPC ß-lactamase enzymes extremely tightly ($K_i < 1$ nM for KPC-2 and KPC-3). The variations among reported crystal structures of the BLIP-0•KPC-2 complex, in particular the position of the F142 loop, support the notion that BLIP-0 binding interactions are highly plastic and adaptable. This helps to explain the wide range of ß-lactamases inhibited by BLIP-0.

BLIP-I

The atomic-resolution structure of BLIP-I from *S. exfoliatus* alone and in complex with TEM-1 was recently reported [37]. As expected, its 38% amino acid identity with BLIP-0 corresponds to close structural resemblance. BLIP-I exhibits the same tandem duplication of structure to form an internal dimer. It is possible to obtain large amounts of mature BLIP-I, fused to a cleavable, N-terminal His_6 tag by recombinant expression in *Escherichia coli.* The protein is produced in an insoluble and inactive form but can be refolded to yield an active BLIP [8]. While the presence of the His_6 tag has not been found to preclude activity, K_i values improve more than 10-fold upon removal of the tag to reach values comparable to native BLIP-I purified from *S. exfoliatus.*

There is experimental evidence for at least one functional epitope shared between BLIP-0 and BLIP-I: the mutation D49A in BLIP-I alters the K_i for TEM-1 to 10 nM, decreasing the binding affinity to the near-identical value for the corresponding BLIP-0 D49A mutant [8]. Otherwise, BLIP-I appears to be quite specific for inhibiting TEM-1 and Bactopenase (a commercial preparation of *B. cereus* ß-lactamase); it has little effect on the limited number of other ß-lactamases tested (Table 1).

Despite close resemblance to BLIP-0 in its uncomplexed state, BLIP-I has noteworthy differences in its mode of binding to TEM-1 [37]. Although the BLIP•TEM-1 interaction surfaces are comparably vast (~1190 $Å^2$ in the BLIP-1•TEM-1 complex compared with ~1280 $Å^2$ in BLIP-0•TEM-1), the binding surface electrostatic charge is predicted to be similar, and the binding affinity and number of contact atoms are equivalent, BLIP-I is shifted by 4.5 Å along the surface of TEM-1 and the majority of individual amino-acid interactions are distinct.

BLIP-II

Recently reported phage display experiments and *in vitro* enzyme inhibition assays show the protein BLIP-II to be a more potent inhibitor than BLIP-0 for several class A beta-lactamases [38] with K_i values in the low picomolar range (Table 1). In many other respects, BLIP-II contrasts with the proteinaceous ß-lactamase inhibition discussed so far. With no significant amino acid sequence identity with the other two BLIPs, the solution of its high-resolution crystal structure unsurprisingly reveals an entirely distinct fold: a funnel-shaped 7-bladed ß-propeller [39]. Bearing overall resemblance to the three-dimensional structure of the human chromatin condensation factor RCC1, each of the seven blades is formed by 3 anti-parallel ß-strands comprised of homologous, tandemly-repeated amino acid segments. The propeller forms a tightly-packed, concave binding surface similar in shape to a satellite dish (Figure 3d). Thus BLIP-II and BLIP-0 both bind TEM-1 using a concave surface and in a similar position (compare Figures 3e, 3b).

Unlike BLIP-0, in the inhibited complex BLIP-II makes no direct contact with TEM-1 active site residues. BLIP-II instead interferes with TEM-1 substrate binding via steric clash. While BLIP-II side chains N50 and D52 insert into the TEM-1 active site pocket without making direct atomic contacts, they nevertheless occupy enough space to preclude substrate binding (Figure 3f). In this same TEM-1 'helix-loop' region, BLIP-II also interacts via an incomplete ring of aromatic residues with a single 'gap' filled by TEM-1 residues E104 and Y105, thus completing a hydrophobic ring that surrounds and may strengthen a central hydrogen bond. A similar binding motif, not discussed further here, is found at the BLIP-0•TEM-1 interface [40]. Thus, despite key structural differences and without apparent common ancestry, BLIP-0 and BLIP-II share a striking degree of functional similarity.

Although the structure of BLIP-II has been solved in complex with TEM-1, that of BLIP-II in apo form has not yet been determined. Thus no information is available on conformational changes that might occur to BLIP-II upon binding, nor is it known to what extent its three-dimensional structure is shaped by interactions with TEM-1.

Early functional studies utilized BLIP-II produced at high levels in *E. coli* as intact pre-protein (including its signal peptide) without a His_6 tag [7]. As with BLIP-I, the recombinant protein was produced in an insoluble, inactive form that could be reactivated by solubilization and refolding. Recombinant pre-proteins frequently retain their signal peptides when accumulated in an insoluble form, so pro-BLIP-II may show ß-lactamase inhibitory activity even with the signal peptide attached, although the size of the recombinant protein product is not clear. BLIP-II was also produced in a soluble and active form in a heterologous *Streptomyces* species.

The mechanism of BLIP-II inhibition (whether it exhibits competitive or non-competitive inhibition kinetics) is somewhat unclear. Variation in the length of time that the ß-lactamase is pre-incubated with inhibitor may be responsible for conflicting reports. One study reports the K_i of BLIP-II with Bactopenase (a *B. cereus* culture filtrate containing a large amount of the class A enzyme) as 274 nM without noting any pre-incubation step [6], while a later study, employing a one-hour pre-incubation of BLIP-II with TEM-1, reports the K_i as 27.2 pM (4 orders of magnitude stronger) and suggests that the slow, tight, Kunitz-type binding requires such an incubation period to achieve full inhibition [39]. In any case, structural data indicate that BLIP-II competitively inhibits TEM-1, and a deacylation-deficient TEM-1 mutant with bound penicillin-G fails to effectively bind BLIP-II. Without a solved BLIP-II•Bactopenase complex in hand, however, the non-competitive inhibition previously observed for that interaction cannot be explained. It is entirely possible that BLIP-II binds Bactopenase without inserting the 'steric clash' residues N50 and D52 and instead causes distortion of the active site and a decreased rate of catalysis. This would result in an apparent decrease in maximum reaction velocity (V_{max}) that cannot be restored even with great excess of substrate (the classical definition of non-competitive inhibition).

The large number of structural and functional studies of BLIP-0 have not been repeated with BLIP-II nor, for that matter, for BLIP-I. Unlike BLIP-I, however, the stark unrelatedness of BLIP-II with BLIP-0 precludes any reasonable inference of its ß-lactamase inhibition profile based on functional studies of BLIP-0.

BLP

Although the natural function of BLP is unclear, its atomic-resolution structure has recently been described and shows it to closely resemble its homologues BLIP-0 and BLIP-I [37]. BLP is very likely to be secreted from the cell. Its signal peptide is predicted with a high degree of confidence by the SignalP server [41] and was able to direct the protein to the periplasm during recombinant *E. coli* expression. In addition, the molecular mass of BLP by SDS-PAGE and mass spectrometry suggests that the mature form is 154 residues in length. This is consistent with electron density observed in the crystal structures at the N-terminus [37].

BLIP-Derived Inhibitors

Efforts to circumvent ß-lactamase activity in the clinic so far take two major approaches. The first is the creation of ß-lactam antibiotics that are resistant to ß-lactamase hydrolysis. The alternative approach is the development of ß-lactamase inhibitors to be co-administered with traditional ß-lactam drugs. Generally, these are poorly hydrolyzed structural analogues of clinically effective ß-lactams, which typically lack antibiotic potency but inactivate ß-lactamases according to branched reaction pathways (see chapters 6 and 9), slowing hydrolysis of the active ß-lactam antibiotic to prolong its activity. For example, the clinically utilized inhibitors clavulanic acid and tazobactam are ß-lactam molecules [42-44]. Both of these approaches, however, have been compromised by mutant ß-lactamases, including the widespread and clinically important TEM-1 [45]. Specific mutations in this enzyme allow cleavage of extended-spectrum (i.e. ß-lactamase resistant) cephalosporins (including the aminothiazole oxime ß-lactam derivatives ceftazidime, cefotaxime, and aztreonam; mutation R164S) and confer resistance to inhibitors like clavulanic acid (R244S). For a time, one study raised the spectre of a double mutant (R164S/R244S) that holds both advantages [46]. While it was later shown that TEM-1 is not yet capable of dual promiscuity and resistance [45, see also chapter 14], it is certainly possible that such compatible mutations may arise and negate the effectiveness of extended-spectrum antibiotic/inhibitor combination therapies.

BLIP-0 Derivatives

Promisingly, it has been shown that TEM-1 inhibition by BLIP-0 is unaffected by the clavulanic acid-resistance mutation R244S and thus does not proceed by the same mechanism as the small-molecule inhibitors. Of course, the macromolecular, proteinaceous nature of BLIP precludes its direct usefulness as a drug: it carries too great a risk of instability and antigenicity for use in the clinic, and is, moreover, far too large to penetrate the outer membrane of gram-negative bacteria. Therefore, a number of BLIP-inspired low molecular mass compounds have been designed to mimic its ß-lactamase inhibitory action, in hopes of ultimately obtaining a new inhibitor for clinical use. There are numerous hurdles involved in translating an effective *in vitro* inhibitor into a drug; there are only three clinical ß-lactamase inhibitors presently in clinical use. However, because of the pressing need for new weapons in the fight against ß-lactam-resistant pathogens, novel ß-lactamase inhibitors showing any reasonable degree of effectiveness have been eagerly welcomed.

There are several examples of proteinaceous inhibitors that have been stripped down to their essential chemical components and reincarnated as potent small-molecule inhibitors [47-49]. Cyclic-peptide inhibitors based on the BLIP-0 domain 1 ß-hairpin loop (A46-Y51) have, unfortunately, not thus far proven effective; their inhibition constants range in the hundreds of µM [25] and have not shown improvement from that of a random fragment peptide (Figure 2, C30–D49) [50]. Another cyclic peptide derived from a region of BLIP-0 optimized for maximum contact with TEM-1 (residues 41-50) gave 2-fold improved inhibition (Figure 2 shows alignments of these peptides with the BLIP-0 sequence). This remains, however, four to six orders of magnitude poorer than inhibition of TEM-1 by various boronic acid (transition-state) inhibitors (see chapter 10), modeled on crystal structures of penicillin-G in the TEM-1 active site [51, 52], and six orders of magnitude poorer than by BLIP itself. Clearly the inhibitory power of the BLIP protein, in part due to whole-molecule charge-charge interactions (involving residues far removed from the active site and even apart from the enzyme-inhibitor interface) that guide docking of the BLIP macromolecule to its target ß-lactamase [19, 21], cannot be easily recapitulated in a peptide excerpt. It should be noted, however, that a BLIP-0-derived peptide may better inhibit ß-lactamases other than TEM-1; while inhibition constants remain poor for the A46-Y51 peptide (hundreds to thousands of µM), it inhibits SHV-1 relatively better than it inhibits TEM-1 (owing to the poor, µM inhibition of SHV-1 by BLIP) and inhibits PC-1 and P99, which BLIP fails to inhibit at all. No further work has been published to date attempting to optimize this particular peptide inhibitor.

Separately, another group has employed yeast-2-hybrid methods to screen a large library of semi-randomized peptides for TEM-1-binding partners [53]. Using the polymerase chain reaction (PCR) to amplify a set of oligonucleotides containing random base pairs at selected positions, peptides of the form $FTIHCX_1X_2X_3AAGDYYX_4X_5X_6X_7X_8X_9GTSF$ were generated, where 'X_n' is any amino acid other than E, K, M, Q, or W; these amino acids appear to have been omitted because of the nature of the genetic code—i.e. to avoid the nonsense codons and to generate as evenly distributed a random sequence as possible. Thus, the 'core' of the semi-random peptides is the A46-Y51 BLIP-0 domain 1 loop, plus BLIP-0 flanking residues IHC and TS (Figure 2). The authors do not elaborate on the reasons for choosing the remainder of the constant residues. With 9 positions of 15 possible amino acids, approximately 3 x 10^{10} ($\approx 15^9$) peptides were theoretically generated. Because of experimental limitations the authors estimate they were only able to screen on the order of 10^7 (<0.1%) of these, although this is still a vast number. Remarkably, only two TEM-1 associating peptides were uncovered, only one of which had the ability to inhibit TEM-1 hydrolysis of ampicillin. The authors do not formally determine K_i but TEM-1 appears to be significantly inhibited at peptide concentrations above ~50 μM.

The discovery of this new BLIP-0-inspired inhibitory peptide is noteworthy because eight of its nine randomly-generated residues (all besides x_5) were not previously predicted to contribute strongly to BLIP-0•TEM-1 binding [16-18, 24, 27]. Furthermore, in the inhibitory peptide position X_5 is occupied by valine instead of tyrosine, the corresponding residue (Y53) in BLIP-0 is critical for TEM-1 binding and conserved in the homologous BLIP-I. For these reasons as well as because no previously described inhibitory peptide incorporates residues beyond Y51 of BLIP-0, this peptide warrants further kinetic and structural study.

BLIP-I: Potential Inhibitors

The high degree of sequence and predicted structural homology of BLIP-I to BLIP-0 strongly suggests that BLIP-I inhibits TEM-1 via a similar protein-protein interaction. This indicates that there is a good chance that BLIP-I domain 1 loop-derived peptides could likewise function as ß-lactamase inhibitors in their own right. Of particular note is the reported ten-fold higher inhibition constant of BLIP-I compared with BLIP-0; while no studies yet report the activity of BLIP-I peptides, it would be interesting to determine whether the partially divergent domain 1 loop sequence (ESGDYA versus AAGDYY) is to any extent responsible for this apparently higher affinity.

BLIP-II: Potential Inhibitors

Since the ß-lactamase inhibitory mechanism of BLIP-II involves no direct contacts with target enzyme active site residues, it seems unlikely that BLIP-II inhibition could be reproduced in a small-molecule inhibitor. Indeed, no published studies to date describe a search for ß-lactamase inhibitory peptides derived from BLIP-II. Furthermore, because of the limited extent to which BLIP-II has been tested against diverse ß-lactamases, it remains unclear whether it is as broad-range an inhibitor as BLIP-0.

Improvements to BLIP-0•TEM-1 Binding

There is evidence that suggests that BLIP-0 inhibition of TEM-1 is not fully optimized in nature. Various mutations to each protein can significantly increase their binding affinity. For example, the amino acid substitutions BLIP-0(Y50A) [17] or TEM-1(E240K) [27] improves the ability of BLIP-0 to inhibit TEM-1 by greater than tenfold. Other, combined mutations have even more dramatic effects: the triple mutant (A113L/T114R/E240K) of TEM-1 is 550 times more susceptible to BLIP-0 inhibition [27], while various single and double mutants to BLIP-0 have the effect of shifting its preference for binding from TEM-1 to other ß-lactamases [18]. Interestingly, a BLIP-0 variant able to bind and inhibit *Staphylococcus* PC1 ß-lactamase has recently been identified by phage display screening of a combinatorial library of BLIP-0 mutants [54]. Thus it seems that BLIP-0 represents a reservoir of adaptability for binding TEM variants as well as other ß-lactamases.

BLIP Roles in Nature

Streptomyces species are well known as producers of degradative exocellular enzymes that process polysaccharides, polypeptides and other macromolecules into soluble, transportable nutrients. As part of their complex interactions with the other microbes that inhabit the competitive soil environment, some species also produce small-molecule and large, proteinaceous inhibitors of these same exoenzymes [55, 56]. Proteinaceous ß-lactamase inhibitors may represent just one member of an array of inhibitors that *Streptomyces* species produces to modulate their interactions with other organisms. Consistent with this theory, all ß-lactamase inhibitory proteins identified to date are products of the genus *Streptomyces* [6, 11]. Recall that the first BLIP to be characterized in detail (BLIP-0) was isolated from *S. clavuligerus* [2]. Since *S. clavuligerus* has the unusual ability to produce ß-lactam metabolites such as cephamycin C and clavulanic acid, production of BLIP-0 might also be related to the production of these ß-lactam metabolites. Clavulanic acid biosynthesis is co-regulated with that of cephamycin C [14], suggesting that clavulanic acid functions in nature as it does in the clinic: to protect the ß-lactam antibiotics produced by *S. clavuligerus* from degradation by ß-lactamases present in the environment. BLIP-0 may provide a second line of defense, working together with clavulanic acid to ensure the broadest protection possible. Certainly the observation that BLIP-0 makes up almost 10% of the total secreted protein of *S. clavuligerus* [2], would indicate such an important role for the protein. However, it must be kept in mind that many ß-lactamases, including the extensively studied TEM-1, are localized to the periplasm of gram-negative bacteria [57] and are thus inaccessible to exocellular BLIPs. Although there are a number of examples of secreted and outer membrane-associated ß-lactamases [58], TEM-1 is very unlikely to be the natural substrate of any BLIP, despite its extremely high binding affinity. In addition, a limited survey of *Streptomyces* species including other cephamycin and clavulanic acid producers, uncovered no other BLIP producers [2] so BLIPs do not appear indispensable to preserving the endogenous ß-lactam stockpile. Conversely, an independent search for BLIP producers reveals only one species, *S. exfoliatus*, that produces two BLIPs (BLIP-I and BLIP-II) but does not produce any ß-lactam metabolites [6]; thus BLIP production is not necessarily linked to production of ß-lactam metabolites.

As part of their complement of degradative exocellular enzymes, many *Streptomyces* species produce ß-lactamases, regardless of their sensitivity to ß-lactam antibiotics [59]. *S. clavuligerus* is included in this group, and so along with its assortment of ß-lactam antibiotics and ß-lactamase inhibitors, *S. clavuligerus* produces at least two ß-lactamase-type proteins. One of these is a Class A ß-lactamase encoded by a gene located at the outermost boundary of the cephamycin arm of the cephamycin-clavulanic acid gene cluster. This ß-lactamase was found to lack detectable activity when assayed in growing or resting cells of *S. clavuligerus* [60]. When expressed in another *Streptomyces* species, the ß-lactamase was sensitive to inhibition by both clavulanic acid and BLIP-0, which would explain the lack of activity in *S. clavuligerus*. However, the enzyme also lacked the highly conserved SDN motif found in most ß-lactamases, and had instead an SDG motif, resulting in intrinsically low activity regardless of the presence of ß-lactamase inhibitors. A second gene encoding a protein with stereotypical ß-lactamase features is located within the clavulanic arm of the gene cluster [61-63]. It includes the SXXK and SDN motifs conserved in ß-lactamases but lacks the KTG motif. The protein has not been investigated for ß-lactamase activity, but it is known to be essential for clavulanic acid biosynthesis since mutation causes loss of production. This calls into question whether either of the above proteins actually functions as a ß-lactamase in *S. clavuligerus*; perhaps instead they act as ß-lactam interacting proteins. For example, the proteins BlaR and MecR from *Bacillus licheniformis* and *Staphylococcus aureus* [64, 65] bear features typical of ß-lactamases but in reality serve as ß-lactam sensing proteins (see chapter 17). These sensors are integral membrane proteins that transduce external information regarding penicillin concentrations into changes in gene expression within the cell. The exocellular nature of most ß-lactamases from *Streptomyces* species is hard to reconcile with signal transducing functions, but if ß-lactam-sensing proteins do exist in *Streptomyces*, BLIPs may interact with them to interfere with or otherwise modulate sensing.

Genetic studies have shown that mutation of bli_{scl} (the gene encoding BLIP-0) has no effect on production of any ß-lactam metabolite by *S. clavuligerus* with the possible exception of a transient delay in production of the non-clavulanic acid clavam compounds [4]. Mutants also appear morphologically normal. To ensure that the simultaneous production of clavulanic acid was not masking a morphological phenotype in the bli_{scl} mutant, a bli_{scl}-*claR* double mutant defective in production of both BLIP-0 and clavulanic acid was prepared. ClaR is a pathway-specific transcriptional regulator that, when mutated, blocks production of clavulanic acid only. Again, no morphological phenotype was associated with the mutations.

In contrast, BLIP-I shows clear evidence of involvement in morphological differentiation [8]. Mutation of *bliA* resulted in a *bld* phenotype indicating a defect in aerial mycelium formation and sporulation. The authors suggest that BLIP-I may function in cell wall biosynthesis by interacting with penicillin binding proteins. While disruption of *bliA* affected morphology, it did not cause a complete loss of proteinaceous ß-lactamase inhibitory activity because of the presence of BLIP-II. Despite its differences from BLIP-I, BLIP-II apparently plays a similar role in morphological differentiation because mutation of BLIP-II likewise gave a *bld* phenotype [40]. The fact that both BLIP-I and BLIP-II appear to be independently required for proper morphology (i.e. neither can compensate for the loss of the other) indicates that they likely function together in a single biochemical pathway related to differentiation. Further mutational analysis supports the involvement of BLIP-II in the condensation of chromosomal DNA during sporulation [66].

Therefore, in the non-ß-lactam producing species *S. exfoliatus*, production of BLIPs seems closely associated with morphogenesis, whereas in *S. clavuligerus*, the morphological role (if any) must be masked by the complexity of the system involving ß-lactam metabolites in addition to BLIP-0.

The observation that production of ß-lactamases causes cell surface alterations suggests that ß-lactamases can serve a biological purpose quite apart from their ability to cleave ß-lactam metabolites [67]. The authors propose that ß-lactamases not only cleave ß-lactam antibiotics but also interact with substituent peptides of nascent peptidoglycan to modulate cell wall biosynthesis. This view is compatible with previous reports that soluble intermediates in cell wall biosynthesis control inducible ß-lactamase production in some species [68]. *Streptomyces* species are unusual amongst bacteria in their ability to produce penicillin-binding proteins as both soluble-exocellular and intrinsic membrane-bound proteins [69, 70]. If membrane-bound penicillin binding proteins work together with their soluble counterparts to shape the peptidoglycan that surrounds cells, and ß-lactamases modulate their activities to influence the extent of cell wall cross-linking, ß-lactamase activity, in turn, may be modified by ß-lactamase inhibitors in those species that produce them, to form a complex network of interacting elements that serve to control normal cell wall biosynthesis. On the other hand, since most *Streptomyces* species form neither proteinaceous nor low molecular mass ß-lactamase inhibitors, these inhibitors may instead be directed against other species to interfere with their cell wall biosynthesis and gain competitive advantage.

Although numerous studies have provided fertile ground for speculation, until more examples of ß-lactamase inhibitory proteins are characterized it is difficult to draw firm conclusions regarding their biological roles.

Finally, the biological function of BLP is unclear. Although its gene *blp* appears to be efficiently transcribed, null mutants in *bli* (the gene encoding BLIP-0) show no residual proteinaceous ß-lactamase inhibitory activity [14]. A double mutant deficient in *bli* and *blp* revealed no further phenotypic effects beyond those associated with the lone *bli* knockout [37]. BLP was also observed to fail to inhibit a small test set of ß-lactamases in vitro [37]. Thus BLP appears to lack the ability to inhibit ß-lactamases, or at minimum is far less promiscuous in its inhibition profile compared with BLIP-0 and BLIP-I. Phage display experiments have suggested that BLIP-0 may bind PBPs [28]; however, gel shift assays failed to detect BLP binding to serine ß-lactamase-like penicillin-binding sensor domains of exogenous PBPs [37].

CONCLUSION

To date, three ß-lactamase inhibitory proteins (BLIPs) and one BLIP-like protein have been identified and isolated from two species of the soil bacteria *Streptomyces*. BLIP-0 and BLIP-I, from *S. clavuligerus* and *S. exfoliatus*, respectively, closely resemble each other in amino acid sequence, structure, and function. BLIP-II was isolated from *S. exfoliatus* and is unrelated in sequence or structure to the other BLIPs. All three are naturally secreted into the extracellular environment, and *in vitro* are able to effectively, selectively, and reversibly inhibit certain ß-lactamases in a 1:1, non-covalent complex. BLP bears close sequence and structural resemblance to BLIP-0 and BLIP-I, yet is not known to inhibit ß-lactamase. While the natural functions of BLIPs remain speculative, BLIP-0 is linked to clavam metabolism and BLIP-I and BLIP-II are known to be required for proper cell wall morphology; they may act in a signaling capacity or to confer competitive advantage in an ecological context.

The BLIP mechanism of ß-lactamase inhibition is distinct from that of the clinical small-molecule inhibitors. Because inhibitor-resistant mutations to the common ß-lactamase TEM-1 confer no resistance to BLIP-0, it is hoped that BLIP derivatives may one day be used in the clinic. Presently, several such synthetic peptides are known to bind and inhibit TEM-1, although with affinities still several of orders of magnitude too poor for prospective use.

A wide variety of protein-protein binding studies employ the BLIP-0•TEM-1 interaction as a model system. Considerable amounts of both structural and functional data are therefore available. Investigations of the BLIPs have added much to the understanding of fundamental protein biochemistry and may provide insight into the development of new ß-lactamase inhibitors. The discovery and characterization of new BLIPs, BLIP properties, and their biological purpose are therefore eagerly anticipated.

REFERENCES

[1] Livermore, D. M. (1995). beta-Lactamases in laboratory and clinical resistance. *Clin Microbiol Rev 8*, 557-84.

[2] Doran, J. L., Leskiw, B. K., Aippersbach, S. & Jensen, S. E. (1990). Isolation and characterization of a beta-lactamase-inhibitory protein from Streptomyces clavuligerus and cloning and analysis of the corresponding gene. *J Bacteriol* 172, 4909-18.

[3] Paradkar, A. S., Petrich, A. K., Leskiw, B. K., Aidoo, K. A. & Jensen, S. E. (1994). Transcriptional analysis and heterologous expression of the gene encoding beta-lactamase inhibitor protein (BLIP) from Streptomyces clavuligerus. *Gene* 144, 31-6.

[4] Thai, W., Paradkar, A. S. & Jensen, S. E. (2001). Construction and analysis of b-lactamase-inhibitory protein (BLIP) non-producer mutants of Streptomyces clavuligerus. *Microbiology* 147, 325-35.

[5] Tahlan, K., Park, H. U., Wong, A., Beatty, P. H. & Jensen, S. E. (2004). Two sets of paralogous genes encode the enzymes involved in the early stages of clavulanic acid and clavam metabolite biosynthesis in Streptomyces clavuligerus. *Antimicrob Agents Chemother* 48, 930-9.

[6] Kim, M. K. & Lee, K. J. (1994). Characteristics of beta-lactamase-inhibiting proteins from Streptomyces exfoliatus SMF19. *Appl Environ Microbiol* 60, 1029-32.

[7] Park, H. U. & Lee, K. J. (1998). Cloning and heterologous expression of the gene for BLIP-II, a beta-lactamase-inhibitory protein from Streptomyces exfoliatus SMF19. *Microbiology* 144 (Pt 8), 2161-7.

[8] Kang, S. G., Park, H. U., Lee, H. S., Kim, H. T. & Lee, K. J. (2000). New beta - lactamase inhibitory protein (BLIP-I) from Streptomyces exfoliatus SMF19 and its roles on the morphological differentiation. *J Biol Chem* 275, 16851-6.

[9] Altschul, S. F., Gish, W., Miller, W., Myers, E. W. & Lipman, D. J. (1990). Basic local alignment search tool. *J Mol Biol* 215, 403-10.

[10] Ono, H., Matsumae, A., Iwai, Y., Nakae, M. & Omura, S. (1973). In vitro and in vivo activity of penicillinase inhibitor KA-107 against Staphylococcus aureus FS-1277. *Antimicrob Agents Chemother* 4, 226-30.

[11] Iwai, Y., Ono, H., Takeshima, H., Yamaguchi, N. & Omura, S. (1973). Screening and isolation of penicillinase inhibitor, KA-107. *Antimicrob Agents Chemother* 4, 222-5.

[12] Hata, T., Omura, S., Iwai, Y., Ono, H. & Takeshima, H. (1972). Studies on penicillinase inhibitors produced by microorganisms. *J Antibiot (Tokyo)* 25, 473-4.

[13] Alexander, D. C. & Jensen, S. E. (1998). Investigation of the Streptomyces clavuligerus cephamycin C gene cluster and its regulation by the CcaR protein. *J Bacteriol* 180, 4068-79.

[14] Perez-Llarena, F. J., Liras, P., Rodriguez-Garcia, A. & Martin, J. F. (1997). A regulatory gene (ccaR) required for cephamycin and clavulanic acid production in Streptomyces clavuligerus: amplification results in overproduction of both beta-lactam compounds. *J Bacteriol* 179, 2053-9.

[15] Strynadka, N. C., Jensen, S. E., Johns, K., Blanchard, H., Page, M., Matagne, A., Frere, J. M. & James, M. N. (1994). Structural and kinetic characterization of a beta-lactamase-inhibitor protein. *Nature* 368, 657-60.

[16] Strynadka, N. C., Jensen, S. E., Alzari, P. M. & James, M. N. (1996). A potent new mode of beta-lactamase inhibition revealed by the 1.7 A X-ray crystallographic structure of the TEM-1-BLIP complex. *Nat Struct Biol* 3, 290-7.

[17] Zhang, Z. & Palzkill, T. (2003). Determinants of binding affinity and specificity for the interaction of TEM-1 and SME-1 beta-lactamase with beta-lactamase inhibitory protein. *J Biol Chem* 278, 45706-12.

[18] Zhang, Z. & Palzkill, T. (2004). Dissecting the protein-protein interface between beta-lactamase inhibitory protein and class A beta-lactamases. *J Biol Chem* 279, 42860-6.

[19] Joughin, B. A., Green, D. F. & Tidor, B. (2005). Action-at-a-distance interactions enhance protein binding affinity. *Protein Sci* 14, 1363-9.

[20] Kozer, N. & Schreiber, G. (2004). Effect of crowding on protein-protein association rates: fundamental differences between low and high mass crowding agents. *J Mol Biol* 336, 763-74.

[21] Selzer, T. & Schreiber, G. (2001). New insights into the mechanism of protein-protein association. *Proteins* 45, 190-8.

[22] Albeck, S., Unger, R. & Schreiber, G. (2000). Evaluation of direct and cooperative contributions towards the strength of buried hydrogen bonds and salt bridges. *J Mol Biol* 298, 503-20.

[23] Albeck, S. & Schreiber, G. (1999). Biophysical characterization of the interaction of the beta-lactamase TEM-1 with its protein inhibitor BLIP. *Biochemistry* 38, 11-21.

[24] Petrosino, J., Rudgers, G., Gilbert, H. & Palzkill, T. (1999). Contributions of aspartate 49 and phenylalanine 142 residues of a tight binding inhibitory protein of beta-lactamases. *J Biol Chem* 274, 2394-400.

[25] Rudgers, G. W., Huang, W. & Palzkill, T. (2001). Binding properties of a peptide derived from beta-lactamase inhibitory protein. *Antimicrob Agents Chemother* 45, 3279-86.

[26] Huang, W., Zhang, Z. & Palzkill, T. (2000). Design of potent beta-lactamase inhibitors by phage display of beta-lactamase inhibitory protein. *J Biol Chem* 275, 14964-8.

[27] Rudgers, G. W. & Palzkill, T. (1999). Identification of residues in beta-lactamase critical for binding beta-lactamase inhibitory protein. *J Biol Chem* 274, 6963-71.

[28] Huang, W., Petrosino, J. & Palzkill, T. (1998). Display of functional beta-lactamase inhibitory protein on the surface of M13 bacteriophage. *Antimicrob Agents Chemother* 42, 2893-7.

[29] Reichmann, D., Rahat, O., Albeck, S., Meged, R., Dym, O. & Schreiber, G. (2005). The modular architecture of protein-protein binding interfaces. *Proc Natl Acad Sci U S* A 102, 57-62.

[30] Reichmann, D., Phillip, Y., Carmi, A. & Schreiber, G. (2008). On the contribution of water-mediated interactions to protein-complex stability. *Biochemistry* 47, 1051-60.

[31] Wang, J., Palzkill, T. & Chow, D.-C. (2008). Structural insight into the kinetics and uCp of interactions between TEM-1 beta-lactamase and BLIP. *J. Biol. Chem.*, M804089200.

[32] Hanes, M. S., Reynolds, K. A., McNamara, C., Ghosh, P., Bonomo, R. A., Kirsch, J. F. & Handel, T. M. (2010). *Specificity and cooperativity at beta-lactamase position 104 in TEM-1/BLIP and SHV-1/BLIP interactions*. Proteins.

[33] Kim, J. L., Nikolov, D. B. & Burley, S. K. (1993). Co-crystal structure of TBP recognizing the minor groove of a TATA element. *Nature* 365, 520-7.

[34] Kim, Y., Geiger, J. H., Hahn, S. & Sigler, P. B. (1993). Crystal structure of a yeast TBP/TATA-box complex. *Nature* 365, 512-20.

[35] Kuzin, A. P., Nukaga, M., Nukaga, Y., Hujer, A. M., Bonomo, R. A. & Knox, J. R. (1999). Structure of the SHV-1 beta-lactamase. *Biochemistry* 38, 5720-7.

[36] Hanes, M. S., Jude, K. M., Berger, J. M., Bonomo, R. A. & Handel, T. M. (2009). Structural and biochemical characterization of the interaction between KPC-2 beta-lactamase and beta-lactamase inhibitor protein. *Biochemistry* 48, 9185-93.

[37] Gretes, M., Lim, D. C., de Castro, L., Jensen, S. E., Kang, S. G., Lee, K. J. & Strynadka, N. C. (2009). Insights into positive and negative requirements for protein-protein interactions by crystallographic analysis of the beta-lactamase inhibitory proteins BLIP, BLIP-I, and BLP. *J Mol Biol* 389, 289-305.

[38] Brown, N. G. & Palzkill, T. (2010). Identification and characterization of beta-lactamase inhibitor protein-II (BLIP-II) interactions with beta-lactamases using phage display. *Protein Eng Des Sel* 23, 469-78.

[39] Lim, D., Park, H. U., de Castro, L., Kang, S. G., Lee, H. S., Jensen, S. E., Lee, K. J. & Strynadka, N. C. (2001). Crystal structure and kinetic analysis of beta-lactamase inhibitor protein-II in complex with TEM-1 beta-lactamase. *Nat Struct Biol* 8, 848-52.

[40] Lim, D., Park, H. U., De Castro, L., Kang, S. G., Lee, H. S., Jensen, S., Lee, K. J. & Strynadka, N. C. (2001). Crystal structure and kinetic analysis of beta-lactamase inhibitor protein-II in complex with TEM-1 beta-lactamase. *Nat Struct Biol* 8, 848-52.

[41] Emanuelsson, O., Brunak, S., von Heijne, G. & Nielsen, H. (2007). Locating proteins in the cell using TargetP, SignalP and related tools. *Nat Protoc* 2, 953-71.

[42] Reading, C. & Cole, M. (1977). Clavulanic acid: a beta-lactamase-inhiting beta-lactam from Streptomyces clavuligerus. *Antimicrob Agents Chemother* 11, 852-7.

[43] Reading, C., Farmer, T. & Cole, M. (1983). The beta-lactamase stability of amoxycillin with the beta-lactamase inhibitor, clavulanic acid. *J Antimicrob Chemother* 11, 27-32.

[44] Bush, K., Macalintal, C., Rasmussen, B. A., Lee, V. J. & Yang, Y. (1993). Kinetic interactions of tazobactam with beta-lactamases from all major structural classes. *Antimicrob Agents Chemother* 37, 851-8.

[45] Schroeder, W. A., Locke, T. R. & Jensen, S. E. (2002). Resistance to beta-lactamase inhibitor protein does not parallel resistance to clavulanic acid in TEM beta-lactamase mutants. *Antimicrob Agents Chemother* 46, 3568-73.

[46] Imtiaz, U., Manavathu, E. K., Mobashery, S. & Lerner, S. A. (1994). Reversal of clavulanate resistance conferred by a Ser-244 mutant of TEM-1 beta-lactamase as a result of a second mutation (Arg to Ser at position 164) that enhances activity against ceftazidime. *Antimicrob Agents Chemother* 38, 1134-9.

[47] Leatherbarrow, R. J. & Salacinski, H. J. (1991). Design of a small peptide-based proteinase inhibitor by modeling the active-site region of barley chymotrypsin inhibitor 2. *Biochemistry* 30, 10717-21.

[48] Li, B., Tom, J. Y., Oare, D., Yen, R., Fairbrother, W. J., Wells, J. A. & Cunningham, B. C. (1995). Minimization of a polypeptide hormone. *Science* 270, 1657-60.

[49] Yanofsky, S. D., Baldwin, D. N., Butler, J. H., Holden, F. R., Jacobs, J. W., Balasubramanian, P., Chinn, J. P., Cwirla, S. E., Peters-Bhatt, E., Whitehorn, E. A., Tate, E. H., Akeson, A., Bowlin, T. L., Dower, W. J. & Barrett, R. W. (1996). High affinity type I interleukin 1 receptor antagonists discovered by screening recombinant peptide libraries. *Proc Natl Acad Sci U S* A 93, 7381-6.

[50] Rudgers, G. W. & Palzkill, T. (2001). Protein minimization by random fragmentation and selection. *Protein Eng* 14, 487-92.

[51] Strynadka, N. C., Martin, R., Jensen, S. E., Gold, M. & Jones, J. B. (1996). Structure-based design of a potent transition state analogue for TEM-1 beta-lactamase. *Nat Struct Biol* 3, 688-95.

[52] Ness, S., Martin, R., Kindler, A. M., Paetzel, M., Gold, M., Jensen, S. E., Jones, J. B. & Strynadka, N. C. (2000). Structure-based design guides the improved efficacy of deacylation transition state analogue inhibitors of TEM-1 beta-Lactamase. *Biochemistry* 39, 5312-21.

[53] Sun, W., Hu, Y., Gong, J., Zhu, C. & Zhu, B. (2005). Identification of beta-lactamase inhibitory peptide using yeast two-hybrid system. *Biochemistry (Mosc)* 70, 753-60.

[54] Yuan, J., Chow, D. C., Huang, W. & Palzkill, T. (2011). Identification of a beta-Lactamase Inhibitory Protein Variant That Is a Potent Inhibitor of Staphylococcus PC1 beta-Lactamase. *J Mol Biol.*

[55] Koller, K. P. & Riess, G. (1989). Heterologous expression of the alpha-amylase inhibitor gene cloned from an amplified genomic sequence of Streptomyces tendae. *J Bacteriol* 171, 4953-7.

[56] Terabe, M., Kojima, S., Taguchi, S., Momose, H. & Miura, K. (1996). New subtilisin-trypsin inhibitors produced by Streptomyces: primary structures and their relationship to other proteinase inhibitors from Streptomyces. *Biochim Biophys Acta* 1292, 233-40.

[57] Farmer, T. H., Degnan, B. A. & Payne, D. J. (1999). Penetration of beta-lactamase inhibitors into the periplasm of gram-negative bacteria. *FEMS Microbiol Lett* 176, 11-5.

[58] Lakaye, B., Dubus, A., Joris, B. & Frere, J. M. (2002). Method for estimation of low outer membrane permeability to beta-lactam antibiotics. *Antimicrob Agents Chemother* 46, 2901-7.

[59] Ogawara, H. (1975). Production and property of beta-lactamases in Streptomyces. *Antimicrob Agents Chemother* 8, 402-8.

[60] Perez-Llarena, F., Martin, J. F., Galleni, M., Coque, J. J., Fuente, J. L., Frere, J. M. & Liras, P. (1997). The bla gene of the cephamycin cluster of Streptomyces clavuligerus encodes a class A beta-lactamase of low enzymatic activity. *J Bacteriol* 179, 6035-40.

[61] Li, R., Khaleeli, N. & Townsend, C. A. (2000). Expansion of the clavulanic acid gene cluster: identification and in vivo functional analysis of three new genes required for biosynthesis of clavulanic acid by Streptomyces clavuligerus. *J Bacteriol* 182, 4087-95.

[62] Mellado, E., Lorenzana, L. M., Rodriguez-Saiz, M., Diez, B., Liras, P. & Barredo, J. L. (2002). The clavulanic acid biosynthetic cluster of Streptomyces clavuligerus: genetic organization of the region upstream of the car gene. *Microbiology* 148, 1427-38.

[63] Jensen, S. E., Paradkar, A. S., Mosher, R. H., Anders, C., Beatty, P. H., Brumlik, M. J., Griffin, A. & Barton, B. (2004). Five additional genes are involved in clavulanic acid biosynthesis in Streptomyces clavuligerus. *Antimicrob Agents* Chemother 48, 192-202.

[64] Hardt, K., Joris, B., Lepage, S., Brasseur, R., Lampen, J. O., Frere, J. M., Fink, A. L. & Ghuysen, J. M. (1997). The penicillin sensory transducer, BlaR, involved in the inducibility of beta-lactamase synthesis in Bacillus licheniformis is embedded in the plasma membrane via a four-alpha-helix bundle. *Mol Microbiol* 23, 935-44.

[65] Zhang, H. Z., Hackbarth, C. J., Chansky, K. M. & Chambers, H. F. (2001). A proteolytic transmembrane signaling pathway and resistance to beta-lactams in staphylococci. *Science* 291, 1962-5.

[66] Kim, E. S., Song, J. Y., Kim, D. W., Ko, E. J., Jensen, S. E. & Lee, K. J. (2008). Involvement of beta-Lactamase Inhibitory Protein, BLIP-II, in Morphological Differentiation of Streptomyces exfoliatus SMF19. *J Microbiol Biotechnol* 18, 1884-9.

[67] Gallant, C. V., Daniels, C., Leung, J. M., Ghosh, A. S., Young, K. D., Kotra, L. P. & Burrows, L. L. (2005). Common beta-lactamases inhibit bacterial biofilm formation. *Mol Microbiol* 58, 1012-24.

[68] Jacobs, C., Frere, J. M. & Normark, S. (1997). Cytosolic intermediates for cell wall biosynthesis and degradation control inducible beta-lactam resistance in gram-negative bacteria. *Cel*l 88, 823-32.

[69] Ogawara, H. & Horikawa, S. (1980). Penicillin-binding proteins of Streptomyces cacaoi, Streptomyces olivaceus, and Streptomyces clavuligerus. *Antimicrob Agents Chemother* 17, 1-7.

[70] Jamin, M., Wilkin, J. M. & Frere, J. M. (1995). Bacterial DD-transpeptidases and penicillin. *Essays Biochem* 29, 1-24.

In: Beta-Lactamases
Editor: Jean-Marie Frère, pp. 315-340
ISBN: 978-1-61324-638-2

Chapter 12

QUANTUM CHEMISTRY APPLIED TO THE STUDY OF THE β-LACTAMASES

Dominique Dehareng and Georges Dive
Centre for Protein Engineering, Institut de Chimie B6a,
Sart Tilman, B-4000, Liège, Belgium.

ABSTRACT

Theoretical models have become necessary to support experimental work. The molecular modelling field covers a wide range of theoretical tools and techniques. In this chapter, the methods of quantum chemistry (QM) are presented and the utilisation of QM in the study of β-lactamase reactivity is discussed in the light of publications by the authors' laboratory or from other sources.

I. INTRODUCTION

Nowadays, it is obvious that theoretical models have become necessary to support experimental works. Almost all the experimental results are explained by models which can be meaningful when the theoretical calculations underlying them are also presented with their results.

In chemistry or biochemistry, this has led to the development of the so-called "molecular modelling" field, which covers a wide range of theoretical tools and techniques. In the particular case of reactivity, the theoretical tool is quantum mechanics (QM) applied to chemistry, i.e. quantum chemistry.

The aim of the chapter is three-fold. First, to present the methods of quantum chemistry. This is developped in Section II. Secondly, to choose some recent articles that use QM to study β-lactamase reactivity and briefly present their results. This is detailed in Section III. Thirdly, in Section IV we analyze two examples studied in our laboratory.

Finally, the Conclusion will highlight some important points about this tool and its good use.

II. Quantum Chemistry and its Methods [1-3]

II.1. The Hartree-Fock Framework: The Main Approximations

Quantum chemistry is the field that uses the mechanics specific to small particles, i.e. quantum mechanics, to describe the molecules, their properties and their reactivities. Due to their small masses, the small particles cannot be described by a well-defined trajectory, like a ball spinning in the roulette wheel or the planets orbiting around the sun. This is due to the fact that their wave character, which is indeed an intrinsic feature of all the particles, be they small or large, becomes as important as their particle character. Consequently, they must be described by a wave function that defines their state. This wave function has no meaning by itself but its squared value represent the probability density that the system have the coordinates or momenta used as arguments. In quantum chemistry, the small particles referred to are the electrons, the nuclei being usually considered as heavy classical particles since they are at least 1830 times heavier than an electron. In quantum mechanics, the physical variables, like the energy, the positions, the momenta, etc.., are associated with an operator (energy operator, position operators,..) which provides the value of the variable after acting on the wave function. For instance, the energy of a stationary system is related to a hamiltonian operator, a tribute to the mathematician William Rowan Hamilton. This hamiltonian is the sum of a kinetic and a potential energy operators, the first term being a function of the momentum operators of the particles while the second usually only depends on their coordinate operators. This is usually written:

$$\mathrm{H}(p,q) = \mathrm{T}(p) + \mathrm{V}(q)$$

H, T and V being respectively the hamiltonian, the kinetic and the potential energy operators, p and q being the vectors whose components are the momenta and the positions of all the particles, electrons and nuclei. If the wave function describing the state of the system is noted $\Psi\ (q)$, the total energy E related with this state is obtained through the famous Schrödinger's equation for stationary systems:

$$\mathrm{H}(p,q)\ \Psi\ (q) = \mathrm{E}\ \Psi\ (q)$$

In quantum chemistry, one does not consider the dynamics of the nuclei and their motions are most often taken account of through the classical image of a trajectory. Thus the motion of the electrons and the nuclei are separated and only the electronic part of the hamiltonian and of the wave function is considered. This approximation, called the Born-Oppenheimer approximation, leads to another equation of the Schrödinger's type:

$$\mathrm{H}^{el}(p^{el},q^{el},Q^{N})\ \Psi^{el}(q^{el},Q^{N}) = \mathrm{E}^{el}(Q^{N})\ \Psi^{el}(q^{el},Q^{N}) \qquad (1)$$

where p^{el},q^{el},Q^{N} are respectively the vectors of the electronic momenta, electronic coordinates and nuclei coordinates, and $\mathrm{E}^{el}(Q^{N})$ is the electronic energy which is no longer a constant but now depends on the (3N-6) internal nuclear coordinates, the 6 degrees of freedom related to the translation and the rotation having been completely separated from the problem; as a matter of fact, the properties of the system are not modified during a translation and/or a rotation according to the three coordinates X,Y,Z. This function is usually referred to as the potential energy surface (PES) because it is in fact the potential energy of the nuclei. They are like slopes and hills on ski tracks but the skiers are the nuclei and they travel in a (3N-6)-dimensional space instead of our 3D one. There are several interesting positions for the skier on the track: the top, from where he will always go down whatever the direction he chooses, the bottom of the slopes, from where he must always provide an effort to go in any direction (unless he pays a ski lift), or the road down to the hotel with the mountain wall on one side and the chasm toward the valley on the other. These special points and paths will be discussed below.

Equation (1) cannot be solved analytically, except for one-electron systems, leading to the hydrogenoïd functions, or orbitals, in the case of atomic ions. Thus, approximations have to be introduced and they concern the way the wave function is expressed. Hartree and Fock have worked out a model, based on the idea that each electron is described as an independent particle feeling the average field of the others. The electronic wave function is then expressed as a product of monoelectronic functions, called the orbitals (atomic or molecular according to the system considered). The first obvious improvement of this wave function was introduced by Slater who expressed it as a determinant built on monoelectronic functions that included a spin function, the spinorbitals. The next approximation, due to Roothaan and Hall, concerns the possible expression of the orbitals. It is called the Linear Combination of Atomic Orbitals, or LCAO, and is based on the following reasoning. Since the orbitals must describe an electron in a molecule, or in an atom, it seems straightforward to consider that their natural building blocks should be the analytically-known hydrogenoïd orbitals, i.e. the functions obtained for one-electron hydrogen-like atomic systems. These orbitals are the well-known 1s, 2s, $2p_x$, $2p_y$, $2p_z$, 3s ... functions. It should be emphasized here that, in spite of the meaning of the acronym LCAO which refers to atomic orbitals, these building blocks are not atomic orbitals, they are just basis functions used to construct the molecular orbitals. They constitute the so-called basis set. All the preceding approximations are summarized in the Appendix.

The framework just described, consisting in the independent electron model and the LCAO approximation, is the Hartree-Fock-Roothaan-Hall framework, most usually referred to in the literature as the Hartree-Fock framework (HF). The most important concepts to keep in mind are the orbitals, or the spinorbitals, the Slater determinant and the basis set functions. The process leading to the knowledge of the orbitals is the resolution of the coupled Fock's equations, which also provide well-known quantities: the orbital energies, that give good approximation of the ionization energies, on the basis of the Koopman's theorem. The Fock's equations are expressed in the basis functions system and provide the LCAO coefficients ($c_{\alpha,i}(Q^N)$ in the Appendix). These equations involve a very large number of 3n-dimensional (n=number of electrons) integrals. The orbitals obtained belong to a representation of the symmetry point group of the molecule and thus determine the symmetry of the related Slater determinant, i.e. of the electronic state. In this paragraph, we will not discuss the virtual orbitals, the overlap, mono and bi-electronic integrals, the spin functions, the different HF frameworks (RHF,UHF,ROHF), or the localized orbitals. All these developments and others can be found in textbooks like those in references 1-3.

In the early years of quantum chemistry, John Pople made a tremendous effort in the development of approximate methods [4], the philosophy of which was based on two axes: the first one was to propose a physically founded way to considerably decrease the number of integrals in the Fock equations, and the second one was to evaluate some of them on the basis of known properties like for instance the ionization energies. Two very famous approximate frameworks are the Pariser-Parr-Pople (PPP)one, concerning only the π electrons, and the Complete Neglect of Differential Overlap (CNDO) one, involving all the valence electrons. These methods constitute the family of so-called semiempirical approaches. Nowadays, the most used method is the Austin Model 1 and has the acronym AM1 [5]. It provides satisfactory qualitative results as to the geometry and the energies, except when hydrogen bonds are concerned, because it does not reproduce the directionality of this peculiar interaction. It has the great advantage to be applicable to large systems (~500-1000 atoms).

II.2. The Basis Set Functions and their Gaussian Expansion

The basis set is qualified as minimal, double-ζ, triple-ζ, .., with polarization and/or diffuse functions, and so on. The meaning of these terms is based on the attribution of a certain number of well-defined basis functions to each element. First of all, a distinction core/valence shells is important in the definition of the basis set, as will be seen below, the core electrons weakly participating to the chemical reactivity. The electrons that must be the best possibly described are the valence ones, that come from the external shells of the constituting atoms. The most obvious way to attribute basis functions to an element is to consider that it will receive as many basis functions as its number of occupied shells in the periodic table. For instance, carbon which has the electronic representation $1s^2$, $2s^2$, $2p^4$, will be attributed five basis functions: 1s, 2s, $2p_x$, $2p_y$, $2p_z$, because the 2p's are equivalent. This is the minimal basis set: hydrogen has one basis function of 1s character, the second row elements have five basis functions (see carbon), the third row elements have 9 basis functions, and so on. However, the fact of choosing atomic functions, optimized for one-electron atoms, to describe molecular systems is equivalent to use bricks to build a doll's house. It lacks the flexibility to involve several molecular details though it will usually lead to qualitatively satisfactory results as to the geometry. This flexibility is included in the basis set by duplicating, or triplicating, each type of basis function. For instance, it includes two 1s-types (noted 1s, 1s'), two 2s-types (2s,2s'), two 2p-types (2p,2p')... functions, in which case it is called double-ζ basis set. If the basis set is triplicated, it is called triple-ζ basis set, and so on. Frequently, the shells related with the core electrons remain single and such basis sets are called valence double(or triple ...)-ζ basis sets. One of these that is very often used today has the acronym 6-31G: the one figure at the left of the minus sign means that the core shells are not duplicated while the two figures at the right of the sign means that the valence shells are duplicated. The flexibility is also brought by including monoelectronic functions that correspond to empty shells in the atom with a high orbital quantum number, thus describing 2p, 3d, .. levels for hydrogen, or 3d, 4f, ...for second row atoms, and so on. This type of functions are called polarization functions and can be represented by an asterisk * in the basis set acronym. One asterisk means inclusion of polarization functions on non-hydrogen atoms, two asterisks refer to inclusion of polarization functions on hydrogen and non-hydrogen atoms (6-31G* or 6-31G**). Finally, the basis set can be improved by adding diffuse functions, i.e. characterized by a small value of the exponential argument. These functions are often s-types and p-types functions and are meant to describe electronically excited states or negative ions. They are represented by a + in the acronym.

The basis function expression initially contains exponential factors. This implies that the above-mentioned integrals must be calculated numerically, which is a very CPU-time-consuming process. John Pople was a pioneer in realizing that the basis functions could be approximated as linear combinations of gaussian functions, that are much easier to handle since many of their product integrals have known values. For instance, in his first attempts to develop minimal basis sets, he proposed his STO-*n*G basis sets, with *n*=3,6, where each basis set function is expanded on *n* gaussian functions. Similarly, the valence double-ζ C-*vw*G involves C gaussian functions for each core shell basis function (1s for instance), *v* gaussians for the first valence shell (2s, 2p,..) and *w* gaussians for the second valence shell (2s', 2p', ..).

As an example, 6-311+G** is a valence triple-ζ basis set with polarization functions on hydrogen and non-hydrogen atoms (2p-type on H and 3d-type on second row atoms) and with diffuse functions on non-hydrogens atoms (sp-type). Each basis function is developed on gaussians, 6 for the core basis functions, 3 for the first, 1 for the second and 1 for the third valence shell.

II.3. The Electronic Correlation

In the Hartree-Fock framework, each electron is described by a one-electron function entering the expression of the total polyelectronic wave function, the Slater determinant, D0 in the following. The influence of the other electrons is taken into account through an average field. Consequently, this framework does not describe correctly the influence of the electrons on one another, i.e. the electronic correlation. This drawback of the method is mitigated by giving the wave function some more flexibility in its description: it is expanded on several Slater determinants $D_{i,a(,j,b)}$ constructed by replacing one or several occupied (i,j,..) orbitals by the virtual (a,b,..) ones, obtained in the LCAO procedure of the HF framework:

$$
\begin{aligned}
\Psi^{el}(q^{el},Q^{N}) \approx\ & u_0\, D0[\varphi_1\sigma_1, \varphi_2\sigma_2, \ldots \varphi_i\sigma_i \ldots \varphi_j\sigma_j \ldots \varphi_n(q_n^{el},Q^{N})\,\sigma_n] + \\
& \sum_{i,a} u_{i,a}\, D_{i,a}[\varphi_1\sigma_1, \varphi_2\sigma_2, \ldots \varphi_a\sigma_i \ldots \varphi_j\sigma_j \ldots \varphi_n(q_n^{el},Q^{N})\,\sigma_n] + \\
& \sum_{i,j,a,b} u_{i,a,j,b}\, D_{i,a,j,b}[\varphi_1\sigma_1, \varphi_2\sigma_2, \ldots \varphi_a\sigma_a \ldots \varphi_b\sigma_b \ldots \varphi_n(q_n^{el},Q^{N})\,\sigma_n] \\
& + \ldots\ldots\ldots\ldots
\end{aligned}
\tag{2}
$$

The coefficients u_t are obtained either variationally, like in the CI (configuration interaction) or QCI (quadratic CI) methods, or perturbationally, like in the Møller-Plesset MPx (x=2,3,4,5) methods. The summations in equation (2) concerns several thousands or even millions configurations. Thus, they are very CPU-time and disk consuming and cannot be applied to large systems (> 10-20 atoms). There are methods that allow to treat intermediate systems (~100 atoms) including electronic correlation: they belong to the density functional theory framework (DFT) [6] and rely on the concept of the density instead of that of the wave function. However, the computation of this density is copied on the HF framework: it remains a simple product of the orbital squared, orbitals that are derived through the Kohn-Sham equations [7], which seems similar to the Fock equations in their writing but whose terms are somewhat more complex and involve the electronic correlation. The terms of these equations that are problematic are the exchange and the correlation terms and there are many different functionals that have appeared in the literature which differ by their exchange and/or correlation functionals. The most known has the acronym B3LYP [8], which consists first of a three term exchange functional due to Becke, one of which is the HF exchange, and second of a Lee-Yang-Parr (LYP) correlation functional.

II.4. The Potential Energy Surface (PES)

The potential energy surface $E^{el}(Q^N)$ is the function that conditions the motion of the nuclei; its partial derivatives according to the nuclear coordinates give the forces acting on them. It is a (3N-6)-dimensional function and its expression or representation is highly dependent on the choice of the nuclear coordinates. Nevertheless, though only (3N-6) linearly independent nuclear coordinates are necessary to describe in a univocal way the molecular system and obtain $E^{el}(Q^N)$, this function can be expressed in a linearly dependent coordinate system, such as the cartesian (X,Y,Z) system or any other one. For small to intermediate systems, the chemist likes to work with his user-chosen *internal coordinates* (bond lengths, valence angles, dihedrals). However, for large systems (>500 atoms), like for instance host-guest complexes, it is much easier and efficient to work with a set of *internal redundant coordinates* [9-11] that can be generated automatically by programs like GAUSSIAN [12], from the cartesian coordinates that are available in the data banks. For instance, in the case of proteins, many 3D structures, determined by crystallography, are available in the Protein Data Bank (PDB) [13] through their cartesian coordinates. These redundant coordinates are chosen according to proximity criteria, e.g. all the distances between nuclei below a certain threshold depending of the van der Waals radii, all the angles between these bonds, and so on.

The characteristics of the large majority of points of the PES vary as a function of the chosen coordinates. Only a few particular points of the PES have constant characteristics no matter what the coordinates are: they are called stationary points and are defined by a zero gradient of $E^{el}(Q^N)$, $\nabla E^{el}(Q^N) = 0$. Let us recall that the gradient is a vector operator the components of which are the partial derivative operators according to the coordinates, here $\nabla = (\partial/\partial Q_1, .., \partial/\partial Q_i, ..., \partial/\partial Q_M)$ if M is the number of nuclear coordinates. The nature of the stationary points is determined thanks to the second derivative matrix, called the hessian matrix, $H = \{\partial^2/\partial^2 Q_i Q_j\}$, and more precisely to its eigenvalues:

$$X^{1} H X = \Lambda \tag{3}$$

where the columns of the X matrix are the eigenvectors of H and Λ is a diagonal matrix the elements of which are the eigenvalues λ_i of H. The components of the eigenvectors are coefficients that weigh the current coordinates Q_i, resulting in new coordinates $\mathcal{Q}_j$

$$\mathcal{Q}_j = \sum_i X_{ij} Q_i, \tag{4}$$

that are linearly independent (or orthogonal) and for which the curvature of the potential energy is known: it is the eigenvalue λ_j. In the particular case of the mass-weighted cartesian coordinates, the hessian matrix is the force constant matrix, the diagonalization of which leads to the frequencies $\nu_i = (2\pi)^{-1}\sqrt{\lambda i}$, and the eigenvectors are the well-known normal modes. If λ_i is negative, the related frequency is imaginary. As a matter of fact, the index of a localized stationary point is determined by calculating the frequencies. The interesting point of the diagonalization of H at a stationary point is that it provides a clear view of the PES topology since it indicates the directions along which the PES rises (all the $\mathcal{Q}_j$ related with λ_j positive) or goes down (all the $\mathcal{Q}_j$ related with λ_j negative). The description of these directions may be complicated since it normally involves all the chosen coordinates Q_i (see equation (4)). However, it often occurs that the coefficients X_{ij} in (4) are large only for a few Q_i, in which case the related coordinate $\mathcal{Q}_j$ is easy to imagine in terms of the user-defined Q_i.

The stationary points are classified according to the number of the hessian negative eigenvalues. If H has only positive eigenvalues, the stationary point is a minimum. In the case of one negative H eigenvalue, the stationary point is called a transition state, with the acronym TS. Otherwise the stationary points are called stationary or critical points of index m, where m is the number of negative H eigenvalues. The transition states are interesting points since they correspond to a region where the energy rises for any displacement away from the stationary point except one: that corresponding to $\mathcal{Q}_1$, associated with the negative eigenvalue λ_1. After two small displacements $-\delta\mathcal{Q}_1$ and $+\delta\mathcal{Q}_1$ along the direction given by the eigenvector $\mathcal{Q}_1$, two paths can be generated by following either the minimum energy or the steepest descent criterion. This latter criterion corresponds to the so-called Intrinsic Reaction Coordinate (IRC) by Fukui [14] and is the most used. The two paths lead to two local minima connected by the initial TS. The reaction coordinate is defined as the coordinate that allows going from one minimum to another and is a line constituted by the two above-mentioned paths constructed from the TS; and from the way it is built, it is also the minimum energy path from one minimum to the other, the TS being the highest point on that path.

II.5. The Geometry: Optimization and Reaction Coordinate

Each point on the PES corresponds to a geometry, i.e. a nuclear arrangement of the molecular system. Only a few peculiar points are interesting on this surface: the stationary points, and among them, the minima and the TS. The problem is thus to localize these points on the PES, i.e. to perform a geometry optimization. If only the minima are to be obtained, rather simple methods based only on the knowledge of the gradient are needed, such as the *steepest descent* method or the *conjugate gradient* method [15]. If stationary points of non-zero index are searched, methods involving the hessian must be used. Actually, the most efficient way to perform a geometry optimization is to start with a calculated hessian and follow a *Quasi Newton Raphson* (QNR) step [15], either in user-defined internal coordinates or in redundant coordinates. For very large systems, the use of a tabulated hessian transformed into the redundant coordinates and the use of a QNR method is most of the time the best compromise.

The definition of a reaction coordinate is somewhat more fuzzy because it can be made either in the framework of quantum chemistry or in that of general chemistry or biochemistry.

For a quantum chemist, the reaction coordinate is the set of geometries corresponding to all the points on the line defined above, i.e. the line made of the two paths connecting the TS to its two related minima. Obviously, only a few points, i.e. a few geometries, are determined on these paths. Thus, the first step in the definition of the reaction coordinate is the localization of the TS. All these geometries are characterized by their coordinates $\{Q_i\}(p)$, p numbering the point of the path. The Q_i's are also called the degrees of freedom (DoF) of the system. The conceptual difficulty with the reaction coordinate is that there is no any simple expression with the $\{Q_i\}(p)$, it can only be written a general expression of the form

$$S = f(\{Q_i(p)\}) \tag{5}$$

which means that the reaction coordinate S is a function of the points p via their coordinates Q_i. Fortunately, in many cases, only a few coordinates present large variations along the reaction coordinate which can thus easily be imagined and represented by the user.

For the organic chemist or biochemist, the concept of reaction coordinate is not necessarily so precise. They usually imagine a pre-defined coordinate, involving one or two internal coordinates or DoF, that the system will follow. In practice, these chosen DoF's are fixed at certain values along the deformation path and all the other DoF's are optimized. On this *a priori* chosen path for which the energy is calculated, the transition point determining the activation barrier is the highest energy point. Usually, such a point is not a stationary point and it is *de facto* higher than the nearest TS. Nevertheless, this method has the advantage that the first step is not the search of a TS and that a simple deformation path can be followed to model the reaction path. This can give good ideas of where the real TS is and provide qualitative results as to the energy barrier.

When the molecular system is large, the calculation of the hessian necessary to converge to a TS is impossible, thus a smaller model is created for which a TS search is possible. In such models, the environment is modelled by separate small entities in non-covalent interaction with one another. For the reacting ones, this is not a problem because the nature of the TS conditions their relative position. As to the others, their optimal positions in the TS and in the minima are generally not related with what should be found in the whole system. This is due to the loss of the covalent constraints with the rest of the system and the too large freedom of these separate entities. To mimic the situation inside the whole system, it is sometimes chosen to work by constraining the separate entities in fixed relative positions. However, without the rest of the environment, such constrained optimizations do not lead to stationary points, thus the number of negative eigenvalues of the hessian is not defined, and the energy of the constraint may be very different for the three points meant to represent the two minima and the TS. Thus, in such constrained optimizations, the energies obtained must be viewed as very qualitative.

Nowadays, to handle large systems, a very interesting approach is used: the so-called mixed QM/MM models, where the reactive centre is modelled at the quantum mechanics level (QM) and the environment is modelled by the classical mechanics formalism through the use of the molecular mechanics (MM) force fields. The quality of the QM/MM approach has been extensively tested by Gherman et al. [16] on a variety of problems, comparing the QM/MM results with both all QM and experimental ones. They can assess that the difference between QM/MM and QM is of the order of 1-2 kcal/mol. Molecular mechanics is a very useful tool to optimize the geometry of huge systems (several thousands of atoms) but is only meant to find local minima. The PES is modelled by very simple functions of the nuclear coordinates constituting the force field. For instance, the deformation of bond lengths can be modelled by a harmonic function:

$$U(R-R_0) = (\tfrac{1}{2})\, k\, (R-R_0)^2$$

k and R_0 being parameters defined according to the nature of the atoms involved. These parameters are tabulated and constitute what is called the force field. The different existing force fields have been parametrized for chemical families. For instance, the AMBER [17] and CHARMM [18] force fields are parametrized for peptides, nucleic acids and carbohydrates. The MM describes the nuclear motions only through a parametrized force field, and it cannot describe electronic changes or rearrangements, i.e. chemical reactions with bond breaking and bond formation.

The mixed approach that will be used in one illustration below is called ONIOM [19] and is implemented in the GAUSSIAN program. It implies two or three different calculation levels related with two or three regions of the whole system studied. Chart 1 illustrates the idea of the method. The smaller system S1 which is the center of the study is considered at the highest calculation level. Its direct environment S2 is concerned by the intermediate calculation level. The rest of the whole system S3 is related with the lowest calculation level. The advantage of the method is that the three calculation levels can be chosen to be all quantum or not. Moreover, only two regions can be considered instead of three. In GAUSSIAN 98 [12a] the electrostatic properties of the MM region were not taken into account in the QM region, i.e., the so-called electronic embedding was not considered. In GAUSSIAN 03 [12b], this feature has been implemented.

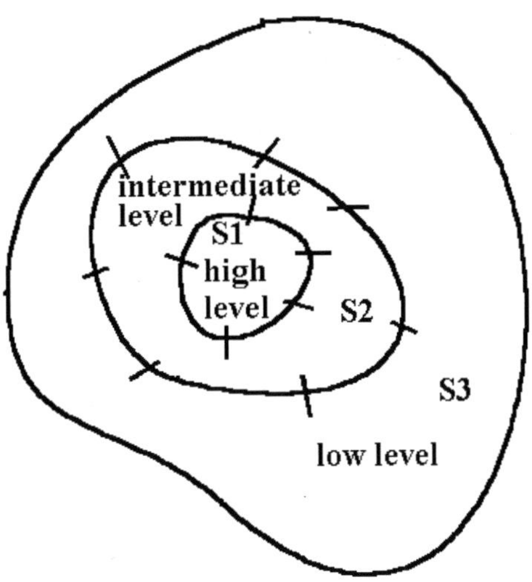

Chart 1.

III. THE MECHANISTIC MODELS AND THE LITERATURE

The review of the literature on the subject will be limited to some examples and is not aimed at being exhaustive. The chosen citations are intended to highlight the difficulty of the task and the variability of the approaches. Moreover, the details of the calculations in the different studies are not mentioned here.

Though there are many studies based on molecular dynamics, used either alone [20,21] or in conjunction with QM/MM calcutations [22-24] but not analyzing the reaction mechanism, we will only focus on studies on the enzymatic reactions based on QM and QM/MM.

The basic line of the reaction pathway leading to the inactivation of β-lactam antibiotics by serine β-lactamases (classes A, C and D) is commonly accepted as a very simple two-stage reaction, from the so-called Henri-Michaelis complex to an acylenzyme, and from the acylenzyme to the open β-lactam and the regenerated enzyme. The first stage is the acylation and the second stage the deacylation. However, the widespread agreement stops there. For instance, two different hypotheses are considered to describe both reactions: they are either described as a one-step concerted process [25-28] or as a two-step process involving a so-called tetrahedral intermediate (TI) [28-31]. Several models as to how the acylation proceeds have been published [25-34], the deacylation having been somewhat less studied [16,35-37] since it was initially considered as a very similar reaction, with a water molecule acting as the nucleophile instead of the catalytic serine. Nevertheless, due to the substantial difference in deacylation rates for acyl-enzyme intermediates in penicillin-binding proteins (PBP) and β-lactamases, this step has been more deeply investigated in the last couple of years. The two reactions are supposed to occur with the help of some residues in the active site that could "activate" the nucleophile (serine or water) (base catalyst) or protonate the leaving group (acid catalyst) [30,34,38]. The chemically and spatially highly conserved motifs in the active site have long been emphasized but their role continues to be controversial. For instance, it is very often supposed that the catalysis is of basic type and the so-called general base (GB) in class A β-lactamases is considered to be either a neutral Lys73 [29] or a water molecule assisted by the anionic Glu166 [31,35,36]. Thus, in the first case, a debate arose as to the pK of this lysine since normally it should be protonated. This is indeed what is suggested by either experimental results [39,40] or theoretical calculations [41,42]. The search of the general base is also largely debated in class C β-lactamases where Tyr150 (Tyr of the second motif YSN) is often modelled as a tyrosinate [16,37] activating the nucleophilic serine. However, this hypothesis seems to have become less likely recently according to experimental evidence indicating the presence of a hydrogen on this tyrosine [43]. It remains that the properties of an alcohol are not the same as those of a phenol.

A point that we would like to emphasize here concerns the constraints introduced in the geometry optimizations of the models. Some authors argue that a model cannot reflect all the constraints existing in the whole system and thus choose to restore them by forbiding some nuclei to move, freezing them at their crystallographic positions for instance. Or in other cases, *a priori* reaction coordinates are chosen within a partially constrained model, like for instance allowing one proton to move between two heavy atoms the distance of which is frozen. This could seem apparently reasonable at first sight but this may lead to erroneous conclusions, like generating an energy barrier for the proton transfer that would not exist if the optimization was unrestrained. In our point of view, a constrained optimization should always be performed in parallel with the unconstrained one for comparison. It should not be forgotten that even if the whole system imposes constraints on the active site, it is also rather mobile and a complete freezing of some points does not finally seem reasonable. Moreover, due to the coupling between the nuclear coordinates, the fact of freezing some of them might have an incidence on the gradient and the hessian elements involving free coordinates coupled with the frozen ones and thus have an effect on the geometry and thus on the energy.

In the following, we will concentrate on the acylation reaction. Two recent reviews on the subject have been written Hermann et al. [31] and López et al. [34]. This matter can be tackled from several points of view, two of which are presented hereafter.

The first one deals with the intrinsic reactivity of the β-lactam ring and a detailed study of its opening upon the action of several nucleophiles has been considered [32-34]. The list of nucleophiles usually considered is mainly the following: OH^-, H_2O, CH_3OH, RNH_2. Moreover, the nucleophiles can be supported by a partner, H_2O or CH_3OH for instance.

The second point of view concerns the event sequence in the acylation. It can be considered as a two-step event [27-31] passing through an intermediate minimum called the tetrahedral intermediate TI and through one or two transition states (TS), or as a one-step concerted process [25-28] passing through only one TS. Depending on this choice, the influence of the environmental residues might be found to be different. This is particularly true as to the so-called general base (GB). The two reaction schemes are presented in Figure 1.

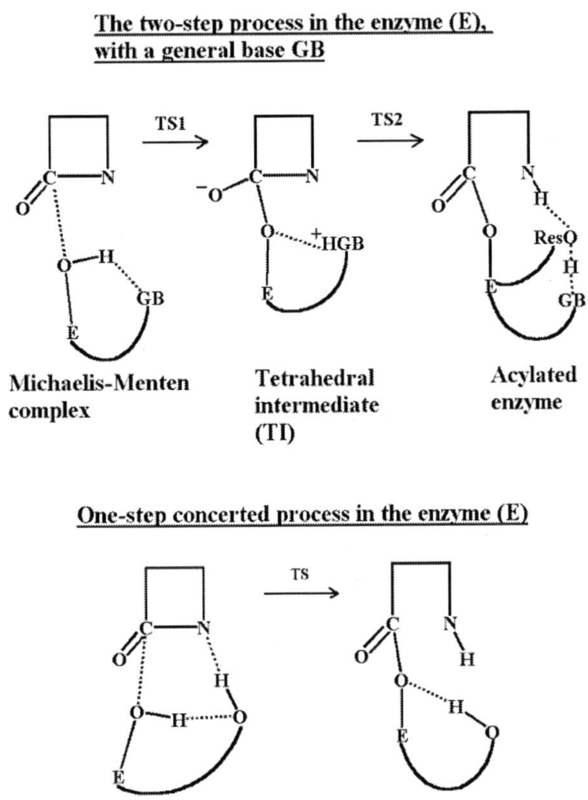

Figure 1. Schematic presentation of the two-step and one-step hypotheses for the acylation mechanism.

The two-step propositions are mainly considering the base-type catalysis and the key factor is thus the determination of the GB that must activate the nucleophilic serine. For Wladkowski et al. [29], the GB is the neutral K73 that abstracts the S70 proton and reaches a transition state (TS1 in Figure 1) around 23 kcal/mol above the energy level of the reactants. The activated S70 attacks the β-lactam carbonyl carbon leading to the TI at 19 kcal/mol. The protonated K73 gives back its proton to S130 (ResOH in Figure 1) which transfers its own to the β-lactam nitrogen with concomitant opening of the ring. The second transition state, TS2, lies around 24 kcal/mol and the acylenzyme is much more stable than the reactants, at –19 kcal/mol. From a purely energetic point of view, the authors found that the oxyanion hole stabilizers had an influence only on the TI energy, not on the TS ones. Without oxyanion hole stabilizer, the energy of the TI was around 25 kcal/mol, i.e. very close to the TS levels. Let us point out that the geometry optimizations were performed under the constraint that some backbone nuclei remain at their crystallographic positions. This role of a neutral K73 is questionable. On the basis of QM/MM energy calculations on snapshots chosen from a MD study on a complex between TEM-1 and benzylpenicillin [22], the ionic pair [K73H^{+}□-E166^{-}] is preferred to the non ionic [K73–E166H]. Moreover, some experimental studies tend to show that K73 is protonated [39,40] as well as theoretical determination of the K73 pKa (~10) by continuum electrostatic calculations [41,42].

Hermann et al. [31] studied the acylation process at the QM/MM level by scanning the energy landscape according to two coordinates meant to represent the hypothetical path from the reactants to the acylenzyme and vice versa. As a matter of fact, they generated their potential energy surface PES in a backward and a forward direction to check the results. Their hypothesis about the GB considered that S70 should be activated by a water molecule assisted by the glutamate E166, K73 being in its protonated state. The activated S70 attacks the carbonyl carbon of the β-lactam to form the TI and a neutral water-E166H pair. Then the protonated K73 gives one proton to S130 which gives its own to the β-lactam nitrogen while the four-membered ring opens. The proton transfer from E166H to the neutral K73 seems to occur spontaneously, i.e. without any energy barrier. The points that the authors called transition states were not defined as is usual in quantum chemistry, through the number of imaginary frequencies at a stationary point, but are taken as the highest energy point on the path connecting the reactants to the TI, or the TI to the acylenzyme. These transition states are probably not stationary points. They obtained only one TS higher in energy than the reactants, the first one, associated with an activation barrier of about 9 kcal/mol. The second TS was found 7 kcal/mol higher than the TI but 8.5 kcal/mol lower than the reactants. The acylenzyme is found 30 kcal/mol lower than the reactants.

In the work by Meroueh et al. [30], the strategy is also based on a 2D scan of the PES at the QM/MM level, leading to the TI starting from the Henri-Michaelis complex with either a zwitterionic pair [K73H^{+}□-E166^{-}] or a non zwitterionic one [K73-□E166H], i.e. considering that the GB is either a water molecule assisted by E166^{-} or the neutral K73. The energy barrier for passing from a zwitterionic to a non zwitterionic pair K73-E166H is equal to 5 kcal/mol, the former being 4 kcal/mol higher in energy than the latter. The influence of the size of the QM region is investigated. The energy barrier for passing from the reactants to the TI is 26 kcal/mol if water-E166^{-} is the GB and 22 kcal/mol if K73 is the GB. The TI does not seem to be a stationary point, just as the TS on these PES. A MD study shows that the TI is subject to conformational changes that orient the pair K73H^{+}-S130 for a proton transfer to the β-lactam nitrogen. This new entity collapses without energy barrier to the acylenzyme, the protonated K73 giving its proton to S130 which gives its own to the β-lactam nitrogen, with the concommitant ring opening. This final process is highly exothermic, by approximately 40 kcal/mol. The authors also performed an analysis of the acid catalysis mechanism where the first step is the protonation of the β-lactam nitrogen by the pair S130-K234 leading to a N-protonated Henri-Michaelis complex which lies 22 kcal/mol higher than the reactants. This complex is then attacked by the S70 assisted by the water-$E166^-$ pair with the opening of the four-membered ring and the formation of the acylenzyme. The energy barrier for the first step is 29 kcal/mol while for the second it amounts to 12 kcal/mol by reference to the N-protonated Michaelis complex, i.e., 34 kcal/mol by reference to the reactants. Thus, the acid catalyzed reaction is not favorable compared with the base catalyzed one. Nevertheless, it seems that none of the so-called transition states are optimized stationary points and the reaction paths proposed are based on *a priori* chosen coordinates which may not reflect the real reaction paths.

In the work by Fujii et al. [35] on the deacylation process, the enzymatic reaction is anticipated to be a substrate-assisted catalysis (SAC) since the carboxylate group on the thiazolidine ring of benzylpenicillin was supposed to play a role in the first step leading to the activation of the water-$E166^-$ to attack the carbonyl of the acylenzyme ester. Their study relied upon QM calculations involving an Onsager continuum solvent model [44], i.e., involving no explicit water molecules, and their geometry optimizations used constraints upon some nuclei positions. This Onsager continuum model considers that the solute lies in a spherical cavity surrounded by a continuum solvent medium characterized by its relative dielectric constant ε, and the solvent response to the solute electric field is proportional to the dipole moment of the latter. The deacylation proceeded in four steps, the highest constrained TS being at about 30 kcal/mol above the reactant acylenzyme energy.

This idea of a SAC is also found in the article of Díaz et al. [27] in their two-step mechanism for the acylation process of TEM1 by benzylpenicillin (BZ). Moreover, these authors compare the two-step mechanism with a one-step concerted one, already proposed in the literature [25,26]: a S70 nucleophilic attack with simultaneous opening of the β-lactam ring and proton transfer from the S130 hydroxyl group to the ring nitrogen. The authors also performed calculations on the hydrolysis of the 2-formamide acetate and the methanolysis of 3α-carboxy penam. They incorporated solvent effects through a continuum model, i.e., with no explicit water molecules. Their optimizations on TEM1-BZ were constrained. In the gas phase, the SAC is preferred for both the amide hydrolysis and the penam methanolysis. However, the solvent effect reverses the trend since the one-step concerted mechanism becomes more favorable in the studied case of the penam methanolysis. In the whole system TEM1-BZ, the favored path is the carboxyl-assisted mechanism, even if the effect of the environment stabilizes preferentially the one-step mechanism.

Wolfe et al. [25] considered gas phase one-step concerted neutral hydrolysis and methanolysis of monocyclic and bicyclic azetidinones, two of which were penams and one of which was bearing a carboxylate group.They considered four-centered TS with either N-protonated or O-protonated conformations (Scheme 1). They also analyzed the catalytic effect of a supplementary water molecule on the reaction. In every case, the N-protonated conformation was lower in energy than the O-protonated one. The presence of an additional water molecule considerably lowers (~10 kcal/mol) the activation barriers.

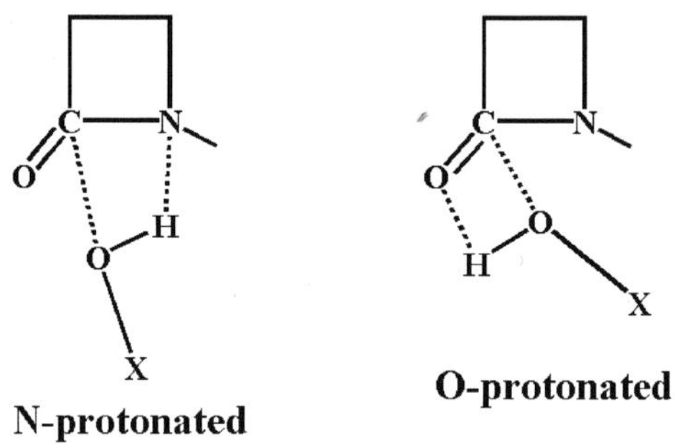

Scheme 1.

Dive and Dehareng [26] also considered gas phase one-step concerted reactions with several combinations of the hydroxyl nucleophile and proton donor, the latter being equivalent to the catalytic water of Wolfe et al.. One monocyclic and five bicyclic azetidinones were studied (four penams and one cephem). The carboxylic head was considered neutral because the stabilizing protein environment was not considered. The different choices of the pair (nucleophile-proton donor) did or did not include one possible oxyanion hole stabilizer. The energy stabilization due to the oxyanion hole stabilizer was found around 4 kcal/mol. It was also pointed out that the conformations of the benzylpenicillin thiazolidine ring, noted PIPCIL and BZPENK, present different reactivities, and that the penams are more reactive than the cephem. It was also found that the geometry of the TS structures fitted very easily into the active site cavity of TEM1, the two serine S70, S130 and the lysine K73 being already well positioned to accomodate the TS.

The intrinsic reactivity of β-lactams was studied by Frau et al. [32] through the nucleophilic attach of the hydroxide anion OH^-. The charge of the nucleophile induced either a C-H or a N-H bond breaking to produce water and a negatively charged β-lactam. This observation compelled the authors to freeze the C-H or N-H breaking bond during the OH^- attack, and resulted in an artificial low energy conformer in the 2D energy profile, followed by a slightly higher energy structure. Anyway, these two conformations were lower in energy than the reactants and the nucleophilic attack proceeded without energy barrier.

Massova and Kollman [33] present a detailed study of the hydrolysis/methanolysis of a variety of amides and β-lactams in order to evaluate the effect of ring strain, substituents and loss of amide resonance on the selected β-lactams. The ΔG/ΔE were evaluated between reaction intermediates, not between TS's. The reaction pathways are studied in the gas phase and in solution through the polarizable continuum model (PCM) [45]. In this continuum model, the solvent is still represented by its dielectric constant but the shape of the solute cavity is defined by the van der Waals radii of the solute nuclei and the solvent response to the electric field of the solute is considered through a charge density on the cavity surface. The first step of the acylation reaction is the formation of the TI and the second step is the acid catalyzed opening of the β-lactam ring. The first step has a positive ΔG which is considered by the authors as the activation barrier associated with the rate-limiting step. Going from an amide to a β-lactam produces an activation energy decrease in solution of about 3.5-4.2 kcal/mol and the inclusion of a second ring brings another energy decrease of 1.6-8.8 kcal/mol. These authors also found that hydroxide attack in the gas phase was occurring without any energy barrier. The second step is found to be very exothermic with negative ΔG. The free energy of the complete acylation reaction is correlated with the barrier height for the first step. They also found that the length of the β-lactam C=O bond correlates with its hydrolytic stability.

López et al. [34] present a review of theoretical studies about β-lactam ring openings through hydrolysis, alcoholysis, aminolysis and ozonolysis, in gas phase as well as in solution. They conclude that this subject is a well understood process and that the intrinsic reactivity of new β-lactam compounds against hydroxyl/alcohol/amine nucleophiles can be routinely accessed using standard computational tools.

To conclude this section, we would like to stress that we agree with López et al. [34] as to the intrinsic reactivity studies and knowledge about β-lactam ring opening. However, as illustrated above by the few chosen examples, such a consensus does not prevail for the analysis of the acylation and deacylation processes. All the investigations present different hypotheses and different ways to approach them. Anyway, two common characteristics arise from the diversity of the analyses. First, it appears that, in the two-step approach, the general base should be the pair water-E166$^-$ and that K73 should be protonated. Secondly, the acylenzyme formation is very exothermic. Beyond that, it seems that the investigation field is still open to debate.

IV. Illustrations on the Use of QM and/or QM/MM in the Investigation of the Enzymatic Reactions

What is obvious from the preceding brief discussion of a few articles on the subject is that the answers obtained are intimately linked with the hypotheses proposed and that they are to be considered as essentially qualitative. This does not alter the great help that theoretical models can provide. As illustrations of this qualitative characteristics and the nevertheless usefulness of the tool, we present two types of studies at the QM or QM/MM level. The first one relates to a comparison of the intrinsic reactivity of one amide, one ester and three β-lactams at the QM level. The second one concerns the protonation state of the lysines involved in the active site, at the QM/MM level.

IV.1. Intrinsic Reactivity of an Amide, an Ester and β-lactams: The Nucleophilic Attack by OH$^-$

The attack by OH$^-$ produces a TI without any energy barrier and its formation energy ΔE_f is calculated as the energy of the TI after subtraction of the sum of the energy of the reactants. The path going from the TI to the final ester passes through a TS. The activation energy ΔE_{act} is calculated as the energy of the TS minus that of the TI. The three stationary points calculated in the present comparison are the initial minimum corresponding to the reactants, the TI and the TS. The β-lactams considered are one monobactam, the azetidin-2-one, and three β-lactams with thiazolidine fused rings in their two conformations noted BZPK and PIPC. The reactants are presented in Scheme 2. The results are shown in Table 1.

Table 1. Energy differences ΔE_f(kJ/mol) between the stationary points TI,TS and the sum of the reactant [ReacI+OH$^-$] energies

Reactant-stationary point	ΔE_f[RHF/ 6-31+G] (ΔE_{act})	ΔE_f[RHF/ 6-31+G*] (ΔE_{act})	ΔE_f[B3LYP/ 6-31+G] (ΔE_{act})	ΔE_f[B3LYP/ 6-31+G*] (ΔE_{act})
Reac1-TI	-62.19	-80.48	-67.04	-84.94
Reac1-TS	62.83 (125.02)	19.78 (100.26)	20.56 (87.60)	-11.38 (73.56)
Reac2-TI	-116.60	-113.00	-118.92	-120.39
Reac2-TS	-69.55 (47.05)	-66.76 (46.24)	-98.96 (19.96)	-99.77 (20.62)
Reac3-TI	-84.45	-102.93	-83.32	-105.81
Reac3-TS	-23.70 (60.75)	-49.38 (53.55)	-50.23 (33.09)	-74.58 (31.23)
Reac4(BZPK)-TI	ND	ND	-127.44	-136.26
Reac4(BZPK)-TS	ND	ND	-123.08 (4.36)	-132.78 (3.48)
Reac4(PIPC)-TI	ND	ND	-125.72	-136.10
Reac4(PIPC)-TS	ND	ND	-117.11 (8.61)	-131.23 (4.87)
Reac5(BZPK)-TI	ND	ND	-152.37	-159.36
Reac5(BZPK)-TS	ND	ND	-151.52 (0.85)	-159.11 (0.25)
Reac5(PIPC)-TI	ND	ND	-151.85	-160.84
Reac5(PIPC)-TS	ND	ND	-143.24 (8.61)	-157.76 (3.08)

ND: not determined.

The reactants are presented in Scheme 2.

For the TS, the activation energies ΔE_{act} by reference to the TI are shown in parentheses.

Including the electronic correlation at the DFT level has a small effect on the TI formation energy ΔE_f but a large one of the activation energy ΔE_{act}: the activation energy decreases significantly (by respectively 22.32 and 37.42 kJ/mol for Reac3/6-31+G* and Reac1/6-31+G).

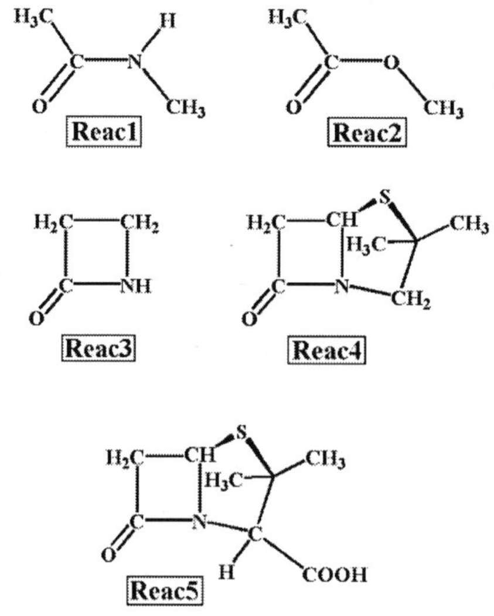

Scheme 2.

Table 2. Scissile C-X (X=N,O) bond length (Å) for the reactant minimum, TI and TS stationary point geometries, obtained at four calculation levels

Reactant-stationary point	RHF/6-31+G	RHF/6-31+G*	B3LYP/6-31+G	B3LYP/6-31+G*
Reac1-min	1.350	1.350	1.368	1.366
Reac1-TI	1.467	1.482	1.485	1.507
Reac1-TS	2.194	2.147	2.220	2.272
Reac2-min	1.345	1.325	1.378	1.354
Reac2-TI	1.462	1.448	1.532	1.507
Reac2-TS	1.995	1.891	2.164	2.053
Reac3-min	1.364	1.356	1.383	1.375
Reac3-TI	1.505	1.509	1.552	1.554
Reac3-TS	1.994	1.964	2.078	2.057
Reac4(BZPK)-min	ND	ND	1.405	1.399
Reac4(BZPK)-TI	ND	ND	1.639	1.638
Reac4(BZPK)-TS	ND	ND	1.874	1.850
Reac4(PIPC)-min	ND	ND	1.403	1.398
Reac4(PIPC)-TI	ND	ND	1.681	1.685
Reac4(PIPC)-TS	ND	ND	2.065	2.023
Reac5(BZPK)-min	ND	ND	1.407	1.401
Reac5(BZPK)-TI	ND	ND	1.674	1.676
Reac5(BZPK)-TS	ND	ND	1.816	1.767
Reac5(PIPC)-min	ND	ND	1.405	1.399
Reac5(PIPC)-TI	ND	ND	1.684	1.701
Reac5(PIPC)-TS	ND	ND	2.121	2.010

On the contrary, the effect of the basis set seems more important on the ΔE_f than on the ΔE_{act}, at least on the absolute energy values, but the relative value variations can be very large for ΔE_{act} too (65-70% for Reac5/PIPC-BZPK).

There does not seem to be a systematic effect of the basis set on the scissile C-X (X=O,N) bond length (Table 2). On the contrary, taking the electronic correlation into account systematically makes the scissile bond length increase.

Thus the energies and geometries can change significantly according to the calculation level. However, the qualitative trends are constant as illustrated below by three characteristics.

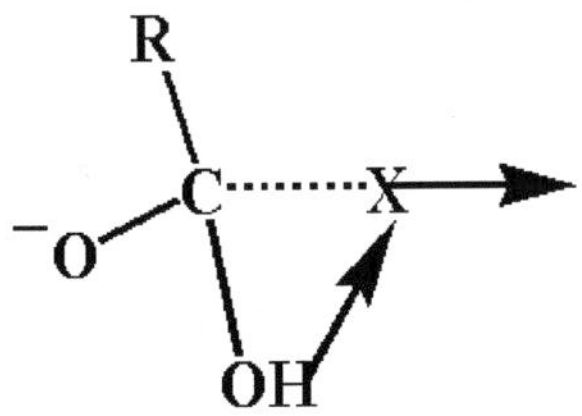

Scheme 3.

Table 3. Intrinsic reactivity classification of the reactants ReacI (Scheme 2), based on the TI energy formation ΔE_f taken from Table 1

	RHF/6-31+G	RHF/6-31+G*	B3LYP/6-31+G	B3LYP/6-31+G*
Most reactive	Reac2 (0.0)	Reac2 (0.0)	Reac5-BZPK (0.0)	Reac5-PIPC (0.0)
	Reac3 (32.15)	Reac3 (10.07)	Reac5-PIPC (0.52)	Reac5-BZPK (1.48)
	Reac1 (54.41)	Reac1 (32.52)	Reac4-BZPK (24.93)	Reac4-BZPK (24.58)
			Reac4-PIPC (26.65)	Reac4-PIPC (24.74)
			Reac2 (33.45)	Reac2 (40.45)
			Reac3 (69.05)	Reac3 (55.03)
Least reactive			Reac1 (85.33)	Reac1 (75.90)

Each column begins by the most reactive species with the $\delta\Delta E_f$ (kJ/mol) in parentheses.

The first one concerns the motion associated with the imaginary frequency of all the TS. Whatever the calculation level, it always corresponds to the bond breaking. However, for the smaller systems (ester, linear amide and monobactam) it also involves the nitrogen protonation by the hydroxyl group (Scheme 3).

The two other characteristics are related to the intrinsic reactivity of the reactants ReacI, first on the basis of the TI formation and secondly on the TS formation.

The attack by OH^- reflects how easily the TI can be formed: the more negative the TI formation energy ΔE_f, the more easily the TI is formed, i.e., the larger the intrinsic reactivity of the reactant. A classification of this easiness is presented in Table 3. Whatever the calculation level, the classification is the same, except the inversion Reac5-BZPK//PIPC but the energy differences (0.52 and 1.48) are too small to be significant. It also appears that the two conformations BZPK and PIPC present the same reactivity upon the hydroxide attack.

Once the TI is formed, the studied reaction model is not completed. There is still a bond to be broken. The easiness of this reaction is monitored by the activation energy ΔE_{act} related to the TS formation: the lower the ΔE_{act}, the more easily the bond is broken. The classification is presented in Table 4. The group Reac1-Reac3 is well separated from Reac4 and Reac5 and is ordered in the same way as what is found with ΔE_f. The difference appears for Reac5 and Reac4 which turn to be very similar according to the ΔE_{act}, which was not the case according to the ΔE_f. The ordering of these four reactants being variable as a function of the basis set used, the only conclusion that can be drawn is that they present very similar reactivities as to the C-N bond breaking.

Table 4. Intrinsic reactivity classification of the reactants ReacI (Scheme 2), based on the TS activation energy ΔE_{act} taken from Table 1

	RHF/6-31+G	RHF/6-31+G*	B3LYP/6-31+G	B3LYP/6-31+G*
Most reactive	Reac2 (0.0)	Reac2 (0.0)	Reac5-BZPK (0.0)	Reac5-BZPK (0.0)
	Reac3 (13.70)	Reac3 (7.31)	Reac4-BZPK (3.51)	Reac5-PIPC (2.83)
	Reac1 (77.97)	Reac1 (54.02)	Reac5-PIPC (7.76)	Reac4-BZPK (3.23)
			Reac4-PIPC (7.76)	Reac4-PIPC (4.62)
			Reac2 (19.11)	Reac2 (20.37)
			Reac3 (32.24)	Reac3 (30.98)
Least reactive			Reac1 (86.75)	Reac1 (73.31)

Each column begins by the most reactive species with the $\delta\Delta E_{act}$ (kJ/mol) in parentheses.

As a conclusion, one can say that the study of intrinsic reactivity through the hydroxide attack remains a qualitative tool that is very useful as a first step in a classification because it provides significant results in a short time. It can be performed as a routine technique. The reactants are presented in Scheme 2. ND = not determined.

IV.2. The Protonation Energies of the Two Active-site Lysines in Serine β-Lactamases

The coordinates of the three chosen β-lactamases TEM1, P99 and OXA10 were taken from the Protein Data Bank [46]. The case of OXA10 is particular since it is well-known [47] that the Lys70 is carboxylated in the optimal pH conditions. However, this enzyme was also considered in its uncarboxylated form in order to have an idea of the energy difference related to this characterictics. No solvent molecules were added to the systems and thus, the side chains of all the Asp, Glu and Lys were considered in their neutralized forms and only the lysines of the two conserved motifs SXXK and KTG were concerned by the protonation process, except for TEM1 where the proximate Glu166 (E166) was also taken into account. All the Arg were taken in their very stable protonated state. For each enzyme, four combinations of Lys protonation states were studied and noted as [$SXXK^0$, K^0TG], [$SXXKH^+$, K^0TG], [$SXXK^0$, KH^+TG], [$SXXKH^+$, KH^+TG], considering moreover E166 in its neutral or anionic states. Thus, sixteen structures were optimized at the molecular mechanics (MM) level using the Amber force field [17]. The dielectric constant was chosen equal to 2 and the convergence test was the maximum force for which the threshold was 0.5 kcal mol^{-1} $Å^{-1}$. The program used was Discover [48]. For every optimized molecule, a hybrid quantum mechanics/molecular mechanics (QM/MM) calculation is performed using the ONIOM method [19] of the GAUSSIAN 98 [12a] program. The high calculation level was the restricted Hartree-Fock (RHF) level using the 6-31G basis set. The intermediate calculation level was the semiempirical AM1 [5]. The low calculation level was the MM UFF one [49].

The High and Intermediate Level REGIONS S1 and S2 are Defined in Table 5

The protonation energies are calculated as the differences between the total energies for the different structures defined as follows.

$$\Delta E1 = E(SXXKH^+,K^0TG) - E(SXXK^0,K^0TG)$$
$$\Delta E2 = E(SXXK^0,KH^+TG) - E(SXXK^0,K^0TG)$$
$$\Delta E3 = E(SXXKH^+,KH^+TG) - E(SXXK^0,KH^+TG)$$
$$\Delta E4 = E(SXXKH^+,KH^+TG) - E(SXXKH^+,K^0TG)$$

ΔE1 corresponds to the protonation energy of the SXXK lysine in an environment where the KTG lysine is in its neutral form, ΔE2 relates to the protonation energy of the KTG lysine in an environment where the SXXK lysine is in its neutral form, ΔE3 is the protonation energy of the SXXK lysine in an environment where the KTG lysine is in its protonated form, ΔE4 corresponds to the protonation energy of the KTG lysine in an environment where the

SXXK lysine is in its protonated form. For TEM1, these four results are considered for both neutral and anionic forms of respectively E166. The results are presented in Table 6.

The penalty on the protonation energy of one lysine due to the protonation of the neighbouring one is respectively about 10 kJ/mol, -90 kJ/mol, 500 kJ/mol, 670 kJ/mol for TEM1(E166neut), TEM1(E166 –1), P99 and OXA10 respectively. For P99 and OXA10, the presence of one charged lysine largely decrease the proton affinity or the basicity of the second lysine. For TEM1, the two lysines behave rather independently as to their protonation tendency. Moreover, when E166 is negatively charged, it appears that the second lysine is slightly more easily protonated when the first one, whatever it may be, is already protonated itself. This result may come from the fact that the geometries are not the same and have been optimized for each structure. Nevertheless, the two results 10 and -90 kJ/mol are much smaller than those concerning P99 and OXA10. Following the conclusions of the experimental [39,40] and theoretical [41,42] works concerning the protonation state of K73 in TEM1, we will consider that both TEM1 lysines are protonated in this case.

As to the protonation trend related with the environment, the comparison of ΔE1 with ΔE2 or ΔE3 with ΔE4 reveals that protonation of the two lysines occurs with equal facility in all the β-lactamases.

As far as the value of the first protonation energy is concerned (see the largest absolute value of ΔE1 or ΔE2), all the lysines appear to present a similar basicity. On the basis of the present results, it is not possible to decide which lysine should be the protonated one. It probably also depends on the interaction with the substrate. On the basis that the two TEM1 lysines are protonated and that ΔE1 and ΔE2, or ΔE3 and ΔE4, are of the same order of magnitude for all β-lactamases, we could make the hypothesis that, for P99 and OXA10, at least one lysine is protonated. In P99, the second one may or may not be protonated but in OXA10, it is most probable that it will not be. This can be put in relation with the fact that K70 is carboxylated in the wild type enzyme.

All these results lead to think that the lysines could have different roles in these enzymes, in relation with their protonation state. The conservation of these residues does not necessarily mean the conservation of their role in the reaction mechanism or even the conservation of the reaction mechanism itself.

Table 5. S1 and S2 definition (see Chart 1). Entire residues are considered except when mentionned by *sch* meaning that only the side chain is chosen, or *bb* meaning that only the backbone is considered

	TEM1	P99	OXA10
S1	K73(*sch*), K234(*sch*), E166 (*sch*)	K67(*sch*), K318(*sch*)	K70(*sch*), K205(*sch*)
S2	S70, K73(*bb*), I127(*sch*), S130-N132, A135, K234(*bb*).	S64, K67(*bb*), Y112(*sch*), Y150-N152, I155(*sch*), G270, K318(*bb*)	S67, K70(*bb*), I112(*sch*), S115-V117, F120(*sch*), K205(*bb*)

The third shell of residues S3 is not listed.

Table 6. Protonation energies of the SxxK and KTG lysines calculated with the ONIOM method

	ΔE1(kJ/mol)	ΔE2(kJ/mol)	ΔE3(kJ/mol)	ΔE4(kJ/mol)
TEM1(E166 neut)	-994	-1317	-985	-1307
TEM1(E166 –1)	-1429	-1299	-1514	-1384
P99	-1572	-1549	-1071	-1048
OXA10	-1304	-1302	-631	-629

CONCLUSIONS

From the above review of both the literature and our personal results, it appears that the QM tool is obviously useful to help in building models and testing them. In the particular case of biochemical reactivity, this technique must be used as a qualitative help since the general trends are well reproduced. It has already proven very useful in predicting relative intrinsic reactivities of compounds. However, this tool cannot provide a quantitative answer to a macroscopic problem like, for instance, that of the acylation or deacylation kinetic models in β-lactamases. As a matter of fact, the results depend namely on the chosen model, on the constraints of the model, the protonation state of the active site.

The future of the method lies in the mixed approaches like the QM/MM ones where the whole system, including the solvent molecules, is taken into account. This could at least allow the user to get rid of many of the constraints and to consider one important point that constitutes the conformational adaptability of the molecular ensemble, with or without a ligand in the active site. Nevertheless, the choice of a model of the concerned chemical reaction will remain a key component independently of the quality of the QM tool itself.

APPENDIX
FROM THE SCHÖDINGER'S EQUATION TO THE HARTREE-FOCK-ROOTHAAN-HALL FRAMEWORK: APPROXIMATIONS

System with N nuclei, n electrons.

1. Separation of the nuclear and electronic motions: the Born-Oppenheimer approximation

$H(\mathbf{p},\mathbf{q})\ \Psi(\mathbf{q}) = E\ \Psi(\mathbf{q})$ to $H^{el}(\mathbf{p}^{el},\mathbf{q}^{el},\mathbf{Q}^{N})\ \Psi^{el}(\mathbf{q}^{el},\mathbf{Q}^{N}) = E^{el}(\mathbf{Q}^{N})\ \Psi^{el}(\mathbf{q}^{el},\mathbf{Q}^{N})$

2. The building blocks of $\Psi^{el}(\mathbf{q}^{el},\mathbf{Q}^{N})$.

2.1. Hartree: the independent electron model, the orbitals and the spinorbitals.

$\Psi^{el}(\mathbf{q}^{el},\mathbf{Q}^{N}) \approx \varphi_1(\mathbf{q}_1^{el},\mathbf{Q}^{N})\ \varphi_2(\mathbf{q}_2^{el},\mathbf{Q}^{N})\ \varphi_3(\mathbf{q}_3^{el},\mathbf{Q}^{N}) \ldots \varphi_n(\mathbf{q}_n^{el},\mathbf{Q}^{N})$

2.2. Fock: coupled equations to which the $\varphi_i(\mathbf{q}_i^{el},\mathbf{Q}^{N})$ obey.

2.3. $\Psi^{el}(\mathbf{q}^{el},\mathbf{Q}^{N})$ must obey the antisymmetry principle for the electrons

=> Slater determinant

$\Psi^{el}(\mathbf{q}^{el},\mathbf{Q}^{N}) \approx D[\varphi_1(\mathbf{q}_1^{el},\mathbf{Q}^{N})\sigma_1, \varphi_2(\mathbf{q}_2^{el},\mathbf{Q}^{N})\sigma_2, \varphi_3(\mathbf{q}_3^{el},\mathbf{Q}^{N})\sigma_3 \ldots \varphi_n(\mathbf{q}_n^{el},\mathbf{Q}^{N})\sigma_n]$

σ_i being the spin function of the ith electron.

3. The building blocks of the $\varphi_i(\mathbf{q}_i^{el}, \mathbf{Q}^N)$: the basis set.

LCAO: $\varphi_i(\mathbf{q}_i^{el}, \mathbf{Q}^N) \approx \sum_{\alpha}^{nf} c_{\alpha,i}(\mathbf{Q}^N)\, \chi_\alpha(\mathbf{q}_i^{el} - \mathbf{Q}_\alpha^N)$, *nf* being the number of basis functions

$\chi_\alpha(\mathbf{q}_i^{el} - \mathbf{Q}_\alpha^N)$

4. The basis functions are replaced by gaussian basis functions.

REFERENCES

[1] W.J. Hehre, L. Radom, P. v.R. Schleyer, J.A. Pople, *Ab initio molecular orbital theory*, Wiley Interscience, New York, 1986.

[2] A. Szabo, N.S. Ostlund, *Modern Quantum Chemistry. Introduction to Advanced Electronic Structure Theory*, Dover publications, New York, 1996.

[3] F. Jensen, *Introduction to computational chemistry*, John Wiley & sons, New York, 1999.

[4] J.A. Pople, D.L. Beveridge, *Approximate molecular orbital theory*, McGraw-Hill Book Company, New York, 1970.

[5] M.J.S. Dewar, E.G. Zoebisch, E.F. Healy, J.J.P. Stewart, "Development and use of quantum mechanical molecular models. 76. AM1: a new general purpose quantum mechanical molecular model." *J. Am. Chem. Soc.*, 1985, *107*, 3902-3909.

[6] R.G. Parr, W. Yang, *Density functional theory of atoms and molecules*, Oxford University Press, New York, 1989.

[7] W. Kohn, L.J. Sham, "Self-Consistent Equations Including Exchange and Correlation Effects", *Phys.Rev.A* 1965, *140(4A)*, 1133-1138.

[8] A.D. Becke, "Density-functional thermochemistry.III. The role of exact exchange", *J. Chem. Phys.* 1993, *98*, 5648-5652.

[9] P. Pulay, G. Fogarasi, "Geometry optimization in redundant internal coordinates", *J.Chem.Phys.* 1992, *95*, 2856-2860.

[10] C. Peng, P.Y. Ayala, H.B. Schlegel, M.J. Frisch, "Using redundant internal coordinates to optimize equilibrium geometries and transition states.", *J.Comput.Chem.* 1996, *17,* 49-56.

[11] K.N. Kudin, G.E. Scuseria, H.B. Schlegel, "A redundant internal coordinate algorithm for optimization of periodic systems", *J.Chem.Phys.* 2001, *114*, 2919-2923.

[12] (a) M.J. Frisch, G.W. Trucks, H.B. Schlegel, G.E. Scuseria, M.A. Robb, J.R. Cheeseman, V.G. Zakrzewski, J.A. Montgomery, Jr., R.E. Stratmann, J.C. Burant, S. Dapprich, J.M. Millam, A.D. Daniels, K.N. Kudin, M.C. Strain, O. Farkas, J. Tomasi, V. Barone, M. Cossi, R. Cammi, B. Mennucci, C. Pomelli, C. Adamo, S. Clifford, J. Ochterski, G.A. Petersson, P.Y. Ayala, Q. Cui, K. Morokuma, D.K. Malick, A.D. Rabuck, K. Raghavachari, J.B. Foresman, J. Cioslowski, J.V. Ortiz, A.G. Baboul, B.B. Stefanov, G. Liu, A. Liashenko, P. Piskorz, I. Komaromi, R. Gomperts, R.L. Martin, D.J. Fox, T. Keith, M.A. Al-Laham, C.Y. Peng, A. Nanayakkara, C. Gonzalez, M. Challacombe, P.M.W. Gill, B. Johnson, W. Chen, M.W. Wong, J.L. Andres, C. Gonzalez, M. Head-Gordon, E.S. Replogle, and J.A. Pople, Gaussian 98, Revision A.7, Gaussian, Inc., Pittsburgh PA, 1998. (www.gaussian.com)

(b) M. J. Frisch, G. W. Trucks, H. B. Schlegel, G. E. Scuseria, M. A. Robb, J. R. Cheeseman, J. A. Montgomery, Jr., T. Vreven, K. N. Kudin, J. C. Burant, J. M. Millam, S. S. Iyengar, J. Tomasi, V. Barone, B. Mennucci, M. Cossi, G. Scalmani, N. Rega, G. A. Petersson, H. Nakatsuji, M. Hada, M. Ehara, K. Toyota, R. Fukuda, J. Hasegawa, M. Ishida, T. Nakajima, Y. Honda, O. Kitao, H. Nakai, M. Klene, X. Li, J. E. Knox, H. P. Hratchian, J. B. Cross, C. Adamo, J. Jaramillo, R. Gomperts, R. E. Stratmann, O. Yazyev, A. J. Austin, R. Cammi, C. Pomelli, J. W. Ochterski, P. Y. Ayala, K. Morokuma, G. A. Voth, P. Salvador, J. J. Dannenberg, V. G. Zakrzewski, S. Dapprich, A. D. Daniels, M. C. Strain, O. Farkas, D. K. Malick, A. D. Rabuck, K. Raghavachari, J. B. Foresman, J. V. Ortiz, Q. Cui, A. G. Baboul, S. Clifford, J. Cioslowski, B. B. Stefanov, G. Liu, A. Liashenko, P. Piskorz, I. Komaromi, R. L. Martin, D. J. Fox, T. Keith, M. A. Al-Laham, C. Y. Peng, A. Nanayakkara, M. Challacombe, P. M. W. Gill, B. Johnson, W. Chen, M. W. Wong, C. Gonzalez, and J. A. Pople, Gaussian 03, Revision B.04, Gaussian, Inc., Pittsburgh PA, 2003. (www.gaussian.com)

[13] The Protein Data Bank, Brookhaven National Laboratory, NY USA. Now maintained by the Research Collaboratory for Structural Bioinformatics (RCSB) Website: http://www.rcsb.org.

[14] K. Fukui, "The path of chemical reactions – the IRC approach", *Acc.Chem.Res.* 1981, *14*, 363-368.

[15] P. Culot, G. Dive, V.H. Nguyen, J.M. Ghuysen, "A quasi-Newton algorithm for first-order saddle-point location", *Theor.Chim.Acta* 1992, *82*, 189-205.

[16] B.F. Gherman, S.D. Goldberg, V.W. Cornish, R.A. Friesner, "Mixed Quantum Mechanical/Molecular Mechanical (QM/MM) Study of the Deacylation Reaction in a Penicillin Binding Protein (PBP) versus in a Class C β-Lactamase", *J.Am.Chem.Soc.* 2004, *126*, 7652-7664.

[17] D.A. Case, T.E. Cheatham, III, T. Darden, H. Gohlke, R. Luo, K.M. Merz, Jr., A. Onufriev, C. Simmerling, B. Wang and R. Woods, "The amber biomolecular simulation programs", *J. Comput. Chem.* 2005, *26*, 1668-1688.

[18] (a) B.R. Brooks, R.E. Bruccoleri, B.D. Olafson, D.J. States, S. Swaminathan, M. Karplus, "CHARMM: a program for macromolecular energy, minimization, and dynamics calculations", *J.Comput.Chem.* 1983, *4*, 187. (b) A.D. MacKerell Jr, B. Brooks, C.L. Brooks III, L. Nilsson, B. Roux, Y. Won, M. Karplus, in *The Encyclopedia of Computational Chemistry*, 1998, *1*, 271. Ed. Schleyer, P.v.R.; et al. Chichester: John Wiley & Sons.

[19] (a) F. Maseras, K. Morokuma, "A new integrated ab initio + molecular mechanics geometry optimization scheme of equilibrium structures and transition states", *J. Comp. Chem.* 1995, *16*, 1170-1179 ; (b) S. Humbel, S. Sieber, K. Morokuma,"The IMOMO(Integrated MO MO) method: integration of different levels of molecular orbitals approximations for geometry optimization of large systems: test for n-butane conformation and SN2 reaction: RCl + Cl□", *J. Chem. Phys.* 1996, *105*, 1959-1967; (c) T. Matsubara, S. Sieber, K. Morokuma, "A test of the new "integrated MO+MM"(IMOMM) method for the conformational energy of ethane and n-butane", *Int. J. Quant. Chem.* 1996, *60*, 1101-1109 ; (d) M. Svensson, S. Humbel, R.D.J. Froese, T. Matsubara, S. Sieber, K. Morokuma, "ONIOM: A Multi-Layered Integrated MO + MM Method for Geometry Optimizations and Single Point Energy Predictions. A Test

for Diels-Alder Reactions and Pt(P(t-Bu)3)2 + H2 Oxidative Addition", *J. Phys. Chem.* 1996, *100*, 19357-19363; (e) S. Dapprich, I. Komaromi, K.S. Byun, K. Morokuma, M.J. Frisch, "A new ONIOM implementation in Gaussian 98. Part I. The calculation of energies, gradients, vibrational frequencies and electric field derivatives", *J. Mol. Struct. (Theochem)* 1999, *461-462*, 1-21.

[20] B. Vilanova, J. Donoso, J. Frau, F. Muñoz, "Kinetic and Molecular-Modelling Studies of Reactions of a Class A β-Lactamase with Compounds Bearing a Methoxy Group on the β-Lactam Ring", *Helv.Chim.Acta* 1999, *82*, 1274-1288.

[21] Y. Fujii, N. Okimoto, M. Hata, T. Narumi, K. Yasuoka, R. Susukita, A. Suenaga, N. Futatsugi, T. Koishi, H. Furusawa, A. Kawai, T. Ebisuzaki, S. Neya, T. Hoshino, "Molecular Dynamics Study on Class A β-Lactamase: Hydrogen Bond Network among the Functional Groups of Penicillin G and Side Chains of the Conserved Residues in the Active Site", *J.Phys.Chem. B* 2003, *107*, 10274-10283.

[22] N. Díaz, T.L. Sordo, K.M. Merz Jr, D. Suárez, "Insights into the Acylation Mechanism of Class A β-Lactamases from Molecular Dynamics Simulations of the TEM-1 Enzyme Complexed with Benzylpenicillin", *J.Am.Chem.Soc.* 2003, *125*, 675-684.

[23] N. Díaz, D. Suárez, K.M. Merz Jr, T.L. Sordo, "Molecular Dynamics Simulations of the TEM-1 β-Lactamase Complexed with Cephalothin", *J.Med.Chem.* 2005, *48*, 780-791.

[24] N. Díaz, D. Suárez, T.L. Sordo, "Molecular Dynamics Simulations of Class C β-Lactamase from *Citrobacter freundii*: Insights into the Base Catalyst for Acylation", *Biochemistry* 2006, *45*, 439-451.

[25] S. Wolfe, C.-K. Kim, K. Yang, "Ab initio molecular orbital calculations on the neutral hydrolysis and methanolysis of azetidinones, including catalysis by water. Relationship to the mechanism of action of β-lactam antibiotics", *Can.J.Chem.* 1994, *72*, 1033-1043.

[26] G. Dive, D. Dehareng, "Serine Peptidase Catalytic Machinery: Cooperative One-Step Mechanism", *Int.J.Quant.Chem.* 1999, *73*, 161-174.

[27] N. Díaz, D. Suárez, T.L. Sordo, K.M. Merz Jr, "Acylation of Class A β-lactamases by Penicillins: A Theoretical Examination of the Role of Serine 130 and the β-lactam Carboxylate Group", *J.Phys.Chem.B* 2001, *105*, 11302-11313.

[28] Y.-Y. Ke, T.-H. Lin, "A theoretical study on the activation of Ser70 in the acylation mechanism of cephalosporin antibiotics", *Biophys.Chem.* 2005, *114*, 103-113.

[29] B.D. Wladkowski, S.A. Chenoweth, J.N. Sanders, M. Krauss, W.J. Stevens, "Acylation of β-Lactams by Class A β-Lactamase: An *ab Initio* Theoretical Study on the Effects of the Oxy-Anion Hole", *J.Am.Chem.Soc.* 1997, *119*, 6423-6431.

[30] S.O. Meroueh, J.F. Fisher, H.B. Schlegel, S. Mobashery, "Ab Initio QM/MM Study of Class A β-Lactamase Acylation: Dual Participation of Glu166 and Lys73 in a Concerted Base Promotion of Ser70", *J.Am.Chem.Soc.* 2005, *127*, 15397-15407.

[31] J.C. Hermann, C. Hensen, L. Ridder, A.J. Mulholland, H.-D. Höltje, "Mechanisms of Antibiotic Resistance: QM/MM Modeling of the Acylation Reaction of a Class A β-Lactamase with Benzylpenicillin", *J.Am.Chem.Soc.* 2005, *127*, 4454-4465.

[32] J. Frau, J. Donoso, F. Muñoz, F. Garcia Blanco, "Semiempirical and ab initio calculations on the alkaline hydrolysis of the β-lactam ring. Influence of the solvent", *J.Mol.Struct.(Theochem)* 1997, *390*, 247-254.

[33] I. Massova, P.A. Kollman, "Quantum Mechanical Study of β-Lactam Reactivity: The Effect of Solvation on Barriers of Reaction and Stability of Transition States and Reaction Intermediates", *J.Phys.Chem. B* 1999, *103*, 8628-8638.

[34] R. López, M.I. Menéndez, N. Díaz, D. Suárez, P. Campomanes, D. Ardura, T.L. Sordo, "Theoretical Studies on the Ring Opening of β-Lactams: Processes in Solution and in Enzymatic Media", *Curr.Org.Chem.* 2006, *10*, 805-821.

[35] Y. Fujii, M. Hata, T. Hoshino, M. Tsuda, "Catalytic Mechanism of Class A β-Lactamase: Role of Lysine 73 and C3-Carboxyl Group of the Substrate Pen G in the Deacylation Step", *J.Phys.Chem. B* 2002, *106*, 9687-9695.

[36] J.C. Hermann, L. Ridder, H.-D. Höltje, A.J. Mulholland, "Molecular mechanisms of antibiotic resistance: QM/MM modelling of deacylation in a class A ββ-lactamase", *Org.Biomol.Chem.* 2006, *4*, 206-210.

[37] M. Hata, Y. Tanaka, Y. Fujii, S. Neya, T. Hoshino, "A Theoretical Study on the Substrate Deacylation Mechanism of Class C β-Lactamase", *J.Phys.Chem. B* 2005, *109*, 16153-16160.

[38] B. Atanasov, D. Mustafi, M.W. Makinen, "Protonation of the β-lactam nitrogen is the trigger event in the catalytic action of class A β-lactamases", *Proc.Nat.Acad.Sci.* 2000, *97*, 3160-3165.

[39] M. Nukaga, K. Mayama, A.M. Hujer, R.A. Bonomo, J.R. Knox, "Ultrahigh Resolution Structure of a Class A β-lactamase: On the Mechanism and Specificity of the Extended-spectrum SHV-2 Enzyme", *J.Mol.Biol.* 2003, *328*, 289-301.

[40] C. Damblon, X. Raquet, L.Y. Lian, J. Lamotte-Brasseur, E. Fonzé, P. Charlier, G.C.K. Roberts, J.-M. Frère, "The catalytic mechanism of β-lactamases: NMR titration of an active-site lysine residue of the TEM-1 enzyme", *Proc.Natl.Acad.Sci.USA* 1996, *93*, 1747-1752.

[41] X. Raquet, V. Lounnas, J. Lamotte-Brasseur, J.-M. Frère, R.C. Wade, "pKa Calculations for class A β-lactamases: methodological and mechanistic implications", *Biophys.J.* 1997, *73*, 2416-2426.

[42] J. Lamotte-Brasseur, V. Lounnas, X. Raquet, R.C. Wade, " pKa Calculations for class A β-lactamases: influence of substrate binding", *Protein Sci.* 1999, *8*, 404-409.

[43] Y. Chen, G. Minasov, T.A. Roth, F. Prati, B.K. Shoichet, "The Deacylation Mechanism of AmpC β-Lactamase at Ultrahigh Resolution", *J.Am.Chem.Soc.* 2006, *128*, 2970-2976.

[44] M.W. Wong, M.J. Frisch, K.B. Wiberg, "Solvent effects. 1. The mediation of electrostatic effects by solvents", *J.Am.Chem.Soc.* 1991, *113*, 4776-4782.

[45] V. Barone, M. Cossi, J. Tomasi, "A new definition of cavities for the computation of solvation free energies by the polarizable continuum model", *J.Chem.Phys.* 1997, *107*, 3210-3221.

[46] 1BTL (TEM1), 1GCE (P99), 1E3U (OXA10)

[47] S. Baurin, L. Vercheval, F. Bouillenne, C. Falzone, A. Brans, L. Jacquamet, J.L. Ferrer, E. Sauvage, D. Dehareng, J.M. Frère, P. Charlier, M. Galleni, F. Kerff, "Critical Role of Tryptophan 154 for the Activity and Stability of class D ß-lactamases", *Biochemistry* 2009, *48*, 11252-11263.

[48] Program Discover, Copyright 1993, Biosym Technologies, Inc., 9685 Scranton Road, San Diego, CA 92121-2777, USA. Now distributed by Accelrys (www.accelrys.com)

[49] A.K. Rappé, C.J. Casewit, K.S. Colwell, W.A. Goddard III, W.M. Skiff, "UFF, a full periodic table force field for molecular mechanics and molecular dynamics simulations", *J.Am.Chem.Soc.* 1992, *114*, 10024-10035.

PART 3: CLINICAL ASPECTS

In: Beta-Lactamases
Editor: Jean-Marie Frère, pp. 343-367
ISBN: 978-1-61324-638-2

Chapter 13

EMERGENCE OF COMPLETELY NEW SEQUENCES

Roger Labia[1,*] and Patrice Nordmann[2]
[1] Université Européenne de Bretagne
Laboratoire Universitaire de Biodiversité et Ecologie Microbienne,
IFR148 ScInBioS, UMT 08.3 PHYSI'Opt, 6 rue de l'Université,
F-29334, Quimper, France.
[2] Service de Bactériologie-Virologie, Hôpital de Bicêtre et Université Paris-Sud, 78,
rue du Général Leclerc, 94275 Le-Kremlin- Bicêtre, France.

ABSTRACT

The action of β-lactamases on β-lactam antibiotics is a major factor of bacterial resistance to these molecules, particularly for Gram-negative organisms. These enzymes are growing considerably in number and variety of catalytic properties associated to phenotypic expression(s). After the introduction of benzylpenicillin in human therapy of infectious diseases, emergence of "completely novel" β-lactamases occurred repetitively and usually step by step. This occurred within the four molecular classes of beta-lactamases defined on the basis of structural and mechanistic properties. In this chapter we report many of the novel sequences identified as naturally occurring or as sources of acquired resistance in Gram-negatives, speculate on the possible origin of "completely novel" β-lactamases and give some biochemical information about the most important leader type enzymes.

INTRODUCTION

The action of β-lactamases on β-lactam antibiotics is a major factor of bacterial resistance to these molecules, particularly for Gram-negative organisms. These enzymes are growing considerably in number and variety of catalytic properties associated to phenotypic expression(s). By 1980, the introduction of the third-generation cephalosporins (e.g. cefotaxime), cephamycins (cefoxitin, moxalactam), carbapenems (imipenem) and

* To whom all correspondence should be addressed, e-mail: labia@univ-brest.fr

monobactams (aztreonam) was followed by a permanent emergence of variants of known enzymes and of "completely novel β-lactamases", which resulted in a rather complex world. Emergence of completely novel enzymes occurred within the four molecular classes of β-lactamases defined on the basis of structural and mechanistic properties Most of these enzymes (but not all) are plasmid-mediated and found in many Gram-negative organisms. Moreover many genes were first identified as silent but some of them are no longer silent! (e.g. *Bacillus anthracis*).

In this chapter we report many of the novel sequences identified as naturally occurring or as sources of acquired resistance in Gram-negatives, speculate on the possible origin of "completely novel" β-lactamases and give some biochemical information about the most important leader type enzymes.

A Few Words about Evolution

It is currently admitted that life appeared on the earth a few billion years ago, probably about 3 billion years, which is a lot of time (Note that a billion years means 1,000,000,000 years). We have only tentative hypotheses concerning the very beginning of life. We know more about what life looked like over the last billion years, mostly through fossils (some of them being of a very high quality, even if close to one billion years old!) and amber inclusions: it seems that one billion years ago life was already highly sophisticated. Surprisingly, some species remained, apparently, unchanged for a few hundred million years, such as *Gingko biloba* (a tree), *Lepisma saccharinum* (an insect) and many others. But this is assumed on the basis of the apparent identity of the present species to those identified in fossils at a macroscopic level. However, phenotypic identity doesn't means genotypic identity.

Conversely, present humans (*Homo sapiens sapiens*) appeared rather recently, about 50,000 years ago (and perhaps much less), whereas the emergence of the very first *Homo* among primates is estimated to be 6 to 8 million years old.

Unfortunately, bacterial fossils do not exist! Yet, even if it is only a credible hypothesis, the very first forms of life were extremely crude but the first organized living organisms were certainly monocellular microorganisms probably more or less similar to present Archaea and / or Bacteria (prokaryotes). Moreover, during a long period, nucleotides and their polymers (e.g. ADN- and ARN-like materials) and amino acids and their polymers (e.g. peptide-, protein- and enzyme-like materials) became progressively stabilized and improved in their structures and functions, in order to yield the highly well-organized presently known corresponding products. This efficiency allowed a highly enhanced evolution of living organisms leading to eukaryotes and pluricellular organisms. For example at the very beginning the number of amino acids was perhaps not 20 (or 21 with seleno cysteine). It seems quite possible to build living organisms on a set of only 16 amino acids.

Reproduction of present living organisms takes place with a very low level of spontaneous mutations. As a counter example, the low fidelity of the HIV reverse transcriptase suggests a rapid evolution of this organism. Similar conclusions may be drawn for many viruses, including the influenza virus.

About two billion years (or perhaps more) ago, some microorganisms started to use peptidoglycan in their envelopes. Peptidoglycan biosynthesis involves transpeptidases and carboxypeptidases which are enzymes inhibited by penicillin. These enzymes are also named "Penicillin-Binding Proteins" or PBPs. Consequently penicillins and β-lactamases rapidly followed peptidoglycan emergence, as chemical weapons in the war between microorganisms. This occurred also for many other "natural" antibiotics and products. Thus β-lactamases evolved over a long period starting about two billion years ago. The corresponding genes remained stored in the genomes of microorganisms. Some of them are presently pathogens for mammalians (e.g. *Klebsiella pneumoniae*…) while others now exhibit low or no pathogenicity for humans or mammalians, with the exception of those with altered immunity.

WHAT IS A COMPLETELY NEW SEQUENCE?

There is no simple answer to this apparently simple question! In terms of residue identity, as 20 amino acids are mostly involved, a 5% value should be observed between two random protein sequences such as computer-generated ones. Nevertheless, as all amino acids are not represented with the same frequencies (e.g. cysteine and tryptophan are present at lower frequencies than other amino acids), purely random identities usually range between 5 and 10%, depending on the size of the proteins. Thus this parameter is probably not an excellent one. Concerning these aspects Higgins and co-workers developed computer programs for evaluation of distances between proteins or genes, which is a complex mathematical or statistical problem [1-5].

Moreover, crystallography may show that proteins with very low levels of identity, if any, may have rather similar foldings and active sites. For example, class A and C β-lactamases (see below) have no significant sequence similarities, but their active sites are rather analogous and distances between some important atoms are often well conserved.

Consequently, in several instances, "new" sequences are probably very old ones which however remained hidden in the chromosome or silent during very long periods. A novel β-lactam pressure may determine if a given hidden or silent gene (and its enzyme product) may be the most efficient against a novel antibiotic menace and, using various genetic strategies (plasmids, transposons, integrons…), jump for instance from the chromosome of often unknown or poorly known non pathogenic organisms to bacteria involved in human pathology.

As a first example, consider the β-lactamase producers *Raoultella planticola* (formerly *Klebsiella planticola*) and *Raoultella terrigena* (formerly *Klebsiella terrigena*): these organisms are respectively associated with botanical and aquatic environments thus of rather low pathogenicity for mammalians. However they are closely related to the *Klebsiella oxytoca* species, which is highly pathogen for humans [6,7]. In this context, Fevre et al. [8] suggested that six identified groups of the *K. oxytoca* OXY β-lactamases evolved from an ancestral enzyme over a period of 100 million years, a very long time ago when compared to the emergence of *Homo sapiens sapiens*. Wild strains of *K. oxytoca* usually produce low levels of an apparent "penicillinase", but overexpression of this enzyme may result in resistance to oximinocephalosporins [9, 10 and see below].

Similarly *Xanthomonas campestris* is a Gram-negative phytopathogenic bacterium often encountered in soil, plants, plant debris, and especially cruciferous plants [11]. In a previous study [12,13] it was found that, among 86 strains of *X. campestris* isolated in Taiwan, 81 were resistant to ampicillin. Antibiotic resistance was transferred by a representative strain through a variety of mechanisms, including transduction, conjugation, or transfection. A β-lactamase was detected in one *X. campestris* strain. The amino acid sequence deduced from the *bla* gene revealed a class A enzyme, which has significant similarity (51.5%) with the L2 serine β-lactamase of *Stenotrophomonas maltophilia* [14], 45% with *Yersinia enterocolitica* and 44.5% with Toho-1 and CTX-M-1 (MEN-1).

Similarly the genus *Erwinia* is classified in the *Enterobacteriaceae*, but most members of this genus characteristically cause diseases in plants [15]. Surprisingly, in an isolate of *Erwinia persicina*, a chromosome-borne class A β-lactamase named ERP-1 was identified. Sequence analysis and kinetic parameters revealed that this plant-associated pathogen produced a class A ESBL that is distantly related to known β-lactamases (50 % amino acid identity with CTX-M-8 and CTX-M-9, see below) [16].

The discovery of a chromosome-mediated oxacillinase (OXA-50) in the genome of *Pseudomonas aeruginosa* PAO1, published in 2000, was a great surprise [17,18]. More recently (2005) a similar observation was made in *Acinetobacter baumannii* [19]. In both cases the β-lactamases seem to play a minor role in bacterial resistance.

Other class A β-lactamases have been described in several Gram-negative genera, such as *Bordetella. Bordetella bronchiseptica* and *Bordetella parapertussis* are closely related Gram-negative rods that may colonize the respiratory tract of mammals. *B. bronchiseptica* causes chronic respiratory infections in a wide range of animals but is rarely implicated as a pathogen in humans, mostly in immunocompromised patients who are often exposed to animals. *B. parapertussis* may cause a mild whooping cough. A narrow-spectrum clavulanic acid-inhibited class A β-lactamase, BOR-1, was identified in a *B. bronchiseptica* clinical isolate. It shared 45% amino acid identity with L-2 from *Stenotrophomonas maltophilia.* An identical β-lactamase gene was found in *B. bronchiseptica* and *B. parapertussis* reference strains that may contribute only in part to their resistance phenotypes [20]

Kluyvera ascorbata [21], *Kluyvera cryocrescens* [22], *Kluyvera georgiana* [23], *Rahnella aquatilis* [24], *Desulfovibrio desulfuricans* [25] and others are β-lactamase producers but they have a low pathogenicity for humans (and mammalians). However *Kluyvera* may be quite important as a reservoir of the emerging CTX-M β-lactamases (see later) that are becoming a serious problem.

These results indicate that identification of chromosome and naturally-encoded β-lactamases may be quite interesting for understanding the spread of what appears to be acquired β-lactamases with often broad-spectrum hydrolytic profiles in clinically-significant bacterial species (*Enterobacteriaceae*, *Pseudomonas aeruginosa* ...).

AMBLER'S CLASSIFICATION OF ß-LACTAMASES

Ambler [26] was a precursor in β-lactamase sequencing and then proposed a sequence-based classification. In this classification the genetic localization (plasmid or chromosome) is not taken into account.

Usually, proteases, peptidases or esterases belong to four families, characterized by the presence in their active site of essential residue(s): i) a serine (e.g. acetylcholine esterase), ii) a cysteine (e.g. papain), iii) two aspartic acids (e.g. pepsin), or iv) a zinc ion (e.g. angiotensin converting enzyme). Surprisingly, β-lactamases, which are β-lactam specific peptidases, belong only to the serine and zinc families.

In one of Ambler's first papers on classification [26], he defined classes A (serine enzymes, initially referring to the *Staphylococcus aureus* penicillinase) and B (metallo enzymes or MBLs, initially from *Bacillus cereus* βII or BcII) and evoked "other classes". Then as the number of sequences increased regularly, Ambler and the β-lactamase community decided to fragment the highly heterogeneous serine β-lactamases into classes A and C. Class C is also known under the name of "cephalosporinases". When PSE-2 (now OXA-10) was sequenced somewhat later, Huovinen and Jacoby [27] proposed to introduce the novel "class D" also identified as "oxacillinases" (OXAs).

The last decades showed that this classification was a good or excellent one. Moreover crystallography of a lot of β-lactamases confirmed this classification. The structures of members of these four classes have been determined. They are monomeric, with the exception of OXAs some of which are dimeric in some conditions and of the L1 MBL which is tetrameric.

Conversely, enzymes from the same class may have very different catalytic properties. As an example, some have an "extended-spectrum", i.e. they hydrolyze β-lactams which are usually resistant to other enzymes (also named parent or ancestral enzymes).

Class A β-lactamases are usually susceptible to inhibition by clavulanic acid and related molecules (sulbactam, tazobactam ...)., whereas enzymes from classes B, C and D are not or poorly inhibited. Chromosomal class A β-lactamases are encountered in *Klebsiella pneumoniae*, *K. oxytoca*, *R. planticola*, *Raoultella ornitholytica*, *Proteus vulgaris*, *Proteus penneri*, *Desulfovibrio desulfuricans, Serratia fonticola*... One will note that some of these enzymes are highly active on third-generation cephalosporins. Others are able to efficiently hydrolyze carbapenems. A standard numbering scheme has been developed to facilitate sequences comparisons [28].

Class B β-lactamases are metallo enzymes with one or two zinc ions in their active sites. Consequently they are inhibited by EDTA and related compounds via zinc depletion. Surprisingly, clavulanic acid is a substrate for these enzymes. Moreover they are able to hydrolyze nearly all β-lactams, including cephamycins and carbapenems, with monobactams as the sole common exceptions. Class B chromosomal enzymes have been described in *Bacteroides* sp., *Stenotrophomonas maltophilia* (formerly *Pseudomonas* then *Xanthomonas maltophilia*), *Aeromonas hydrophila*, *Flavobacterium* sp., *Bacillus cereus*, *B. anthracis* and other organisms. The Liège group (Belgium) developed a classification scheme for class B enzymes [29,30].

In several of these Gram-negative species two (and even three) unrelated broad-spectrum β-lactamases have been identified such as a metallo β-lactamase and a clavulanic acid inhibited ESBL of Ambler's class A. *S. maltophilia* and *Chryseobacterium meningosepticum* [31] are good examples.

Class C enzymes are also known as AmpC or "cephalosporinases" as they confer high levels of resistance to first-generation cephalosporins and, when overexpressed, to second and third generation cephalosporins. They are mostly chromosome-encoded, and usually inducible, but exceptions are known. They are commonly produced by *Escherichia coli*, *Enterobacter cloacae*, *Citrobacter freundii*, *Morganella morganii*, *Pseudomonas aeruginosa*, *Acinetobacter baumannii*...

Class D β-lactamases are markedly active against oxacillin and methicillin, but this gives them no particular clinical relevance since they are encountered only in Gram-negative bacteria which are not susceptible to these antibiotics. Many class D enzymes are plasmid-mediated. Levesque et al. proposed a numbering scheme for class D enzymes [32].

Surprisingly, class D β-lactamases are usually inhibited by chloride ions, but some exceptions are known, whereas those of the other classes are not (i.e. no detectable inhibition is observed with 1M NaCl).

The class D enzymes are naturally occurring in various Gram-negative species such as *P. aeruginosa*, *A. baumannii*, *Shewanella* sp., *Ralstonia* sp., *Aeromonas* sp., *Legionella gormanii*, *Burkholderia pseudomallei*... [18,19,33-36].

It seems likely that enzymes of classes A, C and D are derived from a common ancestral enzyme, probably an ancestral penicillin-binding protein. Conversely class B has a very different origin.

More Recently (1940-1970)

Here "more recently" means the very beginning of the antibiotic area (i.e. from 1940/1945 to the end of the sixties, with the introduction of methicillin, oxacillin, ampicillin, semi-synthetic penicillins, and the first cephalosporins in therapy).

The term "*penicillinase*" was introduced for the first time in 1940 by Abraham and Chain in a paper published even before the first testing of penicillin in humans [37]. This penicillinase was produced by an *Escherichia coli* strain. The authors observed that "An enzyme from bacteria (was) able to destroy penicillin". This enzyme, or a similar one, was sequenced much later in 1981 [38] and is presently assumed to belong to Ambler's class C. Concerning this enzyme, the term penicillinase is no longer used, since members of this class can have a significant catalytic activity towards cephalosporins, monobactams and carbapenems.

When the first penicillins (benzyl penicillin and phenoxymethyl penicillin) were introduced in therapy of infectious diseases, two major facts were observed:

Table 1. The *bla* genes coding for ancestral β-lactamases TEM-1 and TEM-2

	Promoter			Coding region (bases 209-1066)							
	32	162	175	226	317	346	436	604	682	913	925
bla_{TEM-1A}	C	G	A	C	C	A	C	G	T	C	G
bla_{TEM-1B}			G	T			T	T			
bla_{TEM-1C}							T				
bla_{TEM-1D}							T			T	
bla_{TEM-1E}			G	T		G	T	T			
bla_{TEM-1F}		T				G	T		C		A
					Gln-39						
bla_{TEM-2}	T				A	G	T		C		A
					Lys-39						

Note: DNA numbering is according to Sutcliffe [46], and amino acid 39 is numbered according to Ambler [26]. See references [47-50].

1. some organisms were "*naturally*" resistant to these penicillins, such as *E. coli*, *K. pneumoniae* and many others, including most Gram-negative organisms).
2. other organisms, initially very susceptible to these penicillins, such as *S. aureus*, became progressively resistant. This was characterized as "*acquired*" resistance, and at the same time new β-lactamases were often detected.

During this period, the number of totally novel sequences remained limited to the following enzymes probably as investigation techniques were not highly refined:

The Plasmid-Mediated Staphylococcal Penicillinase

This enzyme, which emerged by the end of the forties, belongs to class A. The synthesis of this enzyme is currently inducible. This was the first clinically important β-lactamase as a large number of isolates were producing this enzyme. Presently, the enzyme from the PC1 strain is considered as a reference. This strain is a laboratory mutant which constitutively produces the enzyme [39]. One will note that the origin of *S. aureus* penicillinase is still unknown.

The first studies concerning this enzyme were performed by Richmond and Novick, in 1963 [39-41]. This β-lactamase is one of the first that was sequenced (at the protein level) by Richmond and co-workers [41]. Then, in 1987, Wang and Novick [42] sequenced the structural gene of the β-lactamase coded by the pI258 plasmid, highlighting a signal peptide of 24 amino acids. The mature protein has no tryptophan and no cysteine.

TEM and SHV ß-lactamases

When ampicillin was introduced in therapy by the beginning of the sixties, a lot of quite new enzymes were identified:

The first plasmid-mediated β-lactamase discovered in a Gram-negative organism was the TEM enzyme which was discovered in an *E. coli* strain isolated by Datta and Kontomichalou [43] from a patient hospitalized in Athens at the beginning of the sixties. The producing plasmid was named R6K. Even today we have no idea about the origin of this β-lactamase. The TEM β-lactamase(s) have spread world-wide and are probably the most commonly identified in Gram-negatives. At this time, the TEM enzyme was also called a "broad-spectrum enzyme" as it hydrolyzes cephaloridine, which is not the case for the *S. aureus* penicillinase). This enzyme belongs to class A.

Two different TEM enzymes (TEM-1 and TEM-2) were identified early [44]. They differ by their isoelectric points, respectively 5.4 and 5.6. The TEM-1 β-lactamase produced by plasmid pBR322 was sequenced by Sutcliffe in 1978 [45]. The complete nucleotide sequence of plasmid pBR322 was completed the following year (1979) [46]. However, at the genetic level, sequencing presently shows seven different genes, six coding for TEM-1 and one for TEM-2: see Table 1 [47-50].

The SHV-1 β-lactamase which also emerged in the sixties belongs to class A. TEM-1 and SHV-1 exhibit about 68 % identity and rather similar kinetic properties. It is clear that plasmid-mediated SHV enzyme(s) are derived from a chromosomal enzyme [51,52]. The SHV-1 β-lactamase is chromosomally encoded in the majority of *K. pneumoniae* isolates but is usually plasmid mediated in *E. coli*.

```
                              26▼
TEM-1:   MSIQHFRVALIPFFAAFCLPVFAHPETLVKVKDAEDQLGARVGYIELDLNSGKILESFRPE
SHV-1:     MRYFRLCIISLLATLPLAVHASPQPLEQIKLSESQLSGRVGMIEMDLASGRTLTAWRAD
LEN-1:     MRYVRLCVISLLATLPLVVYAGPQPLEQIKQSESQLSGRVGMVEMDLANGRTLAAWRAD
OHIO-1:    MRYFRLCIISLLATLRLRVHAGPQPLEQIKLSESQLSGSVGMIEMDLARPGTLTAWRAD
OKP-1:     ------------AALPLAVFASPQPLEQVTRSESQLAGRVGYVEMDLASGRTLAAWRAS
Consensus              A         P  L      E QL   VG  E DL     L   R
                                 <-----h1----->  <--s1-->     <s2->

                ▼70▼73                         104▼
TEM-1:    ERFPMMSTFKVLLCGAVLSRVDAGQEQLGRRIHYSQNDLVEYSPVTEKHLTDGMTVRELC
SHV-1:    ERFPMMSTFKVVLCGAVLARVDAGDEQLERKIHYRQQDLVDYSPVSEKHLADGMTVGELC
LEN-1:    ERFPMVSTFKVLLCGAVLARVDAGLEQLDRRIHYRQQDLVDYSPVSEKHLVDGMTIGELC
OHIO-1:   ERFPMMSTFKVVLCGAGLARVDAGDEQLERKIHYRRQDLVDYSPVSEKHLADGMTVGELC
OKP-1:    ERFPLMSTFKVLLCGAVLARVDAGDEQLDRRIRYRQQDLVDYSPVSEKHLADGMTVGELC
Consensus ERFP  STFKV LCGA L RVDAG EQL R I Y   DLV YSPV EKHL DGMTV ELC
               <------h2------->                       <h3>     <---h4

             130▼ ▼132                              ▼166
TEM-1:    SAAITMSDNTAANLLLTTIGGPKELTAFLHNMGDHVTRLDRWEPELNEAIPNDERDTTMP
SHV-1:    AAAITMSDNSAANLLLATVGGPAGLTAFLRQIGDNVTRLDRWETELNEALPGDARDTTTP
LEN-1:    AAAITLSDNSAGNLLLATVGGPAGLTAFLRQIGDNVTRLDRWETALNEALPGDARDTTTP
OHIO-1:   AAAITMSDNSAANLLLPAVGGPAGLTAFLRQIGDNVTRLDRWETELNEALPGDARDTTTA
OKP-1:    AAAITMSDNSAGNLLLKSVGGPAGLTAFLRQIGDNVTRLDRWETEJNEALPGDVRDTTTP
Consensus  AAITMSDN A NLLL   GGP  LTAFL   GD VTRLDRWE ELNEA P D RDTT
          --->    <---h5---->  <---h6--->      <---omega loop---->   <

                                                         234▼    ▼239
TEM-1:    AAMATTLRKLLTGELLTLASRQQLIDWMEADKVAGPLLRSALPAGWFIADKSGAG-ERGS
SHV-1:    ASMAATLRKLLTSQRLSARSQRQLLQWMVDDRVAGPLIRSVLPAGWFIADKTGAG-ERGA
LEN-1:    ASMAATLRKLLTAQHLSARSQQQLLQWMVDDRVAGPLIRAVLPAGWFIADKTGAG-ERGA
OHIO-1:   RSMAATLRKLLTSQRLSARSQRQLLQWMVDDRVAGRLIRSVLPAGWFIADKTGAG-ERGA
OKP-1:    ASMAATLRKLLTSHTLSARSQQQLLQWMVDDQVAGPLIRAVLPAGWFIAEKTGAG-ERGS
Consensus   MA TLRKLLT   L      QL  WM  D VAG L R  LPAGWFIA K GAG-ERG
          ---h8----->       <----h9---->       <h10>   <----s3---->

       244▼        ▼253               275▼▼276
TEM-1:    RGIIAALGP-DGKPSRIVVIYTTGSQATMDERNRQIAEIGASLIKHW
SHV-1:    RGIVALLGP-NNKAERIVVIYLRDTPASMAERNQQIAGIGAALIEHWQR
LEN-1:    RGIVALLGP-DGKPERIVVIYLRDTPASMAERNQHIAGIGQR
OHIO-1:   RGIVALLGP-NNKAERIVVIYLRDTPASMAERNQQIAGIAGALIEHWQR
OKP-1:    RGIVALLGP-NGKAERIVVIYLRDTPASMAERNQQIAR-----------
Consensus RGI A LGP-  K  RIVVIY     A M ERN QIA I
          <--s4-->       <--s5-->     <------h11------>
```

Figure 1. Sequence alignment of five class A β-lactamases. Note the gaps at positions 239 and 253, following Ambler's numbering system [26]. Mature proteins start at residue 26, active-site important residues are located at positions 70 (active serine), 73, 104, 130-132 (SDN loop), 166 (omega loop), 234 and 244. Residues 275 and 276 are involved in resistance to inhibitors (inactivators).

More recently, new chromosomal SHV-related enzymes were discovered with, first OHIO-1, discovered in the state of Ohio, [53], LEN-1 and related enzymes [54] and more recently OKP-1 and related enzymes, standing for ***O**ther **K**. pneumoniae* β-lactamases, [55,56]. Moreover, Grimont's group [57] demonstrated that variants of the *K. pneumoniae* OKP chromosomal β-lactamase are divided into two main groups, OKP-A and OKP-B. If one accepts the hypothesis that *E. coli* and *Salmonella enterica*, diverged 140 million years ago, KpII-A and KpII-B would have diverged 11 to 46 million years ago. Figure 1 shows that alignment of these enzymes with TEM-1 shows strong similarities, mostly in areas close to the active site.

Consequently, it seems credible that the TEM enzymes originated in an unknown enterobacterial species more or less related to *Klebsielleae*.

In 2003, a *K. pneumoniae* isolate was identified that had reduced susceptibility to several expanded-spectrum cephalosporins and imipenem [59]. This isolate produced a chromosome-encoded SHV-type β-lactamase, named SHV-38, which has an alanine to valine substitution in position 146 when compared to SHV-1. The kinetic parameters of the purified SHV-38 and SHV-1 β-lactamases show that the hydrolytic spectrum of SHV-38 is extended to ceftazidime and imipenem. This report is the first example of an SHV-type β-lactamase capable of hydrolyzing imipenem.

Other Enzymes

In the early sixties, Sabath et al [60] observed that *Pseudomonas pyocyanea* produced penicillinase and cephalosporinase activities and demonstrated that a single β-lactamase (from class C) was responsible for the hydrolysis of penicillins and cephalosporins.

The CARBs (class A) and the OXAs (class D) emerged by the beginning of the seventies. At this time, they were also named "PSE" for "*Pseudomonas* Specific Enzymes". Presently the term PSE is more or less abandoned, as these enzymes are commonly found in other bacterial species. As an example, PSE-2 is presently named OXA-10. In 1969, a β-lactamase was identified in a *Pseudomonas aeruginosa* "Dalgleish" strain, and it was noticed that this enzyme was "markedly active against carbenicillin" [61]. This enzyme was named PSE-4, and presently CARB-1. Now about ten CARB enzymes are known, most of them have an arginine in position 234, instead of a lysine in other ß-lactamases. The first class D ß-lactamases (OXAs) were studied by Dale and Smith [62-64].

In 2004 Girlich et al. [18] showed that the $bla_{\mathrm{OXA\text{-}50}}$ gene (formerly known as the PA5514 gene) is an oxacillinase gene identified in silico in the genome of *Pseudomonas aeruginosa* PAO1. By using a mutant strain of *P. aeruginosa* PAO1 in which the bla_{AmpC} cephalosporinase gene was inactivated, the $bla_{\mathrm{OXA\text{-}50}}$ gene was shown to be expressed constitutively in *P. aeruginosa*. This suggests that the *P. aeruginosa* species could be a reservoir for OXA β-lactamases. Other observations confirm this proposal [34].

Also, by the end of the sixties and the beginning of the seventies, it was shown than *Bacillus cereus* was producing twoβ-lactamases: β-lactamase I, a class A enzyme and the class B β-lactamase II, a Zinc-dependent metallo β-lactamase or MBL [65,66].

More recently, it has been shown that *Bacillus anthracis* is also producing two β-lactamases, one class A and one class B [67]. The *B. anthracis* class A (Bla1) and class B (Bla2) β-lactamases share respectively 94% and 92 % amino acid identity with the corresponding *B. cereus* enzymes.

Table 2. Major discoveries in the field of "novel sequences"

Enzyme	Organism	Date of isolation
Class A carbapenemases		
SME-1	*Serratia. marcescens*	1982
IMI-1	*E. cloacae*	1984
NMC-A	*E. cloacae*	1990
KPC-1	*K. pneumoniae*	1996
Class A extended-spectrum		
CTX-M-1, MEN-1	*E. coli*	About 1990
PER-1	*P. aeruginosa*	1991
BES-1	*S. marcescens*	1996
VEB-1		1996
GES-1	*K. pneumoniae*	1998
And others		
Class B metallo enzymes (plasmid)		
IMP-1	*S. marcescens*	1991
VIM-1	*P. aeruginosa*	1997
Class C enzymes (plasmids)		
MIR-1	*K. pneumoniae*	1988
FOX-1	*K. pneumoniae*	1989
CMY-1	*K. pneumoniae*	1989
Class D carbapenemases		
OXA-23	*A. baumannii*	1985
OXA-24	*A. baumannii*	1997
OXA-58	*A. baumannii*	2003

THE MOST RECENT DECADES

After 1980, the so called "third-generation cephalosporins" (including cefepime and cefpirome), monobactams and carbapenems were introduced, starting with cefotaxime. Thus progressively, a very large number of β-lactamases with extended-spectrum or inhibitor-resistance emerged. Moreover as molecular biology techniques improved rapidly, our knowledge concerning these enzymes was considerably enhanced.

Later, after extensive use of the amoxicillin / clavulanate combination, a number of TEM and SHV mutants with extended catalytic activity or reduced susceptibility to inhibitors were identified during the eighties. These enzymes are discussed in chapter 14.

NON-TEM AND NON-SHV EXTENDED-SPECTRUM β-LACTAMASES

a – Class C β-lactamases

Class C or AmpC β-lactamases (also called cephalosporinases) are naturally produced by a variety of enterobacterial species and were identified in strains already isolated before 1920.

A few recent reviews deal with these enzymes [68-70]. The hydrolytic properties of class C β-lactamases are rather similar regardless of their origin.

Chromosome-encoded class C β-lactamases are usually inducible and this mechanism may be rather complex but an *ampC/ampR* tandem is often encountered (see chapter 17 and references 71and 72).

Overexpression may be related to various factors, including mutations in the *ampR* gene. Resistance appeared initially in organisms such as *Enterobacter cloacae, Citrobacter freundii, Serratia marcescens*, and *Pseudomonas aeruginosa* that could, by mutation, overproduce their chromosomal AmpC β-lactamases, thus providing resistance to oximino-cephalosporins, cephamycins and monobactams with relative exceptions for cefepime and cefpirome [73].

Nevertheless, resistance to cefotaxime may be related to overexpression of class C β-lactamases, to plasmid mediated class C β-lactamases or to extended-spectrum class C β-lactamases [74]

The first plasmid-mediated class C β-lactamase was identified in *K. pneumoniae* strains isolated in the Miriam Hospital, Rhode Island, USA before 1990 [75]. The MIR-1 sequence was published later [76]. The DNA sequence from nucleotides 1 to 823 is 98% identical to that from *Pseudomonas pseudoalcaligenes* that resembles an insertion sequence transposase and is also related to ORF341, a component of unidentified function associated with several integrons [77,78].

The amino acid sequence of MIR-1 most closely resembles that of ACT-1 (91% identity), and is closely related to the AmpC enzymes of *E. cloacae* (85 to 86%), *Citrobacter freundii* (74%) and *Escherichia coli* (72%).

A large number of AmpC β-lactamases are known, but they are not all true "novel sequences". They include also mutant enzymes which hydrolyze cefepime and cefpirome.

CTX-M β-lactamases

The CTX-M β-lactamases are now widespread enzymes [79] and their rate of dissemination among bacteria and in most parts of the world has increased dramatically (see also chapter 15).

They were initially identified and characterized by the end the eighties by two groups at the same time, one in France (MEN-1) [80, 81], the other in Germany (CTX-M-1) [82, 83]. In 1990, R. Labia and A. Bauernfeind exchanged their strains (both *E. coli*) and observed that the enzymes were identical. MEN-1 (alias CTX-M-1) shares 72% identity with the chromosome-mediated β-lactamase of *K. oxytoca* D488 [9].

Nevertheless the very first CTX-M enzymes were probably present in South America (Argentina) very early [84]. In 1992, Bauernfeind et al. published a preliminary paper on a CTX-M-2-producing *Salmonella enterica* serovar *typhimurium* strain (pI 7.9) isolated "in the beginning of August 1990" in Argentina without further details. In 1995, Rossi and coworkers described cefotaxime-resistant *Salmonella* spp strains isolated in Argentina in 1991 [85]. They described unidentified β-lactamases of alkaline pI's (7.4, 8.1) which hydrolyze cefotaxime. These enzymes could be CTX-M β-lactamases (ESBLs).

Japan was also an early source of CTX-M enzymes, as suggested by the FEC-1 enzyme which was described in 1988 [86]. This enzyme was produced by a cefotaxime resistant *Escherichia coli* strain isolated from the fecal flora of a laboratory dog which was used for pharmacokinetic studies of β-lactam antibiotics. The sequence of this enzyme [GenBank accession number AB098539] definitely showed that it belongs to the CTX-M family.

Most CTX-M enzymes are associated with an insertion sequence element (IS*Ecp1*) and this was observed with four out of the five bla_{CTX-M} gene clusters (CTX-M-1, -M-2, -M-9, and -M-25 clusters). IS*Ecp1B* has been reported to be associated with and to mobilize the emerging expanded-spectrum β-lactamase bla_{CTX-M} genes in *Enterobacteriaceae*. It contains a gene encoding a 420-amino-acid transposase and carries promoter sequences for high-level expression of the $bla_{CTX-M-14/-18}$, $bla_{CTX-M-17}$, and $bla_{CTX-M-19}$ β-lactamase genes. Recently, we have shown that IS*Ecp1B* is able to mobilize the adjacent $bla_{CTX-M-19}$ gene by a transpositional mechanism in *Escherichia coli* by recognizing a variety of DNA sequences as right inverted repeats (IRR), see [87].

The CTX-M enzymes are also present in environmental species of the genus *Kluyvera*, from which they have been characterized by investigating the mechanisms of natural β-lactam resistance, as shown previously [21-23].

PER β-lactamases

PER stands for *Pseudomonas* Extended Resistance. The PER-1 enzyme was described in 1993 [88, 89]. This enzyme hydrolyzes all β-lactams with the exception of cephamycins and carbapenems. Aztreonam is a poor substrate. PER-1 shares 24% identity with TEM-1.

The X-ray structure of the PER-1 protein, solved at 1.9-Å resolution, reveals that two of the most conserved features in class A ß-lactamases are not present in this enzyme: the fold of the omega-loop and the cis conformation of the peptide bond between residues 166 and 167. The new fold of the omega-loop and the insertion of four residues at the edge of strand S3 generate a broad cavity that may easily accommodate the bulky substituents of cephalosporin substrates [90].

PER-1 is common in Turkey and South East Asia where it was found in various members of the *Enterobacteriaceae* family and *P. aeruginosa* [91].

A PER-2 plasmid-encoded enzyme was identified in a *Salmonella typhimurium* strain isolated in Argentina which confers transferable resistance to oximinocephalosporins, aztreonam, and ceftibuten. The PER-2 β-lactamase is highly susceptible to inhibition by clavulanate. PER-2 is not closely related to TEM or SHV enzymes (25 to 26% identity) but closely related to PER-1 (86.4% identity). PER-2 has been detected only in pathogens (*S. typhimurium*, *E. coli*, *K. pneumoniae*, *Proteus. mirabilis*) isolated from patients in South America [92].

Other class A expanded-spectrum β-lactamases have been identified in Gram-negative organisms such as BES-1 [93], BEL-1 [94], SFO-1 [95], TLA-1 [96] and more recently TLA-2 [97].

Carbapenem Resistance

Carbapenem-hydrolyzing β-lactamases can be Ambler class A enzymes, metallo-β-lactamases or expanded-spectrum oxacillinases.

a-Class A Carbapenem Hydrolyzing Enzymes

Four class A β-lactamases which efficiently hydrolyze carbapenems are known. The first identified enzymes were NMC-A, IMI-1, SME-1, and KPC-1.

NMC-A, which stands for *N*on *M*etallo *C*arbapenemase of class *A*, is chromosome-encoded [98,99]. NMC-A demonstrated unusual strong hydrolytic activity towards imipenem and meropenem. Moreover it hydrolyzed cephamycins with uncommonly high k_{cat} values for these antibiotics [100]. It is susceptible to the action of inhibitors.

NMC-A was produced by a carbapenem-resistant clinical strain of *E. cloacae*, strain NOR-1, isolated in the pus of the fistulized subcutaneous abscess of a patient with multiple wounds at La Pitié-Salpêtrière Hospital in Paris. Before the isolation of the strain, the patient had received a single bolus of intraveneously injected imipenem (500 mg) as he underwent surgery.

Surprisingly, in 1984, a year before approval of imipenem for clinical use in the US, a wound isolate and a bile isolate of *Enterobacter cloacae* were obtained from two patients in a California hospital [101]. These isolates were resistant to imipenem and many β-lactams, but rather susceptible (MICs <4 μg/ml) to cefotaxime, ceftriaxone, ceftazidime, and moxalactam. Both strains produced an apparent TEM-1 β-lactamase and an inducible NMCA-type β-lactamase which was named IMI-1. Purified IMI-1 hydrolyzed imipenem and benzylpenicillin at modest rates, but more slowly than cephaloridine. The enzyme was inhibited by clavulanic acid and tazobactam but not by EDTA. The β-lactamase gene encoding IMI-1 was cloned and sequenced. Sequence analysis identified a class A β-lactamase. The IMI-1 enzyme shares more than 95% identity with NMC-A. DNA sequence analysis also identified a gene upstream that shares >95% identity with *nmcR* and that may encode a regulatory protein. Two other similar enzymes, SME-1 and SME-2, were also identified from two *Serratia marcescens* strains isolated in Great Britain [102].

Later KPC-1 was isolated from a clinical carbapenem-resistant *K. pneumoniae* strain collected in the frame of project ICARE (*I*ntensive *C*are *A*ntimicrobial *R*esistance *E*pidemiology) [103-105]. Analysis of the mechanism(s) of carbapenem resistance suggested that the carbapenem resistance phenotype of the strain was mainly due to the production of KPC-1, a novel class A β-lactamase.

Surprisingly KPC-2 was first found in a *Salmonella enterica* serotype Cubana isolate exhibiting resistance to most β-lactam antibiotics, including oximinocephalosporins and imipenem and isolated from a 4-year-old boy with gastroenteritis in Maryland [106].

The GES-1 β-lactamase produced by *K. pneumoniae* ORI-1 was isolated in 1998 in France from a rectal swab of a 1-month-old girl who was previously hospitalized in Cayenne Hospital, French Guiana. This strain harbored a ca. 140-kb nontransferable plasmid, pTK1, that conferred an extended-spectrum cephalosporin resistance profile antagonized by the addition of clavulanic acid, tazobactam, or imipenem. One will note that this strain is *susceptible* to imipenem. Interestingly, GES-1 possesses cysteine residues at positions 69 and 238, as found in the carbapenem-hydrolyzing enzymes NMC-A, IMI-1, SME-1, and KPC-1 (see later). The kinetic constants of GES-1 do not show any significant imipenem hydrolysis but rather a significant degree of inhibition by imipenem. A possible explanation is that GES-1 may be able to bind imipenem with strong affinity but not to hydrolyze it [107]. Novel GES-type enzymes including GES-2 that is a point mutant derivative of GES-1 may significantly hydrolyze imipenem [108].

NMC-A, IMI-1, SME and KPC enzymes have two cysteines in positions 69 and 238 which probably form a disulfide bridge as shown by the structures of NMC-A [109] and SME-1 [110]. This disulfide bridge seems to be critical for carbapenem hydrolysis [111,112].

b-Class B Carbapenem Hydrolyzing Enzymes

These enzymes are metallo enzymes, with one or two zinc ions in the active site. During a long period, the enzyme from *B. cereus* (BcII), considered as of low clinical importance, was nevertheless well studied. Presently, most metallo-β-lactamase (MBLs) genes are found in environmental species (such as *Bacillus cereus*, some *Aeromonas* species, *Bacteroides fragilis*, *Stenotrophomonas maltophilia*, *Chryseobacterium meningosepticum*, *Chryseobacterium indologenes*, *Legionella gormanii*, *Janthinobacterium lividum*...).

These enzymes usually hydrolyze all β-lactam antibiotics, including cephamycins and carbapenems, but they generally exhibit a poor activity against aztreonam. Thus, many pathogens that produce such enzymes at high levels are resistant to the majority of β-lactam antibiotics, including the carbapenems.

The spreading of metallo-β-lactamases (MBLs) among clinically important human pathogens represents a therapeutic problem of a high and increasing importance. A good knowledge of the distribution of mobile genetic elements could help identifying the original sources of mobile metallo-β-lactamase determinants and increasing our understanding of their intriguing evolutionary history [113, 114].

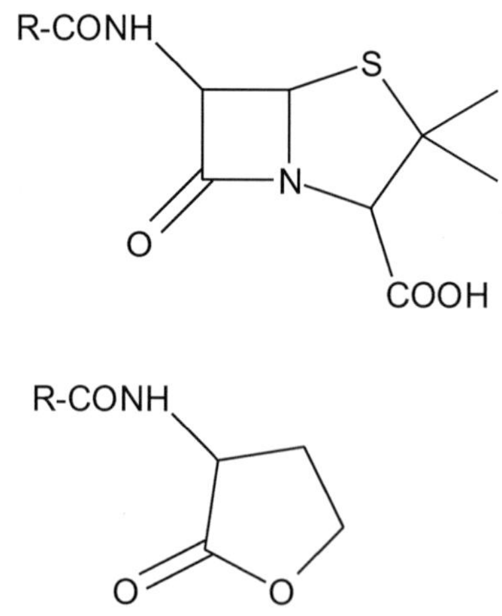

Figure 2. Structures of a penicillin and an acyl homoserine lactone.

Metallo-β-lactamases are clustered into three different subclasses, B1, B2, and B3 [29,30]. Subclass B1 enzymes have been found in strains of *Bacillus (cereus* and *anthracis)*, *Bacteroides*, *Pseudomonas*, *Serratia*, and *Chryseobacterium*, and subclass B3 enzymes have been found in *Stenotrophomonas maltophilia* (L-1); *Janthinobacterium lividum* (THIN-B); *Caulobacter crescentus* (MBL-1), *Legionella gormanii* (FEZ-1), *Chryseobacterium meningosepticum* (GOB-1). Both the B1 and B3 subclasses have broad-spectrum profiles (including penicillins, cephalosphorins, and carbapenems) and require two zinc ions for maximal enzymatic activity in vitro. However, MBLs can display different metal-binding affinities for zinc ions and, at low concentrations (picomolar range) of free zinc appear to be active with a single zinc ion per monomer. Subclass B2 enzymes have a narrow substrate profile, hydrolyzing almost exclusively carbapenems. These enzymes which are are active only in the monozinc form have been identified in *Aeromonas hydrophila* (CphA) and *Serratia fonticola* (SfhI).

A crystal structure of the *B. cereus* MBL enzyme was reported in 1987 [115], but at rather poor resolution (3.5 Å). In 1995, the resolution was improved to 2.5-Å [116]. This was followed by the publication of the structure of the *B. fragilis* CcrA enzyme refined at 1.85-Å resolution [117].

Crystal structures of the two enzymes indicate similar folding patterns for the two enzymes, but only one zinc ionwas identified in the *B. cereus* enzyme, whereas two zinc ions were found in the active site of the *B. fragilis* CcrA3 enzyme. However, the crystallisation conditions were very different, particularly the pH.

The first report of an MBL-producing *Pseudomonas aeruginosa* isolated from a Japanese patient in 1988 was published in 1991 [118]. Presently the occurrence of mobile genetic elements encoding MBLs has extended to many types of Gram-negative organisms distributed throughout the world [118]. The acquired MBLs are mostly reported in *P. aeruginosa* and to a lesser extent in *A. baumannii* and in *Enterobacteriaceae.* They are of the VIM, IMP, SPM, GIM and more recently SIM series.

They have been reported world-wide but with a higher prevalence in Asia, South America and Southern Europe. However, the origin of these resistance determinants is unknown, although an important number of proteins belonging to the MBL superfamily have been identified in microbial genomes. Thus, as suggested by Stoczko et al. in a postgenomic scan of metallo-β-lactamase homologues in *Rhizobacteria* [120] class B enzymes are not related to PBPs.

Many non-β-lactamase related proteins appear to share the metallo-β-lactamase fold (see also chapter 3): N-acyl-homoserine lactone hydrolases (*Arthrobacter*, *Agrobacterium tumefactions*, *Bacillus thuringiensis*): see figure 2, cyclases involved in the polyketide biosynthetic pathway (*Streptomyces arenae*, *Streptomyces coelicolor*), glyoxalase of *Arabidopsis thaliana*, *Saccharomyces cerevisiae*, *Trypanosoma brucei*), cyclic nucleotide phosphodiesterase (*Candida albicans*, *S. cerevisiae*, *Vibrio fischeri*), flavoproteins (*Methanothermobacter thermautotrophicus*, *Moorella thermoacetica*), rubredoxin-oxygen oxidoreductase (*Desulfovibrio gigas*), choline-binding protein (*Streptococcus pneumoniae*), carbofuran hydrolase (*Achromobacter* sp.), methyl parathion hydrolase (*Plesiomonas*).

c-Class D Carbapenem Hydrolyzing Enzymes

Among the class D enzymes, those that may play a significant role for expanding resistance would be oxacillinases with carbapenemase activity identified in *A. baumannii*, *c*arbapenem-*h*ydrolysing class *D* beta-*l*actamases (CHDLs) that are mostly specific to this species [121-124].

These enzymes belong to three unrelated groups of clavulanic acid-resistant β-lactamases, represented by OXA-23, alias ARI-1, [125], OXA-24 [126] and OXA-58 [127,128], that can be either plasmid- or chromosome-encoded. *A. baumannii* also possesses an intrinsic carbapenem-hydrolysing oxacillinase, the expression of which may vary and that may play a role in carbapenem resistance. In addition to β-lactamases, carbapenem resistance in *A. baumannii* may also result from porin or penicillin-binding protein modifications. Several porins, including the 33-kDa CarO protein, that constitute pore channels for influx of carbapenems, might be involved in carbapenem resistance. Surprisingly, the chromosome-encoded class D β-lactamase of *Shewanella oneidensis* is a progenitor of carbapenem-hydrolyzing oxacillinases [129].

It should also be noted that, since nearly all class D enzymes are classified as "OXA", this group is much more heterogeneous than the TEM and SHV families. Although the OXA enzymes certainly share a common, but remote ancestor (but so do the members of the other classes), the limit between "completely new sequences" and "new phenotypes due to amino acid substitutions" (see chapter 14) can be difficult to define.

Concluding Remarks

In the present paper we have not discussed the problem of extended-spectrum β-lactamases, which are derived from "ancestral" or previously-known β-lactamases of classes A, C and D. For example, at the end of 2010, the Jacoby and Bush web site (http://www.lahey.org/studies/) recorded about 180 TEM-derived enzymes and 190 OXAs (see chapter 14 for more details).

In the present paper we described "completely new sequences" following a chronological order. Even if "new sequences" are in fact "old sequences", their emergence is clearly a new problem both for microbiologists and for patients. We did not detail the genetic vehicules that drive the spread of β-lactamases genes especially in Gram-negatives. These vehicules are clearly different from one gene to another: clavulanic acid inhibited ESBLs of the TEM, SHV and PER types are not associated with integrons whereas clavulanic acid inhibited ESBLs of the VEB and GES types and acquired metallo ß-lactamase genes are often associated with these structures.

The origin of acquired β-lactamases has been determined unambiguously in some cases: SHV β-lactamases originate from *K. pneumoniae*, CTX-M from *Kluyvera* species, plasmid-mediated cephalosporinases from several enterobacterial species (*M. morganii*, *Hafnia alvei*...) and OXA-48 from a *Shewanella* species.

The clinical problems due to β-lactamases underline the interest of identifying and surveying novel β-lactamases. For instance, the rapid emergence of CTX-M producers among community acquired enterobacterial species (and mostly in *E. coli*, first human pathogen) and of the MBLs in *P. aeruginosa* are especially pertinent.

Additional examples are presented in chapter 15 where the association of β-lactamases with other resistance determinants is discussed from a clinical point of view. The very recent characterization of NDM-1, a new metallo-β-lactamase is also worth mentioning [130]

REFERENCES

[1] Higgins, D. G. and Sharp, P. M. (1988) CLUSTAL: a package for performing multiple sequence alignments on a microcomputer. *Gene* 73, 237-244.

[2] Higgins, D. G., Bleasby, A. J. and Fuchs, R. (1992) Clustal V: improved software for multiple sequence alignment. *Comp. Appl. In Biosciences (CABIOS)* 8, 189-191.

[3] Thompson, J. D., Higgins, D. G. and Gibson, T. J. (1994) CLUSTAL W: improving the sensitivity of progressive multiple sequence alignment through sequence weighting, positions-specific gap penalties and weight matrix choice. *Nucleic Acids Research* 22, 4673-4680.

[4] Thompson, J. D., Gibson, T. J., Plewniak, F., Jeanmougin, F. and Higgins D. G. (1997) The CLUSTAL_X windows interface: flexible strategies for multiple sequence alignment aided by quality analysis tools. *Nucleic Acids Research* 25, 4876-4882.

[5] Li, K. (2003) ClustalW-MPI: ClustalW analysis using distributed and parallel computing. *Bioinformatics* 19, 1585-1586.

[6] Westbrook, G. L., O'Hara, C. M., Roman, S. B. and Miller J. M. (2000) Incidence and identification of *Klebsiella planticola* in clinical isolates with emphasis on newborns. *J. Clin. Microbiol.* 38, 1495-1497.

[7] Walckenaer, E., Poirel, L., Leflon-Guibout, V., Nordmann, P. and Nicolas-Chanoine, M. H. (2004) Genetic and biochemical characterization of the chromosomal Class A ß-lactamases of *Raoultella* (formerly *Klebsiella*) *planticola* and *Raoultella ornithinolytica*. *Antimicrob. Agents Chemother.* 48, 305-312.

[8] Fevre, C., Jbel, M., Passet, V., Weill, F., Grimont, P. A. D. and Brisse, S. (2005) Six groups of the OXY ß-lactamase evolved over millions of years in *Klebsiella oxytoca.* Antimicrob. *Agents Chemother.* 49, 3453-3462.

[9] Reynaud, A., Peduzzi, J., Barthelemy, M. and Labia, R. (1991) Cefotaxime-hydrolyzing activity of the ß-lactamase of *Klebsiella oxytoca* D488 could be related to a threonine residue at position 140. *FEMS Microbiol. Letters* 81, 185-192.

[10] Fournier, B., Lu C. Y., Lagrange, P. H., Krishnamoorthy, R. and Philippon, A. (1995) Point mutation in the pribnow box, the molecular basis of ß-lactamase overproduction in *Klebsiella oxytoca*. Antimicrob. *Agents Chemother.* 39, 1365-1368.

[11] Williams, P. H. (1980) Black rot: a continuing threat to world crucifers. *Plant Dis.* 64, 736-742.

[12] Weng, S. F., Chen, C. Y., Lee, Y. S., Lin, J. W. and Tseng, Y. H. (1999) Identification of a novel ß-lactamase produced by *Xanthomonas campestris*, a phytopathogenic bacterium. Antimicrob. *Agents Chemother.* 43, 1792-1797.

[13] Weng, S., Lin, J., Chen, C., Chen, Y., Tseng, Y. H. and Tseng, Y. H. (2004) Constitutive expression of a chromosomal class A (BJM Group 2) ß-lactamase in *Xanthomonas campestris*. *Antimicrob. Agents Chemother.* 48, 209-215.

[14] Walsh, T. R., MacGowan, A. P., and Bennett, P. M. (1997) Sequence analysis and enzyme kinetics of the L2 serine ß-lactamase from *Stenotrophomonas maltophilia*. Antimicrob. *Agents Chemother.* 41, 1460–1464.

[15] Starr, M. P. and Chatterjee, A. K. (1972) The genus *Erwinia*: enterobacteria pathogenic to plants and animals. *Ann. Rev. Microbiol.* 26, 389-426.

[16] Vimont, S., Poire, L., Naas, T. and Nordmann P. (2002) Identification of a Chromosome-borne expanded-spectrum class A ß-lactamase from *Erwinia persicina*. Antimicrob. *Agents Chemother.* 46, 3401-3405.

[17] Stover, C. K., Pham, X. Q., Erwin, A. L., Mizoguchi, S. D., Warrener, P., Hickey, M. J., Brinkman, F. S., Hufnagle, W. O., Kowalik, D. J., Lagrou, M., Garber, R. L., Goltry, L., Tolentino, E., Westbrock-Wadman, S., Yuan, Y., Brody, L. L., Coulter, S. N., Folger, K. R., Kas, A., Larbig, K., Lim, R., Smith K., Spencer D., Wong, G. K., Wu, Z., Paulsen, I. T., Reizer, J., Saier, M. H., Hancock, R. E., Lory, S. and Olson, M.V. (2000) Complete genome sequence of *Pseudomonas aeruginosa* PAO1, an opportunistic pathogen. *Nature* 406, 959–964.

[18] Girlich, D., Naas, T. and Nordmann, P. (2004) Biochemical characterization of the naturally occurring oxacillinase OXA-50 of *Pseudomonas aeruginosa. Antimicrob. Agents Chemother.* 48, 2043-2048.

[19] Heritier, C., Poirel, L., Fournier, P., Claverie, J., Raoult, D. and Nordmann, P. (2005) Characterization of the naturally occurring oxacillinase of *Acinetobacter baumannii*. Antimicrob. *Agents Chemother.* 49, 4174-4179.

[20] Lartigue, M., Poirel, L., Fortineau, N. and Nordmann P. (2005) Chromosome-borne class A BOR-1 ß-lactamase of *Bordetella bronchiseptica* and *Bordetella parapertussis*. *Antimicrob. Agents Chemother.* 49, 2565-2567.

[21] Humeniuk, C., Arlet, G., Gauthier, V., Grimont, P., Labia, R. and Philippon, A. (2002) ß-lactamases of *Kluyvera ascorbata*, probable progenitors of some plasmid-encoded CTX-M types. *Antimicrobial Agents Chemother.* 46, 3045-3049..

[22] Decousser, J. W., Poirel, L. and Nordmann, P. (2001) Characterization of a chromosomally encoded extended-spectrum class A ß-lactamase from *Kluyvera cryocrescens*. *Antimicrob. Agents Chemother.* 45, 3595-3598.

[23] Poirel, L., Kampfer, P. and Nordmann, P. (2002) Chromosome-encoded Ambler class A ß-lactamase of *Kluyvera georgiana*, a probable progenitor of a subgroup of CTX-M extended-spectrum ß-lactamases. *Antimicrob. Agents Chemother.* 46, 4038-4040.

[24] Bellais, S., Poirel, L., Fortineau, N., Decousser, J. W. and Nordmann P. (2001) Biochemical-genetic characterization of the chromosomally encoded extended-spectrum class A ß-lactamase from *Rahnella aquatilis*. Antimicrob. *Agents Chemother.* 45, 2965-2968.

[25] Morin, A. S., Poirel, L., Mory, F., Labia, R. and Nordmann P. (2002) Biochemical-genetic analysis and distribution of DES-1, an Ambler class A extended-spectrum ß-lactamase from *Desulfovibrio desulfuricans*. *Antimicrob. Agents Chemother.* 46, 3215-3222.

[26] Ambler, R. P. 1980. The structure of ß-lactamases. *Philos. Trans. R. Soc. Lond.* (Biol.) 289, 321-331.

[27] Huovinen, P., Huovinen, S. and Jacoby, G. A.1988. Sequence of PSE-2 ß-lactamase. Antimicrob. *Agents Chemother.* 32, 134-136.

[28] Ambler, R. P., Coulson, A. F. W., Frère, J. M., Ghuysen, J. M., Joris, B., Forsman, M., Levesque, R.C., Tirabi, G. and Waley, S. G. (1991) A standard numbering scheme for the class A ß-lactamases. *Biochem. J.* 276, 269-270.

[29] Galleni, M., Lamotte-Brasseur, J., Rossolini, G. M., Spencer, J., Dideberg, O. and Frère, J. M. (2001) Standard numbering scheme for class B ß-lactamases. *Antimicrob. Agents Chemother.* 45, 660-663.

[30] Garau, G., Garcia-Saez, I., Bebrone, C., Anne, C., Mercuri, P., Galleni, M., Frère, J.M. and Dideberg, O. (2004) Update of the standard numbering scheme for class B ß-lactamases. Antimicrob. *Agents Chemother.* 48, 2347-2349.

[31] Bellais, S., Aubert, D., Naas, T. and Nordmann, P. (2000) Molecular and biochemical heterogeneity of class B carbapenem-hydrolyzing beta-Lactamases in *Chryseobacterium meningosepticum*. Antimicrob. *Agents Chemother.* 44, 1878-1886.

[32] Sanschagrin, F., Couture, F. and Levesque, R. C. (1995) Primary structure of OXA-3 and phylogeny of oxacillin-hydrolyzing class D beta-lactamases. Antimicrob. *Agents Chemother.* 39, 887-893.

[33] Alksne, L. E. and Rasmussen, B. A. (1997) Expression of the AsbA1, OXA-12, and AsbM1 ß-lactamases in *Aeromonas jandei* AER14 is coordinated by a two-component regulon. *J. Bacteriol.* 179, 2006–2013.

[34] Giuliani, F., Docquier, J., Letizia Riccio M., Pagani L., Rossolini M. G. (2005) OXA-46, a new class D ß-lactamase of narrow substrate specificity encoded by a bla_{VIM-1}-containing integron from a *Pseudomonas aeruginosa* clinical isolate. *Antimicrob. Agents Chemother.* 49, 1973-1980.

[35] Nordmann, P., Poirel, L., Kubina, M., Casetta, A. and Naas, T. (2000) Biochemical-genetic characterization and distribution of OXA-22, a chromosomal and inducible class D ß-lactamase from *Ralstonia* (*Pseudomonas*) *pickettii*. Antimicrob. *Agents Chemother.* 44, 2201–2204.

[36] Rasmussen, B. A., Keeney, D., Yang, Y. and Bush, K. (1994) Cloning and expression of a cloxacillin-hydrolyzing enzyme and a cephalosporinase from *Aeromonas sobria* AER 14M in *Escherichia coli*: requirement for an *E. coli* chromosomal mutation for efficient expression of the class D enzyme. Antimicrob. *Agents Chemother.* 38, 2078–2085.

[37] Abraham, E. P. and Chain, E. (1940) An enzyme from bacteria able to destroy penicillin. *Nature (London)* 146, 837- 837.

[38] Jaurin, B. and Grundstrom, T. (1981) AmpC cephalosporinase of *Escherichia coli* K-12 has a different evolutionary origin from that of ß-lactamases of the penicillinase type *Proc. Natl. Acad. Sci. USA* 78, 4897-4901.

[39] Richmond, M. H. (1963) Purification and properties of the exopenicillinase from *Staphylococcus aureus*. *Biochem. J.* 88, 452-459.

[40] Novick, R. P. (1963) Analysis by transduction of mutations affecting penicillinase formation in *Staphylococcus aureus*. *J. Gen. Microbiol.* 33, 121-136.

[41] Richmond, M. H. (1975) Beta-lactamase (*Staphylococcus aureus*). *Methods Enzymol.*. 43, 664-672.

[42] Wang, P. Z. and Novick, R. P. (1987) Nucleotide sequence and expression of the ß-lactamase gene from *Staphylococcus aureus* pI258 in *Escherichia coli, Bacillus cereus* and *Staphylococcus aureus*. *J. Bacteriol.* 169, 1763-1766.

[43] Datta, N. and Kontomichalou, P. (1965) Penicillinase synthesis controlled by infectious R factors in Enterobacteriaceae. N*ature* 208, 239-241.

[44] Matthew, M., Harris, A. M. Marshall, M.J. and Ross, G. W. (1975) Use of analytical isoelectric focusing for detection and identification of ß-lactamases. *J. Gen. Microbiol.* 88, 169-178.

[45] Sutcliffe, J.G. (1978) Nucleotide sequence of the ampicillin resistance gene of *Escherichia coli* pBR322. *Proc. Nat. Acad. Sci. USA* 75: 3737-3741.

[46] Sutcliffe, J.G. (1979) Complete nucleotide sequence of the *Escherichia coli* plasmid pBR322. C*old Spring Harb. Symp. Quant. Biol.* 43, 77-90.

[47] Leflon-Guibout, V., Heym, B. and Nicolas-Chanoine, M. H. (2000) Updated sequence information and proposed nomenclature for bla_{TEM} genes and their promoters. Antimicrob. *Agents Chemother.* 44, 3232-3234.

[48] Goussard, S., and Courvalin, P. (1991) Sequence of the genes $bla_{T\text{-}1B}$ and $bla_{T\text{-}2}$. Gene 102:71–73.

[49] Goussard, S., and Courvalin, P. (1999) Updated sequence information for TEM ß-lactamase genes. Antimicrob. *Agents Chemother.* 43:367–370.

[50] Leflon-Guibout, V., Speldooren, V., Heym, B. and M.-H. Nicolas-Chanoine M. H. (2000) Epidemiological survey of amoxicillin-clavulanate resistance and corresponding molecular mechanisms in *Escherichia coli* isolates in France: new genetic features of blaTEM genes. Antimicrob. *Agents Chemother.* 44:2709–2714.

[51] Roupas, A. and Pitton, J. S. (1974) R-Factor mediated and chromosomal resistance to ampicillin in *Escherichia coli* Antimicrob. *Agents Chemother.* 5, 186-191.

[52] Babini, G. S. and Livermore, D. M. (2000) Are SHV ß-lactamases universal in *Klebsiella pneumoniae*? *Antimicrob. Agents Chemother.* 44, 2230-2230.

[53] Shlaes, D. M., Medeiros, A. A., Kron, M. A., Currie-McCumber, C., Papa, E. and Vartian, C. V. (1986) Novel plasmid-mediated ß-lactamase in members of the family *Enterobacteriaceae* from Ohio. Antimicrob. *Agents Chemother.* 30, 220-224.

[54] Arakawa, Y., Ohta, M., Kido, M., Fujii, Y., Komatsu, T. and Kato, N. (1986) Close evolutionary relationship between the chromosomally encoded ß-lactamase from gene of *Klebsiella pneumoniae* and the TEM ß-lactamase gene mediated by R plasmids. *FEBS Lett.* 207, 69-74.

[55] Siebor E., Pechinot A., Duez J., Neuwirth C.2005. One new LEN enzyme and two new OKP enzymes in *Klebsiella pneumoniae* clinical isolates and proposed nomenclature for chromosomal ß-lactamases of this species. Antimicrob. *Agents Chemother.* 49, 3097-3098.

[56] Haeggman, S., Lofdahl, S., Paauw, A., Verhoef, J. and Brisse, S. (2004) Diversity and evolution of the class A chromosomal ß-lactamase gene in *Klebsiella pneumoniae.* Antimicrob. *Agents Chemother.* 48, 2400-2408.

[57] Fevre, C., Passet, V., Weill, F., Grimont, P.A.D. and Brisse, S. (2005) Variants of the *Klebsiella pneumoniae* OKP chromosomal ß-lactamase are divided into two main Groups, OKP-A and OKP-B. Antimicrob. *Agents Chemother.* 49, 5149-5152.

[58] Ochman, H. and Wilson A. C. (1987) Evolution in bacteria: evidence for a universal substitution rate in cellular genomes. *J. Mol. Evol.* 26, 74-86.

[59] Poirel, L., Heritier, C., Podglajen, I., Sougakoff, W., Gutmann, L. and Nordmann, P. (2003) Emergence in *Klebsiella pneumoniae* of a chromosome-encoded SHV ß-

lactamase that compromises the efficacy of imipenem. Antimicrob. *Agents Chemother.* 47, 755-758.

[60] Sabath, L. D., Jago M. and Abraham, E. P. (1965) Cephalosporinase and penicillinase activities of a beta-lactamase from *Pseudomonas pyocyanea. Biochem. J.* 96, 739-752.

[61] Newsom, S. W .B., Sykes, R. B. and Richmond, M. H. (1970) Detection of a ß-lactamase markedly active against carbenicillin in a strain of *Pseudomonas aeruginosa. J. Bacteriol.* 101, 1079-1080.

[62] Dale, J. W. and Smith, J. T. (1971) The purification and properties of the ß-lactamase specified by the resistance factor R-1818 in *Escherichia coli* and *Proteus mirabilis. Biochem. J.* 123, 493-500.

[63] Dale, J. W. and Smith, J. T. (1974) R-Factor mediated ß-lactamases that hydrolyze oxacillin: evidence for two distinct groups *J. Bacteriol.* 119, 351-356.

[64] Dale, J. W. and Smith, J. T. (1976) The dimeric nature of an R-factor mediated ß-lactamase. *Biochem. Biophys. Res. Comm.* 68, 1000-1005.

[65] Kuwabara S., Abraham E.P. (1967) Some properties of two exocellular ß-lactamases from *Bacillus cereus* 569/H. *Biochem. J.* 103: 27-30.

[66] Kuwabara S.1970. Purification and properties of two extracellular ß-lactamases from *Bacillus cereus* 569/H. *Biochem. J.* 118: 457-465.

[67] Materon I.C., Queenan A. M., Koehler T. M., Bush K. and Palzkill T. (2003) Biochemical Characterization of ß-lactamases Bla1 and Bla2 from *Bacillus anthracis.* Antimicrob. *Agents Chemother.* 47, 2040-2042.

[68] Philippon, A., Arlet, G. and Jacoby, G. A. (2002) Plasmid-determined AmpC type ß-lactamases. *Antimicrob. Agents Chemother.* 46, 1-11.

[69] Barlow, M. and Hall, B. G. (2002) Origin and evolution of the AmpC ß-lactamases of *Citrobacter freundii. Antimicrob. Agents Chemother.* 46, 1190-1198.

[70] Mammeri, H. and Nordmann, P. (2006) Extended-spectrum cephalosporinases in *Enterobacteriaceae. Anti-Infective Agents in Medicinal Chemistry,* 5, in press.

[71] Poirel, L., Guibert, M., Girlich, D., Naas, T. and Nordmann, P. (1999) Cloning, sequence analyses, expression, and distribution of ampC-ampR from *Morganella morganii* clinical isolates. *Antimicrob. Agents Chemother.* 43, 769-776.

[72] Barnaud, G., Arlet, G., Verdet, C., Gaillot, O., Lagrange, P. H. and Philippon, A. (1998) *Salmonella enteritidis*: ampC plasmid-mediated inducible ß-Lactamase (DHA-1) with an ampR gene from *Morganella morganii.* Antimicrob. *Agents Chemother.* 42, 2352-2358.

[73] Sanders, C. C. (1987) Chromosomal cephalosporinases responsible for multiple resistance to newer ß-lactam antibiotics. *Annu. Rev. Microbiol.* 41, 573-593.

[74] Barnaud, G., Benzerara, Y., Gravisse, J., Raskine, L., Sanson-Le Pors, M.J., Labia, R. and Arlet G. (2004) Selection during cefepime treatment of a new cephalosporinase variant with extended-spectrum resistance to cefepime in an *Enterobacter aerogenes* clinical isolate. Antimicrob. *Agents Chemother.* 48, 1040-1042.

[75] Papanicolaou, G. A., Medeiros, A. A. and Jacoby G. A. (1990) Novel plasmid-mediated ß-lactamase (MIR-1) conferring resistance to oxyimino- and α-methoxy ß-lactams in clinical isolates of *Klebsiella pneumoniae. Antimicrob. Agents Chemother.* 34, 2200-2209.

[76] Jacoby, G. A. and Tran, J. (1999) Sequence of the MIR-1 ß-lactamase gene. Antimicrob. *Agents Chemother.* 43, 2950-2959.

[77] Stokes, H. W., Tomaras, C., Parsons, Y. and Hall, R. M. (1993) The partial 39-conserved segment duplications in the integrons In6 from pSa and In7 from pDGO100 have a common origin. *Plasmid* 30, 39-50.

[78] Valentine, C. R., Heinrich, M. J., Chissoe, S. L. and Roe, B. R. (1994) DNA sequence of direct repeats of the *sulI* gene of plasmid pSa. *Plasmid* 32, 222-227.

[79] Bonnet, R. (2004) Growing group of extended-spectrum ß-lactamases: the CTX-M enzymes. Antimicrob. *Agents Chemother.* 48, 1-14.

[80] Bernard, H., Tancrède, C., Livrelli, V., Morand, A., Barthélémy, M. and Labia, R. (1992) A novel plasmid-mediated extended-spectrum ß-lactamase not derived from TEM- or SHV-type enzymes. *J. Antimicrob. Chemother* 29, 590-592.

[81] Barthélémy, M., Péduzzi, J., Bernard, H., Tancrède, C. and Labia, R. (1992) Close amino acid sequence relationship between the new plasmid-mediated extended-spectrum ß-lactamase MEN-1 and chromosomally encoded enzymes of *Klebsiella oxytoca. Biochim. Biophys. Acta* 1122, 15-22.

[82] Bauernfeind, A., Grimm, H. and Schweighart, S. (1990) A new plasmidic cefotaximase in a clinical isolate of Escherichia coli. *Infectio*n 18, 294-298.

[83] Bauernfeind, A., Stemplinger, I., Jungwirth, R., Ernst, S. and Casellas, J. M. (1996) Sequences of ß-lactamase genes encoding CTX-M-1 (MEN-1) and CTX-M-2 and relationship of their amino acid sequences with those of other [beta]-lactamases. Antimicrob. *Agents Chemother.* 40, 509-513.

[84] Radice, M., Power, P., Di Conza, J., Gutkind, G., Bonnet, R., Sirot, D., Sirot, J. and Labia, R. (2002) Early dissemination of CTX-M-derived enzymes in South America. Antimicrob. *Agents Chemother.* 46, 602-604.

[85] Rossi, A., lopardo, H., Woloj, M., Picandet A. M., Marino, M., Galas, M., Radice, M. and Gutkind, G. (1995) Non-typhoid *Salmonella* spp. resistant to cefotaxime. J. Antimicrob. *Chemother.* 36, 697-702.

[86] Matsumoto, Y., Ikeda, F., Kamimura, T., Yokota, Y. and Mine, Y. (1988) Novel plasmid-mediated ß-lactamase from *Escherichia coli* that inactivates oxyimino-cephalosporins. Antimicrob. *Agents. Chemother.* 32, 1243-1246.

[87] Lartigue M., Poirel L., Aubert D. and Nordmann P. (2006) In vitro analysis of ISEcp1B-mediated mobilization of naturally occurring ß-lactamase gene bla_{CTX-M} of *Kluyvera ascorbata. Antimicrob. Agents Chemother*. 50, 1282-1286.

[88] Nordmann, P., Ronco, E., Naas, T., Duport, C., Michel-Briand, Y. and Labia, R. (1993) Characterization of a novel extended-spectrum ß-lactamase from *Pseudomonas aeruginosa.* Antimicrob. *Agents Chemother.* 37, 962-969.

[89] Nordmann, P. and Nass, T. (1994) Sequence analysis of PER-1 extended-spectrum ß-lactamase from *Pseudomonas aeruginosa* and comprison with class A ß-lactamases. Antimicrob. *Agents Chemother.* 38, 104-114.

[90] Tranier, S., Bouthors, A. T., Maveyraud, L., Guillet, V., Sougakoff, W. and Samama, J. P. (2000) The high resolution crystal structure for class A ß-lactamase PER-1 reveals the bases for its increase in breadth of activity. *J. Biol. Chem.* 275, 28075-28082.

[91] Vahaboglu, H., Öztürk, R., Aygün, G., Cokunkan, F., Yaman, A., Kaygusuz A., Leblebicioglu, H., Balik, I., Aydin, K. and Otkun M. (1997) Widespread detection of PER-1-type Extended-Spectrum ß-lactamases among nosocomial *Acinetobacter* and *Pseudomonas aeruginosa* isolates in Turkey: a nationwide multicenter study. Antimicrob. *Agents Chemother.* 41, 2265-2269.

[92] Bauernfeind, A., Stemplinger, I., Jungwirth, R., Mangold, P., Amann, S., Akalin, E., Ang, O., Bal, C. and Casellas, J. M. (1996) Characterization of ß-lactamase genes *bla*$_{PER-2}$, which encodes an extended-spectrum class A ß-lactamase. *Antimicrob. Agents Chemother.* 40, 616-620.

[93] Bonnet, R., Sampaio, J. L. M., Chanal, C., Sirot, D., De Champs, C., Viallard, J. L., Labia, R. and Sirot, J. (2000) A novel class A extended-spectrum ß-lactamase (BES-1) in *Serratia marcescens* isolated in Brazil. Antimicrob. *Agents Chemother.* 44, 3061-3068.

[94] Poirel, L., Brinas, L., Verlinde, A., Ide L. and Nordmann, P. (2005) BEL-1, a novel clavulanic acid-inhibited extended-spectrum ß-lactamase, and the class 1 integron In120 in *Pseudomonas aeruginosa. Antimicrob. Agents Chemother.* 49, 3743-3748.

[95] Matsumoto, Y. and Inoue, M. (1999) Characterization of SFO-1, a plasmidmediated inducible class A beta-lactamase from *Enterobacter cloacae*. Antimicrob. *Agents Chemother.* 43, 307-313.

[96] Silva, J., Aguilar, C., Ayala, G., Estrada, M. A., Garza-Ramos, U., Lara-Lemus, R. and Ledezma, L. (2000) TLA-1: a new plasmid-mediated extended-spectrum ß-lactamase from *Escherichia coli. Antimicrob. Agents Chemother.* 44, 997-1003.

[97] Girlich, D., Poirel, L., Schluter, A. and Nordmann, P. (2005) TLA-2, a novel Ambler class A expanded-spectrum ß-lactamase. Antimicrob. *Agents Chemother.* 49, 4767-4770.

[98] Nordmann, P., Mariotte, S., Naas, T., Labia, R. and Nicolas, M. H. (1993) Biochemical properties of a carbapenem-hydrolyzing ß-lactamase from *Enterobacter cloacae* and cloning of the gene into *Escherichia coli. Antimicrob. Agents Chemother.* 37, 939-946.

[99] Naas, T. and Nordmann, P. (1994) Analysis of a carbapenem-hydrolyzing class A ß-lactamase from *Enterobacter cloacae* and of its Lys-R-type regulatory protein. P*roc. Natl. Acad. Sci. USA* 91, 7693-7697.

[100] Mariotte-Boyer, S., Nicolas, M. H. and Labia, R. (1996) A kinetic study of NMC-A ß-lactamase, an Ambler class A carbapenemase also hydrolyzing cephamycins. *FEMS Micobiology Lett.* 143, 29-33.

[101] Rasmussen, B. A., Bush, K., Keeney D., Yang Y., Hare, R., O'Gara C. and Medeiros, A. A. (1996) Characterization of IMI-1, a class A carbapenem-hydrolyzing enzyme from *Enterobacter cloacae. Antimicrob. Agents Chemother.* 40, 2080-2086.

[102] Yang, Y., P. Wu, and Livermore, D. M. (1990) Biochemical characterization of a ß-lactamase that hydrolyzes penems and carbapenems from two *Serratia marcescens* isolates. *Antimicrob. Agents Chemother.* 34, 755–758.

[103] Archibald, L., Phillips, L., Monnet, D., McGowan, J. E., Tenover, F. C.and Gaynes, R. (1997) Antimicrobial resistance in isolates from inpatients and outpatients in the United States: increasing importance of the intensive care unit. *Clin. Infect. Dis.* 24, 211–215

[104] Fridkin, S. K., Steward, C. D., Edwards, J. R., Pryor, E. R., McGowan, J. E., Archibald, L. K., Gaynes, R. P. and Tenover, F. C. (1999) Surveillance of antimicrobial use and antimicrobial resistance in United States hospitals : project ICARE phase 2. Project Intensive Care Antimicrobial Resistance Epidemiology (ICARE) hospitals. *Clin. Infect. Dis.* 29, 245–252.

[105] Yigit, H., Queenan, A. M., Anderson, G. J., Domenech-Sanchez, A., Biddle, J. W., Steward, C. D., Alberti, S., Bush K. and Tenover F. C. (2001) Novel carbapenem-

hydrolyzing ß-lactamase, KPC-1, from a carbapenem-resistant strain of *Klebsiella pneumoniae*. *Antimicrob. Agents Chemother.* 45, 1151-1161.

[106] Miriagou, V., Tzouvelekis, L. S., Rossiter, S., Tzelepi, E., Angulo, F. J., Whichard, J. M. (2003) Imipenem resistance in a *Salmonella* clinical strain due to plasmid-mediated class A carbapenemase KPC-2. Antimicrob. Agents Chemother. 47, 1297-1300.

[107] Poirel, L., Le Thomas, I., Naas, T., Karim, A. and Nordmann, P. (2000) Biochemical sequence analyses of GES-1, a novel class A extended-spectrum ß-lactamase, and the class 1 integron In52 from *Klebsiella pneumoniae*. *Antimicrob. Agents Chemother.* 44, 622-632.

[108] Poirel, L., Weldhagen, G.F., Naas, T., De Champs, C., Dove, M.G. and Nordmann, P. (2001) GES-2, a class A ß-lactamase from *Pseudomonas aeruginosa* with increased hydrolysis of imipenem. Antimicrob. *Agents Chemother. 45*, 2598-2603.

[109] Swaren, P., Maveyraud, L., Raquet, X., Cabantous, S., Duez, C., Pedelacq, J. D., Mariotte-Boyer, S., Mourey, L., Labia, R., Nicolas-Chanoine, M. H., Nordmann, P., Frère, J. M. and Samama, J. P. (1998) X-Ray analysis of the NMC-A ß-lactamase at 1.64 A resolution, a class A carbapenemase with broad substrate activity. *J. Biol. Chem.* 41, 26714-26721.

[110] Sougakoff, W., L'Hermite, G., Pernot, L., Naas, T., Guillet, V., Nordmann, P., Jarlier, V. and Delettre, J. (2002) Structure of the imipenem-hydrolyzing class A ß-lactamase SME-1 from *Serratia marcescens*. Acta Crystallogr. *D Biol. Crystallogr.* 58, 267-274.

[111] Majiduddin, F. K. and Palzkill, T. (2003) Amino acid sequence requirements at residues 69 and 238 for the SME-1 ß-lactamase to confer resistance to ß-lactam antibiotics. Antimicrob. *Agents Chemother.* 47, 1062-1067.

[112] Majiduddin, F. K. and Palzkill, T. (2005) Amino acid residues that contribute to substrate specificity of class A ß-lactamase SME-1. Antimicrob. *Agents Chemother.* 49, 3421-3427.

[113] Rawlings, N. D. and Barrett, A. J. (1995) Evolutionary families of metallopeptidases. Methods. *Enzymol.* 248, 183-228.

[114] Rasmussen, B. A. and Bush, K. (1997) Carbapenem-hydrolyzing ß-lactamases. Antimicrob. *Agents Chemother*. 41, 223-232.

[115] Sutton, B. J., Artymiuk, P. J., Cordero-Borboa, A. E., Little, C., Phillips, D. C. and Waley, S. G. (1987) An X-ray-crystallographic study of ß-lactamase II from *Bacillus cereus* at 0.35 nm resolution. *Biochem. J.* 248, 181-188.

[116] Carfi, A., Pares, S., Duée, E., Galleni, M., Duez, C., Frère ,J. M. and Dideberg, O. (1995) The 3-D structure of a zinc metallo-ß-lactamase from *Bacillus cereus* reveals a new type of protein fold. *EMBO J.* 14, 4914-4921.

[117] Concha, N. O., Rasmussen, B. A., Bush, K. and Hertzberg O. (1996) Crystal structure of the wide-spectrum binuclear zinc ß-lactamase from *Bacteroides fragilis*. *Structure 4,* 823-836.

[118] Watanabe, M., Iyobe, S., Inoue, M. and Mitsuhashi, S. (1991) Transferable imipenem resistance in *Pseudomonas aeruginosa*. Antimicrob. Agents Chemother. 35, 147-151.

[119] Walsh, T. R., Toleman, M. A., Poirel, L. and Nordmann, P. (2005) Metallo-ß-lactamases: the quiet before the storm? *Clin. Microbiol. Rev.* 18, 306-325.

[120] Stoczko, M., Frère, J.M., Rossolini, G. M. and Docquier, J. (2006) Postgenomic scan of metallo-ß-lactamase homologues in Rhizobacteria: Identification and characterization

of BJP-1, a subclass B3 ortholog from *Bradyrhizobium japonicum*. Antimicrob. *Agents Chemother.* 50, 1973-1981.

[121] Poirel, L, Nordmann, P. (2006) Carbapenem resistance in *Acinetobacter baumannii*: mechanisms and epidemiology. *Clin. Microbiol. Infect.* 12, 826-836.

[122] Donald, H. M., W. Scaife, S. G. B. Amyes, and H.-K. Young. (2000) Sequence analysis of ARI-1, a novel OXA ß-lactamase, responsible for imipenem resistance in *Acinetobacter baumannii* 6B92. Antimicrob. *Agents Chemother.* 44,196–199.

[123] Poirel, L., Heritier, C., Tolun, V. and Nordmann, P. (2004) Emergence of Oxacillinase-Mediated Resistance to Imipenem in *Klebsiella pneumoniae.* Antimicrob. Agents Chemother. 48, 15-22.

[124] Schneider, I., Queenan M. A. and Bauernfeind, A. (2006) Novel Carbapenem-Hydrolyzing Oxacillinase OXA-62 from Pandoraea pnomenusa. Antimicrob. Agents Chemother. 50, 1330-1335.

[125] Paton, R., Miles, R.S., Hood, J. and Amyes S. G. B. (1993) ARI-1: ß-lactamase-mediated imipenem resistance in *Acinetobacter baumannii. Int. J. Antimicrob. Agents* 2, 81-88.

[126] Bou, G., Oliver, A., Martínez-Beltrán, J. (2000) OXA-24, a novel class D ß-lactamase with carbapenemase activity in an *Acinetobacter baumannii* clinical strain. *Antimicrob. Agents Chemother.* 44, 1556-1561.

[127] Bertini, A., Giordano, A., Varesi, P., Villa, L., Mancini, C. and Carattoli, A. (2006) First report of the carbapenem-hydrolyzing oxacillinase OXA-58 in *Acinetobacter baumannii* isolates in Italy. *Antimicrob. Agents Chemother.* 50, 2268-2269.

[128] Coelho, J., Woodford, N., Afzal-Shah, M. and Livermore, D. (2006) Occurrence of OXA-58-like carbapenemases in *Acinetobacter* spp. collected over 10 years in three continents. Antimicrob. *Agents Chemother.* 50, 756-758.

[129] Poirel, L., Heritier, C. and Nordmann, P. (2004) Chromosome-encoded Ambler class D ß-lactamase of *Shewanella oneidensis* as a progenitor of carbapenem-hydrolyzing oxacillinase. *Antimicrob. Agents Chemother.* 48, 348-351.

[130] Yong, D., Toleman, M.A., Giske, C.G., Cho, H.S., Sundman K., Lee, K. and Walsh, T.R. (2009) Characterization of a new metallo-β-lactamase gene, $bla_{\mathrm{NDM-1}}$, and a novel erythromycin esterase gene carried on a unique genetic structure in *Klebsiella pneumoniae* sequence type 14 from India. Antimicrob. *Agents Chemother.* 53, 5046-5054.

In: Beta-Lactamases
Editor: Jean-Marie Frère, pp. 369-382
ISBN: 978-1-61324-638-2

Chapter 14

Emergence of New Phenotypes Mediated by Amino Acid Substitutions in β-Lactamases

George Jacoby[1] and Karen Bush[2]
[1] Lahey Clinic, Burlington, Massachusetts 01805, U.S.A.
[2] Indiana University, Bloomington IN 47405, U.S.A.

Abstract

Substitutions at particular amino acids allow the common plasmid-mediated TEM, SHV, and OXA enzymes to become extended-spectrum β-lactamases able to hydrolyze oxyimino-β-lactams such as ceftazidime, cefotaxime, ceftriaxone, and aztreonam. Other amino acid changes in the TEM and SHV enzymes are associated with resistance to β-lactamase inhibitors such as clavulanic acid and sulbactam. Alterations at a few sites allow enzymes in the CTX-M family to attack ceftazidime or in the AmpC family to hydrolyze cefepime or cefpirome.

Introduction

Epidemiological studies in the early 1980s demonstrated that the most prevalent β-lactamases in Gram-negative clinical isolates were the plasmid-encoded TEM-1, TEM-2, OXA-1 and SHV-1 enzymes, in addition to the chromosomal AmpC-type cephalosporinases [1-3] Because of their wide distribution throughout the *Enterobacteriaceae*, it is not unexpected to find that these enzymes are the foundations of closely-related β-lactamase families, often differing among themselves by one or more amino acid substitutions.

The TEM β-lactamase has been described as a fully efficient enzyme [4]. Nonetheless the TEM enzyme can tolerate substitutions at 220 of its 263 amino acid residues and still confer resistance to ampicillin [5]. That mutations could alter the substrate spectrum of TEM had been demonstrated in the laboratory by Hall and Knowles in 1976 [6]. The identification of isozymes had thus been recognized in the β-lactamase field, similar to other sets of physiological isozyme families such as alcohol dehydrogenase and cytochrome P450 [7, 8].

How important this phenomenon would become was not appreciated until new β-lactams not significantly hydrolyzed by TEM began to be introduced in the 1980s.

These β-lactamase-stable antibiotics included cephamycins with an α-methoxy side chain such as cefoxitin and cefotetan, oximino-cephalosporins such as cefotaxime, ceftazidime, and ceftriaxone, the oximino-monobactam aztreonam, and carbapenems such as imipenem and meropenem. Moreover, the TEM and SHV enzymes were also challenged by the introduction of the β-lactamase inhibitor combinations during the same time [9]. By combining β-lactamase labile penicillins with enzyme inactivators such as clavulanic acid or sulbactam, the hydrolyzing enzymes were no longer able to confer β-lactam resistance to their producing organisms.

In 1983 *Klebsiella* isolates with plasmid-mediated resistance to cefotaxime were reported from Germany [10] and soon were shown to produce a β-lactamase related to the narrower spectrum SHV-1 [11]. In 1987 *Klebsiella* isolates with transmissible resistance to cefotaxime, ceftazidime, ceftriaxone, and aztreonam emerged in France, where the enzyme proved to be a derivative of TEM [12, 13]. In 1993 a plasmid-encoded extended-spectrum variant of an OXA enzyme was described in *Pseudomonas aeruginosa* from Turkey [14]. Since then the number of expanded-spectrum TEM, SHV, and OXA derivatives defined by unique combinations of amino acid substitutions has proliferated and was joined in 1992 by TEM derivatives, and later, SHV variants, resistant to inhibition by clavulanic acid and other β-lactamase inhibitors [15]. In 1997 a web site was created on the internet to keep track of the growing number of TEM, SHV, OXA, and other variants [16].

Other β-lactamase families populated with a variety of enzymes derived from a common parent include the VIM and IMP metallo-β-lactamases [17, 18]. Serine β-lactamases that hydrolyze carbapenems include closely-related families of enzymes such as the IMI, SME, and KPC β-lactamases [19-21]. A few substitutions that alter the spectrum of AmpC-type enzymes have also been reported. However, most of these families of enzymes confer very similar phenotypic properties to the various producing organisms, and are not representative of the dramatic changes seen with the ESBLs resulting from only a few point mutations.

The TEM Family

Not counting alterations in the leader sequence, changes at 43 of the 263 residues of TEM-1 have been observed in different natural isolates with the ESBL phenotype. The most common sites for change are residues 39, 104, 164, 182, 238, 240, and 265, although other sites have been identified in multiple ESBLs (Table 1).

Residue 39 is the site where TEM-2 differs from TEM-1, so that a Q39K change in other derivatives may only reflect derivation of that enzyme from TEM-2, which has a more active promoter [23], rather than TEM-1. Although the assumption has been made that the two enzymes have interchangeable kinetic parameters [24], a careful comparison of the properties of TEM-1 and its Q39K mutant revealed that the Q39K change conferred a consistent two-fold decrease in susceptibility to ceftazidime without an effect on susceptibility to cefotaxime or aztreonam [25]. A 50% increase in catalytic efficiency for cephaloridine hydrolysis by TEM-2, based on an increased k_{cat} value, was also noted in comparison to TEM-1[25]

Table 1. Common substitutions identified in the naturally-occurring TEM and SHV families of enzymes leading to phenotypic changes in the producing bacteria

Enzyme Family	Amino Acid Positions	Amino Acid Substitutions Identified with Phenotype[a]			
		Wild-type	ESBL	Inhibitor-Resistant	ESBL and Inhibitor Resistant
TEM	39	Q or K	Q or K	Q or K	K
	69	M	--[b]	I, L, V	L
	104	E	K	--	K
	127	I	–	V	–
	130	S	–	G	–
	153	H	R	–	–
	164	R	C, H. S	–	S
	165	W	R	R	–
	182	M	T	T	–
	237	G	T	–	T
	238	G	S	–	S
	240	E	K	–	K
	244	R	–	C, G, H, L, S, T	S
	265	T	M	M	–
	275	R	L	L, Q	–
	276	N	–	D	D
SHV	35	L	Q	–	–
	43	R	S	–	–
	54	G	Del[c]	–	Del
	69	M	–	I	–
	89	E	K	–	–
	129	M	V	–	–
	130	S		–	G
	140	A	R	–	R
	156	G	D	–	–
	173	L	F	–	–
	179	D	A, G, N	–	–
	187	A	T	–	–
	192	K	N	–	N
	193	L	V	–	V
	205	R	L	–	–
	238	G	A, S	–	S
	240	E	K	–	K
	243	A	G	–	–

[a] Substitutions that have been identified in at least two natural isolates.

[b] No change from wild-type.

[c] Del = deletion.

Note that one to five mutations may be associated with individual enzymes ([16] and references therein). Amino acid numbering follows Ambler et al. [22].

Derivatives with single amino acid changes at positions 104 (TEM-17), 164 (TEM-12), and 238 (TEM-19) all have the ESBL phenotype. In terms of enzyme structure, changes at these positions open the active site to accommodate the bulky oximino side chain and may also alter specific electrostatic interactions [26, 27].The M182T change has been shown to stabilize other amino acid substitutions in TEM, thus acting as a missense mutation suppressor [27, 28] but has no effect on substrate spectrum. The T265M change affects neither the substrate range nor the isoelectric point of the enzyme [29]. Alterations at position 240 are to date always associated with changes at other sites in natural isolates, but laboratory constructs with only a E240K change have modest increases in ceftazidime and/or aztreonam resistance [25, 30].

ESBLs have increased k_{cat} and decreased K_m values for oximino-β-lactam substrates compared to TEM-1 and TEM-2 (Table 2). Ceftazidime, aztreonam, and cefotaxime MICs are increased, usually in that order and reflect not only intrinsic enzyme efficiency but also the relative rates at which different β-lactams reach the periplasmic space where the enzyme is located.

Single amino acid changes have less of an effect on enzyme properties than dual alterations. For example organisms producing the TEM-12 enzyme with R164S are resistant only to ceftazidime, but if the R164S mutation is combined with E104K (TEM-26) or with E240K (TEM-10), the efficiency of ceftazidime hydrolysis is increased markedly and resistance to aztreonam becomes manifest (Table 2).

Inhibitor resistance is primarily associated with amino acid substitutions at positions 69, 130, and 244, although other substitutions have been reported (Table 1). Characteristically an *Escherichia coli* strain producing one of these enzymes is resistant to amoxicillin/clavulanate or ticarcillin/clavulanate but frequently susceptible to piperacillin/tazobactam [39]. Most inhibitor resistant TEM enzymes have decreased k_{cat} and increased K_m values for all substrates compared to TEM-1 and higher IC_{50} and K_i values for β-lactamase inhibitors, especially sulbactam [39].

Combinations of alterations producing both the ESBL and inhibitor resistance phenotype are rare but have been observed in natural isolates [40-42]. These enzymes have properties that are less extreme than would be predicted for a pure ESBL or inhibitor-resistant TEM.

Table 2. Properties of selected TEM and SHV ESBLs

β-lactamase	Amino acid alteration					Relative V_{max} [a]			k_{cat} (s^{-1})			K_m (μM)			MIC (μg/ml) in *E. coli*			Ref
	104	164	205	238	240	CTX[b]	CAZ[b]	ATM[b]	CTX	CAZ	ATM	CTX	CAZ	ATM	CTX	CAZ	ATM	
TEM-1	E	R			E	0.07	0.01	<0.14	0.67	0.096	<1.3	1000	300	200	0.125	0.25	0.125	[31-33]
TEM-10	E	S			K	9.3	200	32	2.9	62	10	46	150	28	1	64	32	[33, 34]
TEM-12	E	S			E	4.2	6.7	11	0.2	1.4	1.5	94	130	870	0.5	64	8	[33, 35, 36]
TEM-26	K	S			E	6.3	190	36	1.7	60	12	13	180	37	1	128	16	[33, 34]
SHV-1			R	G	E	<1	<1	<1				350	79	260	0.125	1	0.5	[31, 32, 37]
SHV-2			R	S	E	21	<1	<1				41	24	10	64	32	32	[31, 37, 38]
SHV-3			L	S	E	18	<1	<1							64	32	32	[31, 37]
SHV-4			L	S	K	22	6	<1				25	60	0.5	128	128	256	[31, 37, 38]
SHV-5			R	S	K	16	6	<0.48	24	9	<0.72	5	9.2	NH[c]	64	128	256	[31, 33]

a Relative to Vmax of cephaloridine = 100.

b Antibiotic abbreviations: CTX, cefotaxime; CAZ, ceftazidime; ATM, aztreonam.

c NH – hydrolysis too slow to measure Km accurately.

THE SHV FAMILY

SHV-1 has greater intrinsic hydrolytic activity against oximino-β-lactams than TEM-1 and, when expressed at high level with or without an additional porin defect, can confer low level ceftazidime resistance (MIC 8-16 μg/ml) [32, 43]. In the first SHV ESBLs that were identified, the full ESBL phenotype was associated with amino acid alterations at positions 238 and 240, with single amino acid substitutions at one of these sites being sufficient, although the combination provides an even higher level of resistance [44]. The G238S change in particular has been shown to cause a displacement of the B3 β-strand away from the Ω loop thus expanding the binding site for substrates with a bulky oximino side chain [45]. More recently, SHV enzymes with alterations at many other sites have been reported as ESBLs (Table 1). In particular, a change at position 179 provides enhanced resistance specifically to ceftazidime [46, 47] and alterations at positions 142 and 234 have also been associated with selective ceftazidime resistance [48]. Selective ceftazidime resistance has also been described in SHV-16, which contains an unstable pentapeptide insertion 163DRWET167 corresponding to a tandem duplication in the Ω loop of the enzyme [49]. Selective cefotaxime resistance with a modest decrease in ceftazidime susceptibility is seen in strains with SHV-27 carrying a G156D change [50]. Ceftazidime resistance and reduced susceptibility to imipenem have been found in a strain with SHV-38 carrying an A146V replacement [51]. An isogenic set of strains differing only by single amino acid replacements at various sites in the SHV enzyme has not yet been constructed to validate the role of these various changes. An L35Q change without apparent effect on enzyme function is found in many SHV ESBLs and in the non ESBL SHV-11 from which the others may have evolved. As with TEM-1 and TEM-2 the L35Q change is associated with a stronger promoter [52].

A few inhibitor resistant SHV derivatives have been described, including SHV-10 and SHV-49. SHV-10 differs from ESBL SHV-9 by a single S130G substitution which confers resistance to clavulanic acid and tazobactam but removes activity against oximino-β-lactams [53]. SHV-49 differs from SHV-1 by an M69I substitution, identical to the mutation seen in some TEM inhibitor-resistant β-lactamases, but which reduces affinity for amoxicillin and ticarcillin and enhances resistance to both clavulanic acid and tazobactam [54]. Non-ESBL SHV-26 with A187T is also less susceptible to inhibition by clavulanic acid [55].

THE OXA FAMILY

The OXA family of enzymes is a diverse group some of which provide resistance to oximino-β-lactams or even carbapenems. However, many of these enzymes are widely divergent from each other, more so than members of the TEM and SHV families, with multiple amino acid substitutions required before major functional changes are noted [56, 57]. Thus, the ability to hydrolyze carbapenems resulting in carbapenem resistance is found with OXA enzymes that may exhibit only 39-63% sequence identity with other members of the OXA family.

The largest group is related to OXA-10 (once known as PSE-2). OXA-14 differs from OXA-10 by a G167D change and mediates increased resistance especially to ceftazidime [58]. OXA-11 has an additional G145S change associated with a further increase in ceftazidime and a slight increase in aztreonam resistance [14]. OXA-16 has G167D plus A127T mutations without an additional effect on resistance [59]. ESBL OXA-17 has an N76S alteration, while OXA-13, -19, and –28 differ from OXA-10 by eight amino acids, which differ also in the non-ESBL OXA-35, plus individual changes at residues 76, 145, 164, and 167 responsible for their ESBL phenotype [60-63]. They confer significant resistance to oximino-β-lactams in *P. aeruginosa* but not in *E. coli*, indicating that differences in enzyme synthesis, β-lactam permeability, or efflux mechanisms between the two organisms play an important role in expression of resistance.

OXA-15 is a derivative of OXA-2 with a D150G alteration that is associated with an increase of ceftazidime MIC from 16 to 128 μg/ml in *P. aeruginosa* and from 0.25 to 8 μg/ml in *E. coli* [64]. A similar increase in ceftazidime resistance is seen with OXA-32 which differs from OXA-2 by L169I [65]. In a porin-deficient strain of *K. pneumoniae* OXA-2 itself expresses low level ceftazidime (MIC 16 μg/ml) and ertapenem (MIC 32 μg/ml) resistance [66].

OXA-31 is a derivative of OXA-1 with 4 amino acid differences. In *P. aeruginosa* OXA-31 confers resistance to cefepime but not ceftazidime. In *E. coli* OXA-31 provides resistance neither to cefepime nor to other oximino-β-lactams. Cloned OXA-1 has the same phenotype, making it questionable whether OXA-31 should be termed an ESBL [67].

THE AMPC FAMILY

There have been a few reports of mutant AmpC-type β-lactamases in clinical isolates. (In the following discussion the amino acids have been numbered in mutants of *Enterobacter* spp. and *S. marcescens* to correspond with the standard numbering of *E. cloacae* P99)[68].

Resistance to cefepime in a clinical isolate of *Enterobacter aerogenes* from a patient treated with that antibiotic proved to be due to a L293P amino acid substitution combined with overproduction of the enzyme [69]. Precisely this substitution had previously been obtained by PCR mutagenesis and selection for enhanced cefepime resistance from the related P99 AmpC enzyme of *Enterobacter cloacae* [70], and a nearby V298E change was obtained on selecting for cefpirome resistance with another *E. cloacae* AmpC enzyme [71]. That this region is a hot spot for cefepime resistance was also evident from a clinical isolate of *E. cloacae* with enhanced resistance to cefepime and cefpirome due to a six amino acid deletion of residues 289 to 294 along with enzyme overproduction [72], and from a cefepime and cefpirome resistant clinical isolate of *S. marcescens* with a four amino acid deletion in the same region (residues 293-296)[73]. All these substitutions and deletions are located in the H-10 α-helix close to the C-terminus of the enzyme but distant from its active site [74].

Another *E. cloacae* clinical isolate with enhanced activity toward ceftazidime and other oximino-β-lactams was found to have a three amino acid duplication at positions 211-213, which on structural analysis had disordered and expanded the Ω loop facilitating deacylation of oximino-β -lactam substrates [75-77]. In other clinical isolates of *S. marcescens* ceftazidime resistance was found to be due to E219K [78] or S220Y substitutions [68].

A laboratory mutant of the *S. marcescens* enzyme with enhanced resistance to ceftazidime and cefpirome was found to have a T70I substitution close to the active site serine at position 64 [79]. Analogous clinical isolates have not yet been reported.

Amino acid substitutions distinguish plasmid-mediated AmpC-type enzymes as well, but how much such changes contribute to enzyme specificity has been little explored.

Other β-Lactamases

Many CTX-M enzymes have conspicuously greater activity toward cefotaxime than ceftazidime. Exceptions to this rule include CTX-M enzymes with a P167S or P167T substitution [80, 81] and a larger group with a D240G change [82-84]. The D240G substitution decreases the Km for ceftazidime, apparently by modifying the positioning of the β3 strand during hydrolysis [83]. A recent failure to identify by site-directed mutagenesis single amino acid residues that contribute to the carbapenemase activity of class A β-lactamase SME-1 illustrates that specialized properties like the ability to hydrolyze carbapenems are not the consequence of single amino acid alterations but rather the result of interactions at multiple sites which modify the properties of the enzyme [85].

Conclusion

Because sequencing has become so readily available for the study of β-lactamases, hundreds of new enzyme sequences have been identified over the past two decades. Many of these enzymes are the result of point mutations arising from well-characterized parent enzymes. It is probable that those mutations not leading to phenotypic differences represent the natural background of genetic variants that can be accommodated by this "floppy" enzyme family [86]. However, those mutations that confer resistance to newer β-lactams introduced since 1980 have most likely been selected due to the increase use of β-lactamase-stable antibiotics. As additional new agents enter clinical use, we can only expect to see further scrambling of known amino acid substitutions, as well as the identification of new substituted sites that are allowed by this promiscuous enzyme family.

In the last few years, the number of β-lactamases in the various enzyme families has grown dramatically to over 950 discrete, naturally-occurring enzymes (87). As of October 2010 our web site (http://www.lahey.org/Studies/) listed 183 TEM, 134 SHV, 193 OXA, 103 CTX-M, 59 CMY and 41 other AmpC β-lactamases as well as 28 IMP, 26 VIM, 16 GES and 10 KPC enzymes. The growth and spread of CTX-M variants worldwide has been particularly remarkable (88-94), with the emergence of CTX-M-15 as a predominant ESBL worldwide (95). Also notable is the increase in OXA enzymes, some with ESBL and carbapenemase activities (96-98).

Several excellent reviews of the role of amino acid substitutions in particular families have appeared, including TEM (99) and AmpC (100). The topic is also included in recent reviews of particular β-lactamase types such as AmpC enzymes (101), and ESBLs (102).

REFERENCES

[1] Roy, C, Segura, C, Tirado, M, Reig, R, Hermida, M, Teruel, D, and Foz, A. Frequency of plasmid-determined beta-lactamases in 680 consecutively isolated strains of *Enterobacteriaceae*. *Eur. J. Clin. Microbiol.*, 1985 4, 146-147.

[2] Matthew, M. Plasmid mediated β-lactamases of gram-negative bacteria: properties and distribution. *J. Antimicrob. Chemother.*, 1979 5, 349-358.

[3] Medeiros, AA. ß-lactamases. *Br. Med. Bull.*, 1984 40, 18-27.

[4] Christensen, H, Martin, MT, and Waley, SG. ß-lactamases as fully efficient enzymes. *Biochem. J.*, 1990 266, 853-861.

[5] Huang, W, Petrosino, J, Hirsch, M, Shenkin, PS, and Palzkill, T. Amino acid sequence determinants of ß-lactamase structure and activity. *J. Mol. Biol.*, 1996 258, 688-703.

[6] Hall, A, and Knowles, JR. Directed selective pressure on a ß-lactamase to analyse molecular changes involved in development of enzyme function. *Nature (London)*, 1976 264, 803-804.

[7] Pikkarainen, P, and Raiha, NC. Change in alcohol dehydrogenase isoenzyme pattern during development of human liver. *Nature*, 1969 222, 563-564.

[8] Brosen, K. Recent developments in hepatic drug oxidation. Implications for clinical pharmacokinetics. *Clinical Pharmacokinetics*, 1990 18, 220-239.

[9] Bush, K. ß-lactamase inhibitors from laboratory to clinic. *Clin. Microbiol. Rev.*, 1988 1, 109-123.

[10] Knothe, H, Shah, P, Krcméry, V, Antal, M, and Mitsuhashi, S. Transferable resistance to cefotaxime, cefoxitin, cefamandole and cefuroxime in clinical isolates of *Klebsiella pneumoniae* and *Serratia marcescens*. *Infection*, 1983 11, 315-317.

[11] Kliebe, C, Nies, BA, Meyer, JF, Tolxdorff-Neutzling, RM, and Wiedemann, B. Evolution of plasmid-coded resistance to broad-spectrum cephalosporins. *Antimicrob. Agents Chemother.*, 1985 28, 302-307.

[12] Brun-Buisson, C, Legrand, P, Philippon, A, Montravers, F, Ansquer, M, and Duval, J. Transferable enzymatic resistance to third-generation cephalosporins during nosocomial outbreak of multiresistant *Klebsiella pneumoniae*. *Lancet*, 1987 ii, 302-306.

[13] Sirot, D, Sirot, J, Labia, R, Morand, A, Courvalin, P, Darfeuille-Michaud, A, Perroux, R, and Cluzel, R. Transferable resistance to third-generation cephalosporins in clinical isolates of *Klebsiella pneumoniae*: identification of CTX-1, a novel ß-lactamase. *J. Antimicrob. Chemother.*, 1987 20, 323-334.

[14] Hall, LMC, Livermore, DM, Gur, D, Akova, M, and Akalin, HE. OXA-11, an extended-spectrum variant of OXA-10(PSE-2) ß-lactamase from *Pseudomonas aeruginosa*. *Antimicrob. Agents Chemother.*, 1993 37, 1637-1644.

[15] Vedel, G, Bellaaouaj, A, Gilly, L, Labia, R, Philippon, A, N,vot, P, and Paul, G. Clinical isolates of Escherichia coli producing TRI ß-lactamases: novel TEM-enzymes conferring resistance to ß-lactamase inhibitors. *J. Antimicrob. Chemother.*, 1992 30, 449-462.

[16] Jacoby, G, and Bush, K. *In* "http:// www.lahey.org/studies/".

[17] Pasteran, F, Faccone, D, Petroni, A, Rapoport, M, Galas, M, Vázquez, M, and Procopio, A. Novel variant (*bla*$_{\text{VIM-11}}$) of the metallo-ß-lactamase *bla*$_{\text{VIM}}$ family in a

GES-1 extended-spectrum-ß-lactamase-producing *Pseudomonas aeruginosa* clinical isolate in Argentina. *Antimicrob Agents Chemother*, 2005 49, 474-475.

[18] Mendes, RE, Toleman, MA, Ribeiro, J, Sader, HS, Jones, RN, and Walsh, TR. Integron carrying a novel metallo-ß-lactamase gene, bla_{IMP-16}, and a fused form of aminoglycoside-resistant gene *aac(6')-30/aac(6')-Ib'*: report from the SENTRY Antimicrobial Surveillance Program. *Antimicrob Agents Chemother*, 2004 48, 4693-4702.

[19] Aubron, C, Poirel, L, Ash, RJ, and Nordmann, P. Carbapenemase-producing *Enterobacteriaceae*, U.S. rivers. *Emerg Infect Dis*, 2005 11, 260-264.

[20] Woodford, N, Tierno, PM, Jr., Young, K, Tysall, L, Palepou, MF, Ward, E, Painter, RE, Suber, DF, Shungu, D, Silver, LL, Inglima, K, Kornblum, J, and Livermore, DM. Outbreak of *Klebsiella pneumoniae* producing a new carbapenem-hydrolyzing class A ß-lactamase, KPC-3, in a New York Medical Center. *Antimicrob Agents Chemother*, 2004 48, 4793-4799.

[21] Queenan, AM, Torres-Viera, C, Gold, HS, Carmeli, Y, Eliopoulos, GM, Moellering, RC, Jr., Quinn, JP, Hindler, J, Medeiros, AA, and Bush, K. SME-type carbapenem-hydrolyzing class A ß-lactamases from geographically diverse *Serratia marcescens* strains. *Antimicrob Agents Chemother*, 2000 44, 3035-3039.

[22] Ambler, RP, Coulson, AFW, Frère, J-M, Ghuysen, J-M, Joris, B, Forsman, M, Levesque, RC, Tiraby, G, and Waley, SG. A standard numbering scheme for the class A ß-lactamases. *Biochem. J.*, 1991 276, 269-272.

[23] Lartigue, MF, Leflon-Guibout, V, Poirel, L, Nordmann, P, and Nicolas-Chanoine, MH. Promoters *P3*, *Pa/Pb*, *P4,* and *P5* upstream from bla_{TEM} genes and their relationship to ß-lactam resistance. *Antimicrob Agents Chemother*, 2002 46, 4035-4037.

[24] Matagne, A, and Frere, JM. Contribution of mutant analysis to the understanding of enzyme catalysis: the case of class A beta-lactamases. *Biochimica et Biophysica Acta*, 1995 1246, 109-127.

[25] Blazquez, J, Morosini, MI, Negri, MC, Gonzalez-Leiza, M, and Baquero, F. Single amino acid replacements at positions altered in naturally occurring extended-spectrum TEM ß-lactamases. *Antimicrob Agents Chemother*, 1995 39, 145-149.

[26] Knox, JR. Extended-spectrum and inhibitor-resistant TEM-type ß-lactamases: mutations, specificity, and three-dimensional structure. *Antimicrob. Agents Chemother.*, 1995 39, 2593-2601.

[27] Wang, X, Minasov, G, and Shoichet, BK. Evolution of an antibiotic resistance enzyme constrained by stability and activity trade-offs. *J Mol Biol*, 2002 320, 85-95.

[28] Huang, W, and Palzkill, T. A natural polymorphism in beta-lactamase is a global suppressor. *Proc Natl Acad Sci U S A*, 1997 94, 8801-8806.

[29] Mabilat, C, and Courvalin, P. Development of "oligotyping" for characterization and molecular epidemiology of TEM ß-lactamases in members of the family Enterobacteriaceae. *Antimicrob. Agents Chemother.*, 1990 34, 2210-2216.

[30] Sowek, JA, Singer, SB, Ohringer, S, Malley, MF, Dougherty, TJ, Gougoutas, JZ, and Bush, K. Substitution of lysine at position 104 or 240 of TEM-1pTZ18R ß-lactamase enhances the effect of serine-164 substitution on the hydrolysis or affinity for cephalosporins and the monobactam aztreonam. *Biochemistry*, 1991 30, 3179-3188.

[31] Jacoby, GA, and Carreras, I. Activities of ß-lactam antibiotics against Escherichia coli strains producing extended-spectrum ß-lactamases. *Antimicrob. Agents Chemother.*, 1990 34, 858-862.

[32] Petit, A, Yaghlane-Bouslama, HB, Sofer, L, and Labia, R. Does high level production of SHV-type penicillinase confer resistance to ceftazidime in Enterobacteriaceae? *FEMS Microbiol. Lett.*, 1992 92, 89-94.

[33] Queenan, AM, Foleno, B, Gownley, C, Wira, E, and Bush, K. Effects of inoculum and ß-lactamase activity in AmpC- and extended-spectrum ß-lactamase (ESBL)-producing *Escherichia coli* and *Klebsiella pneumoniae* clinical isolates tested by using NCCLS ESBL methodology. *J Clin Microbiol*, 2004 42, 269-275.

[34] Jacoby, GA. Genetics of extended-spectrum beta-lactamases. *Eur. J. Clin. Microbiol. Infect. Dis.*, 1994 13, Supplement 1, 2-11.

[35] Rice, LB, Willey, SH, Papanicolaou, GA, Medeiros, AA, Eliopoulos, GM, Moellering, RC, Jr., and Jacoby, GA. Outbreak of ceftazidime resistance caused by extended-spectrum ß-lactamases at a Massachusetts chronic-care facility. *Antimicrob. Agents Chemother.*, 1990 34, 2193-2199.

[36] Vakulenko, SB, Taibi-Tronche, P, Toth, M, Massova, I, Lerner, SA, and Mobashery, S. Effects on substrate profile by mutational substitutions at positions 164 and 179 of the class A TEM_{pUC19} ß-lactamase from *Escherichia coli. J Biol Chem*, 1999 274, 23052-23060.

[37] Buré, A, Legrand, P, Arlet, G, Jarlier, V, Paul, G, and Philippon, A. Dissemination in five French hospitals of *Klebsiella pneumoniae* serotype K25 harboring a new transferable enzymatic resistance to third generation cephalosporins and aztreonam. *Eur. J. Clin. Microbiol.*, 1988 7, 780-782.

[38] Sirot, D, Chanal, C, Labia, R, Meyran, M, Sirot, J, and Cluzel, R. Comparative study of five plasmid-mediated ceftazidimases isolated in *Klebsiella pneumoniae. J. Antimicrob. Chemother.*, 1989 24, 509-521.

[39] Chaïbi, EB, Sirot, D, Paul, G, and Labia, R. Inhibitor-resistant TEM ß-lactamases: phenotypic, genetic and biochemical characteristics. *J Antimicrob Chemother*, 1999 43, 447-458.

[40] Perilli, M, Felici, A, Franceschini, N, De Santis, A, Pagani, L, Luzzaro, F, Oratore, A, Rossolini, GM, Knox, JR, and Amicosante, G. Characterization of a new TEM-derived ß-lactamase produced in a *Serratia marcescens* strain. *Antimicrob Agents Chemother*, 1997 41, 2374-2382.

[41] Sirot, D, Recule, C, Chaibi, EB, Bret, L, Croize, J, Chanal-Claris, C, Labia, R, and Sirot, J. A complex mutant of TEM-1 ß-lactamase with mutations encountered in both IRT-4 and extended-spectrum TEM-15, produced by an *Escherichia coli* clinical isolate. *Antimicrob Agents Chemother*, 1997 41, 1322-1325.

[42] Poirel, L, Mammeri, H, and Nordmann, P. TEM-121, a novel complex mutant of TEM-type ß-lactamase from *Enterobacter aerogenes. Antimicrob Agents Chemother*, 2004 48, 4528-4531.

[43] Rice, LB, Carias, LL, Hujer, AM, Bonafede, M, Hutton, R, Hoyen, C, and Bonomo, RA. High-level expression of chromosomally encoded SHV-1 ß-lactamase and an outer membrane protein change confer resistance to ceftazidime and piperacillin-tazobactam in a clinical isolate of *Klebsiella pneumoniae. Antimicrob Agents Chemother*, 2000 44, 362-367.

[44] Randegger, CC, Keller, A, Irla, M, Wada, A, and Hächler, H. Contribution of natural amino acid substitutions in SHV extended-spectrum ß-lactamases to resistance against various ß-lactams. *Antimicrob Agents Chemother*, 2000 44, 2759-2763.

[45] Nukaga, M, Mayama, K, Hujer, AM, Bonomo, RA, and Knox, JR. Ultrahigh resolution structure of a class A ß-lactamase: on the mechanism and specificity of the extended-spectrum SHV-2 enzyme. *J Mol Biol*, 2003 328, 289-301.

[46] Arlet, G, Rouveau, M, and Philippon, A. Substitution of alanine for aspartate at position 179 in the SHV-6 extended-spectrum ß-lactamase. *FEMS Microbiol Lett*, 1997 152, 163-167.

[47] Kurokawa, H, Yagi, T, Shibata, N, Shibayama, K, Kamachi, K, and Arakawa, Y. A new SHV-derived extended-spectrum ß-lactamase (SHV-24) that hydrolyzes ceftazidime through a single-amino-acid substitution (D179G) in the -loop. *Antimicrob Agents Chemother*, 2000 44, 1725-1727.

[48] Mulvey, MR, Bryce, E, Boyd, D, Ofner-Agostini, M, Christianson, S, Simor, AE, and Paton, S. Ambler class A extended-spectrum beta-lactamase-producing *Escherichia coli* and *Klebsiella* spp. in Canadian hospitals. *Antimicrob Agents Chemother*, 2004 48, 1204-1214.

[49] Arpin, C, Labia, R, Andre, C, Frigo, C, El Harrif, Z, and Quentin, C. SHV-16, a ß-lactamase with a pentapeptide duplication in the omega loop. *Antimicrob Agents Chemother*, 2001 45, 2480-2485.

[50] Corkill, JE, Cuevas, LE, Gurgel, RQ, Greensill, J, and Hart, CA. SHV-27, a novel cefotaxime-hydrolysing ß-lactamase, identified in *Klebsiella pneumoniae* isolates from a Brazilian hospital. *J Antimicrob Chemother*, 2001 47, 463-465.

[51] Poirel, L, Heritier, C, Podglajen, I, Sougakoff, W, Gutmann, L, and Nordmann, P. Emergence in *Klebsiella pneumoniae* of a chromosome-encoded SHV ß-lactamase that compromises the efficacy of imipenem. *Antimicrob Agents Chemother*, 2003 47, 755-758.

[52] Podbielski, A, Schönling, J, Melzer, B, and Haase, G. Different promoters of SHV-2 and SHV-2a ß-lactamase lead to diverse levels of cefotaxime resistance in their bacterial producers. *J. Gen. Microbiol.*, 1991 137, 1667-1675.

[53] Prinarakis, EE, Miriagou, V, Tzelepi, E, Gazouli, M, and Tzouvelekis, LS. Emergence of an inhibitor-resistant ß-lactamase (SHV-10) derived from an SHV-5 variant. *Antimicrob Agents Chemother*, 1997 41, 838-840.

[54] Dubois, V, Poirel, L, Arpin, C, Coulange, L, Bebear, C, Nordmann, P, and Quentin, C. SHV-49, a novel inhibitor-resistant ß-lactamase in a clinical isolate of *Klebsiella pneumoniae*. *Antimicrob Agents Chemother*, 2004 48, 4466-4469.

[55] Chang, FY, Siu, LK, Fung, CP, Huang, MH, and Ho, M. Diversity of SHV and TEM ß-lactamases in *Klebsiella pneumoniae*: gene evolution in Northern Taiwan and two novel ß-lactamases, SHV-25 and SHV-26. *Antimicrob. Agents Chemother.*, 2001 45, 2407-2413.

[56] Bou, G, Oliver, A, and Martinez-Beltran, J. OXA-24, a novel class D ß-lactamase with carbapenemase activity in an *Acinetobacter baumannii* clinical strain. *Antimicrob Agents Chemother*, 2000 44, 1556-1561.

[57] Brown, S, Young, HK, and Amyes, SG. Characterisation of OXA-51, a novel class D carbapenemase found in genetically unrelated clinical strains of *Acinetobacter baumannii* from Argentina. *Clin Microbiol Infect*, 2005 11, 15-23.

[58] Danel, F, Hall, LMC, Gur, D, and Livermore, DM. OXA-14, another extended-spectrum variant of OXA-10 (PSE-2) ß-lactamase from *Pseudomonas aeruginosa. Antimicrob. Agents Chemother.*, 1995 39, 1881-1884.

[59] Danel, F, Hall, LM, Gur, D, and Livermore, DM. OXA-16, a further extended-spectrum variant of OXA-10 ß-lactamase, from two *Pseudomonas aeruginosa* isolates. *Antimicrob Agents Chemother*, 1998 42, 3117-3122.

[60] Mugnier, P, Podglajen, I, Goldstein, FW, and Collatz, E. Carbapenems as inhibitors of OXA-13, a novel, integron-encoded ß-lactamase in *Pseudomonas aeruginosa. Microbiology*, 1998 144, 1021-1031.

[61] Mugnier, P, Casin, I, Bouthors, AT, and Collatz, E. Novel OXA-10-derived extended-spectrum ß-lactamases selected in vivo or in vitro. *Antimicrob Agents Chemother*, 1998 42, 3113-3116.

[62] Poirel, L, Girlich, D, Naas, T, and Nordmann, P. OXA-28, an extended-spectrum variant of OXA-10 ß-lactamase from *Pseudomonas aeruginosa* and its plasmid- and integron-located gene. *Antimicrob Agents Chemother*, 2001 45, 447-453.

[63] Aubert, D, Poirel, L, Ben Ali, A, Goldstein, FW, and Nordmann, P. OXA-35 is an OXA-10-related ß-lactamase from *Pseudomonas aeruginosa. J. Antimicrob. Chemother*, 2001 48, 717-721.

[64] Danel, F, Hall, LMC, Gur, D, and Livermore, DM. OXA-15, an extended-spectrum variant of OXA-2 ß-lactamase, isolated from a Pseudomonas aeruginosa strain. *Antimicrob. Agents Chemother.*, 1997 41, 785-790.

[65] Poirel, L, Gerome, P, De Champs, C, Stephanazzi, J, Naas, T, and Nordmann, P. Integron-located *oxa-32* gene cassette encoding an extended-spectrum variant of OXA-2 ß-lactamase from *Pseudomonas aeruginosa. Antimicrob Agents Chemother*, 2002 46, 566-569.

[66] Jacoby, GA, Mills, DM, and Chow, N. Role of ß-lactamases and porins in resistance to ertapenem and other ß-lactams in *Klebsiella pneumoniae. Antimicrob Agents Chemother*, 2004 48, 3203-3206.

[67] Aubert, D, Poirel, L, Chevalier, J, Leotard, S, Pages, JM, and Nordmann, P. Oxacillinase-mediated resistance to cefepime and susceptibility to ceftazidime in *Pseudomonas aeruginosa. Antimicrob Agents Chemother*, 2001 45, 1615-1620.

[68] Hidri, N, Barnaud, G, Decré, D, Cerceau, C, Lalande, V, Petit, JC, Labia, R, and Arlet, G. Resistance to ceftazidime is associated with a S220Y substitution in the omega loop of the AmpC ß-lactamase of a *Serratia marcescens* clinical isolate. *J Antimicrob Chemother*, 2005 55, 496-499.

[69] Barnaud, G, Benzerara, Y, Gravisse, J, Raskine, L, Sanson-Le Pors, MJ, Labia, R, and Arlet, G. Selection during cefepime treatment of a new cephalosporinase variant with extended-spectrum resistance to cefepime in an *Enterobacter aerogenes* clinical isolate. *Antimicrob Agents Chemother*, 2004 48, 1040-1042.

[70] Vakulenko, SB, Golemi, D, Geryk, B, Suvorov, M, Knox, JR, Mobashery, S, and Lerner, SA. Mutational replacement of Leu-293 in the class C *Enterobacter cloacae* P99 ß-lactamase confers increased MIC of cefepime. *Antimicrob Agents Chemother*, 2002 46, 1966-1970.

[71] Morosini, MI, Negri, MC, Shoichet, B, Baquero, MR, Baquero, F, and Blázquez, J. An extended-spectrum AmpC-type ß-lactamase obtained by in vitro antibiotic selection. *FEMS Microbiol. Lett*, 1998 165, 85-90.

[72] Barnaud, G, Labia, R, Raskine, L, Sanson-Le Pors, MJ, Philippon, A, and Arlet, G. Extension of resistance to cefepime and cefpirome associated to a six amino acid deletion in the H-10 helix of the cephalosporinase of an *Enterobacter cloacae* clinical isolate. *FEMS Microbiol. Lett*, 2001 195, 185-190.

[73] Mammeri, H, Poirel, L, Bemer, P, Drugeon, H, and Nordmann, P. Resistance to cefepime and cefpirome due to a 4-amino-acid deletion in the chromosome-encoded AmpC ß-lactamase of a *Serratia marcescens* clinical isolate. *Antimicrob Agents Chemother*, 2004 48, 716-720.

[74] Lobkovsky, E, Moews, PC, Liu, H, Zhao, H, Frere, JM, and Knox, JR. Evolution of an enzyme activity: crystallographic structure at 2-Å resolution of cephalosporinase from the *ampC* gene of *Enterobacter cloacae* P99 and comparison with a class A penicillinase. *Proc Natl Acad Sci U S A*, 1993 90, 11257-11261.

[75] Nukaga, M, Haruta, S, Tanimoto, K, Kogure, K, Taniguchi, K, Tamaki, M, and Sawai, T. Molecular evolution of a class C ß-lactamase extending its substrate specificity. *J. Biol. Chem.*, 1995 270, 5729-5735.

[76] Nukaga, M, Taniguchi, K, Washio, Y, and Sawai, T. Effect of an amino acid insertion into the omega loop region of a class C ß-lactamase on its substrate specificity. *Biochemistry*, 1998 37, 10461-10468.

[77] Crichlow, GV, Kuzin, AP, Nukaga, M, Mayama, K, Sawai, T, and Knox, JR. Structure of the extended-spectrum class C ß-lactamase of *Enterobacter cloacae* GC1, a natural mutant with a tandem tripeptide insertion. *Biochemistry*, 1999 38, 10256-10261.

[78] Matsumura, N, Minami, S, and Mitsuhashi, S. Sequences of homologous ß-lactamases from clinical isolates of *Serratia marcescens* with different substrate specificities. *Antimicrob Agents Chemother*, 1998 42, 176-179.

[79] Raimondi, A, Sisto, F, and Nikaido, H. Mutation in *Serratia marcescens* AmpC ß-lactamase producing high-level resistance to ceftazidime and cefpirome. *Antimicrob Agents Chemother*, 2001 45, 2331-2339.

[80] Kimura, S, Ishiguro, M, Ishii, Y, Alba, J, and Yamaguchi, K. Role of a mutation at position 167 of CTX-M-19 in ceftazidime hydrolysis. *Antimicrob Agents Chemother*, 2004 48, 1454-1460.

[81] Stürenburg, E, Kühn, A, Mack, D, and Laufs, R. A novel extended-spectrum ß-lactamase CTX-M-23 with a P167T substitution in the active-site omega loop associated with ceftazidime resistance. *J Antimicrob Chemother*, 2004 54, 406-409.

[82] Bonnet, R, Dutour, C, Sampaio, JL, Chanal, C, Sirot, D, Labia, R, De Champs, C, and Sirot, J. Novel cefotaximase (CTX-M-16) with increased catalytic efficiency due to substitution Asp-240–>Gly. *Antimicrob Agents Chemother*, 2001 45, 2269-2275.

[83] Bonnet, R, Recule, C, Baraduc, R, Chanal, C, Sirot, D, De Champs, C, and Sirot, J. Effect of D240G substitution in a novel ESBL CTX-M-27. *J Antimicrob Chemother*, 2003 52, 29-35.

[84] Cartelle, M, del Mar Tomas, M, Molina, F, Moure, R, Villanueva, R, and Bou, G. High-level resistance to ceftazidime conferred by a novel enzyme, CTX-M-32, derived from CTX-M-1 through a single Asp240-Gly substitution. *Antimicrob Agents Chemother*, 2004 48, 2308-2313.

[85] Majiduddin, FK, and Palzkill, T. Amino acid residues that contribute to substrate specificity of class A ß-lactamase SME-1. *Antimicrob Agents Chemother*, 2005 49, 3421-3427.

[86] Thatcher, DR. Progress on penicillinase. *Nature*, 1975 255, 526.
[87] Bush K, Jacoby GA. Updated functional classification of ß-lactamases. *Antimicrob Agents Chemother*. 2010;54:969-76.
[88] Bonnet R. Growing group of extended-spectrum ß-lactamases: the CTX-M enzymes. *Antimicrob Agents Chemother.* 2004;48:1-14.
[89] Canton R, Coque TM. The CTX-M ß-lactamase pandemic. *Curr Opin Microbiol.* 2006;9:466-75.
[90] Castanheira M, Mendes RE, Rhomberg PR, Jones RN. Rapid emergence of $bla_{\text{CTX-M}}$ among Enterobacteriaceae in U.S. Medical Centers: molecular evaluation from the MYSTIC Program (2007). *Microb Drug Resist.* 2008;14:211-6.
[91] Pallecchi L, Bartoloni A, Fiorelli C, Mantella A, Di Maggio T, Gamboa H, et al. Rapid dissemination and diversity of CTX-M extended-spectrum ß-lactamase genes in commensal *Escherichia coli* isolates from healthy children from low-resource settings in Latin America. *Antimicrob Agents Chemother.* 2007;51:2720-5.
[92] Peirano G, Pitout JD. Molecular epidemiology of *Escherichia coli* producing CTX-M ß-lactamases: the worldwide emergence of clone ST131 O25:H4. *Int J Antimicrob Agents.* 2010;35:316-21.
[93] Zong Z, Partridge SR, Thomas L, Iredell JR. Dominance of $bla_{\text{CTX-M}}$ within an Australian extended-spectrum ß-lactamase gene pool. *Antimicrob Agents Chemother.* 2008;52:4198-202.
[94] Livermore DM, Canton R, Gniadkowski M, Nordmann P, Rossolini GM, Arlet G, et al. CTX-M: changing the face of ESBLs in Europe. *J Antimicrob Chemother.* 2007;59:165-74.
[95] Nicolas-Chanoine MH, Blanco J, Leflon-Guibout V, Demarty R, Alonso MP, Canica MM, et al. Intercontinental emergence of *Escherichia coli* clone O25:H4-ST131 producing CTX-M-15. *J Antimicrob Chemother.* 2008;61:273-2+81.
[96] Walther-Rasmussen J, Høiby N. OXA-type carbapenemases. *J Antimicrob Chemother.* 2006;57:373-83.
[97] Poirel L, Naas T, Nordmann P. Diversity, epidemiology, and genetics of class D ß-lactamases. *Antimicrob Agents Chemother.* 2010;54:24-38.
[98] Queenan AM, Bush K. Carbapenemases: the versatile ß-lactamases. *Clin Microbiol Rev.* 2007;20:440-58.
[99] Salverda ML, de Visser JA, Barlow M. Natural evolution of TEM-1 ß-lactamase: experimental reconstruction and clinical relevance. *FEMS Microbiol Rev.* 2010;34:1015-36.
[100] Nordmann P, Mammeri H. Extended-spectrum cephalosporinases: structure, detection and epidemiology. *Future Microbiol.* 2007;2:297-307.
[101] Jacoby GA. AmpC ß-lactamases. *Clin Microbiol Rev.* 2009;22:161-82.
[102] Paterson DL, Bonomo RA. Extended-spectrum ß-lactamases: a clinical update. *Clin Microbiol Rev.* 2005;18:657-86.

In: Beta-Lactamases
Editor: Jean-Marie Frère, pp. 383-399
ISBN: 978-1-61324-638-2

Chapter 15

THE CLINICAL PROBLEM OF BETA-LACTAM RESISTANCE IN MULTIDRUG RESISTANT (MDR) *KLEBSIELLA PNEUMONIAE*, *ACINETOBACTER BAUMANNII*, AND *PSEUDOMONAS AERUGINOSA*

Robert A. Bonomo*
Research Service, Section of Infectious Diseases,
and Geriatric Research Education and Clinical Center,
Louis Stokes Cleveland Department of Veterans Affairs Medical Center,
Case Western Reserve University School of Medicine, Cleveland, Ohio, U.S.A.

ABSTRACT

Multidrug resistant (MDR) bacterial strains present a challenging problem for the clinician. They are most taxing when belonging to the *Klebsiella pneumoniae*, *Pseudomonas aeruginosa* and *Acinetobacter baumanii* species. The clinical challenge, which can be very different depending upon the country and even the city, is analysed for each of these pathogens. Special attention is devoted to the KPC, CTX-M (both class A) and NDM-1 (class B) enzymes. In *Pseudomonas*, deregulated overproduction of the chromosomal class C enzyme can also be a major problem.

INTRODUCTION

Although antibiotics are a significant advance in clinical medicine, their efficacy as therapeutic agents is being seriously challenged. Clinicians are now faced with the increasing problem of bacteria resistant to almost all classes of antibiotics. These "multidrug resistant" (MDR) strains are most taxing when encountered among Gram negative bacteria. *Klebsiella pneumoniae*, *Acinetobacter baumannii*, and *Pseudomonas aeruginosa* are the Gram negative bacteria that are being increasinlgy described as MDR (i.e., resistant to three or more classes

* E-mail: robert.bonomo@med.va.gov

of antibiotics). Resistance to beta-lactam antibiotics is most critical in these pathogens, as the efficacy of penicillins, cephalosporins and carbapenems are unparalleled. Resistance mediated by beta-lactamases that can hydrolyze penicillins, cephalosporins and carbapenems forms the basis of this dilemma. Future strategies to combat these significant resistance threats need to employ novel and more potent beta-lactamase inhibitors, better beta-lactam antibiotics, and major infection control efforts. In this chapter we review the problem of MDR in *K. pneumoniae*, *A. baumannii*, and *P. aeruginosa*.

Klebsiella Pneumoniae

Multidrug Resistant (MDR) *Klebsiella pneumoniae*: Resistance to Ceftazidime and Imipenem

The introduction of penicillin into clinical care is regarded as one of the most important advances in clinical medicine [1, 2]. Unfortunately, the indiscriminate use of beta-lactam antibiotics has been greeted with an unprecedented and expanding number of beta-lactamases [3]. At no point was this more critical than when the pharmaceutical companies introduced the advanced generation of cephalosporins. The introduction of the extended-spectrum cephalosporins (ceftazidime, cefotaxime, and cefriaxone) into clinical practice in the early 1980s was regarded as a major addition to our therapeutic armamentarium in the fight against beta-lactamase-mediated bacterial resistance especially against *K. pneumoniae* [3, 4]. *Klebsiella* spp. frequently cause human nosocomial infections. In particular, the medically most important *Klebsiella* species, *Klebsiella pneumoniae*, accounts for a significant proportion of hospital-acquired urinary tract infections, pneumonia, septicemias, and soft tissue infections [5].

Unfortunately, the emergence of *K. pneumoniae* exhibiting resistance to ceftazidime and other cephalosporins seriously compromised the efficacy of these life-saving antibiotics [6]. Single amino acid substitutions in the bla_{TEM} and bla_{SHV} genes (Gly238Ser, Glu240Lys, Arg164Ser, Arg164His, Asp179Asn, and Glu(Asp)104Lys) formed the molecular basis of this novel resistance phenotype [7-11]. The beta-lactamases responsible were designated "extended-spectrum beta-lactamases" (ESBLs). TEM and SHV type ESBLs represented the first example of beta-lactamase mediated resistance to beta-lactam antibiotics that resulted from fundamental changes in the substrate spectra of the enzymes. Their presence in *K. pneumoniae* is significant as this organism is an important nosocomial pathogen, responsible for up to 8% of all healthcare-associated infections [12].

In contrast to TEM and SHV ESBLs, heterogeneous groups of cefotaxime hydrolyzing beta-lactamases are also found in *K. pneumoniae.* These are often referred to as "non-TEM and non-SHV" ESBLs. The most prevalent ESBL beta-lactamase families here are the CTX-M ESBLs [13]. This rapidly growing group presents unique challenges in terms of its genetic context and spead to community pathogens such as *E. coli* [13]. At the present time, the CTX-M family beta-lactamases are the most rapidly spreading family of ESBLs in Europe and North America[14-18]. At this point, there are 5 major groups in this class A beta-lactamase family.

The CTX-M enzymes confer resistance to aminopenicillins (ampicillin or amoxicillin), carboxypenicillins (carbenicillin or ticarcillin), ureidopenicillins (piperacillin), and narrow-spectrum cephalosporins (cephalothin, cephaloridine, and cefuroxime), cefotaxime, aztreonam, ceftriaxone and variable levels of resistance to cefepime and cefpirome. For CTX-M enzymes resistance to ceftazidime is usually minimal. Tazobactam seems to be a better inhibitor for most CTX-M enzymes than clavulanic acid or sulbactam (Toho-1 is the exception), but the details of inhibition by mechanism-based inhibitors are still unknown. Moreover, isolates possessing CTX-M ESBLs are frequently resistant to beta-lactamase inhibitors due to the concomitant presence of other beta-lactamases[19] (i.e., OXA-1 encoded on the same plasmid) [20, 21].

As stated above, the frequency of the presence of the CTX-M family of ESBLs is increasing in Gram-negative enteric bacilli [21]. Among 224 isolates of *K. pneumoniae* acquired from 16 reference laboratories in London and Southeast England during a 3 month period (August to October, 2004), 190 possessed a CTX-M ESBL. This high proportion has significant clinical consequences.

Patients who develop infection by ESBL-producing *K. pneumoniae* are usually seriously ill, have been hospitalized for prolonged periods of time, are being treated using invasive medical devices, have received or are receiving antibiotics (particularly ceftazidime or quinolones), or were former residents of long term care facilities [12, 22-24]. The number of ESBL producing *Enterobacteriacea* found in community isolates (mostly CTX-M ESBLs) is also increasing significantly [25, 26]. Community acquired ESBLs have been found in Spain [27-29], Calgary [30, 31] and France [21, 32-35]. The appearance of ESBL-producing enteric bacilli (particularly the CTX-M ESBLs) in the community and of associated infections is one of the most significant epidemiologic challenges in infectious diseases.

The epidemiology of infections caused by CTX-M carrying Gram negative bacteria such as *K. pneumoniae* is still poorly understood and the role of mobile genetic elements is not yet clearly defined.

As with their hospital counterparts, infections caused by ESBL producers are associated with increased mortality, length of stay and increased cost [28, 36]. Inadequate empirical therapy for serious infections caused by CTX-M ESBL producing organisms is independently associated with increased mortality [37, 38]. Based upon susceptibility testing and extensive clinical experience, carbapenems (imipenem, meropenem, doripenem and ertapenem) became the temporary answer to this escalating resistance phenotype [22, 24, 39].

Unfortunately, resistance to carbapenems has also been recently described in *Klebsiella* spp. [23, 40-45]. The unwelcome emergence of carbapenem hydrolyzing serine beta-lactamases encoded on plasmids in *K. pneumoniae* (KPC-2, and -3, etc.) has drastically altered established treatment paradigms [46-52].

In the years 1999 to 2005, as part of the USA Meropenem Yearly Susceptibility Test Information Collection (MYSTIC) Program, 51 strains of *Enterobactericae* with increased imipenem and meropenem MIC values (≥ 2 μg/mL) were detected [53]. In this study the bla_{KPC} was identified in 28 *K. pneumoniae* isolates from 3 medical centers in the New York City area. Reports from multiple research centers located in New York City are monitoring the dreadful emergence of this phenotype [54]. At present the problem of KPCs is escalating rapidly in Europe, Israel, Northern and Southern America. In Greece the presence of KPCs is at alarming proportions.

Diagnostic Considerations and MDR *K. pneumoniae*

Adding to this complexity, KPC is difficult to detect in the clinical microbiology laboratory. A marked inoculum effect is observed when testing imipenem and meropenem. Agar-based diffusion tests with precisely measured inocula and screening with ertapenem and meropenem are recommended [55-58]. The implementation and development of novel testing systems (Modified Hodge Test, ChromAgar, boronic acid discs, Check-Points microarray, and PCR coupled with electrospray ionization mass spectrometry-PCR/ESI-MS) are also being explored.

It is becoming more recognized that the true prevalence of KPC-producing *Enterobacteriaceae* may be under-appreciated due to difficulties with detection by the clinical laboratories. KPC beta-lactamase production can be missed when laboratories rely solely on automated methods. Some clinical laboratories use the Modified Hodge Test to detect KPC production. The Modified Hodge Test, a phenotypic test, has demonstrated high sensitivity for the detection of carbapenemase activity *in vitro*. In the Hodge Test, a lawn of susceptible *E. coli* (ATCC strain 25922) a carbapenem disc and a streak of the suspected KPC producer are placed upon an agar plate. In addition, limited data suggest that a novel chromogenic agar may help in early identification of KPC-producing strains with a sensitivity and specificity of 100% and 98.4% when compared with PCR. The CHROMagar KPC medium was compared to MacConkey agar with carbapenem discs and PCR for bla_{KPC} gene for rapid detection of carbapenem resistant *K. pneumoniae*. The sensitivity and specificity relative to PCR were 100 % and 98.4% for CHROMagar KPC and 92.7 % and 95.9% for MacConkey [59].

Boronic acid discs (e.g. 3-aminophenylboronic acid) have also been used to detect KPC producers with a sensitivity and specificity of 100%. As a clinically useful test, boronic acid disks promise to accurately differentiate KPC-possessing *K. pneumoniae* isolates [58, 60]. In the assays reported by Tsakris *et al.*, phenylboronic acid is used at a concentration of 20 mg/ml. A similar concentration is used by Doi *et al* [61]. These assays can be performed more readily than others and can add to the current need to detect KPCs.

Most recently, many *Klebisiella* spp. from the subcontinent of India and Pakistan have been found to harbor a novel metallo-beta-lactamase called NDM-1 (New Delhi Metallo-beta-lactamase). The gene harboring the MBL appears to be widely disseminated. NDM-1 attracts significant attention because the genetic environment of $bla_{\text{NDM-1}}$ is readily transferable and its genetic/microbiological background is proving to be more complex and, apparently, more unpredictable than that of KPC. Moreover, the number of patients in whom bacteria possess $bla_{\text{NDM-1}}$ seems to be growing. The gene has already spread to the United Kingdom, the United States, Kenya, Japan, Canada, Belgium, the Netherlands and Australia . A quick review of the original description of $bla_{\text{NDM-1}}$ sets the background for the importance of this dissemination.

Therapeutic Considerations in the Treatment of MDR *K. pneumoniae*

Unfortunately, the commercially available beta-lactam/beta-lactamase inhibitor combinations are not used against ESBL producing *K. pneumoniae*. Despite this, the criterion for diagnosing an ESBL bearing *K. pneumoniae* by the Clinical and Laboratory Standards Institute (CLSI, formerly National Committee for Clinical Laboratory Standards) rests upon the increased susceptibility to a clavulanate/extended spectrum cephalosporin combination (e.g., double disc diffusion method). Published clinical experience with beta-lactam/beta-lactamase inhibitor combinations in the treatment of serious clinical infections due to ESBL producers (TEM, SHV and CTX-M) is limited. A cephalosporin/beta-lactamase inhibitor combination (e.g., cefoperazone/sulbactam) was assessed *in vitro* and *in vivo* on bacteria producing ESBLs in numerous studies (see references in [62]). It is recognized that beta-lactam/beta-lactamase inhibitor combinations are subject to rising MICs as inoculum rises [63]. Although efficacy has been reported in animal models of infection [64], clinical failure was reported in a patient with spontaneous bacterial peritonitis who was awaiting liver transplantation for end-stage liver disease [65]. Hyper-production of non-ESBL-producing beta-lactamases or the combination of beta-lactamase production and porin loss can also lead to a reduction in activity of beta-lactamase inhibitors. Additionally, animal studies have shown beta-lactam/beta-lactamase inhibitor combinations to be less effective than carbapenems against ESBL-producing organisms [66]. Stochastic modeling suggests that some dosing regimens of piperacillin/tazobactam may not have a high probability of achieving pharmacokinetic (PK) and pharmacodynamic (PD) targets which have previously been correlated with clinical success [67]. Thus, clinicians are very cautious about beta-lactam/beta-lactamase inhibitor combinations as suitable first-line therapy for serious infections with ESBL-producing organisms. It is clear that a beta-lactamase inhibitor with the correct PK/PD profile and demonstrating low, nanoMolar affinity against multiple serine beta-lactamase would compare very favorably to carbapenems and be essential in the fight against ESBL *and* carbapenem resistant isolates [68].

Acinetobacter Baumannii

A. Baumannii: A Formidable Challenge for the Clinician

In 2005, a frightening commentary in *Nature* described MDR *A. baumannii* as "a real danger" [69]. Alarming drug resistant phenotypes found in *A. baumannii* include resistance to penicillins (piperacillin and ampicillin), extended-spectrum cephalosporins, beta-lactam/beta-lactamase inhibitor combinations (ampicillin/sulbactam and piperacillin/tazobactam) and carbapenems [70-72]. As noted by the *Infectious Disease Society of America*, the problem raised by *Acinetobacter* spp. is "a prime example of the mismatch between unmet medical need and the current antimicrobial research and development pipeline" [73]

Acinetobacter spp. are Gram negative aerobic bacteria that are coccobacillary in shape and are generally described as aerobic, non-lactose fermenting, non-fastidious, non-motile, catalase-positive and oxidase-negative. Unique among microbial pathogens, the appearance of *Acinetobacter* spp. is highly dependent upon the life-cycle. In the early growth phases, *Acinetobacter* spp. appears rod-shaped. In the stationary phase, *Acinetobacter* spp. acquire a coccobacillary morphology. There are now more than 30 different species in the genus *Acinetobacter* and their classification and identification remains problematic for clinicians.

An examination of the current literature reveals that MDR *Acinetobacter* spp. are emerging in every continent. Certain lineages have achieved "epidemic" status reaching multiple hospitals or countries [74-79]:

To illustrate:

i) the European clones I, II, and III, are widespread in continental Europe and have been detected in the United States[80-86];
ii) a clone with the VEB-1 cephalosporinase spread through northeast France and Belgium in late 2003 to 2004 [87, 88];
iii) a clone with OXA-40 (also known as OXA-24) carbapenemase which is prevalent at numerous hospitals in Spain and Portugal [89];
iv) the southeast (SE) clone has been prevalent in southern England since 2000 [90].

Table 1. Beta-Lactamase genes found in *Acinetobacter* spp

Chromosomal class C[1] cephalosporinase
Class D OXA carbapenemases[2]
bla_{OXA-23}
$bla_{OXA-24/40}$
$bla_{OXA-51/69/66}$
bla_{OXA-58}
$bla_{OXA-143}$
Class D OXA extended spectrum cephalosporinases
bla_{OXA-10}
bla_{OXA-14}
bla_{OXA-17}
Class B Metallo-□-lactamases
bla_{NDM-1}
bla_{VIM}
bla_{IMP}
bla_{GIM}
bla_{SPM}
Class A □-lactamases
bla_{TEM}
bla_{SHV}
bla_{PER}[3]
bla_{VEB}

[1] *Acinetobacter* spp., like other gram-negative organisms, possess chromosomally encoded class C beta-lactamases, the *Acinetobacter-d*erived cephalosporinases (ADCs).

[2] Class D OXA beta-lactamases. Some OXAs are also able to hydrolyze extended-spectrum cephalosporins and inactivate carbapenems.

[3] Class A extended-spectrum beta-lactamases (ESBLs). PER-1, an ESBL, demonstrates high-level activities on penicillins and extended-spectrum cephalosporins.

The spread of *A. baumannii* in the United States is receiving increased attention. Reports in Morbidity and Mortality Weekly Report (MMWR, November 19, 2004) first summarized the extent of this problem in military medical facilities treating service men and women injured in Iraq/Kuwait and Afghanistan. Many of these isolates at Walter Reed Army Medical Center (WRAMC) were resistant to three or more classes of antibiotics. This frightening susceptibility profile left clinicians with extremely few options to treat MDR *A. baumannii* infections. Physicians relied heavily upon colistin [91, 92].

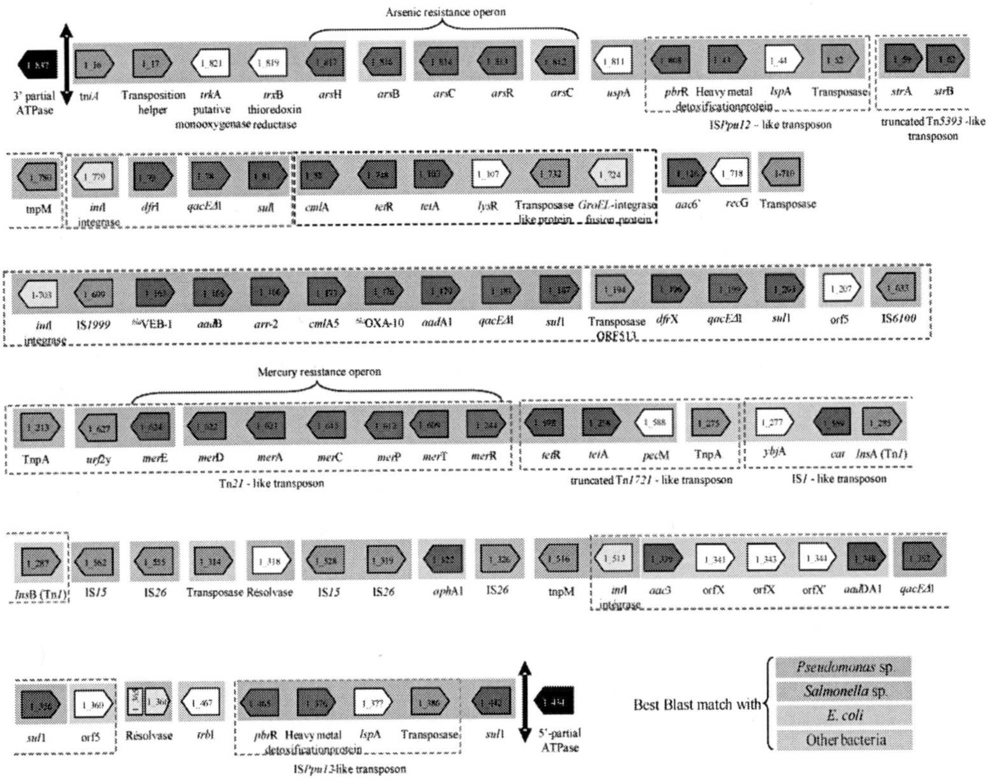

Figure 1. Resistance island of *A. baumanii* (reproduced from ref. 113).

Unfortunately, this reliance is not completely certain. There is still a lot of concern about the real efficiency of colistin. Moreover, the dosing and pharamcokinetics/pharmacodynamics are not completely elucidated [93-96]. Hetero-resistance to colisitin (and other antibiotics) is an unsettling clinical observation that may significantly interfere with one of our truly last line antibiotics [97-99]. Prospective studies using tigecycline remain to be performed. [100-102]

A recent study of 75 unique patient isolates from Walter Reed Army Medical Center in Cleveland revealed that 89.3 % of the isolates were resistant to 3 or more classes of antibiotics [82]. In this study 91% all of the isolates exhibited resistance to ceftazidime and 25% were resistant to meropenem. A genetic survey of the beta-lactamases present in these isolates (9 *bla* primer sets were designed) revealed that the *Acinetobacter* derived cephalosporinase (ADC) and OXA-like carbapenemases are the major beta-lactamases responsible for ceftazidime and imipenem resistance, respectively. In this collection, *bla*$_{OXA-58}$-like and *bla*$_{PER}$-like genes were described for the first time in the U.S. The beta-lactamases of *A. baumannii* also deserve special attention. Except for PER-1 and SHV-ESBLs, the large majority are not inhibited by beta-lactamase inhibitors.

Common infections in which *A. baumannii* is found include ventilator associated pneumonia (VAP), blood stream infections, burn wounds and urinary tract infections [71, 103-105]. *A. baumannii* is responsible for 2%-10% of all Gram-negative bacterial infections in ICUs in Europe and the United States [106]. Despite the importance of *A. baumannii* as a public health problem, several questions remain regarding its virulence [107-111], epidemiology and optimal treatment regimens [50, 71, 112].

The role of resistance gene transfer and expression in *A. baumannii* remains largely unknown. Interestingly, the recent report of the whole genome sequence of *A. baumannii* strain AYE that exhibited an 86-kb genomic region termed a "resistance island" (Figure 1) in which 45 resistance genes are clustered (19 new genes!) suggested a transformable or "switching" genomic structure which may serve as a "hotspot" that could explain the rapid acquisition of resistance markers under antimicrobial pressure [113]. Now, a series of new genomes have been analyzed [109, 114, 115]. In the study by Adams and colleagues the entire MDR phenotype can be explained by the acquisition of discrete resistance determinants distributed throughout the genome. A comparison of closely related MDR and drug-susceptible isolates also suggested that drug efflux may be a less significant contributor to resistance to certain classes of antibiotics than inactivation enzymes are. A resistance island (RI), similar to that found in the Fournier *et al* genomic analysis [113], with a variable composition of resistance determinants interspersed with transposons, integrons, and other mobile genetic elements is a significant but not universal contributor to the MDR phenotype. There are now up to 10 different RIs described. Table 1 summarizes the major resistance determinants found in *A. baumannii.*

Acinetobacter spp. are unique in the bacterial world in their ability to demonstrate resistance to many antibiotics. Some strains of *Acinetobacter* spp. are resistant to all antibiotics and it is these strains that are often responsible for large hospital outbreaks. These MDR strains are also identified as XDR (extreme drug resistant). As is evident, there exists a large collection of beta-lactamase genes, aminoglycoside modifying enzymes, and many efflux pumps.

Unfortunately, resistance to antibiotics also occurs to some of the "last line" agents such as colistin. Colistin (polymyxin E) and it close relative, polymyxin B, are cationic antimicrobial peptides. Colistin is available for administration to patients in two forms (colistin sulfate and colistmethate sodium). Colistmethate sodium is the inactive prodrug form of colistin. Colistmethate sodium is degraded in culture media; therefore colistin sulfate should be used for in vitro testing. It is therefore axiomatic that false susceptibility results can occur due to poor diffusion of colistin sulfate in culture.

Colistin resistant and hetero resistant strains are increasingly identified. The genetic basis of colistin resistance is currently being actively investigated. In fact colistin dependent strains are also reported.

PSEUDOMONAS AERUGINOSA

P. Aeruginosa and Resistance to Beta-Lactams

P. aeruginosa is a versatile and notorious opportunistic nosocomial pathogen [116], ranking among the three most frequently isolated in intensive care units (ICUs) [16]. This pathogen is commonly involved in acute hospital acquired infections, e.g., ventilator-associated pneumonia (VAP) or burn wound infections, conditions associated with high mortality rates [16]. Contemporary isolates exhibit high rates of resistance to beta-lactam antibiotics and are frequently MDR. This problem is further amplified because of the high prevalence of hypermutable strains [117].

The MDR phenotype in *P. aeruginosa* isolates includes resistance to beta-lactams, quinolones, chloramphenicol, aminoglycosides, tetracyclines, polymyxins, and trimethoprim/sulfa [70]. This is mediated by:

1. Alteration in outer membrane proteins (imipenem resistance);
2. Modified penicillin binding proteins (PBPs) (beta-lactams);
3. inactivating enzymes (beta-lactamases and aminoglycoside modifying enzymes);
4. Alteration in enzymes involved in folate metabolism (sulfa resistance);
5. Alterations in membrane composition (colistin resistance);
6. Efflux pumps (constitutive expression of efflux transporters, MexA–MexB–OprM and MexC–MexD–OprJ –multiple agents),
7. Quinolone target site modifications.

Thus, *P .aeruginosa* is the "consummate hospital-acquired pathogen" [118, 119].

P. aeruginosa is naturally resistant to aminopenicillins, cephalothin, cephamycins, and some extended-spectrum cephalosporins such as cefotaxime and moxalactam. Clinical isolates display susceptibility to other beta-lactam agents including carboxypenicillins (carbenicillin, ticarcillin), ureidopenicillins (piperacillin), cephalosporins (ceftazidime), monobactams (aztreonam) and carbapenems (imipenem). Regrettably, resistance development during beta-lactam therapy frequently arises, mediated by the selection of mutations in certain chromosomal genes. This is a frequent problem with major clinical consequences. The most relevant mechanism for development of resistance to antipseudomonal penicillins (such as ticarcillin or piperacillin) and cephalosporins (such as ceftazidime or cefepime) is selection of mutations leading to hyperproduction of the chromosomal cephalosporinase AmpC (also referred to as PDC, *Pseudomonas* Derived Cephalosporinases). *P. aeruginosa* has an inducible, naturally occurring cephalosporinase that confers low-level resistance to aminopenicillins and narrow-spectrum cephalosporins [49, 120-123]. Resistance to extended-spectrum cephalosporins may also arise from acquired beta-lactamases. In the United States, derepression of the chromosomal AmpC beta-lactamases is mainly responsible for reduction in susceptibility to beta-lactam antibiotics. The derepressed mutants demonstrate 'high-level cephalosporinase' phenotype, limiting the use of cephalosporins and penicillins.

As a rule, cefepime maintains activity against AmpC producing *P. aeruginosa* [124]. [122, 125]. There are at least 9 new variants of PDC-1. As stated above, a number of beta-lactamases have been described in *P. aeruginosa,* including ESBLs of class A and class D (e.g., TEM-42, SHV-2, PER-1, VEB, PSE/CARBs, GES, and various OXA-10-type derivative enzymes) and metallo-beta-lactamases (MBLs) of class B (e.g., IMP-1 and the more recently described VIM enzymes). The ESBLs in *P. aeruginosa* (OXA-, TEM-, SHV-, PER-, VEB-type, etc.) can degrade penicillin, expanded-spectrum cephalosporins, and monobactams (but not carbapenems) and are often susceptible to serine-beta-lactamase inhibitors (tazobactam); the OXA enzymes are the major exceptions to this rule. The VIM metallo-beta-lactamases, on the other hand, are notable for their robust carbapenemase activity and can degrade virtually all beta-lactams except monobactams, while being resistant to the currently available inhibitors. Thus, it is acknowledged that the "beta-lactamase threat" in *P. aeruginosa* includes the ubiquitous AmpC cephalosporinase, OXA-10, and IMP/VIM MBLs.

CONCLUSIONS

This review highlights the microbiological and clinical problem of beta-lactam resistance mediated by beta-lactamase enzymes in *K. pneumoniae*, *A. baumannii* and *P. aeruginosa*. The complexity of the beta-lactamase background in these three pathogens will continue to challenge the introduction of each new beta-lactam. Further studies will be needed to understand how these genes arise in these pathogens, how they spread and the best method to control them.

REFERENCES

[1] Donowitz GR, Mandell GL. Drug therapy. Beta-lactam antibiotics (2). *N Engl J Med* 1988 Feb 25;318(8):490-500.

[2] Donowitz GR, Mandell GL. Beta-Lactam antibiotics (1). *N Engl J Med* 1988 Feb 18;318(7):419-26.

[3] Medeiros AA. Evolution and dissemination of beta-lactamases accelerated by generations of beta-lactam antibiotics. *Clin Infect Dis* 1997 Jan;24 Suppl 1:S19-45.

[4] Medeiros AA. Beta-Lactamases: quality and resistance. *Clin Microbiol Infect* 1997 Feb;3 Suppl 4:S2-S9.

[5] Podschun R, Ullmann U. *Klebsiella* spp. as nosocomial pathogens: epidemiology, taxonomy, typing methods, and pathogenicity factors. *Clin Microbiol Rev* 1998 Oct;11(4):589-603.

[6] Knothe H, Shah P, Krcmery V, Antal M, Mitsuhashi S. Transferable resistance to cefotaxime, cefoxitin, cefamandole and cefuroxime in clinical isolates of *Klebsiella pneumoniae* and *Serratia marcescens*. *Infection* 1983 Nov-Dec;11(6):315-7.

[7] Bradford PA. Extended-spectrum beta-lactamases in the 21st century: characterization, epidemiology, and detection of this important resistance threat. Clin Microbiol Rev 2001 Oct;14(4):933-51, table of contents.

[8] Bradford PA. What's New in beta-lactamases? *Curr Infect Dis Rep* 2001 Feb;3(1):13-

[9] Bush K. The evolution of beta-lactamases. *Ciba Found Symp* 1997;207:152-63; discussion 63-6.

[10] Bush K, Jacoby GA, Medeiros AA. A functional classification scheme for beta-lactamases and its correlation with molecular structure. *Antimicrob Agents Chemother* 1995 Jun;39(6):1211-33.

[11] Bush K, Mobashery S. How beta-lactamases have driven pharmaceutical drug discovery. From mechanistic knowledge to clinical circumvention. *Adv Exp Med Biol* 1998;456:71-98.

[12] Paterson DL, Bonomo RA. Extended-spectrum beta-lactamases: a clinical update. *Clin Microbiol Rev* 2005 Oct;18(4):657-86.

[13] Bonnet R. Growing group of extended-spectrum beta-lactamases: the CTX-M enzymes. *Antimicrob Agents Chemother* 2004 Jan;48(1):1-14.

[14] Conceicao T, Brizio A, Duarte A, Lito LM, Cristino JM, Salgado MJ. First description of CTX-M-15-producing *Klebsiella pneumoniae* in Portugal. *Antimicrob Agents Chemother* 2005 Jan;49(1):477-8.

[15] Markovska R, Schneider I, Keuleyan E, Bauernfeind A. Extended-spectrum beta-lactamase (ESBL) CTX-M-15-producing *Escherichia coli* and *Klebsiella pneumoniae* in Sofia, Bulgaria. *Clin Microbiol Infect* 2004 Aug;10(8):752-5.

[16] Vincent JL. Nosocomial infections in adult intensive-care units. *Lancet 2003 Jun* 14;361(9374):2068-77.

[17] Sabate M, Miro E, Navarro F, et al. Beta-lactamases involved in resistance to broad-spectrum cephalosporins in *Escherichia coli* and *Klebsiella* spp. clinical isolates collected between 1994 and 1996, in Barcelona (Spain). *J Antimicrob Chemother* 2002 Jun;49(6):989-97.

[18] Tzouvelekis LS, Tzelepi E, Tassios PT, Legakis NJ. CTX-M-type beta-lactamases: an emerging group of extended-spectrum enzymes. *Int J Antimicrob Agents* 2000 Mar;14(2):137-42.

[19] Mendonca N, Louro D, Castro AP, Diogo J, Canica M. CTX-M-15, OXA-30 and TEM-1-producing *Escherichia coli* in two Portuguese regions. *J Antimicrob Chemother* 2006 May;57(5):1014-6.

[20] Alobwede I, M'Zali FH, Livermore DM, Heritage J, Todd N, Hawkey PM. CTX-M extended-spectrum beta-lactamase arrives in the UK. *J Antimicrob Chemother* 2003 Feb;51(2):470-1.

[21] Livermore DM, Hawkey PM. CTX-M: changing the face of ESBLs in the UK. *J Antimicrob Chemother* 2005 Sep;56(3):451-4.

[22] Paterson DL. Recommendation for treatment of severe infections caused by *Enterobacteriaceae* producing extended-spectrum beta-lactamases (ESBLs). *Clin Microbiol Infect 2*000 Sep;6(9):460-3.

[23] Paterson DL. Resistance in gram-negative bacteria: *Enterobacteriaceae. Am J Infect Control* 2006 Jun;34(5 Suppl 1):S20-8; discussion S64-73.

[24] Paterson DL, Ko WC, Von Gottberg A, et al. International prospective study of *Klebsiella pneumoniae* bacteremia: implications of extended-spectrum beta-lactamase production in nosocomial Infections. *Ann Intern Med* 2004 Jan 6;140(1):26-32.

[25] Azap OK, Arslan H, Serefhanoglu K, et al. Risk factors for extended-spectrum beta-lactamase positivity in uropathogenic *Escherichia coli* isolated from community-acquired urinary tract infections. *Clin Microbiol In*fect 2009 Aug 18.

[26] Calbo E, Romani V, Xercavins M, et al. Risk factors for community-onset urinary tract infections due to *Escherichia coli* harbouring extended-spectrum beta-lactamases. *J Antimicrob Chemother* 2006 Apr;57(4):780-3.

[27] Rodriguez-Bano J, Navarro MD, Romero L, et al. Bacteremia due to extended-spectrum beta -lactamase-producing *Escherichia coli* in the CTX-M era: a new clinical challenge. *Clin Infect Dis* 2006 Dec 1;43(11):1407-14.

[28] Rodriguez-Bano J, Navarro MD, Romero L, et al. Clinical and molecular epidemiology of extended-spectrum beta-lactamase-producing *Escherichia coli* as a cause of nosocomial infection or colonization: implications for control. *Clin Infect Dis* 2006 Jan 1;42(1):37-45.

[29] Rodriguez-Bano J, Picon E, Gijon P, et al. Community-onset bacteremia due to extended-spectrum beta-lactamase-producing *Escherichia col*i: risk factors and prognosis. *Clin Infect Dis* Jan 1;50(1):40-8.

[30] Laupland KB, Church DL, Vidakovich J, Mucenski M, Pitout JD. Community-onset extended-spectrum beta-lactamase (ESBL) producing *Escherichia coli*: importance of international travel. *J Infect* 2008 Dec;57(6):441-8.
[31] Pitout JD, Church DL, Gregson DB, et al. Molecular epidemiology of CTX-M-producing *Escherichia coli* in the Calgary Health Region: emergence of CTX-M-15-producing isolates. Antimicrob Agents Chemother 2007 Apr;51(4):1281-6.
[32] Arpin C, Dubois V, Coulange L, et al. Extended-spectrum beta-lactamase-producing *Enterobacteriaceae* in community and private health care centers. *Antimicrob Agents Chemother* 2003 Nov;47(11):3506-14.
[33] Lartigue MF, Zinsius C, Wenger A, Bille J, Poirel L, Nordmann P. Extended-spectrum beta-lactamases of the CTX-M type now in Switzerland. *Antimicrob Agents Chemother* 2007 Aug;51(8):2855-60.
[34] Naas T, Oxacelay C, Nordmann P. Identification of CTX-M-type extended-spectrum-beta-lactamase genes using real-time PCR and pyrosequencing. *Antimicrob Agents Chemother* 2007 Jan;51(1):223-30.
[35] Zahar JR, Nguile-Makao M, Francais A, et al. Predicting the risk of documented ventilator-associated pneumonia for benchmarking: construction and validation of a score. *Crit Care Med* 2009 Sep;37(9):2545-51.
[36] Helfand MS, Bonomo RA. Extended-spectrum beta-lactamases in multidrug-resistant *Escherichia coli*: changing the therapy for hospital-acquired and community-acquired infections. *Clin Infect Dis* 2006 Dec 1;43(11):1415-6.
[37] Luzzaro F, Vigano EF, Fossati D, et al. Prevalence and drug susceptibility of pathogens causing bloodstream infections in northern Italy: a two-year study in 16 hospitals. *Eur J Clin Microbiol Infect Dis* 2002 Dec;21(12):849-55.
[38] Rodriguez-Bano J, Pascual A. Clinical significance of extended-spectrum beta-lactamases. *Expert Rev Anti Infect Ther* 2008 Oct;6(5):671-83.
[39] Endimiani A, Paterson DL. Optimizing therapy for infections caused by *Enterobacteriaceae* producing extended-spectrum beta-lactamases. *Semin Respir Crit Care Med* 2007 Dec;28(6):646-55.
[40] Bratu S, Mooty M, Nichani S, et al. Emergence of KPC-possessing *Klebsiella pneumoniae* in Brooklyn, New York: epidemiology and recommendations for detection. *Antimicrob Agents Chemother* 2005 Jul;49(7):3018-20.
[41] Bratu S, Tolaney P, Karumudi U, et al. Carbapenemase-producing *Klebsiella pneumoniae* in Brooklyn, NY: molecular epidemiology and in vitro activity of polymyxin B and other agents. *J Antimicrob Chemother* 2005 Jul;56(1):128-32.
[42] Endimiani A, Depasquale JM, Forero S, et al. Emergence of blaKPC-containing *Klebsiella pneumoniae* in a long-term acute care hospital: a new challenge to our healthcare system. *J Antimicrob Chemother* 2009 Nov;64(5):1102-10.
[43] Naas T, Nordmann P, Vedel G, Poyart C. Plasmid-mediated carbapenem-hydrolyzing beta-lactamase KPC in a *Klebsiella pneumoniae* isolate from France. *Antimicrob Agents Chemother* 2005 Oct;49(10):4423-4.
[44] Nordmann P, Poirel L. Emerging carbapenemases in Gram-negative aerobes. *Clin Microbiol Infect* 2002 Jun;8(6):321-31.
[45] Yigit H, Queenan AM, Anderson GJ, et al. Novel carbapenem-hydrolyzing beta-lactamase, KPC-1, from a carbapenem-resistant strain of *Klebsiella pneumoniae*. *Antimicrob Agents Chemother* 2001 Apr;45(4):1151-61.

[46] Daly MW, Riddle DJ, Ledeboer NA, Dunne WM, Ritchie DJ. Tigecycline for treatment of pneumonia and empyema caused by carbapenemase-producing *Klebsiella pneumoniae. Pharmacotherapy* 2007 Jul;27(7):1052-7.

[47] Giamarellou H, Poulakou G. Multidrug-resistant Gram-negative infections: what are the treatment options? *Drugs* 2009;69(14):1879-901.

[48] Livermore DM. The impact of carbapenemases on antimicrobial development and therapy. *Curr Opin Investig Drugs* 2002 Feb;3(2):218-24.

[49] Livermore DM. Has the era of untreatable infections arrived? *J Antimicrob Chemother* 2009 Sep;64 Suppl 1:i29-36.

[50] Perez F, Endimiani A, Hujer KM, Bonomo RA. The continuing challenge of ESBLs. *Curr Opin Pharmacol* 2007 Oct;7(5):459-69.

[51] Pournaras S, Iosifidis E, Roilides E. Advances in antibacterial therapy against emerging bacterial pathogens. *Semin Hematol* 2009 Jul;46(3):198-211.

[52] Weisenberg SA, Morgan DJ, Espinal-Witter R, Larone DH. Clinical outcomes of patients with *Klebsiella pneumoniae* carbapenemase-producing *K. pneumoniae* after treatment with imipenem or meropenem. *Diagn Microbiol Infect Dis* 2009 Jun;64(2):233-5.

[53] Deshpande LM, Rhomberg PR, Sader HS, Jones RN. Emergence of serine carbapenemases (KPC and SME) among clinical strains of *Enterobacteriaceae* isolated in the United States Medical Centers: report from the MYSTIC Program (1999-2005). *Diagn Microbiol Infect Dis* 2006 Dec;56(4):367-72.

[54] Patel G, Huprikar S, Factor SH, Jenkins SG, Calfee DP. Outcomes of carbapenem-resistant *Klebsiella pneumoniae* infection and the impact of antimicrobial and adjunctive therapies. *Infect Control Hosp Epidemiol* 2008 Dec;29(12):1099-106.

[55] Fisher MA, Stamper PD, Hujer KM, et al. Performance of the Phoenix bacterial identification system compared with disc diffusion methods for identifying extended-spectrum beta-lactamase, AmpC and KPC producers. *J Med Microbiol* 2009 Jun;58(Pt 6):774-8.

[56] Goldfarb D, Harvey SB, Jessamine K, Jessamine P, Toye B, Desjardins M. Detection of plasmid-mediated KPC-producing *Klebsiella pneumoniae* in Ottawa, Canada: evidence of intrahospital transmission. J Clin Microbiol 2009 Jun;47(6):1920-2.

[57] Pasteran F, Mendez T, Guerriero L, Rapoport M, Corso A. Sensitive screening tests for suspected class A carbapenemase production in species of *Enterobacteriaceae. J Clin Microbiol* 2009 Jun;47(6):1631-9.

[58] Tsakris A, Kristo I, Poulou A, et al. Evaluation of boronic acid disk tests for differentiating KPC-possessing *Klebsiella pneumoniae* isolates in the clinical laboratory. *J Clin Microbiol* 2009 Feb;47(2):362-7.

[59] Samra Z, Bahar J, Madar-Shapiro L, Aziz N, Israel S, Bishara J. Evaluation of CHROMagar KPC for rapid detection of carbapenem-resistant *Enterobacteriaceae. J Clin Microbiol* 2008 Sep;46(9):3110-1.

[60] Tsakris A, Poulou A, Themeli-Digalaki K, et al. Use of boronic acid disk tests to detect extended- spectrum beta-lactamases in clinical isolates of KPC carbapenemase-possessing *Enterobacteriaceae. J Clin Microbiol* 2009 Nov;47(11):3420-6.

[61] Doi Y, Potoski BA, Adams-Haduch JM, Sidjabat HE, Pasculle AW, Paterson DL. Simple disk-based method for detection of *Klebsiella pneumoniae* carbapenemase-type

beta-lactamase by use of a boronic acid compound. *J Clin Microbiol* 2008 Dec;46(12):4083-6.

[62] Livermore DM, Hope R, Mushtaq S, Warner M. Orthodox and unorthodox clavulanate combinations against extended-spectrum beta-lactamase producers. *Clin Microbiol Infect* 2008 Jan;14 Suppl 1:189-93.

[63] Thomson KS, Moland ES. Cefepime, piperacillin-tazobactam, and the inoculum effect in tests with extended-spectrum beta-lactamase-producing *Enterobacteriaceae. Antimicrob Agents Chemother* 2001 Dec;45(12):3548-54.

[64] Rice LB, Carias LL, Shlaes DM. In vivo efficacies of beta-lactam-beta-lactamase inhibitor combinations against a TEM-26-producing strain of *Klebsiella pneumoniae. Antimicrob Agents Chemother* 1994 Nov;38(11):2663-4.

[65] Paterson DL, Singh N, Gayowski T, Marino IR. Fatal infection due to extended-spectrum beta-lactamase-producing *Escherichia coli*: implications for antibiotic choice for spontaneous bacterial peritonitis. *Clin Infect Dis* 1999 Mar;28(3):683-4.

[66] Karadenizli A, Mutlu B, Okay E, Kolayli F, Vahaboglu H. Piperacillin with and without tazobactam against extended-spectrum beta-lactamase-producing *Pseudomonas aeruginosa* in a rat thigh abscess model. *Chemotherapy* 2001 Jul-Aug;47(4):292-6.

[67] Ambrose PG, Bhavnani SM, Jones RN. Pharmacokinetics-pharmacodynamics of cefepime and piperacillin-tazobactam against *Escherichia coli* and *Klebsiella pneumoniae* strains producing extended-spectrum beta-lactamases: report from the ARREST program. *Antimicrob Agents Chemother* 2003 May;47(5):1643-6.

[68] Petersen PJ, Jones CH, Venkatesan AM, Bradford PA. Efficacy of piperacillin combined with the Penem beta-lactamase inhibitor BLI-489 in murine models of systemic infection. *Antimicrob Agents Chemother* 2009 Apr;53(4):1698-700.

[69] Abbott A. Medics braced for fresh superbug. Nature 2005 Aug 11;436(7052):758.

[70] Bonomo RA, Szabo D. Mechanisms of multidrug resistance in *Acinetobacter* species and *Pseudomonas aeruginosa. Clin Infect* Dis 2006 Sep 1;43 Suppl 2:S49-56.

[71] Perez F, Endimiani A, Bonomo RA. Why are we afraid of *Acinetobacter baumannii*? *Expert Rev Anti Infect Ther* 2008 Jun;6(3):269-71.

[72] Perez F, Hujer AM, Hujer KM, Decker BK, Rather PN, Bonomo RA. Global challenge of multidrug-resistant *Acinetobacter baumannii. Antimicrob Agents Chemother* 2007 Oct;51(10):3471-84.

[73] Talbot GH, Bradley J, Edwards JE, Jr., Gilbert D, Scheld M, Bartlett JG. Bad bugs need drugs: an update on the development pipeline from the Antimicrobial Availability Task Force of the Infectious Diseases Society of America. *Clin Infect Dis* 2006 Mar 1;42(5):657-68.

[74] Marque S, Poirel L, Heritier C, et al. Regional occurrence of plasmid-mediated carbapenem-hydrolyzing oxacillinase OXA-58 in *Acinetobacter* spp. in Europe. *J Clin Microbiol* 2005 Sep;43(9):4885-8.

[75] Seifert H, Dijkshoorn L, Gerner-Smidt P, Pelzer N, Tjernberg I, Vaneechoutte M. Distribution of *Acinetobacter* species on human skin: comparison of phenotypic and genotypic identification methods. *J Clin Microbiol* 1997 Nov;35(11):2819-25.

[76] Shelburne SA, 3rd, Singh KV, White AC, Jr., et al. Sequential outbreaks of infections by distinct *Acinetobacter baumannii* strains in a public teaching hospital in Houston, Texas. *J Clin Microbiol* 2008 Jan;46(1):198-205.

[77] Simhon A, Rahav G, Shazberg G, Block C, Bercovier H, Shapiro M. *Acinetobacter baumannii* at a tertiary-care teaching hospital in Jerusalem, Israel. *J Clin Microbiol* 2001 Jan;39(1):389-91.
[78] van den Broek PJ, van der Reijden TJ, van Strijen E, Helmig-Schurter AV, Bernards AT, Dijkshoorn L. Endemic and epidemic *Acinetobacter* species in a university hospital: an 8-year survey. *J Clin Microbiol* 2009 Nov;47(11):3593-9.
[79] Zhou H, Yang Q, Yu YS, Wei ZQ, Li LJ. Clonal spread of imipenem-resistant *Acinetobacter baumannii* among different cities of China. J Clin Microbiol 2007 Dec;45(12):4054-7.
[80] Coelho JM, Turton JF, Kaufmann ME, et al. Occurrence of carbapenem-resistant *Acinetobacter baumannii* clones at multiple hospitals in London and Southeast England. *J Clin Microbiol* 2006 Oct;44(10):3623-7.
[81] Dijkshoorn L, Aucken H, Gerner-Smidt P, et al. Comparison of outbreak and nonoutbreak *Acinetobacter baumannii* strains by genotypic and phenotypic methods. *J Clin Microbiol* 1996 Jun;34(6):1519-25.
[82] Hujer KM, Hujer AM, Hulten EA, et al. Analysis of antibiotic resistance genes in multidrug-resistant *Acinetobacter* sp. isolates from military and civilian patients treated at the Walter Reed Army Medical Center. *Antimicrob Agents Chemother* 2006 Dec;50(12):4114-23.
[83] Turton JF, Gabriel SN, Valderrey C, Kaufmann ME, Pitt TL. Use of sequence-based typing and multiplex PCR to identify clonal lineages of outbreak strains of *Acinetobacter baumannii. Clin Microbiol Infect* 2007 Aug;13(8):807-15.
[84] Turton JF, Kaufmann ME, Gill MJ, et al. Comparison of *Acinetobacter baumannii* isolates from the United Kingdom and the United States that were associated with repatriated casualties of the Iraq conflict. *J Clin Microbiol* 2006 Jul;44(7):2630-4.
[85] van Dessel H, Dijkshoorn L, van der Reijden T, et al. Identification of a new geographically widespread multiresistant *Acinetobacter baumannii* clone from European hospitals. *Res Microbiol* 2004 Mar;155(2):105-12.
[86] van Dessel H, Kamp-Hopmans TE, Fluit AC, et al. Outbreak of a susceptible strain of *Acinetobacter* species 13 (sensu Tjernberg and Ursing) in an adult neurosurgical intensive care unit. *J Hosp Infect* 2002 Jun;51(2):89-95.
[87] Naas T, Bogaerts P, Bauraing C, Degheldre Y, Glupczynski Y, Nordmann P. Emergence of PER and VEB extended-spectrum beta-lactamases in *Acinetobacter baumannii* in Belgium. *J Antimicrob Chemother* 2006 Jul;58(1):178-82.
[88] Naas T, Kernbaum S, Allali S, Nordmann P. Multidrug-resistant *Acinetobacter baumannii*, Russia. *Emerg Infect Dis* 2007 Apr;13(4):669-71.
[89] Da Silva GJ, Quinteira S, Bertolo E, et al. Long-term dissemination of an OXA-40 carbapenemase-producing *Acinetobacter baumannii* clone in the Iberian Peninsula. *J Antimicrob Chemother* 2004 Jul;54(1):255-8.
[90] Turton JF, Kaufmann ME, Warner M, et al. A prevalent, multiresistant clone of *Acinetobacter baumannii* in Southeast England. *J Hosp Infect* 2004 Nov;58(3):170-9.
[91] Landrum ML, Murray CK. Ventilator associated pneumonia in a military deployed setting: the impact of an aggressive infection control program. *J Trauma* 2008 Feb;64(2 Suppl):S123-7; discussion S7-8.
[92] Murray CK, Hospenthal DR. Treatment of multidrug resistant *Acinetobacter*. *Curr Opin Infect Dis* 2005 Dec;18(6):502-6.

[93] Falagas ME, Grammatikos AP, Michalopoulos A. Potential of old-generation antibiotics to address current need for new antibiotics. *Expert Rev Anti Infect Ther* 2008 Oct;6(5):593-600.

[94] Falagas ME, Kasiakou SK. Toxicity of polymyxins: a systematic review of the evidence from old and recent studies. *Crit Care* 2006 Feb;10(1):R27.

[95] Falagas ME, Rafailidis PI. Nephrotoxicity of colistin: new insight into an old antibiotic. *Clin Infect Dis* 2009 Jun 15;48(12):1729-31.

[96] Falagas ME, Rafailidis PI, Matthaiou DK, Virtzili S, Nikita D, Michalopoulos A. Pandrug-resistant *Klebsiella pneumoniae, Pseudomonas aeruginosa* and *Acinetobacter baumannii* infections: characteristics and outcome in a series of 28 patients. *Int J Antimicrob Agents* 2008 Nov;32(5):450-4.

[97] Li J, Nation RL, Turnidge JD, et al. Colistin: the re-emerging antibiotic for multidrug-resistant Gram-negative bacterial infections. *Lancet Infect Dis* 2006 Sep;6(9):589-601.

[98] Li J, Rayner CR, Nation RL, et al. Heteroresistance to colistin in multidrug-resistant *Acinetobacter baumannii. Antimicrob Agents Chemother* 2006 Sep;50(9):2946-50.

[99] Nation RL, Li J. Colistin in the 21st century. *Curr Opin Infect Dis* 2009 Dec;22(6):535-43.

[100] 100. Bassetti M, Righi E, Esposito S, Petrosillo N, Nicolini L. Drug treatment for multidrug-resistant *Acinetobacter baumannii* infections. *Future Microbiol* 2008 Dec;3(6):649-60.

[101] Poulakou G, Kontopidou FV, Paramythiotou E, et al. Tigecycline in the treatment of infections from multi-drug resistant gram-negative pathogens. *J Infect* 2009 Apr;58(4):273-84.

[102] Wieczorek P, Sacha P, Hauschild T, Zorawski M, Krawczyk M, Tryniszewska E. Multidrug resistant *Acinetobacter baumannii*--the role of AdeABC (RND family) efflux pump in resistance to antibiotics. *Folia Histochem Cytobiol* 2008;46(3):257-67.

[103] Bergogne-Berezin E. The Increasing Role of *Acinetobacter* Species As Nosocomial Pathogens. *Curr Infect Dis Rep* 2001 Oct;3(5):440-4.

[104] Bergogne-Berezin E, Towner KJ. *Acinetobacter* spp. as nosocomial pathogens: microbiological, clinical, and epidemiological features. *Clin Microbiol Rev* 1996 Apr;9(2):148-65.

[105] Towner KJ. Clinical importance and antibiotic resistance of *Acinetobacter* spp. Proceedings of a symposium held on 4-5 November 1996 at Eilat, Israel. *J Med Microbiol* 1997 Sep;46(9):721-46.

[106] Fournier PE, Richet H. The epidemiology and control of *Acinetobacter baumannii* in health care facilities. *Clin Infect Dis* 2006 Mar 1;42(5):692-9.

[107] Choi CH, Hyun SH, Lee JY, et al. *Acinetobacter baumannii* outer membrane protein A targets the nucleus and induces cytotoxicity. *Cell Microbiol* 2008 Feb;10(2):309-19.

[108] Falagas ME, Karveli EA. The changing global epidemiology of *Acinetobacter baumannii* infections: a development with major public health implications. *Clin Microbiol Infect* 2007 Feb;13(2):117-9.

[109] Iacono M, Villa L, Fortini D, et al. Whole-genome pyrosequencing of an epidemic multidrug-resistant *Acinetobacter baumannii* strain belonging to the European clone II group. *Antimicrob Agents Chemother* 2008 Jul;52(7):2616-25.

[110] Joly-Guillou ML. Clinical impact and pathogenicity of *Acinetobacter. Clin Microbiol Infect* 2005 Nov;11(11):868-73.

[111] Robenshtok E, Paul M, Leibovici L, et al. The significance of *Acinetobacter baumannii* bacteraemia compared with *Klebsiella pneumoniae* bacteraemia: risk factors and outcomes. J Hosp Infect 2006 Nov;64(3):282-7.

[112] Villegas MV, Hartstein AI. *Acinetobacter* outbreaks, 1977-2000. *Infect Control Hosp Epidemiol* 2003 Apr;24(4):284-95.

[113] Fournier PE, Vallenet D, Barbe V, et al. Comparative genomics of multidrug resistance in *Acinetobacter baumannii. PLoS Genet* 2006 Jan;2(1):e7.

[114] Adams MD, Goglin K, Molyneaux N, et al. Comparative genome sequence analysis of multidrug-resistant *Acinetobacter baumannii. J Bacteriol* 2008 Dec;190(24):8053-64.

[115] Smith MG, Gianoulis TA, Pukatzki S, et al. New insights into *Acinetobacter baumannii* pathogenesis revealed by high-density pyrosequencing and transposon mutagenesis. *Genes Dev* 2007 Mar 1;21(5):601-14.

[116] Stover CK, Pham XQ, Erwin AL, et al. Complete genome sequence of *Pseudomonas aeruginosa* PAO1, an opportunistic pathogen. *Nature* 2000 Aug 31;406(6799):959-64.

[117] Oliver A, Canton R, Campo P, Baquero F, Blazquez J. High frequency of hypermutable *Pseudomonas aeruginosa* in cystic fibrosis lung infection. *Science* 2000 May 19;288(5469):1251-4.

[118] Bratu S, Quale J, Cebular S, Heddurshetti R, Landman D. Multidrug-resistant *Pseudomonas aeruginosa* in Brooklyn, New York: molecular epidemiology and in vitro activity of polymyxin B. *Eur J Clin Microbiol Infect Dis* 2005 Mar;24(3):196-201.

[119] Quale J, Bratu S, Gupta J, Landman D. Interplay of efflux system, ampC, and oprD expression in carbapenem resistance of *Pseudomonas aeruginosa* clinical isolates. *Antimicrob Agents Chemother* 2006 May;50(5):1633-41.

[120] Arora S, Bal M. AmpC beta-lactamase producing bacterial isolates from Kolkata hospital. *Indian J Med Res* 2005 Sep;122(3):224-33.

[121] Hocquet D, Berthelot P, Roussel-Delvallez M, et al. *Pseudomonas aeruginosa* may accumulate drug resistance mechanisms without losing its ability to cause bloodstream infections. *Antimicrob Agents Chemother* 2007 Oct;51(10):3531-6.

[122] Picao RC, Poirel L, Gales AC, Nordmann P. Diversity of beta-lactamases produced by ceftazidime-resistant *Pseudomonas aeruginosa* isolates causing bloodstream infections in Brazil. *Antimicrob Agents Chemother* 2009 Sep;53(9):3908-13.

[123] Tam VH, Chang KT, Schilling AN, LaRocco MT, Genty LO, Garey KW. Impact of AmpC overexpression on outcomes of patients with *Pseudomonas aeruginosa* bacteremia. *Diagn Microbiol Infect D*is 2009 Mar;63(3):279-85.

[124] Endimiani A, Perez F, Bonomo RA. Cefepime: a reappraisal in an era of increasing antimicrobial resistance. *Expert Rev Anti Infect Ther* 2008 Dec;6(6):805-24.

[125] Rodriguez-Martinez JM, Poirel L, Nordmann P. Molecular epidemiology and mechanisms of carbapenem resistance in *Pseudomonas aeruginosa. Antimicrob Agents Chemother* 2009 Nov;53(11):4783-8.

In: Beta-Lactamases
Editor: Jean-Marie Frère, pp. 401-418
ISBN: 978-1-61324-638-2

Chapter 16

INTERPLAY BETWEEN Β-LACTAMASE ACTIVITY, OUTER MEMBRANE PERMEABILITY AND ACTIVE EFFLUX SYSTEM

Jean-Denis Docquier[1], Gian Maria Rossolini[1], Hiroshi Nikaido[2] and Jean-Marie Frère[3]

[1] Dipartimento di Biotecnologie, Università degli Studi di Siena, 53100 Siena, Italy.
[2] Department of Molecular and Cell Biology, University of California, Berkeley, CA 94720, U.S.A.
[3] CIP, Université de Liège, B 4000 Liège, Belgique.

ABSTRACT

The resistance of a Gram-negative bacterial strain to penicillin is determined by the sensitivity of the essential penicillin-binding proteins (PBPs), the amount and properties of the produced β-lactamase(s), the permeability of the outer membrane and the efficiency of the efflux systems. The interplay between the last three mechanisms is discussed with examples stressing the importance of one or several of them in clinical isolates. Mathematical models are presented. In the relatively simple case of *Escherichia coli* (in which the efflux systems are often negligible when compared to the β-lactamase activity) the MICs can be predicted on the basis of these models. Similarly, correct MIC predictions could be made for an *Enterobacter cloacae* strain which overproduced a class C enzyme.

The production of ß-lactamases is only one of the factors which determine the resistance of a bacterial strain to penicillins. The sensitivity of the target DD-transpeptidases (or PBPs, Penicillin Binding Proteins) is clearly of major importance. A vast number of PBPs have been described and their interactions with β–lactams studied in details. In short (see also the Foreword) active-site serine β-lactamases and PBPs share the kinetic pathways described for the former in Chapter 6 but PBPs consistently exhibit very low k_3 values [1, 2]. Various PBPs exhibit very different sensitivities to the same β-lactam (the k_2/K ratio which characterises the rate of acylation can vary by factors of 100 000 or more) and conversely, the sensitivity of a given PBP to different β-lactams is highly variable [3]. These two observations reflect the different structures of the PBP active sites and their specific interactions with the β-lactam side-chains which result in a more or less favourable positioning of the β-lactam C=O relative to the PBP active-site serine hydroxyl group. To significantly contribute to the resistance mutations which decrease the efficiency of the acylation reaction must preserve the transpeptidase activity of the enzyme. In consequence, it is not surprising to observe that point mutations have seldom been selected and that, when this happens, they never increase the resistance by a large factor. More efficient mechanisms are (i) the importation of a new gene encoding an intrinsically resistant PBP from a related species, as occurs with PBP2a of the Methicillin Resistant *Staphylococcus aureus*, (ii) the manufacturing of mosaic genes containing large DNA pieces characteristic of less sensitive proteins (whose origin is not always clear) as for PBP2x of *Streptococcus pneumoniae* and (iii) the overproduction of a more resistant but less efficient transpeptidase such as PBP5f in *Enterococcus faecium* [4]. It is interesting to note that these resistance mechanisms are mainly encountered in species which do not produce β-lactamases (or do so in very exceptional cases) or have been, for some mysterious reasons, unable to adapt an existing ß-lactamase to hydrolyse new compounds (e.g. the *S. aureus* enzyme) or to import and express a ß-lactamase encoding gene.

Another possibility is to prevent the antibiotic from reaching its target. In Gram-negative bacteria and in mycobacteria, the outer membrane represents a physical obstacle that slows down the rate of diffusion of molecules into the periplasm [5, 6] and active efflux systems can pump compounds with hydrophobic side-chains out from the periplasm [6, 7].

When secreted in the outer medium, the β-lactamase protects both the producing and neighbouring non-producing cells. This is the situation in most Gram-positive strains and in some Gram-negative psychrophilic strains [8], but the latter have never, to our knowledge, been implicated in human infections. Conversely, a case of membrane-bound enzyme has been documented in *Bacillus licheniformis* [9]. In Gram-negatives and mycobacteria, the enzyme is generally confined to the periplasm, which makes it a particularly efficient weapon[4]. Indeed, a small amount of ß-lactamase, or a poorly active one, can strongly increase the MIC if the outer membrane significantly slows down the in-diffusion of the antibiotic or if the latter is pumped out by the efflux system [10].

Zimmermann and Rosselet [11] derived an equation which correlates the MIC with the outer membrane permeability, the sensitivity of the PBPs and the quantity and properties of the periplasmic β-lactamase in the absence of efflux. They mainly used this equation to try to estimate the rates of diffusion through the outer membrane. The same equation was later used by Nikaido and Normark [5], Waley [12] and Frère [13] to predict the MIC values on the basis of the properties of the PBPs, the outer membrane and the β-lactamase.

In this chapter, we present examples where the roles of efflux systems and of modifications of outer membrane permeability have been documented and explain how the mathematical models have evolved to take account of the complexity of *in vivo* situations.

---

[4] In infection foci, some ß-lactamase might also be liberated by cell lysis, but the possible influence of this additional factor is outside the scope of the present analysis. Similarly, some mycobacteria appear to liberate the β-lactamase in the outer medium after several days of growth.

PERMEABILITY AND PORINS

In addition to the production of one or more β-lactamases by Gram-negative opportunistic pathogens, factors affecting the apparent permeability of the outer membrane are now considered of major importance as they can contribute to the development of multi-drug resistance in many pathogens and also play a relevant role in increasing the level of resistance of β-lactamase-producing strains.

The relative permeability of the outer membrane is quite variable according to the organism and represents a key factor explaining the different intrinsic susceptibilities of antimicrobial agents shown by various pathogens [14]. For instance, *Pseudomonas aeruginosa* and *Acinetobacter baumannii* outer membranes display rather low permeabilities when compared to that of *Escherichia coli*, which translate in higher MICs in the former for some β-lactams. This is particularly evident with carbapenems, the last generation of β-lactam antibiotics, which are more active on *Enterobacteriaceae* than on *P. aeruginosa* or *A. baumannii*, thanks to a much lower penetration rate in the latter organisms.

Diffusion through the outer membrane occurs thanks to special channels called porins. These transmembrane proteins are involved in the uptake of various hydrophilic molecules and exhibit a typical β-barrel structure [15]. For instance, the OmpF porin (outer membrane protein F) of *Escherichia coli*, one of the first porins whose crystal structure was obtained, is made of 16 amphiphilic β-strands connected by short turns on the periplasmic side and longer loops on the external side [15]. The substrate specificities of these proteins may be very different, and allow a selection of the transported molecules according to their physico-chemical properties. For instance, the OprD porin of *Pseudomonas aeruginosa*, one of the best studied porins for its relevance in carbapenem uptake (see below), exhibits a strong preference for basic amino acids, such as arginine which has also been shown to control the level of porin production in some organisms [16, 17].

An example of porin specificity is found in *Pseudomonas aeruginosa*, where imipenem, the first carbapenem to be introduced in anti-infective therapy in the mid 80's, enters the periplasm through the OprD2 porin. One of the first mechanisms of imipenem resistance to be identified, besides the production of an acquired carbapenem-hydrolyzing β-lactamase, resulted from the decreased production or alterations of the properties of this porin, achieved by either (i) chromosomal mutations in the *oprD* gene, its promoter region or (ii) chromosomal mutations affecting a regulatory protein (e.g. MexT) or (iii) more complex genetic events, for instance the insertion of mobile genetic elements like insertion sequences that disrupt the promoter region or the structural gene encoding the OprD porin [18]. This mechanism of imipenem resistance is sufficient to shift the MIC values into the resistance range, with obvious clinical consequences. Moreover, chromosomal mutations affecting the level of OprD production appear at rather high frequencies, typically 10^{-7}-10^{-8} in selective environments, and resistance has also been documented to arise during therapy [19]. Although the OprD2 porin also promotes the uptake of other carbapenems (and notably meropenem), porin loss is apparently not sufficient to achieve clinical resistance to these compounds [18].

The contribution of porins in the uptake of β-lactams has also been investigated in other species [20, 21]. In a elegant study, Cornaglia and colleagues showed that both *E. coli* OmpC and OmpF proteins were significantly involved in the uptake of imipenem and meropenem, as the *ompC*$^-$ *ompF*$^-$double mutant showed a 10- to 20-fold decrease in the relative rates of permeation of these agents when compared to the wild-type strain [22]. In more recent studies, various outer membrane proteins of *Acinetobacter baumannii* have been identified for their relevance in the acquisition of carbapenem resistance [23]. Similarly, the loss of an OprD-like protein, of the 29-kDa CarO protein or of the so-called 33- to 36-kDa OMP was recognized as contributing to carbapenem resistance [17, 24, 25].

Efflux Pumps

Bacterial efflux systems are transport proteins, some of which efficiently export antimicrobial compounds. These fall into five major families of transporters, including for instance the ATP-binding cassette or the major facilitator superfamilies [7, 26].

Active efflux was the last antibiotic resistance mechanisms to be discovered. It was originally thought to only work for antimicrobial agents directed against intracellular targets but this view was subsequently challenged by the observation that the outer membrane permeability barrier and the hydrolysis of β-lactams by the periplasmic β-lactamase(s) were not sufficient to explain the resistance levels of certain strains of *Pseudomonas aeruginosa* [27]. Analysis of this phenomenon led to the discovery that some multidrug efflux pumps, namely those belonging to the Resistance-Nodulation-Division (RND) family, could play a role in β-lactam resistance by extruding β-lactam molecules from the periplasmic space [28, 29]. These pumps, for which the best known paradigms are the AcrAB-TolC of *Escherichia coli* and the MexAB-OprM of *P. aeruginosa*, consist of complex multi-subunit structures that span both the cytoplasmic membrane and the outer membrane, and extrude drugs directly into the external medium, rather than into the periplasm [30]. The RND-type efflux system contains three components: a proton-dependent pump located in the cytoplasmic membrane (the RND pump), an outer membrane protein (also called the outer membrane factor and which contains a porin-like β-barrel) and a periplasmic, fusion protein connecting the two transmembrane proteins, the so-called membrane fusion protein [15, 18].

These efflux systems exhibit rather broad substrate profiles and their overproduction causes resistance to a wide range of structurally unrelated antimicrobials such as fluoroquinolones and aminoglycosides whose targets are intracellular. However many β-lactam compounds do not easily cross the cytoplasmic membrane, and therefore stay in the periplasm after diffusing through the outer membrane. Thus it was surprising that efflux pumps, which were earlier thought to capture their substrates only in the cytosol, could contribute to β-lactam resistance. It was proposed that the RND efflux pumps can capture their substrates from the periplasm [28, 30], and this concept fits with the crystallographic structure of AcrB with periplasmic openings called “vestibules” that are thought to supply the substrates with an access to the central pore of the structure [31].

Efflux appears to make a variable contribution to β-lactam resistance depending on one or more of the following conditions.

(a) *Chemical structure of the β-lactam molecule.* Some β-lactams are very good substrates for RND-type efflux systems, as happens with molecules that contain lipophilic substituent(s) [32]; for example, the AcrAB-TolC system increases the cloxacillin MIC more than 128-fold, both in *E. coli* [33] and *Salmonella typhimurium* [32], but produces no or almost no increase in the MICs of hydrophilic β-lactams such as cefazolin and cefmetazole. This observation was earlier thought to support the hypothesis that interaction of the substrate β-lactam with the outer leaflet of the inner membrane is necessary for its capture by the RND pump [28, 30]. However, it now seems more likely to be a reflection of the fact that the binding site within AcrB binds only lipophilic molecules [34]. In any case, efflux was shown to be a major mechanism of intrinsic resistance to penicillins with lipophilic substituents (e. g. oxacillin and cloxacillin) in enterics and other Gram-negative bacteria, while it is overall less important with cephalosporins that have a more hydrophilic nucleus, and generally more hydrophilic side-chains. The apparent K_m values of β-lactams were recently determined, and range from about 1 μM to nearly 300 μM for the AcrB system [35, 36].

(b) *Concomitant production of a β-lactamase for which the β-lactam is a poor substrate.* Although a rapid hydrolysis of the β-lactam by enzymes that are present in the periplasm remains the predominant mechanism for resistance, impermeability and efflux become important mechanisms when the enzymatic hydrolysis is slow and does not overwhelm the whole picture. They cooperate to confer resistance as was predicted by the theoretical analysis which is developed below [10]. This is at least part of the reason why AcrAB-mediated efflux is important for penicillin G in *S. typhimurium*, which does not produce any resident β-lactamase [32], whereas it plays a much lesser role in penicillin G resistance in *E. coli* K-12, with its endogenous AmpC enzyme [33].

(c) *Overall permeability of the outer membrane.* Efflux by RND pumps competes directly with the drug influx across the outer membrane [30]. It is therefore expected that slow influx of the drugs will increase the efficiency of the efflux process. Indeed, the overall contribution of efflux to β-lactam resistance appears to be more important in *P. aeruginosa* [27, 29] with its low permeability outer membrane [37, 38], than in *E. coli*. In a given organism, compounds penetrating more slowly but effluxed at an equal rate will be more affected by efflux. Since porin channels usually allow only a slow diffusion of hydrophobic compounds [39], this may explain the relevant contribution of efflux to intrinsic resistance observed toward β-lactams with lipophilic side-chains. (See the Mathematical Models section).

(d) *Expression level of the efflux system.* The contribution of RND pumps to intrinsic resistance also depends on the expression level. RND pump systems are normally expressed at a basal level (which in some cases is very low), but their expression can be regulated by complex regulatory networks that are still only partially understood [40]. Exposure to environmental stimuli or mutational events that upregulate the expression of an efflux system can be responsible for an increased resistance to the substrates of this efflux system.

The relevance of efflux systems and of their regulation in β-lactam resistance has been best studied in *Pseudomonas aeruginosa*. An analysis of the genome of this opportunistic pathogen revealed the presence of at least 12 RND-type efflux systems, three of which have been demonstrated to extrude β-lactams (see [18] for an excellent review covering the regulatory pathways of the various efflux pumps). The first and perhaps best characterized system is the MexAB-OprM efflux, whose overproduction can confer resistance to a wide range of β-lactams, including carboxypenicillins, monobactams, penems, expanded-spectrum cephalosporins and some carbapenems, but not to imipenem or biapenem [41, 42] in addition to other antimicrobial agents. Other efflux systems, MexCD-OprJ and MexXY-OprM, also act on a wide variety of antibiotics, but regarding β-lactam compounds, they seem to be more specific for fourth-generation cephalosporins (e.g. cefepime) [26, 43].

Efflux pumps also represent relevant resistance mechanisms to β-lactams in other Gram-negative, such as *Enterobacteriaceae* (AcrAB-TolC), *Acinetobacter baumannii* (AdeAB-AdeC), *Campylobacter jejuni* (CmeAB-CmeC), *Stenotrophomonas maltophilia* (SmeAB-SmeC) and *Neisseria gonorrhoeae* (MtrCD-MtrE) [44].

The Case of Mycobacteria

Members of the *Mycobacterium*, *Corynebacterium* and *Nocardia* genera are surrounded by an outer membrane which also defines a periplasmic space but whose chemical composition is quite different from that of the Gram-negative outer membrane. Indeed, they contain mycolic acids, molecules which are unique to these 3 genera. But the general functions and bilayer structure of both types of outer membranes are similar [6, 45]. The mycobacterial outer membrane contains porins which allow the diffusion of hydrophilic molecules and exhibit a β-barrel structure but are nevertheless quite different from their Gram-negative equivalents (6). Moreover, efflux pumps are present which, for instance, contribute to the resistance of *Mycobacterium bovis* to ampicillin [6, 46] and many mycobacteria produce a β-lactamase which most often remains in the periplasm. In consequence, the situation is quite similar to that observed in Gram-negatives and the mathematical models discussed below are valid, unless large amounts of enzyme are released in “older” cultures [47, 48].

Clinical Relevance of the Interplay between Permeability, Efflux and β-lactamases: Some Examples

The impact of efflux on clinical β-lactam resistance depends on the ability to increase MIC values above the clinical breakpoints. When basal expression of the efflux system results in MIC values that are already above the clinical breakpoints (e. g. in the case of AcrAB-TolC and cloxacillin, in *E. coli*), the efflux is responsible for intrinsic resistance. Otherwise, the efflux system can be responsible for acquired resistance upon upregulation of its expression (e. g. in the case of MexAB-OprM and ticarcillin, in *P. aeruginosa*). Acquired β-lactam resistance mediated by upregulation of efflux systems is generally low-level and is not as affected by inoculum-size effect as beta-lactamase-mediated resistance. However, cooperation with the other two mechanisms (beta-lactamases and impermeability) results in increased resistance levels.

In *Enterobacteriaceae* and some non-fermenting pathogens like *P. aeruginosa*, a common mechanism of β-lactam resistance is provided by the derepression of the chromosomal class C β-lactamase and concomitant loss of porins. It is noteworthy that this mechanism of resistance, which might yield high-level resistance to some cephalosporins, does not require the acquisition of additional resistance factors, such as extended-spectrum β-lactamases.

The simultaneous presence of permeability- and efflux-related mechanisms with the production of β-lactamases has been particularly well investigated in the resistance to carbapenems. Resistance to these last generation (and often last-resort) β-lactam compounds is particularly worrisome. Indeed, due to the global spread of multidrug-resistant strains, carbapenems have been increasingly used in the clinical setting, allowing for the selection of carbapenem resistant-strains which become more and more common (carbapenem-resistant strains of the important opportunistic pathogen *Acinetobacter baumannii* increased from 9% in 1995 to 39% in 2004) [49, 50]. Similar trends have been observed with *P. aeruginosa*, while carbapenem resistance in *Enterobacteriaceae*, still less frequent, is becoming an increasingly important issue. More importantly, the recent reports of pan-resistant strains of *A. baumannii* and *Klebsiella pneumoniae*, i.e. strains resistant to the entire range of available antimicrobial agents (see also Chapter 15), represent a potentially dramatic turn back to the pre-antibiotic era [49, 51]. In these strains, β-lactam resistance is often achieved by the production of two or more β-lactamases, porin loss and/or antibiotic efflux.

Another recent example of interplay between permeability defects and β-lactamase production is that of *Enterobacteriaceae* which produce the OXA-48 carbapenemase, an acquired class D carbapenem-hydrolyzing β-lactamase first identified in *Klebsiella pneumoniae* but currently disseminating in other *Enterobacteriaceae* isolates [52, 53]. Among class D β-lactamases, this enzyme exhibits the highest turnover rates for cabapenems [52, 54]. However, strains in which the production of this enzyme was observed are surprisingly associated with extremely variable MIC values for carbapenems, which span from full susceptibility to high-level resistance (imipenem MICs range from 0.5 to 64 μg/ml) [53, 55]. This also results in an important detection issue of this resistance determinants in clinical specimens, which may silently spread in the bacterial population.

The variable susceptibility data obtained with OXA-48-producing isolates indicate that the mere production of the OXA-48 carbapenemase is not *per se* sufficient to provide clinical resistance to carbapenems, unless other porin-related mechanisms are be present. Specifically, the absence of the major outer membrane proteins OmpK35 and OmpK36 in OXA-48-producing *Klebsiella pneumoniae* isolates with high-level of resistance to carbapenems has been documented [52, 56]. Similarly, the absence of OmpC is apparently required for carbapenem resistance in *E. coli* producing the OXA-48 carbapenemase [56].

A similar situation is examplified by metallo-β-lactamase-producing *Enterobacteriaceae*. Metallo-β-lactamases (MBLs) are characterized by a strong catalytic activity on carbapenems (typical k_{cat}/K_m values for these substrates range from 10^5 to 10^6 $M^{-1}.s^{-1}$). These MBLs readily confer carbapenem resistance in hosts like *Pseudomonas aeruginosa*, but usually only confer decreased susceptibility to carbapenems in *Enterobacteriaceae*, especially *E. coli* and *K. pneumoniae* [57]. Again, porin loss has been identified as a critical factor to promote full resistance to the agents [58].

Mathematical Models

1. In the Absence of Efflux

1.1. The Equation of Zimmermannn and Rosselet

The equation of Zimmermann and Rosselet [11] correlates the MIC with the outer membrane permeability, the activity of the β-lactamase and the sensitivity of the PBPs. The rate of diffusion through the outer membrane is

$$v_{in} = k_D \ (I_e - I_p) \tag{1}$$

where k_D is a first order rate constant (in s^{-1}) and I_e and I_p are respectively the external and periplasmic concentrations of the antibiotic (in μM).

The rate of hydrolysis by the β-lactamase is given by the Henri-Michaelis equation

$$v_\beta = \frac{k_{cat} \ E_0 \ I_p}{K_m + I_p} \tag{2}$$

where k_{cat} (in s^{-1}) and K_m (in μM) are the kinetic parameters of the β-lactamase and E_0 (in μM), the concentration of the enzyme in the periplasm.

When $v_{in} = v_\beta$, a steady-state is established and $I_e = I_p + \dfrac{k_{cat} \ E_0 \ I_p}{k_D(K_m + I_p)}$

(3)

When I_p becomes large enough to inactivate a sufficient proportion of the essential PBPs, (I_{pl}), I_e is equal to the MIC so that

$$\mathrm{MIC} = I_{pl} + \frac{k_{cat} \ E_0 \ I_{pl}}{k_d(K_m + I_{pl})} \tag{4}$$

If it is relatively easy to determine the β-lactamase concentration in a defined volume of culture, a difficulty arises in the calculation of E_0 which requires an estimation of the periplasmic volume. Values ranging from 10 to 25% of the total cell volume have been proposed [59]. This problem can be circumvented by replacing k_{cat} E_0 by V (the maximum rate of hydrolysis by the β-lactamase in nmoles of substrate hydrolysed per s and per mg of dry weight) and k_D by P.A where P characterises the rate of diffusion through the outer membrane (in cm s^{-1}) and A is the surface outer membrane area (in cm^2 per mg of dry weight) yielding eq 5

$$\mathrm{MIC} = I_{pl} + \frac{V \ I_{pl}}{PA \ (K_m + I_{pl})} \tag{5}$$

Nikaido and Normark [5] have estimated an A value of 132 cm^2 per mg of dry weight for a Gram-negative rod. Assuming that 1 mg of dry weight corresponds to 4 mg of wet weight, the relationship between k_D and P is k_D = P x 133 000 cm^{-1} if the periplasmic volume is 25% of the total cell volume or k_D = P x 330 000 cm^{-1} if the periplasmic volume is 10% of the total volume. I_{pl} is probably the most difficult factor to estimate.

Table 1. Comparison of the MIC of *E. coli* strain SN03 with the I_{pl} values utilised by Nikaido et Normark (C_{inh}) and calculated by Frère (I_{pl}). All values are in μM [5, 13]

	MIC SN03	C_{inh}	I_{pl}
Cephaloridine	2.5	6	1.2 – 2.4
Cefazolin	5	11	2.5 – 4.7
Cephalothin	2.75	2.75	1 – 2
Cefamandole	0.65	0.63	0.3 – 0.6
Cefaclor	2.5	5	1.2 – 2.4
Cefsulodin	40	9	25 – 35
Cefoxitin	5	16	2.5 – 5
Cefotaxime	0.15	0.125	0.08 – 0.12
Ceftazidime	0.15	0.15	0.08 – 0.15
Penicillin G	5.7	2.3	1.6 – 3.2
Ampicillin	1.4	1.9	0.7 – 1.25
Carbenicillin	21	5.5	10 – 20
Aztreonam	0.15	0.15	0.08 – 0.14

When the acylation rate constant for the most sensitive essential PBPs has been measured with membrane preparations, the variation of active PBP concentration with time can be easily calculated if the synthesis of new PBPs by the cell is neglected. This simplification is certainly valid if I_e is significantly larger than the MIC. However, near the MIC, inactivation of the essential PBP is rather slow, so that its *de novo* synthesis must be taken into account (see point 3 below). In consequence, the best approximation of I_{pl} is given by the MIC of a strain devoid of both β-lactamase and significant efflux.

In *E. coli*, isoxazolylpenicillins such as oxacillin are the only β-lactams for which efflux appears to be a predominant factor. On this basis, equation 5 has been successfully utilised to account for the MIC values of a series of *E. coli* strains producing various quantities of class C or class A β-lactamases [5, 13]. Note that the I_{pl} values utilised by these authors were chosen according to the concentration of antibiotic which reduces the binding of radioactive benzylpenicillin by 50% [5] or by utilising a more complex graphical method [13] but they were never very different from the MIC of strain SN03 which produces a very low level of chromosomal (class C) β-lactamase (Table 1).

By contrast, the contribution of efflux in *Enterobacter cloacae* has been demonstrated by comparing the MICs of *E. coli* SNO3 and *E. cloacae* AD2 (in which the gene encoding the resident class C β-lactamase has been disrupted). Despite the fact that both strains produce a similar set of PBPs with similar kinetic properties, the MICs of AD2 for some compounds were significantly higher than those of SNO3, a result which could not be explained on the basis of the outer membrane permeability but highlighted the presence of efflux for those compounds in AD2 [10]. In the case of deregulated mutants producing large quantities of β-lactamase in a constitutive manner, eq 5 has a somewhat paradoxical consequence. If I_{pl} is much larger than K_m and $\frac{V}{PA}$ much larger than I_{pl},

$$MIC = \frac{V}{PA} \tag{6}$$

and I_{pl} disappears from the equation [10] which means that the sensitivity of the PBP becomes irrelevant. This situation is encountered with clinical isolates overproducing a class C enzyme, a consequence of the rather easy inactivation of the Amp D protein due to point mutations in the corresponding gene. Since the K_m values of these enzymes for compounds such as cefotaxime, imipenem and aztreonam are usually quite low (below 1 μM), the MICs are essentially proportional to the amount of β-lactamase produced.

Table 2. P values in $nm.s^{-1}$

	E. coli [1]	*E. coli* [2]	*E. coli* [3]	*E.aerogenes* [4]	*S.marcescens* [5]	*E. cloacae* [4]	*P. aeruginosa* [4] PAO – 1
Cephaloridine	800	3600	5300[a], 6500[c]		570	200 (86, d)	(11, f)
Cefazolin	600	620	1600[a], 480[c]		370	(35, e)	
Cephalothin	90 – 180	110	290[c]				
Cefamandole	70	11					
Cefaclor	1000	2600					
Cefsulodin	10	300					
Cefoxitin	34	370					
Cefotaxime	65 - 100	180	175[b]	0.32	4.4	0.43 (3, d)	
Ceftazidime	13	96	95[b]			(1, d)	
Cephacetrile		1300*		18		42	(8, f)
Penicillin G	80 – 100	20		0.27		0.7	0.13
Ampicillin	170	980	1500	0.52		1.4	
Carbenicillin	0.06	7.5	11	0.035		0.13	(0.1, i)
Aztreonam	4.3	33	50	0.017		0.032	(0.05)
Imipenem		6800*					137, 18[g]

[1] [13].

[2] [5]; * these values were later determined by the same method although they are not listed in [5].

[3] Alternative values later reported in a) [68]; b) [69]; c) [70].

[4] [10]. The value for cephaloridine was obtained with the Zimmermann and Rosselet method with strain 908R. The values between parentheses are taken from other papers: d) [69]; e) [71]; f) [37]; g) [72]: the first value corresponds to that of the wild-type strain producing porin D1 and the second to that of the mutant lacking this specific channel; i) H. Nikaido, unpublished.

[5] [73].

When a new β-lactamase is identified, the gene is often cloned into a standard strain of *E. coli* and the MICs of the recombinant determined for the β-lactams whose K_m and k_{cat} values are also estimated. It would be interesting to systematically try to correlate the MICs with the enzyme kinetic parameters. Indeed, equation 5 shows that

$$\frac{P\,(MIC - I_{pl})\,(K_m + I_{pl})}{k_{cat}\,I_{pl}}$$

should be independent of the antibiotic and characterises the amount of β-lactamase produced. Incoherent results would suggest errors in the determination of the MIC or of the kinetic parameters or, if a systematic deviation is found for a particular compound, an error in the published value of P.

On the basis of such an analysis, it can be shown for example that the published MIC value for an *E. coli* strain producing an engineered β-lactamase would require a periplasmic concentration of 14 g/ml, a completely unrealistic situation [60].

1.2. The Values of P

The first data were obtained by measuring the rate of hydrolysis of rapidly hydrolysed cephalosporins such as cephacetrile and cephaloridine in cells producing an excess of β-lactamase [61]. However, many of the β-lactams of clinical interest are not readily hydrolysed by commonly-occuring β-lactamases. Thus, Nikaido and Normark used the data obtained by the liposome-swelling method [62] normalised to the P value of cephacetrile obtained in intact cells. However, comparison with the data obtained in intact cells showed that corrections were necessary for compounds with zero and two negative net charges [5]. These values were utilised by the same authors to calculate the MIC values for 4 strains producing increasing quantities of the AmpC β-lactamase and for another one producing the class A TEM-I on the basis of the kinetic parameters of the enzymes. The agreement was generally quite satisfying, which indicated that both the I_{pl} and P values introduced in eq 5 were quite reasonable. Frère [13] later designed a graphical method which allowed an estimation of I_{pl} and P on the basis of Nikaido and Normark's raw data, i.e. the MIC values, the kinetic parameters of the two β-lactamases and the amount and type of enzyme produced by each strain. As shown by Table 1, and as mentioned above, the I_{pl} values so derived were in excellent agreement with the MIC of strain SN03 and the C_{inh} values of Nikaido and Normark. For the P values, there were however some major discrepancies. As shown by Table 2, for three compounds the differences exceeded a factor of 10 (cefsulodin, cefoxitin and most strikingly, carbenicillin). For carbenicillin Nikaido and Normark acknowledged that "the true permeability of the compound could be much lower" since it was near the limit of sensitivity of the liposome swelling method. Alternatively, the value calculated by Frère might be, for some unknown reason, underevaluated although it gives a better agreement with the measured MICs. Note also that some of the Nikaido and Normark's values were later corrected (Table 2), but this did not improve the agreement. Not surprisingly, since they were derived only on the basis of the raw experimental data, with no independent measurement of P, the MIC values calculated by Frère were in somewhat better agreement with the observed ones. The differences might result from minor errors in the MIC determinations (for cefsulodin, the MICs are the same for LA5, LA51 and TE18 despite the fact that the latter produces 100 and 4 times more AmpC β-lactamase than LA5 and LA51, respectively) since it is quite evident that the technique of MIC determination makes it difficult to estimate the values within a factor of less than 2. However, the MICs of Nikaido and Normark's for strain SN03 are in good agreement with those determined by other authors with various *E. coli* strains devoid of β-lactamase. The major problem probably results from the difficulty of obtaining the P values.

With the sole exception of carbenicillin, even the lowest value (4 nm/s for aztreonam) corresponds to a k_D value of at least 0.05 s^{-1} which means that the half equilibration time between the external solution and the periplasm is less than 15 s. This shows that in *E. coli* a modification of permeability is very unlikely to be important, unless combined with efflux or a β-lactamase.

Lakaye et al [63] later proposed a direct method for measuring relatively low P values($\leq$ 40 nm/s) using living cells of *Enterobacter aerogenes*, *Enterobacter cloacae* and *Pseudomonas aeruginosa*. The method rests on the utilisation of a strain which produces a very low amount of β-lactamase (or not at all) but a rather large amount of a very sensitive penicillin-binding protein in the periplasm. The results are shown in Table 2 and were utilised to calculate the MIC values of the β-lactamase overproducing *E. cloacae* strain 908R. The agreement with the measured values was very good [10]. Table 2 also displays a few values obtained by more classical methods. In most cases, the values for both *Enterobacter* species are similar and significantly lower than for *E. coli.* There is again a discrepancy with carbenicillin, the values being lower than those of Nikaido and Normark's [5] but similar to that of Frère's [13]. The values for *P. aeruginosa* are similar to those of *Enterobacter*.

1.3. Presence of Several β-lactamases

When several β-lactamases are present, eq 5 becomes

$$\mathrm{MIC} = I_{pl} + \Sigma \frac{1}{PA} \frac{V\ I_{pl}}{K_m + I_{pl}} \qquad (7)$$

where each term in the sum corresponds to one particular β-lactamase.

1.4. The Viscosity of the Periplasm

The periplasm is a highly concentrated protein solution and it has been suggested that this would strongly decrease the diffusion rate of molecules [59]. Indeed, the diffusion coefficient of fluorescently labelled maltose binding protein (Mr = 40600) has been measured by Brass et al [64] and found to be 1000 fold lower than in water, with a D value of 0.9 x 10^{-10} cm^2s^{-1}. However, the linear mean free path <x> is given by $\sqrt{Dt}$ so that the molecule would migrate about 90 nm in 1 s.

2. The Influence of Efflux

If both a β-lactamase and an efflux system contribute to the MIC, eq 8 applies

$$MIC = I_{pl} + \frac{1}{PA}\left[\frac{V I_{pl}}{K_m + I_{pl}} + \frac{V_e I_{pl}}{K_{me} + I_{pl}}\right] \quad (8)$$

where K_{me} and V_e represent the kinetic parameters of the efflux system. However, it was not possible to determine V_e and K_{me} until recently. This has now been accomplished by measuring P in the presence of a proton conductor that de-energizes the efflux pump, and then by measuring β-lactam hydrolysis in intact cells (in the absence of the proton conductor), where the influx should equal the sum of periplasmic hydrolysis and active efflux from the periplasm [35, 36]. This analysis showed that Hill coefficients larger than 1 were observed for most penicillins and nitrocefin (among cephalosporins). For most cephalosporins, the efflux did not contribute much to resistance, because K_{me} values were so much higher than the values of I_p. Also this situation prevented us to test the validity of eq. 8 [36]. However, with many penicillins, K_{me} values were much lower (close to 1 μM), and eq. 8 (modified for the Hill kinetics) could be tested; it appeared to fit the results rather well. An interesting finding was that cloxacillin and ampicillin had nearly identical values of K_{me}, and V_e values were different only by a factor of 2.4, yet the genetic inactivation of AcrB efflux pump decreased the MIC of the former 256-fold but that of the latter only 2-fold. The analysis showed that this difference is caused mostly by the much lower outer membrane permeability (P) of isoxazolylpenicillins in comparison with ampicillin [35], a conclusion emphasizing the fact that efflux becomes efficient in increasing MIC values only when combined with an effective permeation barrier, a point that can be seen clearly from the form of eq. 8. These results also show that the extent of MIC decrease in pump-deficient mutants, often used as the criterion for the efficacy of efflux in the past, frequently gives very misleading impressions on the function of the pump itself.

3. In a Growing Culture

In a growing culture, new PBPs are continuously synthetised by the cells and a steady state where the proportion of acylated PBP remains constant is reached when the rate of new PBP synthesis is counterbalanced by the antibiotic-induced PBP inactivation [10]. The situation is complicated by the facts that the primary, lethal targets of different antibiotics might be different PBPs and that the exact proportion of each PBP which must remain active to support cell survival and division is not known. Computer-aided simulations then become a very useful tool. Lepage et al [65] could correlate the effects of various β-lactams on growing cultures of a non β-lactamase producing mutant of *Bacillus licheniformis* and showed that PBP1 was the primary target of these antibiotics (but see below). However, the behaviour of other β-lactams such as cloxacillin remained completely unexplained. Lakaye et al [10] have utilised a similar method to simulate the penetration of cefotaxime and aztreonam in the periplasm of growing *E. coli* and *E. cloacae* cells and their inactivation of PBP3. They also measured the proportion of acylated PBPs in the cultures and concluded that about 66% of PBP3 was acylated at a cefotaxime concentration corresponding to the MIC while the other essential PBPs were not significantly acylated under these conditions. However, to yield this proportion of acylated PBP3 at the steady state (which was reached after about 30 min), the k_2/K value which was used in the simulations was about 10% of that measured for membrane preparation of PBP3. This underlined a "crypticity factor", also noticed by Reynolds et al [66] and Lepage et al [65], which might result from a protection of the "working" PBP by its substrates.

This discussion underlines the difficulties which are encountered if one wishes to rationalize the I_{pl} values on the basis of the properties of PBPs.

4. INDUCTION

Equations 5, 7 and 8 are only valid for constitutive producers of β-lactamases. Inducible β-lactamase production significantly complicates the problem and the characteristic differential equations cannot be integrated. Duval et al [67] have tried to simulate the situation in a growing, inducible strain of the Gram positive *Bacillus licheniformis* but the major conclusion was that the sole acylation of the penicillin sensor BlaR (see chapter 17) was not sufficient to induce derepression of the β-lactamase gene. Some alteration of peptidoglycan metabolism also appears to be necessary (see chapter 17).

To our knowledge, no mathematical model has been proposed to rationalize the MIC values of inducible strains.

REFERENCES

[1] Frère, J.M. (1995) Beta-lactamases and bacterial resistance to antibiotics. *Mol Microbiol* 3, 385-95.

[2] Sauvage, E., Kerff, F., Terrak, M., Ayala, J.A., Charlier, P. (2008) The penicillin-binding proteins: structure and role in peptidoglycan biosynthesis. *FEMS Microbiol Rev* 2, 234-58.

[3] Frère, J.M., Joris, B. (1985) Penicillin-sensitive enzymes in peptidoglycan biosynthesis. *Crit Rev Microbiol* 4, 299-396.

[4] Zapun, A., Contreras-Martel, C., Vernet, T. (2008) Penicillin-binding proteins and beta-lactam resistance. *FEMS Microbiol Rev* 2, 361-85.

[5] Nikaido, H., Normark, S. (1987) Sensitivity of *Escherichia coli* to various beta-lactams is determined by the interplay of outer membrane permeability and degradation by periplasmic beta-lactamases: a quantitative predictive treatment. *Mol Microbiol* 1, 29-36.

[6] Niederweis, M., Danilchanka, O., Huff, J., Hoffmann, C., Engelhardt, H. (2010) Mycobacterial outer membranes: in search of proteins. *Trends Microbiol* 3, 109-16.

[7] Dreier, J. (2007) Active efflux in bacteria. In: *Enzyme-mediated resistance to antibiotics*. R. A. Bonomo. ASM Press, Washington DC. 235-264.

[8] Feller, G., Zekhnini, Z., Lamotte-Brasseur, J., Gerday, C. (1997) Enzymes from cold-adapted microorganisms. The class C beta-lactamase from the antarctic psychrophile *Psychrobacter immobilis* A5. *Eur J Biochem* 1, 186-91.

[9] Yamamoto, S., Lampen, J.O. (1976) Purification of plasma membrane penicillinase from *Bacillus licheniformis* 749/C and comparison with exoenzyme. *J Biol Chem* 13, 4095-101.

[10] Lakaye, B., Dubus, A., Lepage, S., Groslambert, S., Frere, J.M. (1999) When drug inactivation renders the target irrelevant to antibiotic resistance: a case story with beta-lactams. *Mol Microbiol* 1, 89-101.

[11] Zimmermann, W., Rosselet, A. (1977) Function of the outer membrane of *Escherichia coli* as a permeability barrier to beta-lactam antibiotics. *Antimicrob Agents Chemother* 3, 368-72.

[12] Waley, S.G. (1987) An explicit model for bacterial resistance: application to beta-lactam antibiotics. *Microbiol Sci* 5, 143-6.

[13] Frère, J.M. (1989) Quantitative relationship between sensitivity to beta-lactam antibiotics and beta-lactamase production in gram-negative bacteria--I. Steady-state treatment. *Biochem Pharmacol* 9, 1415-26.

[14] Nikaido, H. (2003) Molecular basis of bacterial outer membrane permeability revisited. *Microbiol Mol Biol Rev* 4, 593-656.

[15] Cowan, S.W., Schirmer, T., Rummel, G., Steiert, M., Ghosh, R., Pauptit, R.A., Jansonius, J.N., Rosenbusch, J.P. (1992) Crystal structures explain functional properties of two *E. coli* porins. *Nature* 6389, 727-33.

[16] Ochs, M.M., Bains, M., Hancock, R.E. (2000) Role of putative loops 2 and 3 in imipenem passage through the specific porin OprD of *Pseudomonas aeruginosa*. *Antimicrob Agents Chemother* 7, 1983-5.

[17] Dupont, M., Pages, J.M., Lafitte, D., Siroy, A., Bollet, C. (2005) Identification of an OprD homologue in *Acinetobacter baumannii*. *J Proteome Res* 6, 2386-90.

[18] Lister, P.D., Wolter, D.J., Hanson, N.D. (2009) Antibacterial-resistant *Pseudomonas aeruginosa*: clinical impact and complex regulation of chromosomally encoded resistance mechanisms. *Clin Microbiol Rev* 4, 582-610.

[19] Pagani, L., Landini, P., Luzzaro, F., Debiaggi, M., Romero, E. (1990) Emergence of cross-resistance to imipenem and other beta-lactam antibiotics in *Pseudomonas aeruginosa* during therapy. *Microbiologica* 1, 43-53.

[20] Pages, J.M., James, C.E., Winterhalter, M. (2008) The porin and the permeating antibiotic: a selective diffusion barrier in Gram-negative bacteria. *Nat Rev Microbiol* 12, 893-903.

[21] Delcour, A.H. (2009) Outer membrane permeability and antibiotic resistance. *Biochim Biophys Acta* 5, 808-16.

[22] Cornaglia, G., Guan, L., Fontana, R., Satta, G. (1992) Diffusion of meropenem and imipenem through the outer membrane of *Escherichia coli* K-12 and correlation with their antibacterial activities. *Antimicrob Agents Chemother* 9, 1902-8.

[23] Quale, J., Bratu, S., Landman, D., Heddurshetti, R. (2003) Molecular epidemiology and mechanisms of carbapenem resistance in *Acinetobacter baumannii* endemic in New York City. *Clin Infect Dis* 2, 214-20.

[24] Clark, R.B. (1996) Imipenem resistance among *Acinetobacter baumannii*: association with reduced expression of a 33-36 kDa outer membrane protein. *J Antimicrob Chemother* 2, 245-51.

[25] Limansky, A.S., Mussi, M.A., Viale, A.M. (2002) Loss of a 29-kilodalton outer membrane protein in *Acinetobacter baumannii* is associated with imipenem resistance. *J Clin Microbiol* 12, 4776-8.

[26] Poole, K. (2002) Outer membranes and efflux: the path to multidrug resistance in Gram-negative bacteria. *Curr Pharm Biotechnol* 2, 77-98.

[27] Livermore, D.M., Davy, K.W. (1991) Invalidity for *Pseudomonas aeruginosa* of an accepted model of bacterial permeability to beta-lactam antibiotics. *Antimicrob Agents Chemother* 5, 916-21.

[28] Li, X.Z., Ma, D., Livermore, D.M., Nikaido, H. (1994) Role of efflux pump(s) in intrinsic resistance of *Pseudomonas aeruginosa*: active efflux as a contributing factor to beta-lactam resistance. *Antimicrob Agents Chemother* 8, 1742-52.

[29] Li, X.Z., Nikaido, H., Poole, K. (1995) Role of mexA-mexB-oprM in antibiotic efflux in *Pseudomonas aeruginosa. Antimicrob Agents Chemother* 9, 1948-53.

[30] Nikaido, H. (1998) Antibiotic resistance caused by gram-negative multidrug efflux pumps. *Clin Infect Dis* S32-41.

[31] Murakami, S., Nakashima, R., Yamashita, E., Yamaguchi, A. (2002) Crystal structure of bacterial multidrug efflux transporter AcrB. *Nature* 6907, 587-93.

[32] Nikaido, H., Basina, M., Nguyen, V., Rosenberg, E.Y. (1998) Multidrug efflux pump AcrAB of *Salmonella typhimurium* excretes only those beta-lactam antibiotics containing lipophilic side chains. *J Bacteriol* 17, 4686-92.

[33] Mazzariol, A., Cornaglia, G., Nikaido, H. (2000) Contributions of the AmpC beta-lactamase and the AcrAB multidrug efflux system in intrinsic resistance of *Escherichia coli* K-12 to beta-lactams. *Antimicrob Agents Chemother* 5, 1387-90.

[34] Murakami, S., Nakashima, R., Yamashita, E., Matsumoto, T., Yamaguchi, A. (2006) Crystal structures of a multidrug transporter reveal a functionally rotating mechanism. *Nature* 7108, 173-9.

[35] Lim, S.P., Nikaido, H. (2010) Kinetic parameters of efflux of penicillins by the multidrug efflux transporter AcrAB-TolC of *Escherichia coli. Antimicrob Agents Chemother* 5, 1800-6.

[36] Nagano, K., Nikaido, H. (2009) Kinetic behavior of the major multidrug efflux pump AcrB of *Escherichia coli. Proc Natl Acad Sci U S A* 14, 5854-8.

[37] Yoshimura, F., Nikaido, H. (1982) Permeability of *Pseudomonas aeruginosa* outer membrane to hydrophilic solutes. *J Bacteriol* 2, 636-42.

[38] Angus, B.L., Carey, A.M., Caron, D.A., Kropinski, A.M., Hancock, R.E. (1982) Outer membrane permeability in *Pseudomonas aeruginosa*: comparison of a wild-type with an antibiotic-supersusceptible mutant. *Antimicrob Agents Chemother* 2, 299-309.

[39] Schulz, G.E. (1993) Bacterial porins: structure and function. *Curr Opin Cell Biol* 4, 701-7.

[40] Poole, K. (2005) Efflux-mediated antimicrobial resistance. *J Antimicrob Chemother* 1, 20-51.

[41] Poole, K., Krebes, K., McNally, C., Neshat, S. (1993) Multiple antibiotic resistance in *Pseudomonas aeruginosa*: evidence for involvement of an efflux operon. *J Bacteriol* 22, 7363-72.

[42] Okamoto, K., Gotoh, N., Nishino, T. (2002) Alterations of susceptibility of *Pseudomonas aeruginosa* by overproduction of multidrug efflux systems, MexAB-OprM, MexCD-OprJ, and MexXY/OprM to carbapenems: substrate specificities of the efflux systems. *J Infect Chemother* 4, 371-3.

[43] Poole, K., Gotoh, N., Tsujimoto, H., Zhao, Q., Wada, A., Yamasaki, T., Neshat, S., Yamagishi, J., Li, X.Z., Nishino, T. (1996) Overexpression of the mexC-mexD-oprJ efflux operon in nfxB-type multidrug-resistant strains of *Pseudomonas aeruginosa. Mol Microbiol* 4, 713-24.

[44] Poole, K. (2004) Efflux-mediated multiresistance in Gram-negative bacteria. *Clin Microbiol Infect* 1, 12-26.

[45] Hoffmann, C., Leis, A., Niederweis, M., Plitzko, J.M., Engelhardt, H. (2008) Disclosure of the mycobacterial outer membrane: cryo-electron tomography and vitreous sections reveal the lipid bilayer structure. *Proc Natl Acad Sci U S A* 105, 3963-7.

[46] Danilchanka, O., Mailaender, C., Niederweis, M. (2008) Identification of a novel multidrug efflux pump of *Mycobacterium tuberculosis*. *Antimicrob Agents Chemother* 7, 2503-11.

[47] Amicosante, G., Franceschini, N., Segatore, B., Oratore, A., Fattorini, L., Orefici, G., Van Beeumen, J., Frere, J.M. (1990) Characterization of a beta-lactamase produced in *Mycobacterium fortuitum* D316. *Biochem J* 3, 729-34.

[48] Quinting, B., Reyrat, J.M., Monnaie, D., Amicosante, G., Pelicic, V., Gicquel, B., Frere, J.M., Galleni, M. (1997) Contribution of beta-lactamase production to the resistance of mycobacteria to beta-lactam antibiotics. *FEBS Lett* 3, 275-8.

[49] Bonomo, R.A., Szabo, D. (2006) Mechanisms of multidrug resistance in *Acinetobacter species* and *Pseudomonas aeruginosa*. *Clin Infect Dis* S49-56.

[50] Rice, L.B. (2009) The clinical consequences of antimicrobial resistance. *Curr Opin Microbiol* 5, 476-81.

[51] Croft, A.C., D'Antoni, A.V., Terzulli, S.L. (2007) Update on the antibacterial resistance crisis. *Med Sci Monit* 6, RA103-18.

[52] Poirel, L., Heritier, C., Tolun, V., Nordmann, P. (2004) Emergence of oxacillinase-mediated resistance to imipenem in *Klebsiella pneumoniae*. *Antimicrob Agents Chemother* 1, 15-22.

[53] Carrer, A., Poirel, L., Yilmaz, M., Akan, O.A., Feriha, C., Cuzon, G., Matar, G., Honderlick, P., Nordmann, P. (2010) Spread of OXA-48-encoding plasmid in Turkey and beyond. *Antimicrob Agents Chemother* 3, 1369-73.

[54] Docquier, J.D., Calderone, V., De Luca, F., Benvenuti, M., Giuliani, F., Bellucci, L., Tafi, A., Nordmann, P., Botta, M., Rossolini, G.M., et al. (2009) Crystal structure of the OXA-48 beta-lactamase reveals mechanistic diversity among class D carbapenemases. *Chem Biol* 5, 540-7.

[55] Cuzon, G., Naas, T., Bogaerts, P., Glupczynski, Y., Huang, T.D., Nordmann, P. (2008) Plasmid-encoded carbapenem-hydrolyzing beta-lactamase OXA-48 in an imipenem-susceptible *Klebsiella pneumoniae* strain from Belgium. *Antimicrob Agents Chemother* 9, 3463-4.

[56] Gulmez, D., Woodford, N., Palepou, M.F., Mushtaq, S., Metan, G., Yakupogullari, Y., Kocagoz, S., Uzun, O., Hascelik, G., Livermore, D.M. (2008) Carbapenem-resistant *Escherichia coli* and *Klebsiella pneumoniae* isolates from Turkey with OXA-48-like carbapenemases and outer membrane protein loss. *Int J Antimicrob Agents* 6, 523-6.

[57] Vatopoulos, A. (2008) *High rates of metallo-beta-lactamase-producing Klebsiella pneumoniae in Greece--a review of the current evidence.* Euro Surveill 4.

[58] Chen, L.R., Zhou, H.W., Cai, J.C., Zhang, R., Chen, G.X. (2009) Combination of IMP-4 metallo-beta-lactamase production and porin deficiency causes carbapenem resistance in a *Klebsiella oxytoca* clinical isolate. *Diagn Microbiol Infect Dis* 2, 163-7.

[59] Oliver, D. (1996) Periplasm in *Escherichia coli* and *Salmonella*. Editor. ASM Press, Washington DC. 88-103.

[60] Park, H.S., Nam, S.H., Lee, J.K., Yoon, C.N., Mannervik, B., Benkovic, S.J., Kim, H.S. (2006) Design and evolution of new catalytic activity with an existing protein scaffold. *Science* 5760, 535-8.

[61] Nikaido, H., Rosenberg, E.Y., Foulds, J. (1983) Porin channels in *Escherichia coli*: studies with beta-lactams in intact cells. *J Bacteriol* 1, 232-40.

[62] Yoshimura, F., Nikaido, H. (1985) Diffusion of beta-lactam antibiotics through the porin channels of *Escherichia coli* K-12. *Antimicrob Agents Chemother* 1, 84-92.

[63] Lakaye, B., Dubus, A., Joris, B., Frere, J.M. (2002) Method for estimation of low outer membrane permeability to beta-lactam antibiotics. *Antimicrob Agents Chemother* 9, 2901-7.

[64] Brass, J.M., Higgins, C.F., Foley, M., Rugman, P.A., Birmingham, J., Garland, P.B. (1986) Lateral diffusion of proteins in the periplasm of *Escherichia coli*. *J Bacteriol* 3, 787-95.

[65] Lepage, S., Lakaye, B., Galleni, M., Thamm, I., Crine, M., Groslambert, S., Frere, J.M. (1995) Saturation of penicillin-binding protein 1 by beta-lactam antibiotics in growing cells of *Bacillus licheniformis*. *Mol Microbiol* 2, 365-72.

[66] Reynolds, P.E., Shepherd, S.T., Chase, H.A. (1978) Identification of the binding protein which may be the target of penicillin action in *Bacillus megaterium*. *Nature* 5645, 568-70.

[67] Duval, V., Swinnen, M., Lepage, S., Brans, A., Granier, B., Franssen, C., Frere, J.M., Joris, B. (2003) The kinetic properties of the carboxy terminal domain of the *Bacillus licheniformis* 749/I BlaR penicillin-receptor shed a new light on the derepression of beta-lactamase synthesis. *Mol Microbiol* 6, 1553-64.

[68] Liu, W., Nikaido, H. (1991) Contribution of the cell-surface-associated enzyme in the Zimmermann-Rosselet assay of outer membrane permeability of beta-lactam antibiotics. *Antimicrob Agents Chemother* 1, 177-9.

[69] Nikaido, H., Liu, W., Rosenberg, E.Y. (1990) Outer membrane permeability and beta-lactamase stability of dipolar ionic cephalosporins containing methoxyimino substituents. *Antimicrob Agents Chemother* 2, 337-42.

[70] Sen, K., Hellman, J., Nikaido, H. (1988) Porin channels in intact cells of *Escherichia coli* are not affected by Donnan potentials across the outer membrane. *J Biol Chem* 3, 1182-7.

[71] Vu, H., Nikaido, H. (1985) Role of beta-lactam hydrolysis in the mechanism of resistance of a beta-lactamase-constitutive *Enterobacter cloacae* strain to expanded-spectrum beta-lactams. *Antimicrob Agents Chemother* 3, 393-8.

[72] Trias, J., Dufresne, J., Levesque, R.C., Nikaido, H. (1989) Decreased outer membrane permeability in imipenem-resistant mutants of *Pseudomonas aeruginosa*. *Antimicrob Agents Chemother* 8, 1202-6.

[73] Raimondi, A., Sisto, F., Nikaido, H. (2001) Mutation in *Serratia marcescens* AmpC beta-lactamase producing high-level resistance to ceftazidime and cefpirome. *Antimicrob Agents Chemother* 8, 2331-9.

Part 4: B-Lactamases as Model Enzymes and Tools

In: Beta-Lactamases
Editor: Jean-Marie Frère, pp. 421-435
ISBN: 978-1-61324-638-2

Chapter 17

THE INDUCTION OF β-LACTAMASES IN EUBACTERIA

Bernard Joris and Jean Dusart
Centre d'Ingénierie des Protéines
Institut de chimie B6a, Université de Liège, Sart-Tilman,
B4000 Liège, Belgium.

ABSTRACT

The production of a β-lactamase can be induced by the presence of a β-lactam in the growth medium. Three distinct mechanisms have been discovered. In *Bacillus licheniformis* and *Staphylococcus aureus*, a penicillin receptor whose properties are similar to those of a PBP detects the presence of the antibiotic. In *Citrobacter freundii*, the damage caused by β-lactams to the peptidoglycan is detected by an intracellular protein and a similar system might prevail in *Streptomyces cacaoi*. Finally, in *Aeromonas sp* regulation involves a two-component system.

1. INTRODUCTION

In some Gram-positive or -negative bacteria, the presence of a β-lactam antibiotic in the external medium induces the biosynthesis of a β-lactamase that hydrolyses this antibiotic and allows bacterial survival [1]. In *Bacillus licheniformis* and *Staphylococcus aureus*, the mechanism of induction involves similar genes and results in the induction of a class A β-lactamase [2]. This mechanism is also related to that which leads to the induction of the resistant PBB2a in methicillin resistant *S. aureus* (MRSA) [3]. In *Citrobacter freundii*, the presence of penicillin induces the production of a class C β-lactamase. It is one of the two mechanisms observed in Gram-negative bacteria [4]. The second one was described in *Aeromonas sp.* and it is not related to the *C. freundii* one [5-6]. On the contrary, an induction mechanism which seems to be related to the *C. freundii* mechanism was also found in *Streptomyces cacaoi*. Finally, it is reported in literature that the two β-lactamases present in *Bacillus cereus* are inducible and that an extracytoplasmic function sigma factor and its anti-sigma factor control their expression [7].

The physico-chemical properties of the bacterial cytoplasmic membrane and those of the β-lactam antibiotics make the passage of these antibiotics through this membrane very difficult. As a consequence, in all induction phenomena listed above, the different microorganisms have developed systems that detect the antibiotic in the external medium and transmit into the cytoplasm a signal targeting the regulatory protein that controls the expression of the gene coding for the β-lactamase. The following paragraphs give a synthesis of our current knowledge about these induction mechanisms.

2. Induction Mechanism of the BlaP β-lactamase in *Bacillus licheniformis* and of the BlaZ β-lactamase and PBP2a in *Staphylococcus aureus*

2.1. *Bacillus licheniformis*

At least three genes, *blaI*, *blaR*1 and *blaR*2, are involved in the induction of the expression of the *blaP* β-lactamase gene of *B. licheniformis*. The *blaP*, *blaI* and *blaR*1 genes form a divergeon (*bla* divergeon) in which the latter two genes constitute an operon (Figure 1A). The *blaI* gene codes for a cytoplasmic protein of 128 amino acids that binds to the DNA, and the *blaR*1 gene for a membrane protein of 601 amino acids, a sensor receptor able to detect penicillin. The *blaR*2 gene, situated in another locus distant from the *bla* divergeon, remains to be isolated, but it has been postulated that the BlaR2 protein would play a role in the cytoplasmic signal between the BlaR1 sensor and the regulatory protein BlaI [1].

In the absence of inducing antibiotic, BlaI, as a dimer (Kd = 25 μM, [8]), recognises and binds specifically to three inverted repeats (operator regions or BlaI operator) located in the intergenic region between the *blaP* gene and the *blaI-blaR*1 operon. Two of these operator regions are localised in the *blaP* promoter (P*blaP*) and the third one is in the promoter of the operon (P*blaI-R*1) (Figure 1A). When bound to DNA, BlaI acts as a repressor and regulates the expression of *blaP* and of the *blaI-blaR*1 operon. However, the expression of the β-lactamase gene is repressed more tightly than that of the operon. This is easily explainable by the presence of two operator regions in P*blaP* instead of one in P*blaI-R*1, and by a higher affinity of BlaI for the operator regions of P*blaP* [8]. So, in the absence of inducer, "leakage" from the P*blaI-R*1 promoter repressed by BlaI is more important than leakage from P*blaP* and the amount of BlaI is sufficient to maintain the production of β-lactamase at a basal level, and allow the presence of the BlaR1 sensor in the membrane [8]. The tridimensional structure of the DNA-recognising N-terminal domain of BlaI (Bla-NTD, residues 1 to 82) was determined by RMN and showed that this repressor belongs to the winged helix family of repressors. The 2D NMR spectrum of N^{15}-labeled BlaI shows that only the N-terminal domain (1-82) gives a well dispersed picture, whereas the C-terminal domain (83-128, BlaI-CTD) is poorly defined and appears in the same 2D spectrum zone as an unstructured polypeptidic chain [9]. The circular dichroism spectrum of the native protein under the same conditions or of BlaI-CTD purified in the presence of detergent shows that in solution, this domain is well structured. The observation by NMR of a non structured C-terminal domain could be explained by a more dynamic structure of this part of the molecule. Study of the denaturation of BlaI by guadinium chloride (GdnHCl) confirms that BlaI is indeed composed of two domains which are denatured independently and that the C-terminal domain is denatured at a lower concentration of the denaturing agent ($[GdnHCl]_{1/2}$ = 1.75 M) than the N-terminal domain ($[GdnHCl]_{1/2}$ = 4 M). This property is in agreement with a more dynamic C-terminal domain, as suggested by NMR experiments [10]. Another peculiar feature of the dimerisation domain is its importance in the affinity of BlaI for its operator sequence. In the case of purified BlaI-NTD, the affinity for the DNA target is decreased 10,000 fold (Kd = 200 μM) and in the whole BlaI protein, denaturation of the C-terminal domain by GdnHCl leads to the loss of the high affinity for DNA. These results support the importance of the C-terminal domain for the inactivation of BlaI during the induction process [11].

By determining the organisation of BlaR1 in the membrane, it was shown that this protein is constituted of two domains, an N-terminal domain containing four transmembrane segments (BlaR-NTD, residues 1 to 350, the NH_2 terminal end of which is exposed at the outside of the cell) and a C-terminal domain (BlaR-CTD, residues 351-601) localised exclusively outside the cell (Figure 2). The amino acid sequence of BlaR-CTD is homologous to those of the class D β-lactamases. The four transmembrane segments of BlaR-NTD delimit three loops; L1 (9 aa), L2 (63 aa) and L3 (189 aa), L1 and L3 being intracellular, and L2 extracellular. In L3, an H^{212}-E-X-X-H motif has been discovered which is the signature of the zinc proteases of the thermolysin family [12]. Directed mutagenesis and radioactive zinc binding experiments demonstrated that this motif binds a Zn^{2+} ion and that it is important for the mechanism of induction of the β-lactamase (K. Benlafya, S. Berzigotti and B. Joris unpublished data). When the part of the gene coding for BlaR-CTD (separated from BlaR-NTD by genetic engineering) is introduced in *E. coli*, and placed under the control of transcription and export signals proper to this host, the BlaR-CTD domain is produced under a soluble form in the periplasm. Moreover, BlaR-CTD produced and purified from this source is able to recognise and form very stable covalent adducts with β-lactam antibiotics [13]. The topological description of BlaR1 corresponds to the general organisation of a receptor in which BlaR-CTD plays the role of a sensor, able to detect penicillin. The transmembrane segments and the L2 loop constitute the transducing module, the L3 loop acting as a transmitter that amplifies in the cytoplasm the signal given by BlaR-CTD when the latter is acylated by penicillin. Interaction between BlaR-CTD, immobilised in an ELISA test well, and a phage displayed L2 loop was observed in the absence of penicillin. It was also shown that this interaction is weakened in the presence of the inducing antibiotic [14]. The tridimensional structure of BlaR-CTD from *B. licheniformis* has been determined and, as expected, it was found to be very similar to the 3D-structure of a class D β-lactamase (Figure 3, [15]).

Cloning of the *blaP-blaI-blaR*1 divergeon on a *Bacillus sp* plasmid and its transfer into *Bacillus subtilis* 168 (a strain which produces no β-lactamase) have shown that in this recombinant strain, the induction of BlaP β-lactamase is possible. The existence of the *blaR*2 gene being well established in *B. licheniformis*, it must be assumed that this gene is also present in *B. subtilis* 168, and that it exerts a cellular function in all *Bacillus* species. Given that the *B. licheniformis* strain is hardly transformable, all the studies of the induction mechanism of the BlaP β-lactamase were carried out mainly in *B. subtilis* [16].

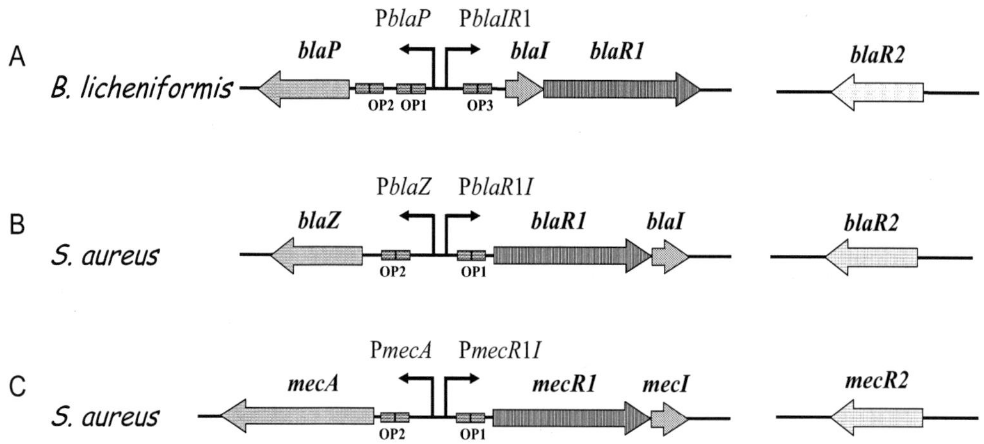

Figure 1. Organisation of the chromosomal genes involved in the induction of the BlaP β-lactamase of *B. licheniformis* (A) and the BlaZ beta-lactamase (*blaZ* gene)/PBP2a (*mecA* gene) in *S. aureus* (B and C). A: in *B. licheniformis*, the expression of the gene coding for the BlaP β-lactamase is regulated by the products of at least three genes: *blaI*, *blaR*1 and *blaR*2. *blaI* and *blaR*1 form an operon. The BlaI protein is a cytoplasmic protein, that binds DNA and recognises specifically three operator regions (OP1, OP2 and OP3), situated in the intergenic region of the *blaP-blaI-blaR*1 divergeon. The *blaR*2 gene is not identified yet and its inactivation gives rise to a non inducible mutant. See text for more details. B and C: on the *S. aureus* chromosome, *blaI/mecI* and *blaR*1/*mecR*1 are organised in an operon and form a divergeon with the gene encoding the β-lactamase or the resistant PBP. In this strain the order of the genes in the operons differs from that in *B. licheniformis*. Two operator sequences (OP1and OP2) recognised by BlaI/MecI are present in the intergenic region between *blaZ/mecA* and *blaR*1/*mecR*1.

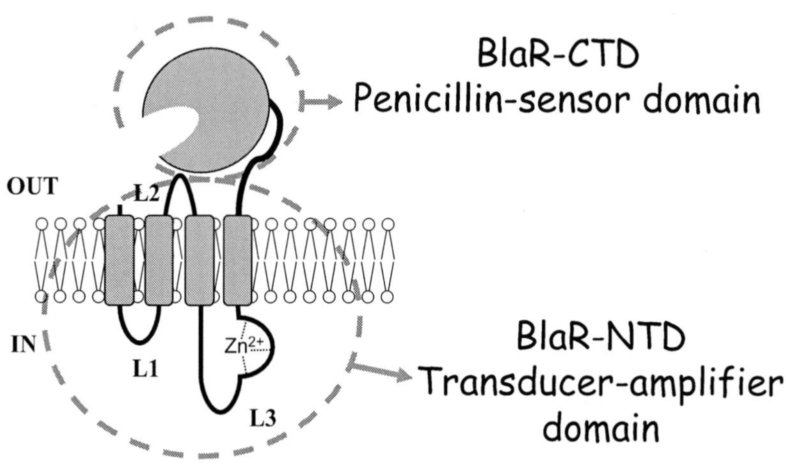

Figure 2. Membrane organisation of the *B. licheniformis* BlaR1 penicillin receptor.
BlaR-CTD acts as a penicillin sensor, whereas BlaR-NTD plays the role of a transducer-transmitter of the signal generated by the acylation of BlaR-CTD in the presence of the inducing antibiotic.
Cylinders : transmembrane segments. L1, L2 and L3: BlaR-NTD loops. See text for more details.

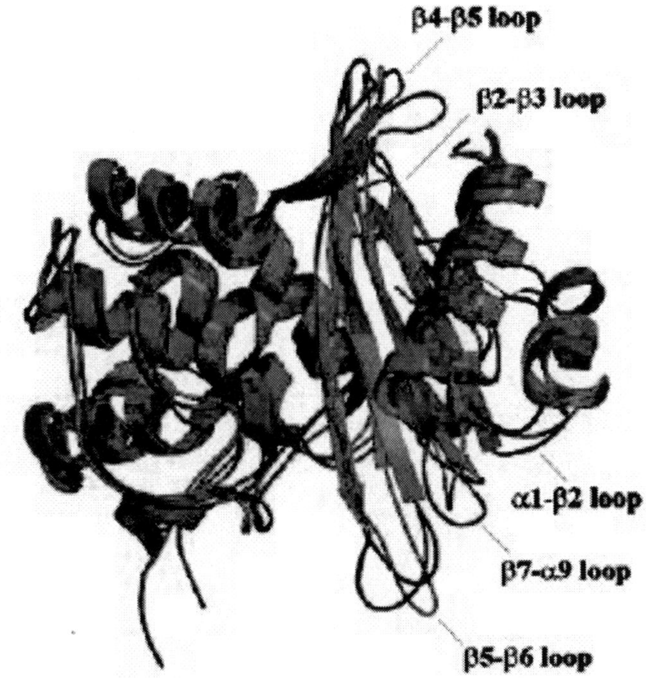

Figure 3. Surimposed ribbon diagrams of BlaR-CTD (magenta) and the class D beta-lactamases OXA-10 (green) and OXA-1 (red)[15].

At the present time, the induction mechanism of the *B. licheniformis* BlaP β-lactamase can be summarized as follows. When a β-lactam antibiotic is present in the external medium, it is detected by BlaR-CTD. The acylation of the active serine of BlaR-CTD results in a change of the interaction between BlaR-CTD and the L2 loop, which would lead to a reorganisation of the transmembrane segments and to a conformational change of the cytoplasmic L3 loop, activating the enzyme activity of this latter. The activation of L3 would be obtained by a cleavage of its polypeptidic chain, between residues R^{293} and R^{294} [20], as in the case of the induction of the BlaZ β-lactamase of *S. aureus* (see point 2.2, Figure 4). It is not clearly established yet if this cleavage is due to an autolysis or to the action of another cytoplasmic protease. This type of enzyme activation is found in several proteases, which are produced first as a proenzyme (for example thermolysin [17]). In the case of the induction of the BlaP β-lactamase in *B. licheniformis*, the activated L3 loop would cleave a still unknown pro-coactivator generating a coactivator of BlaI. When complexed by its coactivator, the latter, would be unable to bind DNA thus allowing the induction of BlaP. The coactivator of BlaI has not been identified yet but its presence has been indirectly demonstrated by following the fate of BlaI during the induction, in both *B. subtilis* and *B. licheniformis*. In the latter strain, BlaI was degraded during the induction, whereas in the former, BlaI is neither proteolysed nor modified, but loses its DNA-binding capacity. These observations in *B. subtilis* are the basis of the assumption that a coactivator is involved in the inactivation of BlaI [18]. The proteolysis of BlaI in the cytoplasm of *B. licheniformis* would be the consequence of the formation of the BlaI-coactivator complex, followed by its cleavage by some cytoplasmic proteases.

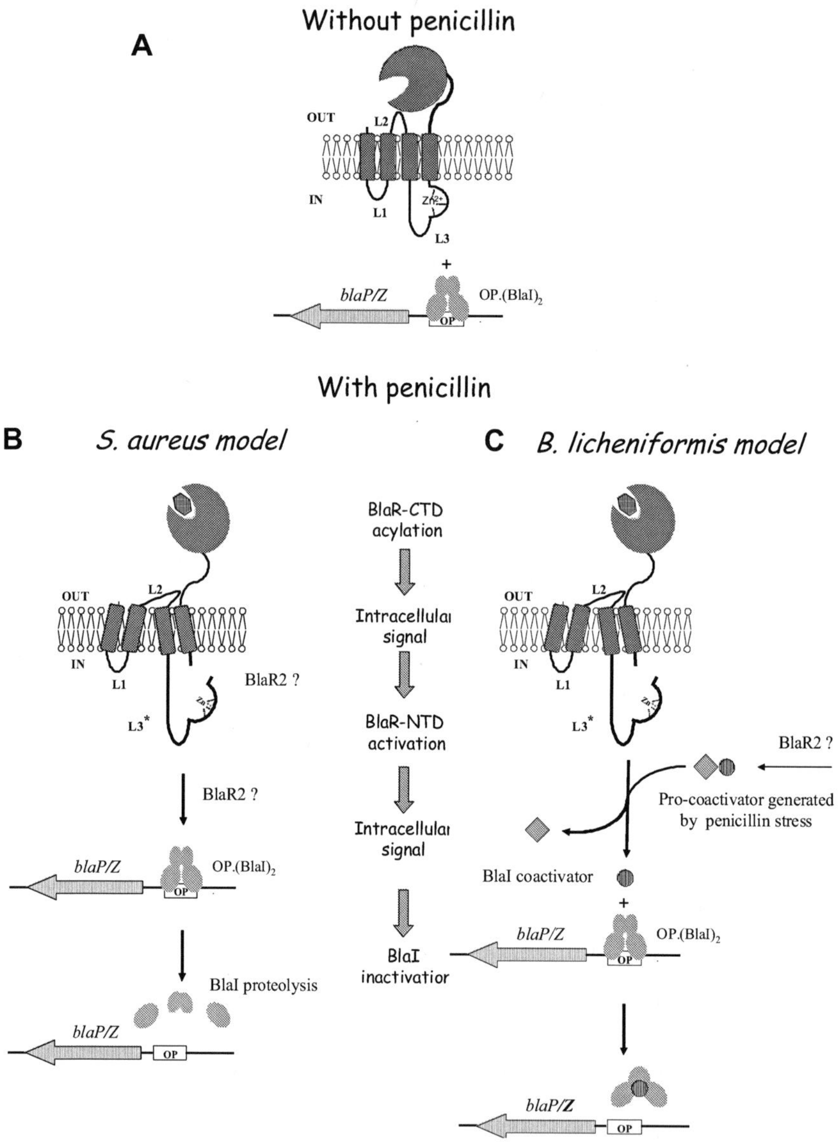

Figure 4. Induction mechanism of the BlaP and BlaZ β-lactamases in *B. licheniformis* and *S. aureus*, respectively. In both microorganisms, products of the *blaI* and *blaR*1 genes involved in the induction of the BlaP and BlaZ β-lactamases are similar and play the same role in the induction mechanism. BlaI is a DNA-binding protein acting as a repressor and BlaR1 is a membrane protein that plays the role of penicillin receptor. A: in absence of penicillin outside the cell B and C: the major difference between the two induction models resides in the mode of inactivation of the BlaI repressor. In *B. licheniformis*, it is proposed that BlaI is inactivated by the linking of a coactivator, whereas in *S. aureus*, the hypothesis is that BlaI is inactivated by proteolysis. In both organisms, the *blaR*2 gene remains unidentified. It is proposed that the BlaR2 protein could play a role in the cleavage of the large cytoplasmic loop of the BlaR receptor when the latter is acylated by the inducer or could alternatively be involved in the production of the proactivator in *B. licheniformis* or in the proteolysis of BlaI in *S. aureus*.

A difficulty in the study of the induction process of a β-lactamase resides in the fact that the inducing antibiotic is also a substrate of the induced enzyme. That is why the inducer disappears more or less rapidly depending on the hydrolytic efficiency of the enzyme towards the inducing antibiotic. Consequently, in order to increase the half-life of the inducer, a poor substrate of the induced β-lactamase can be used as inducer in the study of the induction phenomenon. In the case of the BlaR-CTD receptor, the kinetic data obtained about the acylation of BlaR-CTD show that 7-amino-cephalosporanic acid (7-ACA), a cephalosporin deprived of C7 acyl side chain, is both a good acylating agent for BlaR-CTD and a very poor substrate for the β-lactamase. The analysis of the results obtained by using 7-ACA as a substrate, leads to the conclusion that two conditions are necessary to induce the β-lactamase: firstly, the BlaR receptor must be acylated by the inducer, and secondly, a cellular stress caused by the presence of this antibiotic inducer must be generated (2). At concentrations in 7-ACA allowing the acylation of BlaR (around 2 μM) but having no effect on the growth of *B. licheniformis*, no induction of β-lactamase was observed. When the 7-ACA concentration affects the bacterial growth ([7-ACA] = 18,5 μM), the β-lactamase is induced. The kinetic constants of acylation of the PBP in *B. licheniformis* indicate that at a 7-ACA concentration allowing the induction of the β-lactamase, PBP1, the lethal target of this bacterial strain, is more than 70 % inhibited which would be the cause of the cellular stress. Similar experiments carried out with cephalexin, an antibiotic specific of the PBP3 of *B. licheniformis*, lead to the same conclusion: induction only occurs when PBP1 is acylated. This takes place at a concentration higher than that necessary to inhibit PBP3 [2]. The cellular stress generated by the presence of the inducing antibiotic would result in the production and the accumulation in the bacterial cytoplasm of a molecule which would be the pro-coactivator of BlaI and whose hydrolysis by BlaR1 would yield the coactivator necessary for interfering with the interaction between BlaI and its nucleic operator.

2.2. *Staphylococus aureus*

In *S. aureus*, the products of the *blaI* and *blaR*1 genes involved in the induction of the *blaZ* β-lactamase gene, are homologous to those described in *B. licheniformis,* and exhibit similar properties [3, 19]. The largest genetic difference resides mainly in the organisation of the divergeon. In *S. aureus*, the order of the genes in the operon is inverted compared to that in the *B. licheniformis* operon. Moreover the *S. aureus* intergenic region between *blaZ* and the *blaR*1-*blaI* operon contains two operatory sequences, one in each operatory region, instead of three in *B. licheniformis* (Figure 1B). In *S. aureus*, the presence of the *blaR*2 gene is postulated, but this gene remains to be identified. In the staphylococcal strains harbouring the resistant PBP2a, this protein can also be induced and the genes responsible for this induction are located in the mec divergeon, including *mecA*, *mecI* and *mecR*1, which code for the resistant PBP, the repressor and the penicillin receptor respectively (Figure 1 C). Concerning the induction mechanism in *S. aureus*, Zhang et al. suggested that the inactivation of the BlaI or MecI repressors during induction was due to the cleavage of the peptide bond linking the residues N^{101} and F^{102} (*S. aureus* BlaI numbering), directly or indirectly, by the L3 loop of the activated BlaR1 receptor. These authors have also shown that acylation of BlaR1 by the inducing antibiotic leads to the cleavage of the receptor, between residues R^{293} and R^{294} in the L3 loop (*S. aureus* BlaR1 numbering). This would activate its metalloproteolytic activity (Figure 4, ref 20) . It should be noticed that the proteolysis of the BlaI repressor in *S. aureus* does not exclude the presence of a coactivator, as in *B. licheniformis*. Indeed, proteolysis of the repressor could also result from binding of the coactivator and its release in the cytoplasm, rendering it more susceptible to cytoplasmic proteases. This possibility cannot be excluded because in vitro proteolysis of BlaI by BlaR1 or by a protease activated by the action of the activated receptor has not been demonstrated yet. Moreover, in the tridimensional structures of BlaI and MecI of *S. aureus* the cleavage site is not accessible to the solvent [21]. However, as was shown by the 2D NMR spectrum and chemical denaturation experiments of *B. licheniformis* BlaI, the C-terminal domains of these repressors should be very mobile [9]. This result is in agreement with those obtained by 3D analysis of both *S. aureus* repressors in which two cavities can be observed [22-23]. These cavities display the poorly compacted aspect of the C-terminal domains of both repressors. Do they allow the binding to a coactivator? To answer this question, identification of the coactivator suggested in *B. licheniformis* is needed. So far, four BlaR-CTD tertiary structures are available in the structure banks. The first to be published corresponds to that of the C-terminal domain of the non acylated *B. licheniformis* receptor [15]. The other three correspond to the apo-BlaR-CTD of *S. aureus* and to the same C-terminal domain acylated by penicillin G or by ceftazidine [24-25]. MecR-CTD has been also crystallized and its structures with or without penicillin G or oxacillin have been obtained [26]. As quoted above, the tertiary structures of BlaR/MecR-CTD are very close to those of class D serine β-lactamases (Figure 3). In the active sites of these β-lactamases, a lysine residue is carboxylated by the CO_2 dissolved in the buffer as in the OXA-10 β-lactamase for which this modification is essential for the catalytic activity [27]. For BlaR/MecR-CTD, carboxylation of the corresponding lysine is debated. Indeed, it could not be demonstrated, except in solution and it is absent in the structures obtained by X-ray diffraction. To explain this difference, Birk *et al* propose that in the apo-BlaR-CTD, Lys 426 (*S. aureus* BlaR1 numbering) is carboxylated, and that acylation of BlaR-CTD by the antibiotic leads to its decarboxylation. So, these authors explain why BlaR-CTD catalyses the opening of the β-lactam ring as do the class D β-lactamases and why the formed acylenzyme is stable. Indeed, in the class D β-lactamases, carboxylation of the lysine is essential for the formation and hydrolysis of the acylenzyme [24].

Acylation by penicillin G does not induce important conformation changes in the tertiary structure of the *S. aureus* BlaR/MecR-CTD [24-26], in contrast to the proposal of Golemi *et al* based on circular dichroism studies [28]. These crystallographic results are in agreement with those obtained with *B. licheniformis* BlaR-CTD. In this case, no conformational change could be detected upon acylation of the sensor by penicillin G by circular dichroism or by isotopic hydrogen/deuterium exchange in the protein peptide bonds [14].

3. Induction Mechanism in *Citrobacter Freundii*

In *C. freundii*, expression of the *ampC* gene coding for a class C β-lactamase is both respressed and activated by the product of the *ampR* gene, that plays a repressor-activator role. Comparison of the primary structure of the AmpR protein with the sequences available in protein libraries shows that its N-terminal domain, which recognises DNA, belongs to the LysR regulator family. In the *C. freundii* chromosome, the *ampC* and *ampR* genes are organised as a divergeon (Figure 5). AmpR recognises an operator sequence, located in the intergenic region between *ampC* and *ampR*, and controls expression of both *ampC* and its own gene. Generally, in Gram-negative bacteria possessing an inducible β-lactamase whose gene is controlled by AmpR, the induction mechanism is directly linked to the peptidoglycan

(PG) biosynthesis, to its turnover and to a cellular stress due to the inducing antibiotic. Indeed, AmpR recognises two groups of ligands. The first group consists of the UDP-MurNAc-pentapeptide, the most abundant PG cytoplasmic precursor, the second one contains the 3 anhydro-MurNAc-peptides (tri, tetra and pentapeptides) which originate from peptidoglycan turnover [29]. In the absence of inducer, AmpR which always interacts with DNA, binds the UDP-MurNAc-pentapeptide and the complex acts as a repressor. During growth, a large proportion of pre-existing PG is degraded yielding mainly periplasmic GNAc-anhydro-MurNAc-tetrapeptide which reenters the cytoplasm via the AmpG permease. There, it is degraded to GNAc, anhydro-MurNAc, D-alanine and the LAla—Dglu-γ→m-A_2pm tripeptide by the successive actions of the cytoplasmic NagZ, LdcA and AmpD enzymes (Figure 5). The tripeptide can subsequently be reintroduced into the PG precursor by Mpl. The concentrations of anhydro-MurNAc-tetra- and –tripeptides are never significant in the presence of an active AmpD. In *ampD*⁻ mutants, the anhydro-MurNAc-tripeptide accumulates, binds to AmpR and transforms it into an activator. The anhydro-MurNAc-tetra- and pentapeptides have similar effects on AmpR. It is not known if the UDP-MurNAc-pentapeptide is released from the new complex. Mutations responsible for the disappearance of AmpD activity are rather common and in consequence class C β-lactamase overproducers appear with a high frequency in the hospital setting.

In the presence of an inducing β-lactam, the inactivation of the PBPs disturbs the balance between PG biosynthesis and degradation. By an unknown mechanism, this also increases the activity of the autolytic system and in consequence results in the formation of large amounts of peptidoglycan turnover products. This modifies the quantity and the composition of the mixture of degradation products which reenter the cytoplasm via AmpG [30]. For instance, in the presence of benzylpenicillin, the GNAc-anhydro-MurNAc-pentapeptide becomes the major component [30]. The successive actions of NagZ and AmpD degrade this compound into GNAc, anhydro-MurNAc and pentapeptide and the latter is added to UDP-MurNAc by Mpl [31]. However, AmpD cannot cope with the increased amount of substrate and the anhydro-MurNAc pentapeptide accumulates, binds to AmpR and transforms it into an activator. Interestingly, the anhydro-MurNAc-tetrapeptide never seems to be present in significant quantities. In fact, its reintroduction into the biosynthesis pathway is lethal.

In addition to the induction phenomenon, this model nicely explains why *ampG*⁻ mutants are non-inducible while *ampD*⁻ mutants are constitutive overproducers.

4. Induction Mechanism in *Streptomyces Cacaoi*

In the chromosome of *S. cacaoi*, two genes coding for two class A β-lactamases are present and induced by the same mechanism [1]. The *blaL* gene is located upstream of regulatory *blaA* and *blaB* genes with which it constitutes a divergeon (Figure 6). The *blaU* gene is situated in another chromosomal region. The *blaA* gene codes for a protein whose N-terminal DNA-binding module belongs to the LysR family, as the AmpR protein [32]. However, in the proteins of this family, the C-terminal modules are highly variable, being adapted to the structure of a regulator. Thus, the C-terminal domain of BlaA differs from that of AmpR. The BlaB protein is cytoplasmic and of unknown function. However, BlaB exhibits some sequence similarities with some class A β-lactamases and forms stable acylenzymes with some β-lactam antibiotics and with N-acetyl-L-Lys-D-Ala-D-Ala-OH [33]. So far, the induction mechanism of BlaU and BlaL is not elucidated and the link between the antibiotic present outside the cell and the two regulatory cytoplasmic proteins remains to be identified.

5. Induction Mechanism in *Aeromonas Hydrophila*

In several *Aeromonas* species, the expression of one or more β-lactamases is inducible. In *Aeromonas hydrophila*, three β-lactamases AmpH (class D), CepH (class C) and ImiH (class B) are coded by unlinked genes and are regulated by the same couple of proteins (BlrA and BlrB) belonging to the two component system (TCS) family. These two proteins are coded by an operon localised upstream to the *ampH* gene. BlrB is a membrane protein possessing two transmembrane segments and a cytoplasmic segment containing the histidine kinase signature. BlrA is a cytoplasmic protein which would be phosphorylated by BlrB [5]. As in the three other induction mechanisms, the inducer in *Aeromonas* is also a β-lactam antibiotic. The mechanism by which the presence, of this antibiotic outside the cell is signalled to the cytoplasm remains unknown. BlrA and BlrB are homologous to the TCS CreB and CreC of *Escherichia coli* which is a general metabolic regulator, controlling the expression of many genes, in response to nutrient variations in the external medium. Avison *et al* have shown that the *CreBC* system of *E. coli* can regulate the expression of the *A. hydrophila* β-lactamases [34].

Conclusion

The induction of the AmpC β-lactamase in *C. freundii* is coupled to the peptidoglycan biosynthesis, to the peptidoglycan turnover and to the stress due to the action of the inducing antibiotic on the peptidoglycan biosynthesis. This stress leads to the accumulation in the cytoplasm of peptidoglycan fragments to be recycled and to the activation of the β-lactamase gene transcription. In *B. licheniformis*, we postulate that both a similar cellular stress and the acylation of the BlaR1 receptor are necessary for the induction of the BlaP β-lactamase. These conclusions differ from those proposed in the literature to describe the inactivation of BlaI/MecI repressor, during the induction of the BlaZ β-lactamase and of the resistant PBP2a in *S. aureus*. However, the similarity between the products of the regulatory genes present in both organisms is such that different mechanisms of repressor inactivation are unexpected. Nevertheless, the inactivation mechanism proposed for *B. licheniformis* could also describe the one which prevails in *S. aureus*, if the release of BlaI/MecI from the DNA and its proteolysis by cytoplasmic proteases in *S. aureus* is due to the binding of a coactivator. Identification of the coactivator of BlaI in *B. licheniformis* and the verification that the same coactivator does not exist in *S. aureus* should allow to decide if the two mechanisms are different. In *S. aureus*, it cannot be excluded that expression of the *blaR*2 gene depends on a cellular stress due to the inducing antibiotic.

In the induction mechanism of the *S. cacaoi* BlaU and BlaL β-lactamases, only the cytoplasmic proteins BlaA and BlaB are known, but it is reasonable to suggest that a cellular relay exists, signalling the presence of the antibiotic outside the cell. This cellular relay must be common to some *Streptomyces* species, since the BlaL β-lactamase is inducible in *Streptomyces lividans*, after the plasmid-borne *blaL-blaA-blaB* divergeon has been transformed into this organism [31]. As in *Bacillus*, it is quite possible that this cellular relay depends on a cellular stress due to the presence of the inducing antibiotic in the external medium. This hypothesis is reinforced by the fact that BlaA and AmpR belong to the same family of regulators (LysR family) and that those are generally regulated by a coactivator.

In *A. hydrophilia,* the nature of the signal detected by BlrB when the β-lactam inducer is added in the medium remains unknown. However, it is possible that it is linked to the effect of the antibiotic on the growing cell. So, in addition to the BlrB activation, a cellular stress would be necessary for β-lactamase induction in *A. hydrophilia.*

On the basis of our knowledge on the four induction mechanisms of β-lactamases, it would seem that in each of them, a stress due to the inducing antibiotic is necessary.

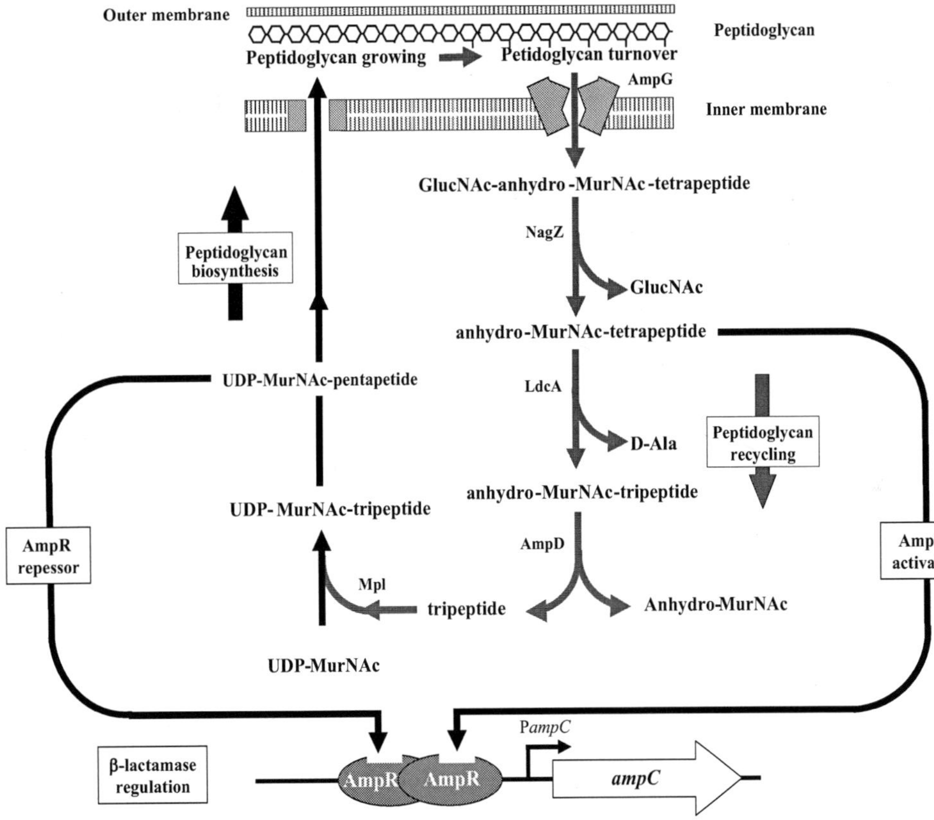

Figure 5. Control of β-lactamase biosynthesis in *C. freundii*: peptidoglycan synthesis (left) and recycling pathways (right). In the absence of inducing β-lactam and with active AmpG and AmpD proteins, the concentrations of anhydro-MurNAc-tetra- and -tripeptides are very low. In *ampD⁻* mutants, the latter accumulates and binds to AmpR transforming it into an activator. In the presence of β-lactams, the GNAc-anhydro-MurNAc-pentapeptide is the main product which reenters the cytoplasm (not shown) It is hydrolysed by NagZ to yield the anhydro-MurNAc-pentapeptide which is a substrate of AmpD (for details, see text) but the activity of the latter is not sufficient to keep the concentration of anhydro-MurNAc-pentapeptide at a low level. In turn, this compound transforms AmpR into an activator. It is easy to understand why *ampG⁻ ampD⁻* double mutants exhibit the phenotype of *ampG⁻* mutants.

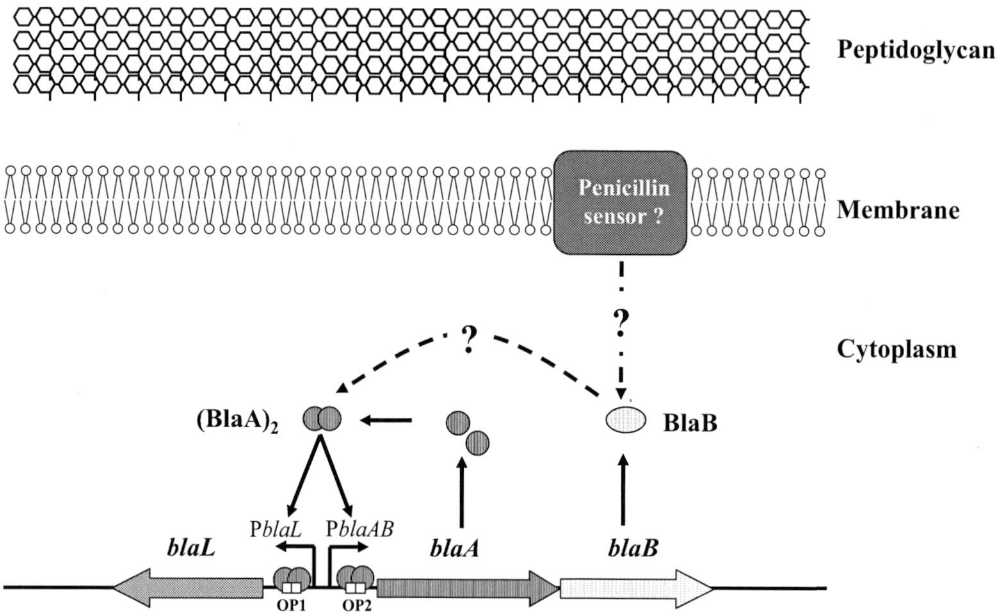

Figure 6. Induction mechanism of the BlaL β-lactamase in *S. cacaoi.*
The *blaL*, *blaA* and *blaB* genes code for the β-lactamase BlaL, the activator-repressor BlaA and the cytoplasmic protein BlaB (of unknown functions), respectively. The mechanism by which the presence of penicillin is detected outside the cell is unknown, as is the nature of the signal transmitted to BlaA. See text for more details.

If this hypothesis is correct, it would be noteworthy that during the evolution, four different mechanisms have been selected to link this cellular stress to the induction of a resistance protein, allowing the survival of the bacterium in the presence of the antibiotic. It might be difficult to verify this hypothesis but it represents an excellent starting point for the elucidation of the still unknown mechanisms.

References

[1] Philippon, A., Dusart, J., Joris, B., and Frere, J. M. (1998) The diversity, structure and regulation of beta-lactamases, *Cell Mol Life Sci 54*, 341-346.

[2] Duval, V., Swinnen, M., Lepage, S., Brans, A., Granier, B., Franssen, C., Frere, J. M., and Joris, B. (2003) The kinetic properties of the carboxy terminal domain of the *Bacillus licheniformis* 749/I BlaR penicillin-receptor shed a new light on the derepression of beta-lactamase synthesis, *Mol Microbiol 48*, 1553-1564.

[3] Mallorqui-Fernandez, G., Marrero, A., Garcia-Pique, S., Garcia-Castellanos, R., and Gomis-Ruth, F. X. (2004) Staphylococcal methicillin resistance: fine focus on folds and functions, *FEMS Microbiol Lett 235*, 1-8.

[4] Jacobs, C., Frere, J. M., and Normark, S. (1997) Cytosolic intermediates for cell wall biosynthesis and degradation control inducible beta-lactam resistance in gram-negative bacteria, *Cell 88*, 823-832.

[5] Avison, M. B., Niumsup, P., Nurmahomed, K., Walsh, T. R., and Bennett, P. M. (2004) Role of the 'cre/blr-tag' DNA sequence in regulation of gene expression by the *Aeromonas hydrophila* beta-lactamase regulator, BlrA, *J Antimicrob Chemother 53*, 197-202.

[6] Segatore, B., Massidda, O., Satta, G., Setacci, D., and Amicosante, G. (1993) High specificity of cphA-encoded metallo-beta-lactamase from *Aeromonas hydrophila* AE036 for carbapenems and its contribution to beta-lactam resistance, *Antimicrob Agents Chemother 37*, 1324-1328.

[7] Ross, C. L., Thomason, K. S., and Koehler, T. M. (2009) An extracytoplasmic function sigma factor controls beta-lactamase gene expression in *Bacillus anthracis* and other *Bacillus cereus* group species, *J Bacteriol 191*, 6683-6693.

[8] Filee, P., Vreuls, C., Herman, R., Thamm, I., Aerts, T., De Deyn, P. P., Frere, J. M., and Joris, B. (2003) Dimerization and DNA binding properties of the *Bacillus licheniformis* 749/I BlaI repressor, *J Biol Chem 278*, 16482-16487.

[9] Van Melckebeke, H. V., Vreuls, C., Gans, P., Filee, P., Llabres, G., Joris, B., and Simorre, J. P. (2003) Solution structural study of BlaI: implications for the repression of genes involved in beta-lactam antibiotic resistance, *J Mol Biol 333*, 711-720.

[10] Vreuls, C., Filee, P., Van Melckebeke, H., Aerts, T., De Deyn, P., Llabres, G., Matagne, A., Simorre, J. P., Frere, J. M., and Joris, B. (2004) Guanidinium chloride denaturation of the dimeric *Bacillus licheniformis* BlaI repressor highlights an independent domain unfolding pathway, *Biochem J 384*, 179-190.

[11] Boudet, J., Duval, V., Van Melckebeke, H., Blackledge, M., Amoroso, A., Joris, B., and Simorre, J. P. (2007) Conformational and thermodynamic changes of the repressor/DNA operator complex upon monomerization shed new light on regulation mechanisms of bacterial resistance against beta-lactam antibiotics, *Nucleic Acids Res 35*, 4384-4395.

[12] Hardt, K., Joris, B., Lepage, S., Brasseur, R., Lampen, J. O., Frere, J. M., Fink, A. L., and Ghuysen, J. M. (1997) The penicillin sensory transducer, BlaR, involved in the inducibility of beta-lactamase synthesis in *Bacillus licheniformis* is embedded in the plasma membrane via a four-alpha-helix bundle, *Mol Microbiol 23*, 935-944.

[13] Joris, B., Ledent, P., Kobayashi, T., Lampen, J. O., and Ghuysen, J. M. (1990) Expression in *Escherichia coli* of the carboxy terminal domain of the BLAR sensory-transducer protein of *Bacillus licheniformis* as a water-soluble Mr 26,000 penicillin-binding protein, *FEMS Microbiol Lett 58*, 107-113.

[14] Hanique, S., Colombo, M. L., Goormaghtigh, E., Soumillion, P., Frere, J. M., and Joris, B. (2004) Evidence of an intramolecular interaction between the two domains of the BlaR1 penicillin receptor during the signal transduction, *J Biol Chem 279*, 14264-14272.

[15] Kerff, F., Charlier, P., Colombo, M. L., Sauvage, E., Brans, A., Frere, J. M., Joris, B., and Fonze, E. (2003) Crystal structure of the sensor domain of the BlaR penicillin receptor from *Bacillus licheniformis*, *Biochemistry 42*, 12835-12843.

[16] Kobayashi, T., Zhu, Y. F., Nicholls, N. J., and Lampen, J. O. (1987) A second regulatory gene, blaR1, encoding a potential penicillin-binding protein required for induction of beta-lactamase in *Bacillus licheniformis*, *J Bacteriol 169*, 3873-3878.

[17] Yasukawa, K., Kusano, M., and Inouye, K. (2007) A new method for the extracellular production of recombinant thermolysin by co-expressing the mature sequence and pro-sequence in *Escherichia coli*, *Protein Eng Des Sel 20*, 375-383.

[18] Filee, P., Benlafya, K., Delmarcelle, M., Moutzourelis, G., Frere, J. M., Brans, A., and Joris, B. (2002) The fate of the BlaI repressor during the induction of the *Bacillus licheniformis* BlaP beta-lactamase, *Mol Microbiol 44*, 685-694.

[19] Fuda, C. C., Fisher, J. F., and Mobashery, S. (2005) Beta-lactam resistance in *Staphylococcus aureus:* the adaptive resistance of a plastic genome, *Cell Mol Life Sci 62*, 2617-2633.

[20] Zhang, H. Z., Hackbarth, C. J., Chansky, K. M., and Chambers, H. F. (2001) A proteolytic transmembrane signaling pathway and resistance to beta-lactams in staphylococci, *Science 291*, 1962-1965.

[21] Garcia-Castellanos, R., Marrero, A., Mallorqui-Fernandez, G., Potempa, J., Coll, M., and Gomis-Ruth, F. X. (2003) Three-dimensional structure of MecI. Molecular basis for transcriptional regulation of staphylococcal methicillin resistance, *J Biol Chem 278*, 39897-39905.

[22] McKinney, T. K., Sharma, V. K., Craig, W. A., and Archer, G. L. (2001) Transcription of the gene mediating methicillin resistance in *Staphylococcus aureus* (mecA) is corepressed but not coinduced by cognate mecA and beta-lactamase regulators, *J Bacteriol 183*, 6862-6868.

[23] Safo, M. K., Zhao, Q., Ko, T. P., Musayev, F. N., Robinson, H., Scarsdale, N., Wang, A. H., and Archer, G. L. (2005) Crystal structures of the BlaI repressor from *Staphylococcus aureus* and its complex with DNA: insights into transcriptional regulation of the bla and mec operons, *J Bacteriol 187*, 1833-1844.

[24] Birck, C., Cha, J. Y., Cross, J., Schulze-Briese, C., Meroueh, S. O., Schlegel, H. B., Mobashery, S., and Samama, J. P. (2004) X-ray crystal structure of the acylated beta-lactam sensor domain of BlaR1 from *Staphylococcus aureus* and the mechanism of receptor activation for signal transduction, *J Am Chem Soc 126*, 13945-13947.

[25] Wilke, M. S., Hills, T. L., Zhang, H. Z., Chambers, H. F., and Strynadka, N. C. (2004) Crystal structures of the Apo and penicillin-acylated forms of the BlaR1 beta-lactam sensor of *Staphylococcus aureus*, *J Biol Chem 279*, 47278-47287.

[26] Marrero, A., Mallorqui-Fernandez, G., Guevara, T., Garcia-Castellanos, R., and Gomis-Ruth, F. X. (2006) Unbound and acylated structures of the MecR1 extracellular antibiotic-sensor domain provide insights into the signal-transduction system that triggers methicillin resistance, *J Mol Biol 361*, 506-521.

[27] Li, J., Cross, J. B., Vreven, T., Meroueh, S. O., Mobashery, S., and Schlegel, H. B. (2005) Lysine carboxylation in proteins: OXA-10 beta-lactamase, *Proteins 61*, 246-257.

[28] Golemi-Kotra, D., Cha, J. Y., Meroueh, S. O., Vakulenko, S. B., and Mobashery, S. (2003) Resistance to beta-lactam antibiotics and its mediation by the sensor domain of the transmembrane BlaR signaling pathway in *Staphylococcus aureus*, *J Biol Chem 278*, 18419-18425.

[29] Park, J. T., and Uehara, T. (2008) How bacteria consume their own exoskeletons (turnover and recycling of cell wall peptidoglycan), *Microbiol Mol Biol Rev 72*, 211-227).

[30] Wiedemann, B., Pfeifle, D., Wiegand, I., and Janas, E. (1998) Beta-lactamase induction and cell wall recycling in gram-negative bacteria, *Drug Resist Updat 1*, 223-226.

[31] Hervé M, Boniface A, Gobec S, Blanot D, Mengin-Lecreulx D.(2007) Biochemical characterization and physiological properties of *Escherichia coli* UDP-N-acetylmuramate:L-alanyl-gamma-D-glutamyl-meso-diaminopimelate ligase*, J Bacteriol* 189, 3987-95.

[32] Lenzini, V. M., Magdalena, J., Fraipont, C., Joris, B., Matagne, A., and Dusart, J. (1992) Induction of a *Streptomyces cacaoi* beta-lactamase gene cloned in *S. lividans*, *Mol Gen Genet 235*, 41-48.

[33] Raskin, C., Gerard, C., Donfut, S., Giannotta, E., Van Driessche, G., Van Beeumen, J., and Dusart, J. (2003) BlaB, a protein involved in the regulation of *Streptomyces cacaoi* beta-lactamases, is a penicillin-binding protein, *Cell Mol Life Sci 60*, 1460-1469.

[34] Avison, M. B., Niumsup, P., Walsh, T. R., and Bennett, P. M. (2000) *Aeromonas hydrophila* AmpH and CepH beta-lactamases: derepressed expression in mutants of *Escherichia coli* lacking creB, *J Antimicrob Chemother 46*, 695-702.

In: Beta-Lactamases
Editor: Jean-Marie Frère, pp. 437-457

ISBN: 978-1-61324-638-2

Chapter 18

RANDOM MUTAGENESIS AS A PROBE OF β-LACTAMASE STRUCTURE, FUNCTION AND EVOLUTION

Jacques Fastrez[1] and Timothy Palzkill[2]

[1] Laboratoire d'Ingénierie des Protéines et des Peptides, Institut des sciences de la Vie, Université catholique de Louvain, Croix du sud 4-5 bte L7.07.06, B1348 Louvain-la-Neuve, Belgium.

[2] Department of Pharmacology, Baylor College of Medicine, One Baylor Plaza, Houston, TX 77030, U.S.A.

ABSTRACT

Because of the ready selection for function, β-lactamases have been intensively studied by random mutagenesis methods. In particular, the TEM-1 β-lactamase has served as a model for methods development for random mutagenesis techniques and as a model for protein engineering and the directed evolution of new enzymatic functions. This chapter reviews the use of random mutagenesis methods for studies of β-lactamase structure, function, evolution and engineering.

INTRODUCTION

Random mutagenesis is a powerful method to address issues of protein structure and function as well for protein engineering applications. The basic approach is to generate random mutations in a gene of interest and to screen the collection for mutants that provide for wild-type function or that exhibit altered function. There are several types of random mutants that can be introduced, such as random point mutations introduced by methods such as error-prone PCR or saturation mutagenesis whereby codons are completely randomized by one of various oligonucleotide directed mutagenesis procedures. The key to the success of any random mutant screen is an efficient way to sort among the large number of mutants for those with the desired function. Ideally, an *in vivo* genetic selection is used to sort through millions of mutants in a single experiment.

A large number of β-lactamases have been identified from both Gram-positive and negative bacteria [1]. This chapter discusses mutagenesis approaches using β-lactamases as model systems. Several enzymes are discussed, however, the TEM-1 β-lactamase has been the subject of intense investigation by both genetic and biochemical approaches and as such is prominent in this chapter. TEM-1 is encoded by the $bla_{\text{TEM-1}}$ gene [2] and is a member of the structurally conserved class A β-lactamases. Epidemiological studies have shown that TEM-1 is the most common plasmid-encoded β-lactamase in Gram-negative bacteria [3]. It is able to efficiently hydrolyze penicillins and many cephalosporins; therefore, it is an important source of bacterial resistance to the β-lactam antibiotics [1, 4].

Two approaches have been used to combat TEM-1 β-lactamase-mediated resistance. First, extended-spectrum cephalosporins such as cefotaxime and ceftazidime were developed, in part, because the TEM-1 β-lactamases is not able to efficiently hydrolyze these antibiotics [5]. A second option has been the co-administration of an inactivator (unsually called inhibitor) of β-lactamase, such as clavulanic acid or sulbactam, with a β-lactam antibiotic such as ampicillin or amoxicillin to overcome β-lactamase-mediated resistance [6, 7]. Clinical use of either of these strategies creates a strong selective pressure for β-lactamase mutations conferring either the ability to cleave extended-spectrum cephalosporins or to decrease the efficiency of the inactivation [8-10]. Several TEM-1 mutations have been discovered in clinical isolates recalcitrant to either of these treatments. These mutations modify TEM-1's substrate profile to generate extended-spectrum TEM-1 β-lactamases (ESBLs) or inhibitor resistant TEM-1 β-lactamases (IRTs) [4, 11, 12].

The clinical importance of TEM-1 and the impact of emergent TEM-1 variants on antibiotic treatment options have made this enzyme the subject of intense research. In addition, the β-lactamase system has served as a model for the development of random mutagenesis and selection approaches because of its widespread presence on recombinant DNA plasmids and the experimental ease of selecting mutants with wild-type or altered functions. These selections typically consist of introducing a plasmid encoding a population of mutagenized β-lactamase genes into *Escherichia coli* and sorting the mutants for function by spreading the transformed bacteria on agar plates containing the β-lactam antibiotic of interest. This chapter reviews the various random mutagenesis approaches that have been employed to study β-lactamase structure and function as well as the use of β-lactamase as a model for the development of directed evolution approaches to protein engineering.

Early Random Mutagenesis Experiments on TEM-1

One of the earliest applications of random mutagenesis to study proteins was performed by Knowles and colleagues in 1976 [13]. For this study, *E. coli* containing a plasmid encoding the TEM-1 β-lactamase was grown in the presence of increasing concentrations of cephalosporin C. As no mutagens were used, the occurrence of mutants depended on the inherent mutation rate of *E. coli*. A mutant was found that provided increased resistance to cephalosporin C compared to *E. coli* encoding the wild type enzyme. Although not reported in the original paper, the mutation was later found to be an alanine to threonine substitution at position 237 in the active site of TEM-1. It is of interest that this substitution was later found in *bla*$_{TEM}$ genes from clinical isolates resistant to extended spectrum cephalosporins (ESBLs).

Several site-directed codon saturation mutagenesis studies took place in the 1980s. These experiments involved randomizing the DNA sequence of a codon to all 19 other amino acids and then assaying each mutant for its effect on enzyme structure and function [14, 15]. The experimental approach was to perform cassette mutagenesis whereby synthetic oligonucleotides that encoded the random sequence nucleotides at the codon of interest were annealed to create double-stranded DNA fragments that contained staggered ends compatible with restriction enzyme sites. The appropriate restriction enzyme sites flanking the codon of interest were present in the plasmid containing the gene to be mutagenized. The plasmid was digested with the restriction enzymes and the synthetic cassette was annealed to replace the original sequence. This method was used to create all 19 amino acid substitutions at residue 71 near the active site of TEM-1 [14]. It was found that position 71 was tolerant to amino acid substitutions and therefore it was not likely to be directly involved in catalysis [14].

Cassette mutagenesis was also used by Oliphant and Struhl [16] to randomly mutagenize multiple codons simultaneously wherein restriction endonuclease sites flanking codons 61 and 77 were engineered into the *bla*$_{TEM-1}$ gene. A double stranded cassette derived from oligonucleotide synthesis of one strand and self-primed synthesis of the second strand was inserted between the engineered sites to replace the 61 to 77 codons with mutagenized codons. The oligonucleotide synthesis was biased to incorporate 20% non-wild type nucleotides at each position. This experimental design resulted in a majority of mutants containing multiple amino acid substitutions [16]. It is interesting that the Met69Leu substitution was identified in a clavulanic acid resistant mutant selected in this study. The Met69Leu substitution has also been identified in *bla*$_{TEM-1}$ mutants isolated from clinical isolates resistant to clavulanate-β-lactam antibiotic combinations. Thus, this was an early prediction of the potential of TEM-1 to acquire inhibitor resistance by mutation [16].

Another early cassette mutagenesis study of the region surrounding serine 70 was performed by Dube and Loeb [17]. In contrast to the moderate mutagenesis levels utilized above, the codons encoding residues 67-69 and 71-72 were completely randomized in a saturation mutagenesis approach. A total of 2 x 10^5 colonies were screened to identify 7 carbenicillin resistant mutants that each contained multiple amino acid substitutions in the 67-72 region. Therefore, despite the fact that the primary mutant library was not large enough to be representative of all possible mutants, a number of functional mutants were found [17]. This was an early indication that TEM-1 β-lactamase is relatively tolerant of amino acid substitutions.

Random Replacement Mutagenesis of TEM-1

Random replacement mutagenesis was developed as a means of performing cassette mutagenesis at a number of positions within a gene [18]. First, a short DNA fragment that contains a *BspMI* recognition site at each end was inserted into the TEM-1 gene. *BspMI* is a type IIs restriction enzyme that recognizes an asymmetric 6 base pair sequence and creates a staggered 4/8 nucleotide base pair (bp) cut downstream of that sequence. Next, the plasmid with the *BspMI* insert was digested to release the linker along with 4 bp's of neighboring plasmid DNA from each side of the linker. This results in a linear plasmid with 4 bp of the

TEM-1 gene deleted from each end. The released DNA fragment was replaced with a linker that contains random sequence DNA. The final result is the randomization of 8 bp of DNA or approximately 3 codons [18]. By generating a set of 19 individual *BspM1* insertions it was possible to randomize 63 codons in 19 random libraries. The random libraries were sorted for functional mutants by introducing each library into *E. coli* by transformation and spreading the transformed cells on agar containing either 10 or 1000 μg/ml of ampicillin. The 10 μg/ml selection was used to identify mutants that retained minimal levels of function while the 1000 □g/ml selection was performed to isolate mutants that possessed near wild type levels of function. DNA sequence analysis of the selected clones was then performed to infer the amino acid sequence requirements for low and high levels of function.

The sequence requirements for function determined in the random replacement experiments were found to vary greatly depending on the position of the substitutions on the enzyme [18]. For example, substitutions at surface positions far from the active site were relatively tolerant to amino acid substitutions, especially for the 10 μg/ml selection. In contrast, positions in the active site were, in general, not tolerant of substitutions regardless of whether the mutants were selected at 10 or 1000 □g/ml ampicillin. The selections at high and low concentrations of ampicillin therefore allowed increased resolution of the importance of residue positions in the enzyme. Essential positions, such as active site positions, exhibited strict sequence requirements at both high and low ampicillin concentrations while moderately important positions tolerated amino acid substitutions at low ampicillin concentrations but not high. Positions unimportant for enzyme structure and function were freely substituted at both high and low ampicillin concentrations [18].

The random replacement mutagenesis libraries described above were also used to examine amino acid sequence requirements around the active site of the TEM-1 β-lactamase with several different substrates [19]. In an initial study, seven random libraries encompassing residues 69-71, 72-74, 103-105, 168-170, 233-235, 235-237 and 238-241 were used to screen for positions that, when substituted, result in increased resistance to the extended spectrum cephalosporin, cefotaxime. Only three of the libraries, 103-105, 168-170 and 238-241, were found to contain mutants with increased cefotaxime resistance [19]. DNA sequencing of β–lactamase gene of the cefotaxime resistant mutants revealed that substitutions of lysine or arginine at position 104, serine at position 238 and small, in-frame deletions in the 168-170 region were responsible for the observed resistance. These results were an early indication that substitutions at positions 104 and 238 would be an important source of ESBLs.

Additional random replacement mutagenesis studies focused on the potential for substitutions in the omega loop in TEM-1 to provide increased hydrolysis of extended spectrum cephalosporins [20, 21]. Systematic randomization of codons and selection of ceftazidime resistant mutants indicated that virtually any substitution within the omega loop results in slightly elevated ceftazidime resistance [20]. There was also a correlation observed between ceftazidime resistance and enzyme stability since more active mutant enzymes were more rapidly proteolyzed in *E. coli* than the wild type β-lactamase as judged by immunoblotting [20, 21]. This was interpreted, although not experimentally proven, as meaning that mutations providing increased mobility of the omega loop provide increased access for ceftazidime and thereby result in increased levels of hydrolysis [20, 21].

The random replacement libraries covering active site residues were also screened for mutants with altered substrate specificity with respect to penicillins and cephalosporins [22].

For these studies, 11 random libraries encompassing 30 residue positions in and near the substrate binding pocket were selected in separate experiments for mutants that exhibited high levels of resistance to either ampicillin or cephaloridine, a second generation cephalosporin. The mutants obtained from each of these selections were then subjected to DNA sequencing and the inferred amino acid sequences were compared for cephaloridine versus ampicillin. The sequence requirements at several positions, including the catalytically critical Ser70, Lys73, Ser130, Glu166 and Lys234 were the same for both substrates in that only the wild type residue was observed among functional mutants [22]. Positions were identified, however, that, when substituted, retained high levels of cephaloridine hydrolysis while exhibiting reduced ampicillin hydrolysis. For example, the Ala237Thr substitution was identified in this screen based on the fact that the threonine mutant was found among cephaloridine resistant mutants but not among ampicillin resistant mutants [22]. Purification and characterization of the Ala237Thr enzyme revealed a two-fold increase in k_{cat}/K_M for cephaloridine hydrolysis and a five-fold reduction in k_{cat}/K_M for ampicillin hydrolysis. The Ala237Thr enzyme thereby hydrolyzes ampicillin and cephaloridine with nearly equal catalytic efficiency while the wild type enzyme displays an 18-fold preference for ampicillin [22]. Therefore, changes at a limited number of positions can have a large effect on the substrate specificity of β-lactamase.

Finally, the random replacement mutagenesis strategy was extended to randomize codons for all of the TEM-1 β-lactamase amino acids. The entire set of 263 amino acids present in the mature enzyme was randomized in a set of 88 random libraries [23]. A total of 78 these libraries were created using a different mutagenesis protocol than the *BspMI* protocol described above. First, a 9 base pair region corresponding to three contiguous codons was chosen for mutagenesis. A β-lactamase insert mutant was then created by using the Kunkel et al. mutagenesis method [24] to insert a unique *Sal* I restriction enzyme site within the 9 base pair region targeted for mutagenesis [23]. Single stranded DNA of the insertion mutant was then created in an *ung- dut-* mutant of *E. coli* [24] and randomization was accomplished by annealing an oligonucleotide designed to replace the 9 base pairs with random sequence DNA [23]. Since randomization is not 100% efficient, some plasmids will remain unmutagenized. If the starting template were the wild type gene rather than the *Sal*I insertion mutant, it would result in an unacceptable level of wild type background in the random library. The presence of the *Sal*I site in the template facilitates the removal of all DNA that was not randomized by digesting the mutagenesis reaction products with the *Sal*I enzyme [23].

The 88 random libraries were sorted for functional mutants by transforming each library into *E. coli* and spreading the transformed cells on agar plates containing 1000 μg/ml ampicillin [23]. This concentration of ampicillin was used to select for phenotypically wild type mutants. In practice, mutants selected from a number of the libraries were tested for ampicillin hydrolysis and it was found that, on average, the mutants possessed 85% of wild-type function [23]. In order to determine the spectrum of amino acid substitutions at each position consistent with function, the DNA sequences of an average of 10 ampicillin resistant mutants were determined for each library. In total, 43 out of the 263 mutated residues were inferred to be essential for TEM-1 structure and function because only the wild-type amino acid was found encoded at these positions among the sequenced mutants [23]. Several of these were critical active site residues such as Ser70, Lys73, Ser130, Glu166 and Lys234. Several positions away from the active site pocket were also identified, however, including several buried tryptophan residues such as Trp165, Trp210, Trp229, and Trp290 [23]. The mutagenesis data was also compared with sequence conservation in the class A β-lactamase family and several positions were identified that did not tolerate substitutions in the mutagenesis studies but are freely substituted among members of the gene family [23]. The substitution of these residues in the gene family may be due to the accumulation of compensatory mutations elsewhere in the other class A enzymes. Finally, the amino acid sequence variation at each position in TEM-1 was quantitated by calculating the effective number of substitutions for each site using information-theoretical entropy [23]. The quantitation allowed a statistical estimate of the correlation between the position of an amino acid on the structure of the enzyme and the ease of substituting the residue. A correlation was observed between surface location and increased tolerance of substitutions. In total, the systematic mutagenesis of TEM-1 β-lactamase has contributed to this enzyme being among the most intensively studied protein structure and function models.

When generating a library by random mutagenesis, a relatively high percentage of mutants can be incompatible with the protein fold, particularly if the number of mutated residues becomes high. To circumvent this problem, Hayes et al [25] have developed a computational screen preliminary to the experimental one, in which the fitness of sequences is scored by using physical potential functions. An extremely efficient search algorithm allowed testing up to 10^{80} sequences in a few hours. Application of this strategy to TEM-1 led to the creation of a library of $\approx 2 \times 10^5$ mutants from which variants exhibiting a 1,280-fold increased resistance against cefotaxime could be selected. All mutations appeared to be novel and different from those found in the naturally evolved mutants. Contrary to the experiments described above, this approach is not designed to mimic natural evolution but to explore more widely the sequence space for compatibility with activity and stability.

Evolution and Prediction of Future Antibiotic Resistance

With the emergence of directed evolution techniques, the prediction of future antibiotic resistance has become more easily feasible. Application of both *in vitro* DNA shuffling [26] or *in vivo* mutagenesis using a hypermutator *E. coli* strain with cefotaxime selection [27] to the TEM-1 β-lactamase resulted in the generation of the same Glu104Lys/ Met182Thr/Gly238Ser combination of mutations. These mutations have been found in the TEM-52 clinical isolate. The Met182Thr substitution had been shown earlier to be a global suppressor of missense mutations that disrupt the folding or stability of the enzyme [28]. The observation of this combination of mutations led to the proposal of tentative pathways of accumulation of mutations, in which stabilizing mutations are brought in before additional mutations responsible for increased resistance can be definitely accepted. While monitoring the trajectory of changes in the TEM β-lactamase gene inserted in the genome of the ϕ6 RNA virus under selection pressure of cefotaxime, Makeyev and Bamford [29] observed an initial explosion in sequence variants with multiple mutations throughout the gene followed by a stabilization to a homogeneous population of sequences containing only 3 mutations, among which the Glu104Lys and Gly238Ser have been seen in clinical isolates. These results suggest how genetic systems may explore adaptative pathways.

In an investigation to better mimic natural evolution, Barlow and Hall [30] created 9 libraries of TEM-1 mutants by error-prone PCR with the Mutazyme polymerase whose mutation spectrum corresponds to the average mutation spectrum of *E. coli.* These libraries were submitted to a complex selection protocol mimicking the clinical selective pressure. In this protocol, the cells were grown in the presence of increasing concentrations of one of four 3rd generation cephalosporins (cefotaxime cefuroxime, ceftazidime) or aztreonam, then on the three others and on ampicillin, pipericillin and cephalothin, the order of the first selection being changed in different experiments. After two rounds of mutagenesis and selection, seven of the nine mutations that have arisen independently multiple times in nature were recovered. These results suggest that it is possible to predict the emergence of future resistant mutants and emphasize the importance of selection schemes. Using the same *in vitro* evolution method, TEM-1 alleles conferring resistance to cefepime, a more recently introduced cephalosporin, were discovered after four rounds. They involve three amino acid substitutions. The results suggest further that several independent evolution pathways are plausible [31]. CMY-2, a class C β-lactamase, was similarly evolved to give rise to alleles whose increased resistance was lower than with TEM-1 [32]. Again, multiple pathways are observed and it does not appear to be any single substitution that is crucial for increased cefepime resistance.

As indicated above, the TEM-1 Met182Thr substitution is known to stabilize the enzyme and thereby compensate for other active site substitutions that change the catalytic parameters of the enzyme but also destabilize it [28, 33]. The Met182Thr substitution was originally identified, after introduction of random point mutations, as a second site suppressor mutation that restored stability to an enzyme containing a substitution in the hydrophobic core. A similar strategy was recently used to identify a number of other TEM-1 substitutions, such as Thr265Met, that act as suppressors of stability defects and are also found in clinical isolates suggesting they play an important role in the evolution of extended-spectrum TEM-1 β-lactamases [34].

Serine β-lactamases are widely believed to have evolved from DD-peptidases through an enormous acceleration of the rate of hydrolysis of the acyl-enzyme formed with β-lactam antibiotics. This evolutionary step has been difficult to reproduce in the laboratory. Peimbert and Segovia [35] have nevertheless succeeded in modifying the catalytic properties of the PBP2x DD-transpeptidase towards a cephalosporinase activity. A combinatorial library was first created, in which 10 sequence positions were mutated or substituted by 2-20 residues. These positions were chosen to increase water access to the active site or to offer a large choice of active site walls shapes for acyl-enzyme orientation. Amplification of the combinatorial library by error-prone PCR allowed the appearance of advantageous mutations at unforeseen positions. Selection on a low concentration of cefotaxime (0.2 μg/ml) afforded a few clones conferring 4- to 10-fold higher resistance than the wild-type enzyme and containing mutations in 6 positions. These were further shuffled and submitted to saturation mutagenesis at 4 positions. One clone featuring only 3 mutations exhibited a 10^5-fold increase in cefotaxime rate of deacylation ($k_{cat} = 1.5 \times 10^{-2}\ s^{-1}$) while retaining the wild-type activity on a thioester mimicking the DD-transpeptidase substrate. Among the 3 mutations, Gly336Ala involves a residue next to the catalytic Ser337 while the other 2 (Phe450Leu and Gln452His) concern the Ω-loop. His452 is proposed to play a role similar to that of Glu166 in class A β-lactamases.

In Vitro Insertion and Deletion Mutagenesis

As a further indication of its role as a model system for mutagenesis studies, new methods for creating insertion and deletion mutants have been tested on TEM-1. For example, Hayes et al. [36] developed pentapeptide scanning mutagenesis using TEM-1 as a model system. Pentapeptide scanning mutagenesis is a method in which a five amino acid cassette is inserted at random into a protein of interest. The method involves generating a set of transposon insertions within the gene of interest. The plasmid containing the transposon insertion is then digested with the restriction enzyme *Kpn* I which removes the majority of the transposon and, upon religation of the plasmid, generates a 15 base pair insertion [36]. For the β-lactamase experiments, a pool of random insertion mutants in TEM-1 were generated and selected on agar plates for mutants with altered substrate specificity towards increased hydrolysis of ceftazidime [36]. Subsequent DNA sequencing of the selected mutants revealed that insertions that increase ceftazidime hydrolysis cluster in the omega loop region of the enzyme. These results demonstrate how an insertion mutant library can be simple in design and construction and yet powerful enough to yield enzymes with altered the catalytic properties [36].

A method for creating in-frame deletions throughout a target gene was also developed and tested on the TEM-1 β-lactamase [37]. This method, named codon-based random deletion mutagenesis (COBARDE), relies on oligonucleotide synthesis using orthogonal protecting groups [37].

For these experiments, a fraction of the oligonucleotides being synthesized are blocked after a desired codon by introducing an alternate protecting group. Synthesis proceeds normally for the oligonucleotides not modified by the alternative protecting group. Upon removal of the alternative protecting group, synthesis reinitiates and thereby introduces a deletion in a fraction of the oligonucleotide pool.

By controlling the deprotection cycles, it is possible to create a family of oligonucleotides containing random deletions of codons [37]. The oligonucleotides are then used for site-directed mutagenesis to create a library of random deletions mutants. This method was performed with the active site omega loop region of TEM-1 as the target and it was found that several different deletions in the region resulted in enzymes providing increased ceftazidime resistance [37]. In contrast, the enzyme was less tolerant of deletions in this region when the mutants were selected for the ability to hydrolyze ampicillin. As with the insertion studies described above, these experiments demonstrated that deletions can alter the catalytic properties of enzymes [37].

Another method for introducing deletions in proteins was developed with TEM-1 as the model system [38, 39]. This method makes use of the mini-Mu transposon to introduce the recognition sequence for the MlyI restriction endonuclease. The Mu transposon has low sequence specificity for insertion and therefore it was used to randomly insert the MlyI site into the TEM-1 β-lactamase gene [38].

Digestion with the MlyI enzyme followed by relegation was performed to create 3 base pair deletions at random positions in the gene. A total of 22 deletions were constructed and it was found that deletions that retained β-lactamase activity occurred in loop regions but also in helices in the enzyme [38].

In Vitro Mutagenesis and Recombination: DNA Shuffling and other Methods

TEM-1 has also served as a model system for the evolution of enzymatic properties. Some of these studies have used combinatorial approaches to address basic questions about how certain properties of enzymes evolve while other studies have focused on the development of mutational techniques that could then be applied to other systems.

Recombination is known to play an important role in the evolution of genomes as well as individual genes. Willem Stemmer used TEM-1 as a model system to develop a powerful *in vitro* mutagenesis method that combines random PCR mutagenesis with recombination [26]. The method, termed DNA Shuffling, involves error-prone PCR amplification of the gene of interest which, in the initial experiment, was that of TEM-1 [26]. The PCR mutagenized bla_{TEM-1} gene was cloned into a plasmid and introduced into *E. coli* and mutants with increased resistance to cefotaxime were selected. The transformants were then pooled and the bla_{TEM-1} genes were amplified. The DNA shuffling occurred when the amplified genes were randomly fragmented with DNAseI and reassembled by multiple cycles of PCR [26]. Reassembly permits multiple independent mutations to be combined into a single clone. Combinations of mutants that resulted in increased cefotaxime resistance were selected on cefotaxime containing media. These experiments demonstrated that a gene can be rapidly evolved *in vitro* by introducing recombination into the mutagenesis and selection method [26]. The experiments also revealed new combinations of β-lactamase mutations that can lead to extended spectrum cephalosporin hydrolysis [26].

The DNA shuffling method outlined above is dependent on sequence identity or similarity between the DNA fragments that are reassembled during the multiple cycles of PCR [40]. Therefore, mutagenesis and recombination mediated by DNA shuffling is limited to closely related genes. Other *in vitro* mutagenesis and recombination methods have been developed to allow recombination between sections of unrelated DNA and thus allow shuffling of unrelated genes. Several of these sequence independent methods have been reported including ITCTHY [41], SCRATCHY [42], and SHIPREC [43]. Another such method, sequence-independent site-directed chimeragenesis (SISDC), utilizes tag DNA sequences that encode the type IIb restriction endonuclease *Bae* I [44]. The tag sequences are placed at analogous positions in two (or more) genes to be recombined to create chimeras. The shuffling of the segments of genes occurs after digestion of the genes containing the tags with the *Bae* I enzyme followed by ligation into full-sized products [44]. The remnants of the tags are removed by the digestion because the *Bae* I restriction enzyme cleaves downstream of its recognition sequence. The SISDC method was used to create gene chimeras of the TEM-1 and PSE-4 β-lactamase genes as a model system [44]. It was found that approximately 6% of the TEM-PSE chimeras were active and chimeras containing few cross-over events, in general, exhibited higher activity.

TEM-1 has also been used as a model to study how sequence independent recombination can be optimized to preferentially create chimeras that fold properly and therefore are potentially functional [45]. A computational method named SCHEMA was developed based on schema in genetic algorithms to compare two structurally homologous proteins to identify fragments that can be recombined without disturbing the three dimensional structure of the resulting chimera [45]. The method works by examining a window of residues for interactions between amino acids that would be disrupted upon chimera formation. The method was used to compare the TEM-1 and PSE-4 β-lactamases to determine positions at which recombination would not disrupt structure. A number of recombination sites were predicted and tested by constructing TEM-1-PSE-4 hybrid enzymes that were found to be functional [45]. These studies have provided important information in understanding minimal units of folding in protein structures as well as an algorithm for designing protein evolution experiments.

In addition, TEM-1 has been used as a target for proof of concept of a codon shuffling approach to modifying protein structure and function [46]. The approach involves the in vitro assembly of synthesized hexamer DNA duplexes that represent all 20 amino acids as codon pairs or "dicodons" [46]. The assembled hexamers are then ligated into a target gene scaffold and the mutant library is selected for the appropriate function. The application of this technique to β-lactamase involved first severely truncating the enzyme to only 86 amino acids and then cloning random hexamers into this truncated version downstream of the conserved Ser70-Lys73 region that was retained [46]. Surprisingly, active versions of the truncated protein were isolated that are only approximately 100 amino acids in size compared to the wild type enzyme which consists of 263 amino acids. These experiments suggest that the structural and functional plasticity of TEM-1 β-lactamase is extreme [46]. It would be of obvious interest for future studies to determine the structure of these truncated enzymes.

ENGINEERING PRACTICAL APPLICATIONS OF β-LACTAMASES

Random mutagenesis has been applied to modify β-lactamases in order to make them better suitable for specific applications. In the Antibody Directed Enzyme Prodrug Therapy (ADEPT), a fusion protein is created between an antibody that recognises specific tumour cells antigens and an enzyme that releases a cytotoxic drug from a prodrug in the tumour vicinity; this minimizes collateral exposure [47]. β-lactamases can activate a prodrug consisting of a cephalosporin bearing a cytotoxic drug like doxorubicin as 3'-substituent (C-Dox). Hydrolysis of the β-lactam ring leads to a secondary elimination reaction releasing the drug [48]. β-lactamases are particularly useful enzymes for ADEPT systems since there will be little interference from mammalian enzymes in activating the prodrug. The class C β-lactamases, being more active on cephalosporins, are more suitable for this task. However, their kinetic parameters are not necessarily ideal for the application. Furthermore, on repeated use, these enzymes will elicit an immune response that will limit their use. Both problems can be tackled using random mutagenesis and a directed evolution approach.

To investigate the effect of mutations in a region of the *Enterobacter cloacae* P99 β-lactamases presumably involved in 3'-substituent recognition and hopefully improve its catalytic activity on C-Dox, Siemers *et al.*, [49] generated a library of random mutants of residues 286-290. Selection of active mutants was performed by plating on cefaloram, a cephalosporin containing an acetate at the 3' position (it could not be performed on C-Dox directly since the prodrug is highly toxic to bacteria). It afforded a variety of mutants, one of which had a modestly increased (1.7-fold) activity on C-Dox. A test of cytotoxicity on adenocarcinoma cells confirmed that this mutant was significantly more potent than the wild-type enzyme in inhibiting thymidine uptake (indicative of DNA synthesis) by cells treated with β-lactamase and C-Dox. Harding *et al.,* [50], on the other hand have reduced the immunogenicity of this enzyme by identifying CD4+ T-cell epitopes and introducing two amino acid changes that resulted in reduced immune cells proliferative response without compromising stability or activity.

DEVELOPMENT OF METHODS FOR *IN VITRO* SELECTION OF FUNCTION

In experiments whose purpose is to analyse the amino acid sequence determinants of function or in directed evolution projects, clones expressing active enzymes can be easily selected by plating on selective medium if the expressed enzyme confers a biological advantage. The requirement for a biological advantage is, however, a significant limitation. To overcome this limitation, two strategies can be envisaged: the development of efficient high throughput screening methods or the development of *in vitro* selection protocols. The ease with which active β-lactamases mutants can be isolated by *in vivo* selection protocols renders these enzymes excellent model systems for the development of efficient *in vitro* selection techniques.

The *in vitro* selection protocols are based on the establishment of a physical link between a gene and its expressed protein. The best known of these techniques is display on filamentous phages, in which the gene encoding the peptide or protein of interest is inserted within the gene of a coat protein, for instance between the sequences encoding the signal peptide and the mature form of protein III of filamentous phages [51]. Upon phage morphogenesis, the protein will be displayed on the capsid with the gene encoded within the phage particle. If the gene encodes an enzyme, a phage-enzyme will have been created, which in most experiments will behave like a free enzyme except that its molecular mass will be that of the phage ($\approx 2 \times 10^6$ Da). This will limit the concentration that can be put in solution.

The phage display technology was first developed to find peptides or proteins (e.g. antibodies) endowed with affinity for a chosen target ligand [51]. Libraries of phages displaying random peptides or protein mutants are submitted to affinity chromatography protocols. However, selection for activity is less straightforward. Two approaches have been explored: affinity chromatography on immobilized transition state analogues and capture by reaction with suicide substrates [52].

The second method was developed with TEM-1 [53]. In this technique, a library of phages-β-lactamases is incubated in the presence of a biotinylated suicide substrate that will react with the active phages and label them. The biotinylated phages are then extracted from the library by binding to streptavidin coated magnetic beads. The bound phages are recovered either by chemical cleavage of a disulfide bond present in the spacer connecting the suicide substrate to biotin or enzymatic cleavage of the peptide connecting the displayed enzyme to the coat protein of the phage. Infection of an *E. coli* culture allows amplification of the selected phage for characterisation or for submission to a new round of selection, usually necessary to extract the desired mutants from the background. By progressive adjustment in the successive rounds of the concentration of the biotinylated label and the time of labeling, the selection pressure can be increased to select the enzymes displaying the best catalytic activities under the chosen selection conditions.

The fact that there is no requirement for a biological advantage and that the labeling conditions can be freely defined to select enzymes with particular properties is clearly an advantage of the technique. It must be realized, however, that there is a hidden *in vivo* selection in the process since clones expressing even slightly toxic phages-enzymes will inevitably disappear from the library. Similarly, unstable enzymes that are largely degraded during expression and consequently displayed at a low level on the phage will not be efficiently selected. The first problem can be solved by application of the same selection protocol to libraries of β-lactamases that are displayed on ribosomes since this display format does not involve living cells at any step [54]. This technique allowed selection of monoclonal antibodies endowed with a β-lactamase activity [55] as well as other enzymes of modified specificity, stereospecificity or stability [56], for a review, see [57].

Application of the technique to libraries of TEM-1 mutants led to the realization that the method had to be altered to avoid selecting mutants that do not turnover. This results from the mechanism of suicide inhibition with serine β-lactamases, which at the level of the acyl-enzyme involves a competition between the normal deacylation pathway versus a series of chemical transformations leading to irreversible inhibition. If, for a particular mutant, the rate of deacylation becomes smaller than the rate of the inhibitory pathway, the enzyme will become efficiently labeled and thus selected. The solution to this problem is to preincubate the library with non-biotinylated substrate to block the slowly deacylating enzymes and then to add the biotinylated suicide substrate to the solution. This modification of the technique has allowed selection of β-lactamase mutants present at low concentration in a library containing mainly slowly deacylating enzymes Alternatively, incubation of the same library with a biotinylated penicillin at low concentration and for short reaction times allowed selection of mutants whose rate of acylation approached that of the wild-type enzyme in absence of Glu 166, a residue thought to play a significant role on the rate of acylation [58].

A selection process by catalytic elution was developed with the metallo-β-lactamase BcII from *Bacillus cereus* [59]. It involved three steps: 1. inactivation of active phage-bound enzymes by metal ion complexation, 2. binding to substrate-coated magnetic beads, 3. release of phages capable of transforming the substrate into product upon zinc salt addition. This selection protocol was applied to extract active phage-bound enzymes from a library of mutants generated by error-prone PCR. The activity of the library was shown to increase 60-fold after two rounds of selection. Randomly picked clones from the second round were shown to have a wild-type like activity.

Metallo-β-Lactamases

Codon randomization and selection techniques have also been applied to the study of metallo-β-lactamases, particularly of IMP-1, a zinc metallo-enzyme encoded by the transferable bla_{IMP-1} gene, which confers resistance to virtually all β-lactam antibiotics including carbapenems. This enzyme contains two zinc atoms in its active site coordinated to six residues, four histidines, one aspartic acid and one cysteine. In a first study, 27 libraries were created, in which individual residues in the active site area were mutated to express all possible amino acid substitutions and selected by disk diffusion tests [60]. Up to 14 positions were observed to be critical for ampicillin hydrolysis since only the wild-type residue provided resistance. Another 6 positions accepted only 2 residues. Among the critical residues are the 6 zinc ligands and a lysine interacting with the carboxyl group of the substrate. Other critical or important residues are in a flexible flap believed to be involved in substrate binding. One asparagine whose side chain is thought to bind the carbonyl oxygen of the substrate was surprisingly found to tolerate mutations [60].

The relative intolerance of the IMP-1 β-lactamase to amino acid substitutions was confirmed in another study in which codon randomization was extended to 29 residues [61]. In this study, it was observed that the hydrolysis of imipenem, cephaloridin and ampicillin has more stringent sequence requirements than the hydrolysis of cefotaxime. *In vitro* evolution experiments on BcII, a metallo-β-lactamase homologous to IMP-1 were performed by DNA shuffling and selection on cephalexin, a cephalosporin not efficiently hydrolysed by BcII [62]. They resulted in an expanded substrate spectrum without compromising hydrolytic efficiency on other substrates. Among the most frequent mutation found, Gly262Ser was involving a residue shown to be a serine for all substrates except cefotaxime by Materon *et al.* [61]. On the other hand, in an investigation whose purpose was to mimic natural evolution and predict the potential of IMP-1 to evolve an increased ability to confer resistance to imipenem, 8 libraries containing $\approx 10^7$ variants were generated by error-prone PCR. This study detected no increased resistance to imipenem and thus concluded that IMP-1 is unlikely to evolve to confer the dreaded increased resistance [63].

The metallo-β-lactamases share the αβ/βα metallohydrolase scaffold and a conserved metal-binding site with glyoxalase II (GlyII), an enzyme involved in the conversion of a cytotoxic 2-oxoaldehyde into a 2-hydroxycarboxylic acid by hydrolysis of the thioester bond of D-lactoyl-glutathione. In an effort to generate a β-lactamase activity within the GlyII scaffold, Park *et al.* [64] adopted a strategy of extensive active site rearrangement, in which after deletion of the C-terminal domain, β-lactamase catalytic elements were introduced by substitution in a GlyII loop, substrate binding elements consisting of predesigned functional loops comprising randomized residues were inserted and random point mutation adjustments were induced by DNA shuffling. Selection on cefotaxime at concentrations progressively increased from 0.2 to 4.5 μg/ml during the construction process allowed isolating a clone that was characterized. Sequencing showed that 81 of the 198 amino acid residues of GlyII were changed while the sequence identity with IMP-1 had almost doubled from 13% to 25%. The resulting enzyme had lost its original activity but catalysed the hydrolysis of cefotaxime with a $(k_{cat}/K_m)^{app}$ of 1.8×10^2 $M^{-1}.s^{-1}$, about four order of magnitude slower than that of IMP-1.

CREATION OF ALLOSTERIC β-LACTAMASES

Two strategies have been explored to transform β-lactamases into molecular sensors. These engineered enzymes would be able to detect the presence of analytes unrelated to the substrate, the response being a change in the specific activity. They could be considered as models of allosterically regulated catalysts.

In the first strategy, an allosteric binding site was planned to be created *de novo* by genetic insertion of random peptide libraries into surface loops, followed by selection and screening of active insertion mutants whose activity is modulated by binding of a target ligand. This strategy was implemented with TEM-1 taking advantage of the fact that it had been functionally displayed on phage [53]. In practice, libraries of 3 to 9 codons encoding random residues were inserted in replacement of 1 to 3 residues of the wild-type sequence between two engineered restriction sites. Alternatively, the libraries of insertion mutants were assembled by PCR using degenerate primers for the regions to be engineered. These constructions were performed in vectors featuring a mutation that inactivates the enzyme to avoid a wild-type background activity in the subsequent selection step. After transformation in *E. coli*, the library was plated on a medium containing 20 μg/ml of ampicillin, which insures that mutants whose activity corresponds to approximately 5% of the wild-type are selected. The tolerance to insertion in the chosen loop was measured by determination of the ratio between the number of clones growing in presence of a mixture of ampicillin and tetracycline *versus* tetracycline alone (the phage vector contained a tetracycline resistance gene). Insertion mutants that bound to a chosen target were then selected by several rounds of incubation with beads on which the target ligand was immobilized, followed by competitive elution. The order in which these two selection steps are performed can be inverted. A final screening step was carried out on the selected clones to find mutants whose activity is modulated by binding of the chosen ligand.

The tolerance to insertion was assessed in 5 surface loops initially chosen on the basis of several criteria including solvent exposure and mobility of the loops, information on the tolerance of mutations and proximity to the active site [65, 66]. Two relatively rigid loops connecting respectively the N-terminal helix (α1) to the first stand (β1) and the last strand (β5) to the C-terminal helix (α11) were observed to accept insertions but with some constraints. Insertions between α1 and β1 were tolerated if cysteine residues were included at both ends of the insert, suggesting the requirement for a stabilising disulfide bridge. Random sequences were also relatively well accommodated within the loop connecting β5 and α11 particularly if the wild-type residue was retained at one of the loops end. Interestingly, we observed that hydrophobic and functional residues were better accepted in the engineered loops of the active mutants than in medium-sized loops of natural proteins. On the other hand, two loops bordering the active site hardly tolerated insertions. Replacement of residues 103-105 or 104-105, in the α domain, by 6 random residues generated libraries containing less than 0.1% of active mutants. Similarly, replacement of residues 238-241 in the α/β domain by random loops of 3-6 residues afforded libraries of very low activity. Mutations of residue 104 and residues 237-240 have been reported to be responsible for an altered specificity in favour of extended-spectrum cephalosporins [19, 67]. Selection of the libraries of 238-241 insertants on cefotaxime or ceftazidime showed that the percentage of mutants active on these two extended-spectrum antibiotics was again very low [66]. Insertions were also attempted at the bottom of the Ω-loop by replacing residues 172-173 by 3-5 residues but without much success [65]. In their pentapeptide scanning mutagenesis experiments, Hayes [36] could nevertheless isolate three mutants featuring insertions between residues 170 and 171 or 171 and 172 whose activity against ampicillin was only slightly reduced (4 to 7-fold) while the activity against ceftazidime was significantly increased (up to 17-fold). It is worth pointing out that in these experiments, no relationship was detected between tolerance to mutations and tolerance to insertion.

The first attempts to generate enzymes whose activity could be modulated by binding a target ligand were designed to include a protuberance that might fit within the cavity of a monoclonal antibody. It was hoped that some of the peptides inserted into permissible sites might constitute mimotopes of epitopes of a chosen protein. Selection from active libraries in which Thr 271 had been replaced by random hexapeptides (sometimes in combination with random mutations of Glu 104 and Tyr 105), by panning on several monoclonal antibodies directed against Prostate Specific Antigen (PSA) afforded enzymes whose activity was modulated by antibody binding. With the best mutant, which was inhibited 5-fold by antibody binding, PSA could be detected in a competitive immunoassay in solution at a concentration of 1 nM [65]. Mutants endowed with a good affinity for other proteins than antibodies could also be extracted from these libraries, although, in this case, the affinity modulation was much less [68].

In an extension of this work, attempts have been made to create more complex binding sites designed to bind small organic ligands or ions by simultaneous insertion of random peptide libraries into three contiguous loops of the β-lactamase three-dimensional structure. These enzymes would contain an allosteric site, which would share some characteristics with binding sites of camelid antibodies featuring only heavy chains and three loops for binding. The libraries were constructed hierarchically. First, 3 libraries in which residues 41-42 between α1 and β1 were replaced by 5-7 random residues between 2 cysteines while residues 238-241 were randomized were constructed. They were selected for affinity: the percentage of active mutants was 5.2%. The active libraries were then combined with the library in which Thr271 had been replaced by 6 random residues and selected for activity (7% active mutants). A library of 2.9×10^8 clones was obtained of which 4.7% were active and 1.0×10^6 contained 2 extended loops and one randomized loop as planned. Altogether, the library of properly constructed active mutants was rather low (0.025%) but the hierarchical construction allowed obtaining a combinatorial library of significant diversity [66]. This library was then selected and screened to find mutants whose activity was modulated by binding of transition metal ions (Ni^{2+}, Cu^{2+} and Zn^{+2}). Several mutants containing at least 4 histidines in the engineered loops were extracted from the library. They bound the transition metal ions with binding constants in the μM range. One mutant was found whose activity was either enhanced (3-fold), inhibited (> 10-fold) or unaffected depending on the metal [69]. Selection on a support on which kanamycin was immobilized allowed isolation of a mutant which, on kanamycin binding, was activated by expulsion of an aminosulfonic acid inhibitor fortuitously bound to the engineered site ($K_d < 10^{-6}$ M). Binding of other aminoglycosides also activated that mutant; the affinity and the activation factor depended on the structure of the ligand [70]. This strategy has allowed the creation of alternative binding sites in the β-lactamase scaffold with a potential for allosteric regulation of the activity.

The second strategy consisted of the creation of protein switches by coupling two normally uncoupled proteins and their function. It involved recombination between two genes encoding, respectively, a module able to detect the presence of a ligand as input signal, and a module giving as output a modulation of an enzymatic activity [71, 72].

In practice, the two partners chosen for combination were the *E. coli* maltose binding protein (MBP), a protein involved in the specific transport of maltodextrins into bacteria, and the TEM-1 β-lactamase. MBP is made up of two domains connected by a hinge; it exists in an open and a closed conformation. Binding of the ligand induces a conformational change that brings the two domains together and closes the binding site. The hypothesis of this work was that in the coupled protein, conformational changes induced in MBP would alter the specific activity of the β-lactamase. Since it is difficult to predict how to couple the two proteins in a way that would couple the functions, Guntas *et al.,* [72] resorted to a random insertion approach. They first created a library of fusion proteins by random insertion of the β-lactamase gene into the *MBP* gene. Clones able to grow on a medium containing ampicillin and maltose were selected then screened in 96-well format for maltose dependence of β-lactamase activity. Two proteins were isolated in which the β-lactamase activity was enhanced in the presence of maltose [71]. A superior switch was isolated from a library in which circularly permuted β-lactamase genes were randomly inserted into the MBP gene. One protein was isolated whose activity was enhanced approximately 20-fold in the presence of maltose (3-fold enhancement in k_{cat} and 8-fold reduction in K_m), while other mono- or di-saccharides had no effect. The β-lactamase was shown by sequencing to be circularly permuted in a loop that precedes a β-sheet lining the active site and to have been inserted at the beginning of an α-helix in the C-terminal part of MBP. The switching activity was sufficient to result in an observable biological effect: maltose dependent resistance to ampicillin [72]. Mutants in the hinge region that have a better maltose affinity, presumably as a consequence of their closing of the apo-MBP, reduced the amplitude of the switching effect [73]. Further exploration of the possibilities of coupling the two proteins allowed isolation of fusion proteins in which the β-lactamase was circularly permuted in the Ω-loop [74]. One of the MBP-β-lactamase switch enzymes displayed a maltose-altering catalytic activity as large as 600-fold. Its activity in the presence of maltose remained, however, approximately 40 times lower than that of the wild-type β-lactamase. In an attempt to generalize the concept, Guntas et al [74] introduced random mutations in five positions of the MBP involved in maltose binding with the goal of creating a sucrose sensitive switch. They indeed isolated by *in vivo* selection and screening a mutant whose activity was increased 80-fold in the presence of sucrose but maltose had a similar switching effect. The strategy developed by Ostermeier and his group has a clear potential for creating molecular sensors for a variety of analytes.

Finally, the use of the mini-Mu transposon to insert MlyI restriction sites and thereby create 3 base pair deletions that was described above [38] has also been used to generate random domain insertions of cytochrome b_{562} within TEM-1 [74]. It was found that several of the insertion variants conferred a switching phenotype on *E. coli* with bacterial resistance levels to ampicillin dependent on heme in the growth medium [75]. It has been recently also been shown *in vitro* that heme binds the cytochrome b_{562} domain in the fusion protein and reduces hydrolysis of the β-lactam nitrocefin upon binding [76]. Therefore, random insertions of cytochrome b_{562} domains via directed evolution is another strategy to create protein switches.

CONCLUSIONS

As seen by the descriptions in this chapter, β-lactamases, and especially the TEM-1 β-lactamase, have served as targets for a large number of *in vitro* random mutagenesis experiments. In fact, the studies described above are only a fraction of the total mutagenesis studies that have been performed on TEM-1. Taken together, these experiments have made TEM-1 one of the most intensively studied of all known proteins. Information is available for the effect of substitutions at each residue position on the structure, stability and activity of the enzyme. In addition, information is available for many of the residues in terms of their impact on the substrate specificity of the enzyme. These data are useful not only in understanding β the TEM-1 structure and function but also as a source of comparative data for understanding the determinants of folding and activity of proteins in general. Because of their ease of selection, β-lactamases have also been extensively used to develop and test new molecular biology techniques. Finally, numerous protein engineering experiments on β-lactamase have served as a model for the development of advanced sensing devices.

REFERENCES

[1] Bush, K., and Jacoby, G.A. (2010). Updated functional classification of beta-lactamases. *Antimicrob. Agents Chemother. 54*, 969-976.

[2] Datta, N., and Kontomichalou, P. (1965). Penicillinase synthesis controlled by infectious R factors in *Enterobacteriacea. Nature 208*, 239-241.

[3] Wiedemann, B., Kliebe, C., and Kresken, M. (1989). The epidemiology of β-lactamases. *J. Antimicrob. Chemother. 24 (Suppl. B)*, 1-22.

[4] Perez, F., Endimiani, A., Hujer, K.M., and Bonomo, R.A. (2007). The continuing challenge of ESBLs. *Curr. Opin. Pharmacol. 7*, 459-469.

[5] Carmine, A.A., Brogden, R.N., Heel, R.C., Speight, T.M., and Avery, G.S. (1983). Cefotaxime. A review of its anti-bacterial activity, pharmacological properties and therapeutic use. *Drugs 25*, 223-265.

[6] Page, M.I., and Laws, A.P. (1998). The mechanism of catalysis and the inhibition of β-lactamases. *Chem. Commun.,* 1609-1617.

[7] Parker, R.H., and Eggleston, M. (1987). β-lactamase inhibitors: another approach to overcoming antimicrobial resistance. *Infect. Control 8*, 36-40.

[8] Jacoby, G.A., and Medeiros, A.A. (1991). More extended-spectrum β-lactamases. *Antimicrob. Agents and Chemother. 35*, 1697-1704.

[9] Medeiros, A.A. (1997). Evolution and dissemination of β-lactamases accelerated by generations of β-lactam antibiotics. *Clin. Infect. Dis. 24*, S19-S45.

[10] Palzkill, T. (1998). β-lactamases are changing their activity spectrums. *ASM News 64*, 90-95.

[11] De Pascale, G., and Wright, G.D. (2010). Antibiotic resistance by enzyme inactivation: from mechanisms to solutions. *ChemBioChem. 11*, 1325-1334.

[12] Petrosino, J., Cantu III, C., and Palzkill, T. (1998). β-lactamases: protein evolution in real time. *Trends Microbiol. 6*, 323-327.

[13] Hall, A., and Knowles, J.R. (1976). Directed selective pressure on a β-lactamase to analyse molecular changes involved in development of enzyme function. *Nature 264*, 803-804.

[14] Schultz, S.C., and Richards, J.H. (1986). Site-saturation studies of β-lactamase: production and characterization of mutant β-lactamases with all possible amino acid substitutions at residue 71. *Proc. Natl. Acad. Sci. USA 83*, 1588-1592.

[15] Healey, W.J., Labgold, M.R., and Richards, J.H. (1989). Substrate specificities in class A beta-lactamases: preference for penams vs. cephems. The role of residue 237. *Proteins: Struc. Funct. Genet. 6*, 275-283.

[16] Oliphant, A.R., and Struhl, K. (1989). An efficient method for generating proteins with altered enzymatic properties: application to β-lactamase. *Proc. Natl. Acad. Sci. USA 86*, 9094-9098.

[17] Dube, D.K., and Loeb, L.A. (1989). Mutants generated by the insertion of random oligonucleotides into the active site of the β-lactamase gene. *Biochemistry 28*, 5703-5707.

[18] Palzkill, T., and Botstein, D. (1992). Probing β-lactamase structure and function using random replacement mutagenesis. *Proteins 14*, 29-44.

[19] Palzkill, T., and Botstein, D. (1992). Identification of amino acid substitutions that alter the substrate specificity of TEM-1 β-lactamase. *J. Bacteriol. 174*, 5237-5243.

[20] Palzkill, T., Le, Q.-Q., Venkatachalam, K.V., LaRocco, M., and Ocera, H. (1994). Evolution of antibiotic resistance: several different amino acid substitutions in an active site loop alter the substrate profile of β-lactamase. *Mol. Microbiol. 12*, 217-229.

[21] Petrosino, J.F., and Palzkill, T. (1996). Systematic mutagenesis of the active site omega loop of TEM-1 β-lactamase. *J. Bacteriol. 178*, 1821-1828.

[22] Cantu III, C., Huang, W.Z., and Palzkill, T. (1997). Cephalosporin substrate specificity determinants of TEM-1 β-lactamase. *J. Biol. Chem. 272*, 29144-29150.

[23] Huang, W.Z., Petrosino, J., Hirsch, M., Shenkin, P.S., and Palzkill, T. (1996). Amino acid sequence determinants of β-lactamase structure and activity. *J. Mol. Biol. 258*, 688-703.

[24] Kunkel, T.A., Roberts, J.D., and Zakour, R.A. (1987). Rapid and efficient site-specific mutagenesis without phenotypic selection. *Methods Enzymol. 154*, 367-382.

[25] Hayes, R.J., Bentzien, J., Ary, M.L., Hwang, M.Y., Jacinto, J.M., Vielmetter, J., Kundu, A., and Dahiyat, B.I. (2002). Combining computational and experimental screening for rapid optimization of protein properties. *Proc. Natl. Acad. Sci.* USA *99,* 15926-15931.

[26] Stemmer, W.P.C. (1994). Rapid evolution of a protein *in vitro* by DNA shuffling. *Nature* (London) *370*, 389-391.

[27] Orencia, M.C., Yoon, J.S., Ness, J.E., Stemmer, W.P., and Stevens, R.C. (2001). Predicting the emergence of antibiotic resistance by directed evolution and structural analysis. *Nat. Struct. Biol. 8*, 238-242.

[28] Huang, W.Z., and Palzkill, T. (1997). A natural polymorphism in β-lactamase is a global suppressor. *Proc. Natl. Acad. Sci. USA 94*, 8801-8806.

[29] Makeyev, E.V., and Bamford, D.H. (2004). Evolutionary potential of an RNA virus. *J. Virol. 78*, 2114-2120.

[30] Barlow, M., and Hall, B.G. (2002). Predicting evolutionary poteintial: In vitro evolutions accurately reproduces natural evolution of the TEM β-lactamase. *Genetics 160*, 823-832.

[31] Barlow, M., and Hall, B.G. (2003). Experimental prediction of the natural evolution of antibiotic resistance. *Genetics 163*, 1237-1241.

[32] Barlow, M., and Hall, B.G. (2003). Experimental prediction of the evolution of cefepime resistance from the CMY-2 AmpC b-lactamase. *Genetics 164*, 23-29.

[33] Wang, X.J., Minasov, G., and Shoichet, B.K. (2002). Evolution of an antibiotic resistance enzyme constrained by stability and activity trade-offs. *J. Mol. Biol. 320*, 85-95.

[34] Brown, N.G., Pennington, J.M., Huang, W.Z., Ayvaz, T., and Palzkill, T. (2010). Multiple global suppressors of protein stability defects facilitate the evolution of extended-spectrum TEM β-lactamases. *J. Mol. Biol. 404*, 832-846.

[35] Peimbert, M., and Segovia, L. (2003). Evolutionary engineering of a β-lactamase activity on a D-Ala D-Ala transpeptidase fold. *Protein Engineer. 16*, 27-35.

[36] Hayes, F., Hallet, B., and Cao, Y. (1997). Insertion mutagenesis as a tool in the modication of protein function. *J. Biol. Chem. 272*, 28833-28836.

[37] Osuna, J., Yanez, J., Soberon, X., and Gaytan, P. (2004). Protein evolution by codon-based random deletions. *Nucleic Acids Res. 32*, e136.

[38] Jones, D.D. (2005). Triplet nucleotide removal at random positions in a target gene: the tolerance of TEM-1 β-lactamase to an amino acid deletion. *Nucl. Acids Res. 33*, e80.

[39] Simm, A.M., Baldwin, A.J., Busse, K., and Jones, D.D. (2007). Investigating protein structural plasticity by surveying the consequence of an amino acid deletion from TEM-1 beta-lactamase. *FEBS Lett. 581*, 3904-3908.

[40] Stemmer, W.P.C. (1994). DNA shuffling by random fragmentation and reassembly: *In vitro* recombination for molecular evolution. *Proc. Natl. Acad. Sci.USA 91*, 10747-10751.

[41] Ostermeier, M., Shim, J.H., and Benkovic, S.J. (1999). A combinatorial approach to hybrid enzymes independent of DNA homology. *Nature Biotechnol. 17*, 1205-1209.

[42] Lutz, S., Ostermeier, M., Moore, G.L., Maranas, C.D., and Benkovic, S.J. (2001). Creating multiple-crossorver DNA libraries independent of sequence identity. *Proc. Natl. Acad. Sci. USA 98*, 11248-11253.

[43] Sieber, V., Martinez, C.A., and Arnold, F.H. (2001). Libraries of hybrid proteins from distantly related sequences. *Nature Biotechnol. 19*, 456-460.

[44] Hiraga, K., and Arnold, F.H. (2003). General method for sequence-independent site-directed chimeragenesis. *J. Mol. Biol. 330*, 287-296.

[45] Voigt, C.A., Martinez, C., Wang, Z.-G., Mayo, S.L., and Arnold, F.H. (2002). Protein building blocks preserved by recombination. *Nat. Struct. Biol. 9*, 553-558.

[46] Chopra, S., and Ranganathan, A. (2003). Protein evolution by "codon shuffling": A novel method for generating highly variant mutant libraries by assembly of hexamer DNA duplexes. *Chemistry & Biology 10*, 917-926.

[47] Bagshawe, K.D. (1987). Antibody directed enzymes revive anticancer prodrugs concept. *Br. J. Cancer 56*, 531-532.

[48] Vrudhula, V.M., Svensson, H.P., Kennedy, K.A., Senter, P.D., and Wallace, P.M. (1993). Antitumor activities of a cephalosporin prodrug in combination with monoclonal antibody-beta-lactamase conjugates. *Bioconjug. Chem. 4*, 334-340.

[49] Siemers, N.O., Yelton, D.E., Bajorath, J., and Senter, P.D. (1996). Modifying the specificity and activity of the *Enterobacter cloacae* P99 b-lactamase by mutagenesis within an M13 phage vector. *Biochemistry 35*, 2104-2111.

[50] Harding, F.A., Liu, A.D., Stickler, M., Razo, O.J., Chin, R., Faravashi, N., Viola, W., Graycar, T., Yeung, V.P., Aehle, W., Meijer, D., Wong, S., Rashid, M.H., Valdes, A.M., and Schellenberger, V.A. (2005). β-lactamase with reduced immunogenicity for the targeted delivery of chemotherapeutics using antibody-directed enzyme prodrug therapy. *Mol. Cancer Ther. 4*, 1791-1800.

[51] Smith, G.P., and Petrenko, V.A. (1997). Phage display. *Chem. Rev. 97*, 391-410.

[52] Soumillion, P., Legendre, D., and Fastrez, J. (2004). Enzyme engineering by phage display. In *Enzyme Functionality, design, engineering and screening*, A. Svendsen, ed. (New York: Marcel Dekker), pp. 391-412.

[53] Soumillion, P., Jespers, L., Bouchet, M., Marchand-Brynaert, J., Winter, G., and Fastrez, J. (1994). Selection of β-lactamase on filamentous bacteriophage by catalytic activity. *J. Mol. Biol. 237*, 415-422.

[54] Amstutz, P., Pelletier, J.N., Guggisberg, A., Jermutus, L., Cesaro-Tadic, S., Zahnd, C., and Plückthun, A. (2002). In vitro selection for catalytic activity with ribosome display. *J. Am. Chem. Soc. 124*, 9396-9403.

[55] Tanaka, F., Almer, H., Lerner, R.A., and Barbas, C.F. (1999). Catalytic single-chain antibodies possessing β-lactamase activity selected from a phage displayed combinatorial library using a mechanism-based inhibitor. *Tetrahedron Letters 40*, 8063-8066.

[56] Droge, M.J., Boersma, Y.L., van Pouderoyen, G., Vrenken, T.E., Ruggeberg, C.J., Reetz, M.T., Dijkstra, B.W., and Quax, W.J. (2006). Directed evolution of Bacillus subtilis lipase A by use of enantiomeric phosphonate inhibitors: crystal structures and phage display selection. *ChemBioChem 7*, 149-157.

[57] Fernandez-Gacio, A., Uguen, M., and Fastrez, J. (2003). Phage display as a tool for the directed evolution of enzymes. *Trends Biotechnol. 21*, 408-414.

[58] Vanwetswinkel, S., Avalle, B., and Fastrez, J. (2000). Selection of β-lactamases and penicillin binding mutants from a library of phage displayed TEM-1 β-lactamase randomly mutated in the active site Ω-loop. *J. Mol. Biol., 295*, 527-540.

[59] Ponsard, I., Galleni, M., Soumillion, P., and Fastrez, J. (2001). Selection of metalloenzymes by catalytic activity using phage display and catalytic elution. *ChemBioChem 2*, 253-259.

[60] Materon, I.C., and Palzkill, T. (2001). Identification of residues critical for metallo-β-lactamase function by codon randomization and selection. *Protein Sci. 10*, 2556-2565.

[61] Materon, I.C., Beharry, Z., Huang, W.Z., Perez, C., and Palzkill, T. (2004). Analysis of the context dependent sequence requirements of active site residues in the metallo-beta-lactamase IMP-1. *J. Mol. Biol. 344*, 653-663.

[62] Tomatis, P.E., Rasia, R.M., Segovia, L., and Vila, A.J. (2005). Mimicking natural evolution in metallo-β-lactamases through second-shell ligand mutations. *Proc. Natl. Acad. Sci.* USA *102*, 13761-13766.

[63] Hall, B.G. (2004). In vitro evolution predicts that the IMP-1 metallo-β-lactamase does not have the potential to evolve increased activity against imipenem. *Antimicrob. Agents Chemother. 48*, 1032-1033.

[64] Park, H.-S., Nam, S.-H., Lee, J.K., Yoon, C.N., Mannervik, B., Benkovic, S.J., and Kim, H.-S. (2006). Design and evolution of new catalytic activity with an existing protein scaffold. *Science 311*, 535-538.

[65] Legendre, D., Soumillion, P., and Fastrez, J. (1999). Engineering a regulatable enzyme for homogeneous immunoassays. *Nat. Biotechnol. 17*, 67-72.

[66] Mathonet, P., Deherve, J., Soumillion, P., and Fastrez, J. (2006). Active TEM-1 β-lactamase mutants with random peptides inserted in three contiguous surface loops. *Protein Sci. 15*, 2323-2334.

[67] Cantu III, C., Huang, W.Z., and Palzkill, T. (1996). Selection and characterization of amino acid substitutions at residues 237-240 of TEM-1 β-lactamase with altered substrate specificity for aztreonam and ceftazidime. *J. Biol. Chem. 271*, 22538-22545.

[68] Legendre, D., Vucic, B., Hougardy, V., Girboux, A.L., Henrioul, C., Van Haute, J., Soumillion, P., and Fastrez, J. (2002). TEM-1 beta-lactamase as a scaffold for protein recognition and assay. *Protein Sci. 11*, 1506-1518.

[69] Mathonet, P., Barrios H., Soumillion, P., and Fastrez, J. (2006). Selection of allosteric β-lactamase mutants featuring an activity regulation by transition metal ions. *Protein Sci. 15*, 2335-2343.

[70] Volkov, A.N., Barrios, H., Mathonet, P., Evrard, C., Ubbink, M., Declercq, J.-P., Soumillion, P., and Fastrez, J. (2011). Engineering an allosteric binding site for aminoglycosides into TEM1-β-lactamase. ChemBioChem 12, 904-913.

[71] Guntas, G., and Ostermeier, M. (2004). Creation of an allosteric enzyme by domain insertion. *J. Mol. Biol. 336*, 263-273.

[72] Guntas, G., Mitchell, S.F., and Ostermeier, M. (2004). A molecular switch created by in vitro recombination of nonhomologous genes. *Chem. Biol. 11*, 1483-1487.

[73] Kim, J.R., and Ostermeier, M. (2006). Modulation of effector affinity by hinge region mutations also modulates switching activity in an engineered allosteric TEM-1 beta-lactamase switch. *Arch. Biochem. Biophys. 446*, 44-51.

[74] Guntas, G., Mansell, T.J., Kim, J.R., and Ostermeier, M. (2005). Directed evolution of protein switches and their application to the creation of ligand-binding proteins. *Proc. Natl. Acad. Sci.* USA *102*, 11224-11229.

[75] Edwards, W.R., Busse, K., Allemann, R.K., and Jones, D.D. (2008). Linking the functions of unrelated proteins using a novel directed evolution domain insertion method. *Nucl. Acids Res. 36*, e78.

[76] Edwards, W.R., Williams, A.J., Morris, J.L., Baldwin, A.J., Allemann, R.K., and Jones, D.D. (2010). Regulation of β-lactamase activity by remote binding of heme: Functional coupling of unrelated proteins through domain insertion. *Biochemistry 49*, 6541-6549.

In: Beta-Lactamases
Editor: Jean-Marie Frère, pp. 459-477

ISBN: 978-1-61324-638-2

Chapter 19

FOLDING AND STABILITY OF CLASS A β-LACTAMASES

***André Matagne*[1] *and Roger H. Pain*[2]**

[1] Laboratoire d'Enzymologie et Repliement des Protéines, Centre d'Ingénierie des Protéines, Université de Liège, Institut de Chimie B6, 4000 Liège, Belgium.

[2] Department of Biotechnology, Jožef Stefan Institute, Jamova 39, 1000 Ljubljana, Slovenia.

ABSTRACT

Class A β-lactamases have proved useful as model proteins in studying a wide variety of aspects of protein folding. We review those features that have shed light on kinetic intermediates that take part in folding, including some insight into the molecular basis of the kinetic and stable molten-globule states that have been identified. The contrast between the folding behaviour of PC1 and the two lactamase mutants, P54 and P2, can be attributed in some detail to changes in molecular conformation. The early, very rapid stages of folding of β-lactamases have been shown to be multiphasic, and an interesting intermediate is described that has non-native contacts involving burial of the C-terminal tryptophan. A further feature is that the region around the disulphide bond in the TEM enzymes is formed very early in the folding reaction. The unusual feature of class A β-lactamases, i.e. a *cis* peptide bond critical in the stereochemistry of the active site that usually, but not always, involves a proline residue, has been shown to be important in accounting for the slow folding reactions.

The effect of substrates on the stabilization of the enzymes and on their reversible deactivation is also reviewed.

INTRODUCTION

Much of the progress in understanding the way in which proteins fold has been marked by intensive studies on model proteins. Proteins such as ribonuclease, lysozyme and bovine pancreatic trypsin inhibitor were chosen for their availability, for particular points of structural interest, and for their solubility and ready reversibility of unfolding in the test tube.

Class A β-lactamases, initially known as penicillinases, constitute one such group and are among the largest proteins studied in terms of their folding properties. The enzyme from *Staphylococcus aureus* PC1, for example, is readily prepared and purified, very soluble and kinetically stable in solution, and can be reversibly folded in high yield, even at high concentrations. The absence of disulphide bonds in some members of the family makes them particularly suitable, since the folding depends solely on noncovalent interactions.

These β-lactamases (257-281 residues; $M_r \approx 29000$) have one particularly interesting property from the folding point of view. Although comprising two 'domains' (Figure 1), these are not the classically defined domains that are joined by just a single chain since, sequence-wise, the α/β domain (111-135 residues) is composed of approximately 45 (35-56) residue N-terminal and 80 (74-89) residue C-terminal extensions to the all-α domain (146-154 residues). The five β-strands of the α/β domain are contributed by the two extensions. One face of the β sheet is packed closely to the all-α domain, while the other is covered by three or four α-helices in contact with the solvent [1]. This raises the question, as yet unanswered, as to whether the protein folds by an initial general collapse, from which state it sorts itself out into the two domains, or whether the all-α domain folds first, forming a template onto which the α/β domain folds. The alternative, traditional view on two domain proteins – that the domains fold separately and then associate – is less easy to envisage in this case. There is a certain structural similarity to hen egg white lysozyme, which features two structural domains, one of which, the α-domain, is also composed of N- and C-terminal extensions to the β-domain. Hydrogen-deuterium exchange studies [2,3] have indicated that, at pH 5.5 and 20°C, the majority (~70%) of the lysozyme refolding molecules acquire native-like protection in the α-domain before a stable β-domain is formed. It is not impossible that the α/β domain of β-lactamase folds first but, since it contains the β-sheet, this is conceptually more difficult to envisage. Experimental verification of this point is eagerly awaited.

We present some of the findings obtained on the stability and folding of a number of β-lactamases, with an indication of how these have contributed to the wider fields of protein stability and folding.

STABILITY

The stability of β-lactamases varies markedly, from *S. aureus* PC1 at 20-30 kJ·mol^{-1} [4-6] to TEM-1 [7] and *Bacillus licheniformis* BS3 [8] at ~45 kJ·mol^{-1}. The thermal unfolding temperatures (T_m) of the enzymes from *B. licheniformis*, strains BS3 and 749/C, are particularly high, with values around 63°C, compared with 50°C and 42°C for the TEM-1 and PC1 enzymes, respectively [9]. The physiological significance of these differences is not understood but the variation has been ascribed to differences in numbers of intramolecular hydrogen bonds and salt bridges, rather than in the global hydrophobicity of the proteins [9].

In a molecular dynamics simulation study of the β-lactamase from *S. aureus*, Beveridge and his colleagues [10] found that the α-helices (mainly in the α domain) are, interestingly, less stable to thermal denaturation than the β structure of the α/β domain, presumably reflecting the greater overall stability of the latter. Approximately 85% of the sheet structure was retained at the end of the simulation but only 35–40% of the helices. This is reminiscent of the values of ~100% and ~60%, respectively, obtained experimentally for a partially denatured state (state H, see below) using circular dichroism [11].

STABILIZATION BY SUBSTRATES

The stabilization of enzymes in the presence of substrates has been recognized for a long time. Quantitative information regarding this phenomenon is, however, sparse, since the enzyme destroys the potential stabilizing agent during the course of the experiments. The stability of the *S. aureus* PC1 β-lactamase was determined in the presence of a number of substrates [12]. Heat-induced inactivation of the enzyme was monitored in real time by following directly the decrease of the rate of substrate hydrolysis. Significant enzyme stabilization was observed in the presence of substrate concentrations above the K_m values for all compounds tested. The effectiveness of the protection was found to be strongly substrate dependent and to correlate well with the calculated energies of activation of the various Henri-Michaelis, non-covalent complexes, which also appeared to be valid for the corresponding covalent (i.e. acyl-enzyme) species.

SUBSTRATE INDUCED DEACTIVATION

Certain poor substrates for β-lactamases, such as methicillin, cloxacillin, quinacillin and nafcillin, reversibly lower the activity of the enzyme towards benzylpenicillin, and Citri termed them A type substrates [13] (reviewed in [14]). The inactivation has been associated with a conformational change in the β-lactamase on the basis of changes in the degree of hydrogen-deuterium exchange [15]. Half the hydrogens probed were affected by methicillin and cloxacillin, but not by penicillin G or cephalosporin C. Further, both inactivation and the subsequent reactivation were prevented by antibodies and also by chemical cross-linking which locked the enzyme into either the active or inactive form [16]. The addition of sulphate, which is known to stabilize, to increase the asymmetry of tyrosine residues, and to slightly increase the helicity of the *S. aureus* enzyme [5], slowed the rate of inactivation [17]. The substrate is covalently attached to the inactivated enzyme [17–19], which is more susceptible to urea denaturation and trypsinolysis [15,20]. This change in conformation is reversed at the conclusion of hydrolysis of the substrate. The inactivated enzyme, separated rapidly at low temperature from excess substrate and product, was examined by far UV circular dichroism and fluorescence emission and found, by these criteria, to be indistinguishable from the native enzyme [20].

Taken together, these effects are consistent with a 'tightening up' and reduction of the dynamics of the enzyme on binding the substrate that is sufficient to affect interaction with antibody, and that prevents the conformational changes necessary for hydrolysis and/or release of the product. The absence of change of fluorescence suggests that the effect is local. The loop that supports some of the active site residues, termed the Ω-loop (Figure 1; see further), interacts with the main body of the enzyme, mainly through hydrogen bonded water molecules, and is thus an attractive candidate region. This would be consistent with the sensitivity of this interaction to thermal denaturation shown in a molecular dynamics simulation [21]. Among other suggested explanations, it has been proposed [20] that the deactivation is due to the formation of two alternative acyl-enzyme complexes of similar free energy, the 'inactive' one either deacylating slowly or undergoing a slow conformational change to the 'native' enzyme, thus enabling deacylation. These two possibilities are kinetically indistinguishable.

Molten Globule-like Intermediate States

The search for partially folded states has formed an important part of experimental studies of protein folding, and a number of β-lactamases were among the earliest proteins to yield interesting information about thermodynamically stable states having conformations intermediate between those of the native and unfolded states ($U \rightleftharpoons I_1 \rightleftharpoons (\ldots) \rightleftharpoons I_n \rightleftharpoons N$) [4, 22–24]. Studies on these and some other proteins formed the basis of the characterization and understanding of the molten globule state [25]. These states, produced originally by reversible chemical denaturation and termed state H (H standing for helical), have been observed by a variety of experimental methods, including optical spectroscopy, NMR, size exclusion chromatography, urea gradient electrophoresis, sedimentation analysis, viscosity and ANS (1-anilino-8-naphthalene sulphonate) binding, and also by molecular dynamics simulation [21].

The molten globule state is characterized by pronounced secondary structure, and differs from the native state mainly in the presence of reduced persistent tertiary structure in at least some regions of the molecule, expressed as increased dynamics of the 'globule', greater accessibility of hydrophobic surfaces and/or ability to intercalate ANS, and a small increase in the hydrodynamic radius. The partially unfolded – or partially folded – state, state H, is thermodynamically stable in the presence of low concentrations of urea or guanidinium chloride denaturants. Intermediate states with similar properties have been shown to be induced by electrostatic destabilization at low pH (A states) and at high pH (B states) in a number of proteins, including β-lactamases [24]. Titration of β-lactamase to pH 2 in the presence of a low concentration of salt results in an acid unfolded state. In the presence of 0.45 M KCl, however, the compact A state is formed, with a far UV CD spectrum closely similar to that of the native protein but with no significant near UV ellipticity. A further probe of this intermediate, using a spin label attached to a cysteine residue introduced by mutation at position 166 [26], exhibited a rotational relaxation identical to that of the probe at neutral pH (i.e. under conditions favouring the native state). This apparent contrast to the results from other probes of tertiary structure probably indicates that β-lactamase molten globules are, like those of other smaller proteins (reviewed in [27]), not expanded homogeneously, but display a combination of natively packed and 'molten' regions. In larger proteins, such as the β-lactamases, this heterogeneity could include segments of structure, such as the Ω-loop (Figure 1; see further), that could be detached from the main structure, thus resulting in a larger hydrodynamic radius.

These molten globule states can be unfolded rapidly to the unfolded state or folded relatively slowly to the native state. Their stability relative to the native state varies from protein to protein, but is approximately half that of the native enzyme. TEM-1 β-lactamase, for example, forms a molten globule state that is 21.7 (± 1.7) $kJ \cdot mol^{-1}$ less stable than the native state and 23.8 (± 0.8) $kJ \cdot mol^{-1}$ more stable than the fully denatured protein [7].

Similarly, in PC1 from *S. aureus,* state H is 10.1 (± 1.5) kJ·mol^{-1} less stable than N and 10.1 (± 1.6) kJ·mol^{-1} more stable than U [5].

Sulphate ions show a remarkable effect on the stability of PC1 [5]. The ellipticity at 220 nm and at 278 nm is significantly increased, suggesting a tightening up of the conformation of the native protein. In the presence of unfolding concentrations of urea, the addition of sulphate progressively reverses the transitions N $\rightleftharpoons$ H and H $\rightleftharpoons$ U . In the presence of 1 M sulphate, both transitions are shifted to higher concentrations of urea by 5.3 M. The fact that the same values of $\Delta G°$ are obtained from these sulphate folding transitions in the presence of urea as from the urea unfolding transitions in the absence of sulphate proves the thermodynamic reversibility in this mixed solvent system. The effect of sulphate on the stability of N relative to H is twice that on the stability of H relative to U, the values of $\Delta G°_{\text{H-N}}$ and $\Delta G°_{\text{U-H}}$ being made more negative by ~41 kJ·mol^{-1} and ~20 kJ·mol^{-1}, respectively, per mol·L^{-1} of sulphate. The increase in stability of H with respect to U is reflected almost entirely by a decrease in the unfolding rate constant – by a factor of 10^{11} – in the presence of 1.4 M sulphate. The difference in the effects of sulphate on the N $\rightleftharpoons$ H and H $\rightleftharpoons$ U transitions indicates that a higher proportion of the non-polar interactions are broken on going from N to H than from H to U. Nandi and Robinson have shown, by model compound studies, that the major effect of sulphate is in salting out non-polar side-chains rather than peptide groups [28,29]. Thus the stabilization of H relative to U is more dependent on the burial of peptide bonds than is the stabilization of N relative to H. This is consistent with the molten globule model for state H described above. This result is further reinforced by analysis of the *m* values calculated in unfolding experiments, with both PC1 (5,6) and TEM-1 (7). This parameter, which account for the dependence of $\Delta G°$ on denaturant concentration ($m = \delta\Delta G°/\delta$ [denaturant]), i.e. for the slope of the transition(s), can be interpreted as a measure of exposure to solvent, of normally buried non-polar side chains, upon unfolding [30]. On this basis, it can be calculated that 60-70% of the hydrophobic residues buried in the native state, are exposed in the H state of PC1 and TEM-1. Note also that for the four β-lactamases (i.e. PC1, TEM-1, BS3 and NMCA) unfolded in guanidinium chloride, *m* values identical within the error limit (34-39 kJ·mol^{-1} [8]) were found, indicating similar exposure of solvent-accessible surface area upon global unfolding of these enzymes.

The TEM-1 β-lactamase is particularly interesting, in that it contains a single disulphide bond which connects Cys77 (α-2 helix) and Cys123 (α-4 helix) in the all-α□domain (Figure 1). This disulphide bridge is not essential for enzyme activity or correct folding [33–37], but its removal destabilizes the protein by *ca.* 14 kJ·mol^{-1} [36,38]. This protein exhibits a thermodynamically stable molten globule state, N $\rightleftharpoons$ H $\rightleftharpoons$ U, whether the disulphide is intact or not [36,39]. Interestingly, the N $\rightleftharpoons$ H transition ($\Delta\Delta G°$ = 3.7 kJ·mol^{-1}) is much less affected by removal of the disulphide than H $\rightleftharpoons$ U ($\Delta\Delta G°$ = 10.4 kJ·mol^{-1}), presumably because the entropy of the U state is much more affected than that of the H state in the presence of the disulphide bridge [36]. The observation that the disulphide bond does not significantly change the stability of N relative to H [36] is in keeping with the fact that it is not necessary for enzymatic activity [34,35].

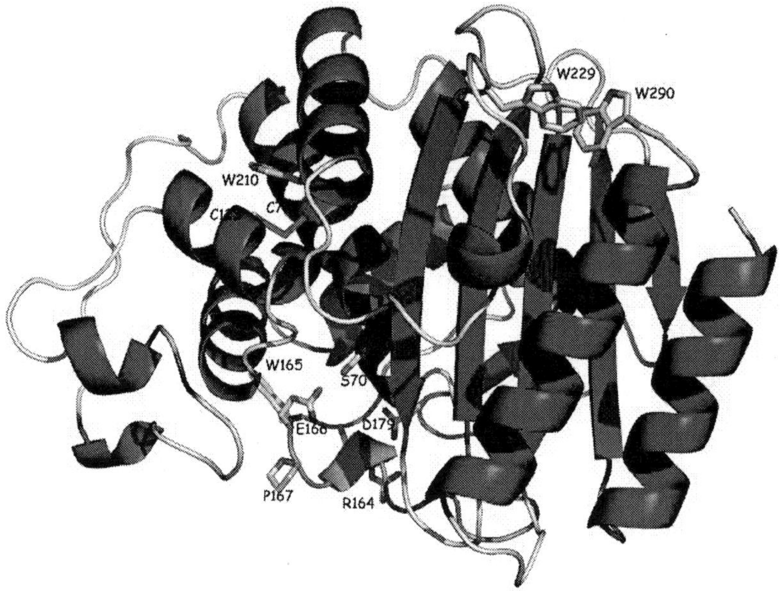

Figure 1. Schematic view of the structure of the TEM-1 β-lactamase [31,32]. The secondary structure elements in the native structure are highlighted red (α-helices) and blue (β-strands). The Ω-loop and the disulphide bonds are shown in green and yellow, respectively. A few important residues (see text) are indicated.

A Kinetic Intermediate in Protein Folding Is Similar to the Thermodynamically Stable Molten Globule

The first experimental evidence for the occurrence of a condensed kinetic intermediate conformational state in protein folding reactions was obtained from urea-gradient gel electrophoresis experiments carried out at low temperature [22,40]. With the PC1 β-lactamase, these experiments (Figure 2) revealed a partially folded species that accumulates rapidly and is transiently populated during refolding from the unfolded (U) to the native (N) state. This intermediate, termed state I, shows a mobility on gel close to that of the native protein and is more compact than the well defined, thermodynamically stable state H, which is also seen on the gels (Figure 2). It thus displays the considerably more compact conformation of the classical molten globule. Although state I is in rapid equilibrium with state H, its interconversion with state N is slow (at temperatures below 10°C, folding takes place over a period of hours), and this constitutes the rate-limiting step of the folding process.

Examination of the refolding kinetics of the PC1 β-lactamase [4,5,25] by a variety of other techniques, including size exclusion chromatography, absorbance and circular dichroism spectroscopy, and enzyme activity measurements, indicated that the condensed state I is formed very rapidly during refolding ($t_{1/2}$ < 100 ms), is enzymatically inactive, exhibits a substantial amount of secondary structure (~70-80 % of the native far UV CD signal) and displays only a few, if any, specific tertiary contacts (confirmed by ^{1}H-NMR experiments, [25]). Furthermore, these experiments showed that the values of the rate constants for the recovery of both specific enzymatic activity and native-like optical properties are identical within the error limit, indicating that folding from I to N is a cooperative two-state process.

Finally, the adsorption of ANS (maximal by *ca* 50 ms, [25]) to state I revealed that the rapid collapse of the protein molecule is characterized by formation of a solvent-accessible hydrophobic core. These distinct features of the kinetic intermediate state I (i.e. compact, high secondary structure content but reduced persistent tertiary interactions, biologically inactive, high affinity for ANS) are typical of the properties of the molten globule species (state H) observed under equilibrium conditions. State I is, however, clearly more compact than state H (Figure 2) and has a slightly different secondary structure (from far UV CD spectra). This is in contrast to α-lactalbumin, in which, on the basis of the similarity of their far UV CD spectra, the H and I states have been reported as being very closely similar [41].

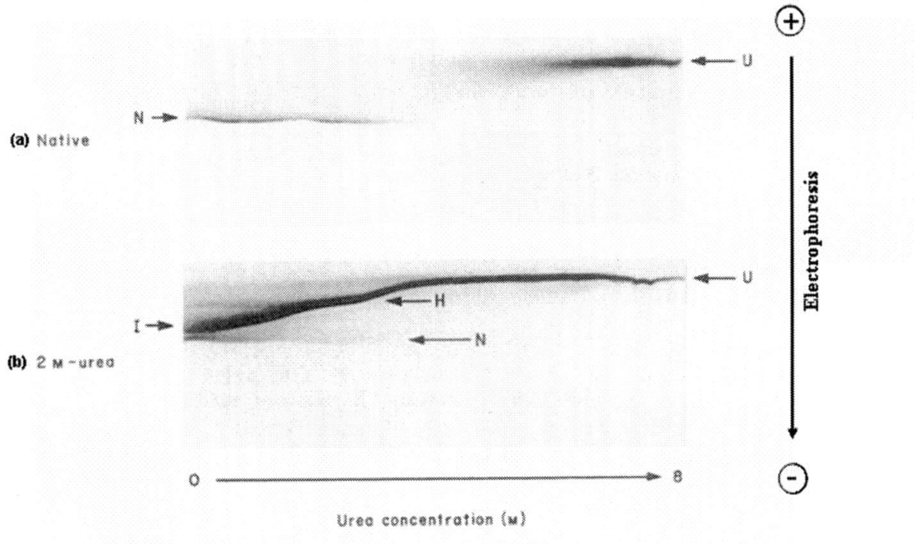

Figure 2. Urea-gradient acrylamide gel electrophoresis of (a) native β-lactamase, (b) β-lactamase in 2 M urea. Each gel contained a horizontal gradient from 0 to 8 M urea, with an inverse gradient of 15 to 11% acrylamide to compensate for the electrophoretic effects of the urea [40]. Samples were layered on the top of the gel and electrophoresed at 2°C for 60 min. (a) At low concentrations of urea, the protein migrates as a horizontal line at a position corresponding to N; only at high concentrations does it unfold sufficiently rapidly, relative to the time of electrophoresis, to run at a position corresponding to U. At intermediate concentrations, where the half time for unfolding is comparable to the duration of electrophoresis, the band of protein is, as expected, smeared out between the positions expected for N and U. (b) When applied in 2 M urea, in which the thermodynamically stable state H predominates, a band is observed in the region of 2 M urea whose mobility is compatible with that of state H [11]. At high concentrations of urea, a band corresponding to U is seen, which is shown by the sharp, continuous line to be in rapid equilibrium with H. At concentrations approaching zero, a sharp band with a mobility corresponding to that of N is separate from that of the slightly more expanded intermediate I, indicating slow interconversion between N and I. The latter intermediate is shown to be distinctly more compact than H, with which it interconverts rapidly. The presence of slowly refolding species migrating more slowly than I at low concentrations of urea is consistent with restrictions on folding resulting from slow proline isomerization. Reprinted from J. Mol. Biol., 137, T. E. Creighton & R. H. Pain, Unfolding and refolding of *Staphylococcus aureus* penicillinase by urea-gradient electrophoresis, pp. 431-436, 1980, with permission from Elsevier.

The fact that molten globule-like kinetic intermediate species have been observed with a few globular proteins displaying different structural properties, led to the proposal that the molten globule state could be a general intermediate in protein folding [25]. This is now generally accepted, at least for proteins with more than a hundred residues.

The possibility that the two partially folded states H and I are on-pathway species (i.e. U $\rightleftharpoons$ H $\rightleftharpoons$ I $\rightleftharpoons$ N) has been discussed [5,42] but not firmly established.

Early Kinetic Events in β-lactamase Folding

Kinetic experiments, in which refolding is triggered by diluting out the chemical denaturant manually (leading to experimental dead-times $\geq$ 5 s), indicate that secondary structure is formed much more rapidly than tertiary structure in class A β-lactamases. Extrapolation of these kinetics of folding to zero time shows that different enzymes of the same class (PC1, TEM-1, NMCA, BS3 and PER-1) all fold to a kinetic molten globule intermediate (I-type), containing at least 70% of the far UV ellipticity of the native protein but with reduced persistent tertiary structure and no catalytic activity [7,8,25].

More recently, stopped-flow experiments have shown that a collapsed state, with substantial amounts of secondary structure, is formed within the dead-time of these rapid-mixing experiments (2-4 ms for fluorescence and 10-15 ms for CD measurements). Thus, with both the PC1 [6] and TEM-1 [43] enzymes, a large proportion (60-70 %) of the native ellipticity in the far UV is acquired during the burst phase (i.e. the dead-time of the stopped-flow experiment). With the TEM-1 β-lactamase, complementary fluorescence methods were used to monitor refolding. These experiments indicate that the burst phase intermediate is characterized by significant enhancement of the intrinsic fluorescence intensity, decreased access of acrylamide to aromatic residues and substantial ANS binding. The maximal binding of ANS actually occurs within the burst phase, a finding consistent with early hydrophobic collapse of the protein molecule. The intrinsic fluorescence measurements, both in the absence (Figure 3a) and presence (Figure 3c) of the quenching agent, show that at least one of the four tryptophan residues encounters a significant change in its environment, resulting in an increase in fluorescence intensity, with lower accessibility to the solvent than in the native protein, in which they are largely exposed. The collapsed intermediate species formed at the onset of the refolding reaction thus contains non-native hydrophobic interactions involving tryptophan residues, which are indicative of a non-hierarchical folding pathway [43]. Examination of the TEM-1 β-lactamase structure [43] and investigation of the folding kinetics of the W290F mutant [44] suggest that the fast early collapse leads to a folding intermediate in which the C-terminal end of the polypeptide chain is trapped in a non-native hydrophobic environment. In this highly fluctuating species, the C-terminal tryptophan could well find a lower energy state by becoming buried in a collapsed and partially folded (molten globule-like) α domain [43]. When refolding of the PC-1 and NMCA β-lactamases is monitored by stopped-flow fluorescence [6,45], no corresponding burst phase is observed. It is nevertheless quite possible that similar non-native interactions occur with these enzymes also, but are not detected, simply because of the absence of any tryptophan residue at their C-termini (Phe in PC1 and Lys in NMCA)

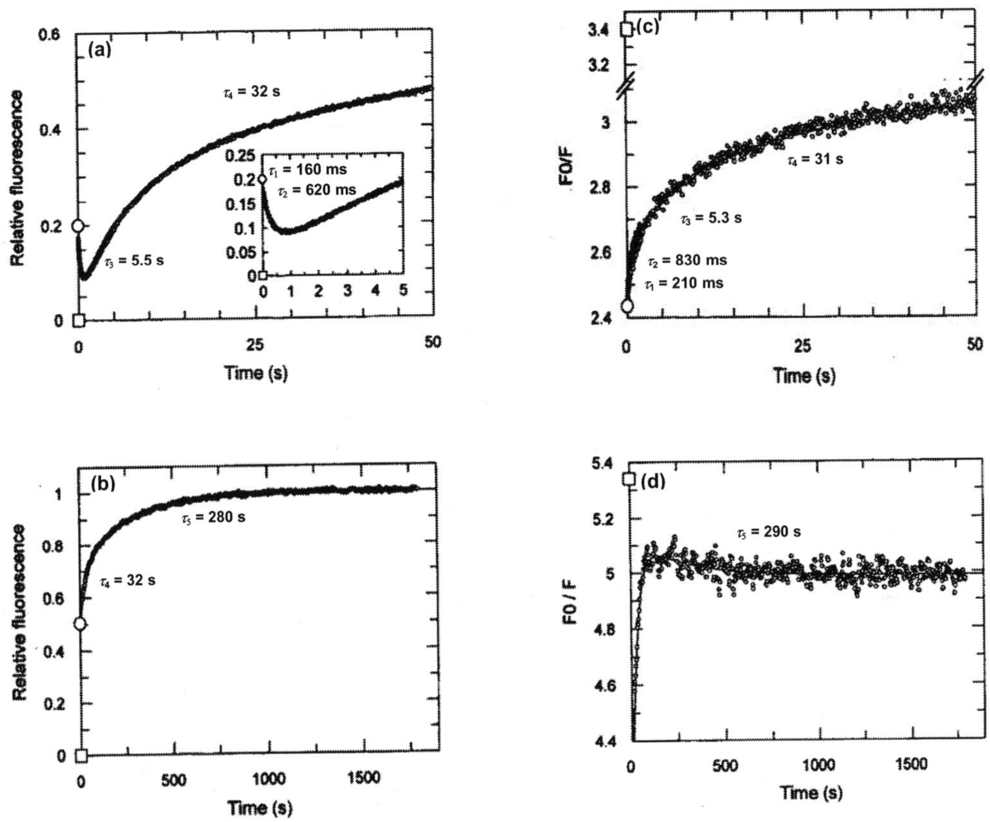

Figure 3. The folding kinetics of TEM-1 β-lactamase at pH 7, 25°C, at a final guanidinium chloride concentration of 0.27 M. Refolding was monitored by intrinsic tryptophan fluorescence in the absence (a, b) and in the presence (c, d) of 0.182 M acrylamide. The inset in (a) shows the first 5 s of the reaction. Quenching of fluorescence was measured by adding acrylamide in the refolding buffer. In this case, parallel folding experiments in the absence of acrylamide allowed F_0/F, i.e. the ratio of fluorescence in the absence and in the presence of quencher, to be calculated as a function of time. The fluorescence emission has been normalized so that the value of the denatured state under refolding conditions is 0 (represented by □) and that of the native state is 1. For acrylamide quenching experiments, F_0/F of the fully unfolded protein was deduced from the fluorescence intensity of the protein in 3.0 M guanidinium chloride, measured either in the presence or in the absence of acrylamide. (a, c) stopped-flow mixing, (b, d) manual mixing. The solid lines represent the fit of a sum of four exponentials (a, c) and of a sum of two exponentials (b, d) to the data. (○) Intrinsic fluorescence (a, b) and F_0/F (c) extrapolated to zero time from the kinetic data. Reprinted from Biochemistry 37, Vanhove, M., Lejeune, A., Guillaume, G., Virden, R., Pain, R.H., Schmid, F.X. and Frère, J.-M., A collapsed intermediate with non-native packing of hydrophobic residues in the folding of TEM-1 β-lactamase, pp. 1941-1950, with permission from the American Chemical Society.

Following the very rapid formation of the collapsed state, a series of distinct kinetic phases are observed with both PC1 and TEM-1 [6,43]. The increase in intrinsic fluorescence of the TEM-1 enzyme (Figure 3a,b) in the burst phase is followed by a rapid decrease in two exponential phases ($\tau_1 \approx 160$ ms and $\tau_2 \approx 620$ ms, with time constant $\tau = 1/k$), before the signal characteristic of the native state is eventually reached in three further phases ($\tau_3 \approx 5.5$ s, $\tau_4 \approx 32$ s and $\tau_5 \approx 280$ s), resulting in the lowest fluorescence intensity at ~800 ms. Based on the study of a mutant TEM-1 enzyme (C77A and C123A) lacking the disulphide bond of the wild type enzyme, Gervasoni and Plückthun [46] assigned this quenching of fluorescence to the location of Trp210 in a native-like environment next to the enzyme disulphide bond (Figure 1). Formation of the native ellipticity (35-40% is lacking after the burst phase) occurs on a 60 s time scale, concomitantly with a strong increase in quenching by acrylamide (Figure 3c,d). These observations can be assigned to the formation of native-like structure in both the all-α and α/β domains, which is accompanied by the expulsion of Trp290 from the α domain core and its exposure to solvent.

The folding kinetics of the TEM-1 β-lactamase (summarized in Table 1) exhibit at least five measurable phases (only four phases were observed with the PC1 enzyme). As discussed below, the last two slow phases, with time constants of 20-30 s and 200-300 s at 25°C, are rate-limited by proline isomerization. The close similarity of the rate constant values measured by complementary techniques shows that each kinetic phase represents a well-defined, cooperative transition.

Slow Phases and XAA-Pro Bond Isomerization

A distinctive feature of class A β-lactamases is the presence of a long Ω-loop of ~16 residues, packed against the main body of the native enzyme, near the active site (Figure 1). Although the loop is not tightly bound to the rest of the protein structure, contacts being mediated mainly by about 10 water molecules, it is stabilized by a salt bridge between residues Arg164 and Asp179. This loop can thus be seen as a ‘flap’ at the bottom of the enzyme active site. Remarkably, it contains a *cis*-peptide bond preceding residue 167 (a proline in most, but not all, enzymes). The precise conformation of the loop and the *cis* configuration of the bond to residue 167 are important for the orientation of Glu166, which is essential for catalytic activity.

The acquisition of enzyme activity of class A β-lactamases (time constant (τ) between 3 and 15 min at 20-25°C), is concomitant with the final step in the folding and is determined by the rate of isomerization of the 166-167 peptide bond to the *cis*-configuration. Thus, when residue 167 is a proline (e.g. TEM-1 and BS3), a classical double-jump experiment [47] shows that both the optical and catalytic properties of the native enzyme are restored within the dead-time of the experiment (5-10 s [7,8]). In this experiment, the protein is fully unfolded for a period of time (5-10 s) too short to allow significant isomerization of proline-containing peptide bonds, and then allowed to refold. Refolding thus proceeds from a population of unfolded molecules in which most of the proline isomers are in the native conformation. In contrast, the double-jump procedure has no effect on the refolding kinetics of the PC1 (Ile167) and NMCA (Leu167) β-lactamases [6,8]. These experiments suggest that the very slow phase observed with the TEM-1 and BS3 enzymes results from isomerization of a peptide bond involving a proline (i.e. Glu166-Pro167), whereas in the case of the PC1 and NMCA enzymes, refolding is rate-limited by isomerization of the same bond, but with no proline involved.

With some enzymes (e.g. TEM-1, NMCA and BS3), a second slow kinetic phase is also observed (τ = 10-40 s at 25°C) that is sensitive to the double-jump assay (i.e. it is not observed after a short incubation time of the protein in the unfolded state [7,8]) and hence probably accounts for a complex process kinetically limited by *cis*/*trans* isomerization of still unassigned proline residues [7,8,48].

Table 1. Kinetic parameters for the folding of TEM-1 β-lactamase at pH 7, 25°C, in 0.273 M guanidinium chloride[a]

probe	burst phase	phase 1	phase 2	phase 3	phase 4	phase 5
Trp fluorescence[(b)] τ (s)	< 0.001 (+)	0.16 (–)	0.62 (–)	5.5 (+)	32 (+)	280 (+)
Acryl quenching[(c)] τ (s)	< 0.001 (–)	0.21 (+)	0.83 (+)	5.3 (+)	31 (+)	290 (–)
ANS fluorescence τ (s)	< 0.001 (+)	0.15 (–)	0.71 (–)	5.6 (–)	30 (–)	290 (–)
Far UV CD[(d)] τ (s)	< 0.003 (+)	0.065 (+)	1.1 (+)	18 (+)	–	–
Activity[(e)] τ (s)	No	No	No	No	20 (+)	200 (+)
Sensitivity to DJ[(f)]	No	No	No	No	Yes	Yes

[a] Data from [7,43].

[b] Intrinsic tryptophan fluorescence.

[c] Quenching of tryptophan fluorescence by acrylamide.

[d] Circular dichroism at 225 nm (these experiments were performed at 21°C).

[e] Recovery of enzymatic activity (these experiments were performed at a final guanidinium chloride concentration of 0.17 M).

[f] Sensitivity to double-jump assay.

The + and – signs denote an increase and a decrease, respectively, of the measured parameter. Note that an increase in fluorescence quenching by acrylamide corresponds to an increase in tryptophan solvent exposure.

In order to investigate further the possible role of the *cis* configuration of the Glu166-Xaa167 peptide bond in determining the kinetics of folding, mutants of the TEM-1 [48], the NMCA [8] and the PC-1 [6] enzymes were constructed in which the residue at position 167 was replaced, giving the P167T, L167P and I167P enzymes, respectively. The catalytic activity of the mutant enzymes was not significantly modified, indicating that in all cases the Glu166-Xaa167 peptide bond, like that in the corresponding wild type enzymes, remains *cis*. Interestingly, with TEM-1 the mutation led to a significant decrease in stability (~7 kJ·mol^{-1}), which is within the expected range of 5-15 kJ·mol^{-1} if the *trans*/*cis* ratio changes from between 2.5:1 and 10:1 (Glu166-Pro167) to between 100:1 and 1000:1 (Glu166-Thr167). With both the PC1 and NMCA enzymes, however, no evidence for enhanced stability could be found on introducing a proline at position 167, suggesting that the favourable contribution of the *cis*/*trans* equilibrium is offset by less favourable packing of the proline side-chain within the folded PC1 and NMCA proteins [6,8]. It is thus likely that the active site of most class A β-lactamases has evolved to accommodate a Pro at position 167, which is thermodynamically favourable to the *cis* conformation adopted in all enzymes but PER-1 (see below).

With the three enzymes, the mutations led to significant changes in the refolding kinetics. With the TEM-1 β-lactamase, the two slower phases ($\tau_1 \approx 20$ s and $\tau_2 \approx 200$ s) observed in the refolding kinetics of the wild type protein were replaced by a single, slower phase ($\tau \approx 360$ s), which was insensitive to the double-jump experiment. This observation points to the *trans* → *cis* isomerization of the Glu166-Thr167 peptide bond as the structural event responsible for the slow monophasic refolding reaction of the mutant. In contrast, substitution of Ile and Leu by Pro in the PC1 and NMCA enzymes, respectively, led to faster folding and sensitivity to the double-jump assay. Results obtained with PC1 brought support to the suggestion of Herzberg and co-workers [49], based on the structure of the PC1 and P54 enzymes (see below), that the *trans* → *cis* isomerization of the Glu166-Ile167 bond should be the rate-limiting step in the conversion of the kinetic intermediate to the native state.

These experiments performed with the three enzymes bearing a mutation at position 167, led to the proposal that the slowest, rate-limiting phase observed in the folding process of class A β-lactamases is due to the *trans* → *cis* isomerization of the Glu166-Xaa167 peptide bond. Further support for this conclusion has been obtained recently [8]. The PER-1 β-lactamase is the only class A enzyme known to display a *trans* configuration of the Glu166-Ala167 peptide bond [50]. With this enzyme, the slow refolding phase ($\tau > 3$ min at 25°C) observed with all other class A β-lactamases studied to date is absent. This observation further reinforces the view that the folding of all class A enzymes containing a *cis* Glu166-Xaa167 peptide bond (except PER-1) is kinetically limited by *trans* → *cis* isomerization of this bond in the Ω-loop.

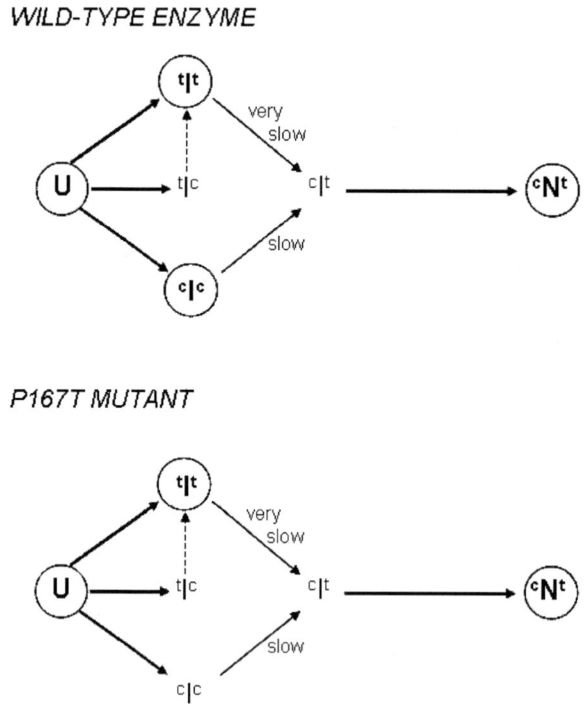

Figure 4. Kinetic models for the folding of wild type and mutant P167T TEM-1 enzymes. The circled symbols in bold type represent the forms which can be present in significant quantities during the folding process. The superscripts preceding and following the capital letter which identifies the state of folding of the protein refer to the conformations, respectively, of the 166-167 peptide bond and the other Xaa-Pro bond(s), whose isomerization accounts for the observed phases of folding. The steps represented by bold arrows are fast compared with the mixing dead time. It remains possible that ${}^{t}I^{c}$ might be formed in a non-negligible proportion but, in this case, the transition ${}^{t}I^{c} \rightarrow {}^{t}I^{t}$ (bold, dashed arrow) should be faster than both ${}^{t}I^{t} \rightarrow {}^{c}I^{t}$ and ${}^{c}I^{c} \rightarrow {}^{c}I^{t}$ [48].

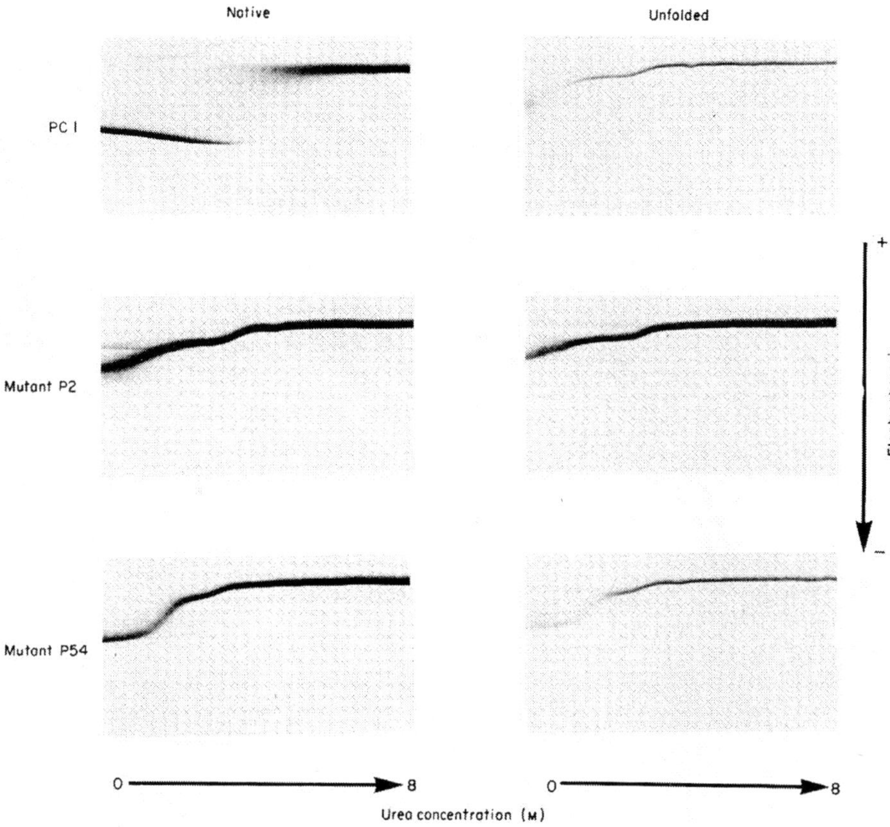

Figure 5. Urea-gradient acrylamide gel electrophoresis of (a) PC1 and (b) P2 and (c) P54 β-lactamase variants. Experimental details as for Figure 2. Each protein was applied (a) dissolved in buffer and (b) in 8 M urea, and electrophoresed for 30 min. (a) Unfolded PC1 folds rapidly to H and to I, but N is not seen clearly due to the folding being slow compared to the time of electrophoresis. (b) and (c) The similarity of the patterns for the two mutants, applied in their folded and unfolded forms, demonstrates rapid equilibration between U and H and between H and the final folded I-like state. Comparison of the patterns for P2 and P54 at very low urea concentrations shows that the former is on the border of being a globular protein. Reprinted from J. Mol. Biol., 185, S. Craig, M. Hollecker, T. E. Creighton and R. H. Pain, Single amino acid mutations block a late step in the folding of β-lactamase from *Staphylococcus aureus*, pp. 681- 687, 1985, with permission from Elsevier.

Finally, as stated above, formation of the native state of TEM-1, monitored by recovery of activity, occurs with biphasic kinetics [7,8,46], indicating that the polypeptide chain folds *via* two parallel pathways. This is not the case with the P167T mutant of TEM-1, which displays only one slow phase. Possible models for the folding kinetics of the wild type and mutant TEM-1 enzymes are given in Figure 4. A recent analysis [8] of the slow refolding phase observed with the various enzymes studied to date, suggest that these models adequately describe the folding of all class A β-lactamases (except PER-1). They differ, however, in the relative population of the three major kinetic species (i.e. ${}^{t}I^{t}$, ${}^{c}I^{c}$, ${}^{t}I^{c}$; note that ${}^{c}I^{t}$ is never observed, since it is readily converted to ${}^{c}N^{t}$). In particular, the absence of a prolyl at position 167 very significantly destabilizes the intermediate species (${}^{c}I^{c}$) with a *cis* peptide bond at position 166-167. In all cases, ${}^{t}I^{t}$ appears to be the most stable species. This incompletely folded species can be viewed as the native structure of the enzyme, lacking only the well-defined native fold of the Ω-loop. This species could reasonably be expected to exhibit a stability similar to that of the native state, whereas in all other significant intermediate species, the wider structural consequences of non-native proline isomers in the main core of the protein would be expected to result in substantially lower stability than the native state.

The Region around the Disulphide Bond Is Native-like at an Early Stage of Folding

As mentioned previously, the enzymes of the TEM family contain a disulphide bond between cysteines 77 and 123 (Figure 1). Walker and Gilbert [37,38] have shown that oxidative refolding of the reduced TEM-1 enzyme involves a competition between disulphide bond formation and folding of the molecule into a compact, catalytically active conformation that buries the two reduced cysteines in the core of the enzyme. Based on decreasing accessibility to *N*-ethyl-maleimide, half of the burial of the thiol groups occurs at a very early stage of folding [38]. In agreement with these reports, other lines of evidence suggest that the region around the disulphide bond is folded in a native-like pattern early during folding: (1) Trp210 comes close to its native topology, next to the disulphide bond, within 1-4 s [43,46] (as noted above, the quenching in fluorescence shown in Figure 3A, with a minimum in intensity at ~800 ms, has been assigned to the location of Trp210 in a native-like topology), and (2) the kinetics of the phases leading to the native conformation are scarcely affected by removal of the disulphide bond [33,36,46]. Interestingly, the bacterial thiol/disulphide oxidoreductase DsbA cannot catalyze formation of the disulphide in the reduced native-like state, and only does so in the H state because of the latter's fast exchange with U [33].

The Staphylococcal P54 Mutant Enzyme

One of the two mutants made by Novick [51], P54, has the aspartate at position 179 in PC1 substituted by an asparagine [52]. This results in an enzyme with substantially reduced activity, reduced intensity in the near UV CD spectrum (that is not increased in the presence of sulphate) and a slightly greater Stokes radius than that of the wild type enzyme [42,49]. In the crystal structure [49], the Ω-loop is disordered, due presumably to the loss of the main stabilizing interactions of Asp179 with the side chain of Arg164 and the main chain nitrogens of Val163 and Arg164. The main body of the enzyme, however, is closely similar to that of PC1, with an rms deviation for all atoms of 0.76 Å and, for main chain atoms, 0.45 Å. The former value reflects primarily surface side-chains, especially lysine, while the only main chain atoms that deviate more than 1 Å from the position of PC1 are located in close proximity to the Ω-loop and are associated with the active site depression.

The kinetics of folding of P54 are dramatically different from those of PC1, in that the slow, rate-determining step is completely lacking. While the fast phases of folding are indistinguishable from those of the PC1 enzyme (Christensen, H. & Pain, R.H., unpublished observations), the slow phase characteristic of PC1 is absent, as monitored by near UV CD. Urea-gradient gel electrophoresis supports these results, showing that the rapidly formed collapsed state H is in fast equilibrium with the more condensed, fully folded P54 protein which, in turn, is less compact than PC1 (Figure 5). Based on this kinetic behaviour, it was suggested that the folded state closely resembles the transient intermediate I observed with PC1. That it is not the same as I is evident from the near UV CD spectrum [42] and subsequently from the crystal structure [49], the main body of the enzyme being folded sufficiently stably to impart optical asymmetry to one or more tyrosines (14 in total), while the disordered Ω-loop accounts for the observed increase in Stokes radius and for the absence of the constraint inducing the Glu166-Ile167 peptide bond into the *cis* configuration. These results support the suggestion [49], strengthened by the kinetic studies on both TEM-1 [48] and PC1 [6], that the slow folding phase of PC1 is due to isomerization of the Glu166-Ile167 peptide bond. Furthermore, this last step in the folding of most β-lactamases has been proposed to be intimately associated with the folding and collapse of the Ω-loop residues on to the main body of the enzyme, into the specific conformation that orients Glu166 (and also probably Asn170, see [53]) properly for catalytic activity [6,54].

The fact that P54, unlike the intermediate I, behaves as a stable globular protein, despite having lost the stabilizing interactions involving Asp179, would appear to be due to the compensating formation of an alternative interaction between the amide nitrogen of Asn179 and the main chain nitrogen of Ala69 [49]. The other Novick mutation of Thr40 to Ile in P2 results in a folded species that is very much on the border of being a stable globular protein [42] (Figure 5). From the structure of PC1 [55], the mutation would be expected to prevent the Ω-loop from 'bedding down' on to the main body of the enzyme. In this case, there are not likely to be any compensating stabilizing interactions.

Perhaps the most interesting feature of the folding of P54 is that the main body of the enzyme, which contains nine proline residues, folds rapidly from the equilibrium unfolded state, within the time scale of manual mixing. This demands further investigation.

ACKNOWLEDGEMENTS

R.H.P. is indebted to Mark Richmond who introduced him to the reversible deactivation of penicillinases, to Richard Ambler who encouraged him in the early days of β-lactamase folding, to Richard Virden as a long-term collaborator, to all those students and postdocs who made life interesting and did the work, to the stimulating friends at Liège, and last, but not least, all those who survived and appeared to enjoy the Holy Island β-lactamase Workshops. Thanks are due to Professor Vito Turk for support in Ljubljana.

A.M. is grateful to Jean-Marie Frère for continual encouragement and support since his very first day in a research lab, and to Chris Dobson who supervised his first step in protein folding research and remains of great help and inspiration.

A.M. and R.H.P. also wish to acknowledge the excellent experimental work done in Liège by Marc Vanhove, Annabelle Lejeune, Julie Vandenameele and Alexandre Di Paolo, as well as the many enlightening discussions we had together. We are grateful to Paulette Charlier for preparing Figure 1.

A.M. was Research Associate of the F.R.S.-FNRS (1998-2010) and was supported in part by a grant from the Fonds de la Recherche Fondamentale et Collective (contract numbers 2.4550.05 and 2.4530.09). Most of the work performed in Liège and reported in this review was supported by the Belgian program of Interuniversity Attraction Poles initiated by the

Federal Office for Scientific Technical and Cultural Affaires (PAI n° P3/19, P4/03, P5/33 and P6/19).

REFERENCES

[1] Herzberg, O. and Moult, J. (1987) Bacterial resistance to β-lactam antibiotics: crystal structure of β-lactamase from *Staphylococcus aureus* PC1 at 2.5 Å resolution. *Science* 236, 694-701.

[2] Radford, S.E., Dobson, C.M. and Evans, P.A. (1992) The folding of hen lysozyme involves partially structured intermediates and multiple pathways. *Nature* 358, 302-307.

[3] Miranker, A., Robinson, C.V., Radford, S.E., Aplin, R.T. and Dobson, C.M. (1993) Detection of transient protein folding populations by mass spectrometry. *Science* 262, 896-900.

[4] Robson, B. and Pain, R.H. (1976) The mechanism of folding of globular proteins. Equilibria and kinetics of conformational transitions of penicillinase from *Staphylococcus aureus* involving a state of intermediate conformation. Biochem. J. 155, 331-344.

[5] Mitchinson, C. and Pain, R.H. (1985) Effects of sulphate and urea on the stability and reversible unfolding of β-lactamase from *Staphylococcus aureus*. Implications for the folding pathway of β-lactamase. *J. Mol. Biol.* 184, 331-342.

[6] Wheeler, K.A., Hawkins, A.R., Pain, R.H. and Virden, R. (1998) The slow step of folding of *Staphylococcus aureus* PC1 β-lactamase involves the collapse of a surface loop rate limited by the trans to cis isomerization of a non-proline peptide bond, Prot. *Struct. Funct. Gen.* 33, 550-557.

[7] Vanhove, M., Raquet, X. and Frère, J.-M. (1995) Investigation of the folding pathway of the TEM-1 β-lactamase. *Prot. Struct. Funct. Gen.* 22, 110-118.

[8] Vandenameele, J., Lejeune, A., Di Paolo, A., Brans, A., Frère, J.-M., Schmid, F.X., and Matagne, A. (2010) Folding of class A β-lactamases is rate-limited by peptide bond isomerization and occurs via parallel pathways, *Biochemistry,* 49, 4264-4275.

[9] Vanhove, M., Houba, S., Lamotte-Brasseur, J. and Frère, J.-M. (1995) Probing the determinants of protein stability: comparison of class A β-lactamases. *Biochem.* J. 308, 859-864.

[10] Vijayakumar, S., Vishveshwara, S., Ravishanker, G. and Beveridge D.L. (1993) Differential stability of β-sheets and a-helices in β-lactamase: a high temperature molecular dynamics study of unfolding intermediates. *Biophys. J.* 65, 2304–2312.

[11] Carrey, E.A. and Pain, R.H. (1978) Conformation of a stable intermediate on the folding pathway of *Staphylococcus aureus* penicillinase. *Biochim. Biophys. Acta,* 533, 12-22.

[12] Lejeune, A., Vanhove, M., Lamotte-Brasseur, J., Pain, R.H., Frère, J.-M. and Matagne, A. (2001) Quantitative analysis of the stabilization by substrate of *Staphylococcus aureus* PC1 β-lactamase. *Chem. Biol.* 8, 831-842.

[13] Citri, N., Samuni, A. and Zyk, N. (1976) Acquisition of substrate-specific parameters during the catalytic reaction of penicillinase. *Proc. Natl. Acad. Sci. U.S.A.* 73, 1048-1052.

[14] Pain R.H. and Virden R. (1979), The structural and conformational basis of β-lactamases. In *β-lactamases* (Hamilton-Miller, J.M.T. and Smith, J.T. eds.), pp. 141-180, Academic Press, New York.

[15] Kiener, P.A. and Waley, S.G. (1977) Substrate-induced deactivation of penicillinases. Studies of β-lactamase I by hydrogen exchange. *Biochem. J.* 165, 279-285.

[16] Carrey, E.A., Virden, R. and Pain, R.H. (1984) The reversible deactivation of β-lactamase from *Staphylococcus aureus* by quinacillin and cephaloridine and its modification by antibodies. *Biochim. Biophys. Acta.* 785, 104-110.

[17] Persaud, K.C., Pain, R.H. and Virden, R. (1986) Reversible deactivation of β-lactamase by quinacillin. Extent of the conformational change in the isolated transitory complex. *Biochem. J.* 237, 723-730.

[18] Virden, R., Bristow, A.F. and Pain, R.H. (1975) The active site of penicillinase from *Staphylococcus aureus* PC1. Isolation of a specific covalent complex with the substrate quinacillin. *Biochem. J.* 149, 397–401.

[19] Kiener, P.A., Knott-Hunziker, V., Petursson, S. and Waley, S.G. (1980) Mechanism of substrate-induced inactivation of β-lactamase I. *Eur. J. Biochem.* 109, 575-80.

[20] Fink, A.L., Behner, K.M. and Tan, A. (1987) Kinetic and structural characterization of reversibly inactivated β-lactamase. *Biochemistry,* 26, 4248-4258.

[21] Roccatano, D., Sbardella, S., Aschi, M., Amicosante, G., Bossa, C., Di Nola, A. and Mazza, F. (2005) Dynamical aspects of TEM-1 β-lactamase probed by molecular dynamics. *J. Comput. Aided Mol. Des.*19, 329-340.

[22] Creighton, T.E. and Pain, R.H. (1980) Unfolding and refolding of *Staphylococcus aureus* penicillinase by urea-gradient electrophoresis. *J. Mol. Biol.* 137, 431-436.

[23] Thomas, R.M., Feeney J., Nicholson R.B., Pain, R.H. and Roberts, G.C.K. (1983) Identification by n.m.r. spectroscopy of a stable intermediate structure in the unfolding of staphylococcal β-lactamase. *Biochem. J.* 215, 525-9.

[24] Goto, Y. and Fink, A.L. (1989), Conformational states of β-lactamase: molten-globule states at acidic and alkaline pH with high salt. *Biochemistry* 28, 945-952.

[25] Ptitsyn, O.B., Pain, R.H., Semisotnov, G.V., Zerovnik, E. and Razgulyaev, O.I. (1990) Evidence for a molten globule state as a general intermediate in protein folding. *FEBS Lett.* 262, 20-24.

[26] Calciano L.J., Escobar, W.A., Millhauser, G.L., Miick, S.M., Rubaloff, J., Todd, A.P. and Fink, A.L. (1993) Side-chain mobility of the β-lactamase A state probed by electron spin resonance spectroscopy. *Biochemistry.* 32, 5644-9.

[27] Kuwajima, K. and Arai, M. (2000) The molten globule state: the physical picture and biological significance. In *Mechanisms of Protein Folding,* 2nd Edn., Pain, R.H., ed. pp. 138-174, Oxford University Press, Oxford.

[28] Nandi, P.K. and Robinson, D.R. (1972) The effects of salts on the free energy of the peptide group, *J. Amer. Chem. Soc.* 94, 1299-1308.

[29] Nandi, P.K. and Robinson, D.R. (1972) The effects of salts on the free energies of nonpolar groups in model peptides, *J. Amer. Chem. Soc.* 94, 1308-1315.

[30] Myers, A. J., Pace, C. N. and Scholtz, J. M. (1995) Denaturant *m* values and heat capacity changes: Relation to changes in accessible surface areas of protein unfolding, *Protein Sci.* 4, 2138-2148.

[31] Strynadka, N.C., Adachi, H., Jensen, S.E., Johns, K., Sielecki, A., Betzel, C., Sutoh, K. and James, M.N. (1992) Molecular structure of the acyl-enzyme intermediate in β-lactam hydrolysis at 1.7 Å resolution, *Nature* 359, 700-705.

[32] Jelsch, C., Lenfant, F., Masson, J.M. and Samama, J.P. (1993) Crystal structure of *Escherichia coli* TEM1 β-lactamase at 1.8 Å resolution, *Prot. Struct. Funct. Gen.* 16, 364-383.

[33] Frech, C., Wunderlich, M., Glockshuber, R. and Schmid, F.X. (1996) Competition between DsbA-mediated oxidation and conformational folding of RTEM1 β-lactamase, *Biochemistry* 35, 11386-11395.

[34] Laminet, A.A. and Plückthun, A. (1989) The precursor of β-lactamase: purification, properties and folding kinetics, *EMBO J.* 8, 1469-1477.

[35] Schultz, S.C., Dalbadie-McFarland, G., Neitzel, J.J. and Richards, J.H. (1987) Stability of wild-type and mutant RTEM-1 β-lactamases: effect of the disulfide bond, *Prot. Struct. Funct. Gen.* 2, 290-297.

[36] Vanhove, M., Guillaume, G., Ledent, P., Richards, J.H., Pain, R.H. and Frère, J.-M. (1997) Kinetic and thermodynamic consequences of the removal of the Cys-77-Cys-123 disulphide bond for the folding of TEM-1 β-lactamase, *Biochem. J.* 321, 413-417.

[37] Walker, K.W. and Gilbert, H.F. (1994) Effect of redox environment on the *in vitro* and *in vivo* folding of RTEM-1 β-lactamase and *Escherichia coli* alkaline phosphatase, *J. Biol. Chem.* 269, 28487-28493.

[38] Walker, K.W. and Gilbert, H.F. (1995) Oxidation of kinetically trapped thiols by protein disulfide isomerase, *Biochemistry* 34, 13642-13650.

[39] Zahn, R. and Plückthun, A. (1994) Thermodynamic partitioning model for hydrophobic binding of polypeptides by GroEL. II. GroEL recognizes thermally unfolded mature β-lactamase. *J. Mol. Biol.* 242, 165-174.

[40] Creighton, T.E. (1980) Kinetic study of protein unfolding and refolding using urea gradient electrophoresis. *J. Mol. Biol.* 137, 61-80.

[41] Ikeguchi, M., Kuwajima, K., Mitani, M. and Sugai, S. (1986) Evidence for identity between the equilibrium unfolding intermediate and a transient folding intermediate: a comparative study of the folding reactions of α-lactalbumin and lysozyme, *Biochemistry* 25, 6965-6972.

[42] Craig, S., Hollecker, M., Creighton, T.E. and Pain, R.H. (1985) Single amino acid mutations block a late step in the folding of β-lactamase from *Staphylococcus aureus*. *J. Mol. Biol.* 185, 681-687.

[43] Vanhove, M., Lejeune, A., Guillaume, G., Virden, R., Pain, R.H., Schmid, F.X. and Frère, J.-M. (1998) A collapsed intermediate with nonnative packing of hydrophobic residues in the folding of TEM-1 β-lactamase, *Biochemistry* 37, 1941-1950.

[44] Lejeune, A., Pain, R.H., Charlier, P., Frère, J.-M. and Matagne, A. (2008) TEM-1 β-lactamase folds in a nonhierarchical manner with transient non-native interactions involving the C-terminal region, *Biochemistry* 47, 1186-1193.

[45] Lejeune, A. (2001) Etude du folding des β-lactamases de classe A, *Ph.D. thesis,* University of Liège.

[46] Gervasoni, P. and Plückthun, A. (1997) Folding intermediates of β-lactamase recognized by GroEL. *FEBS Lett.* 401, 138-142.

[47] Brandts, J.F., Halvorson, H.R. and Brennan, M. (1975) Consideration of the possibility that the slow step in protein denaturation reactions is due to *cis-trans* isomerism of proline residues, *Biochemistry* 14, 4953-4963.

[48] Vanhove, M., Raquet, X., Palzkill, T., Pain, R.H. and Frère, J.-M. (1996) The rate-limiting step in the folding of the *cis*-Pro167Thr mutant of TEM-1 β-lactamase is the *trans* to *cis* isomerization of a non-proline peptide bond. *Prot. Struct. Funct. Gen.* 25, 104-111.

[49] Herzberg, O., Kapadia, G., Blanco, B., Smith, T.S. and Coulson A. (1991) Structural basis for the inactivation of the P54 mutant of β-lactamase from *Staphylococcus aureus* PC1. *Biochemistry* 30, 9503-9509.

[50] Tranier, S., Bouthors, A.T., Maveyraud, L., Guillet, V., Sougakoff, W. and Samama, J.P. (2000) The high resolution crystal structure for class A β-lactamase PER-1 reveals the bases for its increase in breadth of activity. *J. Biol. Chem.* 275, 28075-28082.

[51] Novick, R.P. (1963) Analysis by transduction of mutations affecting penicillinase formation in S*taphylococcus aureus. J. Gen. Microbiol.* 33, 121-136.

[52] Ambler, R.P, (1980) The structure of β-lactamases. *Phil. Trans. Roy. Soc. Lond.* 289, 321-331.

[53] Lamotte-Brasseur, J., Dive, G., Dideberg, O., Charlier, P., Frère, J.-M. and Ghuysen, J.-M. (1991) Mechanism of acyl transfer by the class A serine β-lactamase of *Streptomyces albus G. Biochem. J.* 279, 213-221.

[54] Vanhove, M., Lejeune, A. and Pain, R.H. (1998) β-Lactamases as models for protein-folding studies, *Cell. Mol. Life Sci.* 54, 372-377.

[55] Herzberg, O. (1991) Refined crystal structure of β-lactamase from *Staphylococcus aureus* PC1 at 2.0 Å resolution, *J. Mol. Biol.* 217, 701-719.

In: Beta-Lactamases
Editor: Jean-Marie Frère, pp. 479-503

ISBN: 978-1-61324-638-2

Chapter 20

How to Make Good Use of a "Bad" Enzyme: Utilization of Efficient β-lactamases for the Benefits of Biochemical Research

Marylène Vandevenne[a], Moreno Galleni[a] and Patrice Filée[a,b].
[a] Center for Protein Engineering, Institute of Chemistry, Bât B6a,
University of Liège, B 4000 Sart Tilman, Belgium.
[b] ProGenosis, Boulevard du Rectorat 27b, Bât B22, B 4000 Sart-Tilman, Belgium.

Abstract

The activity of β-lactamases is easy to estimate, they often exhibit very high turn-over numbers and many of them are quite stable. For these reasons they have emerged as promising reporter enzymes to study gene expression and as models to analyse protein-ligand interactions. They also accept the insertion of rather large peptides without losing their activity. This has allowed the development of practically useful technologies which are described in this chapter.

Introduction

β-lactamases have been extensively used as biochemical markers for identification of β-lactam antibiotics which are active against bacterial pathogens. But during the last decade, class A β-lactamases have emerged as promising reporter enzymes to study gene expression and to study protein/ligand interactions. In this context these enzymes are very attractive because they are monomeric, very stable and of small size [1]. Their three-dimensional structures are well-defined and they are considered as very efficient enzymes [2]. Among the various reporter enzymes described in the literature, β-lactamases are particularly interesting because they are not expected to interfere with the function of most of the proteins to which they might be fused and the absence of endogenous counterparts in mammalian cells [3]. They have also been successfully expressed in prokaryotic and eukaryotic cells allowing their utilizations for *in vitro* as well as *in vivo* applications [4]. They can be overexpressed with prokaryotic expression systems so that their production times and costs are drastically reduced [5]. Their success is also partly due to the development of a variety of β-lactam substrates including chromogenic β-lactams for immunoassays, fluorescent membrane-permeant β-lactams which allow cell imagery, flow cytometry and gene expression assay, and non-toxic β-lactam prodrugs for antibody-directed enzyme prodrug therapy [3, 6-8].

Thanks to this uniqueness they have been extensively employed as "reporters" or "sensors" for monitoring various biological processes and interactions. Sensitive and rapid detection of β-lactamase activity in biological samples is thus of importance in both research and clinical applications [3]. This chapter is dedicated to the many practical utilisations of β-lactamases.

1. β-Lactamases as Reporters of Gene Expression

Biological specificity is mediated by the precise and selective regulation of gene expression in response to intrinsic developmental programs and extrinsic signals. To understand the regulation of gene expression, it is essential to use an assay of high sensitivity and fidelity that reports expression at the level of a single cell [9]. Correlation of gene expression with physiological responses and developmental fates would be facilitated by non-destructive transcriptional assays compatible with high-throughput methods such as fluorescence-activated cell sorting (FACS). Such assays are currently based on reporter genes such as those of chloramphenicol acetyltransferase, secreted alkaline phosphatase, β-galactosidase, and firefly luciferase [10] which typically require cell permeabilization or lack single-cell resolution, whereas the green fluorescent protein (GFP) is a relatively insensitive reporter due to the lack of enzymatic amplification [9, 11].

In this context, Zlokarnik and coworkers demonstrated the use of the 29 kDa isoform of the TEM-1 β-lactamase and a membrane-permeant, fluorogenic ester substrate to measure gene expression in single live mammalian cells with high sensitivity and real-time responses. To limit interference from variations in cell size, probe concentration, excitation intensity, and emission sensitivity, they constructed a new substrate molecule presented in figure 1 based on the following strategy: since cleavage of cephalosporins by a β-lactamase triggers the spontaneous elimination of any leaving group in the C3' position [9], they attached two fluorophores with overlapping spectra on the 7' and 3' positions of a cephalosporin. When on the same molecule, these exhibit efficient fluorescence energy transfer (FRET).

The β-lactamase action allows the release of the C3' fluorophore, disrupting FRET and reestablishing fluorescence emission from the 7' donor. For that purpose they used a membrane-permeant ester derivative of a cephalosporin CCF2/AM. The donor fluorophore was 7-hydroxycoumarin and the 3'-acceptor fluorophore was fluorescein. They validated this substrate with several receptor-regulated gene expressions under the control of adequate promoters in living mammalian cells by visualizing the β-lactamase-catalysed hydrolysis of CCF2/AM. This was done in both transitory expression systems and also in stable cell lines.

β-Lactamase reporter gene assays have also been successfully used as readouts for ligand-induced activation of G protein-coupled receptors [13], for selecting high-producing clones in the yeast *Pichia pastoris* [14], to identify inhibitors of the hepatitis C virus (HCV) replication [15], and to monitor spatially restricted patterns of genes expression in the early zebrafish embryos [16].

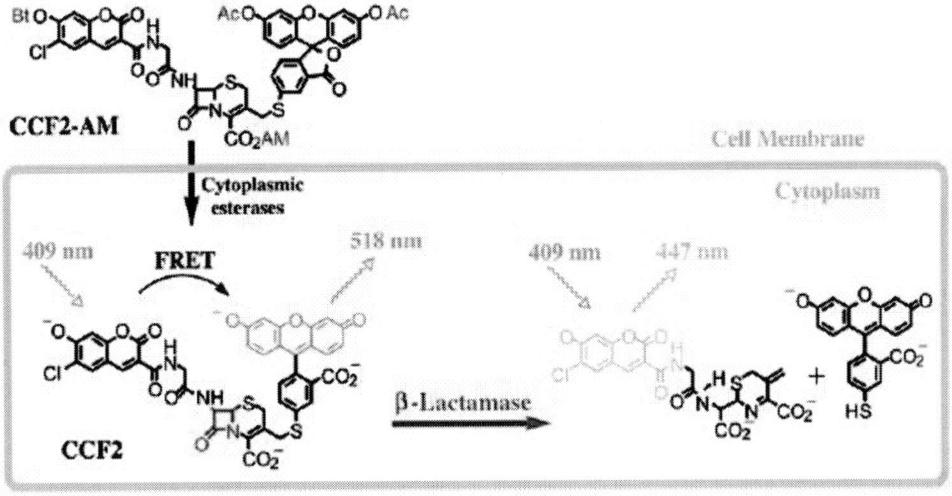

Figure 1. Schematic representation of the CCF2/4-AM substrate molecule loading into cultured cells [12]. The esterified (acetoxymethylated: AM) form of the CCF2/4 substrate, called CCF2/4-AM, diffuses across the cell plasma membrane where endogenous esterases convert it to CCF2/4, thereby trapping it inside the cell. Exciting CCF2/4 at 408 nm leads to efficient FRET from the coumarin moiety to the fluorescein derivative and produces green fluorescence detectable at 530 nm. After CCF2/4 hydrolysis by the β-lactamase, the two fluorophores separate causing loss of FRET. Excitation at 408 nm now results in blue fluorescence detectable at 460 nm.

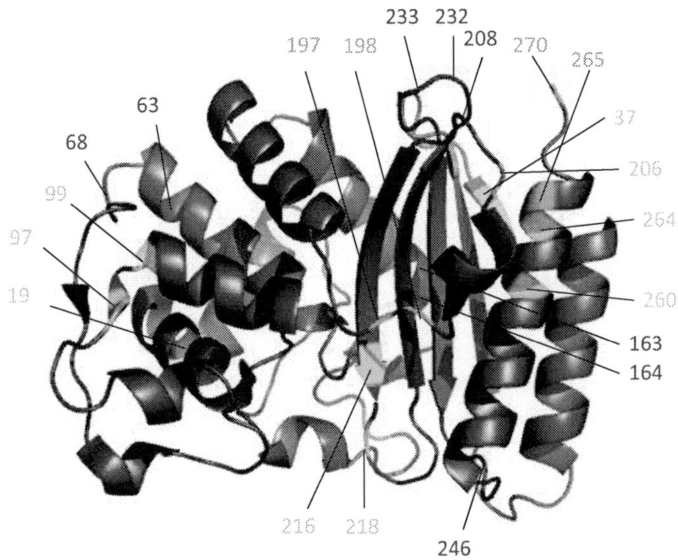

Figure 2. Sites of pentapeptides insertions in the three-dimensional structure of TEM-1 β-lactamase [25]. Numbers identify the insertion positions. Green, cyan and red residues indicate pentapeptide insertions which result in high-level, intermediate-level and loss of ampicillin hydrolysis activity, respectively.

1. β-LACTAMASE AS A REPORTER ACTIVITY

The β-lactamase reporter activity has also been extensively used to study biochemical processes, especially protein/ligand interactions. For this purpose, three different possibilities detailed in this section were developed.

The Use of β-lactamase Fusion Proteins

An interesting application of β-lactamase fusion proteins is the use of this reporter system to study exported proteins. For example McCann and coworkers identified proteins exported by *Mycobacterium tuberculosis* growing in host cells with the assumption that, when used in protein fusions, a β-lactamase can report on the subcellular location of another protein as measured by the protection from β-lactam antibiotics afforded to a mutant strain of *M.tuberculosis* lacking its major β-lactamase [17]. The exported proteins of *M.tuberculosis* have received attention for a long time because of the well-established fact that the majority of bacterial virulence factors and antigens are proteins exported to the bacterial cell wall or secreted [18]. Proteomic and genetic methods were used to identify proteins exported from the cytoplasm especially enzymatic reporter systems [19-21]. Among these reporter systems, β-lactamases are exported proteins that were not initially employed in *M.tuberculosis* because of endogenous β-lactam resistance. But the construction of a β-lactam-sensitive mutant of this bacterium has opened the door to the utilization of β-lactamases as reporters of protein export directly in *M.tuberculosis* with the β-lactam-resistance as a powerful indicator of export. This reporter enzyme presents the advantage of using a selection system instead of a more labor-intensive screening process. This study demonstrated that the TEM-1 β-lactamase lacking its export signal sequence conferred β-lactam resistance when fused to the major *M.tuberculosis* endogenous signal sequences. They further showed that β-lactamase fusion proteins report on protein export even while the bacterium is growing in its host cell namely human macrophages.

In another interesting study, Simonen and coworkers [22] investigated the relationship between folding and secretion competence of a *Saccharomyces cerevisiae* secretory protein: hsp150. They fused the *E.coli* TEM-1 β-lactamase, deprived of its own signal sequence, to the C-termini of various portions of hsp150. The evaluation of the β-lactamase activity of the different chimeric proteins was used to study the effect of the hsp150-carrier sequence on the folding of the β-lactamase portion and the effect of folding on the secretion competence of the fusion proteins. With this reporter system they identified the hsp150 subunits responsible for both the folding and the secretion of the fused protein.

The TEM β-lactamse was also used to probe the topology of membrane proteins [23]. When the β-lactamase gene (devoid of signal peptide) is fused to DNA fragments encoding portions of increasing lengths of the target protein, starting from the N-terminus, the β-lactamase can localise into the periplasm or the cytoplasm. Resistance of single-cell colonies to ampicillin is increased in the first case, not in the second. This technique was used to elucidate the transmembrane organisation of the *Bacillus licheniformis* BlaR sensory transducer (see chapter 17) involved in the induction of β-lactamase in this bacterium [24].

The Use of β-lactamases as Carrier Proteins

Besides the use of β-lactamases as fusion proteins, protein and or peptide insertions were also demonstrated to be efficient tools especially to study protein/ligand interactions. Indeed, class A β-lactamases were demonstrated to accept insertions of both random peptides as well as structured protein/domain. Hallet and coworkers determined the tolerance of the *E.coli* TEM-1 β-lactamase to 23 randomly inserted pentapeptides (Figure 2). They associated the phenotypes of the mutated β-lactamases to the local peptide structures in which the pentapeptide insertions occurred and to their positions in the three-dimensional structure of the enzyme [25].

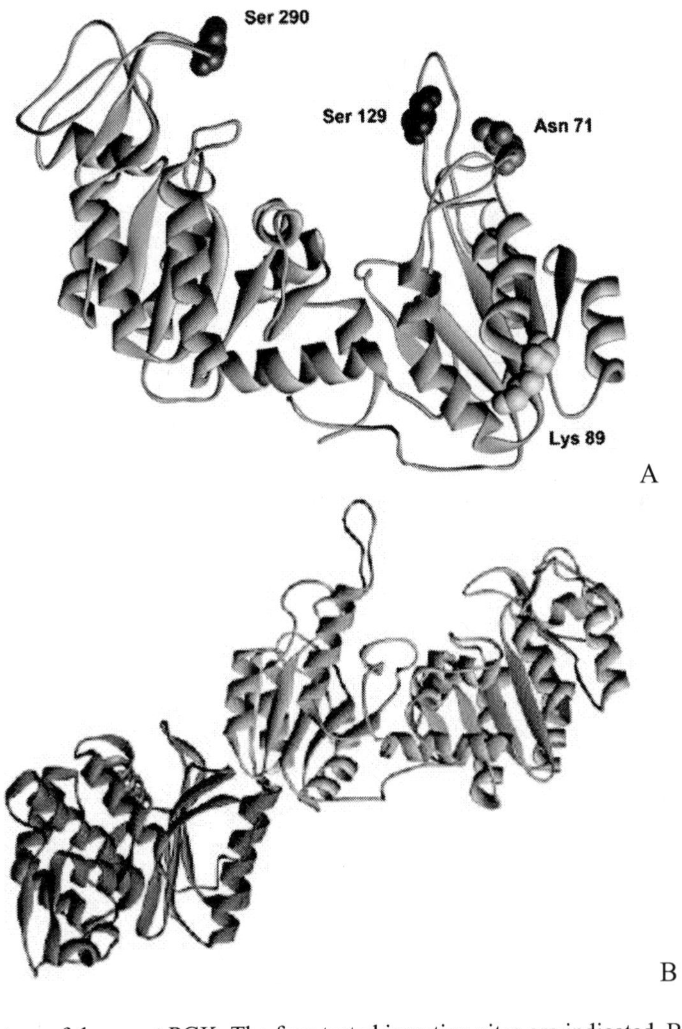

Figure 3. A. Structure of the yeast PGK. The four tested insertion sites are indicated. B. Illustrative scheme of the chimeric protein in which the *E.coli* TEM-1 β-lactamase was inserted into the yeast PGK at position 89 [27].

In the case of complete domains or proteins, the location of the insertion site was demonstrated to strongly influence the tolerance of the inserted polypeptide. For example

Betton and coworkers inserted the TEM-1 β-lactamase into the maltose-binding protein MalE. This resulted in a chimeric protein retaining the activity of both parental proteins [26]. They showed that this bifunctionality was dependent on the insertion site and that the activity of one of the chimeric protein partners could be modulated by an independent ligand specific for the other partner of the macromolecule. Similarly, Collinet and coworkers inserted TEM-1 into four surface loops of the phosphoglycerate kinase (PGK) and showed that the chosen insertion site strongly influenced the tolerance to insertions [27] (Figure 3).

The class A BlaP β-lactamase produced by *Bacillus licheniformis* was used as a carrier protein to study the function of heterologous protein fragments [28]. The chitin-binding domain (ChBD) of the human macrophage chitotriosidase was inserted into a solvent exposed loop of BlaP far away from its active site (position 198) (Figure 4). The product of this construction behaved as a soluble chimeric protein that conserved both the capacity to bind chitin and to hydrolyze β-lactams. Using the β-lactamase reporter activity, the relative association constant (K_r) between the chimeric protein and insoluble chitin was determined. Furthermore, this hybrid protein allowed the detection of chitin in fungal cell wall and to measure the amount of protein bound to various insoluble polysaccharides with a good accuracy. This is a major advantage of this system when compared to the techniques already described in the literature attempting to determine the affinity parameters of carbohydrate binding domains. Indeed, in these studies, protein concentration determination assays or intrinsic fluorescence measurements used to characterize the interaction between the protein domains and the insoluble polysaccharides resulted in large standard deviations [29-30].

The Use of β-lactamase Fragments Complementation

The complementation of β-lactamase fragments fused to interacting polypeptides that regenerate enzymatic activity was also used as a reporter system in assays monitoring inducible protein-protein interactions in eukaryotic cells (Figure 5) [4, 12, 31]. This strategy, called protein fragment complementation assay (PCA), is based on protein interaction assisted folding of rationally designed fragments of enzymes. It has also been used with other reporter proteins such as luciferase, β-galactosidase, the green fluorescent protein and the dihydrofolate reductase. However the advantage of the β-lactamase activity is that no ortholog exists in eukaryotes, and thus PCA based on β-lactamase activity can be used universally in eukaryotic cells and many prokaryotes with no intrinsic background activity. β-lactamase also has the desirable features of enzymatic amplification and facile *in vivo* and *in vitro* assays. Finally as mentioned above, it has the advantage that the versatile membrane-permeable fluorescent β-lactamase substrate, CCF2/AM, is available which allows ratiometric fluorescence detection and thus superior reproducibility and excellent quantification of the results in intact cells [9]. β-lactamase complementation also exhibits an extremely high signal-to noise ratio [31].

The strategy for the selection and design of fragments in the *E.coli* TEM-1 β-lactamase was based on the analysis of its structure and consists in the cleavage of the enzyme between Glu 197 and Leu198, because this site is located on a surface loop opposite to the active site (Figure 6) and produces fragments of similar lengths. It contains no periodic secondary structure, and the fragments can fold independently into native-like structures [4]. This β-lactamase complementation assay has been validated with several known protein-protein interactions, including the homodimerizing GCN4 parallel coiled-coil leucine zipper, the heterodimerizing soluble Bad and truncated (Bcl2T) pair [4] and even the interaction between a transmembrane receptor and cytoplasmic protein as in the case of the epidermal growth factor receptor and the FKBP12 cytoplasmic protein [31].

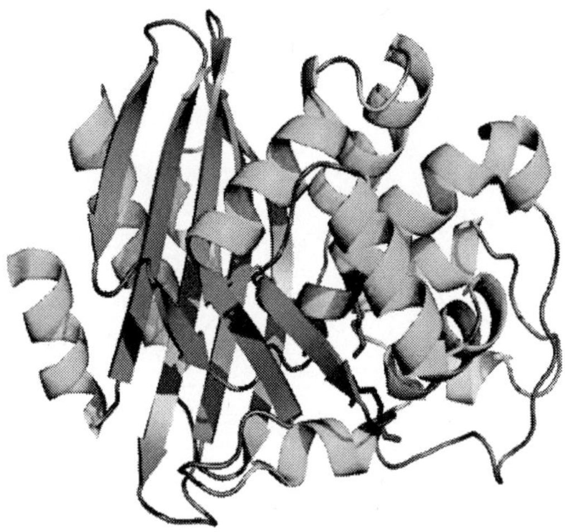

Figure 4. Ribbon diagram of the three dimensional structure of the *Bacillus licheniformis* BlaP β-lactamase. α-helices and β-sheets are represented in blue and orange respectively. The insertion site is indicated by an arrow. It is located in a loop connecting helix 9 to helix 10 between residues Lys 197 and Leu 198. Residues involved in the enzymatic catalysis are shown in red.

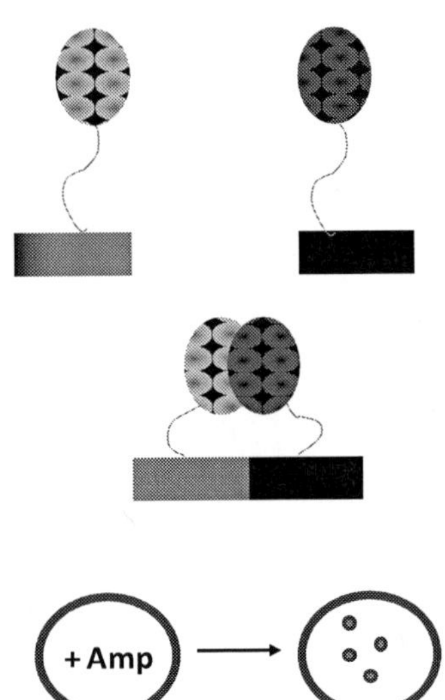

Figure 5. Schematic representation of the ß-lactamase based protein fragment complementation assay (PCA) [32]. The β-lactamase is split into two inactive fragments of similar lengths. Interacting fusion proteins attached to the fragments allow fragment reassembly, which can be detected by growing cells on media containing ampicillin or alternatively by using a chromogenic substrate.

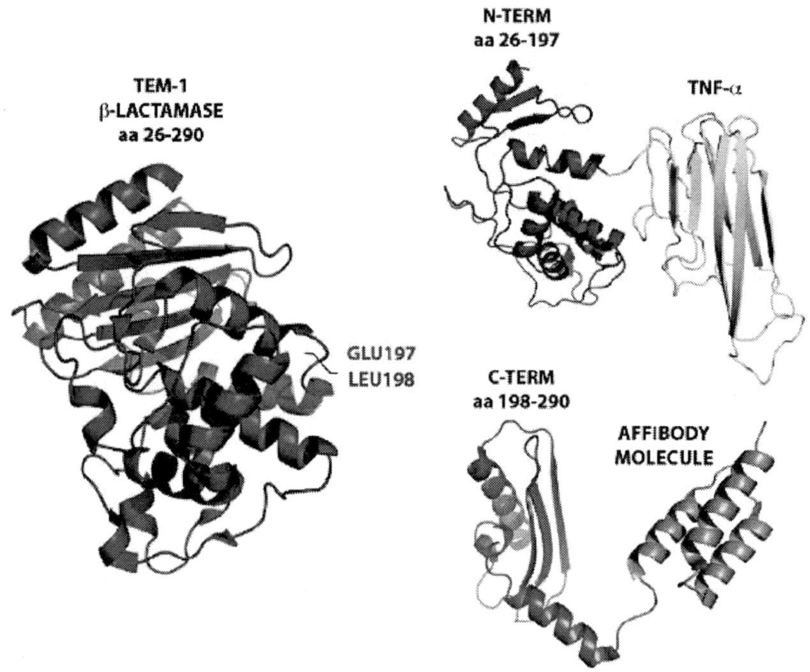

Figure 6. Representation of the *E.coli* TEM-1 β-lactamase with the N- and C- terminal subfragments highlighted in red and blue respectively [33]. An example of two co-expressed constructs whose complementation is monitored in the PCA system is also represented: the TNF-α target (yellow) is fused to the N-terminal fragment and an affibody protein library member (orange) to the C-terminal fragment.

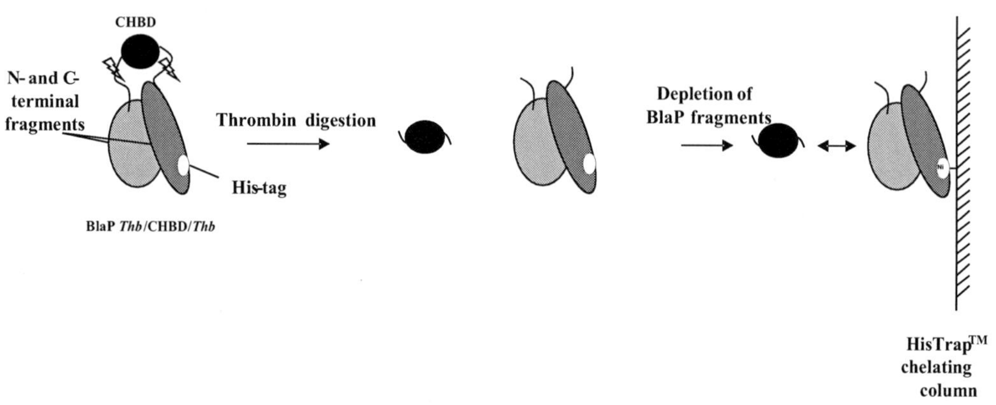

Figure 7. Schematic representation of the 2-step protocol used to prepare purified ChBD. Purified ChBD is collected in the flow through after loading the digested chimeric protein on the HisTrap™ chelating HP [37].

More recently, Löfdah and coworkers applied this TEM-1 β-lactamase based PCA system to select affibody molecules binding the human tumor necrosis factor-α (TNF-α) from a combinatorial library [33]. This system has also been used for the high throughput screening of toll-like receptor signaling inhibitors [34].

3. β-Lactamase as an Alternative for the Production of Difficult to Express Protein Domains

Many *Escherichia coli* expression vectors are available to produce and to purify proteins fused to affinity tags, such as the *Schistosoma japonicum* glutathione S-transferase (GST), the *E.coli* maltose binding protein and thioredoxin [35-36]. All these systems allow the subsequent proteolytic cleavage of the fusion protein to separate the protein of interest from the affinity tags. But a common feature of these systems is the limited utility of the affinity tag in the biochemical characterization of the studied protein. Recently, Vandevenne and coworkers [37] described a system able to elicit the expression of protein fragments inserted into a solvent exposed loop of the *Bacillus licheniformis* BlaP class A β-lactamase. Two thrombine (*thb*) cleavage sites were engineered on both sides of the β-lactamase permissive loop in order to cleave off the exogenous protein fragment from the carrier protein by an original two-step procedure [37]. This system has been used to express the chitin binding domain of the human macrophage chitotriosidase (ChBD) inserted between the two protease cleavage sites. This 73 residues polypeptide contains 3 disulfide bridges reported as essential to its chitin binding activity [38-39]. Recent studies showed that ChBD expression in *E.coli* can be achieved by fusing the domain to the *glutathion-S-transferase* (GST) [39] but with very limited yields which, combined with its low solubility and stability, strongly limited the characterization of ChBD. Upon insertion into the carrier β-lactamase, the resulting chimeric protein named BlaP *Thb*/ChBD/*Thb* was successfully produced in *E.coli* in a soluble form with a very good yield (≈ 20 mg of pure protein per liter of culture) and purified to homogeneity. The separation of ChBD from its carrier protein did alter neither the biochemical activity nor the native disulphide bridge formation of the inserted ChBD. Pure ChBD poplypeptides were prepared based on both the mutual affinity of the N- and C-terminal fragments of BlaP after thrombin cleavage and the presence of a poly-histidine tag at the C-terminal end of BlaP. The two-step purification procedure presented in figure 7 consists in a cleavage of BlaP *Thb*/ChBD/*Thb* with thrombin in solution followed by the elimination of BlaP fragments on a HisTrap Chelating HP column. This step can be repeated until the desired level of purity is obtained. This is a good example of the utilization of the β-lactamase carrier protein to increase the production and the solubility of difficult to express protein fragments and facilitate their purifications and studies.

4. β-Lactamase as a Biosensing Molecule

Biosensors are based on induced conformational changes of proteins to detect small ligands and ions [40]. The proteins selected for biosensor construction are usually allosteric enzymes which upon ligand binding exhibit altered catalytic efficiencies. Some hydrolytic enzymes have been widely used [41-42]. However, despite all the advantages exhibited by β-lactamases, the rational design of biosensors based on these enzymes remains a great challenge since neither substrate binding nor catalysis has been reported to induce large conformational changes in these proteins. The Ω-loop, a loosely packed 17-residue segment (residues 163-179) adjacent to the active site of class A β-lactamases, appears to be flexible and its motion essential to the enzymatic reaction [43]. This Ω-loop contains a key active-site residue Glu-166, whose substitution impairs but still allows acylation of the enzyme by β-lactam antibiotics. By contrast, the deacylation step is much more affected [44]. Moreover, crystal structures of the enzyme show that the Glu-166 side chain is pointing toward the active site. Based on these results Chan and coworkers hypothesized that, if an environment-sensitive fluorescent probe is placed specifically at the 166 position, an enzyme with drastically suppressed hydrolytic activity that can function as a biosensor can be constructed. To demonstrate this concept, they constructed a mutant of the β-lactamase I from *Bacillus cereus* 569/H, E166C, in which the Glu-166 residue on the Ω-loop was replaced by a cysteine residue. The E166C mutant exhibiting a 1800-fold lowered efficiency was labeled with the thiol-specific fluorophore (fluorescein-5-maleimide) to give the labeled E166Cf mutant. As the wild-type β-lactamase I contains no cysteine residue, this single-point mutation allows the specific attachment of a thiol-reactive fluorophore. In aqueous buffers, E166Cf is only weakly fluorescent but when β-lactam antibiotics are added, the fluorescence intensity at 515 nm increases as a function of the antibiotic concentration (Figure 8).

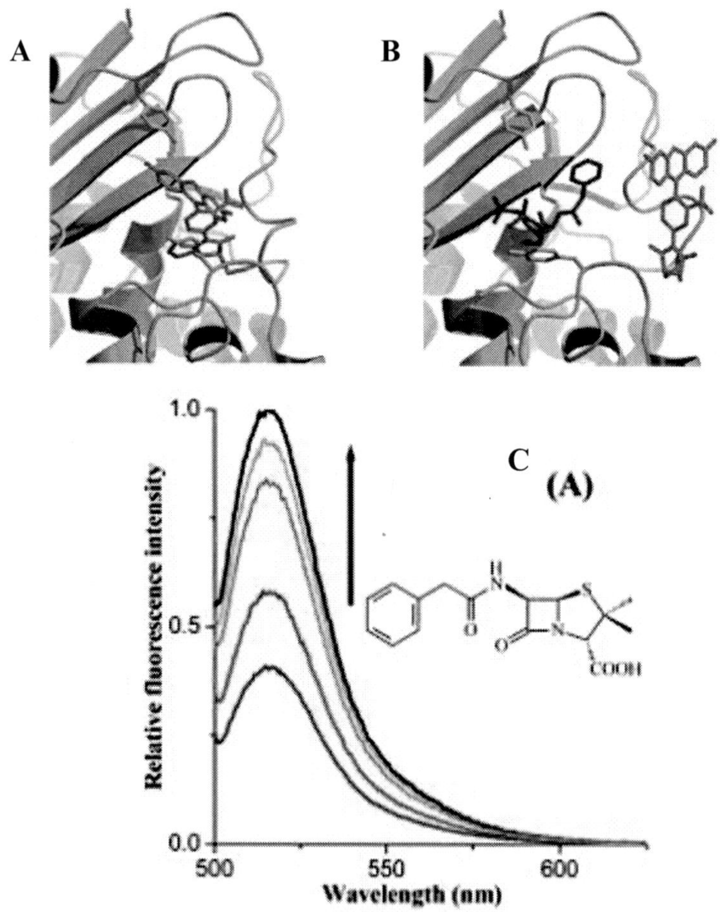

Figure 8. (A) Molecular models of the fluorescein label (green) on E166Cf before and (B) after binding with penicillin G (red). (C) Fluorescence spectra of E166Cf in the presence of penicillin G. Red, green, light blue, and black, after addition of 0.1, 1.0, 10, 100 μM antibiotics respectively [41].

The increase in fluorescence intensity can be observed even when the antibiotic concentration is as low as 0.05 μM, indicating that an extremely sensitive biosensor for β-lactam antibiotics has been obtained [45]. Indeed the detection of antibiotics present at trace levels in food (e.g. milk) is an important application since the ubiquitous presence of antibiotics can have serious health and economical consequences ranging from allergic reactions and the evolution of antibiotic-resistant bacteria [41] to serious problems in the dairy industry.

In another interesting work, Legendre and coworkers attempted to induce conformational changes in β-lactamases to study protein interactions [46]. Their strategy was to chemically modify the TEM-1 β-lactamase by coupling of a hapten molecule in order to create a conjugate whose activity is regulated by the binding of an antihapten antibody. Peptide mimotopes were chosen as haptens. They are nonlinear protein epitopes selected from phage display libraries [47]. These peptides bind to antibodies or to other proteins with high affinity. This technology was used to insert random peptides libraries near the active site of TEM-1 displayed on the phage fd in fusion to the coat protein pIII and to select variants in which the engineered mimotopes serve as binding sites for antibodies directed against antigens of interest. Combination of both elements (insertion of degenerate sequences and *in vivo* selection of active clones) permitted the production of enzyme libraries in which a large variety of antigen epitopes or mimotopes were perfectly integrated within the protein structure. The binding of antibodies to the antigen molecule was demonstrated to affect the β-lactamase activity.

5. The Use of β-Lactamases in Immunoassays

Utilization of enzyme immunoassays (EIAs) is widespread for the detection and/or quantification of small organic analytes (e.g. drugs, hormones) [46] or for the recognition of microbial antigens in clinical specimens. EIA systems utilize enzyme-catalysed reactions to measure the binding of antigens to specific antibodies [48-49]. The performance characteristics of EIAs are thus determined to a great extent by the reaction kinetics of the enzyme-substrate system used as the indicator [50]. Enzymes which are usually used in these assays include alkaline phosphatise [51], horseradish peroxidase [52], and β-galactosidase [53]. These enzymes have been chosen because they can be linked to immunoglobulin molecules to yield stable antibody-enzyme conjugates which retain a substantial portion of the enzymatic and antigen-binding activities of the original reagents [54]. β-lactamases represent another group of enzymes useful for EIA systems and present several advantages. They are easily available and exhibit favourable reaction kinetics with a wide range of available substrates devoid of mutagenic or carcinogenic potential [55-56]. In addition, β-lactamases can be linked to biotin, thus allowing their use in avidin-biotin systems [48, 57-58]. A few examples in which β-lactamases have been successfully implemented into EIAs are described.

β-lactamase enzyme-linked immunosorbent assays (ELISA) for various reproductive hormones have been developed and were found to be quite useful in fertility related problems. This system has now replaced radioimmunoassay (RIAs) which use radioactive labels. The short half-life of I^{125}-labeled hormones and the potential of health hazards in handling and disposing of gamma ray emitters limit the useful lifetime of a kit and increase its cost, thereby hindering its general application [59]. In this context, Shrivastav and co-workers used β-lactamase EIAs to detect prolactin in plasma at concentrations as low as 2,5 ng/ml [60]. This assay has been used to detect and estimate a wide range of β-lactamase labelled hormones and steroids in plasma such as testosterone derivatives [61-62], cortisol [59], human chorionic gonadotropin [63], estradiol, lutropin, follitropin [60] with a high sensitivity and a high specificity. Furthermore a prototype of a ready-to-use β-lactamase based ELISA kit has been developed by Khatkhatay and co-workers [64].

As mentioned above, the noncovalent but exceptionally strong avidin-biotin interaction has been implemented in β-lactamase based immunoassays. In such an indirect procedure, biotin labelled antibodies were used [49] to quantify and specifically detect antigens of the human adenoviruses and the rotavirus [54].

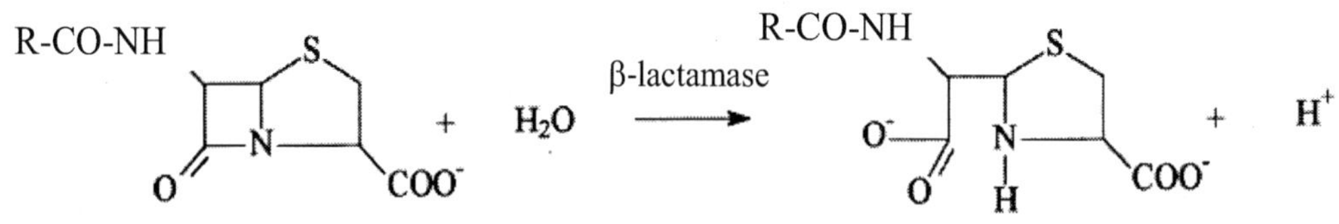

Figure 9. Schematic representation of a β-lactam antibiotic hydrolysis by a β-lactamase accompanied by the release of a proton.

Other interesting β-lactamase-based immunoassays are those implementing the biosensor technology. The enzyme-based biosensors technology relies on two operational mechanisms. The first one involves the catalytic transformation of a non-detectable molecule into a detectable form. The second mechanism involves the detection of molecules that modulate the enzyme activity. Among the various detection methods, the potentiometric measurements rely on the development of charges related to the reporter enzyme activity. Most potentiometric biosensors for detection of environmental analytes have used enzymes that catalyze the appearance or disappearance of protons. For instance, phosphoric and carbamic pesticides can be evaluated through the use of a pH electrode that measures the activity of acetylcholinesterase [65-66]. In this approach, polyalanine (PANI) is an efficient polymer for developing biosensors because its redox potential, conductivity and adsorption wavelength are modified in response to a pH variation [67-69]. In this context, since the hydrolysis of β-lactams generates protons, β-lactamases are good candidates for the development of enzyme-based biosensors (Figure 9).

A good example of application of these biosensors has been described by Sergeyeva and coworkers in which a β-lactamase label-based potentiometric biosensor for the detection of α-2 interferon was used consisting of a highly specific, stable and miniature electrochemical biosensor based on pH-sensitive field effect transistors (pH-FET) [70]. A specific sensing element was built by immobilizing α-2 interferon on the gate of a pH-FET. The interaction between anti-interferon antibodies labeled β-lactamase and interferon-pH-FET (in the presence of a specific enzyme substrate) leads to a local pH-change at the surface of the transducer and produces an electrochemical signal which is proportional to the conjugate concentration.

Gaudin and coworkers also used a similar β-lactamase-based biosensor immunoassay for the screening of penicillins residues in milk [71].

6. IMPLEMENTATION OF β-LACTAMASES INTO THE PHAGE DISPLAY TECHNOLOGY

Phage display is a powerful technology to study protein/ligand interactions. In this technique, peptides, proteins or protein fragments are expressed on the surface of phage particles in fusion with one of their coat proteins [72]. The advantage of this system rests on the physical link between the heterologous polypeptide expressed at the surface of the phage and the corresponding nucleotide sequence cloned into the phagemid genome. The ability of each phage to individually replicate and to display a specific polypeptide on its surface allows a high throughput screening of libraries of variant nucleotide sequences with diversities from 10^6 to 10^{10}.

This technology is of particular interest to study protein interactions and offers many advantages: (i) the peptides, proteins, or protein fragments expressed at the surface of phages particles are solvent-exposed and thus accessible for interaction with their targets; (ii) the bacteriophage can easily be purified; and (iii) there is a physical link between the phenotype and the corresponding genotype. However, phage display presents a few bottlenecks: (i) the lack of an easy and rapid quality control of the representativeness of the initial library; (ii) the lack of a direct assessment of the enrichment level of libraries during the successive rounds of panning without interfering with the phage elution step and finally (iii) the large nucleotidic sequencing efforts necessary to characterize the selected epitope. This can be time-consuming and laborious.

To overcome these difficulties, Legendre and co-workers proposed to combine the phage display technology with the enzymatic activity of the *E.coli* TEM-1 β-lactamase. They identified two loops near the active site of the enzyme in which small peptides could be inserted, replacing one or two codons of the TEM-1 gene [73]. They displayed libraries of hybrid β-lactamases on the surface of phages and selected them for affinity towards specific targets. Allosteric regulations (inhibition or activation) of the enzyme activity were observed for some hybrid proteins upon interaction with their specific targets [46]

This coupling between the phage display technology and the enzymatic β-lactamase activity presents several advantages: (i) it decreases the conformational freedom of the displayed inserted protein fragment; (ii) it protects the inserted polypeptide from proteolytic cleavage; (iii) it allows to directly assess the interaction of the inserted fragments using the β-lactamase as a reporter activity; (iv) the enzymatic activity can also be used to monitor the enrichment of the libraries during the successive selection rounds; and finally (v) it facilitates the determination of the best parameters of the assay as well as the optimization of the experimental procedure.

Recently Filée and co-workers have implemented this coupling between phage display and the so-called "bifunctional hybrid β-lactamase technology" (BHP) to perform epitope mapping. Furthermore, they validated the use of these combined technologies to methods relying on both potentiometric biosensors and DNA-arrays (publication in preparation).

Epitope Mapping

In immunology, epitope mapping is used to study antigen/antibody (B-cell epitope) or antigen/T cell receptor (T cell epitope) interactions, develop immunoassays, identify neutralising sites, produce protective peptide vaccines, develop hypoallergens, investigate the pathogenesis of autoimmune diseases, etc. Chevigné and co-workers demonstrated that using the hybrid β-lactamase technology coupled with phage display could not only identify protein regions involved in protein-protein interactions but also supply useful molecules in which these regions are linked to an enzymatic activity [74]. They validated this approach by preparing an influenza virus hemagglutinin (HA1) gene fragment library inserted into the solvent-exposed permissive loop of the *Bacillus licheniformis* BlaP β-lactamase (position 198). Hybrid β-lactamases, fused to protein pIII of the fdTet phage, were exposed on the surface of phage particles. This library was successfully used to select a linear epitope recognised by a monoclonal antibody. The procedure alternating successive affinity selection of phages and growth of phage infected cells in the presence of β-lactams allows the selection of bifunctional hybrid β-lactamases with high affinity for the target ligand (figure 10). Furthermore they demonstrated that modifying the β-lactam concentration in the culture medium results in a positive selection of hybrid β-lactamases that retain a native-like catalytic efficiency. This is a useful tool to accelerate the convergent selection of bifunctional β-lactamases.

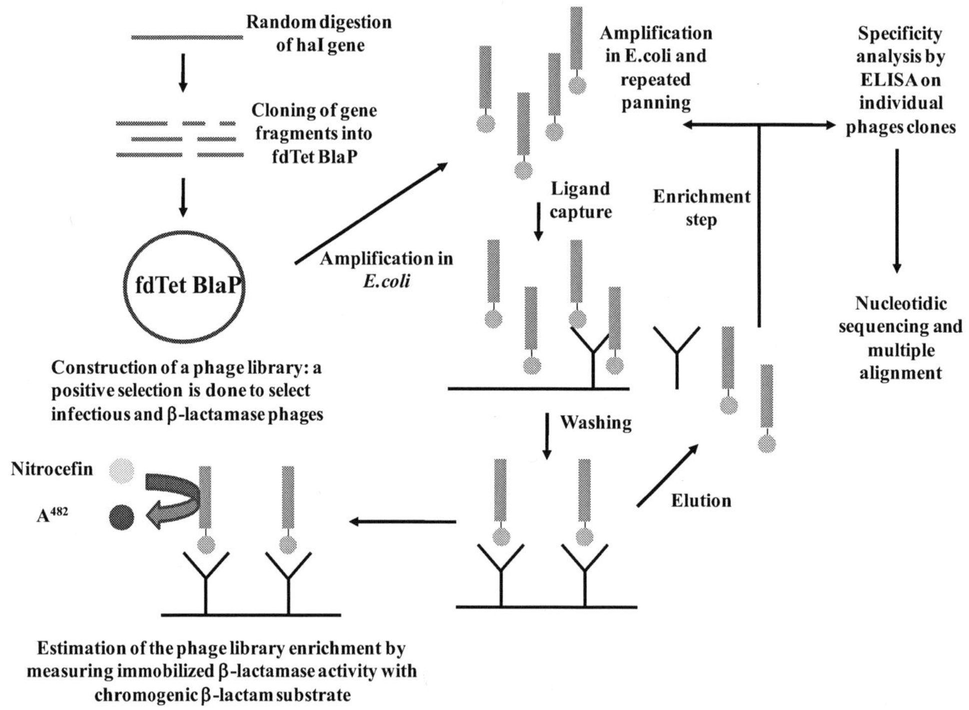

Figure 10. Schematic representation of the epitope mapping procedure [74].

Biosensor

As mentioned above, since the hydrolysis of β-lactams generates protons, β-lactamases are good candidates for the development of enzyme-based biosensors. Recently, Ruth and co-workers combined the phage display-BHP technology with potentiometric biosensor and validated this system for the epitope mapping of the human epidermal growth factor receptor (h*egfr*) (submitted publication). The h*egfr* cDNA was randomly digested and used to generate a random epitope library within the BlaP β-lactamase. This library, exposed on the surface of the fdTet phage, was screened by affinity selection against mouse monoclonal anti-human EGFr using phage display. The biopanning was successfully achieved by coating the anti-EGFr antibodies on a biosensor platform composed of a carbon electrode coated with a thin layer of polyalanine (PANI) (Figure 11).

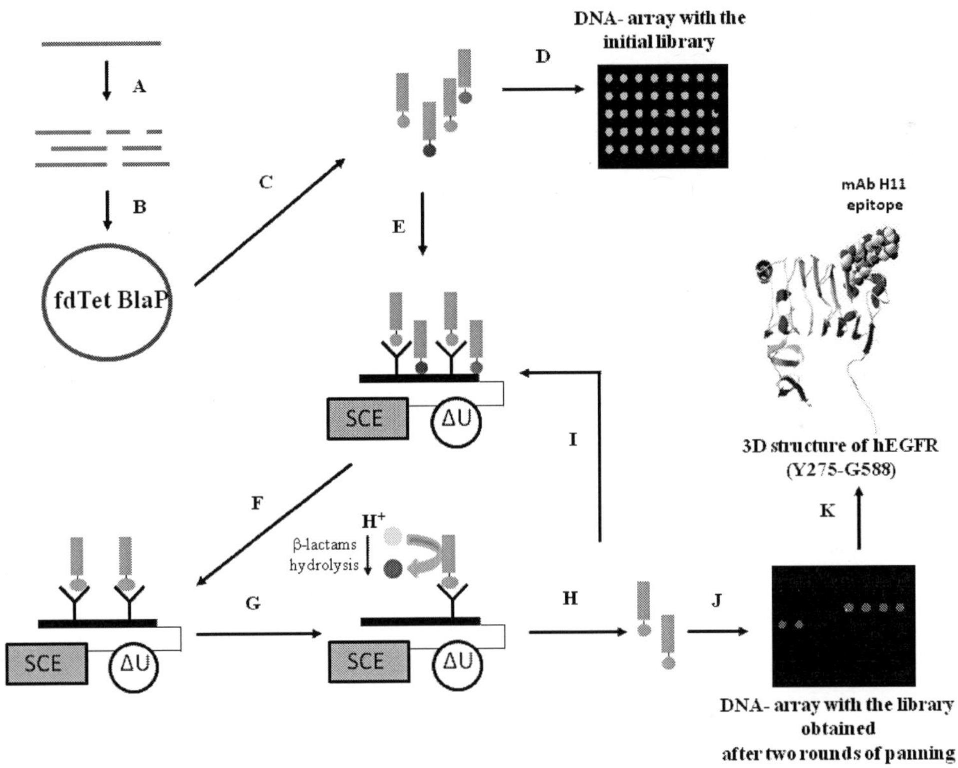

Figure 11. Schematic representation of the epitope mapping procedure combining the combinatorial search for target domains based on hybrid protein synthesis, Phage display, Potentiometric biosensor and DNA-array. SCE: saturated calomel electrode. ΔU: electrical potential variation. A: PCR amplification of the gene coding for the target protein (cDNA encoding the hEGFR) and random digestion of the PCR product with DNaseI. B: cloning of the gene fragments into the fdTet BlaP phagemide. A positive selection is performed to select infectious and β-lactamase positive phages. C: amplification of phage particles in *E. coli*. D: DNA-array analysis to verify the representativeness of the initial library. E: Capture of the phages with specific antibodies (mab H11). F: washing step. G: measurement of the immobilized β-lactamase activity on the potentiometric biosensor using benzylpenicillin as substrate. H: Elution step. I: amplification of eluted phages particles in *E. coli* and performance of a new round of biopanning. J: DNA-array analysis to identify the mAb H11 epitope. K: Assignment of the mAb H11 epitope into the 3D structure of the hEGFR (Y275-G585).

They demonstrated that the release of protons during benzylpenicillin hydrolysis induces a detectable change of electrode potential between the PANI-coated carbon and calomel reference electrodes upon specific binding of various monoclonal antibodies to phages harbouring the antibodies related epitopes. This indicated that biofunctionnalized PANI-electrode constitutes a suitable platform for performing biological screening with the advantage that this type of measurement is less time consuming than the colorimetric assay.

DNA Arrays

The DNA-array technology is used for expression profiling, to monitor expression levels of thousands of genes simultaneously, for comparative genomic hybridization, for identification of protein binding sites or for the identification of single nucleotide polymorphism in the genome of populations. The nucleotidic material spotted on microarrays can be oligonucleotides, cDNA or PCR products. Consequently the broad diversity of nucleotidic material that can be used offers many possibilities of detection in various biological contexts [75-76]. Ruth and co-workers took advantage of the phage-display-BHP technology to develop a new generation of DNA-array (Figure 11). Indeed, in addition to the complementation of the β-lactamase based potentiometric biosensor used to map epitope of the human epidermal growth factor receptor (h*egfr*), they constructed a so called "epitope array" also based on the combination of phage-display-BHP technologies. In practice, the "epitope-array" was created by spotting short adjacent oligonucleotides covering the coding sequence of hEGFR. Labelled probes were prepared from h*egfr* gene fragments cloned into the BlaP gene by PCR amplification incorporating dCTP-cy5 or cy-3. By combining the array results obtained with both the non-coding and coding strands of *hegfr*, they concluded that the different hEGFR protein regions displayed on phages from the initial library entirely covered the protein sequence and gave rise to β-lactamase positive and infectious phages. Moreover, the array experiments performed with phages obtained after two rounds of screening confirmed that fluorescence signals were only observed with oligonucleotides that encoded a protein region which overlapped that mapped with the multiple sequence alignment. Taking advantage of fluorescence intensities obtained on each spotted oligonucleotide, they reduced the epitope to a 28 residues polypeptide. Altogether, this indicates that the implementation of the phage display-BHP technology into DNA array presents the advantage to control the quality of the initial library and can be successfully used to identify the nucleotide sequence that encodes the polypeptide selected during the phage display procedure.

7. The Use of β-Lactamase in DNA/Protein Vaccination

Another interesting application field of β-lactamases is their use as hapten-carrier proteins to develop vaccines against non-immunogenic proteins. A good example is the work described by Ruth and co-workers who developed a DNA vaccination system against the heat-stable enterotoxin STa from *Escherichia coli* [77]. This toxin is a major cause of enteric colibacillosis, characterized by profuse fluid diarrhoea leading to dehydration and even death of newborn calves and piglets [78-79]. Because of its small size, STa is poorly antigenic and not immunogenic in its native form. Several approaches were explored to obtain non-toxic immunogenic molecules for safe vaccine design. Several proteins were chemically coupled to STa in order to increase the immunogenicity of the enterotoxin, including bovine serum albumin [80], the B-subunit of cholera toxin (CTB) [81] and the heat-labile enterotoxin [81-82]. Some hybrid fusion proteins were constructed between STa and the major subunit ClpG of the *E.coli* surface antigen (ClpG-ST) [83], between STa and one or more subunits of the cholera toxin [84-85] and between STa and LT (labile toxin) [86-87]. The OmpC outer-membrane protein and the flagellin have also been used as carriers for STa [88-89]. However, in general, these constructions failed to elicit the production of neutralizing antibodies or retained some degree of STa-associated toxicity, probably because the immunogenic properties of the toxin are influenced by the conformation associated with toxicity. In consequence, Ruth and co-workers decided to insert the STa toxin in the solvent-exposed permissive loop (position 195) of the *E.coli* TEM-1 β-lactamase in order to enhance its immunogenic properties. Furthermore, since the production and purification of hybrid protein vaccines is labour intensive and expensive, they choose to validate this system using the DNA vaccine technology. Indeed, DNA vaccines are stable, easy to produce, and potent tools for inducing both humoral and cell-mediated immune responses [90-91]. Moreover, immune responses induced by DNA vaccines may display a broader epitopic repertoire than immune responses induced by traditional peptide or protein vaccines [92-93]. In this context, Ruth et al genetically inserted the nucleotide sequence of STa into the TEM-1 gene and cloned the constructed hybrid gene into an appropriate plasmid DNA expression vector (V1Jns.tPA vector). The plasmids were tested for their capacity to induce a TEM-1 and STa specific immune response in a mice model. No STa specific IgG could be detected after three DNA immunizations. But a boost with the oligopeptide toxin, which is not immunogenic in its native form, induced high STa specific antibody titres. This demonstrated that immunization with plasmid DNA encoding hybrid hapten-carrier proteins, in this case a β-lactamase with the *E.coli* STa insert, can be used as a priming stimulus to provide an antibody response against a non-immunogenic peptide.

8. β-Lactamases as an Antibody Directed Enzyme in Prodrug Therapy

The toxicities associated with conventional cancer chemotherapy arise primarily from the lack of specificity for tumour cells [94]. Most of the presently available drugs are designed to be selectively toxic for rapidly dividing cells [95]. A variety of approaches are under development to improve the efficency and tumor cell specificity of cancer treatments. Many of these involve monoclonal antibodies and offer attractive means of directing toxic agents such as drugs, radioisotopes, protein cytotoxins, cytokines, effector cells of the immune system and enzymes to tumor cells. In one approach, known as antibody-directed enzyme prodrug therapy (ADEPT) (Figure 12), enzymes are delivered to tumor cells via the recognition of tumor associated antigens (by a specific antibody conjugated to the enzyme) where they catalyse the activation of a pro-drug exactly at the site of tumor [94].

This method is now well advanced, with many examples of systems already tested in preclinical models and some entering clinical trials. Enzymes usually utilized in ADEPT include alkaline phosphatase, carboxypeptidases, β-glucosidases and β-lactamases.

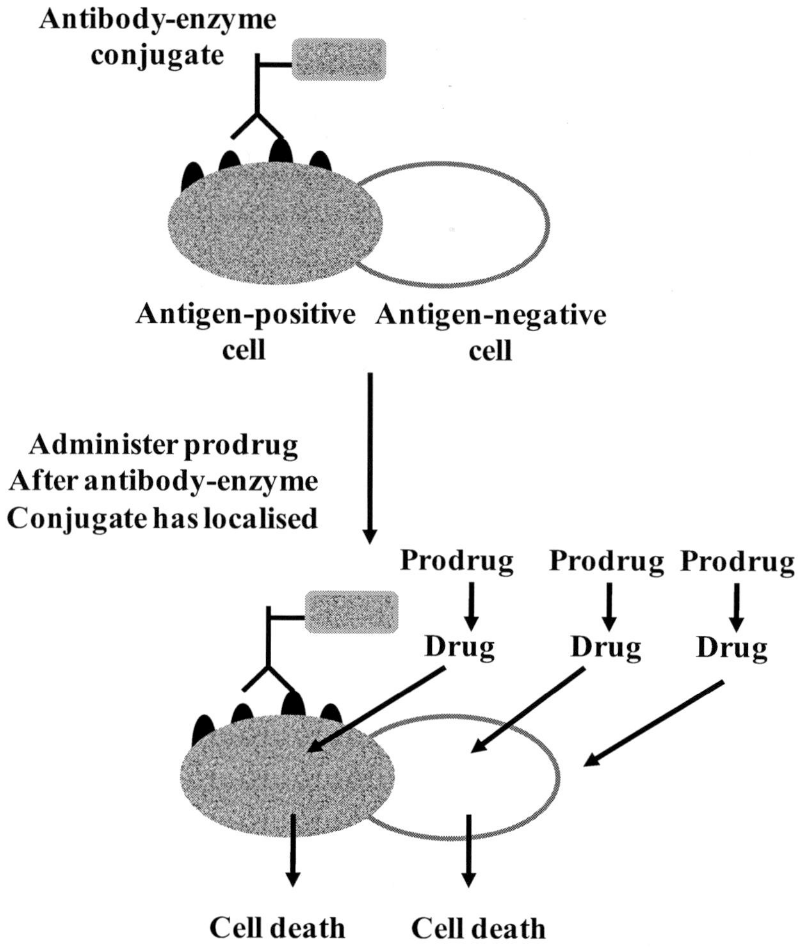

Figure 12. Schematic representation of the antibody-directed enzyme prodrug therapy (ADEPT) in which enzymes are localized to the surface of the target cell where they activate systemically administered prodrugs. The toxic drug can then diffuse inside the tumor [94].

Many β-lactamases hydrolyse a wide range of cephalosporins with 3' leaving groups which are released upon hydrolysis of the β-lactam ring. Various toxic compounds can thus be released from cephalosporin prodrugs, including nitrogen mustard and vinca alkaloid derivatives [96]. Moreover, eukaryotic cells do not produce β-lactamases related enzymes which could cause unwanted activation of prodrugs *in vivo*. A cocktail of different prodrugs can be activated by β-lactamases to release a panel of active drugs which can act independently or synergistically. These reasons explain why β-lactamases are good candidates for this particular application, and examples of their *in vitro* utilization, with the L6 monoclonal antibody, have demonstrated good targeted prodrug activation [97]. The prodrug tested was cephalosporin mustard (CM) which was designed as an anticancer prodrug that can be activated by a site-specific L6 monoclonal antibody-β-lactamase conjugates targeted to antigens present on tumour cell surfaces. The antibody-β-lactamase conjugate catalyzes the release of the active drug phenyl mustard (PDM). Similar antibodies - β□-lactamase conjugates have also been assayed with different prodrugs [98]. More recently,

Roberge and co-workers described the construction of a novel antibody-enzyme fusion protein for use in ADEPT [99]. The fusion protein combines the proven tumour specificity of the CC49 antibody [100] with β-lactamase. But this first generation of fusion proteins is rapidly degraded when expressed in *E.coli*. Optimization of the antibody fragment by combinatorial consensus mutagenesis (CCM), yielded an improved variant exhibiting significant increased stability allowing its use for ADEPT.

REFERENCES

[1] Philippon, A., J. Dusart, B. Joris, and J.M. Frere, *The diversity, structure and regulation of beta-lactamases.* Cell Mol Life Sci, 1998. 54: 341-6.

[2] Matagne, A., A. Dubus, M. Galleni, and J.M. Frere, *The beta-lactamase cycle: a tale of selective pressure and bacterial ingenuity.* Nat Prod Rep, 1999. 16: 1-19.

[3] Naqvi, T. and R. Singh, *A novel chemiluminescent substrate for detecting lactamase.* Mol Biosyst, 2007. 3: 431-8.

[4] Galarneau, A., M. Primeau, L.E. Trudeau, and S.W. Michnick, *Beta-lactamase protein fragment complementation assays as in vivo and in vitro sensors of protein protein interactions.* Nat Biotechnol, 2002. 20: 619-22.

[5] Frate, M.C., E.J. Lietz, J. Santos, J.P. Rossi, A.L. Fink, and M.R. Ermacora, *Export and folding of signal-sequenceless Bacillus licheniformis beta-lactamase in Escherichia coli.* Eur J Biochem, 2000. 267: 3836-47.

[6] Bebrone, C., C. Moali, F. Mahy, S. Rival, J.D. Docquier, G.M. Rossolini, J. Fastrez, R.F. Pratt, J.M. Frere, and M. Galleni, *CENTA as a chromogenic substrate for studying beta-lactamases.* Antimicrob Agents Chemother, 2001. 45: 1868-71.

[7] Vrudhula, V.M., D.E. Kerr, N.O. Siemers, G.M. Dubowchik, and P.D. Senter, *Cephalosporin prodrugs of paclitaxel for immunologically specific activation by L-49-sFv-beta-lactamase fusion protein.* Bioorg Med Chem Lett, 2003. 13: 539-42.

[8] Lee, M., D. Hesek, and S. Mobashery, *A practical synthesis of nitrocefin.* J Org Chem, 2005. 70: 367-9.

[9] Zlokarnik, G., P.A. Negulescu, T.E. Knapp, L. Mere, N. Burres, L. Feng, M. Whitney, K. Roemer, and R.Y. Tsien, *Quantitation of transcription and clonal selection of single living cells with beta-lactamase as reporter.* Science, 1998. 279: 84-8.

[10] Alam, J. and J.L. Cook, *Reporter genes: application to the study of mammalian gene transcription.* Anal Biochem, 1990. 188: 245-54.

[11] Cubitt, A.B., R. Heim, S.R. Adams, A.E. Boyd, L.A. Gross, and R.Y. Tsien, *Understanding, improving and using green fluorescent proteins.* Trends Biochem Sci, 1995. 20: 448-55.

[12] Qureshi, S.A., *Beta-lactamase: an ideal reporter system for monitoring gene expression in live eukaryotic cells.* Biotechniques, 2007. 42: 91-6.

[13] Kunapuli, P., R. Ransom, K.L. Murphy, D. Pettibone, J. Kerby, S. Grimwood, P. Zuck, P. Hodder, R. Lacson, I. Hoffman, J. Inglese, and B. Strulovici, *Development of an intact cell reporter gene beta-lactamase assay for G protein-coupled receptors for high-throughput screening.* Anal Biochem, 2003. 314: 16-29.

[14] Hribar, G., V. Smilovic, A.L. Zupan, and V. Gaberc-Porekar, *Beta-lactamase reporter system for selecting high-producing yeast clones.* Biotechniques, 2008. 44: 477-8, 480, 482 passim.

[15] Zuck, P., E.M. Murray, E. Stec, J.A. Grobler, A.J. Simon, B. Strulovici, J. Inglese, O.A. Flores, and M. Ferrer, *A cell-based beta-lactamase reporter gene assay for the identification of inhibitors of hepatitis C virus replication.* Anal Biochem, 2004. 334: 344-55.

[16] Raz, E., G. Zlokarnik, R.Y. Tsien, and W. Driever, *beta-lactamase as a marker for gene expression in live zebrafish embryos.* Dev Biol, 1998. 203: 290-4.

[17] McCann, J.R., J.A. McDonough, M.S. Pavelka, and M. Braunstein, *Beta-lactamase can function as a reporter of bacterial protein export during Mycobacterium tuberculosis infection of host cells.* Microbiology, 2007. 153: 3350-9.

[18] Finlay, B.B. and S. Falkow, *Common themes in microbial pathogenicity revisited.* Microbiol Mol Biol Rev, 1997. 61: 136-69.

[19] Braunstein, M., T.I. Griffin, J.I. Kriakov, S.T. Friedman, N.D. Grindley, and W.R. Jacobs, Jr., *Identification of genes encoding exported Mycobacterium tuberculosis proteins using a Tn552'phoA in vitro transposition system.* J Bacteriol, 2000. 182: 2732-40.

[20] Chubb, A.J., Z.L. Woodman, F.M. da Silva Tatley, H.J. Hoffmann, R.R. Scholle, and M.R. Ehlers, *Identification of Mycobacterium tuberculosis signal sequences that direct the export of a leaderless beta-lactamase gene product in Escherichia coli.* Microbiology, 1998. 144 (Pt 6): 1619-29.

[21] Downing, K.J., R.A. McAdam, and V. Mizrahi, *Staphylococcus aureus nuclease is a useful secretion reporter for mycobacteria.* Gene, 1999. 239: 293-9.

[22] Simonen, M., E. Jamsa, and M. Makarow, *The role of the carrier protein and disulfide formation in the folding of beta-lactamase fusion proteins in the endoplasmic reticulum of yeast.* J Biol Chem, 1994. 269: 13887-92.

[23] Broome-Smith, J.K., M. Tadayyon, and Y. Zhang, *Beta-lactamase as a probe of membrane protein assembly and protein export.* Mol Microbiol, 1990. 4: 1637-44.

[24] Hardt, K., B. Joris, S. Lepage, R. Brasseur, J.O. Lampen, J.M. Frere, A.L. Fink, and J.M. Ghuysen, *The penicillin sensory transducer, BlaR, involved in the inducibility of beta-lactamase synthesis in Bacillus licheniformis is embedded in the plasma membrane via a four-alpha-helix bundle.* Mol Microbiol, 1997. 23: 935-44.

[25] Hallet, B., D.J. Sherratt, and F. Hayes, *Pentapeptide scanning mutagenesis: random insertion of a variable five amino acid cassette in a target protein.* Nucleic Acids Res, 1997. 25: 1866-7.

[26] Betton, J.M., J.P. Jacob, M. Hofnung, and J.K. Broome-Smith, *Creating a bifunctional protein by insertion of beta-lactamase into the maltodextrin-binding protein.* Nat Biotechnol, 1997. 15: 1276-9.

[27] Collinet, B., M. Herve, F. Pecorari, P. Minard, O. Eder, and M. Desmadril, *Functionally accepted insertions of proteins within protein domains.* J Biol Chem, 2000. 275: 17428-33.

[28] Vandevenne, M., P. Filee, N. Scarafone, B. Cloes, G. Gaspard, N. Yilmaz, M. Dumoulin, J.M. Francois, J.M. Frere, and M. Galleni, *The Bacillus licheniformis BlaP beta-lactamase as a model protein scaffold to study the insertion of protein fragments.* Protein Sci, 2007. 16: 2260-71.

[29] Hashimoto, M., T. Ikegami, S. Seino, N. Ohuchi, H. Fukada, J. Sugiyama, M. Shirakawa, and T. Watanabe, *Expression and characterization of the chitin-binding domain of chitinase A1 from Bacillus circulans WL-12.* J Bacteriol, 2000. 182: 3045-54.

[30] Tomme, P., A.L. Creagh, D.G. Kilburn, and C.A. Haynes, *Interaction of polysaccharides with the N-terminal cellulose-binding domain of Cellulomonas fimi CenC. 1. Binding specificity and calorimetric analysis.* Biochemistry, 1996. 35: 13885-94.

[31] Wehrman, T., B. Kleaveland, J.H. Her, R.F. Balint, and H.M. Blau, *Protein-protein interactions monitored in mammalian cells via complementation of beta -lactamase enzyme fragments.* Proc Natl Acad Sci U S A, 2002. 99: 3469-74.

[32] Barnard, E., McFerran, N.V., Nelson, J. and Timson, D.J., *Detection of Protein-Protein Interactions Using Protein-Fragment Complementation Assays (PCA).* Current Proteomics, 2007. 4: 17-27.

[33] Lofdahl, P.A., O. Nord, L. Janzon, and P.A. Nygren, *Selection of TNF-alpha binding affibody molecules using a beta-lactamase protein fragment complementation assay.* N Biotechnol, 2009. 26: 251-9.

[34] Lee, H.K., S.J. Brown, H. Rosen, and P.S. Tobias, *Application of beta-lactamase enzyme complementation to the high-throughput screening of toll-like receptor signaling inhibitors.* Mol Pharmacol, 2007. 72: 868-75.

[35] LaVallie, E.R. and J.M. McCoy, *Gene fusion expression systems in Escherichia coli.* Curr Opin Biotechnol, 1995. 6: 501-6.

[36] Tsunoda, Y., N. Sakai, K. Kikuchi, S. Katoh, K. Akagi, J. Miura-Ohnuma, Y. Tashiro, K. Murata, N. Shibuya, and E. Katoh, *Improving expression and solubility of rice proteins produced as fusion proteins in Escherichia coli.* Protein Expr Purif, 2005. 42: 268-77.

[37] Vandevenne, M., G. Gaspard, N. Yilmaz, F. Giannotta, J.M. Frere, M. Galleni, and P. Filee, *Rapid and easy development of versatile tools to study protein/ligand interactions.* Protein Eng Des Sel, 2008. 21: 443-51.

[38] Tjoelker, L.W., L. Gosting, S. Frey, C.L. Hunter, H.L. Trong, B. Steiner, H. Brammer, and P.W. Gray, *Structural and functional definition of the human chitinase chitin-binding domain.* J Biol Chem, 2000. 275: 514-20.

[39] Ujita, M., K. Sakai, K. Hamazaki, M. Yoneda, S. Isomura, and A. Hara, *Carbohydrate binding specificity of the recombinant chitin-binding domain of human macrophage chitinase.* Biosci Biotechnol Biochem, 2003. 67: 2402-7.

[40] Doi, N. and H. Yanagawa, *Design of generic biosensors based on green fluorescent proteins with allosteric sites by directed evolution.* FEBS Lett, 1999. 453: 305-7.

[41] Chan, P.H., H.B. Liu, Y.W. Chen, K.C. Chan, C.W. Tsang, Y.C. Leung, and K.Y. Wong, *Rational design of a novel fluorescent biosensor for beta-lactam antibiotics from a class A beta-lactamase.* J Am Chem Soc, 2004. 126: 4074-5.

[42] de Lorimier, R.M., J.J. Smith, M.A. Dwyer, L.L. Looger, K.M. Sali, C.D. Paavola, S.S. Rizk, S. Sadigov, D.W. Conrad, L. Loew, and H.W. Hellinga, *Construction of a fluorescent biosensor family.* Protein Sci, 2002. 11: 2655-75.

[43] Banerjee, S., U. Pieper, G. Kapadia, L.K. Pannell, and O. Herzberg, *Role of the omega-loop in the activity, substrate specificity, and structure of class A beta-lactamase.* Biochemistry, 1998. 37: 3286-96.

[44] Guillaume, G., M. Vanhove, J. Lamotte-Brasseur, P. Ledent, M. Jamin, B. Joris, and J.M. Frere, *Site-directed mutagenesis of glutamate 166 in two beta-lactamases. Kinetic and molecular modeling studies.* J Biol Chem, 1997. 272: 5438-44.

[45] Chan, P.H., P.K. So, D.L. Ma, Y. Zhao, T.S. Lai, W.H. Chung, K.C. Chan, K.F. Yiu, H.W. Chan, F.M. Siu, C.W. Tsang, Y.C. Leung, and K.Y. Wong, *Fluorophore-labeled beta-lactamase as a biosensor for beta-lactam antibiotics: a study of the biosensing process.* J Am Chem Soc, 2008. 130: 6351-61.

[46] Legendre, D., P. Soumillion, and J. Fastrez, *Engineering a regulatable enzyme for homogeneous immunoassays.* Nat Biotechnol, 1999. 17: 67-72.

[47] Parmley, S.F. and G.P. Smith, *Antibody-selectable filamentous fd phage vectors: affinity purification of target genes.* Gene, 1988. 73: 305-18.

[48] Yolken, R.H., S.B. Wee, and M. Van Regenmortel, *The use of beta-lactamase in enzyme immunoassays for detection of microbial antigens.* J Immunol Methods, 1984. 73: 109-23.

[49] Guesdon, J.L., T. Ternynck, and S. Avrameas, *The use of avidin-biotin interaction in immunoenzymatic techniques.* J Histochem Cytochem, 1979. 27: 1131-9.

[50] Wisdom, G.B., *Enzyme-immunoassay.* Clin Chem, 1976. 22: 1243-55.

[51] Yolken, R.H.S., P.J. and Harris, C.C., ed. *Manual of clinical immunology*. Rose and H. Friedman ed., ed. W. American Society for Microbiology, D.C. 1980. 692-699.

[52] Avrameas, S., and Ternynck, T., *Peroxidase-labbeled antibody and Fab conjugates with enhanced intracellular penetration.* Immunochemistry, 1969. 8: 1175-1179.

[53] Imagawa, M., S. Yoshitake, E. Ishikawa, Y. Endo, S. Ohtaki, E. Kano, and Y. Tsunetoshi, *Highly sensitive sandwich enzyme immunoassay of human IgE with beta-D-galactosidase from Escherichia coli.* Clin Chim Acta, 1981. 117: 199-207.

[54] Yolken, R.H. and S.B. Wee, *Enzyme immunoassays in which biotinillated beta-lactamase is used for the detection of microbial antigens.* J Clin Microbiol, 1984. 19: 356-60.

[55] Citri, N., ed. *Penicillinase and other b-lactamases*. Academic Press, Inc., New York ed. The enzymes, ed. P.D. Boyer. Vol. 4. 1971. 23-46.

[56] Voogd, C.E., J.J. Van der Stel, and J.J. Jacobs, *On the mutagenic action of some enzyme immunoassay substrates.* J Immunol Methods, 1980. 36: 55-61.

[57] Yolken, R.H., F.J. Leister, L.S. Whitcomb, and M. Santosham, *Enzyme immunoassays for the detection of bacterial antigens utilizing biotin-labeled antibody and peroxidase biotin--avidin complex.* J Immunol Methods, 1983. 56: 319-27.

[58] Yolken, R.H., *Avidin-biotin radioimmunoassay for human rotavirus.* J Infect Dis, 1983. 148: 942.

[59] Shrivastav, T.G., G.L. Kumari, and P.N. Rao, *Enzyme immunoassay of cortisol in human plasma using penicillinase as label.* Clin Chim Acta, 1988. 174: 83-91.

[60] Shrivastav, T.G., P.K. Pandey, and G.L. Kumari, *Enzyme immunosorbent assay of prolactin with penicillinase as label.* Clin Chem, 1988. 34: 2205-7.

[61] Rassaie, M.J., G.L. Kumari, P.N. Rao, T.G. Shrivastav, and H.P. Pandey, *Influence of different combinations of antibodies and penicillinase-labeled testosterone derivatives on sensitivity and specificity of immunoassays.* Steroids, 1992. 57: 112-8.

[62] He, J., Y. Li, and X. Gao, *[Studies of enzyme-linked immunoassay for testosterone using penicillinase as label].* Hua Xi Yi Ke Da Xue Xue Bao, 1994. 25: 452-5.

[63] Prasad, P.V., S.K. Chaube, T.G. Shrivastav, and G.L. Kumari, *Development of colorimetric enzyme-linked immunosorbent assay for human chorionic gonadotropin.* J Immunoassay Immunochem, 2006. 27: 15-30.

[64] Khatkhatay, M.I., M.P. Desai, G.M. Sankolli, D.K. Pardhe, and U.M. Joshi, *Stability studies of the components of a prototype penicillinase (beta-lactamase)-linked ELISA kit.* J Clin Lab Anal, 1993. 7: 95-9.

[65] Budnikov M.C., a.E., G.A., ed. *Biosensors for direct monitoring of environmental pollutants in field.* Academic publishers, London ed., ed. D. Nikolelis, Krull, U.,Wang J. and Mascini M., Kluwer. Vol. 239. 1998.

[66] Koncki R., H., A., and Glab, S., *Biochemical modifications of membrane ion-selective sensors.* Trends Anal. Chem. , 1997. 16: 528-537.

[67] Karyakin, A.A., Vuki, M., Lukachova, L.V., Karyakina, E., Orlov, L.V., Karpachova, G.P., and Wang, J.W., *Processible polyalanine as an advanced potentiometric pH transducer application to biosensors.* Anal. Chem., 1999. 71: 2534–2540.

[68] Matharu, Z., G. Sumana, S.K. Arya, S.P. Singh, V. Gupta, and B.D. Malhotra, *Polyaniline Langmuir-Blodgett film based cholesterol biosensor.* Langmuir, 2007. 23: 13188-92.

[69] Pandey, P., S.P. Singh, S.K. Arya, A. Sharma, M. Datta, and B.D. Malhotra, *Gold nanoparticle-polyaniline composite films for glucose sensing.* J Nanosci Nanotechnol, 2008. 8: 3158-63.

[70] Sergeyeva, T.A., Soldatkin, A.P., Rachkov, A.E., Tereschenko, S.A., Piletsky, S.A., El skaya, A.V., *b-lactamase label-based potentiometric biosensor for a-2 interferon detection.* Analytica Chimica Acta, 1999. 390: 73-81.

[71] Gaudin, V.a.M., P., *Development of a biosensor-based immunoassay for the screening of penicillins residues in milk.* Food and Agricultural Immunology, 2001. 13: 77-86.

[72] Smith, G.P., *Filamentous fusion phage: novel expression vectors that display cloned antigens on the virion surface.* Science, 1985. 228: 1315-7.

[73] Legendre, D., B. Vucic, V. Hougardy, A.L. Girboux, C. Henrioul, J. Van Haute, P. Soumillion, and J. Fastrez, *TEM-1 beta-lactamase as a scaffold for protein recognition and assay.* Protein Sci, 2002. 11: 1506-18.

[74] Chevigne, A., N. Yilmaz, G. Gaspard, F. Giannotta, J.M. Francois, J.M. Frere, M. Galleni, and P. Filee, *Use of bifunctional hybrid beta-lactamases for epitope mapping and immunoassay development.* J Immunol Methods, 2007. 320: 81-93.

[75] Kulesh, D.A., D.R. Clive, D.S. Zarlenga, and J.J. Greene, *Identification of interferon-modulated proliferation-related cDNA sequences.* Proc Natl Acad Sci U S A, 1987. 84: 8453-7.

[76] Schena, M., D. Shalon, R.W. Davis, and P.O. Brown, *Quantitative monitoring of gene expression patterns with a complementary DNA microarray.* Science, 1995. 270: 467-70.

[77] Ruth, N., J. Mainil, V. Roupie, J.M. Frere, M. Galleni, and K. Huygen, *DNA vaccination for the priming of neutralizing antibodies against non-immunogenic STa enterotoxin from enterotoxigenic Escherichia coli.* Vaccine, 2005. 23: 3618-27.

[78] Holland, R.E., *Some infectious causes of diarrhea in young farm animals.* Clin Microbiol Rev, 1990. 3: 345-75.

[79] Nagy, B. and P.Z. Fekete, *Enterotoxigenic Escherichia coli (ETEC) in farm animals.* Vet Res, 1999. 30: 259-84.

[80] Svennerholm, A.M., M. Wikstrom, M. Lindblad, and J. Holmgren, *Monoclonal antibodies against Escherichia coli heat-stable toxin (STa) and their use in a diagnostic ST ganglioside GM1-enzyme-linked immunosorbent assay.* J Clin Microbiol, 1986. 24: 585-90.

[81] Klipstein, F.A., R.F. Engert, and R.A. Houghten, *Properties of cross-linked toxoid vaccines made with hyperantigenic forms of synthetic Escherichia coli heat-stable toxin.* Infect Immun, 1984. 44: 268-73.

[82] Klipstein, F.A., R.F. Engert, J.D. Clements, and R.A. Houghten, *Vaccine for enterotoxigenic Escherichia coli based on synthetic heat-stable toxin crossed-linked to the B subunit of heat-labile toxin.* J Infect Dis, 1983. 147: 318-26.

[83] Batisson, I. and M. Der Vartanian, *Contribution of defined amino acid residues to the immunogenicity of recombinant Escherichia coli heat-stable enterotoxin fusion proteins.* FEMS Microbiol Lett, 2000. 192: 223-9.

[84] Sanchez, J., A.M. Svennerholm, and J. Holmgren, *Genetic fusion of a non-toxic heat-stable enterotoxin-related decapeptide antigen to cholera toxin B-subunit.* FEBS Lett, 1988. 241: 110-4.

[85] Sanchez, J., R. Argotte, and A. Buelna, *Engineering of cholera toxin A-subunit for carriage of epitopes at its amino end.* FEBS Lett, 1997. 401: 95-7.

[86] Sanchez, J., B.E. Uhlin, T. Grundstrom, J. Holmgren, and T.R. Hirst, *Immunoactive chimeric ST-LT enterotoxins of Escherichia coli generated by in vitro gene fusion.* FEBS Lett, 1986. 208: 194-8.

[87] Aitken, R. and T.R. Hirst, *Recombinant enterotoxins as vaccines against Escherichia coli-mediated diarrhoea.* Vaccine, 1993. 11: 227-33.

[88] Saarilahti, H.T., E.T. Palva, J. Holmgren, and J. Sanchez, *Fusion of genes encoding Escherichia coli heat-stable enterotoxin and outer membrane protein OmpC.* Infect Immun, 1989. 57: 3663-5.

[89] Pereira, C.M., B.E. Guth, M.E. Sbrogio-Almeida, and B.A. Castilho, *Antibody response against Escherichia coli heat-stable enterotoxin expressed as fusions to flagellin.* Microbiology, 2001. 147: 861-7.

[90] Gurunathan, S., D.M. Klinman, and R.A. Seder, *DNA vaccines: immunology, application, and optimization*.* Annu Rev Immunol, 2000. 18: 927-74.

[91] Donnelly, J.J., J.B. Ulmer, and M.A. Liu, *DNA vaccines.* Life Sci, 1997. 60: 163-72.

[92] Denis, O., A. Tanghe, K. Palfliet, F. Jurion, T.P. van den Berg, A. Vanonckelen, J. Ooms, E. Saman, J.B. Ulmer, J. Content, and K. Huygen, *Vaccination with plasmid DNA encoding mycobacterial antigen 85A stimulates a CD4+ and CD8+ T-cell epitopic repertoire broader than that stimulated by Mycobacterium tuberculosis H37Rv infection.* Infect Immun, 1998. 66: 1527-33.

[93] Chen, Y., R.G. Webster, and D.L. Woodland, *Induction of CD8+ T cell responses to dominant and subdominant epitopes and protective immunity to Sendai virus infection by DNA vaccination.* J Immunol, 1998. 160: 2425-32.

[94] Deonarain, M.P. and A.A. Epenetos, *Targeting enzymes for cancer therapy: old enzymes in new roles.* Br J Cancer, 1994. 70: 786-94.

[95] Valeriote, F. and L. van Putten, *Proliferation-dependent cytotoxicity of anticancer agents: a review.* Cancer Res, 1975. 35: 2619-30.

[96] Alexander, R.P., Beeley, N., Odriscoll, M., Oneill, F.P., Pratt, A.T. and Willenbrock, F.W., *Cephalosporin nitrogen mustard carbamate prodrugs.* Tetrahedron. Lett., 1991. 32: 3269-3272.

[97] Svensson, H.P., J.F. Kadow, V.M. Vrudhula, P.M. Wallace, and P.D. Senter, *Monoclonal antibody-beta-lactamase conjugates for the activation of a cephalosporin mustard prodrug.* Bioconjug Chem, 1992. 3: 176-81.

[98] Meyer, D.L., L.N. Jungheim, S.D. Mikolajczyk, T.A. Shepherd, J.J. Starling, and C.N. Ahlem, *Preparation and characterization of a beta-lactamase-Fab' conjugate for the site-specific activation of oncolytic agents.* Bioconjug Chem, 1992. 3: 42-8.

[99] Roberge, M., M. Estabrook, J. Basler, R. Chin, P. Gualfetti, A. Liu, S.B. Wong, M.H. Rashid, T. Graycar, L. Babe, and V. Schellenberger, *Construction and optimization of a CC49-based scFv-beta-lactamase fusion protein for ADEPT.* Protein Eng Des Sel, 2006. 19: 141-5.

[100] Muraro, R., M. Kuroki, D. Wunderlich, D.J. Poole, D. Colcher, A. Thor, J.W. Greiner, J.F. Simpson, A. Molinolo, P. Noguchi, and et al., *Generation and characterization of B72.3 second generation monoclonal antibodies reactive with the tumor-associated glycoprotein 72 antigen.* Cancer Res, 1988. 48: 4588-96.

PART 5: OTHER BETA-LACTAMASES

In: Beta-Lactamases
Editor: Jean-Marie Frère, pp. 507-514

ISBN: 978-1-61324-638-2

Chapter 21

ARE THERE UNKNOWN CLASSES OF β-LACTAMASES? SENSITIVITY OF β-LACTAMS TO NUCLEOPHILIC AGENTS AND OTHER ENZYMES

Stephane Baurin, Xavier Henry,
Nathalie Hoyez and Jean-Marie Frère
Centre d'Ingénierie des Protéines, Université de Liège, Institut de Chimie,
B6, Sart-Tilman, B 4000 Liège, Belgium.

ABSTRACT

With PBPs and active-site serine β-lactamases, β-lactams behave as acylating agents whereas the serine hydroxyl group acts as a nucleophile. In this chapter, we briefly discuss the sensitivity of β-lactams to nucleophilic agents, both small molecules and protein-borne and the β-lactamase-like activities of several other proteins, some of them unrelated to the penicillin-recognising enzymes families.

1. SENSITIVITY TO NUCLEOPHILIC AGENTS

1.1. Hydroxide Ions, Simple Amines and Alcohols

When compared to linear amides, penicillins, cephalosporins and carbapenems exhibit an increased sensitivity to base- and acid-catalysed hydrolysis [1-3]. The stability of a series of β-lactams at high pH has been studied in detail [4-7]. Aminolysis, alcoholysis and thiolysis of the same compounds have also been observed [8, 9]. For instance, in Tris buffer and even at 4°C, benzylpenicillin and nitrocefin progressively lose their potency to acylate penicillin-binding proteins and nitrocefin solutions turn red after a few hours. The values of the second-order rate constants obtained with various compounds upon reaction with OH^- ions, Tris base and methanol, displayed in Table 1, underline the more pronounced reactivities of clavulanic acid and nitrocefin and, to a lesser degree, imipenem. However, the rates of the reactions are clearly dependent upon the exact nature of the nucleophile. Reactions of benzylpenicillin with the thiolate ion and hydroxylamine are significantly faster [9] but cephalosporins seemed to be much less sensitive to thiolysis [10].

1.2. Acylation of Proteins

Not surprisingly, β-lactams are unstable in human and animal sera [11]. The rates of degradation of benzylpenicillin, imipenem, ceftazidime, cefotetan and nitrocefin were similar to those observed at pH 10 or in the presence of 0.1 M Tris base (Hoyez and Frère, unpublished). From the data in Table 1, it seems likely that reactions involve the amino groups of proteins. Indeed, acylation of the ε-amino group of Lys 199 has been demonstrated in serumalbumin [12]. The preference for Lys 199 has been attributed to a lowered pKa (7.9] for this residue and was analysed by theoretical chemistry methods [13]. After the first study [12], a more complete analysis identified 5 other penicilloylated residues in human serumalbumin [14]. Similarly penicilloylated samples were obtained by *in vitro* reaction of benzylpenicillin or isolated from penicillin-treated patients [14, 15]. We monitored the reaction between the fluoresceyl derivative [16] of ampicillin (concentration range 50-150 μM) and 13 μM bovine serumalbumin at pH 7.0. This yielded a second-order rate constant of 0.5 M^{-1} s^{-1}. Interestingly, this value is not very different from that observed for the acylation of a highly resistant *Enterococcus faecium* PBP5fm and benzylpenicillin (about 1,5 M^{-1} s^{-1}) although in the latter case, the reaction is probably with a serine hydroxyl group [17]. Serumalbumin is thus orders of magnitude more reactive than Tris base and insulin and lysozyme (see below). This reactivity can be correlated to allergy phenomena (see for instance refs 18-20).

Table 1. Observed second-order rate constants

β-lactam	Second-order rate constant (M^{-1} h^{-1})		
	OH^- ions	Aminolysis (#)	Alcoholysis
Benzylpenicillin	1200	0.82	2.0 X 10^{-3}
7-Aminocephalosporanic acid	800	0.13	1.3 X 10^{-3}
Nitrocefin	11800*	25	30 X 10^{-3}
Cephaloridine	4300	0.81	2.5 X 10^{-3}
Cefotetan	1300	1.41	5.8 X 10^{-3}
Cefuroxime	1200	0.23	1.0 X 10^{-3}
Ceftazidime	720	0.81	2.5 X 10^{-3}
Cefoxitin	1400	0.31	3.8 X 10^{-3}
Imipenem	5200	4.9	3.3 X 10^{-3}
Clavulanic acid	8900	22	14 X 10^{-3}

adjusted for the concentration of Tris base in the buffer (pK_a = 7.8).
* from reference [2].
SD values were below 10%.

In independent experiments with pure proteins, Corran and Waley [21] showed that benzylpenicillin reacted with the α-amino groups of the N-terminal glycine and the Lys 29 ε-amino group of pig insulin and Lys 116 of hen egg white lysozyme. The second-order rate constants for these reactions are significantly lower (20-fold or more) than that given in Table 1 for Tris base but the values must be corrected for the proportion of free base in the proteins. Interestingly, in the case of lysozyme, the α-amino group of the N-terminal Lys residue did not seem to be penicilloylated. This shows again that the exact nature of the nucleophile and its accessibility are important factors in determining the rates of the reactions. As one would expect, the newly formed amide bonds appeared to be stable. Interestingly, no reaction with serine or threonine hydroxyl groups was observed, in agreement with the low alcoholysis/aminolysis ratio (Table 1).

In regard to the higher reactivity of thiol groups, it may appear surprising that penicilloylation of such groups in proteins has, to our knowledge, never been reported (with the exception of some thiol proteases, see below). This might be due to the facts that free thiols are often buried in extracellular proteins and that penicillins do not easily penetrate into the cytosol.

2. Hydrolysis by other Proteins

2.1. Classical Proteases

It was also interesting to determine if classical proteases could hydrolyse β-lactams. With the five compounds mentioned above, the following enzymes were assayed (Baurin and Frère, unpublished):

- active-site serine enzymes: elastase (from porcine pancreas), trypsin (from porcine pancreas) and chymotrypsin (from bovine pancreas),
- active-site cysteine enzyme: papain (crude papaya extract),
- metallo-enzyme: caboxypeptidase A (from bovine pancreas).

With 1.5 to 5 μM enzymes and over a 90-min incubation, significant hydrolysis was only observed of nitrocefin by elastase and trypsin and of imipenem by trypsin (Table 2) and the steady-state rates were proportional to enzyme concentrations. The other assays did not result in detectable hydrolysis, indicating k_{cat}/K_m values below 1 $M^{-1}s^{-1}$. Even when hydrolysis was recorded, the Enzyme Rate Enhancement Factors [22] remained very low.

Table 2. Enzymatic hydrolysis of nitrocefin and imipenem

	Trypsin			Elastase		
	k_{cat}/K_m ($M^{-1}s^{-1}$)	k_{cat} (s^{-1})	EREF*	k_{cat}/K_m ($M^{-1}s^{-1}$)	k_{cat} (s^{-1})	EREF
Nitrocefin	17	$> 3.4 \times 10^{-3}$	5	8	$> 1.7 \times 10^{-3}$	2.5
Imipenem	15	$> 3 \times 10^{-3}$	10	< 1	–	

* Enzyme rate enhancement factor (22) = (k_{cat}/K_m) / k_{OH^-}

Maybe not surprisingly, nitrocefin and imipenem also rank as the most chemically unstable compounds in the group. For nitrocefin, the k_{cat} values are larger than 0.1 min^{-1} for both elastase and trypsin. This shows that it is dangerous to conclude that a protein exhibits a "poor β-lactamase activity" on the basis of tests performed with this sole compound. It is not easy to rationalise why a positive result was obtained with trypsin and elastase and not with chymotrypsin. The very low values or absence of reaction reported here contrast with the rather high activity ($k_{cat}/K_m = 675\ M^{-1}s^{-1}$) of porcine liver elastase on benzylpenicillin [23].

Some cephalosporins have been selected as rather specific elastase inactivators [24, 25]. The most active compound exhibited a very respectable second-order acylation rate constant of 160000 $M^{-1}s^{-1}$ [20]. The active compounds were usually sulphones, the C4 carboxylic group was esterified and the C7 side chain was α rather than β. Some of the compounds also behaved, although much less efficiently, as inhibitors of chymotrypsin, plasmin and thrombin, but not of cathepsin G and trypsin. Other, similar derivatives of cephalosporins, penicillins and monobactams have been described which inactivate elastase and other active-site serine and cysteine enzymes. The active compounds are always devoid of the typical negative group of antibacterial β-lactams and, for penicillins and cephalosporins, the sulphones are much more active [25]. Interestingly, some monobactams were described as reversible inhibitors of elatase [26, 27] but their activities were due to the fact that they behaved as competitive substrates and the hydrolysed bond was not the β-lactam amide, but an exocyclic ester bond.

2.2. Other Enzymes

Two cysteine-rich *Helicobacter pylori* proteins, HcpA and HcpB have been purified which appear to exhibit very low β-lactamase activities with k_{cat}/K_m values ranging from 10 to 250 $M^{-1}s^{-1}$and k_{cat} values up to 0.02 s^{-1} for the best substrates [28, 29]. They also form covalent complexes with two fluorescent penicillins [28]. They belong to a small family of proteins unique to microorganisms from the *Helicobacter* and *Campylobacter* genera. The 3D structure of HcpB has been determined. There are no similarities with the primary or tertiary structures of the 4 classical classes of β-lactamases.

In the crystals of HcpB, a low-molecular mass compound identified as N-acetylmuramic acid by mass spectrometry was found in the putative penicillin-binding site with partial occupancy. It is noteworthy that the 8 cysteine residues of HcpB are immobilised in disulphide bridges and cannot, in consequence, represent the sites of acylation by penicillins.

The biological function of the Hcp proteins remains unclear. It should however be noted that the K_m values of most of the studied substrates are similar to those of the TEM1/2 enzymes and that the reported activities could correspond to a very low level of contamination (one molecule in 30000 to 100000) by one of these highly efficient β-lactamases.

An even stranger situation occurs with Tp47 (Tpp47 in the first paper) the 47-kDa major lipoprotein immunogen of *Treponema pallidum* [30, 31]. The protein was first characterised as a penicillin-binding protein which could hydrolyse, in a Zn-dependent manner, some ester and thiolester substrates common to β-lactamases and DD- peptidases [30] and later as a β-lactamase [31]. The esterase/thiolesterase activity was however suspected to be an artefact due to aggregation of the protein by Zn^{++} ions [31]. SXXK, TGN and KTG motifs could be identified in the sequence, but no further similarities with β-lactamases or PBPs could be found.

Accordingly, the crystallographic data revealed a novel structure in which the three motifs were located in three different domains "separated by distances greater than 30 Å" and thus did not come together to yield a typical PBP/serine β-lactamase active site [31]. Formation of acylenzymes with various penicillins and cephalosporins was demonstrated by mass spectrometry, but after digestion by trypsin, no peptide fragment containing the penicilloyl moiety could be isolated.

A more detailed kinetic analysis [32] highlighted a non-negligible β-lactamase activity with k_{cat} and k_{cat}/K_m values up 10 s^{-1} and 9000 $mM^{-1}s^{-1}$, respectively, however rapidly quenched by a rather strong product inhibition with K_i values around 6 μM. The k_{cat} values for ampicillin and benzylpenicillin were 25- and 45-fold lower than those previously reported [31] and no activity was found with cephalosporins. Upon prolonged incubation with high concentrations of BOCILLIN FL, a fluorescent penicillin, a covalent reaction took place (with $k_2/K = 0.02\ mM^{-1}s^{-1}$ vs $k_{cat}/K_m = 1000\ mM^{-1}s^{-1}$ for the hydrolysis of the same compound). The authors do not mention if the covalent adduct still exhibits β-lactamase or esterase activities but in contrast to the previous results, they did not observe a Zn-dependence of the reactions. They conclude that the β-lactamase and penicillin-binding activities take place in two different sites.

The physiological functions of Tp47 remain mysterious but the fact that *Escherichia coli* cells expressing this protein exhibit cell wall abnormalities is consistent with a role in peptidoglycan metabolism [30].

2.3. β-lactamase Activity of PBPs

A few PBPs exhibit relatively high k_3 values with some penicillins, for instance *E. coli* PBP5 with benzylpenicillin (0.001 s^{-1}, refs 33 and 34) and *Staphylococcus. aureus* PBP4 with the same substrate (0.01 s^{-1}, ref 35). The W233L mutant of the *Streptomyces* R61 DD-peptidase, engineered for other purposes, unexpectedly exhibited a 300–fold increased k_3 value with benzylpenicillin (from 1.4 x 10^{-4} to 4.6 X 10^{-2} s^{-1}, ref 36). Maybe surprisingly, attempts to engineer β-lactamases from PBPs or DD-peptidases have failed to generate activities higher than those detailed above in this paragraph (see chapter 18).

Transforming a PBP into a β-lactamase is thus not as easy as it might appear on the basis of the structural similarities between the two types of proteins.

In this respect, it is interesting to remember that the highest degree of structural similarity between PBPs and β-lactamases is that between the C-terminal, exocellular domains of the BlaR receptors and class D β-lactamases (37 and chapters 5 and 17). The k_2/K values for the BlaR receptors can be quite high for many penicillins and cephalosporins (up to 1000 $mM^{-1}s^{-1}$, ref 38) but the k_3 values are exceeding low, even by PBP standards (never higher than 10^{-5} s^{-1}).

ACKNOWLEDGMENTS

This work was supported by FRFC grant n° 2.4.508.01.F (FNRS, Brussels) and PAI P5/33 (Politique Scientifique Fédérale, Belgium)

REFERENCES

[1] MI Page. Structure-activity relationships: chemical. In: MI Page, editor. *The chemistry of β-lactams.* London, Glasgow, New-York, Tokyo, Melbourne, *Madras Blackie Academic and Professional,* 1992; 79-100.

[2] M.I. Page (1991). The energetics of intramolecular reactions and enzyme catalysis. *Phil. Trans. R. Soc. Lond., 332*, 149-156.

[3] M.I. Page (1999). The reactivity of β-lactams, the mechanism of catalysis and the inhibition of β-lactamases. *Current Pharmaceutical Design, 5*, 895-913.

[4] J.M. Frère, J.A. Kelly, D. Klein and J.M. Ghuysen (1982). Δ^{2-} and Δ^{3-} cephalosporins, penicillinate and 6-unsubstituted penems. Intrinsic reactivity and interaction with β-lactamases and D-alanyl-D-alanine-cleaving serine peptidases. *Biochem. J.*, 203, 223-224.

[5] J.M. Indelicato and W.L. Wilham (1974). Effect of 6-alpha substitutions in penicillins and 7-beta substitutions in cephalosporins upon beta-lactam reactivity. *J. Med. Chem., 17*, 528-529.

[6] J.M. Indelicato, T.T. Norvilas, R.R. Pfeiffer, W.J. Wheeler and W.L. Wilham (1974). Substituent effects upon the base hydrolysis of penicillins and cephalosporins. Competitive intramolecular nucleophilic amino attacks in cephalosporins. *J. Med. Chem., 17*, 523-527.

[7] T. Yamana and A. Tsuji (1976). Comparative stability of cephalosporins in aqueous solution: kinetics and mechanisms of degradation. *Journal of Pharmaceutical Sciences, 65*, 1563-1574.

[8] C.H. Schneider and A.L. De Weck (1968). Studies on the direct neutral penicilloylation of functional groups occurring on proteins. *Biochim. Biophys. Acta, 168*, 27-35.

[9] K.F. Nakken, L. Eldjarn and A. Pihl (1960). The mechanism of inactivation of penicillin by cysteine and other mercaptoamines. *Biochem. Pharmacol., 3*, 89-100.

[10] E.S. Wagner and M. Gorman (1971). The reaction of cysteine and related compounds with penicillins and cephalosporins. *The Journal of Antibiotics, 24*, 647-650.

[11] C.H. O'Callaghan, A. Morris, S.M. Kirby and A.H. Shingler (1972). Novel method for detection of β-lactamases by using a chromogenic cephalosporin substrate. *Antimicrobial. Agents and Chemotherapy, 1*, 283-288.

[12] M. Yvon and J.M. Wal (1988). Identification of lysine residue 199 of human serum albumin as a binding site for benzylpenicilloyl groups. *FEBS Letters, 239*, 237-240.

[13] N. Diaz, D. Suarez, T.L. Sordo and K.M. Merz Jr (2001). A theoretical study of the aminolysis reaction of lysine 199 of human serum albumin with benzylpenicillin: consequences for immunochemistry of penicillins. *J. Am. Chem. Soc., 123*, 7574-7583.

[14] M. Yvon, P. Anglade and J.M. Wal (1989). Binding of benzyl penicilloyl to human serum albumin. Evidence for a highly reactive region at the junction of domains 1 and 2 of the albumin molecule. *FEBS Letters, 247*, 273-278.

[15] M. Yvon, P. Anglade and J.M. Wal (1990). Identification of the binding sites of benzyl penicilloyl, the allergenic metabolite of penicillin, on the serum albumin molecule. *FEBS Letters, 263*, 237-240.

[16] B. Lakaye, C. Damblon, M. Jamin, M. Galleni, S. Lepage, B. Joris, J. Marchand-Brynaert, C. Frydrych and J.M. Frère (1994). Synthesis, purification and kinetic properties of fluorescein-labelled penicillins. *Biochem.J. 300*, 141-145.

[17] W. Zorzi, X.Y. Zhou, O. Dardenne, J. Lamotte, D. Raze, J. Pierre, L. Gutmann and J. Coyette (1996). Structure of the low-affinity penicillin-binding protein 5 PBP5fm in wild-type and highly penicillin-resistant strains of *Enterococcus faecium. J. Bacteriol., 178*, 4948-4957.

[18] J.W. Coleman (1990). Allergic reactions to drugs: current concepts and problems. *Clinical and Experimental Allergy, 20*, 79-85.

[19] P. Lafaye and C. Lapresie (1988). Fixation of penicilloyl groups to albumin and appearance of anti-penicilloyl antibodies in penicillin-treated patients. *J. Clin. Invest., 82*, 7-12.

[20] J.T. Di Piro, N.F. Adkinson Jr and R.G. Hamilton. Facilitation of penicillin haptenation to serum proteins (1993). *Antimicrobial. Agents and Chemotherapy, 37*, 1463-1467.

[21] P.H. Corran and S.G. Waley (1975). The reaction of penicillin with proteins. *Biochem. J., 149*, 357-364.

[22] A.P. Laws and M.I. Page (1989). The effect of the carboxy group on the chemical and β-lactamase reactivity of β-lactam antibodies. *J. Chem. Soc. Perkin Tans II*, , 1577-1581.

[23] M. Jones and M.I. Page (1991). An esterase with β-lactamase activity. *J. Chem. Soc., Chem. Commun.*, 316-317.

[24] J.B. Doherty, B.M. Ashe, L.W. Argenbright, P.L. Barker, R.J. Bonney, G.O. Chandler, M.E. Dahlgren, C.P.J. Dorn, P.E. Finke, R.A. Firestone, D. Fletcher, W.K. Hagmann, R. Mumford, L. O'Grady, A.L. Maycock, J.M. Pisano, S.K. Shah, K.R. Thompson and M. Zimmerman (1986). Cephalosporin antibiotics can be modified to inhibit human leukocyte elastase. *Nature, 322*, 192-194.

[25] G. Veinberg, M. Vorona, I. Shestakova, I. Kanepe and E. Lukevics (2003). Design of β-lactams with mechanism based nonantibacterial activities. *Current Medicinal Chemistry, 10*, 1741-1757.

[26] S. Gérard, G. Dive, B. Clamot, R. Touillaux and J. Marchand-Brynaert (2002). Synthesis, hydrolysis, biochemical and theoretical evaluation of 1,4-bis(alkoxycarbonyl) azetidin-2-ones as potential elastase inhibitors. *Tetrahedron, 58*, 2423-2433.

[27] S. Gérard, M. Galleni, G. Dive and J. Marchand-Brynaert (2004). Synthesis and evaluation of N1/C4-substituted β-lactams as PPE and HLE inhibitors.*Bioorg. Med. Chem., 12*, 129-138.

[28] P.R.E. Mittl, L. Lüthy, P. Hunziker and M.G. Grütter (2000). The cysteine-rich protein A from *Helicobacter pylori* is a β-lactamase. *J. Biol. Chem., 275*, 17693-17699.

[29] L. Lüthy, M.G. Grütter and P.R.E. Mittl (2002). The crystal structure of *Helicobacter pylori* cysteine-rich protein B reveals a novel fold for a penicillin-binding protein. *J. Biol. Chem., 277*, 10187-10193.

[30] L.M. Weigel, J.D. Radolf and M.V. Norgard (1994). The 47-kDa major lipoprotein immunogeng of *Treponema pallidum* is a penicillin-binding protein with carboxypeptidase activity. *Proc. Natl. Acad. Sci. USA, 91*, 11611-11615.

[31] R.K. Deka, M. Machius, M.V. Norgard and D.R. Tomchick (2002). Crystal structure of the 47-kDa lipoprotein of *Treponema pallidum* reveals a novel penicillin-binding protein. *Proc. Natl. Acad. Sci. USA, 277*, 41857-41864.

[32] J.Y. Cha, A. Ishiwata and S. Mobashery (2004). A novel β-lactamase activity from a penicillin-binding protein of *Treponema pallidum* and why syphilis is still treatable with penicillin. *J. Biol. Chem., 279*, 14917-14921.

[33] H. Amanuma and J.L. Strominger (1980). Purification and properties of penicillin-binding proteins 5 and 6 from *Escherichia coli* membranes. *J. Biol. Chem., 255*, 11173-11180.

[34] R.A. Nicholas and J.L. Strominger (1988). Site-directed mutants of soluble form of penicillin-binding protein 5 from *Escherichia coli* and their catalytic properties. *J. Biol. Chem., 263*, 2034-2040.

[35] D.J. Waxman and J.L. Strominger (1979). Cephalosporin-sensitive penicillin-binding proteins of *Staphylococcus aureus* and *Bacillus subtilis* active in the conversion of [^{14}C]Penicillin G to [^{14}C]Phenylacetylglycine. *J. Biol. Chem., 254*, 12056-12061.

[36] C. Bourguignon-Bellefroid, J.M. Wilkin, B. Joris, R.T. Aplin, C. Houssier, F.G. Prendergast, J. Van Beeumen, J.M. Ghuysen and J.M. Frère (1992). Importance of the two tryptophan residues in the *Streptomyces* R61 exocellular DD-peptidase. *Biochem. J., 282*, 361-367.

[37] F. Kerff, P. Charlier, M.L. Colombo, E. Sauvage, A. Brans, J.M. Frère, B. Joris and E. Fonzé (2003). Crystal structure of the sensor domain of the BlaR penicillin receptor from *Bacillus licheniformis*. *Biochemistry, 42*, 12835-12843.

[38] V. Duval, M. Swinnen, S. Lepage, A. Brans, B. Granier, C. Franssen, J.M. Frère and B. Joris (2003). The kinetic properties of the carboxy terminal domain of the *Bacillus licheniformis* 749/I BlaR penicillin-receptor shed a new light on the derepression of β-lactamase synthesis. *Molecular Microbiology, 48*, 1553-1564.

INDEX

B

C

D

E

F

G

H

I

J

K

L

M

P

S

T

U

V

W

X

Y

Z

β